中华人民共和国进出口税则
(法律文本)

Customs Tariff of Import and Export of the People's Republic of China
(The legal texts)

(2005年)
(Enforced from January 1, 2005)

国务院关税税则委员会办公室
中华人民共和国海关总署关税征收管理司 编
中华人民共和国财政部关税司

Compiled by the Office of Customs Tariff Commission of the State Council
Department of Duty Collection of General Administration of Customs P.R.C.
Tariff Department of Ministry of Finance P.R.C.

中国财政经济出版社
China Financial & Economic Publishing House

图书在版编目（CIP）数据

2005年中华人民共和国进出口税则：法律文本/国务院关税税则委员会办公室，中华人民共和国海关总署关税征收管理司，中华人民共和国财政部关税司编. —北京：中国财政经济出版社，2004.12

ISBN 7-5005-7840-7

Ⅰ.2… Ⅱ.①国… ②海… ③中… Ⅲ.海关税则-中国-2005 Ⅳ.F752.50

中国版本图书馆CIP数据核字（2004）第138626号

中国财政经济出版社出版

URL：http：//www.cfeph.com

E-mail：cfeph@drc.gov.cn

社址：北京海淀区阜成路甲28号 邮政编码：100036

发行处电话：88190406 财经书店电话：64033436

北京人卫印刷厂印刷 各地新华书店经销

889×1194毫米 16开 47印张 1270000字

2005年1月第1版 2005年1月北京第1次印刷

印数：1—10 060 定价：240.00元

ISBN 7-5005-7840-7/F·6886

（图书出现印装问题，本社负责调换）

税率适用说明

1. 最惠国税率

原产于共同适用最惠国待遇条款的世界贸易组织成员的进口货物，原产于与中华人民共和国签订含有相互给予最惠国待遇条款的双边贸易协定的国家或地区的进口货物，以及原产于中华人民共和国境内的进口货物，适用最惠国税率。

2. 协定税率

原产于与中华人民共和国签订含有关税优惠条款的区域性贸易协定的国家或地区的进口货物，适用协定税率。

目前适用协定税率的国家或地区为原产于文莱达鲁萨兰国、柬埔寨王国、印度尼西亚共和国、老挝人民民主共和国、马来西亚、缅甸联邦、新加坡共和国、泰王国、越南社会主义共和国，大韩民国、斯里兰卡民主社会主义共和国、孟加拉人民共和国、印度共和国，巴基斯坦伊斯兰共和国，中华人民共和国香港特别行政区以及中华人民共和国澳门特别行政区的部分进口货物。

3. 特惠税率

原产于与中华人民共和国签订含有特殊关税优惠条款的贸易协定的国家或地区的进口货物，适用特惠税率。

目前适用特惠税率的国家或地区为原产于孟加拉人民共和国、柬埔寨王国、老挝人民民主共和国、缅甸联邦，贝宁共和国、布隆迪共和国、佛得角共和国、中非共和国、科摩罗联盟、刚果民主共和国、吉布提共和国、厄立特里亚国、埃塞俄比亚联邦民主共和国、几内亚共和国、几内亚比绍共和国、莱索托王国、利比里亚共和国、马达加斯加共和国、马里共和国、毛里塔尼亚伊斯兰共和国、莫桑比克共和国、尼日尔共和国、卢旺达共和国、塞拉利昂共和国、苏丹共和国、坦桑尼亚联合共和国、多哥共和国、乌干达共和国、赞比亚共和国的部分进口货物。

4. 普通税率

原产于上述国家或地区以外的国家或地区的进口货物，以及原产地不明的进口货物，适用普通税率。

5. 暂定税率

适用最惠国税率的进口货物有暂定税率的，应当适用暂定税率；适用协定税率、特惠税率的进口货物有暂定税率的，应当从低适用税率；适用普通税率的进口货物，不适用暂定税率。

适用出口税率的出口货物有暂定税率的，应当适用暂定税率。

暂定税率仅在当年有效。

6. 关税减免

特定地区、特定企业或者特定用途的进出口货物减征或者免征关税的，以及其他依法减征或者免征关税的，按照国务院的有关规定执行。

Application of Tariff Rates

1. The Most Favored Nation (MFN) Tariff Rates

The MFN tariff rates shall apply to goods imported from and originating in the members of the World Trade Organization providing that the MFN treatment is reciprocal between the People's Republic of China and these members; or those countries or regions with which the People's Republic of China has concluded bilateral agreements for reciprocal tariff preference; or the custom territory of the People's Republic of China.

2. The Agreement Tariff Rates

The agreement tariff rates shall apply to goods imported from and originating in the countries or regions which join together with the People's Republic of China into regional trade agreements for tariff preferences.

The agreement tariff rates are currently applicable to some goods imported from and originating in Brunei Darussalam, the Kingdom of Cambodia, the Republic of Indonesia, the Lao People's Democratic Republic, Malaysia, the Union of Myanmar, the Republic of Singapore, the Kingdom of Thailand and the Socialist Republic of Viet Nam; the Republic of Korea, the Democratic Socialist Republic of Sri Lanka, the People's Republic of Bangladesh, the Republic of India; Islamic Republic of Pakistan, the Hong Kong Special Administrative Region of the People's Republic of China, the Macao Special Administrative Region of the People's Republic of China.

3. The Special Preferential Tariff Rates

The special preferential tariff rates shall apply to goods imported from and originated in countries or regions that have concluded special tariff preferential agreement with the People's Republic of China.

The special preferential tariff rates are applicable to some goods imported from and originating in the Kingdom of Cambodia, the Lao People's Democratic Republic, the Union of Myanmar, the People' s Republic of Bangladesh; the Republic of Benin, the Republic of Burundi, the Republic of Cape Verde, the Central African Republic, the Vnion of Comoros, the Democratic Republic of Congo, the Republic of Djibouti, the State of Eritrea, the Federal Democratic Republic of Ethiopia, the Republic of Guinea, the Republic of Guinea Bissau, the Kingdom of Lesotho, the Republic of Liberia, the Republic of Madagascar, the Republic of Mali, the Islamic Republic of Mauritania, the Republic of Mozambique, the Republic of Niger, the Republic of Rwanda, the Republic Sierra Leone, the Republic of the Sudan, the Vnited Republic of Tanzania, the Republic of Togo, the Republic of Vganda the Republic of Zambia.

4. The General Tariff Rates

The general tariff rates shall apply to goods imported from and originating in countries or regions with which the People' s Republic of China has concluded no agreements for reciprocal tariff preference, goods with non-origin or no definite origin.

5. The Interim Tariff Rates

The interim import tariff rates have the priority over the MFN tariff rates for application. Goods eligible for interim import tariff rates, when imported under the agreement tariff rates or special preferential tariff rates, will be levied under the lower of the two rates.

The interim export tariff rates have the priority over the export tariff rates for application.

Interim tariff rates are effective in one year from January 1st to December 31st.

6. Exemption and Reduction

Customs duties shall be reduced or exempted in accordance with the provisions set out in the relevant regulations by the State Council on goods imported into or exported out of the designated areas, the designated enterprises, or for designated uses.

总　目　录

Contents

中华人民共和国进口税则 …… (1)

Customs Tariff of Import of the People's Republic of China

中华人民共和国出口税则 …… (615)

Customs Tariff of Export of the People's Republic of China

附表： …… (619)

Appendix

表 1：《中国－东盟全面经济合作框架协议》早期收获安排附加表 …… (621)

Table 1: Annex to Early Havest Arrangement under the Framework on Comprehensive Economic Co-operation Between China and ASEAN

表 2：进口关税特惠税率表 …… (639)

Table 2: Special Preferential Duty Rate on Imported Goods

表 3：中国－巴基斯坦优惠关税安排税目、税率表 …… (673)

Table 3: Duty Rate under the China-Pakistan Preferential Tariff Arrangement

表 4：进口商品从量税、复合税税目税率表 …… (714)

Table 4: Specific and Compound Duty Rate on Imported Goods

表 5：关税配额商品税目、税率表 …… (721)

Table 5: Tariff Quota Rate on Imported Goods

表 6：进口商品暂定税率表 …… (723)

Table 6: Interim Duty Rate on Imported Goods

表 7：出口商品暂定税率表 …… (735)

Table 7: Interim Duty Rate on Exported Goods

表 8：纺织品出口暂定税率表 …… (736)

Table 8: Interim Duty Rate on Exported Textile

表 9：非全税目信息技术产品税率表 …… (743)

Table 9: Duty Rate on Specific Information Technology Products

表 10：中华人民共和国进境物品进口税率表 …… (745)

Table 10: Duty Rate on Inward Articles of the People's Republic of China

中华人民共和国进口税则

(2005 年 1 月 1 日起实施)

Customs Tariff of Import of the People's Republic of China

(Enforced from January 1, 2005)

目　　录

INDEX

归类总规则……………………………………………（11）

第一类　活动物；动物产品

第一章　活动物……………………………………（13）
第二章　肉及食用杂碎……………………………（16）
第三章　鱼、甲壳动物、软体动物及其他水生无脊椎动物……………………（20）
第四章　乳品；蛋品；天然蜂蜜；其他食用动物产品……………………………（29）
第五章　其他动物产品……………………………（33）

第二类　植物产品

第六章　活树及其他活植物；鳞茎、根及类似品；插花及装饰用簇叶………（37）
第七章　食用蔬菜、根及块茎……………………（39）
第八章　食用水果及坚果；柑桔属水果或甜瓜的果皮…………………………（45）
第九章　咖啡、茶、马黛茶及调味香料……（49）
第十章　谷物………………………………………（52）
第十一章　制粉工业产品；麦芽；淀粉；菊粉；面筋……………………………（54）
第十二章　含油子仁及果实；杂项子仁及果实；工业用或药用植物；稻草、秸秆及饲料…………………………（58）
第十三章　虫胶；树胶、树脂及其他植物液、汁…………………………………（64）
第十四章　编结用植物材料；其他植物产品……………………………………（66）

第三类　动、植物油、脂及其分解产品；精制的食用油脂；动、植物蜡

第十五章　动、植物油、脂及其分解产品；精制的食用油脂；动、植物蜡……（68）

GENERAL RULES FOR THE INTERPRETATION OF THE HARMONIZED SYSTEM ……………………（11）

SECTION Ⅰ　LIVE ANIMALS；ANIMAL PRODUCTS

Chapter 1　Live animals ……………………（13）
Chapter 2　Meat and edible meat offal ……………（16）
Chapter 3　Fish and crustaceans，molluscs and other aquatic invertebrates ……………（20）
Chapter 4　Dairy produce；birds'eggs；natural honey；edible products of animal origin，not elsewhere specified or included …………（29）
Chapter 5　Products of animal origin，not elsewhere specified or included ……………（33）

SECTION Ⅱ　VEGETABLE PRODUCTS

Chapter 6　Live trees and other plants；bulbs，roots and the like；cut flowers and ornamental foliage ……………………（37）
Chapter 7　Edible vegetables and certain roots and tubers ……………………（39）
Chapter 8　Edible fruit and nuts；peel of citrus fruit or melons ……………………（45）
Chapter 9　Coffee，tea，mate and spices ……………（49）
Chapter 10　Cereals ……………………（52）
Chapter 11　Products of the milling industry；malt；starches；inulin；wheat gluten ……………（54）
Chapter 12　Oil seeds and oleaginous fruits；miscellaneous grains，seeds and fruit；industrial or medicinal plants；straw and fodder ………（58）
Chapter 13　Lac；gums，resins and other vegetable saps and extracts ……………………（64）
Chapter 14　Vegetable plaiting materials；vegetable products not elsewhere specified or included ……………………（66）

SECTION Ⅲ　ANIMAL OR VEGETABLE FATS AND OILS AND THEIR CLEAVAGE PRODUCTS；PREPARED EDIBLE FATS；ANIMAL OR VEGETABLE WAXES

Chapter 15　Animal or vegetable fats and oils and their cleavage products；prepared edible fats；animal or vegetable waxes ……………………（68）

第四类 食品；饮料、酒及醋；烟草、烟草及烟草代用品的制品

第十六章 肉、鱼、甲壳动物、软体动物及其他水生无脊椎动物的制品……（74）
第十七章 糖及糖食……（77）
第十八章 可可及可可制品……（79）
第十九章 谷物、粮食粉、淀粉或乳的制品；糕饼点心……（81）
第二十章 蔬菜、水果、坚果或植物其他部分的制品……（84）
第二十一章 杂项食品……（90）
第二十二章 饮料、酒及醋……（93）
第二十三章 食品工业的残渣及废料；配制的动物饲料……（96）
第二十四章 烟草、烟草及烟草代用品的制品……（99）

第五类 矿产品

第二十五章 盐；硫磺；泥土及石料；石膏料、石灰及水泥……（100）
第二十六章 矿砂、矿渣及矿灰……（108）
第二十七章 矿物燃料、矿物油及其蒸馏产品；沥青物质；矿物蜡……（112）

第六类 化学工业及其相关工业的产品

第二十八章 无机化学品；贵金属、稀土金属、放射性元素及其同位素的有机及无机化合物……（118）
第二十九章 有机化学品……（134）
第三十章 药品……（166）
第三十一章 肥料……（172）
第三十二章 鞣料浸膏及染料浸膏；鞣酸及其衍生物；染料、颜料及其他着色料；油漆及清漆；油灰及其他类似胶粘剂；墨水、油墨……（176）
第三十三章 精油及香膏；芳香料制品及化妆盥洗品……（182）
第三十四章 肥皂、有机表面活性剂、洗涤剂、润滑剂、人造蜡、调制蜡、光洁剂、蜡烛及类似品、塑型用膏、“牙科用蜡”及牙科用熟石膏制剂……（186）

SECTION IV PREPARED FOODSTUFFS; BEVERAGES, SPIRITS AND VINEGAR, TOBACCO AND MANUFACTURED TOBACCO SUBSTITUTES

Chapter 16 Preparations of meat, of fish or of crustaceans, molluscs or other aquatic invertebrates …… (74)
Chapter 17 Sugars and sugar confectionery …… (77)
Chapter 18 Cocoa and cocoa preparations …… (79)
Chapter 19 Preparations of cereals, flour, starch or milk; pastrycook's products …… (81)
Chapter 20 Preparations of vegetables, fruit, nuts or other parts of plants …… (84)
Chapter 21 Miscellaneous edible preparations …… (90)
Chapter 22 Beverages, spirits and vinegar …… (93)
Chapter 23 Residues and waste from the food industries; prepared animal fodder …… (96)
Chapter 24 Tobacco and manufactured tobacco substitutes …… (99)

SECTION V MINERAL PRODUCTS

Chapter 25 Salt; sulphur; earths and stone; plastering materials, lime and cement …… (100)
Chapter 26 Ores, slag and ash …… (108)
Chapter 27 Mineral fuels, mineral oils and products of their distillation; bituminous substances; mineral waxes …… (112)

SECTION VI PRODUCTS OF THE CHEMICAL OR ALLIED INDUSTRIES

Chapter 28 Inorganic chemicals; organic or inorganic compounds of precious metals, of rare-earth metals of radioactive elements or of isotopes …… (118)
Chapter 29 Organic chemicals …… (134)
Chapter 30 Pharmaceutical products …… (166)
Chapter 31 Fertilizers …… (172)
Chapter 32 Tanning or dyeing extracts; tannins and their derivatives; dyes, pigments and other colouring matter; paints and varnishes; putty and other mastics; inks …… (176)
Chapter 33 Essential oils and resinoids; perfumery, cosmetic or toilet preparations …… (182)
Chapter 34 Soap, organic surface-active agents, washing preparations, lubricating preparations, artificial waxes, prepared waxes, polishing or scouring preparations, candles and similar articles, modelling pastes, “dental waxes” and dental preparations with a basis of plaster …… (186)

第三十五章　蛋白类物质；改性淀粉；胶；酶……………………………………（190）
第三十六章　炸药；烟火制品；火柴；引火合金；易燃材料制品……………（192）
第三十七章　照相及电影用品…………………（194）
第三十八章　杂项化学产品……………………（198）

第七类　塑料及其制品；橡胶及其制品

第三十九章　塑料及其制品……………………（207）
第 四 十 章　橡胶及其制品……………………（220）

第八类　生皮、皮革、毛皮及其制品；鞍具及挽具；旅行用品、手提包及类似容器；动物肠线（蚕胶丝除外）制品

第四十一章　生皮（毛皮除外）及皮革………（229）
第四十二章　皮革制品；鞍具及挽具；旅行用品、手提包及类似容器；动物肠线（蚕胶丝除外）制品………（235）
第四十三章　毛皮、人造毛皮及其制品………（239）

第九类　木及木制品；木炭；软木及软木制品；稻草、秸秆、针茅或其他编结材料制品；篮筐及柳条编结品

第四十四章　木及木制品；木炭………………（242）
第四十五章　软木及软木制品…………………（252）
第四十六章　稻草、秸秆、针茅或其他编结材料制品；篮筐及柳条编结品…………………………………（253）

第十类　木浆及其他纤维状纤维素浆；回收（废碎）纸或纸板；纸、纸板及其制品

第四十七章　木浆及其他纤维状纤维素浆；回收（废碎）纸或纸板…………（255）

Chapter 35 Albuminoidal substances; modified starches; glues; enzymes ………………… (190)
Chapter 36 Explosives; pyrotechnic products; matches; pyrophoric alloys; certain combustible preparations ……………………… (192)
Chapter 37 Photographic or cinematographic goods ……………………………………… (194)
Chapter 38 Miscellaneous chemical products …………… (198)

SECTION Ⅶ PLASTICS AND ARTICLES THEREOF; RUBBER AND ARTICLES THEREOF

Chapter 39 Plastics and articles thereof ………………… (207)
Chapter 40 Rubber and articles thereof ………………… (220)

SECTION Ⅷ RAW HIDES AND SKINS, LEATHER, FURSKINS AND ARTICLES THEREOF; SADDLERY AND HARNESS; TRAVEL GOODS, HANDBAGS AND SIMILAR CONTAINERS; ARTICLES OF ANIMAL GUT (OTHER THAN SILK-WORM GUT)

Chapter 41 Raw hides and skins (other than furskins) and leather ……………………………… (229)
Chapter 42 Articles of leather; saddlery and harness; travel goods, handbags and similar containers; articles of animal gut (other than silk-worm gut) ……………… (235)
Chapter 43 Furskins and artificial fur; manufactures thereof ……………………………… (239)

SECTION Ⅸ WOOD AND ARTICLES OF WOOD; WOOD CHARCOAL; CORK AND ARTICLES OF CORK; MANUFACTURES OF STRAW, OF ESPARTO OR OF OTHER PLAITING MATERIALS; BASKETWARE AND WICKERWORK

Chapter 44 Wood and articles of wood; wood charcoal ……………………………… (242)
Chapter 45 Cork and articles of cork ………………… (252)
Chapter 46 Manufactures of straw, of esparto or of other plaiting materials; basketware and wickerwork ……………………… (253)

SECTION Ⅹ PULP OF WOOD OR OF OTHER FIBROUS CELLULOSIC MATERIAL; RECOVERED (WASTE AND SCRAP) PAPER OR PAPERBOARD; PAPER AND PAPERBOARD AND ARTICLES THEREOF

Chapter 47 Pulp of wood or of other fibrous cellulosic material; recovered (waste and scrap) paper or paperboard ……………………… (255)

第四十八章　纸及纸板；纸浆；纸或纸板制品 …… (257)
第四十九章　书籍、报纸、印刷图画及其他印刷品；手稿、打字稿及设计图纸 …… (271)

第十一类　纺织原料及纺织制品
第五十章　蚕丝 …… (281)
第五十一章　羊毛、动物细毛或粗毛；马毛纱线及其机织物 …… (283)
第五十二章　棉花 …… (286)
第五十三章　其他植物纺织纤维；纸纱线及其机织物 …… (296)
第五十四章　化学纤维长丝 …… (299)
第五十五章　化学纤维短纤 …… (305)
第五十六章　絮胎、毡呢及无纺织物；特种纱线；线、绳、索、缆及其制品 …… (312)
第五十七章　地毯及纺织材料的其他铺地制品 …… (317)
第五十八章　特种机织物；簇绒织物；花边；装饰毯；装饰带；刺绣品 …… (319)
第五十九章　浸渍、涂布、包覆或层压的纺织物；工业用纺织制品 …… (324)
第六十章　针织物及钩编织物 …… (330)
第六十一章　针织或钩编的服装及衣着附件 …… (333)
第六十二章　非针织或非钩编的服装及衣着附件 …… (342)
第六十三章　其他纺织制成品；成套物品；旧衣着及旧纺织品；碎织物 …… (352)

第十二类　鞋、帽、伞、杖、鞭及其零件；已加工的羽毛及其制品；人造花；人发制品
第六十四章　鞋靴、护腿和类似品及其零件 …… (358)
第六十五章　帽类及其零件 …… (362)
第六十六章　雨伞、阳伞、手杖、鞭子、马鞭及其零件 …… (364)
第六十七章　已加工羽毛、羽绒及其制品；人造花；人发制品 …… (365)

Chapter 48　Paper and paperboard; articles of paper pulp, of paper or of paperboard …… (257)
Chapter 49　Printed books, newspapers, pictures and other products of the printing industry; manuscripts, typescripts and plans …… (271)

SECTION XI　TEXTILES AND TEXTILE ARTICLES
Chapter 50　Silk …… (281)
Chapter 51　Wool, fine or coarse animal hair; horsehair yarn and woven fabric …… (283)
Chapter 52　Cotton …… (286)
Chapter 53　Other vegetable textile fibres; paper yarn and woven fabrics of paper yarn …… (296)
Chapter 54　Man-made filaments …… (299)
Chapter 55　Man-made staple fibres …… (305)
Chapter 56　Wadding, felt and nonwovens; special yarns; twine, cordage, ropes and cables and articles thereof …… (312)
Chapter 57　Carpets and other textile floor coverings …… (317)
Chapter 58　Special woven fabrics; tufted textile fabrics; lace; tapestries; trimmings; embroidery …… (319)
Chapter 59　Impregnated, coated, covered or laminated textile fabrics; textile articles of a kind suitable for industrial use …… (324)
Chapter 60　Knitted or crocheted fabrics …… (330)
Chapter 61　Articles of apparel and clothing accessories, knitted or crocheted …… (333)
Chapter 62　Articles of apparel and clothing accessories, not knitted or crocheted …… (342)
Chapter 63　Other made up textile articles; sets; worn clothing and worn textile articles; rags …… (352)

SECTION XII　FOOTWEAR, HEADGEAR, UMBRELLAS, SUN UMBRELLAS, WALKING-STICKS, SEAT-STICKS, WHIPS, RIDING-CROPS AND PARTS THEREOF; PREPARED FEATHERS AND ARTICLES MADE THEREWITH; ARTIFICIAL FLOWERS; ARTICLES OF HUMAN HAIR
Chapter 64　Footwear, gaiters and the like; parts of such articles …… (358)
Chapter 65　Headgear and parts thereof …… (362)
Chapter 66　Umbrellas, sun umbrellas, walking-sticks, seat-sticks, whips, riding-crops and parts thereof …… (364)
Chapter 67　Prepared feathers and down and articles made of feathers or of down; artificial flowers; articles of human hair …… (365)

第十三类　石料、石膏、水泥、石棉、云母及类似材料的制品；陶瓷产品；玻璃及其制品

第六十八章　石料、石膏、水泥、石棉、云母及类似材料的制品……………（367）
第六十九章　陶瓷产品……………………………（373）
第 七 十 章　玻璃及其制品……………………（377）

第十四类　天然或养殖珍珠、宝石或半宝石、贵金属、包贵金属及其制品；仿首饰；硬币

第七十一章　天然或养殖珍珠、宝石或半宝石、贵金属、包贵金属及其制品；仿首饰；硬币………………（384）

第十五类　贱金属及其制品

第七十二章　钢铁………………………………（395）
第七十三章　钢铁制品…………………………（411）
第七十四章　铜及其制品………………………（421）
第七十五章　镍及其制品………………………（429）
第七十六章　铝及其制品………………………（433）
第七十七章　（保留为税则将来所用）
第七十八章　铅及其制品………………………（439）
第七十九章　锌及其制品………………………（442）
第 八 十 章　锡及其制品………………………（445）
第八十一章　其他贱金属、金属陶瓷及 制品……………………………………（449）
第八十二章　贱金属工具、器具、利口器、餐匙、餐叉及其零件……………（452）
第八十三章　贱金属杂项制品…………………（458）

第十六类　机器、机械器具、电气设备及其零件；录音机及放声机、电视图像、声音的录制和重放设备及其零件、附件

第八十四章　核反应堆、锅炉、机器、机械器具及其零件…………………（464）
第八十五章　电机、电气设备及其零件；录音机及放声机、电视图像、声

SECTION XIII　ARTICLES OF STONE, PLASTER, CEMENT, ASBESTOS, MICA OR SIMILAR MATERIALS; CERAMIC PRODUCTS; GLASS AND GLASSWARE

Chapter 68　Articles of stone, plaster, cement, asbestos, mica or similar materials ……………………… (367)
Chapter 69　Caramic products ……………………………… (373)
Chapter 70　Glass and glassware ………………………… (377)

SECTION XIV　NATURAL OR CULTURED PEARLS, PRECIOUS OR SEMI-PRECIOUS STONES, PRECIOUS METALS, METALS CLAD WITH PRECIOUS METAL, AND ARTICLES THEREOF; IMITATION JEWELLERY; COIN

Chapter 71　Natural or cultured pearls, precious or semi-precious stones, precious metals, metals clad with precious metal, and articles thereof; imitation jewellery; coin ………… (384)

SECTION XV　BASE METALS AND ARTICLES OF BASE METAL

Chapter 72　Iron and steel ……………………………… (395)
Chapter 73　Articles of iron or steel ……………………… (411)
Chapter 74　Copper and articles thereof ………………… (421)
Chapter 75　Nickel and articles thereof ………………… (429)
Chapter 76　Aluminium and articles thereof …………… (433)
Chapter 77　(Reserved for possible futuroe use in the Tariff)
Chapter 78　Lead and articles thereof …………………… (439)
Chapter 79　Zinc and articles thereof …………………… (442)
Chapter 80　Tin and articles thereof …………………… (445)
Chapter 81　Other base metals; cermets; artiles thereof ………………………………………… (449)
Chapter 82　Tools, implements, cutlery, spoons and forks, of base metal ………………………… (452)
Chapter 83　Miscellaneous articles of base metal ………… (458)

SECTION XVI　MACHINERY AND MECHANICAL APPLIANCES; ELECTRICAL EQUIPMENT; PARTS THEREOF; SOUND RECORDERS AND REPRODUCERS, TELEVISION IMAGE AND SOUND RECORDERS AND REPRODUCERS, AND PARTS AND ACCESSORIES OF SUCH ARTICLES

Chapter 84　Nuclear reactors, boilers, machinery and mechanical appliances; parts thereof ………………………………………… (464)
Chapter 85　Electrical machinery and equipment and parts thereof; sound recorders and

音的录制和重放设备及其零件、附件……………………………… (514)

reproducers, television image and sound recorders and reproducers, and parts and accessories of such articles …………………………………… (514)

第十七类　车辆、航空器、船舶及有关运输设备

SECTION XⅦ　VEHICLES, AIRCRAFT, VESSELS AND ASSOCIATED TRANSPORT EQUIPMENT

第八十六章　铁道及电车道机车、车辆及其零件；铁道及电车道轨道固定装置及其零件、附件；各种机械（包括电动机械）交通信号设备……………………………… (546)

Chapter 86　Railway or tramway locomotives, rolling-stock and parts thereof; railway or tramway track fixtures and fittings and parts thereof; mechanical (including electro-mechanical) traffic signalling equipment of all kinds ……………………… (546)

第八十七章　车辆及其零件、附件，但铁道及电车道车辆除外……………… (550)

Chapter 87　Vehicles other than railway or tramway rolling-stock, and parts and accessories thereof ……………………………………… (550)

第八十八章　航空器、航天器及其零件……… (563)

Chapter 88　Aircraft, spacecraft and parts thereof ………………………………………………… (563)

第八十九章　船舶及浮动结构体……………… (565)

Chapter 89　Ships, boats and floating structures ……… (565)

第十八类　光学、照相、电影、计量、检验、医疗或外科用仪器及设备、精密仪器及设备；钟表；乐器；上述物品的零件、附件

SECTION XⅧ　OPTICAL, PHOTOGRAPHIC, CINEMATOGRAPHIC, MEASURING, CHECKING, PRECISION, MEDICAL OR SURGICAL INSTRUMENTS AND APPARATUS; CLOCKS AND WATCHES; MUSICAL INSTRUMENTS; PARTS AND ACCESSORIES THEREOF

第 九 十 章　光学、照相、电影、计量、检验、医疗或外科用仪器及设备、精密仪器及设备；上述物品的零件、附件……………………… (568)

Chapter 90　Optical, photographic, cinematographic, measuring, checking, precision, medical or surgical instruments and apparatus; parts and accessories thereof ………………… (568)

第九十一章　钟表及其零件…………………… (586)

第九十二章　乐器及其零件、附件…………… (590)

Chapter 91　Clocks and watches and parts thereof ……… (586)

Chapter 92　Musical instruments; parts and accessories of such articles ……………………… (590)

第十九类　武器、弹药及其零件、附件

SECTION XIX　ARMS AND AMMUNITION; PARTS AND ACCESSORIES THEREOF

第九十三章　武器、弹药及其零件、附件…… (593)

Chapter 93　Arms and ammunition; parts and accessories thereof ………………………… (593)

第二十类　杂项制品

SECTION XX　MISCELLANEOUS MANUFACTURED ARTICLES

第九十四章　家具；寝具、褥垫、弹簧床垫、软座垫及类似的填充制品；未列名灯具及照明装置；发光标志、发光铭牌及类似品；活动房屋……………………………… (596)

Chapter 94　Furniture; bedding, mattresses, mattress supports, cushions and similar stuffed furnishings; lamps and lighting fittings, not elsewhere specified or included; illuminated signs, illuminated name-plates and the like; prefabricated buildings ……… (596)

第九十五章　玩具、游戏品、运动用品及其零件、附件…………………………（601）
第九十六章　杂项制品…………………………（606）

第二十一类　艺术品、收藏品及古物

第九十七章　艺术品、收藏品及古物…………（613）

Chapter 95 Toys, games and sports requisites; parts and accessories thereof ………………………… (601)
Chapter 96 Miscellaneous manufactured articles ………… (606)

SECTION XXI WORKS OF ART, COLLECTORS' PIECES AND ANTIQUES

Chapter 97 Works of art, collectors' pieces and antiques ……………………………………… (613)

归类总规则

货品在本税则目录上的归类，应遵循以下原则：

规则一 类、章及分章的标题，仅为查找方便而设；具有法律效力的归类，应按税目条文和有关类注或章注确定，如税目、类注或章注无其他规定，按以下规则确定。

规则二 （一）税目所列货品，应视为包括该项货品的不完整品或未制成品，只要在进口或出口时该项不完整品或未制成品具有完整品或制成品的基本特征；还应视为包括该项货品的完整品或制成品（或按本款可作为完整品或制成品归类的货品）在进口或出口时的未组装件或拆散件。

（二）税目中所列材料或物质，应视为包括该种材料或物质与其他材料或物质混合或组合的物品。税目所列某种材料或物质构成的货品，应视为包括全部或部分由该种材料或物质构成的货品。由一种以上材料或物质构成的货品，应按规则三归类。

规则三 当货品按规则二（二）或由于其他原因看起来可归入两个或两个以上税目时，应按以下规则归类：

（一）列名比较具体的税目，优先于列名一般的税目。但是，如果两个或两个以上税目都仅述及混合或组合货品所含的某部分材料或物质，或零售的成套货品中的某些货品，即使其中某个税目对该货品描述得更为全面、详细，这些货品在有关税目的列名应视为同样具体。

（二）混合物、不同材料构成或不同部件组成的组合物以及零售的成套货品，如果不能按照规则三（一）归类时，在本款可适用的条件下，应按构成货品基本特征的材料或部件归类。

（三）货品不能按照规则三（一）或（二）归类时，应按号列顺序归入其可归入的最末一个税目。

GENERAL RULES FOR THE INTERPRETATION OF THE HARMONIZED SYSTEM

Classification of goods in the Nomenclature shall be governed by the following Rules:

1. The titles of Sections, Chapters and sub-Chapters are provided for ease of reference only; for legal purposes, classification shall be determined according to the terms of the headings and any relative Section or Chapter Notes and provided such headings or Notes do not otherwise require, according to the following provisions.

2. (a) Any reference in a heading to an article shall be taken to include a reference to that article incomplete or unfinished, provided that, as presented, the incomplete or unfinished article has the essential character of the complete or finished article. It shall also be taken to include a reference to that article complete or finished (or falling to be classified as complete or finished by virtue of this Rule), presented unassembled or disassembled.

   (b) Any reference in a heading to a material or substance shall be taken to include a reference to mixtures or combinations of that material or substance with other materials or substances. Any reference to goods of a given material or substance shall be taken to include a reference to goods consisting wholly or partly of such material or substance. The classification of goods consisting of more than one material or substance shall be according to the principles of Rule 3.

3. When by application of Rule 2 (b) or for any other reason, goods are, *prima facie*, classifiable under two or more headings, classification shall be effected as follows:

   (a) The heading which provides the most specific description shall be preferred to headings providing a more general description. However, when two or more headings each refer to part only of the materials or substances contained in mixed or composite goods or to part only of the items in a set put up for retail sale, those heading are to be regarded as equally specific in relation to those goods, even if one of them gives a more complete or precise description of the goods.

   (b) Mixtures, composite goods consisting of different materials or made up of different components, and goods put up in sets for retail sale, which cannot be classified by reference to Rule 3 (a), shall be classified as they consisted of the material or component which gives them their essential character, insofar as this criterion is applicable.

   (c) When goods cannot be classified by reference to Rule 3 (a) or Rule 3 (b), they shall be classified under the heading which occurs last in numerical order among those which equally merit consideration.

规则四　根据上述规则无法归类的货品，应归入与其最相类似的货品的税目。

4．Goods which cannot be classified in accordance with the above Rules shall be classified under the heading appropriate to the goods to which they are most akin.

规则五　除上述规则外，本规则适用于下列货品的归类：

（一）制成特殊形状仅适用于盛装某个或某套物品并适合长期使用的照像机套、乐器盒、枪套、绘图仪器盒、项链盒及类似容器，如果与所装物品同时进口或出口，并通常与所装物品一同出售的，应与所装物品一并归类。但本款不适用于本身构成整个货品基本特征的容器。

（二）除规则五（一）规定的以外，与所装货品同时进口或出口的包装材料或包装容器，如果通常是用来包装这类货品的，应与所装货品一并归类。但明显可重复使用的包装材料和包装容器可不受本款限制。

5．In addition to the foregoing Rules，the following Rules shall apply in respect of the goods referred to therein：

（a）Camera cases，musical instrument cases，gun cases，drawing instrument cases，necklace cases and similar containers，specially shaped or fitted to contain a specific article or set of articles，suitable for long-term use and presented with the articles for which they are intended，shall be classified with such articles when of a kind normally sold therewith. This Rule does not，however，apply to containers which give the whole its essential character；

（b）Subject to the provisions of Rule 5（a）above，packing materials and packing containers presented with the goods therein shall be classified with the goods if they are of a kind normally used for packing such goods. However，this provision is not binding when such packing materials or packing containers are clearly suitable for repetitive use.

规则六　货品在某一税目项下各子目的法定归类，应按子目条文或有关的子目注释以及以上各条规则来确定，但子目的比较只能在同一数级上进行。除本税则目录条文另有规定的以外，有关的类注、章注也适用于本规则。

6．For legal purposes，the classification of goods in the subheadings of a heading shall be determined according to the terms of those subheadings and any related Subheading Notes and ，*mutatis mutandis*，to the above Rules，on the understanding that only subheadings at the same level are comparable. For the purposes of this Rule the relative Section and Chapter Notes also apply，unless the context otherwise requires.

第一类

活动物；动物产品

注释：

一、本类所称的各属种动物，除条文另有规定的以外，均包括其幼仔在内。

二、除条文另有规定的以外，本目录所称干的产品，均包括经脱水、蒸发或冷冻干燥的产品。

SECTION Ⅰ

LIVE ANIMALS; ANIMAL PRODUCTS

Notes:

1. Any reference in this Section to a particular genus or species of an animal, except where the context otherwise requires, includes a reference to the young of that genus or species.

2. Except where the context otherwise requires, throughout the Nomenclature any reference to "dried" products also covers products which have been dehydrated, evaporated or freeze-dried.

第 一 章
活 动 物

注释：

本章包括所有活动物，但下列各项除外：

一、税号 03.01、03.06 或 03.07 的鱼、甲壳动物、软体动物及其他水生无脊椎动物；

二、税号 30.02 的培养微生物及其他产品；

三、税号 95.08 的动物。

Chapter 1
Live animals

Notes:

This Chapter covers all live animals except:

(a) Fish and crustaceans, molluscs and other aquatic invertebrates, of heading No.03.01, 03.06 or 03.07;

(b) Cultures of micro-organisms and other products of heading No.30.02; and

(c) Animals of heading No.95.08.

序号 No.	税则号列 Tariff Item	货 品 名 称	最惠国税率(%) M.F.N.	协定税率(%) Agreement				普通税率(%) Gen.	Article Description
				*东盟 ASEAN	**曼谷协定 BA	香港 H.K.	澳门 Macao		
	01.01	马、驴、骡：							**Live horses, asses, mules and hinnies:**
		-改良种用：							-Pure-bred breeding:
1	0101.1010	---马	0	0				0	---Horses
2	0101.1020	---驴	0	0				0	---Asses

＊适用本栏东盟协定税率的国家有文莱达鲁萨兰国、柬埔寨王国、印度尼西亚共和国、老挝人民民主共和国、马来西亚、缅甸联邦、新加坡共和国、泰王国、越南社会主义共和国。

The tariff rates in the column"ASEAN"are applicable to Brunei Darussalam, the Kingdom of Cambodia, the Republic of Indonesia, the Lao People's Democratic Republic, Malaysia, the Union of Myanmar, the Republic of Singapore, the Kingdom of Thailand and the Socialist Republic of Viet Nam.

＊＊适用本栏曼谷协定税率的国家有大韩民国、斯里兰卡民主社会主义共和国、孟加拉人民共和国、印度共和国、老挝人民民主共和国。

The tariff rates in the column"BA"are applicable to the Republic of Korea, the Democratic Socialist Republic of Sri Lanka, the People's Republic of Bangladesh, the Republic of India, the Lao People's Democratic Republic.

△ 为拟生产产品。此类产品须经《〈内地与香港关于建立更紧密经贸关系的安排〉补充协议》、《〈内地与澳门关于建立更紧密经贸关系的安排〉补充协议》规定的程序批准后，按照零关税进口。

Products marked"△", release the relevant imports on zero tariff basis in line with Supplementary Protocol to the Mainland and Macao Closer Economic Partnership Arrangement, or Supplementary Protocol to the Mainland and Hong Kong closer Economic Partnership Arrangement.

序号 No.	税则号列 Tariff Item	货品名称	最惠国税率(%) M.F.N.	协定税率(%) Agreement 东盟 ASEAN	曼谷协定 BA	香港 H.K.	澳门 Macao	普通税率(%) Gen.	Article Description
		-其他:							-Other:
3	0101.9010	---马	10	0				30	---Horses
4	0101.9090	---驴、骡	10	0				30	---Asses, mules and hinnies
	01.02	**牛:**							**Live bovine animals:**
5	0102.1000	-改良种用	0	0				0	-Pure-bred breeding
6	0102.9000	-其他	10	0				30	-Other
	01.03	**猪:**							**Live swine:**
7	0103.1000	-改良种用	0	0				0	-Pure-bred breeding
		-其他:							-Other:
		--重量在50公斤以下:							--Weighing less than 50kg:
8	0103.9110	---重量在10公斤以下	10	0				50	---Weighing less than 10kg
9	0103.9120	---重量在10公斤及以上,但在50公斤以下	10	0				50	---Weighing 10kg or more, but less than 50kg
10	0103.9200	--重量在50公斤及以上	10	见表1				50	--Weighing 50kg or more
	01.04	**绵羊、山羊:**							**Live sheep and goats:**
		-绵羊:							-Sheep:
11	0104.1010	---改良种用	0	0				0	---Pure-bred breeding
12	0104.1090	---其他	10	0				50	---Other
		-山羊:							-Goats:
13	0104.2010	---改良种用	0	0				0	---Pure-bred breeding
14	0104.2090	---其他	10	0				50	---Other
	01.05	**家禽,即鸡、鸭、鹅、火鸡及珍珠鸡:**							**Live poultry, that is to say, fowls of the species *Gallus domesticus*, ducks, geese, turkeys and guinea fowls:**
		-重量不超过185克:							-Weighing not more than 185g:
		--鸡:							--Fowls of the species *Gallus domesticus*:
15	0105.1110	---改良种用	0	0				0	---Pure-bred breeding
16	0105.1190	---其他	10	见表1				50	---Other
		--火鸡:							--Turkeys:
17	0105.1210	---改良种用	0	0				0	---Pure-bred breeding
18	0105.1290	---其他	10	0				50	---Other
		--其他:							--Other:
19	0105.1910	---改良种用	0	0				0	---Pure-bred breeding
20	0105.1990	---其他	10	0				50	---Other
		-其他:							-Other:
		--鸡,重量不超过2000克:							--Fowls of the species *Gallus domesticus*, weighing not more than 2000g:
21	0105.9210	---改良种用	0	见表1				0	---Pure-bred breeding
22	0105.9290	---其他	10	见表1				50	---Other
		--鸡,重量超过2000克:							--Fowls of the species *Gallus domesticus*, weighing more than 2000g:
23	0105.9310	---改良种用	0	见表1				0	---Pure-bred breeding
24	0105.9390	---其他	10	见表1				50	---Other
		--其他:							--Other:
25	0105.9910	改良种用	0	见表1				0	---Pure-bred breeding
		---其他:							---Other:

序号 No.	税则号列 Tariff Item	货品名称	最惠国税率(%) M.F.N.	协定税率(%)Agreement 东盟 ASEAN	曼谷协定 BA	香港 H.K.	澳门 Macao	普通税率(%) Gen.	Article Description
26	0105.9991	----鸭	10	见表1				50	----Ducks
27	0105.9992	----鹅	10	见表1				50	----Geese
28	0105.9993	----珍珠鸡	10	见表1				50	----Guinea fowls
29	0105.9994	----火鸡	10	见表1				50	----Turkeys
	01.06	**其他活动物：**							**Other live animals:**
		-哺乳动物：							-Mammals:
		--灵长目：							--Primates:
30	0106.1110	---改良种用	0	见表1				0	---Pure-bred breeding
31	0106.1190	---其他	10	见表1				50	---Other
32	0106.1200	--鲸、海豚及鼠海豚(鲸目哺乳动物)；海牛及儒艮(海牛目哺乳动物)	10	见表1				50	--Whales, dolphins and porpoises (mammals of the order Cetacea); manatees and dugongs (mammals of the order Sirenia)
		--其他：							--Other:
33	0106.1910	---改良种用	0	见表1				0	---Pure-bred breeding
34	0106.1920	---食用	10	见表1				50	---For human consumption
35	0106.1990	---其他	10	见表1				50	---Other
		-爬行动物(包括蛇及龟鳖)：							-Reptiles (including snakes and turtles):
		---改良种用：							---Pure-bred breeding:
36	0106.2011	----鳄鱼苗	0	见表1				0	----Crocodiles for cultivation
37	0106.2019	----其他	0	见表1				0	----Other
38	0106.2020	---食用	10	见表1				50	---For human consumption
39	0106.2090	---其他	10	见表1				50	---Other
		-鸟：							-Birds:
		--猛禽：							--Birds of prey:
40	0106.3110	---改良种用	0	见表1				0	---Pure-bred breeding
41	0106.3190	---其他	10	见表1				50	---Other
		--鹦形目(包括普通鹦鹉、长尾鹦鹉、金刚鹦鹉及美冠鹦鹉)：							--Psittaciformes (including parrots, parakeets, macaws and cockatoos):
42	0106.3210	---改良种用	0	见表1				0	---Pure-bred breeding
43	0106.3290	---其他	10	见表1				50	---Other
		--其他：							--Other:
44	0106.3910	---改良种用	0	见表1				0	---Pure-bred breeding
		---食用：							---For human consumption:
45	0106.3921	----乳鸽	10	见表1				50	----Squabs
46	0106.3922	----鸵鸟	10	见表1				50	----Ostrich
47	0106.3923	----野鸭	10	见表1				50	----Teals
48	0106.3929	----其他	10	见表1				50	----Other
49	0106.3990	---其他	10	见表1				50	Other
		-其他：							-Other:
		---改良种用：							---Pure-bred breeding:
50	0106.9011	----蛙苗	0	见表1				0	----Tadpole and young frogs
51	0106.9019	----其他	0	见表1				0	----Other
52	0106.9020	---食用	10	见表1	9			50	---For human corsumption
53	0106.9090	---其他	10	见表1	9			50	---Other

第二章
肉及食用杂碎

Chapter 2
Meat and edible meat offal

注释：

本章不包括：

一、税号02.01至02.08或02.10的不适合供人食用的产品；

二、动物的肠、膀胱、胃（税号05.04）或动物血（税号05.11、30.02）；

三、税号02.09所列产品以外的动物脂肪（第十五章）。

Notes:

1. This Chapter does not cover:
   (a) Products of the kinds described in headings No.02.01 to 02.08 or 02.10, unfit or unsuitable for human consumption;
   (b) Guts, bladders or stomachs of animals (heading No.05.04) or animal blood (heading No.05.11 or 30.02); or
   (c) Animal fat, other than products of heading No.02.09 (Chapter15).

序号 No.	税则号列 Tariff Item	货品名称	最惠国税率(%) M.F.N.	协定税率(%) Agreement 东盟 ASEAN	曼谷协定 BA	香港 H.K.	澳门 Macao	普通税率(%) Gen.	Article Description
	02.01	**鲜、冷牛肉：**							**Meat of bovine animals, fresh or chilled:**
54	0201.1000	-整头及半头	20	5				70	-Carcasses and half-carcasses
55	0201.2000	-带骨肉	12	5				70	-Other cuts with bone in
56	0201.3000	-去骨肉	12	5				70	-Boneless
	02.02	**冻牛肉：**							**Meat of bovine animals, frozen:**
57	0202.1000	-整头及半头	25	见表1				70	-Carcasses and half-carcasses
58	0202.2000	-带骨肉	12	见表1				70	-Other cuts with bone in
59	0202.3000	-去骨肉	12	见表1				70	-Boneless
	02.03	**鲜、冷、冻猪肉：**							**Meat of swine, fresh, chilled or frozen:**
		-鲜或冷的：							-Fresh or chilled:
		--整头及半头：							--Carcasses and half-carcasses:
60	0203.1110	---乳猪	20	见表1				70	---Sucking pig
61	0203.1190	---其他	20	见表1				70	---Other
62	0203.1200	--带骨的前腿、后腿及其肉块	20	见表1				70	--Hams, shoulders and cuts thereof, with bone in
63	0203.1900	--其他	20	见表1				70	--Other
		-冻的：							-Frozen:
		--整头及半头：							--Carcasses and half-carcasses:
64	0203.2110	---乳猪	12	见表1				70	---Sucking pig
65	0203.2190	---其他	12	见表1				70	---Other
66	0203.2200	--带骨的前腿、后腿及其肉块	12	见表1				70	--Hams, shoulders and cuts thereof, with bone in
67	0203.2900	--其他	12	0				70	--Other
	02.04	**鲜、冷、冻绵羊肉或山羊肉：**							**Meat of sheep or goats, fresh, chilled or frozen:**
68	0204.1000	-鲜或冷的整头及半头羔羊	15	5				70	-Carcasses and half-carcasses of lamb, fresh or chilled
		-其他鲜或冷的绵羊肉：							-Other meat of sheep, fresh or chilled:
69	0204.2100	--整头及半头	23	5				70	--Carcasses and half-carcasses
70	0204.2200	--带骨肉	15	5				70	--Other cuts with bone in

序号 No.	税则号列 Tariff Item	货品名称	最惠国税率(%) M.F.N.	协定税率(%)Agreement 东盟 ASEAN	曼谷协定 BA	香港 H.K.	澳门 Macao	普通税率(%) Gen.	Article Description
71	0204.2300	--去骨肉	15	5				70	--Boneless
72	0204.3000	-冻的整头及半头羔羊	15	5				70	-Carcasses and half-carcasses of lamb, frozen
		-其他冻的绵羊肉:							-Other meat of sheep, frozen:
73	0204.4100	--整头及半头	23	5				70	--Carcasses and half-carcasses
74	0204.4200	--带骨肉	12	0				70	--Other cuts with bone in
75	0204.4300	--去骨肉	15	5				70	--Boneless
76	0204.5000	-山羊肉	20	5				70	-Meat of goats
	02.05	**鲜、冷、冻马、驴、骡肉:**							**Meat of horses, asses, mules or hinnies, fresh, chilled or frozen:**
77	0205.0000	鲜、冷、冻马、驴、骡肉	20	5				70	Meat of horses, asses, mules or hinnies, fresh, chilled or frozen
	02.06	**鲜、冷、冻牛、猪、绵羊、山羊、马、驴、骡的食用杂碎:**							**Edible offal of bovine animals, swine, sheep, goats, horses, asses, mules or hinnies, fresh, chilled or frozen:**
78	0206.1000	-鲜、冷牛杂碎	12	0				70	-Of bovine animals, fresh or chilled
		-冻牛杂碎:							-Of bovine animals, frozen:
79	0206.2100	--舌	12	0				70	--Tongues
80	0206.2200	--肝	12	0				70	--Livers
81	0206.2900	--其他	12	0				70	--Other
82	0206.3000	-鲜、冷猪杂碎	20	5				70	-Of swine, fresh of chilled
		-冻猪杂碎:							-Of swine, frozen:
83	0206.4100	--肝	20	5				70	--Livers
84	0206.4900	--其他	12	0				70	--Other
85	0206.8000	-其他鲜或冷杂碎	20	5				70	-Other, fresh or chilled
86	0206.9000	-其他冻杂碎	18	5				70	-Other, frozen
	02.07	**税号01.05所列家禽的鲜、冷、冻肉及食用杂碎:**							**Meat and edible offal, of the poultry of heading No.01.05, fresh, chilled or frozen:**
		-鸡:							-Of fowls of the species *Gallus domesticus*:
87	0207.1100	--整只,鲜或冷的	20	见表1				70	--Not cut in pieces, fresh or chilled
88	0207.1200	--整只,冻的	见表4	见表1				见表4	--Not cut in pieces, frozen
		--块及杂碎,鲜或冷的:							--Chicken cut and offal, frozen:
		---块:							---Cut:
89	0207.1311	----带骨的	20	见表1				70	----With bone
90	0207.1319	----其他	20	见表1				70	----Other
		---杂碎:							---Offal:
91	0207.1321	----翼(不包括翼尖)	20	见表1				70	----Midjoint wing
92	0207.1329	----其他	20	见表1				70	----Other
		--块及杂碎,冻的:							--Cuts and offal, frozen:
		---块:							---Cut:
93	0207.1411	----带骨的	见表4	见表1				见表4	----With bone
94	0207.1419	----其他	见表4	见表1				见表4	----Other
		---杂碎:							---Offal:

序号 No.	税则号列 Tariff Item	货品名称	最惠国税率(%) M.F.N.	协定税率(%) Agreement 东盟 ASEAN	曼谷协定 BA	香港 H.K.	澳门 Macao	普通税率(%) Gen.	Article Description
95	0207.1421	----翼(不包括翼尖)	见表 4	见表 1				见表 4	----Midjoint wing
96	0207.1422	----鸡爪	见表 4	见表 1				见表 4	----Chicken claw
97	0207.1429	----其他	见表 4	见表 1				见表 4	----Other
		-火鸡:							-Of turkeys:
98	0207.2400	--整只,鲜或冷的	20	见表 1				70	--Not cut in pieces,fresh or chilled
99	0207.2500	--整只,冻的	20	见表 1				70	--Not cut in pieces,frozen
100	0207.2600	--块及杂碎,鲜或冷的	20	见表 1				70	--Cuts and offal,fresh or chilled
101	0207.2700	--块及杂碎,冻的	10	见表 1				70	--Cuts and offal,frozen
		-鸭、鹅或珍珠鸡:							-Of ducks,geese or guinea fowls:
		--整只,鲜或冷的:							--Not cut in pieces,fresh or chilled:
102	0207.3210	---鸭	20	见表 1				70	---Ducks
103	0207.3220	---鹅	20	见表 1				70	---Geese
104	0207.3230	---珍珠鸡	20	见表 1				70	---Guinea fowls
		--整只,冻的:							--Not cut in pieces,frozen:
105	0207.3310	---鸭	20	见表 1				70	---Ducks
106	0207.3320	---鹅	20	见表 1				70	---Geese
107	0207.3330	---珍珠鸡	20	见表 1				70	---Guinea fowls
108	0207.3400	--肥肝,鲜或冷的	20	见表 1				70	--Fatty livers(*foie gras*),fresh or chilled
		--其他,鲜或冷的:							--Other, fresh or chilled:
109	0207.3510	---鸭块及杂碎	20	见表 1				70	---Cuts and offal of ducks
110	0207.3520	---鹅块及杂碎	20	见表 1				70	---Cuts and offal of geese
111	0207.3530	---珍珠鸡块及杂碎	20	见表 1				70	---Cuts and offal of guinea fowls
		--其他,冻的:							--Other,frozen:
112	0207.3610	---鸭块及杂碎	20	5				70	---Cuts and offal of ducks
113	0207.3620	---鹅块及杂碎	20	5				70	---Cuts and offal of geese
114	0207.3630	---珍珠鸡块及杂碎	20	5				70	---Cuts and offal of guinea fowls
	02.08	**其他鲜、冷、冻肉及食用杂碎:**							**Other meat and edible meat offal, fresh,chilled or frozen:**
		-家兔或野兔的:							-Of rabbits or hares:
115	0208.1010	---鲜、冷兔肉,兔头除外	20	5				70	---Meat of rabbits,fresh or chilled,excluding head
116	0208.1020	---冻兔肉,兔头除外	20	5				70	---Meat of rabbits, frozen, excluding head
117	0208.1090	---其他	20	5				70	---Other
118	0208.2000	-田鸡腿	20	5				70	-Frogs' legs
119	0208.3000	-灵长目的	23	5				70	-Of primates
120	0208.4000	-鲸、海豚及鼠海豚(鲸目哺乳动物)的;海牛及儒艮(海牛目哺乳动物)的	23	5				70	-Of whales, dolphins and porpoises (mammals of the order Cetacea); of manatees and dugongs(mammals of the order Sirenia)
121	0208.5000	-爬行动物(包括蛇及龟鳖)的	23	5				70	-Of reptiles(including snakes and turtles)
		-其他:							-Other:
122	0208.9010	---乳鸽的	20	5				70	---Squabs
123	0208.9090	---其他	23	5				70	---Other

序号 No.	税则号列 Tariff Item	货 品 名 称	最惠国税率(%) M.F.N.	协定税率(%) Agreement 东盟 ASEAN	曼谷协定 BA	香港 H.K.	澳门 Macao	普通税率(%) Gen.	Article Description
	02.09	**未炼制或用其他方法提取的不带瘦肉的肥猪肉、猪脂肪及家禽脂肪,鲜、冷、冻、干、熏、盐腌或盐渍的:**							**Pig fat free of lean meat and poultry fat not rendered or otherwise extracted, fresh, chilled, frozen, salted, in brine, dried or smoked:**
124	0209.0000	未炼制或用其他方法提取的不带瘦肉的肥猪肉、猪脂肪及家禽脂肪,鲜、冷、冻、干、熏、盐腌或盐渍的	20	5				70	Pig fat free of lean meat and poultry fat not rendered or otherwise extracted, fresh, chilled, frozen, salted, in brine, dried or smoked
	02.10	**肉及食用杂碎,干、熏、盐腌或盐渍的;可供食用的肉或杂碎的细粉、粗粉:**							**Meat and edible meat offal, salted, in brine, dried or smoked; edible flours and meals of meat or meat offal:**
		-猪肉:							-Meat of swine:
		--带骨的前腿、后腿及其肉块:							--Hams, shoulders and cuts thereof, with bone in:
125	0210.1110	---带骨的腿	25	见表1				80	---Hams and shoulders, with bone in
126	0210.1190	---其他	25	见表1				80	---Other
127	0210.1200	--腹肉(五花肉)	25	见表1				80	--Bellies (streaky) and cuts thereof
128	0210.1900	--其他	25	见表1				80	--Other
129	0210.2000	-牛肉	25	见表1				80	-Meat of bovine animals
		-其他,包括可供食用的肉或杂碎的细粉、粗粉:							- Other, including edible flours and meals of meat and meat offal:
130	0210.9100	--灵长目的	25	见表1				80	--Of primates
131	0210.9200	--鲸、海豚及鼠海豚(鲸目哺乳动物)的;海牛及儒艮(海牛目哺乳动物)的	25	见表1				80	--Of whales, dolphins and porpoises (mammals of the order Cetacea); of manatees and dugongs (mammals of the order Sirenia)
132	0210.9300	--爬行动物(包括蛇及龟鳖)的	25	见表1				80	--Of reptiles (including snakes and turtles)
133	0210.9900	--其他	25	见表1				80	--Other

第 三 章
鱼、甲壳动物、软体动物及其他水生无脊椎动物

Chapter 3
Fish and crustaceans, molluscs and other aquatic invertebrates

注释：

一、本章不包括：

(一) 税目 01.06 的哺乳动物

(二) 税目 01.06 的哺乳动物的肉（税目 02.08 或 02.10）

(三) 因品种或鲜度不适合供人食用的死鱼（包括鱼肝及鱼卵）、死甲壳动物、死软体动物及其他死水生无脊椎动物（第五章）；不适合供人食用的鱼、甲壳动物、软体动物、其他水生无脊椎动物的粉、粒（税号 23.01）；

(四) 鲟鱼子酱及用鱼卵制成的鲟鱼子酱代用品（税号 16.04）。

二、本章所称“团粒”，是指直接挤压或加入少量粘合剂制成的粒状产品。

Notes:

1. This Chapter does not cover:

(a) Mammals of heading No.01.06;

(b) Meat of mammals of heading No.01.06 (heading No.02.08 or 02.10);

(c) Fish (including livers and roes thereof) or crustaceans, molluscs or other aquatic invertebrates, dead and unfit or unsuitable for human consumption by reason of either their species or their condition (Chapter 5); flours, meals or pellets of fish or of crustaceans, molluscs or other aquatic invertebrates, unfit for human consumption (heading No.23.01); or

(d) Caviar or caviar substitutes prepared from fish eggs (heading No.16.04).

2. In this Chapter, the term “pellets” means products which have been agglomerated either directly by compression or by the addition of a small quantity of binder.

序号 No.	税则号列 Tariff Item	货品名称	最惠国税率(%) M.F.N.	协定税率(%)Agreement 东盟 ASEAN	曼谷协定 BA	香港 H.K.	澳门 Macao	普通税率(%) Gen.	Article Description
	03.01	活鱼：							**Live fish:**
134	0301.1000	-观赏鱼	17.5	5				80	-Ornamental fish
		-其他活鱼：							-Other live fish:
		--鳟鱼：							--Trout (*Salmo trutta*, *Oncorhynchus mykiss*, *Oncorhynchus clarki*, *Oncorhynchus aguabonita*, *Oncorhynchus gilae*, *Oncorhynchus apache and Oncorhynchus chrysogaster*):
135	0301.9110	---鱼苗	0	0				0	---Fry
136	0301.9190	---其他	10.5	见表1	8			40	---Other
		--鳗鱼：							--Eels(*Anguilla spp.*):
137	0301.9210	---鱼苗	0	0				0	---Fry
138	0301.9290	---其他	10	0	6.7			40	---Other
		--鲤鱼：							--Carp:
139	0301.9310	---鱼苗	0	见表1				0	---Fry
140	0301.9390	---其他	10.5	见表1	8			40	---Other
		--其他：							--Other:
		---鱼苗：							---Fry:
141	0301.9911	----鲈鱼	0	0				0	----Of perches
142	0301.9912	----鲟鱼	0	0				0	----Of sturgeon
143	0301.9919	----其他	0	0				0	----Other
		---其他：							---Other:
144	0301.9991	----罗非鱼	10.5	0		0		40	----Tilapia

序号 No.	税则号列 Tariff Item	货品名称	最惠国税率(%) M.F.N.	协定税率(%)Agreement 东盟 ASEAN	曼谷协定 BA	香港 H.K.	澳门 Macao	普通税率(%) Gen.	Article Description
145	0301.9992	----鲀	10.5	0	8	0			----Puffer fish
146	0301.9999	----其他	10.5	0	8	0		40	----Other
	03.02	**鲜、冷鱼,但税号 03.04 的鱼片及其他鱼肉除外:**							**Fish, fresh or chilled, excluding fish fillets and other fish meat of heading No.03.04:**
		-鲑鱼,但鱼肝及鱼卵除外:							-Salmonidae, excluding livers and roes:
147	0302.1100	--鳟鱼	12	0				40	--Trout (*Salmo trutta*, *Oncorhynchus mykiss*, *Oncorhynchus clarki*, *Oncorhynchus aguabonita*, *Oncorhynchus gilae*, *Oncorhynchus apache* and *Oncorhynchus chrysogaster*)
		--大马哈鱼、大西洋鲑鱼及多瑙哲罗鱼:							--Pacific salmon, Atlantic salmon, and Danube salmon:
148	0302.1210	---大西洋鲑鱼	10	0				40	---Atlantic salmon
149	0302.1220	---大马哈鱼和多瑙哲罗鱼	10	0				40	---Pacific salmon (*Oncorhynchus nerka*, *Oncorhynchus gorbuscha*, *Oncorhynchus keta*, *Oncorhynchus tschawytscha*, *Oncorhynchus kisutch*, *Oncorhynchus masou* and *Oncorhynchus rhodurus*) and Danube salmon(*Hucho hucho*)
150	0302.1900	--其他	12	0	8			40	--Other
		-比目鱼,但鱼肝及鱼卵除外:							-Flat fish (*Pleuronectidae*, *Bothidae*, *Cynoglossidae*, *soleidae*, *Scophthalmidae and Citharidae*), excluding livers and roes:
151	0302.2100	--庸鲽鱼	12	0	9			40	--Halibut(*Reinhardtius hippoglossoides*, *Hippoglossus hippoglossus*, *Hippoglossus stenolepis*)
152	0302.2200	--鲽鱼	12	0	9			40	--Plaice(*Pleuronectes platessa*)
153	0302.2300	--鳎鱼	12	0	9			40	--Sole(*Solea spp.*)
154	0302.2900	--其他	12	0	9			40	--Other
		-金枪鱼、鲣鱼,但鱼肝及鱼卵除外:							-Tunas (*of the genus Thunnus*), skipjack or stripe-bellied bonito(*Euthynnus* (*Katsuwonus*) *pelamis*), excluding livers and roes:
155	0302.3100	--长鳍金枪鱼	12	0	9			40	--Albacore or longfinned tunas(*Thunnus alalunga*)
156	0302.3200	--黄鳍金枪鱼	12	0	9	0		40	--Yellowfin tunas (*Thunnus albacares*)
157	0302.3300	--鲣鱼	12	0	8			40	--Skipjack or stripe-bellied bonito
158	0302.3400	--大眼金枪鱼	12	0				40	--Bigeye tunas (*Thunnus obesus*)
159	0302.3500	--蓝鳍金枪鱼	12	0				40	--Bluefin tunas (*Thunnus thynnus*)

序号 No.	税则号列 Tariff Item	货品名称	最惠国税率(%) M.F.N.	协定税率(%)Agreement 东盟 ASEAN	曼谷协定 BA	香港 H.K.	澳门 Macao	普通税率(%) Gen.	Article Description
160	0302.3600	--南金枪鱼	12	0				40	--Southern bluefin tunas (*Thunnus maccoyii*)
161	0302.3900	--其他	12	0	8			40	--Other
162	0302.4000	-鲱鱼(大西洋鲱鱼、太平洋鲱鱼),但鱼肝及鱼卵除外	12	0	8			40	-Herrings (*Clupea harengus*, *Clupea pallasii*), excluding livers and roes
163	0302.5000	-鳕鱼,但鱼肝及鱼卵除外	12	0	8			40	- Cod (*Gadus morhua*. *Gadus ogac*, *Gadus macrocephalus*), excluding livers and roes
		-其他鱼,但鱼肝及鱼卵除外:							-Other fish, excluding livers and roes:
164	0302.6100	--沙丁鱼、黍鲱鱼	12	0	8			40	--Sardines (*Sardina pilchardus*, *Sardinops spp.*) sardinella (*Sardinella spp.*), brisling or sprats (*Sprattus sprattus*)
165	0302.6200	--黑线鳕鱼	12	0	8			40	--Haddock (*Melanogrammus aeglefinus*)
166	0302.6300	--绿青鳕鱼	12	0	8			40	--Coalfish(*Pollachius virens*)
167	0302.6400	--鲭鱼	12	0	8			40	--Mackerel (*Scomber scombrus*, *Scomber australasicus*, *Scomber Japonicus*)
168	0302.6500	--角鲨及其他鲨鱼	12	0	9	0		40	--Dogfish and other sharks
169	0302.6600	--鳗鱼	12	0	8	0		40	--Eels(*Anguilla spp.*)
		--其他:							--Other:
170	0302.6910	---带鱼	12	0	8	0		40	---Scabber fish(*Trichurius*)
171	0302.6920	---黄鱼	12	0	8	0		40	---Yellow croaker(*Pseudosicaena*)
172	0302.6930	---鲳鱼	12	0	8	0		40	---Butterfish(*Pamus*)
173	0302.6940	---罗非鱼	12	0				40	---Tilapia
174	0302.6950	---鲀	12	0	8	0			---Puffer fish
175	0302.6990	---其他	12	0	8	0		40	---Other
176	0302.7000	-鱼肝及鱼卵	12	0				50	-Livers and roes
	03.03	**冻鱼,但税号03.04的鱼片及其他鱼肉除外:**							**Fish, frozen, excluding fish fillets and other fish meat of heading No.03.04:**
		-大马哈鱼,但鱼肝及鱼卵除外:							-Pacific salmon (*Oncorhynchus nerka*, *Oncorhynchus gorbuscha*, *Oncorhynchus keta*, *Oncorhynchus tschawytscha*, *Oncorhynchus kisutch*, *Oncorhynchus masou and Oncorhynchus rhodurus*), excluding livers and roes
177	0303.1100	--红大马哈鱼	10	0	6.7			40	--Sockeye salmon (red salmon) (On corhynchus nerka)
178	0303.1900	--其他	10	0	6.7			40	--Other
		-其他鲑鱼,但鱼肝及鱼卵除外:							-Other salmonidae, excluding livers and roes:

序号 No.	税则号列 Tariff Item	货 品 名 称	最惠国税率(%) M.F.N.	协定税率(%)Agreement 东盟 ASEAN	曼谷协定 BA	香港 H.K.	澳门 Macao	普通税率(%) Gen.	Article Description
179	0303.2100	--鳟鱼	12	0				40	--Trout (*Salmo trutta*, *Oncorhynchus mykiss*, *Oncorhynchus clarki*, *Oncorhynchus aguabonita*, *Oncorhynchus gilae*, *Oncorhynchus apache and Oncorhynchus chrysogaster*).
		--大西洋鲑鱼及多瑙哲罗鱼:							--Atlantic salmon (*Salmo salar*), Danube salmon:
180	0303.2210	---大西洋鲑鱼	10	0				40	---Atlantic salmon (*Salmo salar*)
181	0303.2220	---多瑙哲罗鱼	10	0				40	---Danube salmon
182	0303.2900	--其他	10	0	6.7			40	--Other
		-比目鱼,但鱼肝及鱼卵除外:							-Flatfish (*Pleuronectidae*, *Bothidae*, *Cynoglossidae*, *soleidae*, *Scophthalmidae and Citharidae*), excluding livers and roes:
		--庸鲽鱼:							--Halibut (*Reinhardtius hippoglossoides*, *Hippoglossus hippoglossus*, *Hippoglossus stenolepis*)
183	0303.3110	---格陵兰庸鲽鱼	10	0	6.7	0△		40	---Greenland halibut
184	0303.3190	---其他	10	0	6.7	0△		40	---other
185	0303.3200	--鲽鱼	12	0	8			40	--Plaice (*Pleuronectes platessa*)
186	0303.3300	--鳎鱼	12	0	8			40	--Sole(*Solea spp.*)
187	0303.3900	--其他	10	0	8			40	--Other
		-金枪鱼、鲣鱼,但鱼肝及鱼卵除外:							-Tunas (*of the genus Thunnus*), skipjack or stripe-bellied bonito (*Euthynnus (Katsuwonus) pelamis*), excluding livers and roes:
188	0303.4100	--长鳍金枪鱼	12	0	9			40	--Albacore or longfinned tunas (*Thunnus alalunga*)
189	0303.4200	--黄鳍金枪鱼	12	0	9			40	--Yellowfin tunas (*Thunnus albacares*)
190	0303.4300	--鲣鱼	12	0	9			40	--Skipjack or stripe-bellied bonito
191	0303.4400	--大眼金枪鱼	12	0		0		40	--Bigeye tunas (*Thunnus obesus*)
192	0303.4500	--金枪鱼(蓝鳍金枪鱼)	12	0				40	--Bluefin tunas (*Thunnus thynnus*)
193	0303.4600	--南金枪鱼(南方蓝鳍金枪鱼)	12	0				40	--Southern bluefin tunas (*Thunnus maccoyii*)
194	0303.4900	--其他	12	0	9			40	--Other
195	0303.5000	-鲱鱼(大西洋鲱鱼、太平洋鲱鱼),但鱼肝及鱼卵除外	10	0	6.7			40	-Herrings (*Clupea harengus*, *Clupea pallasii*), excluding livers and roes
196	0303.6000	-鳕鱼,但鱼肝及鱼卵除外	10	0	6.7			40	-Cod (*Gadus morhua*, *Gadus ogac*, *Gadus macrocephalus*), excluding livers and roes
		-其他鱼,但鱼肝及鱼卵除外:							-Other fish, excluding livers and roes:

序号 No.	税则号列 Tariff Item	货品名称	最惠国税率(%) M.F.N.	协定税率(%) Agreement 东盟 ASEAN	曼谷协定 BA	香港 H.K.	澳门 Macao	普通税率(%) Gen.	Article Description
197	0303.7100	--沙丁鱼、黍鲱鱼	12	0	8			40	--Sardines (*Sardina pilchardus*, *Sardinops spp.*) sardinella (*Sardinella spp.*), brisling or sprats (*Sprattus sprattus*)
198	0303.7200	--黑线鳕鱼	12	0	8			40	--Haddock (Melanogrammus aeglefinus)
199	0303.7300	--绿青鳕鱼	12	0	8			40	--Coalfish (*Pollachius virens*)
200	0303.7400	--鲭鱼	10	0	6.7			40	--Mackerel (*Scomber scombrus*, *Scomber australasicus*, *Scomber japonicus*)
201	0303.7500	--角鲨及其他鲨鱼	12	0	9			40	--Dogfish and other sharks
202	0303.7600	--鳗鱼	12	0	8	0△		40	--Eels (*Anguilla spp.*)
203	0303.7700	--尖吻鲈鱼	12	0	8	0△		40	--Sea bass (*Dicentrarchus labrax*, *Dicentrarchus punctatus*)
204	0303.7800	--狗鳕鱼(无须鳕鱼、长鳍鳕鱼)	12	0				40	--Hake (*Merluccius spp.* *Urophycis spp.*)
		--其他:							--Other:
205	0303.7910	---带鱼	10	0	6.7	0		40	---Scabber fish (*Trichurius*)
206	0303.7920	---黄鱼	10	0	6.7	0		40	---Yellow croaker (*Pseudosicaena*)
207	0303.7930	---鲳鱼	10	0	6.7	0		40	---Butterfish (*Pampus*)
208	0303.7940	---罗非鱼	10	0				40	---Tilapia
209	0303.7990	---其他	10	0	6.7	0		40	---Other
210	0303.8000	-鱼肝及鱼卵	10	0	9			50	-Livers and roes
	03.04	**鲜、冷、冻鱼片及其他鱼肉(不论是否绞碎):**							**Fish fillets and other fish meat (whether or not minced), fresh, chilled or frozen:**
211	0304.1000	-鲜或冷的	12	5	9	0△		70	-Fresh or chilled
		-冻鱼片:							-Frozen fillets:
212	0304.2010	---罗非鱼的	10	5		0△		70	---Of tilapia
213	0304.2090	---其他鱼的	10	5		0△		70	---Other
214	0304.9000	-其他	10	5				70	-Other
	03.05	**干、盐腌或盐渍的鱼;熏鱼,不论在熏制前或熏制过程中是否烹煮;适合供人食用的鱼的细粉、粗粉及团粒:**							**Fish, dried, salted or in brine; smoked fish, whether or not cooked before or during the smoking process; flours, meals and pellets of fish, fit for human consumption:**
215	0305.1000	-适合供人食用的鱼的细粉、粗粉及团粒	10	5				80	-Flours, meals and pellets of fish, fit for human consumption
216	0305.2000	-干、熏、盐腌或盐渍的鱼肝及鱼卵	10	0				80	-Livers and roes, dried, smoked, salted or in brine
217	0305.3000	-干、盐腌或盐渍的鱼片,但熏制的除外	10	5	7.8			80	-Fish fillets, dried, salted or in brine, but not smoked
		-熏鱼,包括鱼片:							-Smoked fish, including fillets:

序号 No.	税则号列 Tariff Item	货品名称	最惠国税率(%) M.F.N.	协定税率(%) Agreement 东盟 ASEAN	曼谷协定 BA	香港 H.K.	澳门 Macao	普通税率(%) Gen.	Article Description
		--大马哈鱼、大西洋鲑鱼及多瑙哲罗鱼:							--Pacific salmon (*Oncorhynchus nerka*, *Oncorhynchus gorbuscha*, *Oncorhynchus keta*, *Oncorhynchus tschawytscha*, *Oncorhynchus kisutch*, *Oncorhynchus masou* and *Oncorhynchus rhodurus*), Atlantic salmon (*Salmo salar*) and Danube salmon (*Hucho hucho*):
218	0305.4110	---大西洋鲑鱼	14	5				80	---Atlantic salmon
219	0305.4120	---大马哈鱼及多瑙哲罗鱼	14	5				80	---Pacific salmon and Danube salmon
220	0305.4200	--鲱鱼(大西洋鲱鱼、太平洋鲱鱼)	16	5				80	--Herrings (*Clupea harengus*, *Clupea pallasii*)
221	0305.4900	--其他	14	5				80	--Other
		-干鱼,不论是否盐腌,但熏制的除外:							-Dried fish, whether or not salted but not smoked:
222	0305.5100	--鳕鱼	16	5				80	-- Cod (*Gadus morhua*, *Gadus ogac*, *Gadus macrocephalus*)
		--其他:							--Other:
223	0305.5910	---海龙、海马	2	0				20	---Pipefish and hippocampi
224	0305.5920	---鱼翅	15	0			0△	80	---Sharks'fins
225	0305.5990	---其他	16	5				80	---Other
		-盐腌及盐渍的鱼,但干或熏制的除外:							-Fish, salted but not dried nor smoked and fish in brine:
226	0305.6100	--鲱鱼(大西洋鲱鱼、太平洋鲱鱼)	16	5	11.1			80	--Herrings (*Clupea harengus*, *Clupea pallasii*)
227	0305.6200	--鳕鱼	16	5	12			80	--God (*Gadus morhua*, *Gadus ogac*, *Gadus Macrocephalus*)
228	0305.6300	--鳀鱼	16	5	12			80	--Anchovies (*Engraulis spp.*)
		--其他:							--Other:
229	0305.6910	---带鱼	16	5				80	---Scabber fish (*Trichurius*)
230	0305.6920	---黄鱼	16	5				80	---Yellow croaker (*Pseudosicaena*)
231	0305.6930	---鲳鱼	16	5				80	---Butterfish (*Pampus*)
232	0305.6940	---罗非鱼	16	5				80	---Tilapia
233	0305.6990	---其他	16	5				80	---Other
	03.06	**带壳或去壳的甲壳动物,活、鲜、冷、冻、干、盐腌或盐渍的;蒸过或用水煮过的带壳甲壳动物,不论是否冷、冻、干、盐腌或盐渍的;适合供人食用的甲壳动物的细粉、粗粉及团粒:**							**Crustaceans, whether in shell or not, live, fresh, chilled, frozen, dried, salted or in brine; crustaceans, in shell, cooked by steaming or by boiling in water, whether or not chilled, frozen, dried, salted or in brine; flours, meals and pellets of crustaceans, fit for human consumption:**
		-冻的:							-Frozen:
234	0306.1100	--龙虾	10	5				70	--Rock lobster and other sea crawfish (*Palinurus spp.*, *Panulirus spp.*, *Jasus spp.*)

序号 No.	税则号列 Tariff Item	货品名称	最惠国税率(%) M.F.N.	协定税率(%)Agreement				普通税率(%) Gen.	Article Description
				东盟 ASEAN	曼谷协定 BA	香港 H.K.	澳门 Macao		
235	0306.1200	--大螯虾	10	5	7.2			70	--Lobsters (*Homarus spp.*)
		--小虾及对虾：							--Shrimps and prawns:
		---小虾：							---Shrimps:
236	0306.1311	----虾仁	8	0	5	0		70	----Shelled
237	0306.1312	----北方长额虾	5	0		0		70	----Northem pandalus (*Pandalus*)
238	0306.1319	----其他	5	0	4	0		70	----Other
		---对虾：							---Prawns:
239	0306.1321	----虾仁	8	0	5	0		70	----Shelled
240	0306.1329	----其他	5	0	4	0		70	----Other
		--蟹：							--Crabs:
241	0306.1410	---梭子蟹	10	5				70	---Swimming crab
242	0306.1490	---其他	10	5				70	---Other
		--其他，包括适合供人食用的甲壳动物的细粉、粗粉及团粒：							--Other, including flours, meals and pellets of crustaceans, fit for human consumption:
		---淡水小龙虾：							---Freshwater crawfish:
243	0306.1911	虾仁	16	5				70	----Shelled
244	0306.1919	----其他	16	5				70	----Other
245	0306.1990	---其他	16	5				70	---Other
		-未冻的：							-Not frozen:
		--龙虾：							--Rock lobster and other sea crawfish (*Palinurus spp.*, *Panulirus spp.*, *Jasus spp.*):
246	0306.2110	---种苗	0	0				0	---For cultivation
247	0306.2190	---其他	15	5		0		70	---Other
		--大螯虾：							--Lobsters (*Homarus spp.*):
248	0306.2210	---种苗	0	0				0	---For cultivation
249	0306.2290	---其他	15	5		0		70	---Other
		--小虾及对虾：							--Shrimps and prawns:
250	0306.2310	---种苗	0	0				0	---For cultivation
		---其他：							---Other:
251	0306.2391	----鲜、冷对虾	15	5		0		70	----Prawns, fresh or chilled
252	0306.2399	----其他	12	5		0		70	----Other
		--蟹：							--Crabs:
253	0306.2410	---种苗	0	0				0	---For cultivation
		---其他：							---Other:
254	0306.2491	----中华绒螯蟹	14	0				70	----Freshwater crabs, live
255	0306.2492	----梭子蟹	14	5		0		70	----Swimming crab
256	0306.2499	----其他	14	0		0		70	----Other
		--其他，包括适合供人食用的甲壳动物的细粉、粗粉及团粒：							--Other, including flours, meals and pellets of crustaceans, fit for human consumption:
257	0306.2910	---种苗	0	0				0	---For cultivation
258	0306.2990	---其他	14	5		0		70	---Other

序号 No.	税则号列 Tariff Item	货品名称	最惠国税率(%) M.F.N.	协定税率(%) Agreement 东盟 ASEAN	曼谷协定 BA	香港 H.K.	澳门 Macao	普通税率(%) Gen.	Article Description
	03.07	**带壳或去壳的软体动物,活、鲜、冷、冻、干、盐腌或盐渍的;不属于甲壳动物和软体动物的水生无脊椎动物,活、鲜、冷、冻、干、盐腌或盐渍的;适合供人食用的水生无脊椎动物(甲壳动物除外)的细粉、粗粉及团粒:**							**Molluscs, whether in shell or not, live, fresh, chilled, frozen, dried, salted or in brine; aquatic invertebrates other than crustaceans and molluscs, live, fresh, chilled, frozen, dried, salted or in brine; flours, meals and pellets of aquatic invertebrates other than crustaceans, fit for human consumption:**
		-牡蛎(蚝):							-Oysters:
259	0307.1010	---种苗	0	0				0	---For cultivation
260	0307.1090	---其他	14	5		0		70	---Other
		-扇贝,包括海扇:							-Scallops, including queen scallops, of the genera *Pecten*, *Chlamys or Placopecten*:
		--活、鲜或冷的:							--Live, fresh or chilled:
261	0307.2110	---种苗	0	0				0	---For cultivation
262	0307.2190	---其他	14	5		0		70	---Other
263	0307.2900	--其他	14	5				80	--Other
		-贻贝:							-Mussels(*Mytilus spp.*, *Perna spp.*):
		--活、鲜或冷的:							--Live, fresh or chilled:
264	0307.3110	---种苗	0	0				0	---For cultivation
265	0307.3190	---其他	14	5		0		70	---Other
266	0307.3900	--其他	14	5				70	--Other
		-墨鱼及鱿鱼:							-Cuttle fish (*Sepia officinalis*, *Rossia macrosoma*, *Sepiola spp.*) and squid (*Ommastrephes spp.*, *Loligo spp.*, *Nototodarus spp.*, *Sepioteuthis spp.*):
		--活、鲜或冷的:							--Live, fresh or chilled:
267	0307.4110	---种苗	0	0				0	---For cultivation
268	0307.4190	---其他	12	0		0		70	---Other
269	0307.4900	--其他	12	0	10	0		70	--Other
		-章鱼:							-Octopus (*Octopus spp.*):
270	0307.5100	--活、鲜或冷的	17	5		0		70	--Live, fresh or chilled
271	0307.5900	--其他	17	5		0		70	--Other
		-蜗牛及螺:							-Snails, other than sea snails:
272	0307.6010	---种苗	0	0				0	---For cultivation
273	0307.6090	---其他	14	5				70	---Other
		-其他,包括适合供人食用的水生无脊椎动物(甲壳动物除外)的细粉、粗粉及团粒:							-Other, including flours, meals and pellets of aquatic invertebrates other than crustaceans, fit for human consumption:
		--活、鲜或冷的:							--Live, fresh or chilled:

序号 No.	税则号列 Tariff Item	货品名称	最惠国税率(%) M.F.N.	协定税率(%)Agreement				普通税率(%) Gen.	Article Description
				东盟 ASEAN	曼谷协定 BA	香港 H.K.	澳门 Macao		
274	0307.9110	---种苗	0	0				0	---For cultivation
		---其他:							---Other:
275	0307.9191	----鲍鱼	14	5		0		80	----Abalone
276	0307.9192	----沙蚕	14	5		0		70	----Clamworm
277	0307.9199	----其他	14	5		0		70	----Other
		--其他:							--Other:
278	0307.9910	---鲍鱼	10	5		0		80	---Abalone
279	0307.9920	---海参	10	5				80	---Sea cucumbers (*beches-de-mer*)
280	0307.9990	---其他	10	5				70	---Other

第四章
乳品；蛋品；天然蜂蜜；其他食用动物产品

Chapter 4
Dairy produce; birds' eggs; natural honey; edible products of animal origin, not elsewhere specified or included

注释：

一、所称“乳”，是指全脂乳及半脱脂或全脱脂的乳。

二、税号04.05所称：

(一)“黄油”，仅指从乳中提取的天然黄油、乳清黄油及调制黄油（新鲜、加盐或酸败的，包括罐装黄油），按重量计乳脂含量在80%及以上，但不超过95%，乳的无脂固形物最大含量不超过2%，以及水的最大含量不超过16%。黄油中不含添加的乳化剂，但可含有氯化钠、食用色素、中和盐及无害乳酸菌的培养物。

(二)“乳酱”是一种油包水型可涂抹的乳状物，乳脂是该制品所含的唯一脂肪，按重量计其含量在39%及以上，但小于80%。

三、乳清经浓缩并加入乳或乳脂制成的产品，若同时具有下列三种特性，则视为乳酪归入税号04.06：

(一) 按干重计乳脂含量在5%及以上的；

(二) 按重量计干质成分至少为70%，但不超过85%的；

(三) 已成形或可以成形的。

四、本章不包括：

(一) 按重量计乳糖含量（以干燥无水乳糖计）超过95%的乳清制品（税号17.02）；

(二) 白蛋白（包括按重量计干质成分的乳清蛋白含量超过80%的两种或两种以上的乳清蛋白浓缩物）（税号35.02）及球蛋白（税号35.04）。

子目注释：

一、子目号0404.10所称“改性乳清”，是指由乳清成分构成的制品，即全部或部分去除乳糖、蛋白或矿物质的乳清、加入天然乳清成分的乳清及由混入天然乳清成分制成的产品。

Notes:

1. The expression "milk" means full cream milk or partially or completely skimmed milk.

2. For the purposes of heading No.04.05:

(a) The term "butter" means natural butter, whey butter or recombined butter (fresh, salted or rancid, including canned butter) derived exclusively from milk, with a milkfat content of 80% or more but not more than 95% by weight, a maximum milk solids-not-rat content of 2% by weight and a maximum water content of 16% by weight. Butter does not contain emulsifiers, but may contain sodium chloride, food colours, neutralizing salts and cultures of harmless lactic-acid-pro-ducing bacteria.

(b) The expression "dairy spreads" means a spreadable emulsion of the water-in-oil type, containing milkfat as the only fat in the product, with a milkfat content of 39% or more but less than 80% by weight.

3. Products obtained by the concentration of whey and with the addition of milk or milkfat are to be classified as cheese in heading No.04.06. Provided that they have the three following characteristics:

(a) A milkfat content, by weight of the dry matter, of 5% or more;

(b) A dry matter content, by weight, of at least 70% but not exceeding 85%; and

(c) They are moulded or capable of being moulded.

4. This Chapter does not cover:

(a) Products obtained from whey, containing by weight more than 95% lactose, expressed as anhydrous lactose calculated on the dry matter (heading No.17.02); or

(b) Albumins (including concentrates of two or more whey proteins, containing by weight more than 80% whey proteins, calculated on the dry matter) (heading No.35.02) or globulins (heading No.35.04).

Subheading Notes:

1. For the purpose of subheading No.0404.10, the expression "modified whey" means products consisting of whey constituents, i. e., whey from which all or part of the lactose, proteins or minerals have been removed, whey to which natural whey constituents have been added and products obtained

by mixing natural whey constituents.

二、子目号0405.10所称“黄油”，不包括脱水黄油及印度酥油（子目号0405.90）。

2. For the purposes of subheading No.0405.10 the term “butter” does not include dehydrated butter or ghee (subheading No.0405.90).

序号 No.	税则号列 Tariff Item	货品名称	最惠国税率(%) M.F.N.	协定税率(%) Agreement 东盟 ASEAN	曼谷协定 BA	香港 H.K.	澳门 Macao	普通税率(%) Gen.	Article Description
	04.01	**未浓缩及未加糖或其他甜物质的乳及奶油：**							**Milk and cream, not concentrated nor containing added sugar or other sweetening matter:**
281	0401.1000	-按重量计脂肪含量不超过1%	15	5				40	-Of a fat content, by weight, not exceeding 1%
282	0401.2000	-按重量计脂肪含量超过1%，但不超过6%	15	5		0		40	-Of a fat content, by weight, exceeding 1% but not exceeding 6%
283	0401.3000	-按重量计脂肪含量超过6%	15	5				40	-Of a fat content, by weight, exceeding 6%
	04.02	**浓缩、加糖或其他甜物质的乳及奶油：**							**Milk and cream, concentrated or containing added sugar or other sweetening matter:**
284	0402.1000	-粉状、粒状或其他固体形状，按重量计脂肪含量不超过1.5%	10	0		0△		40	-In powder, granules or other solid forms, of a fat content, by weight, not exceeding 1.5%
		-粉状、粒状或其他固体形状，按重量计脂肪含量超过1.5%：							-In powder, granules or other solid forms, of a fat content, by weight, exceeding 1.5%:
285	0402.2100	--未加糖或其他甜物质	10	0		0△		40	--Not containing added sugar or other sweetening matter
286	0402.2900	--其他	10	0		0△		40	--Other
		-其他：							-Other:
287	0402.9100	--未加糖或其他甜物质	10	5				90	--Not containing added sugar or other sweetening matter
288	0402.9900	--其他	10	5				90	--Other
	04.03	**酪乳、结块的乳及奶油、酸乳、酸乳酒及其他发酵或酸化的乳和奶油，不论是否浓缩、加糖、加其他甜物质、加香料、加水果、加坚果或加可可：**							**Buttermilk, curdled milk and cream, yogurt, kephir and other fermented or acidified milk and cream, whether or not concentrated or containing added sugar or other sweetening matter or flavoured or containing added fruit, nuts or cocoa:**
289	0403.1000	-酸乳	10	5		0		90	-Yogurt
290	0403.9000	-其他	20	5		0		90	-Other

序号 No.	税则号列 Tariff Item	货品名称	最惠国税率(%) M.F.N.	协定税率(%) Agreement 东盟 ASEAN	协定税率(%) Agreement 曼谷协定 BA	协定税率(%) Agreement 香港 H.K.	协定税率(%) Agreement 澳门 Macao	普通税率(%) Gen.	Article Description
	04.04	**乳清,不论是否浓缩、加糖或其他甜物质;其他税号未列名的含天然乳的产品,不论是否加糖或其他甜物质:**							**Whey, whether or not concentrated or containing added sugar or other sweetening matter; products consisting of natural milk constituents, whether or not containing added sugar or other sweetening matter, not elsewhere specified or included:**
291	0404.1000	-乳清及改性乳清,不论是否浓缩、加糖或其他甜物质	6	0				30	-Whey and modified whey, whether or not concentrated or containing added sugar or other sweetening matter
292	0404.9000	-其他	20	5				90	-Other
	04.05	**黄油及其他从乳中提取的脂和油;乳酱:**							**Butter and other fats and oils derived from milk; dairy spreads:**
293	0405.1000	-黄油	10	5				90	-Butter
294	0405.2000	-乳酱	10	5	8.1			90	-Dairy spreads
295	0405.9000	-其他	10	5				90	-Other
	04.06	**乳酪及凝乳:**							**Cheese and curd:**
296	0406.1000	-鲜乳酪(未熟化或未固化的),包括乳清乳酪;凝乳	10	5		0		90	-Fresh (unripened or uncured) cheese, including whey cheese, and curd
297	0406.2000	-各种磨碎或粉化的乳酪	12	5				90	-Grated or powdered cheese, of all kinds
298	0406.3000	-经加工的乳酪,但磨碎或粉化的除外	12	5				90	-Processed cheese, not grated or powdered
299	0406.4000	-蓝纹乳酪	15	5				90	-Blue-veined cheese
300	0406.9000	-其他乳酪	12	5				90	-Other cheese
	04.07	**带壳禽蛋,鲜、腌制或煮过的:**							**Birds' eggs, in shell, fresh, preserved or cooked:**
301	0407.0010	---种用蛋	0	见表1				0	---For hatching
		---其他带壳鲜蛋:							---Other, in shell, fresh:
302	0407.0021	----鸡蛋	20	见表1				80	----Of hens
303	0407.0022	----鸭蛋	20	见表1				80	----Of ducks
304	0407.0023	----鹅蛋	20	见表1				80	----Of geese
305	0407.0029	----其他	20	见表1				80	----Other
		---其他:							---Other:
306	0407.0091	----咸蛋	20	见表1				90	----Salted eggs
307	0407.0092	----皮蛋	20	见表1				90	----Lime-preserved eggs
308	0407.0099	----其他	20	见表1				90	----Other
	04.08	**去壳禽蛋及蛋黄,鲜、干、冻、蒸过或水煮、制成型或用其他方法保藏的,不论是否加糖或其他甜物质:**							**Birds' eggs, not in shell, and egg yolks, fresh, dried, cooked by steaming or by boiling in water, moulded, frozen or otherwise preserved, whether or not containing added sugar or other sweetening matter:**
		-蛋黄:							-Egg yolks:

序号 No.	税则号列 Tariff Item	货品名称	最惠国税率(%) M.F.N.	协定税率(%)Agreement				普通税率(%) Gen.	Article Description
				东盟 ASEAN	曼谷协定 BA	香港 H.K.	澳门 Macao		
309	0408.1100	--干的	20	5				90	--Dried
310	0408.1900	--其他	20	5				90	--Other
		-其他:							-Other:
311	0408.9100	--干的	20	5				90	--Dried
312	0408.9900	--其他	20	5		0		90	--Other
	04.09	**天然蜂蜜:**							**Natural honey:**
313	0409.0000	天然蜂蜜	15	5				80	Natural honey
	04.10	**其他税号未列名的食用动物产品:**							**Edible products of animal origin, not elsewhere specified or included:**
314	0410.0010	---燕窝	25	5				80	---Salanganes' nests
315	0410.0020	---鲜蜂王浆	15	5				70	---Pure royal jelley
316	0410.0030	---鲜蜂王浆粉	15	5				70	---Pure royal jelley, in powder
317	0410.0090	---其他	20	5				70	---Other

第 五 章
其他动物产品

Chapter 5
Products of animal origin, not elsewhere specified or included

注释：

一、本章不包括：

（一）食用产品（整个或切块的动物肠、膀胱和胃以及液态或干制的动物血除外）；

（二）生皮或毛皮（第四十一章、第四十三章），但税号 05.05 的货品及税号 05.11 的生皮或毛皮的边角废料仍归入本章；

（三）马毛及废马毛以外的动物纺织原料（第十一类）；

（四）供制帚、制刷用的成束、成簇的材料（税号 96.03）。

二、仅按长度而未按发根和发梢整理的人发，视为未加工品，归入税号 05.01。

三、本目录所称“兽牙”，是指象、河马、海象、一角鲸和野猪的长牙、犀角及其他动物的牙齿。

四、本目录所称“马毛”，是指马科、牛科动物的鬃毛和尾毛。

Notes:

1. This Chapter does not cover:
   (a) Edible products (other than guts, bladders and stomachs of animals, whole and pieces thereof, and animal blood, liquid or dried);
   (b) Hides or skins (including furskins) other than goods of heading No.05.05 and parings and similar waste of raw hides or skins of heading No.05.11 (Chapter 41 or 43);
   (c) Animal textile materials, other than horsehair and horsehair waste (Section XI); or
   (d) Prepared knots or tufts for broom or brush making (heading No.96.03).

2. For the purposes of heading No.05.01, the sorting of human hair by length (provided the root ends and tip ends respectively are not arranged together) shall be deemed not to constitute working.

3. Throughout the Nomenclature, elephant, hippopotamus, walrus, narwhal and wild boar tusks, rhinoceros horns and the teeth of all animals are regarded as "ivory".

4. Throughout the Nomenclature, the expression "horsehair" means hair of the manes or tails of equine or bovine animals.

序号 No.	税则号列 Tariff Item	货品名称	最惠国税率(%) M.F.N.	协定税率(%) Agreement 东盟 ASEAN	曼谷协定 BA	香港 H.K.	澳门 Macao	普通税率(%) Gen.	Article Description
	05.01	**未经加工的人发，不论是否洗涤；废人发：**							**Human hair, unworked, whether or not washed or scoured; waste of human hair:**
318	0501.0000	未经加工的人发，不论是否洗涤；废人发	15	0				90	Human hair, unworked, whether or not washed or scoured; waste of human hair
	05.02	**猪鬃、猪毛；獾毛及其他制刷用兽毛；上述鬃毛的废料：**							**Pigs', hogs' or boars' bristles and hair; badger hair and other brush making hair; waste of such bristles or hair:**
		-猪鬃、猪毛及其废料：							-Pigs', hogs' or boars' bristles and hair and waste thereof:
319	0502.1010	---猪鬃	20	5				90	---Bristles
320	0502.1020	---猪毛	20	5				90	---Hair
321	0502.1030	---废料	20	5				90	---Waste
		-其他：							-Other:

序号 No.	税则号列 Tariff Item	货品名称	最惠国税率(%) M.F.N.	协定税率(%) Agreement 东盟 ASEAN	曼谷协定 BA	香港 H.K.	澳门 Macao	普通税率(%) Gen.	Article Description
		---獾毛及其他制刷用兽毛:							---Badger hair and other brush making hair:
322	0502.9011	----山羊毛	20	5				90	----Goat hair
323	0502.9012	----黄鼠狼尾毛	20	5				90	----Weasel tail hair
324	0502.9019	----其他	20	5				90	----Other
325	0502.9020	---废料	20	5				90	---Waste
	05.03	**马毛及废马毛,不论是否制成有或无衬垫的毛片:**							**Horsehair and horsehair waste, whether or not put up as a layer with or without supporting material:**
326	0503.0010	---马鬃、马尾	15	0				90	---Horse's manes or tails
327	0503.0090	---其他	15	0				90	---Other
	05.04	**整个或切块的动物(鱼除外)的肠、膀胱及胃,鲜、冷、冻、干、熏、盐腌或盐渍的:**							**Guts, bladders and stomachs of animals (other than fish), whole and pieces thereof, fresh, chilled, frozen, salted, in brine, dried or smoked:**
		---肠衣:							---Casings:
328	0504.0011	----盐渍猪肠衣(猪大肠头除外)	20	见表 1				90	----Hog casings, salted
329	0504.0012	----盐渍绵羊肠衣	18	见表 1				90	----Sheep casings, salted
330	0504.0013	----盐渍山羊肠衣	18	见表 1				90	----Goat casings, salted
331	0504.0014	----盐渍猪大肠头	20	见表 1				90	----Hog fat-ends, salted
332	0504.0019	----其他	18	见表 1				90	----Other
		---胃:							---Gizzard:
333	0504.0021	----冷、冻的鸡胗	见表 4	见表 1				见表 4	----Cold, frozen gizzard
334	0504.0029	----其他	20	见表 1				90	----Other
335	0504.0090	---其他	20	见表 1				80	---Other
	05.05	**带有羽毛或羽绒的鸟皮及鸟体其他部分;羽毛及不完整羽毛(不论是否修边)、羽绒,仅经洗涤、消毒或为了保藏而作过处理,但未经进一步加工;羽毛或不完整羽毛的粉末及废料:**							**Skins and other parts of birds, with their feathers or down; feathers and parts of feathers (Whether or not with trimmed edges) and down, not further worked than cleaned, disinfected or treated for preservation; powder and waste of feathers or parts of feathers:**
336	0505.1000	-填充用羽毛;羽绒	10	0	7.5			100	-Feathers of a kind used for stuffing; down
		-其他:							-Other:
337	0505.9010	---羽毛或不完整羽毛的粉末及废料	10	0				35	---Powder and waste of feathers or parts of feathers
338	0505.9090	---其他	10	0				90	---Other
	05.06	**骨及角柱,未经加工或经脱脂、简单整理(但未切割成形)、酸处理或脱胶;上述产品的粉末及废料.**							**Bones and horn-cores, unworked, defatted, simply prepared (but not cut to shape), treated with acid or degelatinized; powder and waste of these products:**

序号 No.	税则号列 Tariff Item	货品名称	最惠国税率(%) M.F.N.	协定税率(%) Agreement 东盟 ASEAN	曼谷协定 BA	香港 H.K.	澳门 Macao	普通税率(%) Gen.	Article Description
339	0506.1000	-经酸处理的骨胶原及骨	12	0				50	-Ossein and bones treated with acid
		-其他:							-Other:
		---骨粉、骨废料:							---Powder and waste of bones:
340	0506.9011	----含牛羊成分的	12	0				35	----Of bovine and sheep
341	0506.9019	----其他	12	0				35	----Other
342	0506.9090	---其他	12	0				50	---Other
	05.07	**兽牙、龟壳、鲸须、鲸须毛、角、鹿角、蹄、甲、爪及喙,未经加工或仅简单整理但未切割成形;上述产品的粉末及废料:**							**Ivory, tortoise-shell, whalebone and whalebone hair, horns, antlers, hooves, nails, claws and beaks, unworked or simply prepared but not cut to shape; powder and waste of these products:**
343	0507.1000	-兽牙;兽牙粉末及废料	10	0				30	-Ivory;ivory powder and waste
		-其他:							-Other:
344	0507.9010	---羚羊角及其粉末和废料	3	0				14	---Antelope horns and powder or waste thereof
345	0507.9020	---鹿茸及其粉末	11.2	0				30	---Pilose antlers and powder thereof
346	0507.9090	---其他	10	0				50	---Other
	05.08	**珊瑚及类似品,未经加工或仅简单整理但未经进一步加工;软体动物壳、甲壳动物壳、棘皮动物壳、墨鱼骨,未经加工或仅简单整理但未切割成形,上述壳、骨的粉末及废料:**							**Coral and similar materials, unworked or simply prepared but not otherwise worked; shells of molluscs, crustaceans or echinoderms and cuttle-bone, unworked or simply prepared but not cut to shape, powder and waste thereof:**
347	0508.0010	---粉末及废料	12	0		0		35	---Powder and waste
348	0508.0090	---其他	12	0		0		50	---Other
	05.09	**动物质天然海绵:**							**Natural sponges of animal origin:**
349	0509.0000	动物质天然海绵	15	0				70	Natural sponges of animal origin
	05.10	**龙涎香、海狸香、灵猫香及麝香;斑蝥;胆汁,不论是否干制;供配制药用的腺体及其他动物产品,鲜、冷、冻或用其他方法暂时保藏的:**							**Ambergris, castoreum, civet and musk; cantharides; bile, whether of not dried; glands and other animal products used in the preparation of pharmaceutical products, fresh, chilled, frozen or otherwise provisionally preserved:**
350	0510.0010	---黄药	3	0				14	---Bezoar
351	0510.0020	---龙涎香、海狸香、灵猫香	7	0				50	---Ambergris, castoreum and civet
352	0510.0030	---麝香	7	0				20	---Musk
353	0510.0040	---斑蝥	7	0				50	---Cantharides
354	0510.0090	---其他	6	0				20	---Other

序号 No.	税则号列 Tariff Item	货品名称	最惠国税率(%) M.F.N.	协定税率(%)Agreement 东盟 ASEAN	曼谷协定 BA	香港 H.K.	澳门 Macao	普通税率(%) Gen.	Article Description
	05.11	**其他税号未列名的动物产品;不适合供人食用的第一章或第三章的死动物:**							**Animal products not elsewhere specified or included; dead animals of Chapter 1 or 3, unfit for human consumption:**
355	0511.1000	-牛的精液	0	0				0	-Bovine semen
		-其他:							-Other:
		--鱼、甲壳动物、软体动物、其他水生无脊椎动物的产品;第三章的死动物:							--Products of fish or crustaceans, molluscs or other aquatic invertebrates; dead animals of Chapter 3:
		---鱼的:							---fish:
356	0511.9111	----受精鱼卵	12	0				35	----fertilized fish eggs
357	0511.9119	---其他	12	0				35	----Other
358	0511.9190	----其他	12	0				35	---Other
		--其他:							--Other:
359	0511.9910	---动物精液(牛的精液除外)	0	0				0	---Animal semen, other than bovine semen
360	0511.9920	---动物胚胎	0	0				0	---Animal embryo
361	0511.9930	---蚕种	0	0				0	---Silkworm graine
362	0511.9990	---其他	12	0				35	---Other

第二类

植 物 产 品

注释：

本类所称“团粒”，是指直接挤压或加入按重量计比例不超过3%的粘合剂制成的粒状产品。

SECTION Ⅱ

VEGETABLE PRODUCTS

Note:

In this Section the term “pellets” means products which have been agglomerated either directly by compression or by the addition of a binder in a proportion not exceeding 3% by weight.

第 六 章
活树及其他活植物；
鳞茎、根及类似品；
插花及装饰用簇叶

注释：

一、除税号06.01的菊苣植物及其根以外，本章只包括通常由苗圃或花店供应为种植或装饰用的活树及其他货品（包括植物秧苗）；但不包括马铃薯、洋葱、青葱、大蒜及其他第七章的产品。

二、税号06.03、06.04的各种货品，包括全部或部分用这些货品制成的花束、花篮、花圈及类似品，不论是否有其他材料制成的附件。但这些货品不包括税号97.01的拼贴画或类似的装饰板。

Chapter 6
Live trees and other plants;
bulbs, roots and the like;
cut flowers and ornamental foliage

Notes:

1. Subject to the second part of heading No.06.01 this Chapter covers only live trees and goods (including seeding vegetables) of a kind commonly supplied by nursery gardeners or florists for planting or for ornamental use; nevertheless it does not include potatoes, onions, shallots, garlic or other products of Chapter 7.

2. Any reference in heading No.06.03 or 06.04 to goods of any kind shall be construed as including a reference to bouquets, floral baskets, wreaths and similar articles made wholly or partly of goods of that kind, account not being taken of accessories of other materials. However, these headings do not include collages or similar decorative plaques of heading No.97.01.

序号 No.	税则号列 Tariff Item	货品名称	最惠国税率(%) M.F.N.	协定税率(%)Agreement 东盟 ASEAN	曼谷协定 BA	香港 H.K.	澳门 Macao	普通税率(%) Gen.	Article Description
	06.01	鳞茎、块茎、块根、球茎、根颈及根茎，休眠、生长或开花的；菊苣植物及其根，但税号12.12的根除外：							**Bulbs, tubers, tuberous roots, corms, crowns and rhizomes, dormant, in growth or in flower; chicory plants and roots other than roots of heading No.12.12:**
		-休眠的鳞茎、块茎、块根、球茎、根颈及根茎：							-Bulbs, tubers, tuberous roots, corms, crowns and rhizomes, dormant:
363	0601.1010	---番红花球茎	4	0				14	---Stigma croci corms
		---其他：							---Other:
364	0601.1091	----种用	0	0				0	----Seed
365	0601.1099	----其他	5	0				40	----Other

序号 No.	税则号列 Tariff Item	货品名称	最惠国税率(%) M.F.N.	协定税率(%) Agreement 东盟 ASEAN	曼谷协定 BA	香港 H.K.	澳门 Macao	普通税率(%) Gen.	Article Description
366	0601.2000	-生长或开花的鳞茎、块茎、块根、球茎、根颈及根茎;菊苣植物及其根	15	0				80	-Bulbs, tubers, tuberous roots, corms, crowns and rhizomes, in growth or in flower; chicory plants and roots
	06.02	**其他活植物(包括其根)、插枝及接穗;蘑菇菌丝:**							**Other live plants (including their roots) cuttings and ships; mushroom spawn:**
367	0602.1000	-无根插枝及接穗	0	0				0	-Unrooted cuttings and slips
		-食用水果或食用坚果的树、灌木,不论是否嫁接:							-Trees, shrubs and bushes, grafted or not, of kinds which bear edible fruit or nuts:
368	0602.2010	---种用苗木	0	0				0	---Seedlings
369	0602.2090	---其他	10	0				80	---Other
		-杜鹃,不论是否嫁接:							-Rhododendrons and azaleas, grafted or not:
370	0602.3010	---种用	0	0				0	---Seedlings
371	0602.3090	---其他	15	0				80	---Other
		-玫瑰,不论是否嫁接:							-Roses, grafted or not:
372	0602.4010	---种用	0	0				0	---Seedlings
373	0602.4090	---其他	15	0				80	---Other
		-其他:							-Other:
374	0602.9010	---蘑菇菌丝	0	0	0			0	---Mushroom spawn
		---其他:							---Other:
375	0602.9091	----种用苗木	0	0				0	----Seedlings
376	0602.9092	----兰花	10	0				80	----Orchid
377	0602.9093	----菊花	10	0				80	----Chrysathemum
378	0602.9094	----百合	10	0				80	----Lily
379	0602.9095	----康乃馨	10	0				80	----Carnation
380	0602.9099	----其他	10	0				80	----Other
	06.03	**制花束或装饰用的插花及花蕾,鲜、干、染色、漂白、浸渍或用其他方法处理的:**							**Cut flowers and flower buds of akind suitable for bouquets or for ornamental purposes, fresh, dried, dyed, bleached, impregnated or otherwise prepared:**
381	0603.1000	-鲜的	10	0				100	-Fresh
382	0603.9000	-其他	23	5				100	-Other
	06.04	**制花束或装饰用的不带花及花蕾的植物枝、叶或其他部分、草、苔藓及地衣,鲜、干、染色、漂白、浸渍或用其他方法处理的:**							**Foliage, branches and other parts of plants, without flowers or flowerbuds, and grasses, mosses and lichens, being goods of a kind suitable for bouquets or for ornamental purposes, fresh, dried, dyed, bleached, impregnated or otherwise prepared:**
383	0604.1000	-苔藓及地衣	23	5				100	-Mosses and lichens
		-其他:							-Other:
384	0604.9100	--鲜的	10	0				100	--Fresh
385	0604.9900	--其他	10	0				100	--Other

第 七 章
食用蔬菜、根及块茎

Chapter 7
Edible vegetables and certain roots and tubers

注释：

一、本章不包括税号 12.14 的草料。

二、税号 07.09、07.10、07.11 及 07.12 所称“蔬菜”，包括食用的蘑菇、块菌、油橄榄、刺山柑、菜葫芦、南瓜、茄子、甜玉米、辣椒、茴香菜、欧芹、细叶芹、龙蒿、水芹、甜茉乔栾那。

三、税号 07.12 包括干制的归入税号 07.01 至 07.11 的各种蔬菜，但下列各项除外：

（一）作蔬菜用的脱荚干豆（税号 07.13）；

（二）税号 11.02 至 11.04 所列形状的甜玉米；

（三）马铃薯细粉、粗粉、粉末、粉片、颗粒及团粒（税号 11.05）；

（四）用税号 07.13 的干豆制成的细粉、粗粉及粉末（税号 11.06）。

四、本章不包括辣椒干及辣椒粉（税号 09.04）。

Notes:

1. This Chapter does not cover forage products of heading No.12.14.

2. In headings No.07.09, 07.10, 07.11 and 07.12 the word “vegetables” includes edible mushrooms, truffles, olives, capers, marrows, pumpkins, aubergines, sweet corn (*Zea may svar. saccharata*), fruits of the genus *Capsicum* or of the genus *Pimenta*, fennel, parsley, chervil, tarragon, cress and sweet marjoram (*Majorana hortensis or Origanum majorana*).

3. Heading No.07.12 covers all dried vegetables of the kinds falling in headings No.07.01 to 07.11, other than:

(a) dried leguminous vegetables, shelled (heading No.07.13);

(b) sweet corn in the forms specified in headings No.11.02 to 11.04;

(c) flour, meal, powder, flakes, granules and pellets of potatoes (heading No.11.05);

(d) flour, meal and powder of the dried leguminous vegetables of heading No.07.13 (heading No.11.06).

4. However, dried or crushed or ground fruits of the genus *Capsicum* or of the genus *Pimenta* are excluded from this Chapter (heading No.09.04).

序号 No.	税则号列 Tariff Item	货品名称	最惠国税率(%) M.F.N.	协定税率(%)Agreement				普通税率(%) Gen.	Article Description
				东盟 ASEAN	曼谷协定 BA	香港 H.K.	澳门 Macao		
	07.01	**鲜或冷藏的马铃薯：**							**Potatoes, fresh or chilled:**
386	0701.1000	-种用	13	0				70	-Seeds
387	0701.9000	-其他	13	0	9			70	-Other
	07.02	**鲜或冷藏的番茄：**							**Tomatoes, fresh or chilled:**
388	0702.0000	鲜或冷藏的番茄	13	见表 1				70	Tomatoes, fresh or chilled
	07.03	**鲜或冷藏的洋葱、青葱、大蒜、韭葱及其他葱属蔬菜：**							**Onions, shallots, garlic, leeks and other alliaceous vegetables, fresh or chilled:**
		-洋葱及青葱：							-Onions and shallots:
389	0703.1010	---洋葱	13	见表 1				70	---Onions
390	0703.1020	---青葱	13	0				70	---Shallots
		-大蒜：							-Garlic:
391	0703.2010	---蒜头	13	见表 1				70	---Garlic bulbs
392	0703.2020	---蒜苔及蒜苗(青蒜)	13	见表 1				70	---Garlic stems, garlic seedlings
393	0703.2090	---其他	13	见表 1				70	---Other

序号 No.	税则号列 Tariff Item	货品名称	最惠国税率(%) M.F.N.	协定税率(%) Agreement 东盟 ASEAN	曼谷协定 BA	香港 H.K.	澳门 Macao	普通税率(%) Gen.	Article Description
		-韭葱及其他葱属蔬菜:							-Leeks and other alliaceous vegetables:
394	0703.9010	---韭葱	13	0				70	---Leeks
395	0703.9020	---大葱	13	0				70	---Scallion
396	0703.9090	---其他	13	0				70	---Other
	07.04	**鲜或冷藏的卷心菜、菜花、球茎甘蓝、羽衣甘蓝及类似的食用芥菜类蔬菜:**							**Cabbages, cauliflowers, kohlrabi, kale and similar edible brassicas, fresh or chilled:**
397	0704.1000	-菜花及硬花甘蓝	10	见表1				70	-Cauliflowers and headed broccoli
398	0704.2000	-抱子甘蓝	13	0				70	-Brussels sprouts
399	0704.9000	-其他	13	见表1				70	-Other
	07.05	**鲜或冷藏的莴苣及菊苣:**							**Lettuce (*lactuca sativa*) and chicory (*Cichorium spp.*), fresh or chilled:**
		-莴苣:							-Lettuce:
400	0705.1100	--结球莴苣(包心生菜)	10	见表1				70	--Cabbage lettuce (head lettuce)
401	0705.1900	--其他	10	见表1				70	--Other
		-菊苣:							-Chicory:
402	0705.2100	--维特罗夫菊苣	13	0				70	--Witloof chicory (*Cichoriym intybus var. foliosum*)
403	0705.2900	--其他	13	0				70	--Other
	07.06	**鲜或冷藏的胡萝卜、萝卜、色拉甜菜根、婆罗门参、块根芹、小萝卜及类似的食用根茎:**							**Carrots, turnips, salad beetroot, salsify, celeriac, radishes and similar edible roots, fresh or chilled:**
404	0706.1000	-胡萝卜及萝卜	13	见表1				70	-Carrots and turnips
405	0706.9000	-其他	13	见表1				70	-Other
	07.07	**鲜或冷藏的黄瓜及小黄瓜:**							**Cucumbers and gherkins, fresh or chilled:**
406	0707.0000	鲜或冷藏的黄瓜及小黄瓜	13	0				70	Cucumbers and gherkins, fresh or chilled
	07.08	**鲜或冷藏的豆类蔬菜,不论是否脱荚:**							**Leguminous vegetables, shelled or unshelled, fresh or chilled:**
407	0708.1000	-豌豆	13	0				70	-Peas (*Pisum sativum*)
408	0708.2000	-豇豆及菜豆	13	见表1				70	-Beans (*Vigna spp.*, *Phaseolus spp.*)
409	0708.9000	-其他豆类蔬菜	13	0				70	-Other leguminous vegetables
	07.09	**鲜或冷藏的其他蔬菜:**							**Other vegetables, fresh or chilled:**
410	0709.1000	-洋蓟	13	0				70	-Globe artichokes
411	0709.2000	-芦笋	13	0				70	-Asparagus
412	0709.3000	-茄子	13	见表1				70	-Aubergines (egg-plants)
413	0709.4000	-芹菜,但块根芹除外	10	0				70	-Celery other than celeriac
		-蘑菇及块菌:							-Mushrooms and truffles:
414	0709.5100	--伞菌属蘑菇	13	见表1				90	--Mushrooms of the genus Agaricus
415	0709.5200	--块菌	13	0				90	--Truffle
		--其他:							--Other:
416	0709.5910	---松茸	13	0				90	---Sungmo
417	0709.5920	---香菇	13	0				90	---Shiitake

序号 No.	税则号列 Tariff Item	货品名称	最惠国税率(%) M.F.N.	协定税率(%) Agreement 东盟 ASEAN	曼谷协定 BA	香港 H.K.	澳门 Macao	普通税率(%) Gen.	Article Description
418	0709.5930	---金针菇	13	0				90	---Winter mushroom
419	0709.5940	---草菇	13	0				90	---Paddy Straw mushroom
420	0709.5950	---口蘑	13	0				90	---Tricholoma mongolicum Imai
421	0709.5990	---其他	13	0				90	---Other
422	0709.6000	-辣椒,包括甜椒	13	见表 1				70	-Fruits of the genus *Capsicum* or of the genus *Pimenta*
423	0709.7000	-菠菜	13	0				70	-Spinach, New Zealand spinach and orache spinach (garden spinach)
		-其他:							-Other:
424	0709.9010	---竹笋	13	见表 1				70	---Bamboo shoots
425	0709.9090	---其他	13	见表 1				70	---Other
	07.10	**冷冻蔬菜(不论是否蒸煮):**							**Vegetables (uncooked or cooked by steaming or boiling in water), frozen:**
426	0710.1000	-马铃薯	13	0				70	-Potatoes
		-豆类蔬菜,不论是否脱荚:							-Leguminous vegetables, shelled or unshelled:
427	0710.2100	--豌豆	13	0				70	--Peas (*Pisum sativum*)
		--豇豆及菜豆:							--Beans (*Vigna spp.*, *Phaseolus spp.*):
428	0710.2210	---红小豆	13	见表 1				70	---Small red (Adzuki) beans (*Phaseolus* or *Vigna angularis*)
429	0710.2290	---其他	13	见表 1				70	---Other
430	0710.2900	--其他	13	见表 1				70	--Other
431	0710.3000	-菠菜	13	0				70	-Spinach, New Zealand spinach and orache spinach (garden spinach)
432	0710.4000	-甜玉米	10	见表 1				70	-Sweet corn
		-其他蔬菜:							-Other vegetables:
433	0710.8010	---松茸	13	见表 1				70	---Sungmo
434	0710.8020	---蒜苔及蒜苗(青蒜)	13	见表 1				70	---Garlic stems, garlic seedlings
435	0710.8030	---蒜头	13	见表 1				70	---Garlic bulbs
436	0710.8090	---其他	13	见表 1				70	---Other
437	0710.9000	-什锦蔬菜	10	见表 1				70	-Mixtures of vegetables
	07.11	**暂时保藏(例如,使用二氧化硫气体、盐水、亚硫酸水或其他防腐液)的蔬菜,但不适于直接食用的:**							**Vegetables provisionally preserved (for example, by sulphur dioxide gas, in brine, in suphur water or in other preservative solutions), but unsuitable in that state for immediate consumption:**
438	0711.2000	-油橄榄	13	0				70	-Olives
439	0711.3000	-刺山柑	13	0				70	-Capers
440	0711.4000	-黄瓜及小黄瓜	13	见表 1				70	-Cucumbers and gherkins
		-蘑菇及块菌:							-Mushrooms and truffles:
		--伞菌属蘑菇:							--Mushrooms of the genus *Agaricus*:
		---盐水的:							---In brine:
441	0711.5112	----白蘑菇	13	0				90	----White mushroom
442	0711.5119	----其他	13	0				90	----Other

序号 No.	税则号列 Tariff Item	货品名称	最惠国税率(%) M.F.N.	协定税率(%) Agreement 东盟 ASEAN	曼谷协定 BA	香港 H.K.	澳门 Macao	普通税率(%) Gen.	Article Description
443	0711.5190	---其他	13	0				90	---Other
		--其他:							--Other:
		---盐水的:							---Inbrine:
444	0711.5911	----松茸	13	0				90	----Sungmo
445	0711.5919	----其他	13	0				90	----Other
446	0711.5990	---其他	13	0				90	---Other
		-其他蔬菜;什锦蔬菜:							-Other:
		---盐水的:							---Inbrine:
447	0711.9031	----竹笋	13	见表1				70	----Bamboo shoots
448	0711.9034	----大蒜	13	见表1				70	----Garlic
449	0711.9039	----其他	13	见表1				70	----Other
450	0711.9090	---其他	13	见表1				90	---Other
	07.12	**干蔬菜,整个、切块、切片、破碎或制成粉状,但未经进一步加工的:**							**Dried vegetables, whole, cut, sliced, broken or in powder, but not further prepared:**
451	0712.2000	-洋葱	13	见表1				80	-Onions
		-蘑菇、木耳、银耳及块菌:							-Mushrooms, wood ears (*Auricularia spp.*), jelly fungi (*Tremella spp.*) and truffles:
452	0712.3100	--伞菌属蘑菇	13	见表1	9			80	--Mushrooms of the genus *Agaricus*
453	0712.3200	--木耳	13	0				100	--Wood ears (*Auricularia spp.*)
454	0712.3300	--银耳	13	0				90	--Jelly fungi (*Tremella spp.*)
		--其他:							--Other:
455	0712.3910	---香菇	13	见表1	9				---Shiitake
456	0712.3920	---金针菇	13	见表1	9				---Winter mushroom
457	0712.3930	---草菇	13	见表1	9				---Paddy straw mushroom
458	0712.3940	---口蘑	13	见表1	9				---Tricholoma mongolicum Imai
459	0712.3950	---牛肝菌	13	见表1	9				---Boletus;Porcini
460	0712.3990	---其他	13	见表1	9				---Other
		-其他蔬菜;什锦蔬菜:							-Other vegetables; mixtures of vegetables:
461	0712.9010	---笋干丝	13	见表1				80	---Bamboo shoots
462	0712.9020	---紫萁(薇菜干)	13	见表1				80	---Osmund
463	0712.9030	---金针菜(黄花菜)	13	见表1				80	---Day lily flowers
464	0712.9040	---蕨菜	13	见表1				80	---Wild brake
465	0712.9050	---大蒜	13	见表1				80	---Garlic
466	0712.9060	---甜椒	13	见表1				80	---Capsicum annuum Var. grossum
467	0712.9090	---其他	13	见表1				80	---Other
	07.13	**脱荚的干豆,不论是否去皮或分瓣:**							**Dried leguminous vegetables, shelled, whether or not skinned or split:**
		-豌豆:							-Peas (*Pisum sativum*):
468	0713.1010	---种用	0	0				0	---Seed
469	0713.1090	---其他	5	0				20	---Other
		-鹰嘴豆:							-Chickpeas (garbanzos):
470	0713.2010	---种用	0	0				0	---Seed
471	0713.2090	---其他	7	0				20	---Other

序号 No.	税则号列 Tariff Item	货品名称	最惠国税率(%) M.F.N.	协定税率(%) Agreement 东盟 ASEAN	曼谷协定 BA	香港 H.K.	澳门 Macao	普通税率(%) Gen.	Article Description
		-豇豆及菜豆:							-Beans (*Vigna spp.*, *Phaseolus spp.*):
		--绿豆:							--Beans of the species *Vigna mungo* (*L.*) *Hepper or Vigna radiata* (*L.*) *Wilczek*:
472	0713.3110	---种用	0	见表1				0	---Seed
473	0713.3190	---其他	3	见表1				11	---Other
		--赤豆:							--Small red (Adzuki) beans (*Phaseolus* or *Vigna angularis*):
		---种用:							---Seed:
474	0713.3211	----红小豆	0	见表1				0	----Small red (Adzuri) beans
475	0713.3219	----其他	0	见表1				0	----Other
476	0713.3290	---其他	3	见表1				14	---Other
		--芸豆:							--Kidney beans, including white pea beans (*Phaseolus vulgaris*):
477	0713.3310	---种用	0	见表1				0	---Seed
478	0713.3390	---其他	7.5	见表1				20	---Other
479	0713.3900	--其他	7	见表1				20	--Other
		-扁豆:							-Lentils:
480	0713.4010	---种用	0	0				0	---Seed
481	0713.4090	---其他	7	0				20	---Other
		-蚕豆:							-Broad beans (*Vicia faba var. Major*) and horse beans (*Vicia faba var. equina*, *Vicia faba var. minor*):
482	0713.5010	---种用	0	0				0	---Seed
483	0713.5090	---其他	7	0				20	---Other
		-其他:							-Other:
484	0713.9010	---种用干豆	0	见表1				0	---Seed
485	0713.9090	---其他	7	见表1				20	---Other
	07.14	**鲜、冷、冻或干的木薯、竹芋、兰科植物块茎、菊芋、甘薯及含有高淀粉或菊粉的类似根茎,不论是否切片或制成团粒;西谷茎髓:**							**Manioc, arrowroot, salep, Jerusalem artichokes, sweet potatoes and similar roots and tubers with high starch or inulin content, fresh, chilled, frozen or dried, whether or not sliced or in the form of pellets; sago pith:**
		-木薯:							-Manioc(cassava):
486	0714.1010	---鲜的	10	见表1				30	---Fresh
487	0714.1020	---干的	5	见表1				30	---Dried
488	0714.1030	---冷或冻的	10	见表1				80	---Chilled or frozen
		-甘薯:							-Sweet potatoes:
		---鲜的:							-Fresh:
489	0714.2011	----种用	0	见表1				50	----For cultivatron
490	0714.2019	----其他	13	见表1				50	----Other
491	0714.2020	---干的	13	见表1				50	---Dried
492	0714.2030	---冷或冻的	13	见表1				80	---Chilled or frozen
		-其他:							-Other:

序号 No.	税则号列 Tariff Item	货品名称	最惠国税率(%) M.F.N.	协定税率(%)Agreement 东盟 ASEAN	曼谷协定 BA	香港 H.K.	澳门 Macao	普通税率(%) Gen.	Article Description
493	0714.9010	---荸荠	13	见表1				50	---Water chestnut
		---藕：							---Lotus (*Nelumbo nucifera*) rootstock：
494	0714.9021	----种用	0	见表1				0	----For cultivation
495	0714.9029	----其他	13	见表1				50	----Other
496	0714.9090	---其他	13	见表1				50	---Other

第八章
食用水果及坚果；柑桔属水果或甜瓜的果皮

注释：

一、本章不包括非供食用的坚果或水果。

二、冷藏的水果和坚果应按相应的鲜果税号归类。

三、本章的干果可以部分复水或为下列目的进行其他处理：

(一) 为保藏或保持其稳定性（例如，经适度热处理或硫化处理、添加山梨酸或山梨酸钾）；

(二) 为改进或保持其外观（例如，添加植物油或少量葡萄糖浆）。

但必须保持干果的特征。

Chapter 8
Edible fruit and nuts; peel of citrus fruit or melons

Notes:

1. This Chapter does not cover inedible nuts or fruits.

2. Chilled fruits and nuts are to be classified in the same headings as the corresponding fresh fruits and nuts.

3. Dried fruit or dried nuts of this Chapter may be partially rehydrated, or treated for the following purposes:

(a) For additional preservation or stabilization (e. g., by moderate heat treatment, sulphuring, the addition of sorbic acid or potassium sorbate).

(b) To improve or maintain their appearance (e. g., by the addition of vegetable oil or small quantities of glucose syrup).

Provided that they retain the character of dried fruit or dried nuts.

序号 No.	税则号列 Tariff Item	货品名称	最惠国税率(%) M.F.N.	协定税率(%) Agreement 东盟 ASEAN	曼谷协定 BA	香港 H.K.	澳门 Macao	普通税率(%) Gen.	Article Description
	08.01	**鲜或干的椰子、巴西果及腰果，不论是否去壳或去皮：**							**Coconuts, Brazil nuts and cashew nuts, fresh or dried, whether or not shelled or peeled:**
		-椰子：							-Coconuts:
497	0801.1100	--干的	12	0	10.5			80	--Desiccated
		--其他：							--Other:
498	0801.1910	---种用	0	见表1	0			0	---seedlings
499	0801.1990	---其他	12	见表1	10.5			80	---Other
		-巴西果：							-Brazil nuts:
500	0801.2100	--未去壳	10	0				80	--In shell
501	0801.2200	--去壳	10	0				80	--Shelled
		-腰果：							-Cashew nuts:
502	0801.3100	--未去壳	20	见表1				70	--In shell
503	0801.3200	--去壳	10	见表1				70	--Shelled
	08.02	**鲜或干的其他坚果，不论是否去壳或去皮：**							**Other nuts, fresh or dried, whether or not shelled or peeled:**
		-巴旦杏：							-Almonds:
504	0802.1100	--未去壳	24	见表1				70	--In shell
505	0802.1200	--去壳	10	见表1				70	--Shelled
		-榛子：							-Hazelnuts or filberts(*Corylus spp*):
506	0802.2100	--未去壳	25	见表1				70	--In shell
507	0802.2200	--去壳	10	见表1				70	--Shelled
		-核桃：							-Walnuts:
508	0802.3100	--未去壳	25	见表1				70	--In shell

序号 No.	税则号列 Tariff Item	货品名称	最惠国税率(%) M.F.N.	协定税率(%) Agreement 东盟 ASEAN	曼谷协定 BA	香港 H.K.	澳门 Macao	普通税率(%) Gen.	Article Description
509	0802.3200	--去壳	20	见表 1				70	--Shelled
		-栗子:							-Chestnuts(*Castanea spp.*):
510	0802.4010	---板栗	25	见表 1				70	---Chestnuts
511	0802.4090	---其他	25	见表 1				70	---Other
512	0802.5000	-阿月浑子果(开心果)	10	见表 1				70	-Pistachios
		-其他:							-Other:
513	0802.9010	---槟榔	10	见表 1				30	---Betel nuts
514	0802.9020	---白果	25	见表 1				70	---Gingko nuts
515	0802.9030	---松子仁	25	见表 1				70	---Pine-nuts, shelled
		---夏威夷果:							---Hawaiian nuts:
516	0802.9041	----种用	0	见表 1				70	----Seed
517	0802.9049	----其他	24	见表 1				70	----Other
518	0802.9090	---其他	24	见表 1				70	---Other
	08.03	**鲜或干的香蕉,包括芭蕉:**							**Bananas, including plantains, fresh or dried:**
519	0803.0000	鲜或干的香蕉,包括芭蕉	10	0	6.9			40	Bananas, including plantains, fresh or dried
	08.04	**鲜或干的椰枣、无花果、菠萝、鳄梨、番石榴、芒果及山竹果:**							**Dates, figs, pineapples, avocados, guavas, mangoes and mangosteens, fresh or dried:**
520	0804.1000	-椰枣	15	0				40	-Dates
521	0804.2000	-无花果	30	见表 1				70	-Figs
522	0804.3000	-菠萝	12	见表 1	7.9			80	-Pineapples
523	0804.4000	-鳄梨	25	见表 1				80	-Avocados
		-番石榴、芒果及山竹果:							-Guavas, mangoes and mangosteens:
524	0804.5010	---番石榴	15	见表 1	10.6			80	---Guavas
525	0804.5020	---芒果	15	见表 1	10.6			80	---Mangoes
526	0804.5030	---山竹果	15	见表 1	10.6			80	---Mangosteens
	08.05	**鲜或干的柑桔属水果:**							**Citrus fruit, fresh or dried:**
527	0805.1000	-橙	11	见表 1				100	-Oranges
		-柑桔;杂交柑桔:							-Mandarins (including tangerines and satsumas); clementines, wilkings and similar citrus hybrids:
528	0805.2010	---蕉柑	12	见表 1				100	---Chiao-Kan
529	0805.2020	---阔叶柑橘	12	见表 1				100	---Latifolia Citrus
530	0805.2090	---其他	12	见表 1				100	---Other
531	0805.4000	-柚	12	见表 1				100	-Grapefruit
532	0805.5000	-柠檬及酸橙	11	见表 1				100	-Lemons (*Citrus limon*, *Citrus limonum*) and limes (*Citrus aurantifolia*)
533	0805.9000	-其他	30	见表 1				100	-Other
	08.06	**鲜或干的葡萄:**							**Grapes, fresh or dried:**
534	0806.1000	-鲜的	13	见表 1				80	-Fresh

序号 No.	税则号列 Tariff Item	货品名称	最惠国税率(%) M.F.N.	协定税率(%) Agreement 东盟 ASEAN	曼谷协定 BA	香港 H.K.	澳门 Macao	普通税率(%) Gen.	Article Description
535	0806.2000	-干的	10	见表1				80	-Dried
	08.07	**鲜的甜瓜(包括西瓜)及木瓜:**							**Melons (including watermelons) and papaws(papayas), fresh:**
		-甜瓜,包括西瓜:							-Melons(including watermelons):
536	0807.1100	--西瓜	25	见表1				70	--Watermelons
		--其他:							--Other:
537	0807.1910	---哈蜜瓜	12	见表1				70	---Hami melons
538	0807.1920	---罗马甜瓜及加勒比甜瓜	12	见表1				70	---Cantaloupe and Galia melons
539	0807.1990	--其他	12	见表1				70	---Other
540	0807.2000	-木瓜	25	见表1				70	-Papaws(papayas)
	08.08	**鲜的苹果、梨及榅桲:**							**Apples, pears and quinces, fresh:**
541	0808.1000	-苹果	10	0				100	-Apples
		-梨及榅桲:							-Pears and quinces:
		---梨:							---Pears:
542	0808.2012	----鸭梨及雪梨	12	见表1	10			100	----Ya pears, Hsueh pears
543	0808.2013	----香梨	12	见表1	10			100	----Xiang pears
544	0808.2019	----其他	10	0				100	----Other
545	0808.2020	---榅桲	16	见表1				100	---Quinces
	08.09	**鲜的杏、樱桃、桃(包括油桃)、梅及李:**							**Apricots, cherries, peaches (including nectarines), plums and sloes, fresh:**
546	0809.1000	-杏	25	见表1				70	-Apricots
547	0809.2000	-樱桃	10	0				70	-Cherries
548	0809.3000	-桃,包括油桃	10	0				70	-Peaches, including nectarines
549	0809.4000	-梅及李	10	见表1				70	-Plums and sloes
	08.10	**其他鲜果:**							**Other fruit, fresh:**
550	0810.1000	-草莓	21.3	见表1				80	-Strawberries
551	0810.2000	-木莓、黑莓、桑椹及罗甘莓	25	见表1				80	-Raspberries, blackberries, mulberries and loganberries
552	0810.3000	-黑、白或红的醋栗(加仑子)及鹅莓	25	见表1				80	-Black, white or red currants and gooseberries
553	0810.4000	-蔓越桔及越桔	30	见表1				80	-Cranberries, bilberries and other fruits of the genus *Vaccinium*
554	0810.5000	-猕猴桃	20	见表1	16.5			80	-Kiwifruit
555	0810.6000	-榴莲	20	见表1				80	-Durian
		-其他:							-Other:
556	0810.9010	---荔枝	30	见表1	20			80	---Lychee
557	0810.9030	---龙眼	12	见表1				80	---Longan
558	0810.9040	---红毛丹	20	见表1				80	---Rambutan
559	0810.9050	---番荔枝	20	见表1				80	---Sugar apple
560	0810.9060	---杨桃	20	见表1				80	---Carambola
561	0810.9070	---莲雾	20	见表1	16.4			80	---Wax apple
562	0810.9080	---火龙果	20	见表1	16.4			80	----Dragon fruit

序号 No.	税则号列 Tariff Item	货品名称	最惠国税率(%) M.F.N.	协定税率(%) Agreement 东盟 ASEAN	曼谷协定 BA	香港 H.K.	澳门 Macao	普通税率(%) Gen.	Article Description
563	0810.9090	---其他	20	见表1	16.4			80	---Other
	08.11	**冷冻水果及坚果,不论是否蒸煮、加糖或其他甜物质:**							**Fruit and nuts, uncooked or cooked by steaming or boiling in water, frozen, whether or not containing added sugar or other sweetening matter:**
564	0811.1000	-草莓	30	见表1				80	-Strawberries
565	0811.2000	-木莓、黑莓、桑椹、罗甘莓、黑、白或红的醋栗(加仑子)及鹅莓	30	见表1				80	-Raspberries, blackberries, mulberries, loganberries, black, white or red currants and gooseberries
		-其他:							-Other:
566	0811.9010	---栗子,未去壳	30	见表1				80	---Chestnuts, in shell
567	0811.9090	---其他	30	见表1				80	---Other
	08.12	**暂时保藏(例如,使用二氧化硫气体、盐水、亚硫酸水或其他防腐液)的水果及坚果,但不适于直接食用的:**							**Fruit and nuts, provisionally preserved (for example, by sulphur dioxide gas, in brine, in sulphur water or in other preservative solutions), but unsuitable in that state for immediate consumption:**
568	0812.1000	-樱桃	30	见表1				80	-Cherries
569	0812.9000	-其他	29.5	见表1				80	-Other
	08.13	**税号08.01至08.06以外的干果;本章的什锦坚果或干果:**							**Fruit, dried, other than that of headings No. 08.01 to 08.06; mixtures of nuts or dried fruits of this Chapter:**
570	0813.1000	-杏	25	见表1				70	-Apricots
571	0813.2000	-梅及李	25	见表1				70	-Prunes
572	0813.3000	-苹果	25	见表1				70	-Apples
		-其他干果:							-Other fruit:
573	0813.4010	---龙眼干、肉	20	见表1				70	---Longans and longan pulps
574	0813.4020	---柿饼	25	见表1				70	---Persimmons
575	0813.4030	---红枣	25	见表1				70	---Red jujubes
576	0813.4040	---荔枝干	25	见表1				70	---Preserved litchi
577	0813.4090	---其他	25	见表1				70	---Other
578	0813.5000	-本章的什锦坚果或干果	18	见表1				70	-Mixtures of nuts or dried fruits of this Chapter
	08.14	**柑桔属水果或甜瓜(包括西瓜)的果皮,鲜、冻、干或用盐水、亚硫酸水或其他防腐液暂时保藏的:**							**Peel of citrus fruitor melons (including watermelons), fresh, frozen, dried or provisionally preserved in brine, in sulphur water or in other preservative solutions:**
579	0814.0000	柑桔属水果或甜瓜(包括西瓜)的果皮,鲜、冻、干或用盐水、亚硫酸水或其他防腐液暂时保藏的	25	见表1				70	Peel of citrus fruit or melons(including watermelons), fresh, frozen, dried or provisionally preserved in brine, in sulphur water or in other preservative solutions

第 九 章
咖啡、茶、马黛茶及调味香料

Chapter 9
Coffee, tea, mate and spices

注释：

一、税号 09.04 至 09.10 所列产品的混合物，应按下列规定归类：

（一）同一税号的两种或两种以上产品的混合物仍应归入该税号；

（二）不同税号的两种或两种以上产品的混合物应归入税号 09.10。

税号 09.04 至 09.10 的产品（或上述（一）或（二）项的混合物）如添加了其他物质，只要所得的混合物保持了原产品的基本特性，其归类应不受影响。基本特性已经改变的，则不应归入本章；构成混合调味品的，应归入税号 21.03。

二、本章不包括荜澄茄椒或税号 12.11 的其他产品。

Notes:

1. Mixtures of the products of headings No.09.04 to 09.10 are to be classified as follows:

(a) Mixtures of two or more of the products of the same heading are to be classified in that heading;

(b) Mixtures of two or more of the products of different headings are to be classified in heading No.09.10.

The addition of other substances to the products of headings No.09.04 to 09.10 (or to the mixtures referred to in paragraph (a) or (b) above) shall not affect their classification provided the resulting mixtures retain the essential character of the goods of those headings. Otherwise such mixtures are not classified in this Chapter; those constituting mixed condiments or mixed seasonings are classified in heading No.21.03.

2. This Chapter does not cover Cubeb pepper (*Piper cubeba*) or other products of heading No.12.11.

序号 No.	税则号列 Tariff Item	货品名称	最惠国税率(%) M.F.N.	协定税率(%) Agreement				普通税率(%) Gen.	Article Description
				东盟 ASEAN	曼谷协定 BA	香港 H.K.	澳门 Macao		
	09.01	**咖啡，不论是否焙炒或浸除咖啡碱；咖啡豆荚及咖啡豆皮；含咖啡的咖啡代用品：**							**Coffee, whether or not roasted or decaffeinated; coffee husks and skins; coffee substitutes containing coffee in any proportion:**
		-未焙炒的咖啡：							-Coffee, not roasted:
580	0901.1100	--未浸除咖啡碱	8					50	--Not decaffeinated
581	0901.1200	--已浸除咖啡碱	8					50	--Decaffeinated
		-已焙炒的咖啡：							-Coffee, roasted:
582	0901.2100	--未浸除咖啡碱	15			0	0	80	--Not decaffeinated
583	0901.2200	--已浸除咖啡碱	15	见表 1			0	80	--Decaffeinated
		-其他：							-Other:
584	0901.9010	---咖啡豆荚及咖啡豆皮	10					30	---Coffee husks and skins
585	0901.9020	---含咖啡的咖啡代用品	30					80	---Coffee substitutes containing coffee
	09.02	**茶，不论是否加香料：**							**Tea, whether or not flavoured:**
		-绿茶（未发酵），内包装每件净重不超过 3 公斤：							-Green tea (not fermented) in immediate packings of a content not exceeding 3kg:
586	0902.1010	---花茶	15		12.5			100	---Flavoured
587	0902.1090	---其他	15		12.5			100	---Other
		-其他绿茶（未发酵）：							-Other green tea (not fermented):
588	0902.2010	---花茶	15		12.5			100	---Flavoured

序号 No.	税则号列 Tariff Item	货品名称	最惠国税率(%) M.F.N.	协定税率(%) Agreement 东盟 ASEAN	曼谷协定 BA	香港 H.K.	澳门 Macao	普通税率(%) Gen.	Article Description
589	0902.2090	---其他	15		12.5			100	---Other
		-红茶(已发酵)及半发酵茶,内包装每件净重不超过3公斤:							-Black tea(fermented) and partly fermented tea, in immediate packings of a content not exceeding 3kg:
590	0902.3010	---乌龙茶	15		12.5	0		100	---Oolong tea
591	0902.3020	---普洱茶	15			0		100	---Pu-er tea
592	0902.3090	---其他	15		12.5	0		100	---Other
		-其他红茶(已发酵)及半发酵茶:							-Other black tea (fermented) and other partly fermented tea:
593	0902.4010	---乌龙茶	15		12.5	0		100	---Oolong tea
594	0902.4020	---普洱茶	15			0		100	---Pu-er tea
595	0902.4090	---其他	15		12.5	0		100	---Other
	09.03	**马黛茶:**							**Mate:**
596	0903.0000	马黛茶	10					100	Mate
	09.04	**胡椒;辣椒干及辣椒粉:**							**Pepper of the genus *Piper*; dried or crushed or ground fruits of the genus *Capsicum or of the genus Pimenta*:**
		-胡椒:							-Pepper:
597	0904.1100	--未磨	20					70	--Neither crushed nor ground
598	0904.1200	--已磨	20					70	--Crushed or ground
		-辣椒干及辣椒粉:							-Fruits of the genus *Capsicum* or of the genus *Pimenta*, dried or crushed or ground:
599	0904.2010	---辣椒干	20					70	---Neither crushed nor ground
600	0904.2020	---辣椒粉	20					70	---Crushed or ground
	09.05	**香子兰豆:**							**Vanilla:**
601	0905.0000	香子兰豆	15					50	Vanilla
	09.06	**肉桂及肉桂花:**							**Cinnamon and cinnamon-tree flowers:**
602	0906.1000	-未磨	5					50	-Neither crushed nor ground
603	0906.2000	-已磨	15					50	-Crushed or ground
	09.07	**丁香(母丁香、公丁香及丁香梗):**							**Cloves (whole fruit, cloves and stems):**
604	0907.0000	丁香(母丁香、公丁香及丁香梗)	3					14	Cloves(whole fruit, cloves and stems)
	09.08	**肉豆蔻、肉豆蔻衣及豆蔻:**							**Nutmeg, mace and cardamoms:**
605	0908.1000	-肉豆蔻	8					30	-Nutmeg
606	0908.2000	-肉豆蔻衣	8					30	-Mace
607	0908.3000	-豆蔻	3					14	-Cardamoms
	09.09	**茴芹子、八角茴香、小茴香子、芫荽子、枯茗子及贳蒿子;杜松果:**							**Seeds of anise, badian, fennel, coriander, cumin or caraway; juniper berries:**
		-茴芹子或八角茴香:							-Seeds of anise or badian:
608	0909.1010	---八角茴香	20					90	---Star aniseed
609	0909.1090	---其他	15					50	---Other

序号 No.	税则号列 Tariff Item	货品名称	最惠国税率(%) M.F.N.	协定税率(%) Agreement				普通税率(%) Gen.	Article Description
				东盟 ASEAN	曼谷协定 BA	香港 H.K.	澳门 Macao		
610	0909.2000	-芫荽子	15					50	-Seeds of coriander
611	0909.3000	-枯茗子	15					50	-Seeds of cumin
612	0909.4000	-贳蒿子	15					50	-Seeds of caraway
613	0909.5000	-小茴香子;杜松果	15					50	-Seeds of fennel;juniper berries
	09.10	**姜、番红花、姜黄、麝香草、月桂叶、咖喱及其他调味香料:**							**Ginger, saffron, turmeric (curcuma), thyme, bay leaves, curry and other spices:**
614	0910.1000	-姜	15					50	-Ginger
615	0910.2000	-番红花	2					14	-Saffron
616	0910.3000	-姜黄	15					50	-Turmeric(curcuma)
617	0910.4000	-麝香草;月桂叶	15					50	-Thyme;bay leaves
618	0910.5000	-咖喱	15					50	-Curry
		-其他调味香料:							-Other spices:
619	0910.9100	--本章注释一(二)所述的混合物	15					50	--Mixtures referred to in Note 1(b) to this Chapter
620	0910.9900	--其他	15			0		50	--Other

第十章 谷物

Chapter 10 Cereals

注释：

一、（一）本章各税号所列产品必须带有谷粒，不论是否成穗或带杆。

（二）本章不包括已去壳或经其他加工的谷物。但去壳、碾磨、磨光、上光、半熟或破碎的稻米仍应归入税号10.06。

二、税号10.05不包括甜玉米（第七章）。

子目注释：

所称"硬粒小麦"，是指硬粒小麦属的小麦及以该属具有相同染色体数目（28）的小麦种间杂交所得的小麦。

Notes:

1. (a) The products specified in the headings of this Chapter are to be classified in those headings only if grains are present, whether or not in the ear or on the stalk.

(b) This Chapter does not cover grains which have been hulled or otherwise worked. However, rice, husked, milled, polished, glazed, parboiled or broken remains classified in heading No.10.06.

2. Heading No.10.05 does not cover sweet corn (Chapter 7).

Subheading Note:

The term "durum wheat" means wheat of the *Triticum durum* species and the hybrids derived from the inter-specific crossing of *Triticum durum* which have the same number (28) of chromosomes as that species.

序号 No.	税则号列 Tariff Item	货品名称	最惠国税率(%) M.F.N.	协定税率(%) Agreement 东盟 ASEAN	曼谷协定 BA	香港 H.K.	澳门 Macao	普通税率(%) Gen.	Article Description
	10.01	**小麦及混合麦：**							**Wheat and maslin:**
621	1001.1000	-硬粒小麦	见表5					见表5	-Durum wheat
		-其他：							-Other:
622	1001.9010	---种用	见表5					见表5	---Seed
623	1001.9090	---其他	见表5					见表5	---Other
	10.02	**黑麦：**							**Rye:**
624	1002.0010	---种用	0					0	---Seed
625	1002.0090	---其他	3					8	---Other
	10.03	**大麦：**							**Barley:**
626	1003.0010	---种用	0					160	---Seed
627	1003.0090	---其他	3					160	---Other
	10.04	**燕麦：**							**Oats:**
628	1004.0010	---种用	0					0	---Seed
629	1004.0090	---其他	2					8	---Other
	10.05	**玉米：**							**Maize(corn):**
630	1005.1000	-种用	见表5					见表5	-Seed
631	1005.9000	-其他	见表5					见表5	-Other
	10.06	**稻谷、大米：**							**Rice:**
		-稻谷：							-Rice in husk (paddy or rough):
		---种用：							---Seed:
632	1006.1011	----籼米	见表5					见表5	----Long grain
633	1006.1019	----其他	见表5					见表5	----Other
		---其他：							---Other:
634	1006.1091	----籼米	见表5					见表5	----Long grain
635	1006.1099	----其他	见表5					见表5	----Other

序号 No.	税则号列 Tariff Item	货品名称	最惠国税率(%) M.F.N.	协定税率(%) Agreement 东盟 ASEAN	曼谷协定 BA	香港 H.K.	澳门 Macao	普通税率(%) Gen.	Article Description
		-糙米:							-Husked (brown) rice:
636	1006.2010	---籼米	见表5					见表5	---Long grain
637	1006.2090	---其他	见表5					见表5	---Other
		-精米,不论是否磨光或上光:							-Semi-missed or wholly missed rice, whether or not polished or glazed:
638	1006.3010	---籼米	见表5					见表5	---Long grain
639	1006.3090	---其他	见表5					见表5	---Other
		-碎米:							-Broken rice:
640	1006.4010	---籼米	见表5					见表5	---Long grain
641	1006.4090	---其他	见表5					见表5	---Other
	10.07	**食用高粱:**							**Grain sorghum:**
642	1007.0010	---种用	0					0	---Seed
643	1007.0090	---其他	2					8	---Other
	10.08	**荞麦、谷子及加那利草子;其他谷物:**							**Buckwheat, millet and canary seed; other cereals:**
644	1008.1000	-荞麦	2					8	-Buckwheat
645	1008.2000	-谷子	2					8	-Millet
646	1008.3000	-加那利草子	2					8	-Canary seed
		-其他谷物:							-Other cereals:
647	1008.9010	---种用	0					0	---Seed
648	1008.9090	---其他	3					8	---Other

第十一章
制粉工业产品；麦芽；淀粉；菊粉；面筋

Chapter 11
Products of the milling industry; malt; starches; inulin; wheat gluten

注释：

一、本章不包括：

（一）作为咖啡代用品的焙制麦芽（税号09.01或21.01）；

（二）税号19.01的经制作的细粉、粗粒、粗粉或淀粉；

（三）税号19.04的玉米片及其他产品；

（四）税号20.01、20.04或20.05的经制作或保藏的蔬菜；

（五）药品（第三十章）；

（六）具有芳香料制品或化妆盥洗品性质的淀粉（第三十三章）。

二、（一）下表所列谷物碾磨产品按干制品重量计如果同时符合以下两个条件，应归入本章；但是，整粒、滚压、制片或磨碎的谷物胚芽均应归入税目11.04：

1. 淀粉含量（按修订的尤艾斯旋光法测定）超过表列第（2）栏的比例；

2. 灰分含量（除去任何添加的矿物质）不超过表列第（3）栏的比例。

否则，应归入税号23.02。

（二）符合上述规定归入本章的产品，如果用表列第（4）或第（5）栏规定孔径的金属丝网筛过筛，其通过率按重量计不低于表列比例的，应归入税号11.01或11.02。

否则，应归入税号11.03或11.04。

Notes:

1. This Chapter does not cover:

(a) Roasted malt put up as coffee substitutes (heading No.09.01 or 21.01);

(b) Prepared flours, groats, meals or starches of heading No.19.01;

(c) Corn flakes or other products of heading No.19.04;

(d) Vegetables, prepared or preserved, of heading No.20.01, 20.04 or 20.05;

(e) Pharmaceutical products (Chapter 30); or

(f) Starches having the character of perfumery, cosmetic or toilet preparations (Chapter 33).

2. (A) Products from the milling of the cereals listed in the table below fall in this Chapter if they have, by weight on the dry product (However, germ of cereals, whole, rolled, flaked or ground is always classified in heading No.11.04):

(a) a starch content (determined by the modified Ewers polarimetric method) exceeding that indicated in Column (2); and

(b) an ash content (after deduction of any added minerals) not exceeding that indicated in Column (3).

Otherwise, they fall in heading No.23.02.

(B) Products falling in this Chapter under the above provisions shall be classified in heading No.11.01 or 11.02 if the percentage passing through a woven metal wire cloth sieve with the aperture indicated in Column (4) or (5) is not less, by weight, than that shown against the cereal concerned.

Otherwise, they fall in heading No.11.03 or 11.04.

谷　　物 (1)	淀粉含量 (2)	灰分含量 (3)	通过下列孔径筛子的比率	
			315 微米 (4)	500 微米 (5)
小麦及黑麦	45%	2.5%	80%	—
大　　麦	45%	3%	80%	—
燕　　麦	45%	5%	80%	—
玉米及高粱	45%	2%	—	90%
大　　米	45%	1.6%	80%	—
荞　　麦	45%	4%	80%	—

Cereal (1)	Starch content (2)	Ash content (3)	Rate of passage through a sieve with an aperture of	
			315 micrometres (microns) (4)	500 micrometres (microns) (5)
Wheat and rye	45%	2.5%	80%	-
Barley	45%	3%	80%	-
Oats	45%	5%	80%	-
Maize (corn) and grain Sorghum	45%	2%	-	90%
Rice	45%	1.6%	80%	-
Buckwheat	45%	4%	80%	-

三、税号 11.03 所称“粗粒”及“粗粉”，是指谷物经碾碎所得的下列产品：

（一）玉米产品，用 2 毫米孔径的金属丝网筛过筛，通过率按重量计不低于 95% 的；

（二）其他谷物产品，用 1.25 毫米孔径的金属丝网筛过筛，通过率按重量计不低于 95% 的。

3. For the purposes of heading No. 11.03, the terms “groats” and “meal” mean products obtained by the fragmentation of cereal grains, of which:

(a) in the case of maize (corn) products, at least 95% by weight passes through a woven metal wire cloth sieve with an aperture of 2mm.

(b) in the case of other cereal products, at least 95% by weight passes through a woven metal wire cloth sieve with an aperture of 1.25mm.

序号 No.	税则号列 Tariff Item	货品名称	最惠国税率(%) M.F.N.	协定税率(%) Agreement 东盟 ASEAN	曼谷协定 BA	香港 H.K.	澳门 Macao	普通税率(%) Gen.	Article Description
	11.01	**小麦或混合麦的细粉：**							**Wheat or maslin flour:**
649	1101.0000	小麦或混合麦的细粉	见表 5					见表 5	Wheat or maslin flour
	11.02	**其他谷物细粉，但小麦或混合麦的细粉除外：**							**Cereal flours other than of wheat or maslin:**
650	1102.1000	-黑麦细粉	5					14	-Rye flour
651	1102.2000	-玉米细粉	见表 5					见表 5	-Maize(corn)flour
		-大米细粉：							-Rice flour:
652	1102.3010	---籼米的	见表 5					见表 5	---Of long grain
653	1102.3090	---其他	见表 5					见表 5	---Other
654	1102.9000	-其他	5					14	-Other
	11.03	**谷物的粗粒、粗粉及团粒：**							**Cereal groats, meal and pellets:**
		-粗粒及粗粉：							-Groats and meal:
655	1103.1100	--小麦的	见表 5					见表 5	--Of wheat
656	1103.1300	--玉米的	见表 5					见表 5	--Of maize(corn)
		--其他：							--Other:
657	1103.1910	---燕麦的	5					14	---Of oats
		---大米的：							---Of rice:

序号 No.	税则号列 Tariff Item	货品名称	最惠国税率(%) M.F.N.	协定税率(%) Agreement 东盟 ASEAN	曼谷协定 BA	香港 H.K.	澳门 Macao	普通税率(%) Gen.	Article Description
658	1103.1921	----籼米的	见表5					见表5	----Of long grain
659	1103.1929	----其他	见表5					见表5	----Other
660	1103.1990	---其他	5					14	---Other
		-团粒：							-Pellets：
661	1103.2010	---小麦的	见表5					见表5	--Of wheat
662	1103.2090	---其他	20					50	--Of other cereals
	11.04	**经其他加工的谷物(例如，去壳、滚压、制片、制成粒状、切片或粗磨)，但税号10.06的稻谷、大米除外；谷物胚芽，整粒、滚压、制片或磨碎的：**							**Cereal grains otherwise worked (for example, hulled, rolled, flaked, pearled, sliced or kibbled), except rice of heading No.10.06; germ of cereals, whole, rolled, flaked or ground:**
		-滚压或制片的谷物：							-Rolled or flaked grains：
663	1104.1200	--燕麦的	20					50	--Of oats
		--其他：							--Other：
664	1104.1910	---大麦的	20					50	---Of barley
665	1104.1990	---其他	20					50	---Other
		-经其他加工的谷物(例如，去壳、制成粒状、切片或粗磨)：							-Other worked grains (for example, hulled, pearled, sliced or kibbled)：
666	1104.2200	--燕麦的	20					50	--Of oats
667	1104.2300	--玉米的	见表5					见表5	--Of maize(corn)
		--其他：							--Other：
668	1104.2910	---大麦的	65					114	---Of barley
669	1104.2990	---其他	20					50	---Other
670	1104.3000	-谷物胚芽，整粒、滚压、制片或磨碎的	20					50	-Germ of cereals, whole, rolled, flaked or ground
	11.05	**马铃薯的细粉、粗粉、粉末、粉片、颗粒及团粒：**							**Flour, meal, powder, flakes, granules and pellets of potatoes:**
671	1105.1000	-细粉、粗粉及粉末	15					50	-Flour, meal and powder
672	1105.2000	-粉片、颗粒及团粒	15					50	-Flakes, granules and pellets
	11.06	**用税号07.13的干豆或税号07.14的西谷茎髓及植物根茎、块茎制成的细粉、粗粉及粉末；用第八章的产品制成的细粉、粗粉及粉末：**							**Flour, meal and powder of the dried leguminous vegetables of heading No.07.13, of sago or of roots or tubers of heading No.07.14; or of the products of Chapter 8:**
673	1106.1000	-用税号07.13的干豆制成的	10					30	-Of the dried leguminous vegetables of heading No.07.13
674	1106.2000	-用税号07.14的西谷茎髓及植物根茎、块茎制成的	20					50	-Of sago or of roots or tubers of heading No.07.14
675	1106.3000	-用第八章的产品制成的	20					80	-Of the products of Chapter 8
	11.07	**麦芽，不论是否焙制：**							**Malt, whether or not roasted:**

序号 No.	税则号列 Tariff Item	货品名称	最惠国税率(%) M.F.N.	协定税率(%)Agreement 东盟 ASEAN	曼谷协定 BA	香港 H.K.	澳门 Macao	普通税率(%) Gen.	Article Description
676	1107.1000	-未焙制	10					50	-Not roasted
677	1107.2000	-已焙制	10					50	-Roasted
	11.08	**淀粉;菊粉:**							**Starches;inulin:**
		-淀粉:							-Starches:
678	1108.1100	--小麦淀粉	20					50	--Wheat starch
679	1108.1200	--玉米淀粉	20					50	--Maize(corn)starch
680	1108.1300	--马铃薯淀粉	15					50	--Potato starch
681	1108.1400	--木薯淀粉	10					50	--Manioc (cassava)starch
682	1108.1900	--其他	20					50	--Other starches
683	1108.2000	-菊粉	20					50	-Inulin
	11.09	**面筋,不论是否干制:**							**Wheat gluten,whether or not dried:**
684	1109.0000	面筋,不论是否干制	18					80	Wheat gluten,whether or not dried

第十二章
含油子仁及果实；杂项子仁及果实；工业用或药用植物；稻草、秸秆及饲料

注释：

一、税号 12.07 主要包括油棕果及油棕仁、棉子、蓖麻子、芝麻、芥子、红花子、罂粟子、牛油树果。但不包括税号 08.01 或 08.02 的产品及油橄榄（第七章或第二十章）。

二、税号 12.08 不仅包括未脱脂的细粉和粗粉，而且包括部分或全部脱脂以及用其本身的油料全部或部分复脂的细粉和粗粉。但不包括税号 23.04 至 23.06 的残渣。

三、甜菜子、草子及其他草本植物种子、观赏用花的种子、蔬菜种子、林木种子、果树种子、巢菜子（蚕豆除外）、羽扇豆属植物种子，可一律视为种植用种子，归入税号 12.09。

但下列各项即使作种子用，也不归入税号 12.09：

（一）豆类蔬菜或甜玉米（第七章）；

（二）第九章的调味香料及其他产品；

（三）谷物（第十章）；

（四）税号 12.01 至 12.07 或 12.11 的产品。

四、税号 12.11 主要包括下列植物或这些植物的某部分：

罗勒、琉璃苣、人参、海索草、甘草、薄荷、迷迭香、芸香、鼠尾草及苦艾。

但税号 12.11 不包括：

（一）第三十章的药品；

（二）第三十三章的芳香料制品及化妆盥洗品；

（三）税号 38.08 的杀虫剂、杀菌剂、除草剂、消毒剂及类似产品。

五、税号 12.12 的“海草及其他藻类”不包括：

（一）税号 21.02 的已死的单细胞微生物；

（二）税号 30.02 的培养微生物；

（三）税号 31.01 或 31.05 的肥料。

Chapter 12
Oil seeds and oleaginous fruits; miscellaneous grains, seeds and fruit; industrial or medicinal plants; straw and fodder

Notes:

1. Heading No.12.07 applies, *inter alia*, to palm nuts and kernels, cotton seeds, castor oil seeds, sesamum seeds, mustard seeds, safflower seeds, poppy seeds and shea nuts (karite nuts). It does not apply to products of heading No.08.01 or 08.02 or to olives (Chapter 7 or Chapter 20).

2. Heading No.12.08 applies not only to non-defatted flours and meals but also to flours and meals which have been defatted or partially defatted and wholly or partially refatted with their original oils. It does not, however, apply to residues of headings No.23.04 to 23.06.

3. For the purposes of heading No.12.09, beet seeds, grass and other herbage seeds, seeds of ornamental flowers, vegetable seeds, seeds of forest trees, seeds of fruit trees, seeds of vetches (other than those of the species *Vicia faba*) or of lupines are to be regarded as “seeds of a kind used for sowing”.

   Heading No.12.09 does not, however, apply to the following even if for sowing:

   (a) Laguminous vegetables or sweet corn (Chapter 7);

   (b) Spices or other products of Chapter 9;

   (c) Cereals (Chapter 10); or

   (d) Products of headings No.12.01 to 12.07 or 12.11.

4. Heading No.12.11 applies, *inter alia*, to the following plants or parts thereof:

   Basil, borage, ginseng, hyssop, liquorice, all species of mint, rosemary, rue, sage and wormwood.

   Heading No.12.11 does not, however, apply to:

   (a) Medicaments of Chapter 30;

   (b) Perfumery, cosmetic or toilet preparations of Chapter 33; or

   (c) Insecticides, fungicides, herbicides, disinfectants or similar products of heading No.38.08.

5. For the purposes of heading No.12.12, the term “seaweeds and other algae” does not include:

   (a) Dead single-cell micro-organisms of heading No.21.02;

   (b) Cultures of micro-organisms of heading No.30.02; or

   (c) Fertilizers of heading No.31.01 or 31.05.

子目注释：

子目 1205.10 所称“低芥子酸油菜子”，是指所获取的固定油中芥子酸含量按重量计低于 2%，以及所得的固体成分每克葡萄糖苷酸（酯）含量低于 30 微摩尔的油菜子。

Subheading Note:

For the purposes of subheading No.1205.10, the expression “low erucic acid rape or colza seeds” means rape of colza seeds yielding a fixed oil which has an erucic acid content of less than 2% by weight and yielding a solid component which contains less than 30 micromoles of glucosinolates per gram.

序号 No.	税则号列 Tariff Item	货品名称	最惠国税率(%) M.F.N.	协定税率(%)Agreement				普通税率(%) Gen.	Article Description
				东盟 ASEAN	曼谷协定 BA	香港 H.K.	澳门 Macao		
	12.01	**大豆,不论是否破碎:**							**Soya beans, whether or not broken:**
685	1201.0010	---种用	0					180	---Seed
		---其他:							---Other:
686	1201.0091	----黄大豆	3					180	----Yellow soya beans
687	1201.0092	----黑大豆	3					180	----Black soya beans
688	1201.0093	----青大豆	3					180	----Green soya beans
689	1201.0099	----其他	3					180	----Other
	12.02	**未焙炒或未烹煮的花生,不论是否去壳或破碎:**							**Ground-nuts, not roasted or otherwise cooked, whether or not shelled or broken:**
		-未去壳:							-In shell:
690	1202.1010	---种用	0					0	---Seed
691	1202.1090	---其他	15					70	---Other
692	1202.2000	-去壳,不论是否破碎	15					70	-Shelled, whether or not broken
	12.03	**干椰子肉:**							**Copra:**
693	1203.0000	干椰子肉	15					30	Copra
	12.04	**亚麻子,不论是否破碎:**							**Linseed, whether or not broken:**
694	1204.0000	亚麻子,不论是否破碎	15					70	Linseed, whether or not broken
	12.05	**油菜子,不论是否破碎:**							**Rape or colza seeds, whether or not broken:**
		-低芥子酸油菜籽:							-Low erucic acid rape or colza seeds:
695	1205.1010	---种用	0					80	---Seed
696	1205.1090	---其他	9					80	---Other
		-其他:							-Other:
697	1205.9010	---种用	0					80	---Seed
698	1205.9090	---其他	9					80	---Other
	12.06	**葵花子,不论是否破碎:**							**Sunflower seeds, whether or not broken:**
699	1206.0010	---种用	0					0	---Seed
700	1206.0090	---其他	15					70	---Other
	12.07	**其他含油子仁及果实,不论是否破碎:**							**Other oil seeds and oleaginous fruits, whether or not broken:**
		-油棕果及油棕仁:							-Palm nuts and kernels:
701	1207.1010	---种用	0					0	---Seeds for cultivation
702	1207.1090	---其他	10					70	---Other
		-棉子:							-Cotton seeds:
703	1207.2010	---种用	0					0	---Seeds for cultivation
704	1207.2090	---其他	15					70	---Other
		-蓖麻子:							-Castor seeds:

序号 No.	税则号列 Tariff Item	货品名称	最惠国税率(%) M.F.N.	协定税率(%)Agreement 东盟 ASEAN	曼谷协定 BA	香港 H.K.	澳门 Macao	普通税率(%) Gen.	Article Description
705	1207.3010	---种用	0					0	---Seeds for cultivation
706	1207.3090	---其他	15					70	---Other
		-芝麻:							-Sesamum seeds:
707	1207.4010	---种用	0		0			0	---Seeds for cultivation
708	1207.4090	---其他	10		9			70	---Other
		-芥子:							-Mustard seeds:
709	1207.5010	---种用	0					0	---Seeds for cultivation
710	1207.5090	---其他	15					70	---Other
		-红花子:							-Safflower seeds:
711	1207.6010	---种用	0					0	---Seeds for cultivation
712	1207.6090	---其他	20					70	---Other
		-其他:							-Other:
713	1207.9100	--罂粟子	20					70	--Poppy seeds
		--其他:							--Other:
714	1207.9910	---种用	0					0	---Seed
		---其他:							---Other:
715	1207.9991	----牛油树果	20					70	----Shea nuts (karite nuts)
716	1207.9999	----其他	10					70	----Other
	12.08	**含油子仁或果实的细粉及粗粉,但芥子粉除外:**							**Flours and meals of oil seeds or oleaginous fruits, other than those of mustard:**
717	1208.1000	-大豆粉	9					70	-Of soya beans
718	1208.9000	-其他	15					80	-Other
	12.09	**种植用的种子、果实及孢子:**							**Seeds, fruit and spores, of a kind used for sowing:**
719	1209.1000	-糖甜菜子	0					0	-Sugar beet seed
		-饲料植物种子:							-Seeds of forage plants:
720	1209.2100	--紫苜蓿子	0					0	--Lucerne(alfalfa)seed
721	1209.2200	--三叶草子	0					0	--Clover(*Trifolium spp.*) seed
722	1209.2300	--羊茅子	0					0	--Fescue seed
723	1209.2400	--草地早熟禾子	0					0	--Kentucky blue grass (*Poa pratensis L.*) seed
724	1209.2500	--黑麦草种子	0					0	--Rye grass (*Lolium multiflorum Lam.*, *Lolium perenne L.*) seed
725	1209.2600	--梯牧草种子	0					0	--Timothy grass seed
		--其他:							--Other:
726	1209.2910	---甜菜籽,糖甜菜籽除外	0					0	---Sugar beet seed
727	1209.2990	---其他	0					0	---Other
728	1209.3000	-草本花卉植物种子	0					0	Seeds of herbaceous plants cultivated principally for their flowers
		-其他:							-Other:
729	1209.9100	--蔬菜种子	0					0	--Vegetable seeds
		--其他:							--Other:
730	1209.9910	---西瓜子	0					0	---Seeds of water melons
731	1209.9920	---甜瓜子	0					0	---Seeds of melons

序号 No.	税则号列 Tariff Item	货品名称	最惠国税率(%) M.F.N.	协定税率(%) Agreement 东盟 ASEAN	曼谷协定 BA	香港 H.K.	澳门 Macao	普通税率(%) Gen.	Article Description
732	1209.9990	---其他	0					0	---Other
	12.10	**鲜或干的啤酒花,不论是否研磨或制成团粒;蛇麻腺:**							**Hop cones, fresh or dried, whethr or not ground, powdered or in the form of pellets; lupulin:**
733	1210.1000	-啤酒花,未经研磨也未制成团粒	20					50	-Hop cones, neither ground nor powdered nor in the form of pellets
734	1210.2000	-啤酒花,经研磨或制成团粒;蛇麻腺	10					50	-Hop cones, ground, powdered or in the form of pellets; lupulin
	12.11	**主要用作香料、药料、杀虫、杀菌或类似用途的植物或这些植物的某部分(包括子仁及果实),鲜或干的,不论是否切割、压碎或研磨成粉:**							**Plants and parts of plants** (including seeds and fruits)**, of a kind used primarily in perfumery, in pharmacy or for insecticidal, fungicidal or similar purposes, fresh or dried, whether or not cut, crushed or powdered:**
		-甘草:							-Liquorice roots:
735	1211.1010	---新疆胀果甘草	6					30	---Zhang Guo licorice(*Glycyrrhiza uralenisis*)
736	1211.1090	---其他	6					30	---Other
		-人参:							-Ginseng roots:
737	1211.2010	---西洋参	8.6					70	---American ginseng
738	1211.2020	---野山参(西洋参除外)	20		16.4			90	---Wild ginseng(other than American ginseng)
		---其他:							---Other:
739	1211.2091	----鲜的	20			0△		50	----fresh
740	1211.2099	----其他	20			0△		50	----Other
741	1211.3000	-古柯叶	9					50	-Coca leaf
742	1211.4000	-罂粟杆	9					50	-Poppy straw
		-其他:							-Other:
		---主要用作药料的植物及其某部分:							---Of a kind used primarily in pharmacy:
743	1211.9011	----当归	6					30	----Radix angelicae sinensis
744	1211.9012	----田七	6					20	----Radix pseudoginseng
745	1211.9013	----党参	6					20	----Radix codonopsitis
746	1211.9014	----黄连	6					20	----Rhizoma coptidis
747	1211.9015	----菊花	6					20	----Flos chrysanthemi
748	1211.9016	----冬虫夏草	6					20	----Cordyceps sinensis
749	1211.9017	----贝母	6					20	----Bulbs fritillariae thunbergii
750	1211.9018	----川芎	6					20	----Rhizoma ligustici
751	1211.9019	----半夏	6					20	----Rhizoma pinelliae
752	1211.9021	----白芍	6					20	----Radix paeoniae lactiflorae
753	1211.9022	----天麻	6					20	----Rhizoma gastrodiae
754	1211.9023	----黄芪	6					30	----Radix astragali
755	1211.9024	----大黄、籽黄	6					20	----Rhubarb
756	1211.9025	----白术	6					20	----Rhizoma atractylodis macrocephalae

序号 No.	税则号列 Tariff Item	货品名称	最惠国税率(%) M.F.N.	协定税率(%) Agreement 东盟 ASEAN	曼谷协定 BA	香港 H.K.	澳门 Macao	普通税率(%) Gen.	Article Description
757	1211.9026	----地黄	6					20	----Radix rehmanniae
758	1211.9027	----槐米	6					20	----Flos sophorae
759	1211.9028	----杜仲	6					20	----Cortex eucommiae
760	1211.9029	----茯苓	6					20	----Poria
761	1211.9031	----枸杞	6					30	----Fructus lycii
762	1211.9032	----大海子	6					20	----Bantaroi seeds
763	1211.9033	----沉香	3					20	----Aloes wood
764	1211.9034	----沙参	6					20	----Adenophora axilliflora
765	1211.9035	----青蒿	6					20	----Southernwood
766	1211.9039	----其他	6					20	----Other
767	1211.9050	---主要用作香料的植物及其某部分	8					50	---Of a kind used primarily in perfumery
		---其他：							---Other：
768	1211.9091	----鱼藤根、除虫菊	3					11	----Derris roots and pyrethrum
769	1211.9099	----其他	9					30	----Other
	12.12	**鲜、冷、冻或干的刺槐豆、海草及其他藻类、甜菜及甘蔗，不论是否碾磨；主要供人食用的其他税号未列名的果核、果仁及植物产品(包括未焙制的菊苣根)：**							**Locust beans, seaweeds and other algae, sugar beet and sugar cane, fresh, chilled, frozen or dried, whether or not ground; fruit stones and kernels and other vegetable products** (including unroasted chicory roots of the variety *Cichorium intybus sativum*) **of a kind used primarily for human consumption, not elsewhere specified or included:**
770	1212.1000	-刺槐豆，包括刺槐豆子	20					70	-Locust beans, including locust bean seeds
		-海草及其他藻类：							-Seaweeds and other algae：
771	1212.2010	---海带	20					70	---Sea tangle
772	1212.2020	---发菜	20		16.2			70	---Black moss
		---裙带菜：							---Pinnatifida：
773	1212.2031	----干的	15					70	----Dried
774	1212.2032	----鲜的	15					70	----Fresh
775	1212.2039	----其他	15					70	----Other
		---紫菜：							---Laver
776	1212.2041	----干的	15					70	----Dried
777	1212.2042	----鲜的	15					70	----Fresh
778	1212.2049	----其他	15					70	----Other
779	1212.2050	---马尾藻	15					70	---Sagarsso weed
780	1212.2090	---其他	15		13.1			70	---Other
		-杏、桃(包括油桃)、梅或李的核及核仁：							-Apricot, peach (including nectarine) or plum stones and kernels：
		---杏仁：							---Apricot kernels：
781	1212.3011	----苦杏仁	20					80	----Bitter
782	1212.3012	----甜杏仁	20					80	----Sweet

序号 No.	税则号列 Tariff Item	货品名称	最惠国税率(%) M.F.N.	协定税率(%)Agreement 东盟 ASEAN	曼谷协定 BA	香港 H.K.	澳门 Macao	普通税率(%) Gen.	Article Description
783	1212.3090	---其他	20					80	---Other
		-其他:							-Other:
784	1212.9100	--甜菜	20					70	--Sugar beet
		--其他:							--Other:
		---种用:							---For autivatioy:
785	1212.9953	----甘蔗	0					0	----Surgar cane
786	1212.9959	----其他	0					0	----Other
		---其他:							---Other:
787	1212.9991	----黑瓜子	20					80	----Black watermelon seeds
788	1212.9992	----红瓜子	20					80	----Red watermelon seeds
789	1212.9993	----白瓜子	20					80	----Pumpkin seeds
790	1212.9994	----莲子	20					80	----Lotus seeds (*Semen Nelumbinis*)
791	1212.9995	----甘蔗	20					70	----Sugar cane
792	1212.9999	----其他	30					70	----Other
	12.13	**未经处理的谷类植物的茎、秆及谷壳,不论是否切碎、碾磨、挤压或制成团粒:**							**Cereal straw and husks, unprepared, whether or not chopped, ground, pressed or in the form of pellets:**
		未经处理的谷类植物的茎、秆及谷壳,不论是否切碎、碾磨、挤压或制成团粒:							Cereal straw and husks, unprepared, whether or not chopped, ground, pressed or in the form of pellets:
793	1213.0010	---未经处理的稻草的茎、杆	12					35	---Cereal straw, unprepared
794	1213.0090	---其他	12					35	---Other
	12.14	**芜菁甘蓝、饲料甜菜、饲料用根、干草、紫苜蓿、三叶草、驴喜豆、饲料羽衣甘蓝、羽扇豆、巢菜及类似饲料,不论是否制成团粒:**							**Swedes, mangolds, fodder roots, hay, lucerne(alfalfa), clover, sainfoin, forage kale, lupines, vetches and similar forage products, whether or not in the form of pellets:**
795	1214.1000	-紫苜蓿粗粉及团粒	5					35	-Lucerne(alfalfa) meal and pellets
796	1214.9000	-其他	9					35	-Other

第十三章
虫胶；树胶、树脂及其他植物液、汁

Chapter 13
Lac; gums, resins and other vegetable saps and extracts

注释：

税号 13.02 主要包括甘草、除虫菊、啤酒花、芦荟的浸膏及鸦片，但不包括：

一、按重量计蔗糖含量在 10% 以上或制成糖食的甘草浸膏（税号 17.04）；

二、麦芽膏（税号 19.01）；

三、咖啡精、茶精、马黛茶精（税号 21.01）；

四、构成含酒精饮料的植物汁、液（第二十二章）；

五、樟脑、甘草甜及税号 29.14 或 29.38 的其他产品；

六、罂粟杆浓缩物，按重量计生物碱含量不低于 50%（税目 29.39）；

七、税号 30.03 或 30.04 的药品及税号 30.06 的血型试剂；

八、鞣料或染料的浸膏（税号 32.01 或 32.03）；

九、精油、浸膏、净油、香膏、提取的油树脂或精油的水馏液及水溶液；饮料制造业用的以芳香物质为基料的制剂（第三十三章）；

十、天然橡胶、巴拉塔胶、古塔波胶、银胶菊胶、糖胶树胶或类似的天然树胶（税号 40.01）。

本国注释：

子目号 1302.1100 的鸦片，我国禁止进口。

Notes:

Heading No.13.02 applies, *inter alia*, to liquorice extract and extract of pyrethrum, extract of hops, extract of aloes and opium. The heading does not apply to:

1. Liquorice extract containing more than 10% by weight of sucrose or put up as confectionery (heading No.17.04);
2. Malt extract (heading No.19.01);
3. Extracts of coffee, tea or mate (heading No.21.01);
4. Vegetable saps or extracts constituting alcoholic beverages (Chapter 22);
5. Camphor, glycyrrhizin or other products of heading No.29.14 or 29.38;
6. Concentrates of poppy straw containing not less than 50% by weight of alkaloids (heading 29.39);.
7. Medicaments of heading No.30.03 or 30.04 or bloodgrouping reagents (heading No.30.06);
8. Tanning or dyeing extracts (heading No.32.01 or 32.03);
9. Essential oils, concretes, absolutes, resinoids, extracted oleoresins, aqueous distillates or aqueous solutions of essential oils or preparations based on odoriferous substances of a kind used for the manufacture of beverages (Chapter 33); or
10. Natural rubber, balata, gutta-percha, guayule, chicle or similar natural gums (heading No.40.01).

National note:

Opium of Subheading No.1302.1100 is subject to import ban.

序号 No.	税则号列 Tariff Item	货品名称	最惠国税率(%) M.F.N.	协定税率(%) Agreement 东盟 ASEAN	曼谷协定 BA	香港 H.K.	澳门 Macao	普通税率(%) Gen.	Article Description
	13.01	**虫胶；天然树胶、树脂、树胶脂及油树脂（例如，香树脂）：**							**Lac; natural gums, resins, gum-resins and oleoresins**(for example, balsams):
797	1301.1000	-虫胶	15					40	-Lac
798	1301.2000	-阿拉伯胶	15					40	-Gum Arabic
		-其他：							-Other:
799	1301.9010	---胶黄耆树胶（卡喇杆胶）	15					40	---Gum tragacanth
800	1301.9020	---乳香、没药及血竭	3					17	---Olibanum, myrrh and dragon's blood
801	1301.9030	---阿魏	3					17	---Asafoetida
802	1301.9040	---松脂	15					45	---Pine-resin

序号 No.	税则号列 Tariff Item	货品名称	最惠国税率(%) M.F.N.	协定税率(%)Agreement 东盟 ASEAN	曼谷协定 BA	香港 H.K.	澳门 Macao	普通税率(%) Gen.	Article Description
803	1301.9090	---其他	15					45	---Other
	13.02	**植物液汁及浸膏;果胶、果胶酸盐及果胶酸酯;从植物产品制得的琼脂、其他胶液及增稠剂,不论是否改性:**							**Vegetable saps and extracts; pectic substances, pectinates and pectates; agar-agar and other mucilages and thickeners, whether or not modified, derived from vegetable products:**
		-植物液汁及浸膏:							-Vegetable saps and extracts:
804	1302.1100	--鸦片	0					0	--Opium
805	1302.1200	--甘草的	6					20	--Of liquorice
806	1302.1300	--啤酒花的	10					80	--Of hops
807	1302.1400	--除虫菊的或含鱼藤酮植物根茎的	3					11	--Of pyrethrum or of the roots of plants containing rotenone
		--其他:							--Other:
808	1302.1910	---生漆	20					90	---Crude lacquer
809	1302.1920	---印楝素	3						---Azadirachtin
810	1302.1990	---其他	20		15			80	---Other
811	1302.2000	-果胶、果胶酸盐及果胶酸酯	20					80	-Pectic substances, pectinates and pectates
		-从植物产品制得的胶液及增稠剂,不论是否改性:							-Mucilages and thickeners, whether or not modified, derived from vegetable products:
812	1302.3100	--琼脂	10					80	--Agar-agar
813	1302.3200	--从刺槐豆、刺槐豆子或瓜尔豆制得的胶液及增稠剂,不论是否改性	15		10			80	--Mucilages and thickeners, whether or not modified, derived from locust beans, locust bean seeds or guar seeds:
814	1302.3900	--其他	15					80	--Other

第十四章
编结用植物材料；
其他植物产品

Chapter 14
Vegetable plaiting materials; vegetable products not elsewhere specified or included

注释：

一、本章不包括归入第十一类的下列产品：

主要供纺织用的植物材料或植物纤维，不论其加工程度如何；或经过处理使其只能作为纺织原料用的其他植物材料。

二、税号14.01主要包括竹（不论是否劈开、纵锯、切段、圆端、漂白、磨光、染色或进行不燃处理）、劈开的柳条、芦苇及类似品和藤心、藤丝、藤片。但不包括木片条（税号44.04）。

三、税号14.02不包括木丝（税号44.05）。

四、税号14.03不包括供制帚、制刷用成束、成簇的材料（税号96.03）。

Notes:

1. This Chapter does not cover the following products which are to be classified in Section Ⅺ:
Vegetable materials or fibres of vegetable materials of a kind used primarily in the manufacture of textiles, however prepared, or other vegetable materials which have undergone treatment so as to render them suitable for use only as textile materials.

2. Heading No.14.01 applies, *inter alia*, to bamboos (whether or not split, sawn lengthwise, cut to length, rounded at the ends, bleached, rendered non-inflammable, polished or dyed), split osier, reeds and the like, to rattan cores and to drawn or split rattans. The heading does not apply to chipwood (heading No.44.04).

3. Heading No.14.02 does not apply to wood wool (heading No.44.05).

4. Heading No.14.03 does not apply to prepared knots or tufts for broom or brush making (heading No.96.03).

序号 No.	税则号列 Tariff Item	货品名称	最惠国税率(%) M.F.N.	协定税率(%) Agreement 东盟 ASEAN	曼谷协定 BA	香港 H.K.	澳门 Macao	普通税率(%) Gen.	Article Description
	14.01	**主要作编结用的植物材料(例如,竹、藤、芦苇、灯芯草、柳条、酒椰叶,已净、漂白或染色的谷类植物的茎秆,椴树皮):**							**Vegetable materials of a kind used primarily for plaiting (for example, bamboos, rattans, reeds, rushes, osier, raffia, cleaned, bleached or dyed cereal straw, and lime bark):**
815	1401.1000	-竹	10					70	-Bamboos
816	1401.2000	-藤	10					35	-Rattans
		-其他:							-Other:
817	1401.9010	---谷类植物的茎秆(麦秸除外)	10					70	---Cereal straw (other than wheat straw)
818	1401.9020	---芦苇	10					70	---Reeds
819	1401.9030	---灯芯草属	10					70	---Rushes
820	1401.9090	---其他	10					70	---Other
	14.02	**主要作填充或衬垫用的植物材料(例如,木棉、植物毛及大叶藻),不论是否制成有或无支承材料的层片:**							**Vegetable materials of a kind used primarily as stuffing or as padding (for example, kapok, vegetable hair and eelgrass), whether or not put up as a layer with or without supporting material:**

序号 No.	税则号列 Tariff Item	货品名称	最惠国税率(%) M.F.N.	协定税率(%) Agreement 东盟 ASEAN	曼谷协定 BA	香港 H.K.	澳门 Macao	普通税率(%) Gen.	Article Description
821	1402.0000	主要作填充或衬垫用的植物材料(例如,木棉、植物毛及大叶藻),不论是否制成有或无支承材料的层片	15					70	Vegetable materials of a kind used primarily as stuffing or as padding(for example, kapok, vegetable hair and eelgrass), whether or not put up as a layer with or without supporting material
	14.03	**主要供制帚、制刷用的植物材料(例如,帚用高粱、纤维榈纤维、匍匐须芒草及龙舌兰纤维),不论是否成绞或成捆:**							**Vegetable materials of a kind used primarily in brooms or in brushes**(for example, broomcorn, piassave, couch-grass and istle), **whether or not in hanks or bundles:**
822	1403.0000	主要供制帚、制刷用的植物材料(例如,帚用高粱、纤维榈纤维、匍匐须芒草及龙舌兰纤维),不论是否成绞或成捆:	15					70	Vegetable materials of a kind used primarily in brooms or in brushes(for example, broomcorn, piassave, couch-grass and istle), whether or not in hanks or bundles:
	14.04	**其他税号未列名的植物产品:**							**Vegetable products not elsewhere specified or included:**
823	1404.1000	-主要供染料、鞣料用的植物原料	5		4.3			45	-Raw vegetable materials of a kind used primarily in dyeing or tanning
824	1404.2000	-棉短绒	4					30	-Cotton linters
825	1404.9000	-其他	15					70	-Other

第三类

动、植物油、脂及其分解产品；精制的食用油脂；动、植物蜡

第十五章

动、植物油、脂及其分解产品；精制的食用油脂；动、植物蜡

注释：

一、本章不包括：

（一）税号 02.09 的猪脂肪及家禽脂肪；

（二）可可脂、可可油（税号 18.04）；

（三）按重量计税号 04.05 所列产品的含量超过 15%的食品（通常归入第二十一章）；

（四）税号 23.01 的油渣或税号 23.04 至 23.06 的残渣；

（五）第六类的脂肪酸、精制蜡、药品、油漆、清漆、肥皂、芳香料制品、化妆盥洗品、磺化油及其他货品；

（六）从油类提取的油膏（税号 40.02）。

二、税号 15.09 不包括用溶剂提取的橄榄油（税号 15.10）。

三、税号 15.18 不包括变性的油、脂及其分离品，这些货品应归入其相应的未变性油、脂及其分离品的税号。

四、皂料、油脚、硬脂沥青、甘油沥青及羊毛脂残渣，归入税号 15.22。

子目注释：

子目 1514.11 及 1514.19 所称“低芥子酸菜子油”，是指按重量计芥子酸含量低于 2%的固定油

SECTION Ⅲ

ANIMAL OR VEGETABLE FATS AND OILS AND THEIR CLEAVAGE PRODUCTS; PREPARED EDIBLE FATS; ANIMAL OR VEGETABLE WAXES

Chapter 15

Animal or vegetable fats and oils and their cleavage products; prepared edible fats; animal or vegetable waxes

Notes:

1. This Chapter does not cover:

(a) Pig fat or poultry fat of heading No.02.09;

(b) Cocoa butter, fat or oil (heading No.18.04);

(c) Edible preparations containing by weight more than 15% of the products of heading No.04.05 (generally Chapter 21);

(d) Greaves (heading No.23.01) or residues of headings No.23.04 to 23.06;

(e) Fatty acids, prepared waxes, medicaments, paints, varnishes, soap, perfumery, cosmetic or toilet preparations, sulphonated oils or other goods of Section Ⅵ; or

(f) Factice derived from oils (heading No.40.02).

2. Heading No.15.09 does not apply to oils obtained from olives by solvent extraction (heading No.15.10).

3. Heading No.15.18 does not cover fats or oils or their fractions, merely denatured, which are to be classified in the heading appropriate to the corresponding undenatured fats and oils and their fractions.

4. Soap -stocks, oil foots and dregs, stearin pitch, glycerol pitch and wool grease residues fall in heading No.15.22.

Subheading Note:

For the purposes of subheadings 1514.11 and 1514.19, the expression “low erucic acid rape of colza oil” means the fixed oil which has an erucic acid content of less than 2% by weight.

序号 No.	税则号列 Tariff Item	货品名称	最惠国税率(%) M.F.N.	协定税率(%)Agreement 东盟 ASEAN	曼谷协定 BA	香港 H.K.	澳门 Macao	普通税率(%) Gen.	Article Description
	15.01	**猪脂肪(包括已炼制的猪油)及家禽脂肪,但税号02.09及15.03的货品除外:**							**Pig fat (including lard) and poultry fat, other than that of heading No.02.09 or 15.03:**
826	1501.0000	猪脂肪(包括已炼制的猪油)及家禽脂肪,但税号02.09及15.03的货品除外	10					35	Pig fat(including lard)and poultry fat, other than that of heading No. 02.09 or 15.03
	15.02	**牛、羊脂肪,但税号15.03的货品除外:**							**Fats of bovine animals, sheep or goats, other than those of heading No. 15.03:**
827	1502.0010	---未炼制的牛、羊脂肪	8					70	---Raw
828	1502.0090	---其他	8					30	---Other
	15.03	**猪油硬脂、液体猪油、油硬脂、食用或非食用脂油,未经乳化、混合或其他方法制作:**							**Lard stearin, lard oil, oleostearin, oleooil and tallow oil, not emulsified or mixed or otherwise prepared:**
829	1503.0000	猪油硬脂、液体猪油、油硬脂、食用或非食用脂油,未经乳化、混合或其他方法制作	10					30	Lard stearin, lard oil, oleostearin, oleooil and tallow oil, not emulsified or mixed or otherwise prepared
	15.04	**鱼或海生哺乳动物的油、脂及其分离品,不论是否精制,但未经化学改性:**							**Fats and oils and their fractions, of fish or marine mammals, whether or not refined, but not chemically modified:**
830	1504.1000	-鱼肝油及其分离品	12					30	-Fish-liver oils and their fractions
831	1504.2000	-除鱼肝油以外的鱼油、脂及其分离品	12					50	-Fats and oils and their fractions, of fish, other than liver oils
832	1504.3000	-海生哺乳动物的油、脂及其分离品	14.4					50	-Fats and oils and their fractions, of marine mammals
	15.05	**羊毛脂及从羊毛脂制得的脂肪物质(包括纯净的羊毛脂):**							**Wool grease and fatty substances derived therefrom(including lanolin):**
833	1505.0000	羊毛脂及从羊毛脂制得的脂肪物质(包括纯净的羊毛脂):	20					70	Wool grease and fatty substances derived therefrom(including lanolin):
	15.06	**其他动物油、脂及其分离品,不论是否精制,但未经化学改性:**							**Other animal fats and oils and their fractions, whether or not refined, but not chemically modified:**
834	1506.0000	其他动物油、脂及其分离品,不论是否精制,但未经化学改性	20					70	Other animal fats and oils and their fractions, whether or not refined, but not chemically modified
	15.07	**豆油及其分离品,不论是否精制,但未经化学改性:**							**Soya-bean oil and its fractions, whether or not refined, but not chemically modified:**

序号 No.	税则号列 Tariff Item	货品名称	最惠国税率(%) M.F.N.	协定税率(%) Agreement 东盟 ASEAN	曼谷协定 BA	香港 H.K.	澳门 Macao	普通税率(%) Gen.	Article Description
835	1507.1000	-初榨的,不论是否脱胶	见表5					见表5	-Crude oil whether or not degummed
836	1507.9000	-其他	见表5					见表5	-Other
	15.08	**花生油及其分离品,不论是否精制,但未经化学改性:**							**Ground-nut oil and its fractions, whether or not refined, but not chemically modified:**
837	1508.1000	-初榨的	10					100	-Crude oil
838	1508.9000	-其他	10					100	-Other
	15.09	**油橄榄油及其分离品,不论是否精制,但未经化学改性:**							**Olive oil and its fractions, whether or not refined, but not chemically modified:**
839	1509.1000	-初榨的	10					30	-Virgin
840	1509.9000	-其他	10					30	-Other
	15.10	**其他橄榄油及其分离品,不论是否精制,但未经化学改性,包括掺有税号15.09的油或分离品的混合物:**							**Other oils and their fractions, obtained solely from olives, whether or not refined, but not chemically modified, including blends of these oils or fractions with oils or fractions of heading No. 15.09:**
841	1510.0000	其他橄榄油及其分离品,不论是否精制,但未经化学改性,包括掺有税号15.09的油或分离品的混合物	10					30	Other oils and their fractions, obtained solely from olives, whether or not refined, but not chemically modified, including blends of these oils or fractions with oils or fractions of heading No. 15.09
	15.11	**棕榈油及其分离品,不论是否精制,但未经化学改性:**							**Palm oil and its fractions, whether or not refined, but not chemically modified:**
842	1511.1000	-初榨的	见表5					见表5	-Crude oil
		--其他:							--Other:
843	1511.9010	---棕榈液油(熔点19-24度)	见表5					见表5	---Palm olein
844	1511.9020	---棕榈硬脂(熔点44-56度)	8					60	---Palm stearin
845	1511.9090	---其他	见表5					见表5	---Other
	15.12	**葵花油、红花油或棉子油及其分离品,不论是否精制,但未经化学改性:**							**Sunflower-seed, safflower or cotton-seed oil and fractions thereof, whether or not refined, but not chemically modified:**
		-葵花油或红花油及其分离品:							-Sunflower-seed or safflower oil and fractions thereof:
846	1512.1100	--初榨的	9.1					160	--Crude oil
847	1512.1900	--其他	9.1					160	--Other
		-棉子油及其分离品:							-Cotton-seed oil and its fractions:
848	1512.2100	--初榨的,不论是否去除棉子酚	10					70	--Crude oil, whether or not gossypol has been removed
849	1512.2900	--其他	10					70	--Other

序号 No.	税则号列 Tariff Item	货品名称	最惠国税率(%) M.F.N.	协定税率(%) Agreement 东盟 ASEAN	曼谷协定 BA	香港 H.K.	澳门 Macao	普通税率(%) Gen.	Article Description
	15.13	**椰子油、棕榈仁油或巴巴苏棕榈果油及其分离品,不论是否精制,但未经化学改性:**							**Coconut (copra), palm kernel or babassu oil and fractions thereof, whether or not refined, but not chemically modified:**
		-椰子油及其分离品:							-Coconut (copra) oil and its fractions:
850	1513.1100	--初榨的	9.5	见表 1				40	--Crude oil
851	1513.1900	--其他	9.5	见表 1				40	--Other
		-棕榈仁油或巴巴苏棕榈果油及其分离品:							-Palm kernel or babassu oil and fractions thereof:
852	1513.2100	--初榨的	10	见表 1				40	--Crude oil
853	1513.2900	--其他	10	见表 1				40	--Other
	15.14	**菜子油或芥子油及其分离品,不论是否精制,但未经化学改性:**							**Rape, colza or mustard oil and fractions thereof, whether or not refined, but not chemically modified:**
		-低芥子酸菜子油及其分离品:							-Low erucic acid rape of colza oil and its fractions:
854	1514.1100	--初榨的	见表 5					见表 5	--Crude oil
855	1514.1900	--其他	见表 5					见表 5	--Other
		-其他:							-Other:
		--初榨的:							--Crude oil:
856	1514.9110	---菜子油	见表 5					见表 5	---Rape oil
857	1514.9190	---芥子油	见表 5					见表 5	---Mustard oil
858	1514.9900	--其他	见表 5					见表 5	--Other
	15.15	**其他固定植物油、脂(包括希蒙得木油)及其分离品,不论是否精制,但未经化学改性:**							**Other fixed vegetable fats and oils (including jojoba oil) and fractions thereof, whether or not refined, but not chemically modified:**
		-亚麻子油及其分离品:							-Linseed oil and its fractions:
859	1515.1100	--初榨的	15					30	--Crude oil
860	1515.1900	--其他	15					30	--Other
		-玉米油及其分离品:							-Maize (corn) oil and its fractions:
861	1515.2100	--初榨的	10					160	--Crude oil
862	1515.2900	--其他	10					160	--Other
863	1515.3000	-蓖麻油及其分离品	10					70	-Castor oil and its fractions
864	1515.4000	-桐油及其分离品	20					70	-Tung oil and its fractions
865	1515.5000	-芝麻油及其分离品	12					20	-Sesame oil and its fractions
		-其他:							-Other:
866	1515.9010	---希蒙得木油及其分离品	20					70	---Jojoba oil and its fractions
867	1515.9090	---其他	20					70	---Other
	15.16	**动、植物油、脂及其分离品,全部或部分氢化、相互酯化、再酯化或反油酸化,不论是否精制,但未经进一步加工:**							**Animal or vegetable fats and oil and fractions thereof, partly or wholly hydrogenated, inter-esterified, reesterified or elaidinized, whether or not refined, but not further prepared:**

序号 No.	税则号列 Tariff Item	货品名称	最惠国税率(%) M.F.N.	协定税率(%)Agreement 东盟 ASEAN	曼谷协定 BA	香港 H.K.	澳门 Macao	普通税率(%) Gen.	Article Description
868	1516.1000	-动物油、脂及其分离品	5					70	-Animal fats and oils and fractions thereof
869	1516.2000	-植物油、脂及其分离品	25	见表1				70	-Vegetable fats and oils and fractions thereof
	15.17	**人造黄油;本章各种动、植物油、脂及其分离品混合制成的食用油、脂或制品,但税号15.16的食用油、脂及其分离品除外:**							**Margarine; edible mixtures or preparations of animal or vegetable fats or oils or of fractions of different fats or oils of this Chapter, other than edible fats or oils or their fractions of heading No.15.16**
870	1517.1000	-人造黄油,但不包括液态的	30					80	-Margarine, excluding liquid margarine
871	1517.9000	-其他	25	见表1				70	-Other
	15.18	**动、植物油、脂及其分离品,经过熟炼、氧化、脱水、硫化、吹制或在真空、惰性气体中加热聚合及用其他化学方法改性的,但税号15.16的产品除外;本章各种油、脂及其分离品混合制成的其他税号未列名的非食用油、脂或制品:**							**Animal or vegetable fats and oils and fractions thereof, boiled, oxidized, dehydrated, sulphurized, blown, polymerized by heat in vacuum or in inert gas or otherwise chemically modified, excluding those of heading No.15.16; inedible mixtures or preparations of animal or vegetable fats or oils or of fractions of different fats or oils of this Chapter, not elsewhere specified or included:**
872	1518.0000	动、植物油、脂及其分离品,经过熟炼、氧化、脱水、硫化、吹制或在真空、惰性气体中加热聚合及用其他化学方法改性的,但税号15.16的产品除外;本章各种油、脂及其分离品混合制成的其他税号未列名的非食用油、脂或制品	10					70	Animal or vegetable fats and oils and fractions thereof, boiled, oxidized, dehydrated, sulphurized, blown, polymerized by heat in vacuum or in inert gas or otherwise chemically modified, excluding those of heading No.15.16; inedible mixtures or preparations of animal or vegetable fats or oils or of fractions of different fats or oils of this Chapter, not elsewhere specified or included
	15.20	**粗甘油;甘油水及甘油碱液:**							**Glycerol, crude; glycerol waters and glycerol lyes:**
873	1520.0000	粗甘油;甘油水及甘油碱液	20					50	Glycerol, crude; glycerol waters and glycerol lyes
	15.21	**植物蜡(甘油三酯除外)、蜂蜡、其他虫蜡及鲸蜡,不论是否精制或着色:**							**Vegetable waxes (other than triglycerides), beeswax, other insect waxes and spermaceti, whether or not refined or coloured:**
874	1521.1000	-植物蜡	20					80	-Vegetable waxes
875	1521.9000	-其他	20					80	-Other

序号 No.	税则号列 Tariff Item	货品名称	最惠国税率(%) M.F.N.	协定税率(%)Agreement 东盟 ASEAN	曼谷协定 BA	香港 H.K.	澳门 Macao	普通税率(%) Gen.	Article Description
	15.22	**油鞣回收脂;加工处理油脂物质及动、植物蜡所剩的残渣:**							**Degras; residues resulting from the treatment of fatty substances of animal or vegetable waxes:**
876	1522.0000	油鞣回收脂;加工处理油脂物质及动、植物蜡所剩的残渣	20					50	Degras; residues resulting from the treatment of fatty substances of animal or vegetable waxes

第四类

食品；饮料、酒及醋；烟草、烟草及烟草代用品的制品

SECTION Ⅳ

PREPARED FOODSTUFFS; BEVERAGES, SPIRITS AND VINEGAR; TOBACCO AND MANUFACTURED TOBACCO SUBSTITUTES

注释：

本类所称“团粒”，是指直接挤压或加入按重量计比例不超过3%的粘合剂制成的粒状产品。

Note:

In this Section the term “pellets” means products which have been agglomerated either directly by compression or by the addition of a binder in a proportion not exceeding 3% by weight.

第十六章
肉、鱼、甲壳动物、软体动物及其他水生无脊椎动物的制品

Chapter 16
Preparations of meat, of fish or of crustaceans, molluscs or other aquatic invertebrates

注释：

一、本章不包括用第二章、第三章及税号05.04所列方法制作或保藏的肉、食用杂碎、鱼、甲壳动物、软体动物或其他水生无脊椎动物。

二、本章的食品按重量计必须含有20%以上的香肠、肉、食用杂碎、动物血、鱼、甲壳动物、软体动物或其他水生无脊椎动物及其混合物。对于含有两种或两种以上前述产品的食品，则应按其中重量最大的产品归入第十六章的相应税号。但本条规定不适用于税号19.02的包馅食品和税号21.03及21.04的食品。

Notes:

1. This Chapter does not cover meat, meat offal, fish, crustaceans, molluscs or other aquatic invertebrates, prepared or preserved by the processes specified in Chapter 2 or 3 or heading No.05.04.

2. Food preparations fall in this Chapter provided that they contain more than 20% by weight of sausage, meat, meat offal, blood, fish or crustaceans, molluscs or other aquatic invertebrates, or any combination thereof. In cases where the preparation contains two or more of the products mentioned above, it is classified in the heading of Chapter 16 corresponding to the component or components which predominate by weight. These provisions do not apply to the stuffed products of heading No.19.02 or to the preparations of heading No.21.03 or 21.04.

子目注释：

一、子目号1602.10的“均化食品”，是指用肉、食用杂碎或动物血经精细均化制成供婴幼儿食用或营养用的零售包装食品（每件净重不超过250克）。为了调味、保藏或其他目的，均化食品中可以加入少量其他配料，还可以含有少量可见的肉粒或食用杂碎粒。归类时该子目优先于税号16.02的其他子目。

Subheading Notes:

1. For the purposes of subheading No.1602.10, the expression “homogenized preparations” means preparations of meat, meat offal or blood, finely homogenized, put up for retail sale as infant food or for dietetic purposes, in containers of a net weight content not exceeding 250g. For the application of this definition no account is to be taken of small quantities of any ingredients which may have been added to the preparation for seasoning, preservation or other purposes. These preparations may contain a small quantity of visible pieces of meat or meat offal. This subheading takes precedence over all other subheadings of heading No.16.02.

二、税号16.04或16.05项下各子目所列的是鱼及甲壳动物的俗名，它们与第三章中相同名称的鱼及甲壳动物种类范围相同。

2. The fish and crustaceans specified in the subheahings of heading No.16.04 or 16.05 under their common names only, are of the same species as those mentioned in Chapter 3 under the same name.

序号 No.	税则号列 Tariff Item	货品名称	最惠国税率(%) M.F.N.	协定税率(%)Agreement 东盟 ASEAN	曼谷协定 BA	香港 H.K.	澳门 Macao	普通税率(%) Gen.	Article Description
	16.01	**肉、食用杂碎或动物血制成的香肠及类似产品;用香肠制成的食品:**							**Sausages and similar products, of meat, meat offal or blood; food preparations based on these products:**
877	1601.0000	肉、食用杂碎或动物血制成的香肠及类似产品;用香肠制成的食品	15					90	Sausages and similar products, of meat, meat offal or blood; food preparations based on these products
	16.02	**其他方法制作或保藏的肉、食用杂碎或动物血:**							**Other Prepared or preserved meat, meat offal or blood:**
878	1602.1000	-均化食品	15					90	-Homogenized preparations
879	1602.2000	-动物肝	15					90	-Of liver of any animal
		-税号01.05的家禽的:							-Of poultry of heading No.01.05:
880	1602.3100	--火鸡的	15					90	--Of turkeys
		--鸡的:							--Of fowls of the species *Gallus domesticus*:
881	1602.3210	---罐头	15					90	---In airtight containers
		---其他:	15					90	---Other
882	1602.3291	----鸡胸肉	15			0		90	----chicken breast filets
883	1602.3292	----鸡腿肉	15			0		90	----chicken leg meat
884	1602.3299	----其他	15			0		90	----other
		--其他:							--Other:
885	1602.3910	---罐头	15					90	---In airtight containers
		---其他:	15					90	---Other:
886	1602.3991	----鸭的	15					90	----Of duck
887	1602.3999	----其他	15					90	----Other
		-猪的:							-Of swine:
888	1602.4100	--后腿及其肉块	15					90	--Hams and cuts thereof
899	1602.4200	--前腿及其肉块	15					90	--Shoulders and cuts thereof
		--其他,包括混合的肉:							--Other, including mixtures:
890	1602.4910	---罐头	15					90	In airtight containers
891	1602.4990	---其他	15			0		90	---Other
		-牛的:							-Of bovine animals:
892	1602.5010	---罐头	12					90	---In airtight containers
893	1602.5090	---其他	12			0		90	---Other
		-其他,包括动物血的食品:							-Other, including preparations of blood of any animal:
894	1602.9010	---罐头	15					90	---In airtight containers
895	1602.9090	---其他	15					90	---Other
	16.03	**肉、鱼、甲壳动物、软体动物或其他水生无脊椎动物的精及汁:**							**Extracts and juices of meat, fish or crustaceans, molluscs or other aquatic invertebrates:**

序号 No.	税则号列 Tariff Item	货品名称	最惠国税率(%) M.F.N.	协定税率(%) Agreement 东盟 ASEAN	曼谷协定 BA	香港 H.K.	澳门 Macao	普通税率(%) Gen.	Article Description
896	1603.0000	肉、鱼、甲壳动物、软体动物或其他水生无脊椎动物的精及汁	23.3			0△		90	Extracts and juices of meat, fish or crustaceans, molluscs or other aquatic invertebrates
	16.04	**制作或保藏的鱼;鲟鱼子酱及鱼卵制的鲟鱼子酱代用品:**							**Prepared or preserved fish; caviar and caviar substitutes prepared from fish eggs:**
		-鱼,整条或切块,但未绞碎:							-Fish, whole or in pieces, but not minced:
		--鲑鱼:							--Salmon:
897	1604.1110	---大西洋鲑鱼	12					90	---Atalantic salmon
898	1604.1190	---其他	12					90	---Other
899	1604.1200	--鲱鱼	12					90	--Herrings
900	1604.1300	--沙丁鱼、黍鲱鱼	5					90	--Sardines, sardinella and brisling or sprats
901	1604.1400	--金枪鱼、鲣鱼	5			0		90	--Tunas, skipjack and bonito (*Sarda spp.*)
902	1604.1500	--鲭鱼	12					90	--Mackerel
903	1604.1600	--鳀鱼	12					90	--Anchovies
		--其他:							--Other:
904	1604.1910	---河鳗鱼	12		9.9	0		90	---River eel
905	1604.1920	---罗非鱼	12			0		90	---Tilapia
906	1604.1990	---其他	12		9.9	0		90	---Other
		-其他制作或保藏的鱼:							-Other prepared or preserved fish:
		---罐头:							---In airtight containers:
907	1604.2011	----鱼翅	12		9.9	0	0△	90	----shark fin
908	1604.2019	----其他	12		9.9	0		90	----other
		---其他:							---Other:
909	1604.2091	----鱼翅	12		9.9	0	0△	90	----shark fin
910	1604.2099	----其他:	12		9.9	0		90	----other
911	1604.3000	-鲟鱼子酱及鲟鱼子酱代用品	12					90	-Caviar and caviar substitutes
	16.05	**制作或保藏的甲壳动物、软体动物及其他水生无脊椎动物:**							**Crustaceans, molluscs and other aquatic invertebrates, prepared or preserved:**
912	1605.1000	-蟹	5			0		90	-Crab
913	1605.2000	-小虾及对虾	5			0		90	-Shrimps and prawns
914	1605.3000	-龙虾	5			0		90	-Lobster
		-其他甲壳动物:							-Other crustaceans:
		---淡水小龙虾:							---Freshwater crawfish:
915	1605.4011	----虾仁	5					90	----Shelled
916	1605.4019	----其他	5					90	----Other
917	1605.4090	---其他	5					90	---Other
		-其他:							-Other:
918	1605.9010	---海蜇	15			0		90	---Jelly fish
919	1605.9090	---其他	5		3.9	0		90	---Other

第十七章
糖及糖食

Chapter 17
Sugars and sugar confectionery

注释：

本章不包括：

一、含有可可的糖食（税号 18.06）；

二、税号 29.40 的化学纯糖（蔗糖、乳糖、麦芽糖、葡萄糖及果糖除外）及其他产品；

三、第三十章的药品及其他产品。

Notes:

This Chapter does not cover:

1. Sugar confectionery containing cocoa (heading No. 18.06);

2. Chemically pure sugars (other than sucrose, lactose, maltose, glucose and fructose) or other products of heading No. 29.40; or

3. Medicaments or other products of Chapter 30.

子目注释：

子目号 1701.11 及 1701.12 所称"原糖"，是指按重量计干燥状态的蔗糖含量低于旋光读数 99.5°的糖。

Subheading Note:

For the purposes of subheadings No. 1701.11 and 1701.12, "raw sugar" means sugar whose content of sucrose by weight, in the dry state, corresponds to a polarimeter reading of less than 99.5°.

序号 No.	税则号列 Tariff Item	货品名称	最惠国税率(%) M.F.N.	协定税率(%) Agreement 东盟 ASEAN	曼谷协定 BA	香港 H.K.	澳门 Macao	普通税率(%) Gen.	Article Description
	17.01	**固体甘蔗糖、甜菜糖及化学纯蔗糖：**							**Cane or beet sugar and chemically pure sucrose, in solid form:**
		-未加香料或着色剂的原糖：							-Raw sugar not containing added flavouring or colouring matter:
920	1701.1100	--甘蔗糖	见表 5					见表 5	--Cane sugar
921	1701.1200	--甜菜糖	见表 5					见表 5	--Beet sugar
		-其他：							-Other:
922	1701.9100	--加有香料或着色剂	见表 5					见表 5	--Containing added flavouring or colouring matter
		--其他：							--Other:
923	1701.9910	---砂糖	见表 5					见表 5	---Granulated sugar
924	1701.9920	---绵白糖	见表 5					见表 5	---Superfine sugar
925	1701.9990	---其他	见表 5					见表 5	---Other
	17.02	**其他固体糖，包括化学纯乳糖、麦芽糖、葡萄糖及果糖；未加香料或着色剂的糖浆；人造蜜，不论是否掺有天然蜂蜜；焦糖：**							**Other sugars, including chemically pure lactose, maltose, glucose and fructose, in solid form; sugar syrups not containing added flavouring or colouring matter; artificial honey, whether or not mixed with natural honey; caramel:**
		-乳糖及乳糖浆：							-Lactose and lactose syrup:
926	1702.1100	--按重量计干燥无水乳糖含量在 99% 及以上	10					80	--Containing by weight 99% or more lactose, expressed as anhydrous lactose, calculated on the dry matter
927	1702.1900	--其他	10					80	--Other
928	1702.2000	-槭糖及槭糖浆	30					80	-Maple sugar and maple syrup

序号 No.	税则号列 Tariff Item	货品名称	最惠国税率(%) M.F.N.	协定税率(%)Agreement 东盟 ASEAN	曼谷协定 BA	香港 H.K.	澳门 Macao	普通税率(%) Gen.	Article Description
929	1702.3000	-葡萄糖及葡萄糖浆,不含果糖或按重量计干燥状态的果糖含量在20%以下	30					80	-Glucose and glucose syrup, not containing fructose or containing in the dry state less than 20% by weight of fructose
930	1702.4000	-葡萄糖及葡萄糖浆,按重量计干燥状态的果糖含量在20%及以上,但在50%以下	30					80	-Glucose and glucose syrup, containing in the dry state at least 20% but less than 50% by weight of fructose
931	1702.5000	-化学纯果糖	30					80	-Chemically pure fructose
932	1702.6000	-其他果糖及果糖浆,按重量计干燥状态的果糖含量在50%以上	30					80	-Other fructose and fructose syrup, containing in the dry state more than 50% by weight of fructose
933	1702.9000	-其他,包括转化糖	30			0		80	-Other, including invert sugar
	17.03	**制糖后所剩的糖蜜:**							**Molasses resulting from the extraction or refining or sugar:**
934	1703.1000	-甘蔗糖蜜	8					50	-Cane molasses
935	1703.9000	-其他	8					50	-Other
	17.04	**不含可可的糖食(包括白巧克力):**							**Sugar confectionery (including white chocolate), not containing cocoa:**
936	1704.1000	-口香糖,不论是否裹糖	12		9.5			50	-Chewing gum, whether or not sugar-coated
937	1704.9000	-其他	10		8.2	0	0	50	-Other

第十八章
可可及可可制品

Chapter 18
Cocoa and cocoa preparations

注释：

一、本章不包括税号 04.03、19.01、19.04、19.05、21.05、22.02、22.08、30.03、30.04 的制品。

二、税号 18.06 包括含有可可的糖食及注释一以外的其他含可可的食品。

Notes:

1. This Chapter does not cover the preparations of heading No.04.03, 19.01, 19.04, 19.05, 21.05, 22.02, 22.08, 30.03, or 30.04.

2. Heading No.18.06 includes sugar confectionery containing cocoa and, subject to Note 1 to this Chapter, other food preparations containing cocoa.

序号 No.	税则号列 Tariff Item	货品名称	最惠国税率(%) M.F.N.	协定税率(%)Agreement 东盟 ASEAN	曼谷协定 BA	香港 H.K.	澳门 Macao	普通税率(%) Gen.	Article Description
	18.01	**整颗或破碎的可可豆，生的或焙炒的：**							**Cocoa beans, whole or broken, raw or roasted:**
938	1801.0000	整颗或破碎的可可豆，生的或焙炒的	8	见表 1				30	Cocoa beans, whole or broken, raw or roasted
	18.02	**可可荚、壳、皮及废料：**							**Cocoa shells, husks, skins and other cocoa waste:**
939	1802.0000	可可荚、壳、皮及废料	10					30	Cocoa shells, husks, skins and other cocoa waste
	18.03	**可可膏，不论是否脱脂：**							**Cocoa paste, whether or not defatted:**
940	1803.1000	-未脱脂	10	见表 1				30	-Not defatted
941	1803.2000	-全脱脂或部分脱脂	10	见表 1				30	-Wholly or partly defatted
	18.04	**可可脂、可可油：**							**Cocoa butter, fat and oil:**
942	1804.0000	可可脂、可可油	22	见表 1				70	Cocoa butter, fat and oil
	18.05	**未加糖或其他甜物质的可可粉：**							**Cocoa powder, not containing added sugar or other sweetening matter:**
943	1805.0000	未加糖或其他甜物质的可可粉	15	见表 1				40	Cocoa powder, not containing added sugar or other sweetening matter:
	18.06	**巧克力及其他含可可的食品：**							**Chocolate and other food preparations containing cocoa:**
944	1806.1000	-加糖或其他甜物质的可可粉	10	见表 1				50	-Cocoa powder, containing added sugar or other sweetening matter
945	1806.2000	-其他重量超过 2 公斤的块状或条状含可可食品，或液状、膏状、粉状、粒状或其他散装形状的含可可食品，容器包装或内包装每件净重超过 2 公斤的	10		7.7			50	-Other preparations in blocks, slabs or bars weighing more than 2kg or in liquid, paste, powder, granular or other bulk form in containers or immediate packings, of a content exceeding 2kg

序号 No.	税则号列 Tariff Item	货 品 名 称	最惠国税率(%) M.F.N.	协定税率(%)Agreement 东盟 ASEAN	曼谷协定 BA	香港 H.K.	澳门 Macao	普通税率(%) Gen.	Article Description
-		-其他块状或条状的含可可食品:							-Other, in blocks, slabs or bars:
946	1806.3100	--夹心	8		6.4	0		50	--Filled
947	1806.3200	--不夹心	10		7.7	0		50	--Not filled
948	1806.9000	-其他	8		6.4	0		50	-Other

第十九章
谷物、粮食粉、淀粉或乳的制品；糕饼点心

注释：

一、本章不包括：

（一）按重量计含香肠、肉、食用杂碎、动物血、鱼、甲壳动物，软体动物、其他水生无脊椎动物及其混合物超过20%的食品（第十六章），但税号19.02的包馅食品除外；

（二）用粮食粉或淀粉制的专作动物饲料用的饼干及其他制品（税号23.09）；

（三）第三十章的药品及其他产品。

二、税目19.01所称

（一）“粗粒”是指第十一章谷物的粗粒；

（二）“细粉”及“粗粉”，是指：

1. 第十一章谷物的细粉及粗粉；

2. 其他章植物的细粉、粗粉及粉末，但不包括干蔬菜、马铃薯和干豆类的细粉、粗粉及粉末（应分别归入税目07.12、11.05和11.06）。

三、税号19.04不包括按重量计全脱脂可可含量超过6%或裹巧克力的食品以及其他含可可食品（税号18.06）。

四、税号19.04所称“其他方法制作的”，是指制作或加工程度超过第十章或第十一章各税号或注释所规定范围的。

Chapter 19
Preparations of cereals, flour, starch or milk; pastrycook's products

Notes:

1. This Chapter does not cover:

(a) Except in the case of stuffed products of heading No.19.02, food preparations containing more than 20% by weight of sausage, meat, meat offal, blood, fish or crustaceans, molluscs or other aquatic invertebrates, or any combination thereof (Chapter 16);

(b) Biscuits or other articles made from flour or from starch, specially prepared for use in animal feeding (heading No.23.09); or

(c) Medicaments or other products of Chapter 30.

2. For the purposes of heading No.19.01:

(a) The term "groats" means cereal groats of Chapter 11;

(b) The terms "flour" and "meal" mean:

(1) Cereal flour and meal of Chapter 11, and

(2) Flour, meal and powder of vegetable origin of any Chapter, other than flour, meal or powder of dried vegetables (heading No.07.12), of potatoes (heading No.11.05) or of dried leguminous vegetables (heading No.11.06).

3. Heading No.19.04 does not cover preparations containing more than 6% by weight of cocoa calculated on a totally defatted basis or coated with chocolate or other food preparations containing cocoa (heading No.18.06).

4. For the purposes of heading No.19.04, the expression "otherwise prepared" means prepared or processed to an extent beyond that provided for in the headings of or Notes to Chapter 10 or 11.

序号 No.	税则号列 Tariff Item	货品名称	最惠国税率(%) M.F.N.	协定税率(%) Agreement 东盟 ASEAN	曼谷协定 BA	香港 H.K.	澳门 Macao	普通税率(%) Gen.	Article Description
	19.01	**麦精;细粉、粗粒、粗粉、淀粉或麦精制的其他税目未列名的食品,不含可可或按重量计全脱脂可可含量低于40%;税目04.01至04.04所列货品制的其他税目未列名的食品,不含可可或按重量计全脱脂可可含量低于5%:**							**Malt extract; food preparations of flour, groats, meal, starch or malt extract, not containing cocoa or containing less than 40% by weight of cocoa calculated on a totally defatted basis, not elsewhere specified or included; food preparations of goods of headings No.04.01 to 04.04, not containing cocoa or containing less than 5% by weight of cocoa calculated on a totally defatted basis, not elsewhere specified or included:**
949	1901.1000	-供婴幼儿食用的零售包装食品	15					40	-Preparations for infant use, put up for retail sale
950	1901.2000	-供烘焙税号19.05所列面包糕饼用的调制品及面团	25			0	0	80	-Mixes and doughs for the preparation of bakers' wares of heading No. 19.05
951	1901.9000	-其他	10			0	0	80	-Other
	19.02	**面食,不论是否煮熟、包馅(肉馅或其他馅)或其他方法制作,例如,通心粉、面条、汤团、馄饨、饺子、奶油面卷;古斯古斯面食,不论是否制作:**							**Pasta, whether or not cooked or stuffed (with meat or other substances) or otherwise prepared, such as spaghetti, macaroni, noodles, lasagne, gnocchi, ravioli, cannelloni; couscous, whether or not prepared:**
		-生的面食,未包馅或未经其他方法制作:							-Uncooked pasta, not stuffed or otherwise prepared:
952	1902.1100	--含蛋	15				0	80	--Containing eggs
953	1902.1900	--其他	15			0	0	80	--Other
954	1902.2000	-包馅面食,不论是否烹煮或经其他方法制作	15			0		80	-Stuffed pasta, whether or not cooked or otherwise prepared
		-其他面食:							-Other pasta:
955	1902.3010	---米粉干	15				0	80	---Rice vermicelli, cooked
956	1902.3020	---粉丝	15					80	---Bean vermicelli, cooked
957	1902.3030	---即食或快熟面条	15		13.1	0	0	80	---Instant noodle
958	1902.3090	---其他	15		13.1		0	80	---Other
959	1902.4000	-古斯古斯面食	25					80	-Couscous
	19.03	**珍粉及淀粉制成的珍粉代用品,片、粒、珠、粉或类似形状的:**							**Tapioca and substitutes therefor prepared from starch, in the form of flakes, grains, pearls, siftings or in similar forms:**
960	1903.0000	珍粉及淀粉制成的珍粉代用品,片、粒、珠、粉或类似形状的	15					80	Tapioca and substitutes therefor prepared from starch, in the form of flakes, grains, pearls, siftings or in similar forms

序号 No.	税则号列 Tariff Item	货品名称	最惠国税率(%) M.F.N.	协定税率(%)Agreement				普通税率(%) Gen.	Article Description
				东盟 ASEAN	曼谷协定 BA	香港 H.K.	澳门 Macao		
	19.04	**谷物或谷物产品经膨化或烘炒制成的食品(例如,玉米片);其他税号未列名的预煮或经其他方法制作的谷粒(玉米除外)、谷物片或经其他加工的谷粒(细粉及粗粉除外):**							**Prepared foods obtained by the swelling or roasting of cereals or cereal products (for example, corn flakes); cereals (other than maize (corn)) in grain form or in the form of flakes or other worked grains (except flour, groats and meal), pre-cooked or otherwise prepared, not elsewhere specified or included.**
961	1904.1000	-谷物或谷物产品经膨化或烘炒制成的食品	25					80	-Prepared foods obtained by the swelling or roasting of cereals or cereal products
962	1904.2000	-未烘炒谷物片制成的食品及未烘炒的谷物片与烘炒的谷物片或膨化的谷物混合制成的食品	30					80	-Prepared foods obtained from unroasted cereal flakes or from mixtures of unroasted cereal flakes and roasted cereal flakes or swelled cereals
963	1904.3000	-碾碎的干小麦	30					80	-Bulgur wheet
964	1904.9000	-其他	30					80	-Other
	19.05	**面包、糕点、饼干及其他烘焙糕饼,不论是否含可可;圣餐饼、装药空囊、封碱、糯米纸及类似制品:**							**Bread, pastry, cakes, biscuits and other bakers' wares, whether or not containing cocoa; communion wafers, empty cachets of a kind suitable for pharmaceutical use, sealing wafers, rice paper and similar products:**
965	1905.1000	-黑麦脆面包片	20					80	-Crispbread
966	1905.2000	-姜饼及类似品	20					80	-Gingerbread and the like
		-甜饼干;华夫饼干及圣餐饼							-Sweet biscuits; waffles and wafers
967	1905.3100	--甜饼干	15		12.4	0	0	80	--Sweet biscuits
968	1905.3200	--华夫饼干及圣餐饼	15		12.4	0	0	80	--Waffles and wafers
969	1905.4000	-面包干、吐司及类似的烤面包	20				0	80	-Rusks, toasted bread and similar toasted products
970	1905.9000	-其他	20		17.1	0	0	80	-Other

第二十章
蔬菜、水果、坚果或植物其他部分的制品

注释：

一、本章不包括：

(一) 用第七章、第八章或第十一章所列方法制作或保藏的蔬菜、水果或坚果；

(二) 按重量计含香肠、肉、食用杂碎、动物血、鱼、甲壳动物、软体动物、其他水生无脊椎动物及其混合物超过20%的食品（第十六章）；

(三) 税号21.04的均化混合食品。

二、税号20.07及20.08不包括制成糖食的果冻、果膏、糖衣杏仁或类似品（税号17.04）及巧克力糖食（税号18.06）。

三、税号20.01、20.04及20.05仅酌情包括用本章注释一（一）以外的方法制作或保藏的第七章或税号11.05、11.06的产品（第八章产品的细粉、粗粉除外）。

四、干重量在7%及以上的番茄汁归入税号20.02。

五、税目20.07所称“烹煮的”是指，在常压或减压下，通过减少水分或其他方法增加产品粘稠度的热处理。

原注释五变更为注释六。

六、税号20.09所称“未发酵及未加酒精的水果汁”，是指按容量计酒精浓度（标准见第二十二章注释二）不超过0.5%的水果汁。

子目注释：

一、子目号2005.10所称“均化蔬菜”，是指蔬菜经精细均化制成供婴幼儿食用或营养用的零售包装食品（每件净重不超过250克）。为了调味、保藏或其他目的，均化蔬菜中可以加入少量其他配料，还可以含有少量可见的蔬菜粒。归类时，子目号2005.10优先于税号20.05的其他子目。

Chapter 20
Preparations of vegetables, fruit, nuts or other parts of plants

Notes:

1. This Chapter does not cover:

(a) Vegetables, fruit or nuts, prepared or preserved by the processes specified in Chapter 7, 8 or 11;

(b) Food preparations containing more than 20% by weight of sausage, meat, meat offal, blood, fish or crustaceans, molluscs or other aquatic invertebrates, or any combination thereof (Chapter 16); or

(c) Homogenized composite food preparations of heading No.21.04.

2. Headings No.20.07 and 20.08 do not apply to fruit jellies, fruit pastes, sugar-coated almonds or the like in the form of sugar confectionery (heading No.17.04) or chocolate confectionery (heading No.18.06).

3. Headings No.20.01, 20.04 and 20.05 cover, as the case may be, only those products of Chapter 7 or of heading No.11.05 or 11.06 (other than flour, meal and powder of the products of Chapter 8) which have been prepared or preserved by processes other than those referred to in Note 1 (a).

4. Tomato juice, the dry weight content of which is 7% or more, is to be classified in heading No.20.02.

5. For the purposes of heading No.20.07, the expression "obtained by cooking" means obtained by heat treatment at atmospheric pressure or under reduced pressure to increase the viscosity of a product through reduction of water content or other means.

6. For the purposes of heading No.20.09, the expression "juices, unfermented and not containing added spirit" means juices of an alcoholic strength by volume (see Note 2 to Chapter 22) not exceeding 0.5% vol.

Subheading Notes:

1. For the purposes of subheading No.2005.10, the expression "homogenized vegetables" means preparations of vegetables, finely homogenized, put up for retail sale as infant food or for dietetic purposes, in containers of a net weight content not exceeding 250g. For the application of this definition no account is to be taken of small quantities of any ingredients which may have been added to the preparation for seasoning, preparation or other purposes. These preparations may contain a small quantity of visible pieces of vegetables. Subheading No.2005.10 takes precedence over all other subheadings of heading No.20.05.

二、子目号 2007.10 所称“均化食品”，是指果实经精细均化制成供婴幼儿食用或营养用的零售包装食品（每件净重不超过 250 克）。为了调味、保藏或其他目的，均化食品中可以加入少量其他配料，还可以含有少量可见的果粒。归类时，子目号 2007.10 优先于税号 20.07 的其他子目。

2. For the purposes of subheading No.2007.10, the expression "homogenized preparations" means preparations of fruit, finely homogenized, put up for retail sale as infant food or for dietetic purposes, in containers of a net weight content not exceeding 250g. For the application of this definition no account is to be taken of small quantities of any ingredients which may have been added to the preparation for seasoning, preservation or other purposes. These preparations may contain a small quantity of visible pieces of fruit. Subheading No.2007.10 takes precedence over all other subheadings of heading No.20.07.

三、子目 2009.12、2009.12、2009.31、2009.41、2009.61 及 2009.71 所称“白利糖度值”，是指直接从白利糖度计读取的度数或在 20℃ 时从折射计读取的以蔗糖百分比含量计的折射率，在其他温度下读取的数值应折算为 20℃ 时的折射率。

3. For the purposes of subheadings No.2009.12, 2009.21, 2009.31, 2009.41, 2009.61 and 2009.71, the expression "Brix value" means the direct reading of degrees Brix obtained from a Brix hydrometer of refractive index expressed in terms of percentage sucrose content obtained from a refractometer, at a temperature of 20℃ or corrected for 20℃ if the reading is made at a different temperature.

序号 No.	税则号列 Tariff Item	货品名称	最惠国税率(%) M.F.N.	协定税率(%)Agreement 东盟 ASEAN	曼谷协定 BA	香港 H.K.	澳门 Macao	普通税率(%) Gen.	Article Description
	20.01	**蔬菜、水果、坚果及植物的其他食用部分，用醋或醋酸制作或保藏的：**							**Vegetagbles, fruit, nuts and other edible parts of plants, prepared or preserved by vinegar or acetic acid:**
971	2001.1000	-黄瓜及小黄瓜	25					70	-Cucumbers and gherkins
		-其他：							-Other:
972	2001.9010	---大蒜	25					70	---Garlic
973	2001.9090	---其他	25					70	---Other
	20.02	**番茄，用醋或醋酸以外的其他方法制作或保藏的：**							**Tomatoes prepared or preserved otherwise than by vinegar or acetic acid:**
		-番茄，整个或切片：							-Tomatoes, whole or in pieces:
974	2002.1010	---罐头	19					80	---In airtight containers
975	2002.1090	---其他	25					70	---Other
		-其他：							-Other:
976	2002.9010	---番茄酱罐头	20					80	---Tomato paste, in airtight containers
977	2002.9090	---其他	18					70	---Other
	20.03	**蘑菇及块菌，用醋或醋酸以外的其他方法制作或保藏的：**							**Mushrooms and truffles, prepared or preserved otherwise than by vinegar or acetic acid:**
		-伞菌属蘑菇：							-Mushrooms:
		---罐头：							---In airtight containers:
978	2003.1011	----小白蘑菇	25					90	----Small white agaric
979	2003.1019	----其他	25					90	----Other
980	2003.1090	---其他	25					90	---Other
981	2003.2000	-块菌	25					90	-Truffles
		-其他：							-Other:
982	2003.9010	---罐头	25					90	---In airtight containers.

序号 No.	税则号列 Tariff Item	货品名称	最惠国税率(%) M.F.N.	协定税率(%) Agreement 东盟 ASEAN	曼谷协定 BA	香港 H.K.	澳门 Macao	普通税率(%) Gen.	Article Description
983	2003.9090	---其他	25					90	---Other
	20.04	**其他冷冻蔬菜,用醋或醋酸以外的其他方法制作或保藏的,但税号20.06的产品除外:**							**Other vegetables prepared or preserved otherwise than by vinegar or acetic acid, other than products of heading No.20.06:**
984	2004.1000	-马铃薯	13					70	-Potatoes
985	2004.9000	-其他蔬菜及什锦蔬菜	25					70	-Other vegetables and mixtures of vegetables
	20.05	**其他未冷冻蔬菜,用醋或醋酸以外的其他方法制作或保藏的,但税号20.06的产品除外:**							**Other vegetables prepared or preserved otherwise than by vinegar or acetic acid, not frozen, other than products of heading No.20.06:**
986	2005.1000	-均化蔬菜	25					70	-Homogenized vegetables
987	2005.2000	-马铃薯	15					70	-Potatoes
988	2005.4000	-豌豆	25					70	-Peas(*Pisum sativum*)
		-豇豆及菜豆:							-Beans(*Vigna spp.*, *phaseolus spp.*):
		--脱荚的:							--Beans,shelled:
989	2005.5110	---罐头	25			0		80	---In airtight containers
990	2005.5190	---其他	25			0		70	---Other
		--其他:							--Other:
991	2005.5910	---罐头	25					80	---In airtight containers
992	2005.5990	---其他	25					70	---Other
		-芦笋:							-Asparagus:
993	2005.6010	---罐头	25					80	---In airtight containers
994	2005.6090	---其他	25					70	---Other
995	2005.7000	-油橄榄	10					70	-Olives
996	2005.8000	-甜玉米	10					80	-Sweet corn(*Zea mays var. saccharata*)
		-其他蔬菜及什锦蔬菜:							-Other vegetables and mixtures of vegetables:
997	2005.9010	---清水马蹄罐头	25					80	---Water chestnut,in airtight containers
998	2005.9020	---蚕豆罐头	25					80	---Broad beans,in airtight containers
		---竹笋罐头:							---Bamboo shoots, in airtight containers:
999	2005.9031	----水煮竹笋,容积在8公升及以上	25					80	----Boiled Bamboo shoots in airtight containrts, volume exceeding 8 liters
1000	2005.9039	----其他	25					80	----Other
1001	2005.9040	---榨菜	25					70	---Hot pickled mustard tubers
1002	2005.9050	---咸蕨菜	25					70	---Chueh tsai (fiddle-head),salted
1003	2005.9060	---咸藠头	25					70	---Scallion,salted
		---其他:							---Other:
1004	2005.9091	----罐头	25					70	----In airtight containers
1005	2005.9092	----赤豆馅	25					70	----Red bean paste

序号 No.	税则号列 Tariff Item	货品名称	最惠国税率(%) M.F.N.	协定税率(%)Agreement 东盟 ASEAN	曼谷协定 BA	香港 H.K.	澳门 Macao	普通税率(%) Gen.	Article Description
1006	2005.9099	----其他	25					70	----Other
	20.06	**糖渍蔬菜、水果、坚果、果皮及植物的其他部分(沥干、糖渍或裹糖的):**							**Vegetables, fruit, nuts, fruit-peel and other parts of plants, preserved by sugar (drained, glacé or crystallized):**
1007	2006.0010	---蜜枣	30			0		90	---Preserved jujubes
1008	2006.0020	---橄榄	30			0		90	---Preserved olives
1009	2006.0090	---其他	30			0		90	---Other
	20.07	**烹煮的果酱、果冻、柑桔酱、果泥及果膏,不论是否加糖或其他甜物质:**							**Jams, fruit jellies, marmalades, fruit or nut purée and fruit or nut pastes, being cooked preparations, whether or not containing added sugar or other sweetening matter:**
1010	2007.1000	-均化食品	30					80	-Homogenized preparations
		-其他:							-Other:
1011	2007.9100	--柑桔属水果的	30					80	--Citrus fruit
		--其他:							--Other:
1012	2007.9910	---罐头	5					80	---In airtight containers
1013	2007.9990	---其他	5					80	---Other
	20.08	**用其他方法制作或保藏的其他税号未列名水果、坚果及植物的其他食用部分,不论是否加酒、加糖或其他甜物质:**							**Fruit, nuts and other edible parts of plants, otherwise prepared, or preserved, whether or not containing added sugar or other sweetening matter or spirit, not elsewhere specified or included:**
		-坚果、花生及其他子仁,不论是否混合:							-Nuts, ground-nuts and other seeds, whether or not mixed together:
		--花生:							--Ground-nuts:
1014	2008.1110	---花生米罐头	30			0		90	---Ground-nut kernels, in airtight containers
1015	2008.1120	---烘焙花生	30			0		80	---Roasted ground-nuts
1016	2008.1130	---花生酱	30					90	---Ground-nut butter
1017	2008.1190	---其他	30			0		80	---Other
		--其他,包括什锦坚果及其他子仁:							--Other, including mixtures:
1018	2008.1910	---核桃仁罐头	20			0		90	---Walnut meats, in airtight containers
1019	2008.1920	---其他果仁罐头	13			0		90	---Other nuts, in airtight containers
		---其他:							---Other:
1020	2008.1991	----栗仁	10			0			----Chestnut seed
1021	2008.1999	----其他	10			0			----Other
		-菠萝:							-Pineapples:
1022	2008.2010	---罐头	15					90	---In airtight containers
1023	2008.2090	---其他	15					80	---Other
		-柑桔属水果:							-Citrus fruit:
1024	2008.3010	---罐头	20					90	---In airtight containers
1025	2008.3090	---其他	20					80	---Other

序号 No.	税则号列 Tariff Item	货 品 名 称	最惠国税率(%) M.F.N.	协定税率(%)Agreement				普通税率(%) Gen.	Article Description
				东盟 ASEAN	曼谷协定 BA	香港 H.K.	澳门 Macao		
		-梨:							-Pears:
1026	2008.4010	---罐头	20					90	---In airtight containers
1027	2008.4090	---其他	20					80	---Other
1028	2008.5000	-杏	20					90	-Apricots
1029	2008.6000	-樱桃	20					90	-Cherries
		-桃:							-Peaches:
1030	2008.7010	---罐头	10					90	---In airtight containers
1031	2008.7090	---其他	20					80	---Other
1032	2008.8000	-草莓	15					90	-Strawberries
		-其他,包括子目号2008.19以外的什锦果实:							-Other, including mixtures other than those of subheading No. 2008.19:
1033	2008.9100	--棕榈芯	5					80	--Palm hearts
1034	2008.9200	--什锦果实	10					80	--Mixtures
		--其他:							--Other:
1035	2008.9910	---荔枝罐头	20			0		90	---Lychee can
1036	2008.9920	---龙眼罐头	15			0		80	---Longan can
1037	2008.9990	---其他	15			0	0	80	---Other
	20.09	**未发酵及未加酒精的水果汁(包括酿酒葡萄汁)、蔬菜汁,不论是否加糖或其他甜物质:**							**Fruit juices(including grape must)and vegetable juices, unfermented and not containing added spirit, whether or not containing added sugar or other sweetening matter:**
		-橙汁:							-Orange juice:
1038	2009.1100	--冷冻的	7.5			0		90	--Frozen
1039	2009.1200	--非冷冻的,白利糖度值不超过20	30			0		90	--Not frozen, of a Brix value not exceeding 20
1040	2009.1900	--其他	30					90	--Other
		-柚汁:							-Grapefruit juice:
1041	2009.2100	--白利糖度值不超过20的	15			0		90	--Of a Brix value not exceeding 20
1042	2009.2900	--其他	15					90	--Other
		-其他未混合的柑桔属水果汁:							-Juice of any other single citrus fruit:
		--白利糖度值不超过20的:							--Of a Brix value not exceeding 20:
1043	2009.3110	---柠檬汁	18		16.8			90	---Lemon juice
1044	2009.3190	---其他	18		16.8			90	---Other
		--其他:							--Other:
1045	2009.3910	---柠檬汁	18		16.8			90	---Lemon juice
1046	2009.3990	---其他	18		16.8			90	---Other
		-菠萝汁:							-Pineapple juice:
1047	2009.4100	--白利糖度值不超过20	10			0△		90	--Of a Brix value not exceeding 20

序号 No.	税则号列 Tariff Item	货品名称	最惠国税率(%) M.F.N.	协定税率(%)Agreement 东盟 ASEAN	曼谷协定 BA	香港 H.K.	澳门 Macao	普通税率(%) Gen.	Article Description
1048	2009.4900	其他	10					90	--Other
1049	2009.5000	-番茄汁	30					80	--Tomato juice
		-葡萄汁,包括酿酒葡萄汁:							--Grape juice(including grape must):
1050	2009.6100	--白利糖度值不超过20的	20			0		90	--Of a Brix value not exceeding 30
1051	2009.6900	--其他	20					90	--Other
		-苹果汁:							-Apple juice:
1052	2009.7100	--白利糖度值不超过20的	20			0		90	--Of a Brix value not exceeding 20
1053	2009.7900	--其他	20					90	--Other
		-其他未混合的水果汁或蔬菜汁:							-Juice of any other single fruit or vegetable:
		---水果汁:							---Fruit juice:
1054	2009.8011	----椰子汁	10		9			90	----Coconut juice
1055	2009.8012	----芒果汁	20		17.4			90	----Mango juice
1056	2009.8013	----西番莲果汁	20		17.4			90	----Passion-fruit juice
1057	2009.8014	----番石榴果汁	20		17.4			90	----Guva juice
1058	2009.8019	----其他	20		17.4			90	----Other
1059	2009.8020	---蔬菜汁	20					80	---Vegetable juice
		-混合汁:							-Mixtures of juices:
1060	2009.9010	---水果汁	20		17.4	0		90	---Of fruit juices
1061	2009.9090	---其他	20			0		80	---Other

第二十一章
杂项食品

Chapter 21
Miscellaneous edible preparations

注释：

一、本章不包括：

（一）税号07.12的什锦蔬菜；

（二）含咖啡的焙炒咖啡代用品（税号09.01）；

（三）加香料的茶（税号09.02）；

（四）税号09.04至09.10的调味香料或其他产品；

（五）按重量计含香肠、肉、食用杂碎、动物血、鱼、甲壳动物、软体动物、其他水生无脊椎动物及其混合物超过20%的食品（第十六章），但税号21.03或21.04的产品除外；

（六）税号30.03或30.04的药用酵母及其他产品；

（七）税号35.07的酶制品。

二、上述注释一（二）所述咖啡代用品的精汁归入税号21.01。

三、税号21.04所称"均化混合食品"，是指两种或两种以上的基本配料，例如，肉、鱼、蔬菜或果实等，经精细均化制成供婴幼儿食用或营养用的零售包装食品（每件净重不超过250克）。为了调味、保藏或其他目的，可以加入少量其他配料，还可以含有少量可见的小块配料。

Notes:

1. This Chapter does not cover:
   (a) Mixed vegetables of heading No.07.12;
   (b) Roasted coffee substitutes containing coffee in any proportion (heading No.09.01);
   (c) Flavoured tea (heading No.09.02);
   (d) Spices or other products of headings No. 09.04 to 09.10;
   (e) Food preparations, other than the products described in heading No.21.03 or 21.04, containing more than 20% by weight of sausage, meat, meat offal, blood, fish or crustaceans, molluscs or other aquatic invertebrates, or any combination thereof (Chapter 16);
   (f) Yeast put up as a medicament or other products of heading No.30.03 or 30.04; or
   (g) Prepared enzymes of heading No.35.07.

2. Extracts of the substitutes referred to in Note 1 (b) above are to be classified in heading No.21.01.

3. For the purposes of heading No.21.04, the expression "homogenized composite food preparations" means preparations consisting of a finely homogenized mixture of two or more basic ingredients such as meat, fish, vegetables or fruit, put up for retail sale as infant food or for dietetic purposes, in containers of a net weight content not exceeding 250g. For the application of this definition, no account is to be taken of small quantities of any ingredients which may be added to the mixture for seasoning, preservation or other purposes. Such preparations may contain a small quantity of visible pieces of ingredients.

序号 No.	税则号列 Tariff Item	货品名称	最惠国税率(%) M.F.N.	协定税率(%) Agreement 东盟 ASEAN	曼谷协定 BA	香港 H.K.	澳门 Macao	普通税率(%) Gen.	Article Description
	21.01	咖啡、茶、马黛茶的浓缩精汁及以其为基本成分或以咖啡、茶、马黛茶为基本成分的制品；烘焙菊苣和其他烘焙咖啡代用品及其浓缩精汁：							Extracts, essences and concentrates, of coffee, tea or maté and preparations with a basis of these products or with a basis of coffee, tea or maté; roasted chicory and other roasted coffee substitutes, and extracts, essences and concentrates thereof:

序号 No.	税则号列 Tariff Item	货品名称	最惠国税率(%) M.F.N.	协定税率(%) Agreement 东盟 ASEAN	曼谷协定 BA	香港 H.K.	澳门 Macao	普通税率(%) Gen.	Article Description
		-咖啡浓缩精汁及以其为基本成分或以咖啡为基本成分的制品:							-Extracts, essences and concentrates of coffee, and preparations with a basis of these extracts, essences or concentrates or with a basis of coffee:
1062	2101.1100	--浓缩精汁	17			0	0	130	--Extracts, essences and concentrates
1063	2101.1200	--以浓缩精汁或咖啡为基本成分的制品	30			0		130	Preparations with a basis of extracts, essences or concentrates or with a basis of coffee
1064	2101.2000	-茶、马黛茶浓缩精汁及以其为基本成分或以茶、马黛茶为基本成分的制品	32			0		130	-Extracts, essences and concentrates, of tea or maté, and preparations with a basis of these extracts, essences or concentrates or with a basis of tea or mate
1065	2101.3000	-烘焙菊苣和其他烘焙咖啡代用品及其浓缩精汁	32					130	-Roasted chicory and other roasted coffee substitutes, and extracts, essences and concentrates thereof
	21.02	**酵母(活性或非活性);已死的其他单细胞微生物(不包括税号30.02的疫苗);发酵粉:**							**Yeasts (active or inactive); other singlecell micro-organisms, dead (but not including vaccines of heading No. 30.02); prepared baking powders:**
1066	2102.1000	-活性酵母	25					80	-Active yeasts
1067	2102.2000	-非活性酵母;已死的其他单细胞微生物	25					70	-Inactive yeasts; other single-cell microorganisms, dead
1068	2102.3000	-发酵粉	25					70	-Prepared baking powders
	21.03	**调味汁及其制品;混合调味品;芥子粉及其调制品:**							**Sauces and preparations therefor; mixed condiments and mixes seasonings; mustard flour and meal and prepared mustard:**
1069	2103.1000	-酱油	28			0		90	-Soya sauce
1070	2103.2000	-番茄沙司及其他番茄调味汁	15			0		90	-Tomato ketchup and other tomato sauces
1071	2103.3000	-芥子粉及其调制品	15					70	-Mustard flour and meal and prepared mustard
		-其他:							-Other:
1072	2103.9010	---味精	21		18.2	0		130	---Gourmet powder
1073	2103.9020	---别特酒(Aromatic bitters),按体积计酒精含量44.2%-49.2%,按重量计含1.5%-6%的香料、各种配料以及4%-10%的糖	21			0		90	---Aromatic bitters, 44.2% - 49.2% of which is alcoholic strength by volume, 1.5% - 6% of which is spiles and various ingredients by weight and 4% - 10% of which is sugar by weight

序号 No.	税则号列 Tariff Item	货品名称	最惠国税率(%) M.F.N.	协定税率(%) Agreement 东盟 ASEAN	曼谷协定 BA	香港 H.K.	澳门 Macao	普通税率(%) Gen.	Article Description
1074	2103.9090	---其他	21		18.4	0	0	90	---Other
	21.04	**汤料及其制品;均化混合食品:**							**Soups and broths and preparations therefor; homogenized composite food preparations:**
1075	2104.1000	-汤料及其制品	15			0		90	-Soups and broths and preparations therefor
1076	2104.2000	-均化混合食品	32					90	-Homogenized composite food preparations
	21.05	**冰淇淋及其他冰制食品,不论是否含可可:**							**Ice cream and other edible ice, whether or not containing cocoa:**
1077	2105.0000	冰淇淋及其他冰制食品,不论是否含可可	19			0	0	90	Ice cream and other edible ice, whether or not containing cocoa
	21.06	**其他税号未列名的食品:**							**Food preparations not elsewhere specified or included:**
1078	2106.1000	-浓缩蛋白质及人造蛋白物质	10				0	90	-Protein concentrates and textured protein substances
		-其他:							-Other:
1079	2106.9010	---制造碳酸饮料的浓缩物	35			0	0	100	---Beverage bases
1080	2106.9020	---制造饮料用的复合酒精制品	20			0	0	180	---Compound alcoholic preparations of a kind used for the manufacture of beverages
1081	2106.9030	---蜂王浆制剂	3		3	0		80	---Royal jelly, put up as tonic essences
		---其他:							---Other:
1082	2106.9091	----调味紫菜	20		18.4	0	0	90	----Seasoned laver
1083	2106.9099	----其他	20		18.4	0	0	90	----Other

第二十二章
饮料、酒及醋

Chapter 22
Beverages, spirits and vinegar

注释：

一、本章不包括：

（一）本章的产品（税号 22.09 的货品除外）经配制后，用于烹饪而不适于作为饮料的制品（通常归入税号 21.03）；

（二）海水（税号 25.01）；

（三）蒸馏水、导电水及类似的纯净水（税号 28.51）；

（四）按重量计浓度超过 10% 的醋酸（税号 29.15）；

（五）税号 30.03 或 30.04 的药品；

（六）芳香料制品及盥洗品（第三十三章）。

二、本章及第二十章和第二十一章所称“按容量计酒精浓度”，应是温度在 20℃ 时测得的浓度。

三、税号 22.02 所称“无酒精饮料”，是指按容量计酒精浓度不超过 0.5% 的饮料。含酒精饮料应分别归入税号 22.03 至 22.06 或税号 22.08。

子目注释：

子目号 2204.10 所称“汽酒”，是指温度在 20℃ 时装在密封容器中超过大气压力 3 巴及以上的酒。

Notes:

1. This Chapter does not cover:

(a) Products of this Chapter (other than those of heading No.22.09) prepared for culinary purposes and thereby rendered unsuitable for consumption as beverages (generally heading No.21.03);

(b) Sea water (heading No.25.01);

(c) Distilled or conductivity water or water of similar purity (heading No.28.51);

(d) Acetic acid of a concentration exceeding 10% by weight of acetic acid (heading No.29.15);

(e) Medicaments of heading No.30.03 or 30.04; or

(f) Perfumery or toilet preparations (Chapter 33).

2. For the purposes of this Chapter and of Chapter 20 and 21, the "alcoholic strength by volume" shall be determined at a temperature of 20℃.

3. For the purposes of heading No.22.02, the term "non-alcoholic beverages" means beverages of an alcoholic strength by volume not exceeding 0.5% vol. Alcoholic beverages are classified in headings No.22.03 to 22.06 or heading No.22.08 as appropriate.

Subheading Note:

For the purposes of subheading No.2204.10, the expression "sparkling wine" means wine which, when kept at a temperature of 20℃ in closed containers, has an excess pressure of not less than 3 bars.

序号 No.	税则号列 Tariff Item	货品名称	最惠国税率(%) M.F.N.	协定税率(%) Agreement 东盟 ASEAN	曼谷协定 BA	香港 H.K.	澳门 Macao	普通税率(%) Gen.	Article Description
	22.01	**未加糖或其他甜物质及未加味的水，包括天然或人造矿泉水及汽水；冰及雪：**							**Waters, including natural or artificial mineral waters and aerated waters, not containing added sugar or other sweetening matter or flavoured; ice and snow:**
		-矿泉水及汽水：							-Mineral waters and aerated waters:
1084	2201.1010	---矿泉水	20			0		90	---Mineral waters
1085	2201.1020	---汽水	20					90	---Aerated waters
		-其他：							-Other:
1086	2201.9010	---天然水	10					30	---Natural waters
1087	2201.9090	---其他	10					30	---Other

序号 No.	税则号列 Tariff Item	货品名称	最惠国税率(%) M.F.N.	协定税率(%) Agreement 东盟 ASEAN	曼谷协定 BA	香港 H.K.	澳门 Macao	普通税率(%) Gen.	Article Description
	22.02	**加味、加糖或其他甜物质的水,包括矿泉水及汽水,其他无酒精饮料,但不包括税号20.09的水果汁或蔬菜汁:**							**Waters, including mineral waters and aerated waters, containing added sugar or other sweetening matter or flavoured, and other non-alcoholic beverages, not including fruit or vegetable juices of heading No.20.09:**
1088	2202.1000	-加味、加糖或其他甜物质的水,包括矿泉水及汽水	20			0		100	-Waters, including mineral waters and aerated waters, containing added sugar or other sweetening matter or flavoured
1089	2202.9000	-其他	35		29.5	0		100	-Other
	22.03	**麦芽酿造的啤酒:**							**Beer made from malt:**
1090	2203.0000	麦芽酿造的啤酒	见表 4		见尾注			见表 4	Beer made from malt
	22.04	**鲜葡萄酿造的酒,包括加酒精的;税号20.09以外的酿酒葡萄汁:**							**Wine of fresh grapes, including fortified wines; grape must other than that of heading No.20.09:**
1091	2204.1000	-汽酒	14					180	-Sparkling wine
		-其他酒;加酒精抑制发酵的酿酒葡萄汁:							-Other wine; grape must with fermentation prevented or arrested by the addition of alcohol:
1092	2204.2100	--装入 2 升及以下容器的	14					180	--In containers holding 2L or less
1093	2204.2900	--其他	20					180	--Other
1094	2204.3000	-其他酿酒葡萄汁	30					90	-Other grape must
	22.05	**味美思酒及其他加植物或香料的用鲜葡萄酿造的酒:**							**Vermouth and other wine of fresh grapes flavoured with plants or aromatic substances:**
1095	2205.1000	-装入 2 升及以下容器的	65					180	-In containers holding 2L or less
1096	2205.9000	-其他	65					180	-Other
	22.06	**其他发酵饮料(例如,苹果酒、梨酒、蜂蜜酒);其他税号未列名的发酵饮料的混合物及发酵饮料与无酒精饮料的混合物:**							**Other fermented beverages (for example, cider, perry, mead); mixtures of fermented beverages and mixtures of fermented beverages and non-alcoholic beverages, not elsewhere specified or included:**
1097	2206.0000	其他发酵饮料(例如,苹果酒、梨酒、蜂蜜酒);其他税号未列名的发酵饮料的混合物及发酵饮料与无酒精饮料的混合物	51.4			0△	0	180	Other fermented beverages (for example, cider, perry, mead); mixtures of fermented beverages and mixtures of fermented beverages and non-alcoholic beverages, not elsewhere specified or included

序号 No.	税则号列 Tariff Item	货品名称	最惠国税率(%) M.F.N.	协定税率(%) Agreement 东盟 ASEAN	曼谷协定 BA	香港 H.K.	澳门 Macao	普通税率(%) Gen.	Article Description
	22.07	**未改性乙醇,按容量计酒精浓度在80%及以上;任何浓度的改性乙醇及其他酒精:**							**Undenatured ethyl alcohol of an alcoholic strength by volume of 80% vol or higher; ethyl alcohol and other spirits, denatured, of any strength:**
1098	2207.1000	-未改性乙醇,按容量计酒精浓度在80%及以上	40				0	100	-Undenatured ethyl alcohol of an alcoholic strength by volume of 80% vol or higher
1099	2207.2000	-任何浓度的改性乙醇及其他酒精	30					80	-Ethyl alcohol and other spirits, denatured, of any strength
	22.08	**未改性乙醇,按容量计酒精浓度在80%以下;蒸馏酒、利口酒及其他酒精饮料:**							**Undenaturated ethyl alcohol of an alcoholic strength by volume of less than 80% vol; spirits, liqueurs and other spirituous beverages:**
1100	2208.2000	-蒸馏葡萄酒制得的烈性酒	10					180	-Spirits obtained by distilling grape wine or grape marc
1101	2208.3000	-威士忌酒	10					180	-Whiskies
1102	2208.4000	-朗姆酒及其他甘蔗蒸馏酒	10					180	-Rum and tafia
1103	2208.5000	-杜松子酒	10					180	-Gin and geneva
1104	2208.6000	-伏特加酒	10		8.9			180	-Vodka
1105	2208.7000	-利口酒及柯迪尔酒	10		8.9			180	-Liqueurs and cordials
		-其他:							-Other:
1106	2208.9010	---龙舌兰酒	10		8.9			180	---Tequila, Mezcal
1107	2208.9090	---其他	10		8.9			180	---Other
	22.09	**醋及用醋酸制得的醋代用品:**							**Vinegar and substitutes for vinegar obtained from acetic acid:**
1108	2209.0000	醋及用醋酸制得的醋代用品	20			0		70	Vinegar and substitutes for vinegar obtained from acetic acid

注:① 22030000 麦芽酿造的啤酒的协定税率:0元/升。

Note: ①Agreement specific tariff rate of heading No. 22030000 "beer made from malt": 0RMB/litre.

第二十三章
食品工业的残渣及废料；配制的动物饲料

Chapter 23
Residues and waste from the food industries; prepared animal fodder

注释：

税号 23.09 包括其他税号未列号的配制动物饲料，这些饲料是由动、植物原料加工而成的，并且已改变了原料的基本特性，但加工过程中的植物废料、植物残渣及副产品除外。

Note:

Heading No.23.09 includes products of a kind used in animal feeding, not elsewhere specified or included, obtained by processing vegetable or animal materials to such an extent that they have lost the essential characteristics of the original material, other than vegetable waste, vegetable residues and by-products of such processing.

子目注释：

子目 2306.41 所称"低芥子酸油菜籽"，是指第十二章子目注释一所定义的油菜子。

Subheading Note:

For the purposes of subheading No.2306.41, the expression "low erucic acid rape or colza seeds" means seeds as defined in Subheading Note 1 to Chapter 12.

序号 No.	税则号列 Tariff Item	货品名称	最惠国税率(%) M.F.N.	协定税率(%)Agreement 东盟 ASEAN	曼谷协定 BA	香港 H.K.	澳门 Macao	普通税率(%) Gen.	Article Description
	23.01	**不适于供人食用的肉、杂碎、鱼、甲壳动物、软体动物或其他水生无脊椎动物的渣粉及团粒；油渣：**							**Flours, meals and pellets, of meat or meat offal, of fish or of crustaceans, molluscs or other aquatic invertebrates, unfit, for human consumption; greaves:**
		-肉、杂碎的渣粉及团粒；油渣：							-Flours, meals and pellets, of meat or meat offal; greaves:
		---肉骨粉：							---Flours and meals, of meat bones:
1109	2301.1011	----含牛羊成分的	2					11	----Of bovine and sheep
1110	2301.1019	----其他	2					11	----Other
1111	2301.1020	---油渣	5					50	---Greaves
1112	2301.1090	---其他	5					30	---Other
		-鱼、甲壳动物、软体动物或其他水生无脊椎动物的渣粉及团粒：							-Flours, meals and pellets, of fish or of crustaceans, molluscs or other aquatic invertebrates:
1113	2301.2010	---饲料用鱼粉	2			0△		11	---Flours and meals of fish, of a kind used in animal feeding
1114	2301.2090	---其他	5			0△		30	---Other
	23.02	**谷物或豆类植物在筛、碾或其他加工过程中所产生的糠、麸及其他残渣，不论是否制成团粒：**							**Bran, sharps and other residues, whether or not in the form of pellets, derived from the sifting, milling or other working of cereals or of leguminous plants:**
1115	2302.1000	-玉米的	5					30	-Of maize(corn)
1116	2302.2000	-稻米的	5					30	-Of rice
1117	2302.3000	-小麦的	3					30	-Of wheat
1118	2302.4000	-其他谷物的	5					30	-Of other cereals
1119	2302.5000	-豆类植物的	5			0		30	-Of leguminous plants

序号 No.	税则号列 Tariff Item	货品名称	最惠国税率(%) M.F.N.	协定税率(%)Agreement 东盟 ASEAN	曼谷协定 BA	香港 H.K.	澳门 Macao	普通税率(%) Gen.	Article Description
	23.03	**制造淀粉过程中的残渣及类似的残渣,甜菜渣、甘蔗渣及制糖过程中的其他残渣,酿造及蒸馏过程中的糟粕及残渣,不论是否制成团粒:**							**Residues of starch manufacture and similar residues, beet-pulp, bagasses and other waste of sugar manufacture, brewing or distilling dregs and waste, whether or not in the form of pellets:**
1120	2303.1000	-制造淀粉过程中的残渣及类似的残渣	5					30	-Residues of starch manufacture and similar residues
1121	2303.2000	-甜菜渣、甘蔗渣及制糖过程中的其他残渣	5					30	-Beet-pulp, bagasses and other waste of sugar manufacture
1122	2303.3000	-酿造及蒸馏过程中的糟粕及残渣	5					30	-Brewing or distilling dregs and waste
	23.04	**提炼豆油所得的油渣饼及其他固体残渣,不论是否碾磨或制成团粒:**							**Oil-cake and other solid residues, whether or not ground or in the form of pellets, resulting from the extraction of soyabean oil:**
1123	2304.0010	---油渣饼	5					30	---Oil-cake
1124	2304.0090	---其他	5					30	---Other
	23.05	**提炼花生油所得的油渣饼及其他固体残渣,不论是否碾磨或制成团粒:**							**Oil-cake and other solid residues, whether or not ground or in the form of pellets, resulting from the extraction of groundnutoil:**
1125	2305.0000	提炼花生油所得的油渣饼及其他固体残渣,不论是否碾磨或制成团粒	5					30	Oil-cake and other solid residues, whether or not ground or in the form of pellets, resulting from the extraction of groundnut oil
	23.06	**税号 23.04 或 23.05 以外的提炼植物油脂所得的油渣饼及其他固体残渣,不论是否碾磨或制成团粒:**							**Oil-Cake and other solid residues, whether or not ground or in the form of pellets, resulting from the extraction of vegetable fats or oils, other than those of heading No. 23.04 or 23.05:**
1126	2306.1000	-棉子的	5					30	-Of cotton seeds
1127	2306.2000	-亚麻子的	5					30	-Of linseed
1128	2306.3000	-葵花子的	5					30	-Of sunflower seeds
		-油菜子的							-Of rape or colza seeds:
1129	2306.4100	--低芥子酸的	5					30	--Of low erucic acid rape or colza seeds
1130	2306.4900	--其他	5					30	--Other
1131	2306.5000	-椰子或干椰肉的	5					30	-Of coconut or copra
1132	2306.6000	-油棕果或油棕仁的	5					30	-Of palm nuts or kernels
1133	2306.7000	-玉米胚芽的	5					30	-Of maize(corn) germ
1134	2306.9000	-其他	5					30	-Other
	23.07	**葡萄酒渣;粗酒石:**							**Wine lees; argol:**
1135	2307.0000	葡萄酒渣;粗酒石	5					30	Wine lees; argol

序号 No.	税则号列 Tariff Item	货品名称	最惠国税率(%) M.F.N.	协定税率(%) Agreement 东盟 ASEAN	曼谷协定 BA	香港 H.K.	澳门 Macao	普通税率(%) Gen.	Article Description
	23.08	**动物饲料用的其他税号未列名的植物原料、废料、残渣及副产品，不论是否制成团粒：**							**Vegetable materials and vegetable waste, vegetable residues and by-products, whether or not in the form of pellets, of a kind used in animal feeding, not elsewhere specified or included:**
1136	2308.0000	动物饲料用的其他税号未列名的植物原料、废料、残渣及副产品，不论是否制成团粒	5					35	Vegetable materials and vegetable waste, vegetable residues and by-products, whether or not in the form of pellets, of a kind used in animal feeding, not elsewhere specified or included
	23.09	**配制的动物饲料：**							**Preparations of a kind used in animal feeding:**
		-零售包装的狗食或猫食：							-Dog or cat food, put up for retail sale:
1137	2309.1010	---罐头	15					90	---In airtight containers
1138	2309.1090	---其他	15					90	---Other
		-其他：							-Other:
1139	2309.9010	---制成的饲料添加剂	5					14	---Preparations for use in making the complete feeds or supplementary feeds
1140	2309.9090	---其他	6.7					14	---Other

第二十四章
烟草及烟草代用品的制品

Chapter 24
Tobacco and manufactured tobacco substitutes

注释：

本章不包括药用卷烟（第三十章）。

Note:

This Chapter does not cover medicinal cigarettes (Chapter 30).

序号 No.	税则号列 Tariff Item	货品名称	最惠国税率(%) M.F.N.	协定税率(%) Agreement 东盟 ASEAN	曼谷协定 BA	香港 H.K.	澳门 Macao	普通税率(%) Gen.	Article Description
	24.01	**烟草；烟草废料：**							**Unmanufactured tobacco; tobacco refuse:**
		-未去梗的烟草：							-Tobacco, not stemmed/stripped:
1141	2401.1010	---烤烟	10		9.4			70	---Flue-cured
1142	2401.1090	---其他	10					70	---Other
		-部分或全部去梗的烟草：							-Tobacco, partly or wholly stemmed/stripped:
1143	2401.2010	---烤烟	10					70	---Flue-cured
1144	2401.2090	---其他	10					70	---Other
1145	2401.3000	-烟草废料	10					70	-Tobacco refuse
	24.02	**烟草或烟草代用品制成的雪茄烟及卷烟：**							**Cigars, cheroots, cigarillos and cigarettes, of tobacco or of tobacco substitutes:**
1146	2402.1000	-烟草制的雪茄烟	25					180	-Cigars, cheroots and cigarillos, containing tobacco
1147	2402.2000	-烟草制的卷烟	25					180	-Cigarettes containing tobacco
1148	2402.9000	-其他	25					180	-Other
	24.03	**其他烟草及烟草代用品的制品；“均化”或“再造”烟草；烟草精汁：**							**Other manufactured tobacco and manufactured tobacco substitutes; “homogenized” or “reconstituted” tobacco; tobacco extracts and essences:**
1149	2403.1000	-供吸用的烟草，不论是否含有任何比例的烟草代用品	57		50			180	-Smoking tobacco, whether or not containing tobacco substitutes in any proportion
		-其他：							-Other:
1150	2403.9100	--“均化”或“再造”烟草	57					180	--“Homogenized” or “reconstituted” tobacco
1151	2403.9900	--其他	57					180	--Other

第五类

矿　产　品

第二十五章

盐；硫磺；泥土及石料；石膏料、石灰及水泥

注释：

一、除条文及注释四另有规定的以外，本章各税号只包括原产状态的矿产品，或只经过洗涤（包括用化学物质清除杂质而未改变产品结构的）、破碎、磨碎、研粉、淘洗、筛分以及用浮选、磁选和其他机械物理方法（不包括结晶法）精选过的货品，但不得经过焙烧、煅烧、混合或超过税目所列的加工范围。

本章产品可含有添加的抗尘剂，但所加剂料并不使原产品改变其一般用途而适合于某些特殊用途。

二、本章不包括：

（一）升华硫磺、沉淀硫磺及胶态硫磺（税号28.02）；

（二）土色料，按重量计三氧化二铁含量在70%及以上（税号28.21）；

（三）第三十章的药品及其他产品；

（四）芳香料制品及化妆盥洗品（第三十三章）；

（五）长方砌石、路缘石、扁平石（税号68.01）、镶嵌石或类似石料（税号68.02）及铺屋顶、饰墙面或防潮用的板岩（税号68.03）；

（六）宝石或半宝石（税号71.02或71.03）；

（七）每颗重量不低于2.5克的氯化钠或氧化镁培养晶体（光学元件除外）（税号38.24）；氯化钠或氧化镁制的光学元件（税号90.01）；

（八）台球用粉块（税号95.04）；

（九）书写或绘画用粉笔及裁缝划粉（税号96.09）。

SECTION V

MINERAL PRODUCTS

Chapter 25

Salt; sulphur; earths and stone; plastering materials, lime and cement

Notes:

1. Except where their context or Note 4 to this Chapter otherwise requires, the headings of this Chapter cover only products which are in the crude state or which have been washed(even with chemical substances eliminating the impurities without changing the structure of the product), crushed, ground, powdered, levigated, sifted, screened, concentrated by flotation, magnetic separation or other mechanical or physical processes (except crystallization), but not products which have been roasted, calcined, obtained by mixing or subjected to processing be yond that mentioned in each heading.

   The products of this Chapter may contain an added antidusting agent, provided that such addition does not render the product particularly suitable for specific use rather than for general use.

2. This Chapter does not cover:

   (a) Sublimed sulphur, precipitated sulphur or colloidal sulphur (heading No.28.02);

   (b) Earth colours containing 70% or more by weight of combined iron evaluated as Fe_2O_3 (heading No.28.21);

   (c) Medicaments or other products of Chapter 30;

   (d) Perfumery, cosmetic or toilet preparations (Chapter 33);

   (e) Setts, curbstones or flagstones (heading No.68.01); mosaic cubes or the like (heading No.68.02); roofing, facing or damp course slates (heading No.68.03);

   (f) Precious or semi-precious stones (heading No.71.02 or 71.03);

   (g) Cultured crystals (other than optical elements) weighing not less than 2.5g each, of sodium chloride or of magnesium oxide, of heading No.38.24; optical elements of sodium chloride or of magnesium oxide (heading No.90.01);

   (h) Billiard chalks (heading No.95.04); or

   (i) Writing or drawing chalks or tailors' chalks (heading No, 96.09).

三、既可归入税号25.17又可归入本章其他税号的产品，应归入税号25.17。

四、税号25.30主要包括：未膨胀的蛭石、珍珠岩及绿泥石；不论是否煅烧或混合的土色料；天然云母氧化铁；海泡石（不论是否磨光成块）；琥珀；模制后未经进一步加工的片、条、杆或类似形状的粘聚海泡石及粘聚琥珀；黑玉；菱锶矿（不论是否煅烧），但不包括氧化锶；破碎陶器；砖或混凝土的碎块。

3. Any products classifiable in heading No.25.17 and any other heading of this Chapter are to be classified in heading No.25.17.

4. Heading No.25.30 applies, *inter alia*, to: vermiculite, perlite and chlorites, unexpanded; earth colours, whether or not calcined or mixed together; natural micaceous iron oxides; meerschaum (whether or not in polished pieces); amber; agglomerated meerschaum and agglomerated amber, in plates, rods, sticks or similar forms, not worked after moulding; jet; strontianite (whether or not calcined), other than strontium oxide; broken pieces of pottery, brick or concrete.

序号 No.	税则号列 Tariff Item	货品名称	最惠国税率(%) M.F.N.	协定税率(%) Agreement 东盟 ASEAN	曼谷协定 BA	香港 H.K.	澳门 Macao	普通税率(%) Gen.	Article Description
	25.01	**盐(包括精制盐及变性盐)及纯氯化钠，不论是否为水溶液，也不论是否添加抗结块剂或松散剂；海水：**							**Salt (including table salt and denatured salt) and pure sodium chloride, whether or not in aqueous solution or containing added anticaking or free-flowing agents; sea water:**
		---盐：							---Salt:
1152	2501.0011	----食用盐	0					0	----Edible salt
1153	2501.0019	----其他	0					0	----Other
1154	2501.0020	---纯氯化钠	3					35	---Pure sodium chloride
1155	2501.0030	---海水	0					0	---Sea water
	25.02	**未焙烧的黄铁矿：**							**Unroasted iron pyrites:**
1156	2502.0000	未焙烧的黄铁矿	3					20	Unroasted iron pyrite
	25.03	**各种硫磺，但升华硫磺、沉淀硫磺及胶态硫磺除外：**							**Sulphur of all kinds, other than sublimed sulphur, precipitated sulphur and colloidal sulphur:**
1157	2503.0000	各种硫磺，但升华硫磺、沉淀硫磺及胶态硫磺除外	3					17	Sulphur of all kinds, other than sublimed sulphur, precipitated sulphur and colloidal sulphur
	25.04	**天然石墨：**							**Natural graphite:**
		-粉末或粉片：							-In powder or in flakes:
1158	2504.1010	---磷片	3					30	---In flakes
1159	2504.1090	---其他	3					30	---Other
1160	2504.9000	-其他	3					30	-Other
	25.05	**各种天然砂，不论是否着色，但第二十六章的含金属矿砂除外：**							**Natural sands of all kinds, whether or not coloured, other than metal-bearing sands of Chapter 26:**
1161	2505.1000	-硅砂及石英砂	3					40	-Silica sands and quartz sands
1162	2505.9000	-其他	3					40	-Other
	25.06	**石英(天然砂除外)；石英岩，不论是否粗加修整或仅用锯或其他方法切割成矩形(包括正方形)的板、块：**							**Quartz (other than natural sands); quartzite, whether or not roughly trimmed or merely cut, by sawing or otherwise, into blocks or slabs of a rectangular (including square) shape:**

序号 No.	税则号列 Tariff Item	货品名称	最惠国税率(%) M.F.N.	协定税率(%) Agreement 东盟 ASEAN	曼谷协定 BA	香港 H.K.	澳门 Macao	普通税率(%) Gen.	Article Description
1163	2506.1000	-石英	3					40	-Quartz
		-石英岩:							-Quartzite:
1164	2506.2100	--原状或粗加修整	3					40	--Crude or roughly trimmed
1165	2506.2900	--其他	3					40	--Other
	25.07	**高岭土及类似土,不论是否煅烧:**							**Kaolin and other kaolinic clays, whether or not calcined:**
		高岭土及类似土,不论是否煅烧:							Kaolin and other kaolinic clays, whether or not calcined:
1166	2507.0010	---高岭土	3					50	---Kaolin
1167	2507.0090	---其他	3					50	---Other
	25.08	**其他粘土(不包括税号68.06的膨胀粘土)、红柱石、蓝晶石及硅线石,不论是否煅烧;富铝红柱石;火泥及第纳斯土:**							**Other clays (not including expanded clays of heading No.68.06), andalusite, kyanite and sillimanite, whether or not calcined; mullite; chamotte or dinas earths:**
1168	2508.1000	-膨润土	3					50	-Bentonite
1169	2508.2000	-脱色土及漂白土	3					50	-Decolourizing earths and fuller's earth
1170	2508.3000	-耐火粘土	3					20	-Fire-clay
1171	2508.4000	-其他粘土	3					50	-Other clays
1172	2508.5000	-红柱石、蓝晶石及硅线石	3					40	-Andalusite, kyanite and sillimanite
1173	2508.6000	-富铝红柱石	3					40	-Mullite
1174	2508.7000	-火泥及第纳斯土	3					20	-Chamotte or dinas earths
	25.09	**白垩:**							**Chalk:**
1175	2509.0000	白垩	3					45	Chalk
	25.10	**天然磷酸钙、天然磷酸铝钙及磷酸盐白垩:**							**Natural calcium phosphates, natural aluminium calcium phosphates and phosphatic chalk:**
		-未碾磨:							-Unground:
1176	2510.1010	---磷灰石	3					11	---Apatite
1177	2510.1090	---其他	3					20	---Other
		-已碾磨:							-Ground:
1178	2510.2010	---磷灰石	3					11	---Apatite
1179	2510.2090	---其他	3					20	---Other
	25.11	**天然硫酸钡(重晶石);天然碳酸钡(毒重石),不论是否煅烧,但税号28.16的氧化钡除外:**							**Natural barium sulphate (barytes); natural barium carbonate (witherite), whether or not calcined, other than barium oxide of heading No.28.16:**
1180	2511.1000	-天然硫酸钡(重晶石)	3					45	-Natural barium sulphate (barytes)

序号 No.	税则号列 Tariff Item	货品名称	最惠国税率(%) M.F.N.	协定税率(%)Agreement 东盟 ASEAN	曼谷协定 BA	香港 H.K.	澳门 Macao	普通税率(%) Gen.	Article Description
1181	2511.2000	-天然碳酸钡(毒重石)	3					45	-Natural barium carbonate(witherite)
	25.12	**硅质化石粗粉(例如各种硅藻土)及类似的硅质土,不论是否煅烧,其表观比重不超过1:**							**Siliceous fossil meals (for example, kieselguhr, tripolite and diatomite) and similar siliceous earths, whether or not calcined, of an apparent specific gravity of 1 or less:**
		硅质化石粗粉(例如各种硅藻土)及类似的硅质土,不论是否煅烧,其表观比重不超过1:							Siliceous fossil meals (for example, kieselguhr, tripolite and diatomite) and similar siliceous earths, whether or not calcined, of an apparent specific gravity of 1 or less:
1182	2512.0010	---硅藻土	3					40	---Kieselguhr
1183	2512.0090	---其他	3					40	---Other
	25.13	**浮石;刚玉岩;天然刚玉砂;天然石榴石及其他天然磨料,不论是否热处理:**							**Pumice stone; emery; natural corundum, natural garnet and other natural abrasives, whether or not heat-treated:**
		-浮石:							-Pumice stone:
1184	2513.1100	--原状或不规则碎块,包括破碎的在内	3					35	--Crude or in irregular pieces, including crushed pumice("bimskies")
1185	2513.1900	--其他	3					35	--Other
1186	2513.2000	-刚玉岩、天然刚玉砂、天然石榴石及其他天然磨料	3					17	-Emery, natural corundum, narural garnet and other natural abrasives
	25.14	**板岩,不论是否粗加修整或仅用锯或其他方法切割成矩形(包括正方形)的板、块:**							**Slate, whether or not roughly trimmed or merely cut, by sawing or otherwise, into blocks or slabs of a rectangular (including square) shape:**
1187	2514.0000	板岩,不论是否粗加修整或仅用锯或其他方法切割成矩形(包括正方形)的板、块	3					50	Slate, whether or not roughly trimmed or merely cut, by sawing or otherwise, into blocks or slabs of a rectangular (including square) shape
	25.15	**大理石、石灰华及其他石灰质碑用或建筑用石,表观比重为2.5及以上,蜡石,不论是否粗加修整或仅用锯或其他方法切割成矩形(包括正方形)的板、块:**							**Marble, travertine, ecaussine and other calcareous monumental or building stone of an apparent specific gravity of 2.5 or more, and alabaster, whether or not roughly trimmed or merely cut, by sawing or otherwise, into blocks or slabs of a rectangular (including square) shape:**
		-大理石及石灰华:							-Marble and travertine:
1188	2515.1100	--原状或粗加修整	4					80	--Crude or roughly trimmed

序号 No.	税则号列 Tariff Item	货品名称	最惠国税率(%) M.F.N.	协定税率(%) Agreement 东盟 ASEAN	曼谷协定 BA	香港 H.K.	澳门 Macao	普通税率(%) Gen.	Article Description
1189	2515.1200	--用锯或其他方法切割成矩形,包括正方形	4					80	--Merely cut, by sawing or otherwise, into blocks or slabs of a rectangular (including square) shape
1190	2515.2000	-其他石灰质碑用或建筑用石;蜡石	3					50	-Ecaussine and other calcareous monumental or building stone; alabaster
	25.16	**花岗岩、斑岩、玄武岩、砂岩以及其他碑用或建筑用石,不论是否粗加修整或仅用锯或其他方法切割成矩形(包括正方形)的板、块:**							**Granite, porphyry, basalt, sandstone and other monumental or building stone, whether or not roughly trimmed or merely cut, by sawing or otherwise, into blocks or slabs of a rectangular (including square) shape:**
		-花岗岩:							-Granite:
1191	2516.1100	--原状或粗加修整	4					50	--Crude or roughly trimmed
1192	2516.1200	--仅用锯或其他方法切割成矩形,包括正方形	4					50	--Merely cut, by sawing or otherwise, into blocks or slabs of a rectangular (including square) shape
		-砂岩:							-Sandstone:
1193	2516.2100	--原状或粗加修整	3					50	--Crude or roughly trimmed
1194	2516.2200	--仅用锯或其他方法切割成矩形,包括正方形	3					50	--Merely cut, by sawing or otherwise, into blocks or slabs of a rectangular (including square) shape
1195	2516.9000	-其他碑用或建筑用石	3					50	-Other monumental or building stone
	25.17	**通常作混凝土粒料、铺路、铁道路基或其他路基用的卵石、砾石及碎石,圆石子及燧石,不论是否热处理;矿渣、浮渣及类似的工业残渣,不论是否混有本税号第一部分所列的材料;沥青碎石;税号25.15、25.16所列各种石料的碎粒、碎屑及粉末,不论是否热处理:**							**Pebbles, gravel, broken or crushed stone, of a kind commonly used for concrete aggregates, for road metalling or for railway or other ballast, shingle and flint, whether or not heat-treated; macadam of slag, dross or similar industrial waste, whether or not incorporating the materials cited in the first part of the heading; tarred macadam; granules, chippings and powder, of stones of heading No. 25.15 or 25.16, whether or not heat-treated:**
1196	2517.1000	-通常作混凝土粒料、铺路、铁道路基或其他路基用的卵石、砾石及碎石,圆石子及燧石,不论是否热处理	4					50	-Pebbles, gravel, broken or crushed stone, of a kind commonly used for concrete aggregates, for road metalling or for railway or other ballast, shingle and flint, whether or not bead-treated

序号 No.	税则号列 Tariff Item	货品名称	最惠国税率(%) M.F.N.	协定税率(%) Agreement 东盟 ASEAN	曼谷协定 BA	香港 H.K.	澳门 Macao	普通税率(%) Gen.	Article Description
1197	2517.2000	-矿渣、浮渣及类似的工业残渣,不论是否混有子目号 2517.10 所列的材料	3					50	-Macadam of slag, dross or similar industrial waste, whether or not incorporating the materials cited in subheading No.2517.10
1198	2517.3000	-沥青碎石	3					50	-Tarred macadam
		-税号 25.15 及 25.16 所列各种石料的碎粒、碎屑及粉末,不论是否热处理:							-Granules, chippings and powder, of stones of heading No.25.15 or 25.16, whether or not heat-treated:
1199	2517.4100	--大理石的	3					50	--Of marble
1200	2517.4900	--其他	3					50	--Other
	25.18	**白云石,不论是否煅烧或烧结、粗加修整或仅用锯或其他方法切割成矩形(包括正方形)的板、块;夯混白云石:**							**Dolomite, whether or not calcined; including dolomite roughly trimmed or merely cut, by sawing or otherwise, into blocks or slabs of a rectangular (including square) shape; dolomite ramming mix:**
1201	2518.1000	-未煅烧或烧结的白云石	3					40	-Dolomite, not calcinecd or sintered
1202	2518.2000	-已煅烧或烧结的白云石	3					40	-Calcined or sintered dolomite
1203	2518.3000	-夯混白云石	3					40	-Dolomite ramming mix
	25.19	**天然碳酸镁(菱镁矿);熔凝镁氧矿;烧结镁氧矿,不论烧结前是否加入少量其他氧化物;其他氧化镁,不论是否纯净:**							**Natural magnesium carbonate (magnesite); fused magnesia; dead-burned (sintered) magnesia, whether or not containing small quantities of other oxides added before sintering; other magnesium oxide, whether or not pure:**
1204	2519.1000	-天然碳酸镁(菱镁矿)	3					40	-Natural magnesium carbonate (magnesite)
		-其他:							-Other:
1205	2519.9010	---熔凝镁氧矿	3					40	---Fused magnesia
1206	2519.9020	---烧结镁氧矿(重烧镁)	3					40	---Dead-burned (sintered) magnesia
1207	2519.9030	---碱烧镁(轻烧镁)	3					40	---Light-burned magnesia
		---其他:							---Other:
1208	2519.9091	----化学纯氧化镁	3					35	----Magnesium oxide, chemically pure
1209	2519.9099	----其他	3					40	----Other
	25.20	**生石膏;硬石膏;熟石膏(由煅烧的生石膏或硫酸钙构成),不论是否着色,也不论是否带有少量促凝剂或缓凝剂:**							**Gypsum; anhydrite; plasters (consisting of calcined gypsum or calcium sulphate) whether or not coloured, with or without small quantities of accelerators or retarders:**

序号 No.	税则号列 Tariff Item	货品名称	最惠国税率(%) M.F.N.	协定税率(%) Agreement				普通税率(%) Gen.	Article Description
				东盟 ASEAN	曼谷协定 BA	香港 H.K.	澳门 Macao		
1210	2520.1000	-生石膏;硬石膏	5					80	-Gypsum;anhydrite
		-熟石膏:							-Plasters:
1211	2520.2010	---牙科用	5					40	---For dental use
1212	2520.2090	---其他	5					80	---Other
	25.21	**石灰石助熔剂;通常用于制造石灰或水泥的石灰石及其他钙质石:**							**Limestone flux; limestone and other calcareous stone, of a kind used for the manufacture of lime or cement:**
1213	2521.0000	石灰石助熔剂;通常用于制造石灰或水泥的石灰石及其他钙质石	5			0		50	Limestone flux; limestone and other calcareous stone, of a kind used for the manufacture of lime or cement
	25.22	**生石灰、熟石灰及水硬石灰,但税号28.25的氧化钙及氢氧化钙除外:**							**Quicklime, slaked lime and hydraulic lime, other than calcium oxide and hydroxide of heading No.28.25:**
1214	2522.1000	-生石灰	5					80	-Quicklime
1215	2522.2000	-熟石灰	5					80	-Slaked lime
1216	2522.3000	-水硬石灰	5					80	-Hydraulic lime
	25.23	**硅酸盐水泥、矾土水泥、矿渣水泥、富硫酸盐水泥及类似的水凝水泥,不论是否着色,包括水泥熟料:**							**Portland cement, aluminous cement, slag cement, supersulphate cement and similar hydraulic cements, whether or not coloured or in the form of clinkers:**
1217	2523.1000	-水泥熟料	8					30	-Cement clinkers
		-硅酸盐水泥:							-Portland cement:
1218	2523.2100	--白水泥,不论是否人工着色	6					30	--White cement, whether or not artificially coloured
1219	2523.2900	--其他	8			0	0	30	--Other
1220	2523.3000	-矾土水泥	6					30	-Aluminous cement
1221	2523.9000	-其他水凝水泥	8					30	-Other hydraulic cements
	25.24	**石棉:**							**Asbestos:**
1222	2524.0010	---长纤维石棉	5					30	---Asbestos of long staple
1223	2524.0090	---其他	5					35	---Other
	25.25	**云母,包括云母片;云母废料:**							**Mica, including splittings; mica waste:**
1224	2525.1000	-原状云母及劈开的云母片	5					30	-Crude mica and mica rifted into sheets or splittings
1225	2525.2000	-云母粉	5					30	-Mica powder
1226	2525.3000	-云母废料	5					30	-Mica waste
	25.26	**天然冻石,不论是否粗加修整或仅用锯或其他方法切割成矩形(包括正方形)的板、块;滑石:**							**Natural steatite, whether or not roughly trimmed or merely cut, by sawing or other-wise, into blocks or slabs or a rectangular (including square) shape; talc:**
		-未破碎及未研粉:							-Not crushed, not powdered:

序号 No.	税则号列 Tariff Item	货品名称	最惠国税率(%) M.F.N.	协定税率(%)Agreement 东盟 ASEAN	曼谷协定 BA	香港 H.K.	澳门 Macao	普通税率(%) Gen.	Article Description
1227	2526.1010	---冻石	3					50	---Natural steatite
1228	2526.1020	---滑石	3					50	---Talc
		-已破碎或已研粉:							-Crushed or powdered:
1229	2526.2010	---冻石	3					50	---Natural steatite
1230	2526.2020	---滑石	3					50	---Talc
	25.28	**天然硼酸盐及其精矿(不论是否煅烧),但不包括从天然盐水析离的硼酸盐;天然粗硼酸,含硼酸干重不超过85%:**							**Natural borates and concentrates thereof(whether or not calcined), but not including borates separated from natural brine; natural boric acid containing not more than 85% of H_3BO_3 calulated on the dry weight:**
1231	2528.1000	-天然硼砂及其精矿(不论是否煅烧)	3					30	-Natural sodium borates and concentrates thereof (whether or not calcined)
1232	2528.9000	-其他	5					30	-Other
	25.29	**长石;白榴石;霞石及霞石正长岩;萤石(氟石):**							**Felspar; leucite; nepheline and nepheline syenite; fluorspar:**
1233	2529.1000	-长石	3					50	-Felspar
		-萤石:							-Fluorspar:
1234	2529.2100	--按重量计氟化钙含量在97%及以下	3					50	--Containing by weight 97% or less of calcium fluoride
1235	2529.2200	--按重量计氟化钙含量在97%以上	3					50	--Containing by weight more than 97% of calcium fluoride
1236	2529.3000	-白榴石;霞石及霞石正长岩	5					50	-Leucite; nepheline and nepheline syenite
	25.30	**其他税号未列名的矿产品:**							**Mineral substances not elsewhere specified or included:**
		-未膨胀的蛭石、珍珠岩及绿泥石:							-Vermiculite, perlite and chlorites, unexpanded:
1237	2530.1010	---绿泥石	5					30	---Chlorites
1238	2530.1020	---未膨胀的蛭石和珍珠岩	5					30	---Vermiculite, perlite unexpanded
1239	2530.2000	-硫镁矾矿及泻盐矿(天然硫酸镁)	3					30	-Kieserite, epsomite (natural magnesium sulphates)
		-其他:							-Other:
1240	2530.9010	---矿物性药材	3					30	---Mineral medicinal substances
1241	2530.9020	---稀土金属矿	0					0	---Ores of rare earth metals
		---其他:							---Other:
1242	2530.9091	----硅灰石	3					50	----Wollastonite
1243	2530.9099	----其它	3					50	----Other

第二十六章
矿砂、矿渣及矿灰

注释：

一、本章不包括：

（一）供铺路用的矿渣及类似的工业废渣（税号25.17）；

（二）天然碳酸镁（菱镁矿），不论是否煅烧（税号25.19）；

（三）储油罐的淤渣，主要成分为税号27.10项下的油品；

（四）第三十一章的碱性熔渣；

（五）矿物棉（税号68.06）；

（六）贵金属或包贵金属的废碎料；主要用于回收贵金属的含贵金属或贵金属化合物的其他废碎料（税号71.12）；

（七）通过熔炼所产生的铜锍、镍锍或钴锍（第十五类）。

二、税号26.01至26.17所称“矿砂”，是指冶金工业中提炼汞、税号28.44的金属以及第十四类、第十五类金属的矿物，即使这些矿物不用于冶金工业，也包括在内。但税号26.01至26.17不包括不是以冶金工业正常加工方法处理的各种矿物。

三、税号26.20只适用于：（一）在工业上提炼金属或作为生产金属化合物基本原料的矿渣或矿灰，但焚化城市垃圾所产生的灰、渣除外（税目26.21）；（二）含有砷的矿灰、残渣，不论其是否含有金属，用于提取或生产砷、金属及其化合物。

子目注释：

一、子目2620.21所称“含铅汽油的淤渣及含铅抗震化合物的淤渣”，是指取自含铅汽油及含铅抗震化合物（例如，四乙铅）储罐的淤渣，主要含有铅、铅化合物以及铁的氧化物。

二、含有砷、汞、铊及其混合物的矿灰及残渣，用

Chapter 26
Ores, slag and ash

Notes:

1. This Chapter does not cover:

(a) Slag or similar industrial waste prepared as macadam (heading No.25.17);

(b) Natural magnesium carbonate (magnesite), whether or not calcined (heading No.25.19);

(c) Sludges from the storage tanks of petroleum oils, consisting mainly of such oils (heading 27.10);

(d) Basic slag of Chapter 31;

(e) Slag wool, rock wool or similar mineral wools (heading No.68.06);

(f) Waste or scrap of precious metal or of metal clad with precious metal; other waste or scrap containing precious metal or precious metal compounds, of a kind used principally for the recovery of precious metal (heading No.71.12); or

(g) Copper, nickel or cobalt mattes produced by any process of smelting (Section XV).

2. For the purposes of headings Nos.26.01 to 26.17, the term "ores" means minerals of mineralogical species actually used in the metallurgical industry for the extraction of mercury, of the metals of heading No.28.44 or of the metals of Section XIV or XV, even if they are intended for non-metallurgical purposes. Headings Nos.26.01 to 26.17 do not, however, include minerals which have been submitted to processes not normal to the metallurgical industry.

3. Heading No.26.20 applies only to: (a) ash and residues of a kind used in industry either for the extraction of metals or as a basis for the manufacture of chemical compounds of metals, excluding ash and residues from the incineration of municipul waste (hending 26.21); and (b) Ash and residues containing arsenic, whether or not containing metals, of a kind used either for the extraction of arsenic or metals or for the manufacture of their chemical compounds.

Subheading Notes:

1. For the purposes of subheading 2620.21, "leaded gasoline sludges and leaded anti-knock compound sludges" mean sludges obtained from storage tanks of leaded gasoline and leaded anti-knock compounds (for example, tetraethyl lead), and consisting essentially of lead, lead compounds and iron oxide.

2. Ash and residues containing arsenic, mercury, thallium or

于提取或生产砷、汞、铊及其化合物，归入子目 2620.60。

their mixtures, of a kind used for the extraction of arsenic or those metals classified in subheading 2620.60."

序号 No.	税则号列 Tariff Item	货品名称	最惠国税率(%) M.F.N.	协定税率(%) Agreement 东盟 ASEAN	曼谷协定 BA	香港 H.K.	澳门 Macao	普通税率(%) Gen.	Article Description
	26.01	**铁矿砂及其精矿,包括焙烧黄铁矿:**							**Iron ores and concentrates, including roasted iron pyrites:**
		-铁矿砂及其精矿,但焙烧黄铁矿除外:							-Iron ores and concentrates, other than roasted iron pyrites:
1244	2601.1100	--未烧结	0					0	--Non-agglomerated
1245	2601.1200	--已烧结	0					0	--Agglomerated
1246	2601.2000	-焙烧黄铁矿	0					0	-Roasted iron pyrites
	26.02	**锰矿砂及其精矿,包括以干重计含锰量在20%及以上的锰铁矿及其精矿:**							**Manganese ores and concentrates, including ferruginous manganese ores and concentrates with a manganese content of 20% or more, calculated on the dry weight:**
1247	2602.0000	锰矿砂及其精矿,包括以干重计含锰量在20%及以上的锰铁矿及其精矿	0					0	Manganese ores and concentrates, including ferruginous manganese ores and concentrates with a manganese content of 20% or more, calculated on the dry weight
	26.03	**铜矿砂及其精矿:**							**Copper ores and concentrates:**
1248	2603.0000	铜矿砂及其精矿	0					0	Copper ores and concentrates
	26.04	**镍矿砂及其精矿:**							**Nickel ores and concentrates:**
1249	2604.0000	镍矿砂及其精矿	0					0	Nickel ores and concentrates
	26.05	**钴矿砂及其精矿:**							**Cobalt ores and concentrates:**
1250	2605.0000	钴矿砂及其精矿	0					0	Cobalt ores and concentrates
	26.06	**铝矿砂及其精矿:**							**Aluminium ores and concentrate:**
1251	2606.0000	铝矿砂及其精矿	0					0	Aluminium ores and concentrate
	26.07	**铅矿砂及其精矿:**							**Lead ores and concentrates:**
1252	2607.0000	铅矿砂及其精矿	0					0	Lead ores and concentrates
	26.08	**锌矿砂及其精矿:**							**Zinc ores and concentrates:**
1253	2608.0000	锌矿砂及其精矿	0					0	Zinc ores and concentrates
	26.09	**锡矿砂及其精矿:**							**Tin ores and concentrates:**
1254	2609.0000	锡矿砂及其精矿	0					0	Tin ores and concentrates
	26.10	**铬矿砂及其精矿:**							**Chromium ores and concentrates:**
1255	2610.0000	铬矿砂及其精矿	0					0	Chromium ores and concentrates
	26.11	**钨矿砂及其精矿:**							**Tungsten ores and concentrates:**
1256	2611.0000	钨矿砂及其精矿	0					0	Tungsten ores and concentrates
	26.12	**铀或钍矿砂及其精矿:**							**Uranium or thorium ores and concentrates:**
1257	2612.1000	-铀矿砂及其精矿	0					0	-Uranium ores and concentrates
1258	2612.2000	-钍矿砂及其精矿	0					0	-Thorium ores and concentrates
	26.13	**钼矿砂及其精矿:**							**Molybdenum ores and concentrates:**
1259	2613.1000	-已焙烧	0					0	-Roasted
1260	2613.9000	-其他	0					0	-Other

序号 No.	税则号列 Tariff Item	货品名称	最惠国税率(%) M.F.N.	协定税率(%) Agreement 东盟 ASEAN	曼谷协定 BA	香港 H.K.	澳门 Macao	普通税率(%) Gen.	Article Description
	26.14	**钛矿砂及其精矿:**							**Titanium ores and concentrates:**
1261	2614.0000	钛矿砂及其精矿	0					0	Titanium ores and concentrates
	26.15	**铌、钽、钒或锆矿砂及其精矿:**							**Niobium, tantalum, vanadium or zirconium ores and concentrates:**
1262	2615.1000	-锆矿砂及其精矿	0					0	-Zirconium ores and concentrates
		-其他:							-Other:
1263	2615.9010	---水合钽铌原料(钽铌矿富集物)	0					0	---(Hydrated Tantalum/Niobium materials or enriched materials from Tantalum/Niobium Ore)
1264	2615.9090	---其他	0					0	---Other
	26.16	**贵金属矿砂及其精矿:**							**Precious metal ores and concentrates:**
1265	2616.1000	-银矿砂及其精矿	0					0	-Silver ores and concentrates
1266	2616.9000	-其他	0					0	-Other
	26.17	**其他矿砂及其精矿:**							**Other ores and concentrates:**
		-锑矿砂及其精矿:							-Antimony ores and concentrates:
1267	2617.1010	---生锑(锑精矿,选矿产品)	0					0	---Crude antimony(Antimony concentrates which are mineral products)
1268	2617.1090	---其他	0					0	---Other
		-其他:							-Other:
1269	2617.9010	---朱砂(辰砂)	3					14	---Cinnabar
1270	2617.9090	---其他	0					0	---Other
	26.18	**冶炼钢铁所产生的粒状熔渣(熔渣砂):**							**Granulated slag(slag sand) from the manufacture of iron or steel:**
1271	2618.0010	---主要含锰	4					35	---Containing mainly Manganese
1272	2618.0090	---其他	4					35	---Other
	26.19	**冶炼钢铁所产生的熔渣、浮渣(粒状熔渣除外)、氧化皮及其他废料:**							**Slag, dross (other than granulated slag), scalings and other waste from the manufacture of iron or steel:**
1273	2619.0000	冶炼钢铁所产生的熔渣、浮渣(粒状熔渣除外)、氧化皮及其他废料	4					35	Slag, dross (other than granulated slag), scalings and other waste from the manufacture of iron or steel
	26.20	**含有砷、金属及其化合物的矿灰及残渣(冶炼钢铁所产生的灰、渣除外):**							**Ash and residues(other than from the manufacture of iron or steel). containing arsenic, metals or their compounds:**
		-主要含锌:							-Containing mainly zinc:
1274	2620.1100	--含硬锌	4					35	--Hard zinc spelter
1275	2620.1900	--其他	4					35	--Other
		-主要含铅:							-Containing mainly lead:
1276	2620.2100	--含铅汽油的淤渣及含铅抗震化合物的淤渣	4					35	--Leaded gasoline sludges and leaded anti-knock compound sludges

序号 No.	税则号列 Tariff Item	货品名称	最惠国税率(%) M.F.N.	协定税率(%) Agreement 东盟 ASEAN	曼谷协定 BA	香港 H.K.	澳门 Macao	普通税率(%) Gen.	Article Description
1277	2620.2900	--其他	4					35	--Other
1278	2620.3000	-主要含铜	4					35	-Containing mainly copper
1279	2620.4000	-主要含铝	4					35	-Containing mainly aluminium
1280	2620.6000	-含砷、汞、铊及其混合物,用于提取或生产砷、汞、铊及其化合物	4					35	-Containing arsenic, mercury, thallium or their mixtures, of a kind used for the extraction of arsenic or those metals or for the manufacture of their chemical compounds.
		-其他:							-Other:
1281	2620.9100	--含锑、铍、镉、铬及其混合物	4					35	--Containing antimony, beryllinm, cadmium, chromium or their mixtures
		--其他:							--Other:
1282	2620.9910	---主要含钨	4					35	---Containing mainly tungsten
1283	2620.9990	---其他	4					35	---Other
	26.21	**其他矿渣及矿灰,包括海藻灰(海草灰);焚化城市垃圾所产生的灰、渣:**							**Other slag and ash, including seaweed ash(kelp); ash and residues from the incineration of municipal waste:**
1284	2621.1000	-焚化城市垃圾所产生的灰、渣	4					35	-Ash and residues from the incineration of municipal waste
1285	2621.9000	-其他	4			0		35	-Other

第二十七章
矿物燃料、矿物油及其蒸馏产品;沥青物质;矿物蜡

Chapter 27
Mineral fuels, mineral oils and products of their distillation; bituminous substances; mineral waxes

注释:

一、本章不包括:

(一)单独的已有化学定义的有机化合物,但纯甲烷及纯丙烷应归入税号27.11;

(二)税号30.03及30.04的药品;

(三)税号33.01、33.02及38.05的不饱和烃混合物。

二、税号27.10所称"石油及从沥青矿物提取的油类",不仅包括石油、从沥青矿物提取的油及类似油,还包括那些用任何方法提取的主要含有不饱和烃混合物的油,但其非芳族成分的重量必须超过芳族成分。然而,它不包括温度在300℃时,压力转为1013毫巴后减压蒸馏出的液体合成聚烯烃以体积计小于60%的货品(第三十九章)。

三、税目27.10所称"废油",是指主要含石油及从沥青矿物提取的油类(参见本章注释二)的废油,不论其是否与水混合。它们包括:

(一)不再适于作为原产品使用的废油(例如,用过的润滑油、液压油及变压器油);

(二)石油储罐的淤渣油,主要含有石油及高浓度的在生产原产品时使用的添加剂(例如,化学品);

(三)水乳浊液状或与水混合的废油,例如,浮油、清洗油罐所得的油或机械加工中已用过的切削油。

子目注释:

一、子目号2701.11所称"无烟煤",是指含挥发物(以干燥、无矿物质计)不超过14%的煤。

二、子目号2701.12所称"烟煤",是指含挥发物(以干燥、无矿物质计)超过14%,并且热值(以潮湿、无矿物质计)等于或大于5833大卡

Notes:

1. This Chapter does not cover:
   (a) Separate chemically defined organic compounds, other than pure methane and propane which are to be classified in heading No.27.11;
   (b) Medicaments of heading No.30.03 or 30.04; or
   (c) Mixed unsaturated hydrocarbons of heading No.33.01, 33.02 or 38.05

2. References in heading No.27.10 to "petroleum oils and oils obtained from bituminous minerals" include not only petroleum oils and oils obtained from bituminous minerals but also similar oils, as well as those consisting mainly of mixed unsaturated hydrocarbons, obtained by any process, provided that the weight of the non-aromatic constituents exceeds that of the aromatic constituents. However, the references do not include liquid synthetic polyolefins of which less than 60% by volume distils at 300℃, after conversion to 1,013 millibars when a reduced-pressure distillation method is used (Chapter 39).

3. For the purposes of heading 27.10, "waste oils" means waste containing mainly petroleum oils and oils obtained from bituminous minerals (as described in Note 2 to this Chapter), whether or not mixed with water. These include:
   (a) Such oils no longer fit for use as primary products (for example, used lubricating oils, used hydraulic oils and used transformer oils);
   (b) Sludge oils from the storage tanks of petroleum oils, mainly containing such oils and a high concentration of additives (for example, chemicals) used in the manufacture of the primary products; and
   (c) Such oils in the form of emulsions in water or mixtures with water, such as those resulting from oil spills, storage tank washings, or from the use of cutting oils for machining operations.

Subheading Notes:

1. For the purposes of subheading No.2701.11, "anthracite" means coal having a volatile matter limit (on a dry, mineral-matter-free basis) not exceeding 14%.

2. For the purposes of subheading No. 2701. 12, "bituminous coal" means coal having a volatile matter limit (on a dry, mineral-matter-free basis) exceeding 14% and a calorific val-

/公斤的煤。

ue limit (on a moist, mineral-matter-free basis) equal to or greater than 5,833 kcal/kg.

三、子目号 2707.10、2707.20、2707.30、2707.40、2707.60 所称“粗苯”、“粗甲苯”、“粗二甲苯”、“萘”、“酚”，是分别指按重量计苯、甲苯、二甲苯、萘、酚的含量在 50% 以上的产品。

3. For the purposes of subheadings Nos.2707.10, 2707.20, 2707.30, 2707.40 and 2707.60, the terms “benzole”, “toluole”, “xylole”, “naphthalene” and “phenols” apply to products which contain more than 50% by weight of benzene, toluene, xylene, naphthalene or phenols, respectively.

四、子目号 2710.11 所称“轻油及其制品”，是指 210℃（美国材料试验学会 D86 法）以下馏出物体积比不小于 90%（包括损失部分）。

4. For the purposes of subheading 2710.11, “light oils and preparations” are those of which 90% or more by volume (including losses) distil at 210℃ (ASTM D 86 method).

序号 No.	税则号列 Tariff Item	货品名称	最惠国税率(%) M.F.N.	协定税率(%) Agreement 东盟 ASEAN	曼谷协定 BA	香港 H.K.	澳门 Macao	普通税率(%) Gen.	Article Description
	27.01	**煤;煤砖、煤球及用煤制成的类似固体燃料:**							**Coal; briquettes, ovoids and similar solid fuels manufactured from coal:**
		-煤,不论是否粉化,但未制成型:							-Coal, whether or not pulverized, but not agglomerated:
1286	2701.1100	--无烟煤	3	见表 1 Table 1				20	--Anthracite
		--烟煤:							--Bituminous coal:
1287	2701.1210	---炼焦煤	3					20	---Coking coal
1288	2701.1290	---其他	6					20	---Other
1289	2701.1900	--其他煤	5					20	--Other coal
1290	2701.2000	-煤砖、煤球及用煤制成的类似固体燃料	5					50	-Briquettes,ovoids and similar solid fuels manufactured from coal
	27.02	**褐煤,不论是否制成型,但不包括黑玉:**							**Lignite,whether or not agglomerated, excluding jet:**
1291	2702.1000	-褐煤,不论是否粉化,但未制成型	3					20	-Lignite, whether or not pulverized, but not agglomerated
1292	2702.2000	-制成型的褐煤	3					20	-Agglomerated lignite
	27.03	**泥煤(包括肥料用泥煤),不论是否制成型:**							**Peat(including peat litter), whether or not agglomerated:**
1293	2703.0000	泥煤(包括肥料用泥煤),不论是否制成型	5					20	Peat(including peat litter),whether or not agglomerated
	27.04	**煤、褐煤或泥煤制成的焦炭及半焦炭,不论是否制成型;甑炭:**							**Coke and semi-coke of coal,of lignite or of peat,whether or not agglomerated;retort carbon:**
1294	2704.0010	---焦炭及半焦炭	5	见表 1				11	---Coke and semi-coke
1295	2704.0090	---其他	5					11	---Other
	27.05	**煤气、水煤气、炉煤气及类似气体,但石油气及其他烃类气除外:**							**Coal gas,water gas,producer gas and similar gases, other than petroleum gases and other gaseous hydrocarbons:**

序号 No.	税则号列 Tariff Item	货品名称	最惠国税率(%) M.F.N.	协定税率(%) Agreement 东盟 ASEAN	曼谷协定 BA	香港 H.K.	澳门 Macao	普通税率(%) Gen.	Article Description
1296	2705.0000	煤气、水煤气、炉煤气及类似气体,但石油气及其他烃类气除外	5					20	Coal gas, water gas, producer gas and similar gases, other than petroleum gases and other gaseous hydrocarbons
	27.06	**从煤、褐煤或泥煤蒸馏所得的焦油及其他矿物焦油,不论是否脱水或部分蒸馏,包括再造焦油:**							**Tar distilled from coal, from lignite or from peat, and other mineral tars, whether or not dehydrated or partially distilled, including reconstituted tare:**
1297	2706.0000	从煤、褐煤或泥煤蒸馏所得的焦油及其他矿物焦油,不论是否脱水或部分蒸馏,包括再造焦油	6					30	Tar distilled from coal, from lignite or from peat, and other mineral tars, whether or not dehydrated or partially distilled, including reconstituted tars
	27.07	**蒸馏高温煤焦油所得的油类及其他产品;芳族成分重量超过非芳族成分的类似产品:**							**Oils and other products of the distillation of high temperature coal tar; similar products in which the weight of the aromatic constituents exceeds that of the non-aromatic constituents:**
1298	2707.1000	-粗苯	6					20	-Benzole
1299	2707.2000	-粗甲苯	6					30	-Toluole
1300	2707.3000	-粗二甲苯	6					20	-Xylole
1301	2707.4000	-萘	7		6			30	-Naphthalene
1302	2707.5000	-其他芳烃混合物,温度在250℃时蒸馏出的芳烃混合物含量以体积计(包括损耗)在65%及以上(以美国标准试验法D86为准)	7					30	-Other aromatic hydrocarbon mixtures of which 65% or more by volume (including losses) distils at 250℃ by the ASTM D 86 method
1303	2707.6000	-酚	7					30	-Phenols
		-其他:							-Other:
1304	2707.9100	--杂酚油	7					30	--Creosote oils
1305	2707.9900	--其他	7					30	--Other
	27.08	**从煤焦油或其他矿物焦油所得的沥青及沥青焦:**							**Pitch and pitch coke, obtained from coal tar or from other mineral tars:**
1306	2708.1000	-沥青	7					35	-Pitch
1307	2708.2000	-沥青焦	6					11	-Pitch coke
	27.09	**石油原油及从沥青矿物提取的原油:**							**Petroleum oils and oils obtained from bituminous minerals, crude:**
1308	2709.0000	石油原油及从沥青矿物提取的原油	见表4					见表4	Petroleum oils and oils obtained from bituminous minerals, crude

序号 No.	税则号列 Tariff Item	货品名称	最惠国税率(%) M.F.N.	协定税率(%)Agreement				普通税率(%) Gen.	Article Description
				东盟 ASEAN	曼谷协定 BA	香港 H.K.	澳门 Macao		
	27.10	**石油及从沥青矿物提取的油类,但原油除外;以上述油为基本成分(按重量计不低于70%)的其他税号未列名制品;废油:**							**Petroleum oils and oils obtained from bituminous minerals, other than crude; preparations not elsewhere specified or included, containing by weight 70% or more of petroleum oils or of oils obtained from bituminous minerals, these oils being the basic constituents of the preparations; waste oils:**
		-石油及从沥青矿物提取的油类,(但原油除外);以上述油为基本成分(按重量计不低于70%)的其他税目未列名制品,但废油除外:							-Petroleum oils and oils obtained from bituminous minerals (other than crude) and preparations not elsewhere specified or included, containing by weight 70% or more of petroleum oils or of oils obtained from bituminous minerals, these oils being the basic constituents of the preparations, other than waste oils:
		--轻油及其制品:							--Light oils and preparations:
1309	2710.1110	---车用汽油及航空汽油	5					14	---Mortor gasoline, aviation gasoline
1310	2710.1120	---石脑油	6					20	---Naphtha
1311	2710.1130	---橡胶溶剂油,油漆溶剂油、抽提溶剂油	6					30	---Rubber solvent, paint solvent, extractive solvent
		---其他:						20	---Other:
1312	2710.1191	----壬烯	9					20	----Nonene
1313	2710.1199	----其他	9						----Other
		--其他:							--Other:
		---煤油馏分:							---Kerosene distillages:
1314	2710.1911	----航空煤油	9					14	----Aviation kerosene
1315	2710.1912	----灯用煤油	9					14	----Lamp-kerosene
1316	2710.1919	----其他	6					20	----Other
		---柴油及其他燃料油:							---Diesel oils and other fuel oils:
1317	2710.1921	----轻柴油	6					11	----Light diesel oil
1318	2710.1922	----5—7号燃料油	6					20	----Fuel oils No.5～No.7
1319	2710.1929	----其他	6					20	----Other
		---润滑油、润滑脂及其他重油:							---Lubricating oils, lubricating greases and other heavy oils:
1320	2710.1991	----润滑油	6			0	0△	17	----Lubricating grease
1321	2710.1992	----润滑脂	6					17	----Lubricating oils
1322	2710.1993	----润滑油基础油	6					17	----Basic oils for lubricating oils
1323	2710.1994	----液体石蜡和重质液体石蜡	6					20	----Liquid paraffin and heavy liquid paraffin
1324	2710.1999	----其他	6					20	----Other

序号 No.	税则号列 Tariff Item	货品名称	最惠国税率(%) M.F.N.	协定税率(%) Agreement 东盟 ASEAN	曼谷协定 BA	香港 H.K.	澳门 Macao	普通税率(%) Gen.	Article Description
		-废油:							-Waste oils:
1325	2710.9100	--含多氯联苯(PCBs)、多氯三联(PCTs)或多溴联苯(PBBs)的	6					20	--Containing poly chlorinated biphenyls (PCBs), polychlorinated terphenyls(PCTs) or polybrominated biphenyls(PBBs)
1326	2710.9900	--其他	6					20	--Other
	27.11	**石油气及其他烃类气:**							**Petroleum gases and other gaseous hydrocarbons:**
		-液化的:							-Liquefied:
1327	2711.1100	--天然气	0					20	--Natural gas
1328	2711.1200	--丙烷	5					20	--Propane
		--丁烷:							--Butanes:
1329	2711.1310	---直接灌注香烟打火机及类似打火器用,其包装容器的容积超过300立方厘米	11					80	---Liquid or liquefied-gas fuels in containers of a kind used for filling or refilling cigarette or similar lighters and of a capacity exceeding 300cm^3
1330	2711.1390	---其他	5					20	---Other
1331	2711.1400	--乙烯、丙烯、丁烯及丁二烯	5					20	--Ethylene, propylene, butylene and butadiene
		--其他:							--Other:
1332	2711.1910	---直接灌注香烟打火机及类似打火器用的燃料,其包装容器的容积超过300立方厘米	10					80	---Liquid or liquefied-gas fuels in containers of a kind used for filling or refilling cigarette or similar lighters and of a capacity exceeding 300cm^3
1333	2711.1990	---其他	3					20	---Other
		-气态的:							-In gaseous state:
1334	2711.2100	--天然气	0					20	--Natural gas
1335	2711.2900	--其他	6					20	--Other
	27.12	**凡士林;石蜡、微晶石蜡、疏松石蜡、地蜡、褐煤蜡、泥煤蜡、其他矿物蜡及用合成或其他方法制得的类似产品,不论是否着色:**							**Petroleum jelly; paraffin wax, microcrystalline petroleum wax, slack wax, ozokerite, lignite wax, peat wax, other mineral waxes, and similar products obtained by synthesis or by other processes, whether or not coloured:**
1336	2712.1000	-凡士林	8					45	-Petroleum jelly
1337	2712.2000	-石蜡,按重量计含油量小于0.75%	8					45	-Paraffin wax containing by weight less than 0.75% of oil
		-其他:							-Other:
1338	2712.9010	---微晶石蜡	8					45	---Microcrystalline petroleum wax
1339	2712.9090	---其他	8					45	---Other
	27.13	**石油焦、石油沥青及其他石油或从沥青矿物提取的油类的残渣:**							**Petroleum coke, Petroleum bitumen and other residues of petroleum oils or of oils obtained from bituminous minerals:**

序号 No.	税则号列 Tariff Item	货品名称	最惠国税率(%) M.F.N.	协定税率(%)Agreement 东盟 ASEAN	曼谷协定 BA	香港 H.K.	澳门 Macao	普通税率(%) Gen.	Article Description
		-石油焦:							-Petroleum coke:
		--未煅烧:							--Not calcined:
1340	2713.1110	---硫的重量百分比小于3%的	3					11	---Containing by weight less than 3% of Sulphur
1341	2713.1190	---其他	3					11	---Other
		--已煅烧:							--Calcined:
1342	2713.1210	---硫的重量百分比小于0.8%的	3					11	---Containing by weight less than 0.8% of Sulphur
1343	2713.1290	---其他	3					11	---Other
1344	2713.2000	-石油沥青	8					35	-Petroleum bitumen
1345	2713.9000	-其他石油或从沥青矿物提取的油类的残渣	6					35	-Other residues of petroleum oils or of oils obtained from bituminous minerals
	27.14	**天然沥青(地沥青)、沥青页岩、油页岩及焦油砂;沥青岩:**							**Bitumen and asphalt, natural; bituminous or oil shale and tar sands; asphaltites and asphaltic rocks:**
1346	2714.1000	-沥青页岩、油页岩及焦油砂	6					20	-Bituminous or oil shale and tar sands
		-其他:							-Other:
1347	2714.9010	---天然沥青(地沥青)	8					35	---Natural bitumen and asphalt
1348	2714.9020	---乳化沥青	0					20	---Emulsified bitumen and asphalt
1349	2714.9090	---其他	3					20	---Other
	27.15	**以天然沥青(地沥青)、石油沥青、矿物焦油或矿物焦油沥青为基本成分的沥青混合物(例如,沥青胶粘剂、稀释沥青):**							**Bituminous mixtures based on natural asphalt, on natural bitumen, on petroleum bitumen, on mineral tar or on mineral tar pitch (for example, bituminous mastics, cut-backs):**
1350	2715.0000	以天然沥青(地沥青)、石油沥青、矿物焦油或矿物焦油沥青为基本成分的沥青混合物(例如,沥青胶粘剂、稀释沥青)	8					35	Bituminous mixtures based on natura asphalt, on natural bitumen, on petroleum bitumen, on mineral tar or on mineral tar pitch (for example, bituminous mastics, cut-backs)
	27.16	**电力:**							**Electrical energy:**
1351	2716.0000	电力	0					8	Electrical energy

第六类

化学工业及其相关工业的产品

注释：

一、

（一）凡符合税号 28.44 或 28.45 规定的货品（放射性矿砂除外），应分别归入这两个税号而不归入本目录的其他税号。

（二）除上述（一）款另有规定的以外，凡符合税号 28.43 或 28.46 规定的货品，应分别归入这两个税号而不归入本类的其他税号。

二、除上述注释一另有规定的以外，凡由于按一定剂量或作为零售包装而可归入税号 30.04、30.05、30.06、32.12、33.03、33.04、33.05、33.06、33.07、35.06、37.07 或 38.08 的货品，应分别归入以上税号，而不归入本目录的其他税号。

三、由两种或两种以上单独成分配套的货品，其部分或全部成分属于本类范围以内，混合后则构成第六类或第七类的货品，应按混合后产品归入相应的税号，但其组成成分必须同时符合下列条件：

（一）其包装形式足以表明这些成分不需经过改装就可一起使用的；

（二）一起进口或出口的；

（三）这些成分的属性及相互比例足以表明是相互配用的。

SECTION Ⅵ

PRODUCTS OF THE CHEMICAL OR ALLIED INDUSTRIES

Notes:

1. (a) Goods (other than radioactive ores) answering to a description in heading No.28.44 or 28.45 are to be classified in those headings and in no other heading of the Nomenclature.

   (b) Subject to paragraph (a) above, goods answering to a description in heading No.28.43 or 28.46 are to be classified in those headings and in no other heading of this Seciton.

2. Subject to Note 1 above, goods classifiable in heading No.30.04, 30.05, 30.06, 32.12, 33.03, 33.04, 33.05, 33.06, 33.07, 35.06, 37.07, or 38.08 by reason of being put up in measured doses or for retail sale are to be classified in those headings and in no other headings of the Nomenclature.

3. Goods put up in sets consisting of two or more separate constituents, some or all of which fall in this Section and are intended to be mixed together to obtain a product of Section Ⅵ or Ⅶ, are to be classified in the heading appropriate to that product, provided that the constituents are:

   (a) having regard to the manner in which they are put up, clearly identifiable as being intended to be used together without first being repacked;

   (b) presented together; and

   (c) identifiable, whether by their nature or by the relative proportions in which they are present, as being complementary one to another.

第二十八章

无机化学品；贵金属、稀土金属、放射性元素及其同位素的有机及无机化合物

注释：

一、除条文另有规定的以外，本章各税号只适用于：

（一）单独的化学元素及单独的已有化学定义

Chapter 28

Inorganic chemicals; organic or inorganic compounds of precious metals, of rare-earth metals of radioactive elements or of isotopes

Notes:

1. Except where the context otherwise requires, the headings of this Chapter apply only to:

   (a) Separate chemical elements and separate chemically de-

的化合物，不论是否含有杂质；

（二）上述（一）款产品的水溶液；

（三）溶于其他溶剂的上述（一）款产品，但该产品处于溶液状态只是为了安全或运输所采取的正常必要方法，其所用溶剂并不使该产品改变其一般用途而适合于某些特殊用途；

（四）为了保存或运输需要，加入稳定剂（包括抗结块剂）的上述（一）、（二）、（三）款产品；

（五）为了便于识别或安全起见，加入抗尘剂或着色剂的上述（一）、（二）、（三）、（四）款产品，但所加剂料并不使原产品改变其一般用途而适合于某些特殊用途。

二、除以有机物质稳定的连二亚硫酸盐及次硫酸盐（税号 28.31），无机碱的碳酸盐及过碳酸盐（税号 28.36），无机碱的氰化物、氧氰化物及氰络合物（税号 28.37），无机碱的雷酸盐、氰酸盐及硫氰酸盐（税号 28.38），税号 28.43 至 28.46 的有机产品，以及碳化物（税号（28.49）之外，本章仅包括下列碳化合物：

（一）碳的氧化物，氰化氢及雷酸、异氰酸、硫氰酸及其他简单或络合氰酸（税号 28.11）；

（二）碳的卤氧化物（税号 28.12）；

（三）二硫化碳（税号 28.13）；

（四）硫代碳酸盐、硒代碳酸盐、碲代碳酸盐、硒代氰酸盐、碲代氰酸盐、四氰硫基二氨基络酸盐及其他无机碱络合氰酸盐（税号 28.42）；

（五）用尿素固化的过氧化氢（税号 28.47）、氧硫化碳、硫代羰基卤化物、氰、卤化氰、氨基氰及其金属衍生物（税号 28.51），不论是否纯净，但氰氨化钙除外（第三十一章）。

三、除第六类注释一另有规定的以外，本章不包括：

（一）氯化钠或氧化镁（不论是否纯净）及第五类的其他产品；

（二）上述注释二所述以外的有机－无机化合

fined compounds, whether or not containing impurities;

(b) The products mentioned in (a) above dissolved in water;

(c) The products mentioned in (a) above dissolved in other solvents, provided that the solution constitutes a normal and necessary method of putting up these products adopted solely for reasons of safety or for transport, and that the solvent does not render the product particularly suitable for specific use rather than for general use;

(d) The products mentioned in (a), (b) or (c) above with an added stabilizer (including an anti-caking agent necessary for their preservation or transport;

(e) The products mentioned in (a), (b), (c) or (d) above with an added antidusting agent or a colouring substance added to facilitate their identification or for safety reasons, provided that the additions do not render the product particularly suitable for specific use rather than for general use.

2. In addition to dithionites and sulphoxylates, stabilized with organic substances (heading No.28.31), carbonates and peroxocarbonates of inorganic bases (heading No.28.36), cyanides, cyanide oxides and complex cyanides of inorganic bases (heading No.28.37), fulminates, cyanates and thiocyanates, of inorganic bases (heading No.28.38), organic products included in headings Nos.28.43 to 28.46 and carbides (heading No.28.49), only the following compounds of carbon are to the classified in this Chapter:

(a) Oxides of carbon, hydrogen cyanide and fulminic, isocyanic, thiocyanic and other simple or complex cyanogen acids (heading No.28.11);

(b) Halide oxides of carbon (heading No.28.12);

(c) Carbon disulphide (heading No.28.13);

(d) Thiocarbonates, selenocarbonates, tellurocarbonates, selenocyanates, tellurocyanates, tetrathiocyanato diamminochromates (reineckates) and other complex cyanates, of inorganic bases (heading No.28.42);

(e) Hydrogen peroxide, soildified with urea (heading No.28.47), carbon oxysulphide, thiocarbonyl halides, cyanogen, cyanogen halides and cyanamide and its metal derivatives (heading No.28.51) other than calcium cyanamide, whether or not pure (Chapter31).

3. Subject to the provisions of Note 1 to Section Ⅵ, this Chapter does not cover:

(a) Sodium chloride or magnesium oxide, whether or not pure, or other products of Section Ⅴ;

(b) Organo-inorganic compounds other than those men-

物；

（三）第三十一章注释二、三、四或五所述的产品；

（四）税号32.06的用作发光剂的无机产品；税目32.07的搪瓷玻璃料及其他玻璃，呈粉、粒或粉片状的；

（五）人造石墨（税号38.01）；税号38.13的灭火器的装配药及已装药的灭火弹；税号38.24的零售包装的除墨剂；税号38.24的每颗重量不少于2.5克的碱金属或碱土金属卤化物的培养晶体（光学元件除外）；

（六）宝石或半宝石（天然、合成或再造）及这些宝石、半宝石的粉末（税号71.02至71.05），第七十一章的贵金属及贵金属合金；

（七）第十五类的金属（不论是否纯净）、金属合金或金属陶瓷，包括硬质合金（与金属烧结的金属碳化物）；

（八）光学元件，例如用碱金属或碱土金属卤化物制成的（税号90.01）。

四、由本章第二分章的非金属酸和第四分章的金属酸所构成的已有化学定义的络酸，应归入税号28.11。

五、税号28.26至28.42只适用于金属盐、铵盐及过氧酸盐。除条文另有规定的以外，复盐及络盐应归入税号28.42。

六、税号28.44只适用于：

（一）锝（原子序数43）、钷（原子序数61）、钋（原子序数84）及原子序数大于84的所有化学元素；

（二）天然或人造放射性同位素（包括第十四类及第十五类的贵金属和贱金属的放射性同位素），不论是否混合；

（三）上述元素或同位素的无机或有机化合物，不论是否已有化学定义或是否混合；

（四）含有上述元素或同位素及其无机或有机化合物并且具有某种放射性强度超过74贝克勒尔/克（0.002微居里/克）的合金、分散体（包括金属陶瓷）、陶瓷产品及混合物；

（五）核反应堆已耗尽（已辐照）的燃料元件

tioned in Note 2 above;

(c) Products mentioned in Note 2, 3, 4 or 5 to Chapter 31;

(d) Inorganic products of a kind used as luminophores, of heading No.32.06; glass frit and other glass in the form of powder, granules or flakes, of heading 32.07;

(e) Artificial graphite (heading No.38.01); products put up as charges for fire-extinguishers or put up in fireextinguishing grenades, of heading No.38.13; ink removers put up in packings for retail sale, of heading No.38.24, cultured crystals (other than optical elements) weighing not less than 2.5g each, of the halides of the alkali or alkaline-earth metals, of heading No.38.24;

(f) Precious or semi-precious stones (natural, synthetic or reconstructed) or dust or powder of such stones (headings Nos.71.02 to 71.05), or precious metals or precious metal alloys of Chapter 71;

(g) The metals, whether or not pure, metal alloys or cermets, including sintered metal carbides (metal carbides sintered with a metal), of Section ⅩⅤ; or

(h) Optical elements, for example, of the halides of the alkali or alkaline-earth metals (heading No.90.01).

4. Chemically defined complex acids consisting of a nonmetalacid of sub-Chapter Ⅱ and a metal acid of sub-Chapter Ⅳ are to be classified in heading No.28.11.

5. Headings Nos.28.26 to 28.42 apply only to metal or ammonium salts or peroxysalts. Except where the context otherwise requires, double or complex salts are to be classified in heading No.28.42.

6. Heading No.28.44 applies only to:

(a) Technetium (atomic No.43), promethium (atomic No.61), polonium (atomic No.84) and all elements with an atomic number greater than 84;

(b) Natural or artificial radioactive isotopes (including those of the precious metals or of the base metals of Sections ⅩⅣ and ⅩⅤ), whether or not mixed together;

(c) Compounds, inorganic or organic, of these elements or isotopes, whether or not chemically defined, whether or not mixed together;

(d) Alloys, dispersions (inlhuding cermets), ceramic products and mixtures containing these elements or isotopes or inorganic or organic compounds thereof and having a specific radioactivity exceeding 74 Bq/g (0.002μ Ci/g);

(e) Spent (irradiated) fuel elements (cartridges) of nu-

（释热元件）；

（六）放射性的残渣，不论是否有用。

税号28.44、28.45及本注释所称“同位素”，是指：

1. 单独的核素，但不包括自然界中以单一同位素状态存在的核素；
2. 同一元素的同位素混合物，其中一种或几种同位素已被浓缩，即人工地改变了该元素同位素的自然构成。

clear reactors;

(f) Radioactive residues whether or not usable.

The term "isotopes", for the purposes of this Note and of the wording of headings Nos.28.44 and 28.45, refers to:

-individual nuclides, excluding, however, those existing in nature in the monoisotopic state;

-mixtures of isotopes of one and the same element, enriched in one or several of the said isotopes, that is, elements of which the natural isotopic composition has been artificially modified.

七、税号28.48包括按重量计含磷量超过15%的磷化铜（磷铜）。

7. Heading No.28.48 includes copper phosphide (phosphor copper) containing more than 15% by weight of phosphorus.

八、经掺杂用于电子工业的化学元素（例如，硅、硒），如果拉制后未经加工或呈圆筒形、棒形，应归入本章；如果已切成圆片、薄片或类似形状，则归入税号38.18。

8. Chemical elements (for example, silicon and selenium) doped for use in electronics are to be classified in this Chapter, provided that they are in forms unworked as drawn, or in the form of cylinders or rods. When cut in the form of discs, wafers or similar forms, they fall in heading No.38.18.

序号 No.	税则号列 Tariff Item	货品名称	最惠国税率(%) M.F.N.	协定税率(%)Agreement 东盟 ASEAN	曼谷协定 BA	香港 H.K.	澳门 Macao	普通税率(%) Gen.	Article Description
		第一分章 化学元素							Ⅰ. CHEMICAL ELEMENTS
	28.01	**氟、氯、溴及碘：**							**Fluorine, chlorine, bromine and iodine:**
1352	2801.1000	-氯	5.5					80	-Chlorine
1353	2801.2000	-碘	5.5					30	-Iodine
		-氟；溴：							-Fluorine; bromine:
1354	2801.3010	---氟	5.5					30	---Fluorine
1355	2801.3020	---溴	5.5					30	---Bromine
	28.02	**升华硫磺、沉淀硫磺；胶态硫磺：**							**Sulphur, sublimed or precipitated; colloidal sulphur:**
1356	2802.0000	升华硫磺、沉淀硫磺；胶态硫磺	5.5					17	Sulphur, sublimed or precipitated; colloidal sulphur
	28.03	**碳（碳黑及其他税号未列名的其他形态的碳）：**							**Carbon (carbon blacks and other forms of carbon not elsewhere specified or included):**
1357	2803.0000	碳（碳黑及其他税号未列名的其他形态的碳）	5.5					35	Carbon (carbon blacks and other forms of carbon not elsewhere specified or included)
	28.04	**氢、稀有气体及其他非金属：**							**Hydrogen, rare gases and other non-met-als:**
1358	2804.1000	-氢	5.5			0		30	-Hydrogen
		-稀有气体：							-Rare gases:
1359	2804.2100	--氩	5.5			0		30	--Argon
1360	2804.2900	--其他	5.5			0		30	--Other

序号 No.	税则号列 Tariff Item	货品名称	最惠国税率(%) M.F.N.	协定税率(%) Agreement 东盟 ASEAN	曼谷协定 BA	香港 H.K.	澳门 Macao	普通税率(%) Gen.	Article Description
1361	2804.3000	-氮	5.5			0		30	-Nitrogen
1362	2804.4000	-氧	5.5			0		80	-Oxygen
1363	2804.5000	-硼;碲	5.5					17	-Boron;tellurium
		-硅:							-Silicon:
		--按重量计含硅量不少于99.99%:							--Containing by weight not less than 99.99% of silicon:
1364	2804.6110	---经掺杂用于电子工业的直径在7.5厘米及以上的单晶硅棒	4					11	---Monocrystals doped for use in electronics, in the form of cylinders or rods, 7.5cm or more in diameter
1365	2804.6120	---经掺杂用于电子工业的其他单晶硅棒	4					17	---Other monocrystals doped for use in electronics, in the form of cylinders or rods
1366	2804.6190	---其他	4					30	---Other
1367	2804.6900	--其他	4					30	--Other
		-磷:							-Phosphorus:
1368	2804.7010	---黄磷(白磷)	5.5					30	---Yellow phosphorus(white phosphorus)
1369	2804.7090	---其他	5.5					30	---Other
1370	2804.8000	-砷	5.5					30	-Arsenic
		-硒:							-Selenium:
1371	2804.9010	---经掺杂用于电子工业的晶体棒	4					17	---Crystals doped for use in electronics, in the form of cylinders or rods
1372	2804.9090	---其他	5.5					30	---Other
	28.05	**碱金属、碱土金属;稀土金属、钪及钇,不论是否相互混合或相互熔合;汞:**							**Alkali or alkaline-earth metals; rare-earth metals, scandium and yttrium, whether or not intermixed or interalloyed; mercury:**
		-碱金属及碱土金属:							-Alkali metals or alkaline-earth metals:
1373	2805.1100	--钠	5.5					30	--Sodium
1374	2805.1200	--钙	5.5					30	--Calcium
1375	2805.1900	--其他	5.5					30	--Other
		-稀土金属、钪及钇,不论是否相互混合或相互熔合:							-Rare-earth metals, scandium and yttrium, whether or not intermixed or interalloyed:
		---稀土金属、钪及钇,未相互混合或相互熔合:							---Not intermixed or interalloyed:
1376	2805.3011	----钕	5.5					30	----Neodymium
1377	2805.3012	----镝	5.5					30	----Dysprosium
1378	2805.3019	----其他	5.5					30	----Other
		---稀土金属、钪及钇,相互混合或相互熔合:							---Inuermixed or interalloyed:
1379	2805.3021	----电池级	5.5					30	----Battery grade
1380	2805.3029	----其他	5.5					30	----Other

序号 No.	税则号列 Tariff Item	货品名称	最惠国税率(%) M.F.N.	协定税率(%)Agreement 东盟 ASEAN	曼谷协定 BA	香港 H.K.	澳门 Macao	普通税率(%) Gen.	Article Description
1381	2805.4000	-汞	5.5					17	-Mercury
		第二分章　无机酸及非金属　无机氧化物							Ⅱ. INORGANIC ACIDS AND INORGANIC OXYGEN COMPOUNDS OF NON-METALS
	28.06	**氯化氢(盐酸);氯磺酸:**							**Hydrogen chloride (hydrochloric acid); chorosulphuric acid:**
1382	2806.1000	-氯化氢(盐酸)	5.5					80	-Hydrogen chloride(hydrochloric acid)
1383	2806.2000	-氯磺酸	5.5					40	-Chlorosulphuric acid
	28.07	**硫酸;发烟硫酸:**							**Sulphuric acid; oleum:**
1384	2807.0000	硫酸;发烟硫酸	5.5					35	Sulphuric acid; oleum
	28.08	**硝酸;磺硝酸:**							**Nitric acid; sulphonitric acids:**
1385	2808.0000	硝酸;磺硝酸	5.5					40	Nitric acid; sulphonitric acids
	28.09	**五氧化二磷;磷酸;多磷酸,不论是否已有化学定义:**							**Diphosphorus pentaoxide; phosphoric acid; polyphosphoric acids, whethet or not chemically defined:**
1386	2809.1000	-五氧化二磷	1					8	-Diphosphorus pentaoxide
		-磷酸及多磷酸:							-Phosphoric acid and polyphosphoric acids:
1387	2809.2010	---磷酸及偏磷酸、焦磷酸	1					8	---Phosphoric acid, metaphosphoric acid and pyrophosphoric acid
1388	2809.2090	---其他	5.5					35	---Other
	28.10	**硼的氧化物;硼酸:**							**Oxides of boron; boric acids:**
1389	2810.0010	---硼的氧化物	5.5					30	---Oxides of boron
1390	2810.0020	---硼酸	5.5					30	---Boric acids
	28.11	**其他无机酸及非金属无机氧化物:**							**Other inorganic acids and other inorganic oxygen compounds of non-metals:**
		-其他无机酸:							-Other inorganic acids:
1391	2811.1100	--(氟化氢氢氟酸)	5.5					35	--Hydrofluoric acid
		--其他:							--Other:
1392	2811.1910	---氢氰酸	5.5					35	---Hydrocyanic acid
1393	2811.1990	---其他	5.5					35	---Other
		-其他非金属无机氧化物:							-Other inorganic oxygen compounds of nonmetals:
1394	2811.2100	--二氧化碳	5.5			0		30	--Carbon dioxide
1395	2811.2200	--二氧化硅	5.5					30	--Silicon dioxide
1396	2811.2300	--二氧化硫	5.5					30	--Sulphur dioxide
1397	2811.2900	--其他	5.5			0		30	--Other
		第三分章　非金属卤化物及硫化物							Ⅲ. HALOGEN OR SULPHUR COMPOUNDS OF NON-METALS
	28.12	**非金属卤化物及卤氧化物:**							**Halides and halide oxides of non-metals:**
		-氯化物及氯氧化物:							-Chlorides and chloride oxides:
1398	2812.1010	---氯化亚砜	5.5					30	---Sulphoxide chloride

序号 No.	税则号列 Tariff Item	货品名称	最惠国税率(%) M.F.N.	协定税率(%) Agreement 东盟 ASEAN	曼谷协定 BA	香港 H.K.	澳门 Macao	普通税率(%) Gen.	Article Description
1399	2812.1020	---氧氯化磷(磷酰氯;三氯氧磷)	5.5					30	---Phosphorus oxychloride(phosphoryl monochloride, phosphorus oxytrichloride)
1400	2812.1030	---碳酰二氯(光气)	5.5					30	---Carbonyl dichloride(phosgene)
		---非金属氯化物:							---Other chlorides of non-metals:
1401	2812.1041	----一氯化硫(氯化硫)	5.5					30	----Sulfur monochloride
1402	2812.1042	----二氯化硫	5.5					30	----Sulfur dichloride
1403	2812.1043	----三氯化磷	5.5					30	----Phosphorus trichloride
1404	2812.1044	----三氯化砷	5.5					30	----Arsenic trichloride
1405	2812.1045	----五氯化磷	5.5					30	----Phosphorus pentachloride
1406	2812.1049	---其他	5.5					30	----Other
1407	2812.1090	---其他	5.5					30	---Other
1408	2812.9000	-其他	5.5					30	-Other
	28.13	**非金属硫化物;商品三硫化二磷:**							**Sulphides of non-metals; commercial phosphorus trisulphides:**
1409	2813.1000	-二硫化碳	5.5					30	-Carbon disulphide
1410	2813.9000	-其他	5.5					30	-Other
		第四分章 无机碱和金属氧化物、氢氧化物及过氧化物							Ⅳ. INORGANIC BASES AND OXIDES, HYDROXIDES AND PEROXIDES OF METALS
	28.14	**氨及氨水:**							**Ammonia, anhydrous or in aqueous solution:**
1411	2814.1000	-氨	5.5					35	-Anhydrous ammonia
1412	2814.2000	-氨水	5.5					35	-Ammonia in aqueous solution
	28.15	**氢氧化钠(烧碱);氢氧化钾(苛性钾);过氧化钠及过氧化钾:**							**Sodium hydroxide (caustic soda); potassium hydroxide(caustic potash); peroxides or sodium or postassium:**
		-氢氧化钠(烧碱):							-Sodium hydroxide(caustic soda):
1413	2815.1100	--固体	10		8			35	--Solid
1414	2815.1200	--水溶液(氢氧化钠浓溶液及液体烧碱)	8					35	--In aqueous solution(soda lye or liquid soda)
1415	2815.2000	-氢氧化钾(苛性钾)	5.5					30	-Potassium hydroxide(caustic potash)
1416	2815.3000	-过氧化钠及过氧化钾	5.5					30	-Peroxides of sodium or potassium
	28.16	**氢氧化镁及过氧化镁;锶或钡的氧化物、氢氧化物及过氧化物:**							**Hydroxide and peroxide of magnesium; oxides, hydroxides and peroxides, of strontium or barium:**
1417	2816.1000	-氢氧化镁及过氧化镁	5.5					30	-Hydroxide and peroxide of magnesium
1418	2816.4000	-锶或钡的氧化物、氢氧化物及过氧化物	5.5					30	-Oxides, hydroxides and peroxides, of strontium or barium
	28.17	**氧化锌及过氧化锌:**							**Zinc oxide; Zinc peroxide:**
1419	2817.0010	---氧化锌	5.5					40	---Zinc oxide
1420	2817.0090	---过氧化锌	5.5					30	---Zinc peroxide
	28.18	**人造刚玉,不论是否已有化学定义;氧化铝;氢氧化铝:**							**Artificial corundum, whether or not chemically defined; aluminium oxide; aluminium hydroxide:**

序号 No.	税则号列 Tariff Item	货品名称	最惠国税率(%) M.F.N.	协定税率(%)Agreement 东盟 ASEAN	曼谷协定 BA	香港 H.K.	澳门 Macao	普通税率(%) Gen.	Article Description
1421	2818.1000	-人造刚玉,不论是否已有化学定义	5.5					20	-Artificial corundum, whether or not chemically defined
1422	2818.2000	-氧化铝,但人造刚玉除外	8					30	-Aluminium oxide, other than artificial corundum
1423	2818.3000	-氢氧化铝	5.5					30	-Aluminium hydroxide
	28.19	**铬的氧化物及氢氧化物:**							**Chromium oxides and hydroxides:**
1424	2819.1000	-三氧化铬	5.5					20	-Chromium trioxide
1425	2819.9000	-其他	5.5					30	-Other
	28.20	**锰的氧化物:**							**Manganese oxide:**
1426	2820.1000	-二氧化锰	5.5					40	-Manganese dioxide
1427	2820.9000	-其他	5.5					30	-Other
	28.21	**铁的氧化物及氢氧化物;土色料,按重量计三氧化二铁含量在70%及以上:**							**Iron oxides and hydroxides; earth colours containing 70% or more by weight of combined iron evaluated as Fe_2O_3:**
1428	2821.1000	-铁的氧化物及氢氧化物	5.5					30	-Iron oxides and hydroxides
1429	2821.2000	-土色料	5.5					45	-Earth colours
	28.22	**钴的氧化物及氢氧化物;商品氧化钴:**							**Cobalt oxides and hydroxides; commercial cobalt oxides:**
		钴的氧化物及氢氧化物;商品氧化钴:							Cobalt oxides and hydroxides; commercial cobalt oxides:
1430	2822.0010	---四氧化三钴	5.5					30	---Cobalt tetroxide
1431	2822.0090	---其他	5.5					30	---Other
	28.23	**钛的氧化物:**							**Titanium oxides:**
1432	2823.0000	钛的氧化物	5.5					30	Titanium oxides
	28.24	**铅的氧化物;铅丹及铅橙:**							**Lead oxides; red lead and orange lead:**
1433	2824.1000	-一氧化铅(铅黄、黄丹)	5.5					30	-Lead monoxide(litharge, massicot)
1434	2824.2000	-铅丹及铅橙	5.5					45	-Red lead and orange lead
1435	2824.9000	-其他	5.5					30	-Other
	28.25	**肼(联氨)、胲(羟胺)及其无机盐;其他无机碱;其他金属氧化物、氢氧化物及过氧化物:**							**Hydrazine and hydroxylamine and their inorganic salts; other inorganic bases; othermetal oxides, hydroxides and peroxides:**
		-肼(联氨)、胲(羟胺)及其无机盐:							-Hydrazine and hydroxylamine and their inorganic salts:
1436	2825.1010	---水合肼	5.5					30	---Hydrazine hydrate
1437	2825.1090	---其他	5.5					30	---Other
		-锂的氧化物及氢氧化物:							-Lithium oxide and hydroxide:
1438	2825.2010	---氢氧化锂	5.5					30	---Lithium hydroxide
1439	2825.2090	---其他	5.5					30	---Other

序号 No.	税则号列 Tariff Item	货品名称	最惠国税率(%) M.F.N.	协定税率(%) Agreement 东盟 ASEAN	曼谷协定 BA	香港 H.K.	澳门 Macao	普通税率(%) Gen.	Article Description
		-钒的氧化物及氢氧化物:							-Vanadium oxides and hydroxides:
1440	2825.3010	---五氧化二钒	5.5					30	---Divanadium pentaoxide
1441	2825.3090	---其他	5.5					30	---Other
1442	2825.4000	-镍的氧化物及氢氧化物	5.5					30	-Nickel oxides and hydroxides
1443	2825.5000	-铜的氧化物及氢氧化物	5.5					30	-Copper oxides and hydroxides
1444	2825.6000	-锗的氧化物及二氧化锆	5.5					30	-Germanium oxides and zirconium dioxide
1445	2825.7000	-钼的氧化物及氢氧化物	5.5					30	-Molybdenum oxides and hydroxides
1446	2825.8000	-锑的氧化物	5.5					30	-Antimony oxides
		-其他:							-Other:
		---钨的氧化物及氢氧化物:							---Tungsten oxides and hydroxides:
1447	2825.9011	----钨酸	5.5					30	----Tungstic acid
1448	2825.9012	----三氧化钨	5.5					30	----Tungstic oxide
1449	2825.9019	----其他	5.5					30	----Other
1450	2825.9090	---其他	5.5					30	---Other
		第五分章 无机酸盐、无机过氧酸盐及金属酸盐、金属过氧酸盐							Ⅴ. SALTS AND PEROXYSALTS, OFINORGANIC ACIDS AND METALS
	28.26	**氟化物;氟硅酸盐、氟铝酸盐及其他氟络盐:**							**Fluorides; fluorosilicates, fluoroaluminates and other complex fluorine salts:**
		-氟化物:							-Fluorides:
1451	2826.1100	--氟化铵及氟化钠	5.5					30	--Of ammonium or of sodium
1452	2826.1200	--氟化铝	5.5					30	--Of aluminium
1453	2826.1900	--其他	5.5					30	--Other
1454	2826.2000	-氟硅酸钠及氟硅酸钾	5.5					30	-Fluorosilicates of sodium or of potassium
1455	2826.3000	-六氟铝酸钠(人造冰晶石)	5.5					30	-Sodium hexafluoroaluminate(synthetic cryolite)
1456	2826.9000	-其他	5.5					30	-Other
	28.27	**氯化物、氯氧化物及氢氧基氯化物;溴化物及溴氧化物;碘化物及碘氧化物:**							**Chlorides, chloride oxides and chloride hydroxides; bromides and bromide oxides; iodides and iodide oxides:**
		-氯化铵:							-Ammonium chloride:
1457	2827.1010	---肥料用	4					11	---For use as fertilizer
1458	2827.1090	---其他	5.5					30	---Other
1459	2827.2000	-氯化钙	5.5					50	-Calcium chloride
		-其他氯化物:							-Other chlorides:
1460	2827.3100	--氯化镁	5.5					30	--Of magnesium
1461	2827.3200	--氯化铝	5.5					30	--Of aluminium

序号 No.	税则号列 Tariff Item	货品名称	最惠国税率(%) M.F.N.	协定税率(%) Agreement 东盟 ASEAN	曼谷协定 BA	香港 H.K.	澳门 Macao	普通税率(%) Gen.	Article Description
1462	2827.3300	--铁的氯化物	5.5					30	--Of iron
1463	2827.3400	--氯化钴	5.5					30	--Of cobalt
1464	2827.3500	--氯化镍	5.5					30	--Of nickel
1465	2827.3600	--氯化锌	5.5					50	--Of zinc
		--其他:							--Other:
1466	2827.3910	---氯化锂	5.5					30	---Lithium chloride
1467	2827.3920	---氯化钡	5.5					30	---Barium chleride
1468	2827.3990	---其他	5.5					30	---Other
		-氯氧化物及氢氧基氯化物:							-Chloride oxides and chloride hydroxides:
1469	2827.4100	--铜的氯氧化物及氢氧基氯化物	5.5					30	--Of copper
1470	2827.4900	--其他	5.5					30	--Other
		-溴化物及溴氧化物:							-Bromides and bromide oxides:
1471	2827.5100	--溴化钠及溴化钾	5.5					30	--Bromides of sodium or of potassium
1472	2827.5900	--其他	5.5					30	--Other
1473	2827.6000	-碘化物及碘氧化物	5.5					30	-Iodides and iodide oxides
	28.28	**次氯酸盐;商品次氯酸钙;亚氯酸盐;次溴酸盐:**							**Hypochlorites; commercial calcium hypochlorite; chlorites; hypobromites:**
1474	2828.1000	-商品次氯酸钙及其他钙的次氯酸盐	12					80	-Commercial calcium hypochlorite and other calcium hypochlorites
1475	2828.9000	-其他	5.5					30	-Other
	28.29	**氯酸盐及高氯酸盐;溴酸盐及过溴酸盐;碘酸盐及高碘酸盐:**							**Chlorates and perchlorates; bromates and perbromates; iodates and periodates:**
		-氯酸盐:							-Chlorates:
1476	2829.1100	--氯酸钠	12					30	--Of sodium
		--其他:							--Other:
1477	2829.1910	---氯酸钾(洋硝)	5.5					20	---Potassium chlorate
1478	2829.1990	---其他	5.5					30	---Other
1479	2829.9000	-其他	5.5					30	-Other
	28.30	**硫化物;多硫化物,无论是否已有化学定义:**							**Sulphides; polysulphides, whether or not chemically defined:**
		-钠的硫化物:							-Sodium sulphides:
1480	2830.1010	---硫化钠	5.5					40	---Sodium sulphide
1481	2830.1090	---其他	5.5					30	---Other
1482	2830.2000	-硫化锌	5.5					30	-Zinc sulphide
1483	2830.3000	-硫化镉	5.5					30	-Cadmium sulphide
		-其他:							-Other:
1484	2830.9010	---硫化汞	5.5					45	---Mercury sulphide
1485	2830.9020	---硫化锑	5.5					45	---Antimony sulphide
1486	2830.9030	---硫化钴	5.5					30	---Cobalt sulphide
1487	2830.9090	---其他	5.5					30	---Other

序号 No.	税则号列 Tariff Item	货品名称	最惠国税率(%) M.F.N.	协定税率(%)Agreement 东盟 ASEAN	曼谷协定 BA	香港 H.K.	澳门 Macao	普通税率(%) Gen.	Article Description
	28.31	**连二亚硫酸盐及次硫酸盐:**							**Dithionites and sulphoxylates:**
1488	2831.1000	-钠的连二亚硫酸盐及次硫酸盐	5.5					30	-Of sodium
1489	2831.9000	-其他	5.5					30	-Other
	28.32	**亚硫酸盐;硫代硫酸盐:**							**Sulphites;thiosulphates:**
1490	2832.1000	-钠的亚硫酸盐	5.5					30	-Sodium sulphites
1491	2832.2000	-其他亚硫酸盐	5.5					30	-Other sulphites
1492	2832.3000	-硫代硫酸盐	5.5					30	-Thiosulphates
	28.33	**硫酸盐;矾;过硫酸盐:**							**Sulphates; alums; peroxosulphates (persul- phates):**
		-钠的硫酸盐:							-Sodium sulphates:
1493	2833.1100	--硫酸钠	5.5					40	--Disodium sulphate
1494	2833.1900	--其他	5.5					30	--Other
		-其他硫酸盐:							-Other sulphates:
1495	2833.2100	--硫酸镁	5.5					30	--Of magnesium
1496	2833.2200	--硫酸铝	5.5					30	--Of aluminium
1497	2833.2300	--铬的硫酸盐	5.5					30	--Of chromium
1498	2833.2400	--镍的硫酸盐	5.5					30	--Of nickel
1499	2833.2500	--铜的硫酸盐	5.5					30	--Of copper
1500	2833.2600	--硫酸锌	5.5					30	--Of zinc
1501	2833.2700	--硫酸钡	5.5					30	--Of barium
		--其他:							--Other:
1502	2833.2910	---硫酸亚铁	5.5					45	---Ferrous sulphate
1503	2833.2990	---其他	5.5					30	---Other
		-矾:							-Alums:
1504	2833.3010	---钾铝矾	5.5					45	---Potassium aluminum sulfate
1505	2833.3090	---其他	5.5					30	---Other
1506	2833.4000	-过硫酸盐	5.5					30	-Peroxosulphates(persulphates)
	28.34	**亚硝酸盐;硝酸盐:**							**Nitrites;nitrates:**
1507	2834.1000	-亚硝酸盐	5.5					30	-Nitrites
		-硝酸盐:							-Nitrates:
		--硝酸钾:							--Of potassium:
1508	2834.2110	---肥料用	4					11	---For use as fertilizer
1509	2834.2190	---其他	5.5					30	---Other
		--其他:							--Other:
1510	2834.2910	---硝酸钴	5.5					30	---Of cobalt
1511	2834.2990	---其他	5.5					30	---Other
	28.35	**次磷酸盐、亚磷酸盐、磷酸盐及多磷酸盐,无论是否已有化学定义:**							**Phosphinates (hypophosphites), phosphonates(phosphites), phosphates and polyphosphates, whether or not chemically defined:**
1512	2835.1000	-次磷酸盐及亚磷酸盐	5.5					20	-Phosphinates (hypophosphites) and phosphonates(phosphites)

序号 No.	税则号列 Tariff Item	货品名称	最惠国税率(%) M.F.N.	协定税率(%)Agreement 东盟 ASEAN	曼谷协定 BA	香港 H.K.	澳门 Macao	普通税率(%) Gen.	Article Description
		-磷酸盐:							-Phosphates:
1513	2835.2200	--磷酸一钠及磷酸二钠	5.5					20	--Of mono-or disodium
1514	2835.2300	--磷酸三钠	5.5					20	--Of trisodium
1515	2835.2400	--钾的磷酸盐	5.5					20	--Of potassium
1516	2835.2500	--正磷酸氢钙(磷酸二钙)	5.5					20	--Calcium hydrogenorthophosphate (dicalcium phosphate)
1517	2835.2600	--其他磷酸钙	5.5					20	--Other phosphates of calcium
1518	2835.2900	--其他	5.5					20	--Other
		-多磷酸盐:							-Polyphosphates:
1519	2835.3100	--三磷酸钠(三聚磷酸钠)	5.5					20	--Sodium triphosphate(sodium tripoly-phosphate)
1520	2835.3900	--其他	5.5					20	--Other
	28.36	**碳酸盐;过碳酸盐;含氨基甲酸铵的商品碳酸铵:**							**Carbonates; peroxocarbonates (per-carbonates); commercial ammonium carbonate containing ammonium car-bamate:**
1521	2836.1000	-商品碳酸铵及其他铵的碳酸盐	5.5					30	-Commercial ammonium carbonate and other ammonium carbonates
1522	2836.2000	-碳酸钠(纯碱)	5.5					35	-Disodium carbonate
1523	2836.3000	-碳酸氢钠(小苏打)	5.5					45	-Sodium hydrogencarbonate (sodium bicar bonate)
1524	2836.4000	-钾的碳酸盐	5.5					30	-Potassium carbonates
1525	2836.5000	-碳酸钙	5.5					45	-Calcium carbonate
1526	2836.6000	-碳酸钡	5.5					40	-Barium carbonate
1527	2836.7000	-铅的碳酸盐	5.5					30	-Lead carbonates
		-其他:							-Other:
1528	2836.9100	--锂的碳酸盐	5.5					30	--Lithium carbonates
1529	2836.9200	--锶的碳酸盐	5.5					30	--Strontium carbonate
		--其他:							--Other:
1530	2836.9910	---碳酸镁	5.5					45	---Magnesium carbonate
1531	2836.9930	---碳酸钴	5.5					30	---Cobalt carbonate
1532	2836.9990	---其他	5.5					30	---Other
	28.37	**氰化物、氧氰化物及氰络合物:**							**Cyanides, cyanide oxides and complex cyanides:**
		-氰化物及氧氰化物:							-Cyanides and cyanide oxides:
		--氰化钠及氧氰化钠:							--Of sodium:
1533	2837.1110	---氰化钠	5.5					20	---Sodium cyanide
1534	2837.1120	---氧氰化钠	5.5		5			30	---Sodium cyanide oxide
		--其他:							--Other:
1535	2837.1910	---氰化钾	5.5					20	---Potassium cyanide
1536	2837.1990	---其他	5.5					30	---Other
1537	2837.2000	-氰络合物	5.5					30	-Complex cyanides
	28.38	**雷酸盐、氰酸盐及硫氰酸盐:**							**Fulminates, cyanates and thio-cyanates:**
1538	2838.0000	雷酸盐、氰酸盐及硫氰酸盐	5.5					30	Fulminates, cyanates and thiocyanates

序号 No.	税则号列 Tariff Item	货品名称	最惠国税率(%) M.F.N.	协定税率(%)Agreement 东盟 ASEAN	曼谷协定 BA	香港 H.K.	澳门 Macao	普通税率(%) Gen.	Article Description
	28.39	**硅酸盐;商品碱金属硅酸盐:**							**Silicates;commercial alkali metal silicates:**
		-钠盐:							-Of sodium:
1539	2839.1100	--偏硅酸钠	5.5					40	--Sodium metasilicates
1540	2839.1900	--其他	5.5					30	--Other
1541	2839.2000	-钾盐	5.5					30	-Of potassium
1542	2839.9000	-其他	5.5					30	-Other
	28.40	**硼酸盐及过硼酸盐:**							**Borates;peroxoborates(perborates):**
		-四硼酸钠(精炼硼砂):							-Disodium tetraborate(refined borax):
1543	2840.1100	--无水四硼酸钠	5.5					20	--Anhydrous
1544	2840.1900	--其他	5.5					20	--Other
1545	2840.2000	-其他硼酸盐	5.5					30	-Other borates
1546	2840.3000	-过硼酸盐	5.5					30	-Peroxoborates(perborates)
	28.41	**金属酸盐及过金属酸盐:**							**Salts of oxometallic or peroxometallic acids:**
1547	2841.1000	-铝酸盐	5.5					30	-Aluminates
1548	2841.2000	-锌的铬酸盐及铅的铬酸盐	5.5					40	-Chromates of zinc or of lead
1549	2841.3000	-重铬酸钠	5.5					20	-Sodium dichromate
1550	2841.5000	-其他铬酸盐及重铬酸盐;过铬酸盐	5.5					30	-Other chromates and dichromates; peroxochromates
		-亚锰酸盐、锰酸盐及高锰酸盐:							-Manganites, manganates and permanganates:
1551	2841.6100	--高锰酸钾	5.5					30	--Potassium permanganate
1552	2841.6900	--其他	5.5					30	--Other
		-钼酸盐:							-Molybdates:
1553	2841.7010	---钼酸铵	5.5					30	---Ammonium molybdates
1554	2841.7090	---其他	5.5					30	---Other
		-钨酸盐:							-Tungstates(wolframates):
1555	2841.8010	---仲钨酸铵	5.5					30	---Ammonium paratungstate
1556	2841.8020	---钨酸钠	5.5					30	---Sodium tungstate
1557	2841.8030	---钨酸钙	5.5					30	---Calcium wolframate
1558	2841.8040	---偏钨酸铵	5.5					30	---Ammonium metatungstate
1559	2841.8090	---其他	5.5					30	---Other
1560	2841.9000	-其他	5.5					30	-Other
	28.42	**其他无机酸盐及过氧酸盐(包括不论是否已有化学定义的硅酸盐),但迭氮化物除外:**							**Other Salts of inorganic acids or peroxoacids (including aluminosilicates whether or not chemically defined), other than azides:**
1561	2842.1000	-硅酸复盐及硅酸络盐,(包括不论是否已有化学定义的硅铝酸盐)	5.5			0		30	-Double or complex silicates,including aluminosilicates whether or not chemically defined
1562	2842.9000	-其他	5.5					30	-Other
		第六分章 杂项产品							Ⅳ. MISCELLANEOUS

序号 No.	税则号列 Tariff Item	货品名称	最惠国税率(%) M.F.N.	协定税率(%) Agreement 东盟 ASEAN	曼谷协定 BA	香港 H.K.	澳门 Macao	普通税率(%) Gen.	Article Description
	28.43	**胶态贵金属;贵金属的无机或有机化合物,不论是否已有化学定义;贵金属汞齐:**							**Colloidal precious metals; inorganic or organic compounds of precious metals, whether or not chemically defined; amalgams of precious metals:**
1563	2843.1000	-胶态贵金属	5.5					30	-Colloidal precious metals
		-银化合物:							-Silver compounds:
1564	2843.2100	--硝酸银	5.5			0△		30	--Silver nitrate
1565	2843.2900	--其他	5.5			0		30	--Other
1566	2843.3000	-金化合物	5.5			0		30	-Gold compounds
1567	2843.9000	-其他贵金属化合物;贵金属汞齐	5.5			0		30	-Other compounds; amalgams
	28.44	**放射性化学元素及放射性同位素(包括可裂变或可转换的化学元素及同位素)及其化合物;含上述产品的混合物及残渣:**							**Radioactive chemical elements and radioactive isotopes (including the fissile or fertile chemical elements and isotopes) and their compounds; mixtures and residues containing these products:**
1568	2844.1000	-天然铀及其化合物;含天然铀或天然铀化合物的合金、分散体(包括金属陶瓷)、陶瓷产品及混合物	5.5					30	-Natural uranium and its compounds; aloys, dispersions (including cermets), ceramic products and mixtures containing natural uranium or natural uranium compouds
1569	2844.2000	-U_{235}浓缩铀及其化合物;钚及其化合物;含U_{235}浓缩铀、钚或它们的化合物的合金、分散体(包括金属陶瓷)、陶瓷产品及混合物	5.5					30	-Uranium enriched in U_{235} and its compounds; plutonium and its compounds; alloys dispersion (including cermets), ceramic products and mixtures containing uranium enriched in U_{235}, plutonium or compounds of these products
1570	2844.3000	-U_{235}贫化铀及其化合物;钍及其化合物;含U_{235}贫化铀、钍或它们的化合物的合金、分散体(包括金属陶瓷)、陶瓷产品及混合物	5.5					30	-Uranium depleted in U_{235} and its compounds; thorium and its compounds; alloys, dispersions (including cermets), ceramic products and mixtures containing uranium depleted in U_{235}, thorium or compounds of these products
		-除子目号 2844.10、2844.20 及 2844.30 以外的放射性元素、同位素及其化合物;含这些元素、同位素及其化合物的合金、分散体(包括金属陶瓷)、陶瓷产品及混合物:							-Radioactive elements and isotopes and compounds other than those of subheading No.2844.10, 2844.20 or 2844.30; alloys, dispersions (including cermets), ceramic products and mixtures containing these elements, isotopes or compounds; radioactive residues:
1571	2844.4010	---镭及镭盐	4					14	---Radium and its salts

序号 No.	税则号列 Tariff Item	货品名称	最惠国税率(%) M.F.N.	协定税率(%)Agreement 东盟 ASEAN	曼谷协定 BA	香港 H.K.	澳门 Macao	普通税率(%) Gen.	Article Description
1572	2844.4020	---钴及钴盐	4					14	---Cobalt and its salts
1573	2844.4090	---其他	5.5					30	---Other
1574	2844.5000	-核反应堆已耗尽(已辐照)的燃料元件(释热元件)	5.5					30	-Spent(irradiated) fuel elements(cartridges)of nuclear reactors
	28.45	**税号28.44以外的同位素;这些同位素的无机或有机化合物,不论是否已有化学定义:**							**Isotopes other than those of heading No.28.44; compounds, inorganic or organic, of such isotopes, whether or not chemically defined:**
1575	2845.1000	-重水(氧化氘)	5.5					30	-Heavy water(deuterium oxide)
1576	2845.9000	-其他	5.5					30	-Other
	28.46	**稀土金属、钇、钪及其混合物的无机或有机化合物:**							**Compounds, inorganic or organic, of rareearth metals, of yttrium or of scandium or of mixtures of these metals:**
		-铈的化合物:							-Ceric compounds:
1577	2846.1010	---氧化铈	5.5					30	---Cerium oxide
1578	2846.1020	---氢氧化铈	5.5					30	---Cerium hydroxide
1579	2846.1030	---碳酸铈	5.5					30	---Cerium carbonate
1580	2846.1090	---其他	5.5					30	---Other
		-其他:							-Other:
		---氧化稀土(氧化铈除外):							---Rare-earth oxides(other than cerium oxide):
1581	2846.9011	----氧化钇	5.5					30	----Yttrium oxide
1582	2846.9012	----氧化镧	5.5					30	----Lanthanum oxide
1583	2846.9013	----氧化钕	5.5					30	----Neodymium oxide
1584	2846.9014	----氧化铕	5.5					30	----Eurapium oxide
1585	2846.9019	----其他	5.5					30	----Other
		---氯化稀土:							---Rare-earth chlorides
1586	2846.9028	----混合氯化稀土	5.5					30	----Mixture of rare-earth chlorides
1587	2846.9029	----其他	5.5					30	----Other
1588	2846.9030	---氟化稀土	5.5					30	---Rare-earth fluorides
		---碳酸稀土:							---Rare-earth carbonate:
1589	2846.9048	----混合碳酸稀土	5.5					30	----Mixture of rare-earth carbonate
1590	2846.9049	----其他	5.5					30	----Other
1591	2846.9090	---其他	5.5					30	---Other
	28.47	**过氧化氢,不论是否用尿素固化:**							**Hydrogen peroxide, whether or not solidified with urea:**
1592	2847.0000	过氧化氢,不论是否用尿素固化	5.5					30	Hydrogen peroxide, whether or not solidified with urea
	28.48	**磷化物,不论是否已有化学定义,但不包括磷铁:**							**Phosphides, whether or not chemically defined, excluding ferrophosphorus:**
1593	2848.0000	磷化物,不论是否已有化学定义,但不包括磷铁	5.5					20	Phosphides, whether or not chemically defined, excluding ferrophophorus

序号 No.	税则号列 Tariff Item	货品名称	最惠国税率(%) M.F.N.	协定税率(%) Agreement 东盟 ASEAN	曼谷协定 BA	香港 H.K.	澳门 Macao	普通税率(%) Gen.	Article Description
	28.49	**碳化物,不论是否已有化学定义:**							**Carbides, whether or not chemically defined:**
1594	2849.1000	-碳化钙	5.5					45	-Of calcium
1595	2849.2000	-碳化硅	5.5					30	-Of silicon
		-其他:							-Other:
1596	2849.9010	---碳化硼	5.5					30	---Of boron
1597	2849.9020	---碳化钨	5.5					30	---Of tungsten
1598	2849.9090	---其他	5.5					30	---Other
	28.50	**氢化物、氮化物、迭氮化物、硅化物及硼化物,不论是否已有化学定义,但可归入税号28.49的碳化物除外:**							**Hydrides, nitrides, azides, silicides and borides, whether or not chemically defined, other than compounds which are also carbides of heading No.28.49:**
1599	2850.0000	氢化物、氮化物、迭氮化物、硅化物及硼化物,不论是否已有化学定义,但可归入税号28.49的碳化物除外	5.5			0		30	Hydrides, nitrides, azides, silicides and borides, whether or not chemically defined, other than compounds which are also carbides of heading No.28.49
	28.51	**其他无机化合物(包括蒸馏水、导电水及类似的纯净水);液态空气(不论是否除去稀有气体);压缩空气;汞齐,但贵金属汞齐除外:**							**Other inorganic compounds (including distilled or conductivity water and water of similar purity); liquid air (whether or not rare gases have been removed); compressed air; amalgams, other than amalgams of precious metals:**
1600	2851.0010	---饮用蒸馏水	5.5			0		70	---Distilled water for human consumption
1601	2851.0020	---氯化氰	5.5					30	---Chlorocyanogen, cyanogen chloride
1602	2851.0090	---其他	5.5			0		30	---Other

第二十九章
有机化学品

注释：

一、除条文另有规定的以外，本章各税号只适用于：

(一) 单独的已有化学定义的有机化合物，不论是否含有杂质；

(二) 同一有机化合物的两种或两种以上异构体的混合物（不论是否含有杂质），但无环烃异构体的混合物（立体异构体除外），不论是否饱和，应归入第二十七章；

(三) 税号29.36至29.39的产品，税号29.40的糖醚、糖缩醛、糖酯及其盐类和税号29.41的产品，不论是否已有化学定义；

(四) 上述（一）、（二）、（三）款产品的水溶液；

(五) 溶于其他溶剂的上述（一）、（二）、（三）款的产品，但该产品处于溶液状态只是为了安全或运输所采取的正常必要方法，其所用溶剂并不使该产品改变其一般用途而适合于某些特殊用途；

(六) 为了保存或运输的需要，加入稳定剂（包括抗结块剂）的上述（一）、（二）、（三）、（四）、（五）各款产品；

(七) 为了便于识别或安全起见，加入抗尘剂、着色剂或气味剂的上述（一）、（二）、（三）、（四）、（五）、（六）各款产品，但所加剂料并不使原产品改变其一般用途而适合于某些特殊用途；

(八) 为生产偶氮染料而稀释至标准浓度的下列产品：重氮盐，用于重氮盐、可重氮化的胺及其盐类的偶合剂。

二、本章不包括：

(一) 税号15.04的货品及税号15.20的粗甘油；

(二) 乙醇（税号22.07或22.08）；

(三) 甲烷及丙烷（税号27.11）；

(四) 第二十八章注释二所述的碳化合物；

(五) 尿素（税号31.02或31.05）；

Chapter 29
Organic chemicals

Notes:

1. Except where the context otherwise requires, the headings of this Chapter apply only to:

(a) Separate chemically defined organic compounds, whether or not containing impurities;

(b) Mixtures of two or more isomers of the same organic compound (whether ornot containing impurities), except mixtures of acyclic hydrocarbon isomers (other than stereoisomers), whether or not saturated (Chapter 27);

(c) The products of headings 29.36 to 29.39 or the sugar ethers, sugar acetals and sugar esters, and their salts, of heading 29.40, or the products of heading 29.41, whether or not chemically defined;

(d) The products mentioned in (a), (b) or (c) above dissolved in water;

(e) The products mentioned in (a), (b) or (c) above dissolved in other slovents provided that the solution constitutes a normal and necessary method of putting up these products adopted solely for reasons of safety or for transport and that the solvent does not render the product particularly suitable for specific use rather than for general use;

(f) The products mentioned in (a), (b), (c), (d) or (e) above with an added stabilizer (including an anti-caking agent) necessary for their preservation or transport;

(g) The products mentioned in (a), (b), (c), (d), (e) or (f) above with an added anti-dusting agent or a colouring or odoriferous substance added to facilitate their identification or for safety reasons, provided that the additions do not render the product particularly suitable for specific use rather than for general use;

(h) The following products, diluted to standard strengths, for the production of azo dyes: diazonium salts, couplers used for these salts and diazotisable amines and their salts.

2. This Chapter does not cover:

(a) Goods of heading No.15.04 or crude glycerol of heading No.15.20;

(b) Ethyl alcohol (heading No.22.07 or 22.08);

(c) Methane or propane (headng No.27.11);

(d) The compounds of carbon mentioned in Note 2 to Chapter 28;

(e) Urea (heading No.31.02 or 31.05);

（六）植物性或动物性着色料（税号 32.03）、合成有机着色料、用作萤光增白剂或发光体的合成有机产品（税号 32.04）及零售包装的染料或其他着色料（税号 32.12）；

（七）酶（税号 35.07）；

（八）聚乙醛、六亚甲基四胺（乌洛托品）及类似物质，制成片、条或类似形状作为燃料用的，以及包装容器的容积不超过 300 立方厘米的直接灌注香烟打火机及类似打火器用的液体燃料或液化气体燃料（税号 36.06）；

（九）灭火器的装配药及已装药的灭火弹（税号 38.13）；零售包装的除墨剂（税号 38.24）；

（十）光学元件，例如用酒石酸乙二胺制成的（税号 90.01）。

三、可以归入本章两个或两个以上税号的货品，应归入有关税号中的最后一个税号。

四、税号 29.04 至 29.06、29.08 至 29.11 及 29.13 至 29.20 的卤化、磺化、硝化或亚硝化衍生物均包括复合衍生物，例如，卤磺化、卤硝化、磺硝化及卤磺硝化衍生物。

硝基及亚硝基不作为税号 29.29 的含氮基官能团。

税号 29.11、29.12、29.14、29.18 及 29.22 所称“含氧基”，仅限于税号 29.05 至 29.20 的各种含氧基（其特征为有机含氧基）。

五、（一）本章第一分章至第七分章的酸基有机化合物与这些分章的有机化合物构成的酯应归入有关税号中的最后一个税号；

（二）乙醇与本章第一分章至第七分章的酸基有机化合物所构成的酯，应按有关酸基化合物归类。

（三）除第六类注释一及第二十八章注释二另有规定的以外：

1. 第一分章至第十分章及税号 29.42 的有机化合物的无机盐，例如，含酸基、酚基或烯醇基的化合物及有机碱的无机盐，应归

(f) Colouring matter of vegetable or animal origin (heading No.32.03), synthetic organic colouring matter, synthetic organic products of a kind used as fluores cent brightening agents or as luminophores (heading No.32.04) or dyes or other colouring matter put up in forms or packings for retail sale (heading No.32.12);

(g) Enzymes (heading No.35.07);

(h) Metaldehyde, hexamethylenetetramine or similar substances, put up in forms (for example, tablets, sticks or similar forms) for use as fuels, or liquid or liquefied-gas fuels in containers of a kind used for filling or refilling cigarette or similar lighters and of a capacity not exceeding 300cm^3 (heading No.36.06);

(i) Products put up as charges for fire-extinguishers or put up in fire-extinguishing grenades, of heading No.38.13; ink removers put up in packings for retail sale, of heading No.38.24; or

(j) Optical elements, for example, of ethylenediamine tartrate (heading No.90.01)

3. Goods which could be included in two or more of the headings of this Chapter are to be classified in that one of those headings which occurs last in numerical order.

4. In headings Nos.29.04 to 29.06, 29.08 to 29.11 and 29.13 to 29.20, any reference to halogenated, sulphonated, nitrated or nitrosated derivatives includes a reference to compound derivatives, such as sulphohalogenated, nitrohalogenated, nitrosulphonated or nitrosulphohalogenated derivatives.

Nitro or nitroso groups are not to be taken as “nitrogenfunctions” for the purpose of heading No.29.29.

For the purposes of headings Nos.29.11, 29.12, 29.14, 29.18 and 29.22, “oxygen-function” is to be restricted to the functions (the characteristic organic oxygen-containing groups) referred to in headings Nos.29.05 to 29.20.

5. (a) The esters of acid-function organic compounds of sub-Chapters Ⅰ to Ⅶ with organic compounds of these sub-Chapters are to be classified with that compound which is classified in the heading which occurs last in numerical order in these sub-Chapters.

(b) Esters of ethyl alcohol with acid-function organic compounds of sub-Chapters Ⅰ to Ⅶ are to be classified in the same heading as the corresponding acidfunction compounds.

(c) Subject to Note 1 to Section Ⅵ and Note 2 to Chapter 28:

(1) Inorganic salts of organic compounds such as acid, phenol or enol-function compounds or organic bases, of sub-Chapters Ⅰ to Ⅹ or heading No.29.42, are to

入相应的有机化合物的税号；

2. 第一分章至第十分章及税号 29.42 的有机化合物之间生成的盐，应按生成该盐的碱或酸（包括酚基或烯醇基化合物）归入本章有关税号中的最后一个税号。

（四）除乙醇外，金属醇化物应按相应的醇归类（税号 29.05）。

（五）羧酸酰卤化物应按相应的酸归类。

六、税号 29.30 及 29.31 的化合物是指有机化合物，其分子中除含氢、氧或氮原子外，还含有与碳原子直接连接的其他非金属或金属原子（例如，硫、砷、汞或铅）。

税号 29.30（有机硫化合物）及税号 29.31（其他有机—无机化合物）不包括某些磺化或卤化衍生物（含复合衍生物）。这些衍生物分子中除氢、氧、氮之外，只有具有磺化或卤化衍生物（或复合衍生物）性质的硫原子或卤素原子与碳原子直接连接。

七、税号 29.32、29.33 及 29.34 不包括三节环环氧化物、过氧化酮、醛或硫醛的环聚合物、多元羧酸酐、多元醇或酚与多元酸构成的环酯及多元酸酰亚胺。

本条规定只适用于由本条所列环化功能形成环内杂原子的化合物。

八、税目 29.37 所称：

（一）“激素”包括激素释放因子、激素刺激和释放因子、激素抑制剂以及激素抗体；

（二）“主要起激素作用”，不仅适用于激素衍生物以及主要起激素作用的结构类似物，也适用于在本税目所列产品合成过程中主要用作中间体的激素衍生物以及结构类似物。

子目注释：

属于本章任一税号项下的一种（组）化合物的衍生物，如果该税号其他子目未明确将其包括在内，而且有

be classified in the heading appropriate to the organic compound; and

(2) Salts formed between organic compounds of sub-Chapters Ⅰ to Ⅹ or heading No.29.42 are to be classified in the heading appropriate to the base or to the acid (including phenol or enol-function compounds) from which they are formed, whichever occurs last in numerical order in the Chapter.

(d) Metal alcoholates are to be classified in the same heading as the corresponding alcohols except in the case of ethanol (heading No.29.05).

(e) Halides of carboxylic acids are to be classified in the same heading as the corresponding acids.

6. The compounds of headings Nos.29.30 and 29.31 are organic compounds the molecules of which contain, in addition to atoms of hydrogen, oxygen or nitrogen, atoms of other non-metals of or metals (such as sulphur, arsenic, mercury or lead) directly linked to carbon atoms.

Heading No.29.30 (organo-sulphur compounds) and heading No.29.31 (other organo-inorganic compounds) do not include sulphonated or halogenated derivatives (in cluding compound derivatives) which, apart from hydrogen, oxygen and nitrogen, only have directly linked to carbon the atoms of sulphur or of a halogen which give them their nature of sulphonated or halogenated derivatives (or compound derivatives).

7. Headings Nos.29.32, 29.33 and 29.34 do not include epoxides with a three-membered ring, ketone peroxides, cyclic polymers of aldehydes or of thioaldehydes, anhydrides of polybasic carboxylic acids, cyclic esters of polyhydric alcohols or phenols with polybasic acids, and imides of polybasic acids.

These provisions apply only when the ring-position hetero-atoms are those resulting solely form the cyclising function or functions here listed.

8. For the purposes of heading 29.37:

(a) the term "hormones" includes hormone-releasing or hormone-stimulating factors, hormone inhibitors and hormone antagonists (anti-hormones);

(b) the expression "used primarily as hormones" applies not only to hormone derivatives and structural analogues used primarily for their hormonal effect, but also to those derivatives and structural analogues used primarily as intermediates in the synthesis of products of this heading.

Subheading Note:

Within any one heading of this Chapter, derivatives of a chemical compound (or group of chemical compounds) are to be clas-

关的子目中又无列名为“其他”的子目，则应与该种(组)化合物归入同一子目。

sified in the same subheading as that compound (or group of compounds) provided that they are not more specifically covered by any other subheading and that there is no residual subheading named “other” in the series of subheadings concerned.

序号 No.	税则号列 Tariff Item	货品名称	最惠国税率(%) M.F.N.	协定税率(%) Agreement 东盟 ASEAN	曼谷协定 BA	香港 H.K.	澳门 Macao	普通税率(%) Gen.	Article Description
		第一分章　烃类及其卤化、磺化、硝化或亚硝化衍生物							Ⅰ. HYDROCARBONS AND THEIR HALOGENATED, SULPHONATED, NITRATED OR NITROSATED DERIVATIVES
	29.01	**无环烃:**							**Acyclic hydrocarbons:**
1603	2901.1000	-饱和	2					30	-Saturated
		-不饱和:							-Unsaturated:
1604	2901.2100	--乙烯	2					20	--Ethylene
1605	2901.2200	--丙烯	2					20	--Propene(propylene)
1606	2901.2300	--丁烯及其异构体	2					20	--Butene(butylene)and isomers thereof
1607	2901.2400	--1,3-丁二烯及异戊二烯	2					20	--Buta-1,3-diene and isoprene
		--其他:							--Other:
1608	2901.2910	---异戊烯	2					30	---Isopentene
1609	2901.2920	---乙炔	2					45	---Acetylene
1610	2901.2990	---其他	2					30	---Other
	29.02	**环烃:**							**Cyclic hydrocarbons:**
		-环烷烃、环烯及环萜烯:							-Cyclanes,cyclenes and cycloterpenes:
1611	2902.1100	--环己烷	2					30	--Cyclohexane
1612	2902.1900	--其他	2					30	--Other
1613	2902.2000	-苯	2					20	-Benzene
1614	2902.3000	-甲苯	2					30	-Toluene
		-二甲苯:							-Xylenes:
1615	2902.4100	--邻二甲苯	2					20	--*o*-Xylene
1616	2902.4200	--间二甲苯	2					20	--*m*-Xylene
1617	2902.4300	--对二甲苯	2					20	--*p*-Xylene
1618	2902.4400	--混合二甲苯异构体	2					20	--Mixed xylene isomers
1619	2902.5000	-苯乙烯	2		1.4			30	-Styrene
1620	2902.6000	-乙苯	2					30	-Ethylbenzene
1621	2902.7000	-异丙基苯	2					30	-Cumene
		-其他:							-Other:
1622	2902.9010	---四氢萘	2					11	---Tetrahydronaphthalene(tetralin)
1623	2902.9020	---精萘	2					35	---Naphthalene
1624	2902.9030	---十二烷基苯	2					30	---Dodecylbenzene
1625	2902.9090	---其他	2					30	---Other
	29.03	**烃的卤化衍生物:**							**Halogenated derivatives of hydrocarbons:**
		-无环烃的饱和氯化衍生物:							-Saturated chlorinated derivatives of acyclic hydrocarbons:

序号 No.	税则号列 Tariff Item	货品名称	最惠国税率(%) M.F.N.	协定税率(%) Agreement 东盟 ASEAN	曼谷协定 BA	香港 H.K.	澳门 Macao	普通税率(%) Gen.	Article Description
1626	2903.1100	--一氯甲烷及氯乙烷	5.5					30	--Chloromethane(methyl chloride) and chloroethane(ethyl chloride)
1627	2903.1200	--二氯甲烷	8					30	--Dichloromethane (methylene chloride)
1628	2903.1300	--氯仿(三氯甲烷)	10		9			30	--Chloroform(trichloromethane)
1629	2903.1400	--四氯化碳	8					30	--Carbon tetrachloride
1630	2903.1500	--1,2-二氯乙烷	5.5					30	--1,2-Dichloroethane(ethylene dichloride)
		--其他:							--Other:
1631	2903.1910	---1,1,1-三氯乙烷(甲基氯仿)	8					30	---1, 1, 1-Trichloroethane (methylchloro form)
1632	2903.1990	---其他	5.5					30	---Other
		-无环烃的不饱和氯化衍生物:							-Unsaturated chlorinated derivatives of acyclic hydrocarbons:
1633	2903.2100	--氯乙烯	5.5					30	--Vinyl chloride(chloroethylene)
1634	2903.2200	--三氯乙烯	8					30	--Trichloroethylene
1635	2903.2300	--四氯乙烯(全氯乙烯)	5.5					30	--Tetrachloroethylene (perchloroethylene)
1636	2903.2900	--其他	5.5					30	--Other
		-无环烃的氟化、溴化或碘化衍生物:							-Fluorinated, brominated or iodinated derivatives of acyclic hydrocarbons:
1637	2903.3010	---1,1,3,3,3-五氟-2-三氟甲基-1-丙烯(全氟异丁烯;八氟异丁烯)	5.5					30	---1, 1, 3, 3, 3-Pentafluro-2-triflurometliyl-1-propene (Perfluorolisobutylene, isobutylene octafluoride)
1638	2903.3090	---其他	5.5					30	---Other
		-含有两种或两种以上不同卤素的无环烃卤化衍生物:							-Halogenated derivatives of acyclic hydrocarbons containing two or more different halogens:
1639	2903.4100	--三氯氟甲烷	5.5					30	--Trichlorofluoromethane
1640	2903.4200	--二氯二氟甲烷	5.5					30	--Dichlorodifluoromethane
1641	2903.4300	--三氯三氟乙烷	5.5					30	--Trichlorotrifluoroethanes
1642	2903.4400	--二氯四氟乙烷及氯五氟乙烷	5.5					30	--Dichlorotetrafluoroethanes and chloropenta-fluoroethane
		--其他仅含氟和氯的全卤化衍生物:							--Other derivatives perhalogenated only with fluorine and chlorine:
1643	2903.4510	---氯三氟甲烷	5.5					30	---Chlorotrifluoromethane
1644	2903.4520	---五氯氟乙烷	5.5					30	---Pentachlorofluoroethane
1645	2903.4530	---四氯二氟乙烷	5.5					30	---Tetrachlorodifluoroethane
1646	2903.4540	---七氯氟丙烷	5.5					30	---Heptachlorofluoropropane
1647	2903.4550	---六氯二氟丙烷	5.5					30	---Hexachlorodifluoropropane
1648	2903.4560	---五氯三氟丙烷	5.5					30	---Pentachlorotrifluoropropane
1649	2903.4570	---四氯四氟丙烷	5.5					30	---Tetrachlorotetrafluoropropane
1650	2903.4580	---三氯五氟丙烷	5.5					30	---Trichloropentafluoropropane
		---其他:							---Other:
1651	2903.4591	----二氯六氟丙烷	5.5					30	----Dichlorohexafluoropropane

序号 No.	税则号列 Tariff Item	货品名称	最惠国税率(%) M.F.N.	协定税率(%)Agreement				普通税率(%) Gen.	Article Description
				东盟 ASEAN	曼谷协定 BA	香港 H.K.	澳门 Macao		
1652	2903.4592	----氯七氟丙烷	5.5					30	----Chloroheptafluoropropane
1653	2903.4599	----其他	5.5					30	----Other
1654	2903.4600	--溴氯二氟甲烷、溴三氟甲烷及二溴四氟乙烷	5.5					30	--Bromochlorodifluoromethane, bromotrifluoromethane and dibromotetraf luoroethane
1655	2903.4700	--其他全卤化衍生物	5.5					30	-Other perhalogenated derivatives
		--其他：							--Other：
1656	2903.4910	---其他仅含氟和氯的甲烷、乙烷及丙烷的卤化衍生物	5.5					30	---Other methane,ethane and propane halogenated derivatives only with fluorine and chlorine
1657	2903.4920	---其他仅含氟和溴的甲烷、乙烷及丙烷的卤化衍生物	5.5					30	---Other methane,ethane and propane halogenated derivatives with fluorine and bromine
1658	2903.4990	---其他	5.5					30	---Other
		-环烷烃、环烯烃或萜烯烃的卤化衍生物：							-Halogenated derivatives of cyclanic, cyclenic or cycloterpenic hydrocarbons：
1659	2903.5100	--1,2,3,4,5,6-六氯环己烷	5.5					30	--1, 2, 3, 4, 5, 6-Hexachlorocyclohexane
1660	2903.5900	--其他	5.5					30	--Other
		-芳烃卤化衍生物：							-Halogenated derivatives of aromatic hydrocarbons：
		--氯苯、邻二氯苯及对二氯苯：							--Chlorobenzene, *o*-dichlorobenzene and *p*-dichlorobenzene：
1661	2903.6110	---邻二氯苯	5.5					30	---*o*-Dichlorobenzene
1662	2903.6190	---其他	5.5					30	---Other
1663	2903.6200	--六氯苯及滴滴涕(1,1,1-三氯-2,2-双对氯苯乙烷)	5.5					30	--Hexachlorobenzene and DDT[1,1,1-trichloro-2,2-bis(*p*-chlorophenyl) ethane]
		--其他：							--Other：
1664	2903.6910	---对氯甲苯	5.5					30	---*p*-Chlorotoluene
1665	2903.6920	---3,4-二氯三氟甲苯	5.5					30	---3,4-Dichlorotrifluoride toluene
1666	2903.6990	---其他	5.5					30	---Other
	29.04	**烃的磺化、硝化或亚硝化衍生物,不论是否卤化：**							**Sulphonated, nitrated or nitrosated derivatives of hydrocarbons, whether or not halogenated：**
1667	2904.1000	-仅含磺基的衍生物及其盐和乙酯	5.5			0		30	-Derivatives containing only sulpho groups,their salts and ethyl esters
		-仅含硝基或亚硝基的衍生物：							-Derivatives containing only nitro or only nitroso groups：
1668	2904.2010	---硝基苯	5.5					20	---Nitrobenzene
1669	2904.2020	---硝基甲苯	5.5					30	---Nitrotoluene and nitrochlorobenzene
1670	2904.2030	---二硝基甲苯	5.5					20	---Dinitrotoluene and dinitrochlorobenzene
1671	2904.2040	---三硝基甲苯(TNT)	5.5					40	---Trinitrotoluene
1672	2904.2090	---其他	5.5					30	---Other

序号 No.	税则号列 Tariff Item	货品名称	最惠国税率(%) M.F.N.	协定税率(%) Agreement 东盟 ASEAN	曼谷协定 BA	香港 H.K.	澳门 Macao	普通税率(%) Gen.	Article Description
		-其他:							-Other:
1673	2904.9010	---硝基氯化苯	5.5					30	---Nitrochlorobenzene
1674	2904.9020	---二硝基氯化苯	5.5					20	---Dinitrochlorobenzene
1675	2904.9030	---三氯硝基甲烷(氯化苦;硝基氯仿)	5.5					30	---Trichloronitromethane (chloropicrin, nitrochloroform)
1676	2904.9090	---其他	5.5					30	---Other
		第二分章　醇类及其卤化、磺化、硝化或亚硝化衍生物							Ⅱ. ALCOHOLS AND THEIR HALOGENATED, SULPHONATED, NITRATED OR NITROSATED DERIVATIVES
	29.05	**无环醇及其卤化、磺化、硝化或亚硝化衍生物:**							**Acyclic alcohols and their halogenated, sulphonated, nitrated or nitrosated derivatives:**
		-饱和一元醇:							-Saturated monohydric alcohols:
1677	2905.1100	--甲醇	5.5					30	--Methanol(methyl alcohol)
		--丙醇及异丙醇:							--Propan-1-ol (propyl alcohol) and propan-2-ol(isopropyl alcohol):
1678	2905.1210	---丙醇	5.5					30	---Propan-1-ol(propyl alcohol)
1679	2905.1220	---异丙醇	5.5					30	---Propan-2-ol(isopropyl alcohol)
1680	2905.1300	--正丁醇	5.5		5.5			30	--Butan-1-ol(*n*-butyl alcohol)
1681	2905.1400	--其他丁醇	5.5					30	--Other butanols
1682	2905.1500	--戊醇及其异构体	5.5					30	--Pentanol(amyl alcohol) and isomers thereof
1683	2905.1600	--辛醇及其异构体	5.5					30	--Octanol (octyl alcohol) and isomers thereof
1684	2905.1700	--十二醇、十六醇及十八醇	7					30	--Dodecan-1-ol(lauryl alcohol), hexadecan-1-ol (cetyl alcohol) and octadecan-1-ol(stearyl alcohol)
		--其他:							--Other:
1685	2905.1910	---3,3-二甲基丁-2-醇(频哪基醇)	5.5					30	---3,3-Dimethyl-2-butanol (pinacolyl alcohol)
1686	2905.1990	---其他	5.5					30	---Other
		-不饱和一元醇:							-Unsaturated monohydric alcohols:
		--无环萜烯醇:							--Acyclic terpene alcohols:
1687	2905.2210	---香叶醇、橙花醇(3,7-二甲基-2,6-辛二烯-1-醇)	5.5					30	---Geraniol, nerol(cis-3,7-Dimethyl-2,6-octadien-1-ol)
1688	2905.2220	---香茅醇(3,7-二甲基-6-辛烯-1-醇)	5.5					30	---Citronellol(3,7-Dimethyl-6-octen-1-ol)
1689	2905.2230	---芳樟醇	5.5					30	---Linalool
1690	2905.2290	---其他	5.5					30	---Other
1691	2905.2900	--其他	5.5					30	--Other
		-二元醇:							-Diols:
1692	2905.3100	--1,2-乙二醇	5.5					30	--Ethylene glycol(ethanediol)
1693	2905.3200	--丙二醇	5.5					30	--Propylene glycol(propane-1,2-diol)

序号 No.	税则号列 Tariff Item	货品名称	最惠国税率(%) M.F.N.	协定税率(%)Agreement 东盟 ASEAN	曼谷协定 BA	香港 H.K.	澳门 Macao	普通税率(%) Gen.	Article Description
		--其他:							--Other:
1694	2905.3910	---2,5-二甲基已二醇	4					11	---2,5-dimethyl hexandiol
1695	2905.3990	---其他	5.5					30	---Other
		-其他多元醇:							-Other polyhydric alcohols:
1696	2905.4100	--2-乙基-2-(羟甲基)丙烷-1,3-二醇(三羟甲基丙烷)	5.5					30	--2-Ethyl-2-(hydroxymethyl) propane-1,3-diol(trimethylolpropane)
1697	2905.4200	--季戊四醇	5.5					30	--Pentaerythritol
1698	2905.4300	--甘露糖醇	8					30	--Mannitol
1699	2905.4400	--山梨醇	14					40	--D-glucitol(sorbitol)
1700	2905.4500	--丙三醇(甘油)	14					50	--Glycerol
1701	2905.4900	--其他	5.5					30	--Other
		-无环醇的卤化、磺化、硝化或亚硝化衍生物:							-Halogenated, sulphonated, nitrated or nitrosated derivatives of acyclic alcohols:
1702	2905.5100	--乙氯维诺(INN)	5.5					30	--Ethchlorvynol(INN)
1703	2905.5900	--其他	5.5					30	--Other
	29.06	**环醇及其卤化、磺化、硝化或亚硝化衍生物:**							**Cyclic alcohols and their halogenated, sulphonated, nitrated or nitrosated derivatives:**
		-环烷醇、环烯醇及环萜烯醇:							-Cyclanic, cyclenic or cycloterpenic:
1704	2906.1100	--薄荷醇	5					70	--Menthol
1705	2906.1200	--环己醇、甲基环己醇及二甲基环己醇	5.5					30	--Cyclohexanol, methylcyclohexanols and dimethylcyclohexanols
		--固醇及肌醇:							--Sterols and inositols:
1706	2906.1310	---固醇	5.5					30	---Sterol
1707	2906.1320	---肌醇	5.5					30	---Inositol
1708	2906.1400	--萜品醇	5.5					30	--Terpineols
1709	2906.1900	--其他	5.5					30	--Other
		-芳香醇:							-Aromatic:
1710	2906.2100	--苄醇	5					30	--Benzyl alcohol
1711	2906.2900	--其他	5.5					30	--Other
		第三分章 酚、酚醇及其卤化、磺化、硝化或亚硝化衍生物							Ⅲ. PHENOLS, PHENOL-ALCOHOLS, AND THEIR HALOGENATED, SULPHONATED, NITRATED OR NITROSATED DERIVATIVES
	29.07	**酚;酚醇:**							**Phenols; phenol-alcohols:**
		-一元酚:							-Monophenols:
		--苯酚及其盐:							--Phenol (hydroxybenzene) and its salts:
1712	2907.1110	---苯酚	5.5		5.5			30	---Phenol
1713	2907.1190	---其他	5.5		5.5			30	---Other

序号 No.	税则号列 Tariff Item	货品名称	最惠国税率(%) M.F.N.	协定税率(%) Agreement 东盟 ASEAN	曼谷协定 BA	香港 H.K.	澳门 Macao	普通税率(%) Gen.	Article Description
		--甲酚及其盐:							--Cresol and its salts:
		---甲酚:							---Cresol:
1714	2907.1211	----间甲酚	5.5					30	----*m*-Cresol
1715	2907.1212	----邻甲酚	5.5					30	----*o*-Cresol
1716	2907.1219	----其他	5.5					30	----Other
1717	2907.1290	---其他	5.5					30	---Other
		--辛基酚、壬基酚及其异构体以及它们的盐:							--Octylphenol, nonylphenol and their isomers; salts thereof:
1718	2907.1310	---壬基酚	5.5					30	---Nonylphenol
1719	2907.1390	---其他	5.5					30	---Other
1720	2907.1400	--二甲苯酚及其盐	5.5					30	--Xylenols and their salts
		--萘酚及其盐:							--Naphthols and their salts:
1721	2907.1510	---2-萘酚(β-萘酚)	5.5					30	---2-Naphthols(β-naphthol)
1722	2907.1590	---其他	5.5					30	---Other
		--其他:							--Other:
1723	2907.1910	---邻仲丁基酚、邻异丙基酚	4					11	---*o*-Sec-butyl phenol, *o*-isopropyl phenol
1724	2907.1990	---其他	5.5					30	---Other
		-多元酚;酚醇:							-Polyphenols; phenol-alcohols:
1725	2907.2100	--间苯二酚及其盐	5.5					30	--*m*-Dihydroxybenzene(resorcinol) and its salts
		--对苯二酚及其盐:							--*p*-Dihydroxybenzene (hydroquinone) and its salts:
1726	2907.2210	---对苯二酚	5.5					30	---Hydroquinone
1727	2907.2290	---其他	5.5					30	---Other
1728	2907.2300	--4,4′-异亚丙基联苯酚(双酚A,二苯基酚丙烷)及其盐	5.5		5.5			30	--4,4′-Isopropylidenediphenol (bisphenol A, diphenylolpropane) and its salts
		--其他:							--Other:
1729	2907.2910	---邻苯二酚	4					11	---*o*-Dihydroxybenzene(catechol, pyrocatechol)
1730	2907.2990	---其他	5.5					30	---Other
	29.08	**酚及酚醇的卤化、磺化、硝化或亚硝化衍生物:**							**Halogenated, sulphonated, nitrated or nitrosated derivatives of phenols or phenolalcohols:**
		-仅含卤素取代基的衍生物及其盐:							-Derivatives containing only halogen substituents and their salts:
1731	2908.1010	---对氯苯酚	4					11	---*p*-Chlorophenol
1732	2908.1090	---其他	5.5					30	---Other
1733	2908.2000	-仅含磺基的衍生物及其盐和脂	5.5					30	-Derivatives containing only sulpho groups, their salts and esters
		-其他:							-Other:

序号 No.	税则号列 Tariff Item	货品名称	最惠国税率(%) M.F.N.	协定税率(%) Agreement 东盟 ASEAN	曼谷协定 BA	香港 H.K.	澳门 Macao	普通税率(%) Gen.	Article Description
1734	2908.9010	---对硝基酚、对硝基酚钠	5.5					30	---*p*-Nitrophenol, sodium *p*-nitro-phenolate
1735	2908.9090	---其他	5.5					30	---Other
		第四分章 醚、过氧化醇、过氧化醚、过氧化酮、三节环环氧化物、缩醛及半缩醛及其卤化、磺化、硝化或亚硝化衍生物							Ⅳ. ETHERS, ALCOHOL PEROXIDES, ETHER PEROXIDES, KETONE PEROXIDES, EPOXIDES WITH A THREE-MEMBERED RING, ACETALS AND HEMIACETALS, AND THEIR HALOGENATED, SULPHONATED, NITRATED OR NITROSATED DERIVATIVES
	29.09	**醚、醚醇、醚酚、醚醇酚、过氧化醇、过氧化醚、过氧化酮(不论是否已有化学定义)及其卤化、磺化、硝化或亚硝化衍生物:**							**Ethers, ether-alcohols, ether-phenols, ether-alcohol-phenols, alcohol peroxides, ether peroxides, ketone peroxides (whether or not chemically defined), and their halogenated, sulphonated, nitrated or nitrosated derivatives:**
		-无环醚及其卤化、磺化、硝化或亚硝化衍生物:							-Acyclic ethers and their halogenated, sulphonated, nitrated or nitrosated derivatives:
1736	2909.1100	--乙醚	5.5					30	--Diethyl ether
1737	2909.1900	--其他	5.5		5.5			30	--Other
		-环烷醚、环烯醚或环萜烯醚及其卤化、磺化、硝化或亚硝化衍生物:							-Cyclanic, cyclenic or cycloterpenic ethers and their halogenated, sulphonated, nitrated or nitrosated derivatives:
		---环萜烯醚:							---Cycloterpenic ethers:
1738	2909.2081	----1,8-桉树脑	5.5					30	----Eucalgptol
1739	2909.2089	----其他	5.5					30	----Other
1740	2909.2090	---其他	5.5					30	---Other
1741	2909.3000	-芳香醚及其卤化、磺化、硝化或亚硝化衍生物	5.5					30	-Aromatic ethers and their halogenated, sulphonated, nitrated or nitrosated derivatives
		-醚醇及其卤化、磺化、硝化或亚硝化衍生物:							-Ether-alcohols and their halogenated, sulphonated, nitrated or nitrosated derivatives:
1742	2909.4100	--2,2′-氧联二乙醇(二甘醇)	5.5					30	--2, 2′-Oxydiethanol (diethylene glycol, digol)
1743	2909.4200	--乙二醇或二甘醇的单甲醚	5.5					30	--Monomethyl ethers of ethylene glycol or of diethylene glycol
1744	2909.4300	--乙二醇或二甘醇的单丁醚	5.5				0	30	--Monobutyl ethers of ethylene glycol or of diethylene glycol
1745	2909.4400	--乙二醇或二丁醇的其他单烷基醚	5.5					30	--Other monoalkylethers of ethylene glycol or of diethylene glycol

序号 No.	税则号列 Tariff Item	货品名称	最惠国税率(%) M.F.N.	协定税率(%) Agreement 东盟 ASEAN	曼谷协定 BA	香港 H.K.	澳门 Macao	普通税率(%) Gen.	Article Description
		--其他:							--Other:
1746	2909.4910	---间苯氧基苄醇	4					11	---*m*-Phenoxy benzalcohol
1747	2909.4990	---其他	5.5					30	---Other
1748	2909.5000	-醚酚、醚醇酚及其卤化、磺化、硝化或亚硝化衍生物	5.5					30	-Ether-phenols, ether-alcohol-phenols and their halogenated, sulphonated, nitrated or nitrosated derivatives
1749	2909.6000	-过氧化醇、过氧化醚、过氧化酮及其卤化、磺化、硝化或亚硝化衍生物	5.5					30	-Alcohol peroxides, ether peroxides, ketone peroxides and their halogenated, sulphonated, nitrated or nitrosated derivatives
	29.10	**三节环环氧化物、环氧醇、环氧酚、环氧醚及其卤化、磺化、硝化或亚硝化衍生物:**							**Epoxides, epoxyalcohols, epoxyphenols and epoxyethers, with a three-membered ring, and their halogenated, sulphonated, nitrated or nitrosated derivatives:**
1750	2910.1000	-环氧乙烷(氧化乙烯)	5.5					30	-Oxirane(ethylene oxide)
1751	2910.2000	-甲基环氧乙烷(氧化丙烯)	5.5					30	-Methyloxirane(propylene oxide)
1752	2910.3000	-1-氯-2,3-环氧丙烷(表氯醇)	5.5					30	-1-Chloro-2, 3-epoxypropane (epichlorohydrin)
1753	2910.9000	-其他	5.5					30	-Other
	29.11	**缩醛及半缩醛,不论是否含有其他含氧基,及其卤化、磺化、硝化或亚硝化衍生物:**							**Acetals and hemiacetals, whether or not with other oxygen function, and their halogenated, sulphonated, nitrated or nitrosated derivatives:**
1754	2911.0000	缩醛及半缩醛,不论是否含有其他含氧基,及其卤化、磺化、硝化或亚硝化衍生物	5.5					30	Acetals and hemiacetals, whether or not with other oxygen function, and their halogenated, sulphonated, nitrated or nitrosated derivatives
		第五分章 醛基化合物							Ⅴ. ALDEHYDE-FUNCTION COMPO-UNDS
	29.12	**醛,不论是否含有其他含氧基;环聚醛;多聚甲醛:**							**Aldehydes, whether or not with other oxygen function; cyclic polymers of aldehydes; paraformaldehyde:**
		-不含其他含氧基的无环醛:							-Acyclic aldehydes without other oxygen function:
1755	2912.1100	--甲醛	5.5					30	--Methanal(formaldehyde)
1756	2912.1200	--乙醛	5.5					30	--Ethanal(acetaldehyde)
1757	2912.1300	--丁醛	5.5					30	--Butanal (butyraldehyde, normal isomer)
1758	2912.1900	--其他	5.5					30	--Other
		-不含其他含氧基的环醛:							-Cyclic aldehydes without other oxygen function:
1759	2912.2100	--苯甲醛	5.5					30	--Benzaldehyde

序号 No.	税则号列 Tariff Item	货品名称	最惠国税率(%) M.F.N.	协定税率(%)Agreement 东盟 ASEAN	曼谷协定 BA	香港 H.K.	澳门 Macao	普通税率(%) Gen.	Article Description
		--其他：							--Other：
1760	2912.2910	---铃兰醛(对叔丁基-α-甲基-氧化肉桂醛)	5.5					30	---Lilial(p-tert-butyl-a-methyl-oxocinnamaldehyde)
1761	2912.2990	---其他	5.5					30	---Other
1762	2912.3000	-醛醇	5.5					30	-Aldehyde-alcohols
		-醛醚、醛酚及含其他含氧基的醛：							-Aldehyde-ethers, aldehyde-phenols and aldehydes with other oxygen function：
1763	2912.4100	--香草醛(3-甲氧基-4-羟基苯甲醛)	5.5					30	--Vanillin (4-hydroxy-3-methoxybenzaldehyde)
1764	2912.4200	--乙基香草醛(3-乙氧基-4-羟基苯甲醛)	5.5					30	--Ethylvanillin (3-ethoxy-4-hydroxyben-zaldehyde)
1765	2912.4900	--其他	5.5					30	--Other
1766	2912.5000	-环聚醛	5.5					30	-Cyclic polymers of aldehydes
1767	2912.6000	-多聚甲醛	5.5					30	-Paraformaldehyde
	29.13	**税号29.12所列产品的卤化、磺化、硝化或亚硝化衍生物：**							**Halogenated, sulphonated, nitrated or nitrosated derivatives of products of heading No.29.12：**
1768	2913.0000	税号29.12所列产品的卤化、磺化、硝化或亚硝化衍生物	5.5					30	Halogenated, sulphonated, nitrated or nitrosated derivatives of products of heading No.29.12
		第六分章 酮基化合物及醌基化合物							Ⅵ. KETONE-FUNCTION COMPOUNDS AND QUINONE-FUNCTION COMPOUNDS
	29.14	**酮及醌,不论是否含有其他含氧基,及其卤化、磺化、硝化或亚硝化衍生物：**							**Ketones and quinones, whether or not with other oxygen function, and their halogenated, sulphonated, nitrated or nitrosated derivatives：**
		-不含其他含氧基的无环酮：							-Acyclic ketones without other oxygen function：
1769	2914.1100	--丙酮	5.5					20	--Acetone
1770	2914.1200	--丁酮[甲基乙基(甲)酮]	5.5					30	--Butanone(methyl ethyl ketone)
1771	2914.1300	--4-甲基-2-戊酮[甲基异丁基(甲)酮]	5.5					30	--4-Methyl-2-pentanone (isobutyl methyl ketone)
1772	2914.1900	--其他	5.5					30	--Other
		-不含其他含氧基的环烷酮、环烯酮或环萜烯酮：							-Cyclanic, cyclenic or cycloterpenic ketones without other oxygen function：
1773	2914.2100	--樟脑	5.5					40	--Camphor
1774	2914.2200	--环己酮及甲基环己酮	5.5					30	--Cyclohexanone and methylcyclohexanone
1775	2914.2300	--芷香酮及甲基芷香酮	5.5					30	--Ionones and methylionones
1776	2914.2900	--其他	5.5					30	--Other

序号 No.	税则号列 Tariff Item	货品名称	最惠国税率(%) M.F.N.	协定税率(%) Agreement 东盟 ASEAN	曼谷协定 BA	香港 H.K.	澳门 Macao	普通税率(%) Gen.	Article Description
		-不含其他含氧基的芳香酮:							-Aromatic ketones without other oxygen function:
1777	2914.3100	--苯丙酮(苯基丙-2-酮)	5.5					30	--Propiophenone(phenyl propan-2-one)
		--其他:							--Other:
1778	2914.3910	---苯乙酮	4					11	---Acetophenone
1779	2914.3990	---其他	5.5					30	---Other
1780	2914.4000	-酮醇及酮醛	5.5					30	-Ketone-alcohols and ketone-aldehydes
		-酮酚及含有其他含氧基的酮:							-Ketone-phenols and ketones with other oxygen function:
		---酮酚:							---Ketene – alcohols:
1781	2914.5011	----覆盆子酮	5.5					30	----Raspberry ketone
1782	2914.5019	----其他	5.5					30	----Other
1783	2914.5090	---其他	5.5					30	---Other
		-醌:							-Quinones:
1784	2914.6100	--蒽醌	5.5					30	--Anthraquinone
1785	2914.6900	--其他	5.5					30	--Other
1786	2914.7000	-卤化、磺化、硝化或亚硝化衍生物	5.5					30	-Halogenated, sulphonated, nitrated or nitrosated derivatives
		第七分章 羧酸及其酸酐、酰卤化物、过氧化物和过氧酸以及它们的卤化、磺化、硝化或亚硝化衍生物							Ⅶ. CARBOXYLIC ACIDS AND THEIR ANHYDRIDES, HALIDES, PEROXIDES AND PEROXYACIDS AND THEIR HALOGENATED, SULPHONATED, NITRATED OR NITROSATED DERIVATIVES
	29.15	**饱和无环一元羧酸及其酸酐、酰卤化物、过氧化物和过氧酸以及它们的卤化、磺化、硝化或亚硝化衍生物:**							**Saturated acyclic monocarboxylic acids and their anhydrides, halides, peroxides and peroxyacids; their halogenated, sulphonated, nitrated or nitrosated derivatives:**
		-甲酸及其盐和酯:							-Formic acid, its salts and esters:
1787	2915.1100	--甲酸	5.5					40	--Formic acid
1788	2915.1200	--甲酸盐	5.5					30	--Salts of formic acid
1789	2915.1300	--甲酸酯	5.5					30	--Esters of formic acid
		-乙酸及其盐;乙酸酐:							-Acetic acid and its salts; acetic anhydride:
		--乙酸:							--Acetic acid:
1790	2915.2110	---冰乙酸	5.5		5.5			30	---Acetic acid, glacial
1791	2915.2190	---其他	5.5		5.5			50	---Other
1792	2915.2200	--乙酸钠	5.5					50	--Sodium acetate
1793	2915.2300	--钴的乙酸盐	5.5					50	--Cobalt acetates
1794	2915.2400	--乙酸酐	5.5					50	--Acetic anhydride
1795	2915.2900	--其他	5.5					50	--Other
		-乙酸酯:							-Esters of acetic acid:
1796	2915.3100	--乙酸乙酯	5.5					30	--Ethyl acetate
1797	2915.3200	--乙酸乙烯酯	5.5					30	--Vinyl acetate
1798	2915.3300	--乙酸(正)丁酯	5.5					30	--*n*-Butyl acetate

序号 No.	税则号列 Tariff Item	货 品 名 称	最惠国税 率 (%) M.F.N.	协定税率(%)Agreement				普通税率 (%) Gen.	Article Description
				东盟 ASEAN	曼谷协定 BA	香港 H.K.	澳门 Macao		
1799	2915.3400	--乙酸异丁酯	5.5					30	--Isobutyl acetate
1800	2915.3500	--乙酸-2-乙氧基乙酯	5.5					30	--2-Ethoxyethyl acetate
1801	2915.3900	--其他	5.5					30	--Other
1802	2915.4000	-一氯代乙酸、二氯乙酸或三氯乙酸及其盐和酯	5.5					30	-Mono-, di- or trichloroacetic acids, their salts and esters
		-丙酸及其盐和酯:							-Propionic acid, its salts and esters:
1803	2915.5010	---丙酸	5.5					30	---Propionic acid
1804	2915.5090	---其他	5.5					30	---Other
1805	2915.6000	-丁酸、戊酸及其盐和酯	5.5					30	-Butanoic acids, pentanoic acids, their salts and esters
		-棕榈酸、硬脂酸及其盐和酯:							-Palmitic acid, stearic acid, their salts and esters:
1806	2915.7010	---硬脂酸	7					50	---Stearic acid
1807	2915.7090	---其他	5.5					30	---Other
1808	2915.9000	-其他	5.5					30	-Other
	29.16	**不饱和无环一元羧酸、环一元羧酸及其酸酐、酰卤化物、过氧化物和过氧酸以及它们的卤化、磺化、硝化或亚硝化衍生物:**							**Unsaturated acyclic monocarboxylic acids, cyclic monocarboxylic acids, their anhydrides, halides, peroxides and peroxyacids; their halogenated, sulphonated, nitrated or nitrosated derivatives:**
		-不饱和无环一元羧酸及其酸酐、酰卤化物、过氧化物和过氧酸以及它们的衍生物:							-Unsaturated acyclic monocarboxylic acids, their anhydrides, halides, peroxides, peroxyacids and their derivatives:
1809	2916.1100	--丙烯酸及其盐	6.5					30	--Acrylic acid and its salts
1810	2916.1200	--丙烯酸酯	6.5		6.5			30	--Esters of acrylic acid
1811	2916.1300	--甲基丙烯酸及其盐	6.5					80	--Methacrylic acid and its salts
1812	2916.1400	--甲基丙烯酸酯	6.5					80	--Esters of methacrylic acid
1813	2916.1500	--油酸、亚油酸或亚麻酸及其盐和酯	6.5					30	--Oleic, linoleic or linolenic acids, their salts and esters
1814	2916.1900	--其他	6.5					30	--Other
		-环烷一元羧酸、环烯一元羧酸或环萜烯一元羧酸及其酸酐、酰卤化物、过氧化物和过氧酸以及它们的衍生物:							-Cyclanic, cyclenic or cycloterpenic monocarboxylic acids, their anhydrides, halides, peroxides, peroxyacids and their derivatives:
1815	2916.2010	---二溴菊酸、DV 菊酸甲酯	4					11	---Dibromochrysanthemicacid, DVchrysanthemimono carboxylate
1816	2916.2090	---其他	6.5					30	---Other
		-芳香一元羧酸及其酸酐、酰卤化物、过氧化物和过氧酸以及它们的衍生物:							-Aromatic monocarboxylic acids, their anhydrides, halides, peroxides, peroxyacids and their derivatives:
1817	2916.3100	--苯甲酸及其盐和酯	6.5					30	--Benzoic acid, its salts and esters

序号 No.	税则号列 Tariff Item	货品名称	最惠国税率(%) M.F.N.	协定税率(%) Agreement 东盟 ASEAN	曼谷协定 BA	香港 H.K.	澳门 Macao	普通税率(%) Gen.	Article Description
1818	2916.3200	--过氧化苯甲酰及苯甲酰氯	6.5					30	--Benzoyl peroxide and benzoyl chloride
1819	2916.3400	--苯乙酸及其盐	6.5					30	--Phenylacetic acid and its salts
1820	2916.3500	--苯乙酸酯	6.5					30	--Esters of phenylacetic acid
		--其他:							--Other:
1821	2916.3910	---邻甲基苯甲酸	6.5					30	---*m* -Methylbenzoic acid
1822	2916.3920	---布洛芬	6.5					30	---Brufen(Ibuprofen)
1823	2916.3990	---其他	6.5					30	---Other
	29.17	**多元羧酸及其酸酐、酰卤化物、过氧化物和过氧酸以及它们的卤化、磺化、硝化或亚硝化衍生物:**							**Polycarboxylic acids, their anhydrides, halides, peroxides and peroxyacids; their halogenated, sulphonated, nitrated or nitrosated derivatives:**
		-无环多元羧酸及其酸酐、酰卤化物、过氧化物和过氧酸以及它们的衍生物:							-Acyclic polycarboxylic acids, their anhydrides, halides, peroxides, peroxyacids and their derivatives:
		--草酸及其盐和酯:							--Oxalic acid, its salts and esters:
1824	2917.1110	---草酸	6.5					40	---Oxalic acid
1825	2917.1120	---草酸钴	9					30	---Cobalt oxalate
1826	2917.1190	---其他	6.5					30	---Other
1827	2917.1200	--己二酸及其盐和酯	6.5					30	--Adipic acid, its salts and esters
		--壬二酸、癸二酸及其盐和酯:							--Azelaic acid, sebacic acid, their salts and esters:
1828	2917.1310	---癸二酸及其盐和酯	6.5					30	---Sebacic acid, its salts and esters
1829	2917.1390	---其他	6.5					30	---Other
1830	2917.1400	--马来酐	6.5					30	--Maleic anhydride
1831	2917.1900	--其他	6.5					30	--Other
		-环烷多元羧酸、环烯多元羧酸、环萜烯多元羧酸及其酸酐、酰卤化物、过氧化物和过氧酸以及它们的衍生物:							-Cyclanic, cyclenic or cycloterpenic polycarboxylic acids, their anhydrides, halides, peroxides, peroxyacids and their derivatives:
1832	2917.2010	---四氢苯酐	4					11	---Tetrahydro benzoic anhydride
1833	2917.2090	---其他	6.5					30	---Other
		-芳香多元羧酸及其酸酐、酰卤化物、过氧化物和过氧酸以及它们的衍生物:							-Aromatic polycarboxylic acids, their anhydrides, halides, peroxides, peroxyacids and their derivatives:
1834	2917.3100	--邻苯二甲酸二丁酯	6.5					30	--Dibutyl orthophthalates
1835	2917.3200	--邻苯二甲酸二辛酯	6.5		6.5			30	--Dioctyl orthophthalates
1836	2917.3300	--邻苯二甲酸二壬酯及邻苯二甲酸二癸酯	6.5					30	--Dinonyl or didecyl orthophthalates
1837	2917.3400	--其他邻苯二甲酸酯	6.5					30	--Other esters of orthophthalic acid
1838	2917.3500	--邻苯二甲酸酐	6.5		6.5			30	--Phthalic anhydride
		--对苯二甲酸及其盐:							--Terephthalic acid and its salts:
1839	2917.3610	---对苯二甲酸	9.7		9.1			30	---Terephthalic acid

序号 No.	税则号列 Tariff Item	货品名称	最惠国税率(%) M.F.N.	协定税率(%)Agreement 东盟 ASEAN	曼谷协定 BA	香港 H.K.	澳门 Macao	普通税率(%) Gen.	Article Description
1840	2917.3690	---其他	6.5		6.5			30	---Other
1841	2917.3700	--对苯二甲酸二甲酯	6.5					30	--Dimethyl terephthalate
1842	2917.3900	--其他	6.5					30	--Other
	29.18	**含附加含氧基的羧酸及其酸酐、酰卤化物、过氧化物和过氧酸以及它们的卤化、磺化、硝化或亚硝化衍生物：**							**Carboxylic acids with additional oxygen function and their anhydrides, halides, peroxides and peroxyacids; their halogenated, sulphonated, nitrated or nitrosated derivatives:**
		-含醇基但不含其他含氧基的羧酸及其酸酐、酰卤化物、过氧化物和过氧酸以及它们的衍生物：							-Carboxylic acids with alcohol function but without other oxygen function, their anhydrides, halides, peroxides, peroxyacids and their derivatives:
1843	2918.1100	--乳酸及其盐和酯	6.5					30	--Lactic acid, its salts and esters
1844	2918.1200	--酒石酸	6.5					35	--Tartaric acid
1845	2918.1300	--酒石酸盐及酒石酸酯	6.5					30	--Salts and esters of tartaric acid
1846	2918.1400	--柠檬酸	6.5					35	--Citric acid
1847	2918.1500	--柠檬酸盐及柠檬酸酯	6.5					30	--Salts and esters of citric acid
1848	2918.1600	--葡糖酸及其盐和酯	6.5					30	--Gluconic acid, its salts and esters
		--其他：							--Other:
1849	2918.1910	---2,2-二苯基-2-羟基乙酸(二苯羟乙酸；二苯乙醇酸)	6.5					30	---2, 2-Diphenyl-2-hydroxyacetic acid (ahydroxydiphenylacetic acid, benzilic acid)
1850	2918.1990	---其他	6.5					30	---Other
		-含酚基但不含其他含氧基的羧酸及其酸酐、酰卤化物、过氧化物和过氧酸以及它们的衍生物：							-Carboxylic acids with phenol function but without other oxygen function, their anhydrides, halides, peroxides, peroxyacids and their derivatives:
		--水杨酸及其盐：							--Salicylic acid and its salts:
1851	2918.2110	---水杨酸、水杨酸钠	6.5					20	---Salicylic acid and sodium salicylate
1852	2918.2190	---其他	6.5					30	---Other
		--邻乙酰水杨酸及其盐和酯：							--*o* -Acetylsalicylic acid, its salts and esters:
1853	2918.2210	---邻乙酰水杨酸(阿斯匹林)	6					20	---Acetylsalicylic acid
1854	2918.2290	---其他	6.5					30	---Other
1855	2918.2300	--水杨酸的其他酯及其盐	6.5					30	--Other esters of salicylic acid and their salts
1856	2918.2900	--其他	6.5					30	--Other
1857	2918.3000	-含醛基或酮基但不含其他含氧基的羧酸及其酸酐、酰卤化物、过氧化物和过氧酸以及它们的衍生物	6.5					30	-Carboxylic acids with aldehyde or ketone function but without other oxygen function, their anhydrides, halides, peroxides, peroxyacids and their derivatives

序号 No.	税则号列 Tariff Item	货品名称	最惠国税率(%) M.F.N.	协定税率(%) Agreement 东盟 ASEAN	曼谷协定 BA	香港 H.K.	澳门 Macao	普通税率(%) Gen.	Article Description
1858	2918.9000	-其他	6.5					30	-Other
		第八分章　非金属无机酸酯及其盐以及它们的卤化、磺化、硝化或亚硝化衍生物							Ⅷ. ESTERS OF INORGANIC ACIDS OF NON-METALS AND THEIR SALTS, AND THEIR HALOGENATED, SULPHONATED, NITRATED OR NITROSATED DERIVATIVES
	29.19	**磷酸酯及其盐，包括乳磷酸盐，以及它们的卤化、磺化、硝化或亚硝化衍生物：**							**Phosphoric esters and their salts, including lactophosphates; their halogenated, sulphonated, nitrated or nitrosated derivatives:**
1859	2919.0000	磷酸酯及其盐，包括乳磷酸盐，以及它们的卤化、磺化、硝化或亚硝化衍生物	6.5					30	Phosphoric esters and their salts, including lactophosphates; their halogenated, sulphonated , nitrated or nitrosated derivatives
	29.20	**其他非金属无机酸酯**(不包括卤化氢的酯)**及其盐以及它们的卤化、磺化、硝化或亚硝化衍生物：**							**Esters of other inorganic acids of non-metals** (excluding esters of hydrogen halides) **and their salts; their halogenated, sulphonated, nitrated or nitrosated derivatives:**
1860	2920.1000	-硫代磷酸酯及其盐以及它们的卤化、磺化、硝化或亚硝化衍生物	6.5					30	-Thiophosphoric esters (phosphorothioates) and their salts; their halogenated, sulphonated, nitrated or nitrosated derivatives
		-其他：							-Other:
		---亚磷酸酯：							---Phosphites
1861	2920.9011	----亚磷酸三甲酯	6.5					30	----Trimethyl phosphite
1862	2920.9012	----亚磷酸三乙酯	6.5					30	----Trimethyl phosphite
1863	2920.9013	----亚磷酸二甲酯	6.5					30	----Dimethyl phosphite
1864	2920.9014	----亚磷酸二乙酯	6.5					30	----Diethyl phosphite
1865	2920.9019	----其他	6.5					30	----Other
1866	2920.9090	---其他	6.5					30	---Other
		第九分章　含氮基化合物							Ⅸ. NITROGEN-FUNCTION COMPOUNDS
	29.21	**氨基化合物：**							**Amine-function compounds:**
		-无环单胺及其衍生物以及它们的盐：							-Acyclic monoamines and their derivatives; salts thereof:
1867	2921.1100	--甲胺、二甲胺或三甲胺及其盐	6.5					30	--Methylamine, di-or trimethylamine and their salts
1868	2921.1200	--二乙胺及其盐	6.5					30	--Diethylamine and its salts
		--其他：							--Other:
1869	2921.1910	---二正丙胺	4					11	---Di-*n*-propylamine
1870	2921.1920	---异丙胺	6.5					30	---Isopropyl amine
1871	2921.1930	---N，N一二(2-氯乙基)乙胺	6.5					30	---N,N-Bis(2-chloroethyl)ethylamine

序号 No.	税则号列 Tariff Item	货品名称	最惠国税率(%) M.F.N.	协定税率(%)Agreement 东盟 ASEAN	曼谷协定 BA	香港 H.K.	澳门 Macao	普通税率(%) Gen.	Article Description
1872	2921.1940	---N,N一二(2-氯乙基)甲胺	6.5					30	---N, N-Bis (2-chloroethyl) methylamine
1873	2921.1950	---三(2-氯乙基)胺	6.5					30	---Tri-(2-chloroethyl)amine
1874	2921.1960	---二烷(甲、乙、正丙或异丙)氨基乙基-2-氯及其质子化盐	6.5					30	---N,N-Dialkyl(Me,Et,n-Pr or i-Pr) aminoethyl-2-chlorides and corresponking protonated salts
1875	2921.1990	---其他	6.5					30	---Other
		-无环多胺及其衍生物以及它们的盐:							-Acyclic polyamines and their derivatives;salts thereof:
		--乙二胺及其盐:							--Ethylenediamine and its salts:
1876	2921.2110	---乙二胺	6.5					30	---Ethylenediamine
1877	2921.2190	---其他	6.5					30	---Other
		--六亚甲基二胺及其盐:							--Hexamethylenediamine and its salts:
1878	2921.2210	---己二酸己二胺盐(尼龙66盐)	6.5					20	---Nylon-66 salt
1879	2921.2290	---其他	6.5					30	---Other
1880	2921.2900	--其他	6.5					30	--Other
1881	2921.3000	-环烷单胺或多胺、环烯单胺或多胺、环萜烯单胺或多胺及其衍生物以及它们的盐	6.5					30	-Cyclanic, cyclenic or cycloterpenic monoor polyamines, and their derivatives;salts thereof
		-芳香单胺及其衍生物以及它们的盐:							-Aromatic monoamines and their derivatives;salts thereof:
		--苯胺及其盐:							--Aniline and its salts:
1882	2921.4110	---苯胺	6.5					20	---Aniline
1883	2921.4190	---其他	6.5					30	---Other
1884	2921.4200	--苯胺衍生物及其盐	6.5					30	--Aniline derivatives and their salts
1885	2921.4300	--甲苯胺及其衍生物以及它们的盐	6.5					30	--Toluidines and their derivatives;salts thereof
1886	2921.4400	--二苯胺及其衍生物以及它们的盐	6.5					30	--Diphenylamine and its derivatives; salts thereof
1887	2921.4500	--1-萘胺(α-萘胺)、2-萘胺(β-萘胺)及其衍生物以及它们的盐	6.5					30	--1-Naphthylamine (α-naphthylamine),2-naphthylamine(β-naphthylamine) and their derivatives; salts thereof
1888	2921.4600	--安非他明(INN)、苄非他明(INN)、右苯丙胺(INN)、乙非他明、芬坎法明(INN)、利非他明、左苯丙胺(INN)、美芬雷司(INN)、苯丁胺(INN)以及它们的盐	6.5					30	--Amfetamine (INN), benzfetamine (INN), dexamfetamine(INN), etilamfetamine (INN), fencamfamin (INN), lefetamine(INN), levamfetamine(INN), mefenorex(INN) and phentermine(INN); salts thereof

序号 No.	税则号列 Tariff Item	货品名称	最惠国税率(%) M.F.N.	协定税率(%)Agreement 东盟 ASEAN	曼谷协定 BA	香港 H.K.	澳门 Macao	普通税率(%) Gen.	Article Description
		--其他:							--Other:
1889	2921.4910	---对异丙基苯胺	4					11	---*p*-Isopropyl-aniline
1890	2921.4920	---二甲基苯胺	6.5					20	---Dimethylanilines
1891	2921.4930	---2,6-甲基乙基苯胺	4					11	---2,6-Methyl ethyl aniline
1892	2921.4940	---2,6-二乙基苯胺	6.5					20	---2,6-Diethylaniline
1893	2921.4990	---其他	6.5					30	---Other
		-芳香多胺及其衍生物以及它们的盐:							-Aromatic polyamines and their derivatives;salts thereof:
		--邻-、间-、对-苯二胺、二氨基甲苯及其衍生物以及它们的盐:							--*o*-, *m*-, *p*-Phenylenediamine, diaminotoluenes,and their derivatives; salts thereof:
1894	2921.5110	---邻苯二胺	4					11	---*o*-Phenylenediamine
1895	2921.5190	---其他	6.5					30	---Other
1896	2921.5900	--其他	6.5					30	--Other
	29.22	**含氧基氨基化合物:**							**Oxygen-function amino-compounds:**
		-氨基醇(但含有一种以上含氧基的除外)及其醚和酯,以及它们的盐:							-Amino-alcohols,other than those containing more than one kind of oxygen function,their ethers and esters;salts thereof:
1897	2922.1100	--单乙醇胺及其盐	6.5					30	--Monoethanolamine and its salts
1898	2922.1200	--二乙醇胺及其盐	6.5					30	--Diethanolamine and its salts
		--三乙醇胺及其盐:							--Triethanolamine and its salts:
1899	2922.1310	---三乙醇胺	6.5					30	---Triethanolamine
1900	2922.1320	---三乙醇胺盐	6.5					30	---Salts of triethanolamine
1901	2922.1400	--右丙氧吩(INN)及其盐	6.5					30	--Dextropropoxyphene (INN) and its saltsZ
		--其他:							--Other:
1902	2922.1910	---乙胺丁醇	6.5					30	---Ethylamino butanol(Ethambutol)
		---二烷(甲、乙、正丙或异丙)氨基乙-2-醇及其质子化盐:							---N,N-Dialkyl-(Me,Et,n-Pr or i-Pr) aminoethane-2-ols and cornesponding protonated salts:
1903	2922.1921	----二甲氨基乙醇及其质子化盐	6.5					30	----N, N-Dimethylaminoethanol and corresponding protonated salts
1904	2922.1922	----二乙氨基乙醇及其质子化盐	6.5					30	----N,N-Diethylaminoethanol and corresponding protonated salts
1905	2922.1929	----其他	6.5					30	----Other
1906	2922.1930	---乙基二乙醇胺	6.5					30	---Ethyldiethanolamine
1907	2922.1940	---甲基二乙醇胺	6.5					30	---Methyldiethanolamine
1908	2922.1990	---其他	6.5					30	---Other
		-氨基萘酚和其他氨基酚(但含有一种以上含氧基的除外)及其醚和酯,以及它们的盐:							-Amino-naphthols and amino-phenols, other than those containing more than one kind of oxygen function, their ethers and esters;salts thereof:
1909	2922.2100	--氨基羟基萘磺酸及其盐	6.5					30	--Aminohydroxynaphthalenesulphonic acid and their salts
1910	2922.2200	--茴香胺、二茴香胺、氨基苯乙醚及其盐	6.5					30	--Anisidine, dianisidine, phenetidine and their salts

序号 No.	税则号列 Tariff Item	货品名称	最惠国税率(%) M.F.N.	协定税率(%)Agreement 东盟 ASEAN	曼谷协定 BA	香港 H.K.	澳门 Macao	普通税率(%) Gen.	Article Description
1911	2922.2900	--其他	6.5					30	--Other
		-氨基醛、氨基酮和氨基醌,但含有一种以上含氧基的除外,以及它们的盐:							-Amino-aldehydes, amino-ketones and amino-quinones, other than those containing more than one kind of oxygen function; salts thereof:
1912	2922.3100	--安非拉酮、美沙酮和去甲美沙酮以及它们的盐	6.5					30	--Amfepramone (INN), methadone (INN) and normethadone (INN); salts thereof
1913	2922.3900	--其他	6.5					30	--Other
		-氨基酸及其酯,但含有一种以上含氧基的除外,以及它们的盐:							-Amino-acids, other than those containing more than one kind of oxygen function, and their esters; salts thereof:
		--赖氨酸及其酯以及它们的盐:							--Lysine and its esters; salts thereof:
1914	2922.4110	---赖氨酸	5					20	---Lysine
1915	2922.4190	---其他	6					30	---Other
		--谷氨酸及其盐:							--Glutamic acid and its salts:
1916	2922.4210	---谷氨酸	10		8.6			90	---Glutamic acid
1917	2922.4220	---谷氨酸钠	10			0		130	---Sodium glutamate
1918	2922.4290	---其他	6.5					30	---Other
		--氨基酸(但含有一种以上含氧基的除外)及其酯,以及它们的盐:							--Anthranilic acid and its salts:
1919	2922.4310	---邻氨基苯甲酸(氨茴酸)	6.5					20	---Anthranilic acid
1920	2922.4390	---其他	6.5					30	---Other
1921	2922.4400	--替利定(INN)及其盐	6.5					30	--Tilidine(INN)and its salts
		--其他:							--Other:
1922	2922.4910	---其他氨基酸	6.5					20	---Amino acids
		---其他:							---Other:
1923	2922.4991	----普鲁卡因	6					20	----Procaine
1924	2922.4999	----其他	6.5					30	----Other
1925	2922.5000	-氨基醇酚、氨基酸酚及其他含氧基氨基化合物	6.5					30	-Amino-alcohol-phenols, amino-acid-phenols and other amino-compounds with oxygen function
	29.23	**季铵盐及季铵碱;卵磷脂及其他磷氨基类脂,无论是否已有化学定义:**							**Quaternary ammonium salts and hydroxides; lecithins and other phosphoaminolipids, whether or not chemically defined:**
1926	2923.1000	-胆碱及其盐	6.5					30	-Choline and its salts
1927	2923.2000	-卵磷脂及其他磷氨基类脂	6.5					30	-Lecithins and other phosphoaminolipids
1928	2923.9000	-其他	6.5					30	-Other

序号 No.	税则号列 Tariff Item	货 品 名 称	最惠国税率(%) M.F.N.	协定税率(%) Agreement 东盟 ASEAN	曼谷协定 BA	香港 H.K.	澳门 Macao	普通税率(%) Gen.	Article Description
	29.24	**羧基酰胺基化合物;碳酸酰胺基化合物:**							**Carboxyamide-function compounds; amidefunction compounds of carbonic acid:**
		-无环酰胺(包括无环氨基甲酸酯)及其衍生物以及它们的盐:							-Acyclic amides (including acyclic carbamates) and their derivatives; salts thereof:
1929	2924.1100	--甲丙氨酯(INN)	6.5		6.5			30	--Meprobamate(INN)
		--其他(INN):							--Other:
1930	2924.1910	---二甲基甲酰胺	6.5		6.5			30	---N,N-dimethylformamide
1931	2924.1990	---其他	6.5		6.5			30	---Other
		-环酰胺(包括环氨基甲酸酯)及其衍生物以及它们的盐:							-Cyclic amides(including cyclic carbamates) and their derivatives; salts thereof:
1932	2924.2100	--酰脲及其衍生物以及它们的盐	6.5					30	--Ureides and their derivatives; salts thereof
1933	2924.2300	--2-乙酰氨基苯甲酸(N-乙酰邻氨基苯甲酸)及其盐	6.5					30	--2-Acetamidobenzoic acid (N-acetylanthranilic acid)and its salts
1934	2924.2400	--炔已蚁胺(INN)	6.5					30	--Ethinamate(INN)
		--其他:							--Other:
1935	2924.2910	---对乙酰氨基苯乙醚(非那西丁)	6					20	---Phenacetin
1936	2924.2920	---对乙酰氨基酚(扑热息痛)	6					30	---*p*-Acetaminophenol(paracetanol)
1937	2924.2990	---其他	6.5					30	---Other
	29.25	**羧基酰亚胺化合物(包括糖精及其盐)及亚胺基化合物:**							**Carboxyimide-function compounds** (including saccharin and its salts) **and imine-function compounds:**
		-酰亚胺及其衍生物以及它们的盐:							-Imides and their derivatives; salts thereof:
1938	2925.1100	--糖精及其盐	9					90	--Saccharin and its salts
1939	2925.1200	--格鲁米特(INN)	6.5					30	--Glutethimide(INN)
1940	2925.1900	--其他	6.5					30	--Other
1941	2925.2000	-亚胺及其衍生物以及它们的盐	6.5					30	-Imines and their derivatives; salts thereof
	29.26	**腈基化合物:**							**Nitrile-function compounds:**
1942	2926.1000	-丙烯腈	6.5					30	-Acrylonitrile
1943	2926.2000	-1-氰基胍(双氰胺)	6.5					30	-1-cyanoguanidine(dicyandiamide)
1944	2926.3000	-芬普雷司(INN)及其盐;美沙酮中间体(4-氰基-2-二甲氨基-4,4-二苯基丁烷)	6.5					30	-Fenproporex (INN) and its salts; methadone (INN) intermediate (4-cyano-2-dimethylamino-4, 4-diphenylbutane)
		-其他:							-Other:
1945	2926.9010	---对氯氰苄	4					11	---*p*-Chlorobenzyl cyanide

序号 No.	税则号列 Tariff Item	货品名称	最惠国税率(%) M.F.N.	协定税率(%) Agreement 东盟 ASEAN	曼谷协定 BA	香港 H.K.	澳门 Macao	普通税率(%) Gen.	Article Description
1946	2926.9020	---间苯二甲腈	6.5					30	---*m*-Phthalonitrile
1947	2926.9090	---其他	6.5					30	---Other
	29.27	**重氮化合物、偶氮化合物及氧化偶氮化合物:**							**Diazo-,azo-.or azoxy-compounds:**
1948	2927.0000	重氮化合物、偶氮化合物及氧化偶氮化合物	6.5				0△	30	Diazo-,azo-.or azoxy-compounds
	29.28	**肼(联氨)及胲(羟胺)的有机衍生物:**							**Organic derivatives of hydrazine or of hydroxylamine:**
1949	2928.0000	肼(联氨)及胲(羟胺)的有机衍生物	6.5					20	Organic derivatives of hydrazine or of hydroxylamine
	29.29	**其他含氮基化合物:**							**Compounds with other nitrogen function:**
1950	2929.1010	---2,4-和2,6-甲苯二异氰酸酯混合物(甲苯二异氰酸酯TDI)	6.5		6.5			30	---Toluene diisocyanate
1951	2929.1020	---二甲苯二异氰酸酯(TODI)	6.5		6.5			30	---*o*-Xylene diisocyanate
1952	2929.1030	---二苯基甲烷二异氰酸酯(纯MDI)	6.5		6.5			30	---Diphenylmethane diisocyanate
1953	2929.1040	---六亚甲基二异氰酸酯	6.5		6.5			30	---Hexamethelene diisocyanate
1954	2929.1090	---其他	6.5		6.5			30	---Other
		-其他:							-Other:
1955	2929.9010	---环已基氨基磺酸钠(甜蜜素)	9					90	---Sodium cyclamate
1956	2929.9020	---二烷(甲、乙、正丙或异丙)氨基膦酰二卤	6.5					30	---N,N-Dialkyl(Me,Et,n-Pr or i-Pr) phosphoramidic dihalides
1957	2929.9030	---二烷(甲、乙、正丙或异丙)氨基膦酸二烷(甲、乙、正丙或异丙)酯	6.5					30	---Dialkyl(Me,Et,n-Pr or i-Pr)N,N-dialkyl(Me,Et,n-Pr or i-Pr)-phosphoramidates
1958	2929.9090	---其他	6.5					30	---Other
		第十分章　有机—无机化合物、杂环化合物、核酸及其盐以及磺(酰)胺							X. ORGANO-INORGANIC COMPOU-NDS, HETEROCYCLIC COMPOUNDS, NUCLEIC ACIDS AND THEIR SALTS, AND SULPHONAMIDES
	29.30	**有机硫化合物:**							**Organo-sulphur compounds:**
1959	2930.1000	-二硫代碳酸酯(或盐)〔黄原酸酯(或盐)〕	6.5					30	-Dithiocarbonates(xanthates)
1960	2930.2000	-硫代氨基甲酸盐(或酯)及二硫代氨基甲酸盐	6.5					30	-Thiocarbamates and dithiocarbamates

序号 No.	税则号列 Tariff Item	货品名称	最惠国税率(%) M.F.N.	协定税率(%)Agreement 东盟 ASEAN	曼谷协定 BA	香港 H.K.	澳门 Macao	普通税率(%) Gen.	Article Description
1961	2930.3000	-一硫化二烃氨基硫羰、二硫化二烃氨基硫羰及四硫化二烃氨基硫羰	6.5					30	-Thiuram mono-,di or tetrasulphide
1962	2930.4000	-甲硫氨酸(蛋氨酸)	6.5					30	-Methionine
		-其他：							-Other：
1963	2930.9010	---双巯丙氨酸(胱氨酸)	6.5					30	---Cystine
1964	2930.9090	---其他	6.5					30	---Other
	29.31	**其他有机—无机化合物：**							**Other organo-inorganic compounds：**
1965	2931.0000	其他有机—无机化合物	6.5					30	Other organo-inorganic compounds
	29.32	**仅含有氧杂原子的杂环化合物：**							**Heterocyclic compounds with oxygen hetero-atom(s)only：**
		-结构上含有一个非稠合呋喃环(不论是否氢化)的化合物：							-Compounds containing an unfused furan ring(whether or not hydrogenated)in the structure：
1966	2932.1100	--四氢呋喃	6					20	--Tetrahydrofuran
1967	2932.1200	--2-糠醛	6					20	--2-Furaldehyde(furfuraldehyde)
1968	2932.1300	--糠醇及四氢糠醇	6					20	--Furfuryl alcohol and tetrahydrofurfuryl alcohol
1969	2932.1900	--其他	6.5					20	--Other
		-内酯：							-Lactones：
1970	2932.2100	--香豆素、甲基香豆素及乙基香豆素	6.5					20	--Coumarin, methylcoumarins and ethylcoumarins
1971	2932.2900	--其他内酯	6.5					20	--Other lactones
		-其他：							-Other：
1972	2932.9100	--4-丙烯基-1,2-亚甲二氧基苯(异黄樟脑)	6.5					20	--Isosafrole
1973	2932.9200	--1-(1,3-苯并二噁茂-5-基)丙烷-2-酮	6.5					20	--1-(1, 3-Benzodioxol-5-yl) propan-2-one
1974	2932.9300	--3,4-亚甲二氧基苯甲醛(胡椒醛)	6.5					20	--Piperonal
1975	2932.9400	--4-烯丙基-1,2-亚甲二氧基苯(黄樟脑)	6.5					20	--Safrole
1976	2932.9500	--四氢大麻酚(所有的异构体)	6.5					20	--Tetrahydrocannabinols(all is omers)
		--其他：							--Other：
1977	2932.9910	---7-羟基苯并呋喃(呋喃酚)	4					11	---Furan phenol
1978	2932.9920	---2,2′-双甲氧羰基-4,4′-双甲氧基-5,6,5′,6′-双亚甲二氧基联苯(联苯双酯)	6.5					20	---Bifendate
1979	2932.9990	---其他	6.5					20	---Other

序号 No.	税则号列 Tariff Item	货品名称	最惠国税率(%) M.F.N.	协定税率(%) Agreement 东盟 ASEAN	曼谷协定 BA	香港 H.K.	澳门 Macao	普通税率(%) Gen.	Article Description
	29.33	**仅含有氮杂原子的杂环化合物:**							**Heterocyclic compounds with nitrogen hetero-atom(s) only:**
		-结构上含有一个非稠合吡唑环(不论是否氢化)的化合物:							-Compounds containing an unfused pyrazole ring(whether or not hydrogenated)in the structure:
1980	2933.1100	--二甲基苯基吡唑酮(安替比林)及其衍生物	6.5		6			20	--Phenazone(antipyrin)and its derivatives
		--其他:							--Other:
1981	2933.1920	---安乃近	6					20	---Analgin
1982	2933.1990	---其他	6.5					20	---Other
		-结构上含有一个非稠合咪唑环(不论是否氢化)的化合物:							-Compounds containing an unfused imidazole ring(whether or not hydrogenated)in the structure:
1983	2933.2100	--乙内酰脲及其衍生物	6.5					30	--Hydantoin and its derivatives
1984	2933.2900	--其他	6.5					20	--Other
		-结构上含有一个非稠合吡啶环(不论是否氢化)的化合物:							-Compounds containing an unfused pyridine ring(whether or not hydrogenated)in the structure:
1985	2933.3100	--吡啶及其盐	6					20	--Pyridine and its salts
		--六氢吡啶(哌啶)及其盐:							--Piperidine and its salts:
1986	2933.3210	---六氢吡啶(哌啶)	4					11	---Hexahydropyridine(piperidine)
1987	2933.3220	---六氢吡啶(哌啶)盐	6.5					20	---Isoniazidum
1988	2933.3300	--阿芬太尼(INN),阿尼利定(INN),苯氰米特(INN),溴西泮(INN),地芬诺新(INN),地芬诺酯(INN),地匹哌酮(INN),芬太尼(INN),凯托米酮(INN),哌醋甲酯(INN),喷他左辛(INN),哌替啶(INN),哌替啶中间体A(INN),苯环利定(INN),苯哌利定(INN),哌苯甲醇(INN),哌氰米特(INN),哌丙吡胺(INN)和三甲利定(INN)以及它们的盐	6.5					20	--Alfentanil(INN),anileridine(INN),bezitramide(INN),bromazepam(INN),difenoxin(INN),diphenoxylate(INN),dipipanone(INN),fentanyl(INN),ketobemidone(INN),methylphenidate(INN),pentazocine(INN),pethidine(INN),pethidine(INN),intermediate A,phencyclidine(INN)(PCP),phenoperidine(INN),pipradrol(INN),piritramide(INN),propiram(INN) and trimeperidine(INN);salts thereof
		--其他:							--Other:

序号 No.	税则号列 Tariff Item	货品名称	最惠国税率(%) M.F.N.	协定税率(%) Agreement 东盟 ASEAN	曼谷协定 BA	香港 H.K.	澳门 Macao	普通税率(%) Gen.	Article Description
1989	2933.3910	---二苯乙醇酸-3-奎宁环脂	6.5					20	---Benzilic acid-3-quinuclidinate
1990	2933.3920	---奎宁环-3-醇	6.5					20	---Quinuclidine-3-ol
1991	2933.3990	---其他	6.5					20	---Other
		-结构上含有一个喹啉或异喹啉环系(不论是否氢化)的化合物,但未经进一步稠合的:							-Compounds containing a quinoline or isoquinoline ring-system (whether or not hydrogenated), not further fused:
1992	2933.4100	--左非诺(INN)及其盐	6.5					20	--Levorpharol(INN)and its sacts
		--其他:							--Other:
1993	2933.4910	---环丙氟哌酸	6.5		6			20	---Ciprofloxacin
1994	2933.4990	---其他	6.5					20	---Other
		-结构上含有一个嘧啶环(不论是否氢化)或哌嗪环的化合物:							-Compounds containing a pyrimidine ring(whether or not hydrogenated)or piperazine ring in the structure:
1995	2933.5200	--丙二酰脲(巴比土酸)及其盐	6.5					20	--Malonylurea (barbituric acid) and its salts
1996	2933.5300	--阿洛巴比妥(INN),异戊巴比妥(INN),巴比妥(INN),布他比妥(INN),正丁巴比妥(INN),环己巴比妥(INN),甲苯巴比妥(INN),戊巴比妥(INN),苯巴比妥(INN),仲丁巴比妥(INN),司可巴比妥(INN),和乙烯比妥(INN)以及它们的盐	6.5					20	--Allobarbital (INN), amobarbital (INN), barbital (INN), butalbital (INN), butobarbital, cyclobarbital (INN), methylphenobarbital(INN), pentobarbital (INN), phenobarbital (INN), secbutabarbital(INN), secobarbital(INN) and vinylbital(INN); salts thereof
1997	2933.5400	--其他丙二酰脲(巴比土酸)的衍生物以及它们的盐	6.5					20	--Other derivatives of malonylurea (barbituric acid); salts thereof
1998	2933.5500	--氯普唑仑(INN),甲氯喹酮(INN),甲喹酮(INN)和齐培丙醇(INN)以及它们的盐	6.5					20	--Loprazolam (INN), mecloqualone (INN), methaqualone (INN) and zipeprol(INN); salts thereof
1999	2933.5900	--其他	6.5					20	--Other
		-结构上含有一个非稠合三嗪环(不论是否氢化)的化合物:							-Compounds containing an unfused triazine ring (whether or not hydrogenated)in the structure:
2000	2933.6100	--三聚氰胺(蜜胺)	6.5					20	--Melamine
		--其他:							--Other:
2001	2933.6910	---三聚氰氯	6				0	20	---Cyanuric chloride

序号 No.	税则号列 Tariff Item	货品名称	最惠国税率(%) M.F.N.	协定税率(%) Agreement 东盟 ASEAN	曼谷协定 BA	香港 H.K.	澳门 Macao	普通税率(%) Gen.	Article Description
		---异氰脲酸氯化衍生物:							---Chloroisocyanurate:
2002	2933.6921	----二氯异氰脲酸	6.5				0	20	----Dichloroisooyanurate acid
2003	2933.6922	----三氯异氰脲酸	6.5				0	20	----Trichloroisocyanurate acid
2004	2933.6929	----其他	6.5				0	20	----Other
2005	2933.6990	---其他	6.5				0	20	---Other
		-内酰胺:							-Lactams:
2006	2933.7100	--6-己内酰胺	9					35	--6-Hexanolactam (epsilon-caprolactam)
2007	2933.7200	--氯巴占(INN)和甲乙哌酮(INN)	9					15	--Clobazam (INN) and methyprylon (INN)
2008	2933.7900	--其他内酰胺	9					20	--Other lactams
2009	2933.9100	--阿普唑仑(INN),卡马西泮(INN),氯氮卓(INN),氯硝西泮(INN),氯拉卓酸,地洛西泮(INN),地西泮(INN),艾司唑仑(INN),氯氟卓乙酯(INN),氟地西泮(INN),氟硝西泮(INN),氟西泮(INN),哈拉西泮(INN),劳拉西泮(INN),氯甲西泮(INN),马吲哚(INN),美达西泮(INN),咪达唑仑(INN),硝甲西泮(INN),去甲西泮(INN),奥沙西泮(INN),匹那西泮(INN),普拉西泮(INN),吡咯戊酮(INN),替马西泮(INN),四氢西泮(INN)和三唑仑(INN)以及它们的盐	6.5					20	--Alprazolam (INN), camazepam (INN), chlordiazepoxide (INN), clonazepam(INN), clorazepate, delorazepam(INN), diazepam(INN), estazolam (INN), ethyl loflazepate (INN), fludiazepam (INN), flunitrazepam(INN), flurazepam(INN), halazepam(INN), lorazepam(INN), lormetazepam (INN), mazindol (INN), medazepam(INN), midazolam(INN), nimetazepam(INN), nitrazepam(INN), nordazepam(INN), oxazepam(INN), pinazepam(INN), prazepam (INN), pyrovalerone (INN), temazepam (INN), tetrazepam (INN) and triazolam (INN); salts thereof
2010	2933.9900	--其他	6.5					20	--Other
	29.34	**核酸及其盐;无论是否已有化学定义;其他杂环化合物:**							**Nucleic acids and their salts; whether or not chemically defined; Other heterocyclic compounds:**
2011	2934.1000	-结构上含有一个非稠合噻唑环(不论是否氢化)的化合物	6.5		6.5			20	-Compounds containing an unfused thiazole ring (whether or not hydrogenated) in the structure

序号 No.	税则号列 Tariff Item	货品名称	最惠国税率(%) M.F.N.	协定税率(%) Agreement 东盟 ASEAN	曼谷协定 BA	香港 H.K.	澳门 Macao	普通税率(%) Gen.	Article Description
2012	2934.2000	-结构上含有一个苯并噻唑环系(不论是否氢化)的化合物,但未经进一步稠合的	6.5					20	-Compounds containing in the structure a benzothiazole ring-system (whether or not hydrogenated), not further fused
2013	2934.3000	-结构上含有一个吩噻嗪环系(不论是否氢化)的化合物,但未经进一步稠合的	6.5					20	-Compounds containing in the structure a phenothiazine ring-system (whether or not hydrogenated), not further fused
		-其他:							-Other:
2014	2934.9100	--阿米雷司(INN),溴替唑仑(INN),氯噻西泮(INN),氯恶唑仑(INN),右吗拉胺(INN),卤恶唑仑(INN),凯他唑仑(INN),美索卡(INN),恶唑仑(INN),匹莫林(INN),苯巴曲嗪(INN),芬美曲嗪(INN)和舒芬太尼(INN)及它们的盐	6.5					20	--Aminorex(INN), brotizolam(INN), clotiazepam (INN), cloxazolam (INN), dextromoramide (INN), haloxazolam (INN), ketazolam (INN), mesocarb (INN), oxazolam (INN), pemoline(INN), phendimetrazine(INN), phenmetrazine(INN) and sufentanil(INN); salts thereof
		--其他:							--Other:
2015	2934.9910	---磺内酯及磺内酰胺	6.5		6.5			30	---Sultones and sultams
2016	2934.9920	---呋喃唑酮	6					20	---Furazolidone
2017	2934.9930	---核酸及其盐	6.5					35	---Nucleic acids and their salts
2018	2934.9990	---其他	6.5					20	---Other
	29.35	**磺(酰)胺:**							**Sulphonamides:**
2019	2935.0010	---磺胺嘧啶	6.5					35	---Sulphadiazine
2020	2935.0020	---磺胺双甲基嘧啶	6.5					35	---Sulfadimidine
2021	2935.0030	---磺胺甲噁唑	6.5					35	---Sulfamethoxazole
2022	2935.0090	---其他	6.5					35	---Other
		第十一分章 维生素原、维生素及激素							Ⅺ. PROVITAMINS, VITAMINS AND HORMONES
	29.36	**天然或合成再制的维生素原和维生素**(包括天然浓缩物)**及其主要用作维生素的衍生物,上述产品的混合物,不论是否溶于溶剂:**							**Provitamins and vitamins, natural or reproduced by synthesis** (including natural concentrates), **derivatives thereof used primarily as vitamins, and intermixtures of the foregoing, whether or not in any solvent:**
2023	2936.1000	-未混合的维生素原	4					20	-Provitamins, unmixed
		-未混合的维生素及其衍生物:							-Vitamins and their derivatives, unmixed:
2024	2936.2100	--维生素A及其衍生物	4					20	--Vitamins A and their derivatives
2025	2936.2200	--维生素 B_1 及其衍生物	4					20	--Vitamin B_1 and its derivatives

序号 No.	税则号列 Tariff Item	货品名称	最惠国税率(%) M.F.N.	协定税率(%)Agreement 东盟 ASEAN	曼谷协定 BA	香港 H.K.	澳门 Macao	普通税率(%) Gen.	Article Description
2026	2936.2300	--维生素 B_2 及其衍生物	4					20	--Vitamin B_2 and its derivatives
2027	2936.2400	--D或DL-泛酸(维生素 B_3 或维生素 B_5)及其衍生物	4					20	--D-or DL-Pantothenic acid (Vitamin B_3 or Vitamin B_5) and its derivatives
2028	2936.2500	--维生素 B_6 及其衍生物	4					20	--Vitamin B_6 and its derivatives
2029	2936.2600	--维生素 B_{12} 及其衍生物	4					20	--Vitamin B_{12} and its derivatives
2030	2936.2700	--维生素C及其衍生物	4					20	--Vitamin C and its derivatives
2031	2936.2800	--维生素E及其衍生物	4					20	--Vitamin E and its derivatives
2032	2936.2900	--其他维生素及其衍生物	4					20	--Other vitamins and their derivatives
2033	2936.9000	-其他,包括天然浓缩物	4					20	-Other, including natural concentrates
	29.37	**天然或合成再制的激素、前列腺素、血栓烷、血细胞三烯及其衍生物和结构类似物,包括主要用作激素的改性链多肽:**							**Hormones, prostaglandins, thromboxanes and leukotrienes, natural or reproduced by synthesis; derivatives and structural analogues thereof, including chain modified polypeptides, used primarily as hormones:**
		-多肽激素、蛋白激素、糖蛋白激素及其衍生物和结构类似物:							-Polypeptide hormones, protein hormones and glycoprotein hormones, their derivatives and structural analogues:
2034	2937.1100	--生长激素及其衍生物和结构类似物	4					20	--Somatotropin, its derivatives and structural analogues
2035	2937.1200	--胰岛素及其盐	4					20	--Insulin and its salts
2036	2937.1900	--其他	4					20	--Other
		-甾族激素及其衍生物和结构类似物:							-Steroidal hormones, their derivatives and structural analogues:
2037	2937.2100	--可的松、氢化可的松、脱氢可的松及脱氢皮质醇	4					20	--Cortisone, hydrocortisone, prednisone (dehydrocortisone) and prednisolone (dehydrocortisone)
		--皮质甾类激素的卤化衍生物:							--Halogenated derivatives of corticosteroidal hormones:
2038	2937.2210	---地塞米松	4					30	---Dexamethasone
2039	2937.2290	---其他	4					30	---Other
2040	2937.2300	--雌(甾)激素和孕激素	4					30	--Oestrogens and progestogens
2041	2937.2900	--其他	4					30	--Other
		-儿茶酚胺激素及其衍生物和结构类似物:							-Catecholamine hormones, their derivatives and structural analogues:
2042	2937.3100	--肾上腺素	4					30	--Epinephrine
2043	2937.3900	--其他	4					30	--Other
2044	2937.4000	-氨基酸衍生物	4					30	-Amino-acid derivatives

序号 No.	税则号列 Tariff Item	货品名称	最惠国税率(%) M.F.N.	协定税率(%) Agreement 东盟 ASEAN	曼谷协定 BA	香港 H.K.	澳门 Macao	普通税率(%) Gen.	Article Description
2045	2937.5000	-前列腺素、血栓烷和白细胞三烯及其衍生物和结构类似物	4					30	-Prostaglandins, thromboxanes and leukotrienes, their derivatives and structural analogues
2046	2937.9000	-其他	4					30	-Other
		第十二分章 天然或合成再制的苷(配糖物)、植物碱及其盐、醚、酯和其他衍生物							Ⅻ. GLYCOSIDES AND VEGETABLE ALKALOIDS, NATURAL OR REPRODUCED BY SYNTHESIS, AND THEIR SALTS, ETHERS, ESTERS AND OTHER DERIVATIVES
	29.38	**天然或合成再制的苷(配糖物)及其盐、醚、酯和其他衍生物:**							**Glycosides, natural or reproduced by synthesis, and their salts, ethers, esters and other derivatives:**
2047	2938.1000	-芸香苷及其衍生物	6.5					20	-Rutoside(rutin) and its derivatives
2048	2938.9000	-其他	6.5					20	-Other
	29.39	**天然或合成再制的生物碱及其盐、醚、酯和其他衍生物:**							**Vegetable alkaloids, natural or reproduced by synthesis, and their salts, ethers, esters and other derivatives:**
		-鸦片碱及其衍生物以及它们的盐:							-Alkaloids of opium and their derivatives; salts thereof:
2049	2939.1100	--罂粟秆浓缩物、丁丙诺啡(INN)、可待因、双氢可待因(INN)、乙基吗啡、埃托啡(INN)、海洛因、氢可酮(INN)、氢吗啡酮(INN)、吗啡、尼可吗啡(INN)、羟考酮(INN)、羟吗啡酮(INN)、福尔可定(INN)、醋氢可酮(INN)和蒂巴因,以及它们的盐	4					50	--Concentrates of poppy straw; buprenorphine(INN), codeine, dihydrocodeine(INN), ethylmorphine, etorphine(INN), heroin, hydrocodone (INN), hydromorphone(INN), morphine, nicomorphine (INN), oxycodone(INN), oxymorphone(INN), pholcodine(INN), thebacon (INN) and thebaine; salts thereof
2050	2939.1900	--其他	4					50	--Other
		-金鸡纳生物碱及其衍生物以及它们的盐:							-Alkaloids of cinchona and their derivatives; salts thereof:
2051	2939.2100	--奎宁及其盐	4					30	--Quinine and its salts
2052	2939.2900	--其他	4					20	--Other
2053	2939.3000	-咖啡因及其盐	4					20	-Caffeine and its salts
		-麻黄碱类及其盐:							-Ephedrines and their salts:
2054	2939.4100	--麻黄碱及其盐	4					20	--Ephedrine and its salts
2055	2939.4200	--假麻黄碱及其盐	4					20	--Pseudoephedrine(INN) and its salts
2056	2939.4300	--d-去甲假麻黄碱(INN)及其盐	4					20	--d-Norpseudoephedrine and its salts

序号 No.	税则号列 Tariff Item	货品名称	最惠国税率(%) M.F.N.	协定税率(%) Agreement 东盟 ASEAN	曼谷协定 BA	香港 H.K.	澳门 Macao	普通税率(%) Gen.	Article Description
2057	2939.4900	--其他	4					20	--Other
		-茶碱和氨茶碱及其衍生物以及它们的盐：							-Theophylline and aminophylline (theophylline-ethylenediamine) and their derivatives; salts thereof:
2058	2939.5100	--芬乙茶碱(INN)及其盐	4					20	--Fenetylline(INN)and its salts
2059	2939.5900	--其他	4					20	--Other
		-麦角生物碱及其衍生物以及它们的盐：							-Alkaloids of rye ergot and their derivatives; salts thereof:
2060	2939.6100	--麦角新碱及其盐	4					20	--Ergometrine(INN)and its salts
2061	2939.6200	--麦角胺及其盐	4					20	--Ergotamine(INN)and its salts
2062	2939.6300	--麦角酸及其盐	4					20	--Lysergic acid and its salts
2063	2939.6900	--其他	4					20	--Other
		--可卡因，芽子碱、左甲苯丙胺、去氧麻黄碱(INN)，去氧麻黄碱外消旋体，它们的盐、酯及其他衍生物：							--Cocaine, ecgonine, levometamfetamine, metamfetamine (INN), metamfetamine racemate; salts, esters and other derivatives thereof:
2064	2939.9110	---可卡因及其盐	4					20	---Cocaine and its salts
2065	2939.9190	---其他：	4					20	---Other:
		-其他：							-Other:
2066	2939.9910	---烟碱及其盐	4					20	---Nicotine and its salts
2067	2939.9920	---番木鳖碱(士的年)及其盐	4					17	---Strychnine and its salts
2068	2939.9990	---其他	4					20	---Other
		第十三分章　其他有机化合物							XIII. OTHER ORGANIC COMPOUNDS
	29.40	**化学纯糖，但蔗糖、乳糖、麦芽糖、葡萄糖及果糖除外；糖醚、糖缩醛、糖酯及其盐，但不包括税号29.37、29.38及29.39的产品：**							**Sugars, chemically pure, other than sucrose, lactose, maltose, glucose and fructose; sugar ethers, sugar acetals and sugar esters, and their salts, other than products of heading No. 29.37, 29.38 or 29.39:**
2069	2940.0000	化学纯糖，但蔗糖、乳糖、麦芽糖、葡萄糖及果糖除外；糖醚、糖缩醛、糖酯及其盐，但不包括税号29.37、29.38及29.39的产品	6					30	Sugars, chemically pure, other than sucrose, lactose, maltose, glucose and fructose; sugar ethers and sugar esters, and their salts, other than products of heading No. 29.37, 29.38 or 29.39
	29.41	**抗菌素：**							**Antibiotics:**
		-青霉素和具有青霉烷酸结构的青霉素衍生物及其盐：							-Penicillins and their derivatives with a penicillanic acid structure; salts thereof:
		---氨苄青霉素及其盐：							---Ampicillin and its salts:

序号 No.	税则号列 Tariff Item	货品名称	最惠国税率(%) M.F.N.	协定税率(%)Agreement				普通税率(%) Gen.	Article Description
				东盟 ASEAN	曼谷协定 BA	香港 H.K.	澳门 Macao		
2070	2941.1011	----氨苄青霉素	6		5			20	----Ampicillin
2071	2941.1012	----氨苄青霉素三水酸	6		5			20	----Ampicillin trihydrate
2072	2941.1019	----其他	6		5			20	----Other
		---其他：							---Other：
2073	2941.1091	----羟氨苄青霉素	4		4			20	----Amoxycillin
2074	2941.1092	----羟氨苄青霉素三水酸	4		4			20	----Amoxycillin trihydrate
2075	2941.1093	----6 氨基青霉烷酸(6APA)	4		4			20	----6-Aminopenicillanic acid
2076	2941.1094	----青霉素 V	4		4			20	----Penicillin V
2077	2941.1095	----磺苄青霉素	4		4			20	----Sulfobenzylpenicillin
2078	2941.1096	----邻氯青霉素	4		4			20	----Cloxacillin
2079	2941.1099	----其他	4		4			20	----Other
2080	2941.2000	-链霉素及其衍生物以及它们的盐	4		4			20	-Streptomycins and their derivatives; salts thereof
		-四环素及其衍生物以及它们的盐：							-Tetracyclines and their derivatives; salts thereof：
		---四环素及其盐：							---Tetracyclines and their salts
2081	2941.3011	----四环素	4		4		0	20	----Tetracyclines
2082	2941.3012	----四环素盐	4		4		0	20	----Salts of tetracyclines
2083	2941.3020	---四环素衍生物及其盐	4		4		0	20	---Tetracyclines derivatives and their salts
2084	2941.4000	-氯霉素及其衍生物以及它们的盐	4		4			20	-Chloramphenicol and its derivatives; salts thereof
2085	2941.5000	-红霉素及其衍生物以及它们的盐	4		4		0	20	-Erythromycin and its derivatives;salts thereof
		-其他：							-Other：
2086	2941.9010	---庆大霉素及其衍生物以及它们的盐	4		4			20	---Gentamycin and its derivatives;salts thereof
2087	2941.9020	---卡那霉素及其衍生物以及它们的盐	4		4			20	---Kanamycin and its derivatives; salts thereof
2088	2941.9030	---利福平及其衍生物以及它们的盐	4		4			20	---Rifampicin(RFP);salts thereof
2089	2941.9040	---林可霉素及其衍生物以及它们的盐	4		4			20	---Lincomycin and its derivatives; salts thereof
		---先锋霉素及其衍生物以及它们的盐：							---Cephamycin and its derivatives;salts thereof：
2090	2941.9051	----7 氨基头孢烷酸、7 氨基脱乙酰氧基头孢烷酸	6		5			20	----7-aminocephalosporianic acid, 7-aminodeacetoxycefanoic acid
2091	2941.9052	----头孢氨苄及其盐	6		5			20	----Cefalexin and its salts
2092	2941.9053	----头孢唑啉及其盐	6		5			20	----Cefazolin and its salts
2093	2941.9054	----头孢拉啶及其盐	6		5			20	----Cefradine and its salts
2094	2941.9055	----头孢三嗪(头孢曲松)及其盐	6		5			20	----Ceftriaxone and its salts
2095	2941.9056	----头孢哌酮及其盐	6		5			20	----Cefoperazone and its salts

序号 No.	税则号列 Tariff Item	货品名称	最惠国税率(%) M.F.N.	协定税率(%) Agreement 东盟 ASEAN	曼谷协定 BA	香港 H.K.	澳门 Macao	普通税率(%) Gen.	Article Description
2096	2941.9057	----头孢噻肟及其盐	6		5			20	----Cefotaxime and its salts
2097	2941.9058	----头孢克罗及其盐	6		5			20	----Cefaclor and its salts
2098	2941.9059	----其他	6		5			20	----Other
2099	2941.9060	---麦迪霉素及其衍生物以及它们的盐	6		5			20	---Midecamycin and its derivatives; salts thereof
2100	2941.9070	---乙酰螺旋霉素及其衍生物以及它们的盐	4		4			20	---Acetyl-spiramycin and its derivatives;salts thereof
2101	2941.9090	---其他	6		5			20	---Other
	29.42	**其他有机化合物:**							**Other organic compounds:**
2102	2942.0000	其他有机化合物	6.5					30	Other organic compounds

第三十章
药　　品

注释：

一、本章不包括：

（一）食品及饮料（例如，营养品、糖尿病食品、强化食品、保健食品、滋补饮料及矿泉水）（第四类），但不包括供静脉摄入用的滋养品；

（二）经特殊煅烧或精细研磨的牙科用熟石膏（税号 25.20）；

（三）适合医药用的精油水馏液及水溶液（税号 33.01）；

（四）税号 33.03 至 33.07 的制品，不论是否具有治疗及预防疾病的作用；

（五）加有药料的肥皂及税号 34.01 的其他产品；

（六）以熟石膏为基本成分的牙科用制品（税号 34.07）；

（七）不作治疗及预防疾病用的血清蛋白（税号 35.02）。

二、税号 30.02 所称的“修饰免疫制品”，仅适用于单系抗体（MABs）、抗体片段、抗体及抗体片段缀合物。

三、税号 30.03 及 30.04 以及本章注释四(四)所述的非混合产品及混合产品，按下列规定处理：

（一）非混合产品：

1. 溶于水的非混合产品；
2. 第二十八章及第二十九章的所有货品；
3. 税号 13.02 的单一植物浸膏，只经标定或溶于溶剂的。

（二）混合产品：

1. 胶体溶液及悬浮液（胶态硫磺除外）；
2. 从植物性混合物加工所得的植物浸膏；
3. 蒸发天然矿质水所得的盐及浓缩物。

四、税号 30.06 仅适用于下列物品（这些物品只能归入税号 30.06 而不得归入本目录其他税号）：

（一）无菌外科肠线、类似的无菌缝合材料及外伤用的无菌粘合胶布；

（二）无菌昆布及无菌昆布塞条；

（三）外科及牙科用无菌吸收性止血材料；

（四）用于病人的 X 光检查造影剂及其他诊断

Chapter 30
Pharmaceutical products

Notes:

1. This Chapter does not cover:

(a) Foods or beverages (such as dietetic, diabetic or fortified foods, food supplements, tonic beverages and mineral waters) other than nutritional preparations for intraverwas administration (Section Ⅳ);

(b) Plasters specially calcined or finely ground for use in dentistry (heading No.25.20);

(c) Aqueous distillates or aqueous solutions of essential oils, suitable for medicinal uses (heading No.33.01);

(d) Preparations of headings Nos.33.03 to 33.07, even if they have therapeutic or prophylactic properties;

(e) Soap or other products of heading No.34.01 containing added medicaments;

(f) Preparations with a basis of plaster for use in dentistry (heading No.34.07); or

(g) Blood albumin not prepared for therapeutic or prophylactic uses (heading No.35.02).

2. For the purposes of heading No.30.02, the expression “modified immunological products” applies only to monoclonal antibodies (MABs), antibody fragments, antibody conjugates and antibody fragment conjugates.

3. For the purposes of headings Nos.30.03 and 30.04 and of Note 4 (d) to this Chapter, the following are to be treated:

(a) As unmixed products:

(1) Unmixed products dissolved in water;

(2) All goods of Chapter 28 or 29; and

(3) Simple vegetable extracts of heading No.13.02, merely standardised or dissolved in any solvent;

(b) As products which have been mixed:

(1) Colloidal solutions and suspensions (other than colloidal sulphur);

(2) Vegetable extracts obtained by the treatment of mixtures of vegetable materials; and

(3) Salts and concentrates obtained by evaporating natural mineral waters.

4. Heading No.30.06 applies only to the following, which are to be classified in that heading and in no other heading of the Nomenclature:

(a) Sterile surgical catgut, similar sterile suture materials and sterile tissue adhesives for surgical wound closure;

(b) Sterile laminaria and sterile laminaria tents;

(c) Sterile absorbable surgical or dental haemostatics;

(d) Opacifying preparations for X-ray examinations and diag-

试剂，这些药剂是由单一产品配定剂量或由两种以上成分混合而成的；	nostic reagents designed to be administered to the patient, being unmixed products put up in measured doses or products consisting of two or more ingredients which have been mixed together for such uses;
（五）血型试剂；	(e) Blood-grouping reagents;
（六）牙科粘固剂及其他牙科填料；骨骼粘固剂；	(f) Dental cements and other dental fillings; bone reconstruction cements;
（七）急救药箱、药包；	(g) First-aid boxes and kits;
（八）以激素、税目 29.37 的其他产品或杀精子剂为基本成分的化学避孕药物。	(h) Chemical contraceptive preparations based on hormones, on other products ot heading 2G37 or on spermicides.
（九）专用于人类或作兽药用的凝胶制品，作为外科手术或体检时躯体部位的润滑剂，或者作为躯体和医疗器械之间的偶合剂。	(i) Gel preparations designed to be used in human or veterinary medicine as a lubricant for parts of the body for surgical operations or physical examinations or as a coupling agent between the body and medical instruments, and.
（十）废药物即那些因超过有效保存期等原因而不适于作原用途的药品。	(j) Waste pharmaceuticals, that is, pharmaceutical products which are unfit for their original intended purpose due to, for example, expiry of shelf life.

序号 No.	税则号列 Tariff Item	货品名称	最惠国税率(%) M.F.N.	协定税率(%)Agreement 东盟 ASEAN	曼谷协定 BA	香港 H.K.	澳门 Macao	普通税率(%) Gen.	Article Description
	30.01	**已干燥的器官疗法用腺体及其他器官，不论是否制成粉末；器官疗法用腺体、其他器官及其分泌物的提取物；肝素及其盐；其他供治疗或预防疾病用的其他税号未列名的人体或动物制品：**							**Glands and other organs for organo-therapeutic uses, dried, whether or not powdered; extracts of glands or other organs or of their secretions for organo-therapeutic uses; heparin and its salts; other human or animal substances prepared for therapeutic or prophylactic uses, not elsewhere specified or included:**
2103	3001.1000	-已干燥的腺体及其他器官，不论是否制成粉末	3					30	-Glands and other organs, dried, whether or not powdered
2104	3001.2000	-腺体、其他器官及其分泌物的提取物	3					30	Extracts of glands or other organs or of their secretions
		-其他：							-Other:
2105	3001.9010	---肝素及其盐	3					30	---Heparin and its salts
2106	3001.9090	---其他	3					30	---Other
	30.02	**人血；治病、防病或诊断用的动物血制品；抗血清、其他血份及修饰免疫制品，不论是否通过生物工艺加工制得；疫苗、毒素、培养微生物（不包括酵母）及类似产品：**							**Human blood; animal blood prepared for therapeutic, prophylactic or diagnostic uses; antisera and other blood fractions and modified immunological products, whether or not obtained by means of biotechnological processes; vaccines, toxins, cultures of micro-organisms (excluding yeasts) and similar products:**

序号 No.	税则号列 Tariff Item	货品名称	最惠国税率(%) M.F.N.	协定税率(%) Agreement 东盟 ASEAN	曼谷协定 BA	香港 H.K.	澳门 Macao	普通税率(%) Gen.	Article Description
2107	3002.1000	-抗血清、其他血份及修饰免疫制品,不论是否通过生物工艺加工制得	3					20	-Antisera and other blood fractions and modified immunological products, whether or not obtained by means of biotechnological processes
2108	3002.2000	-人用疫苗	3					20	-Vaccines for human medicine
2109	3002.3000	-兽用疫苗	3					20	-Vaccines for veterinary medicine
		-其他:							-Other:
2110	3002.9010	---石房蛤毒素	3					20	---Saxitoxin
2111	3002.9020	---蓖麻毒素	3					20	---Ricitoxin
2112	3002.9030	---细菌及病毒	3					20	---Bacteria and Virus
2113	3002.9040	---遗传物质和基因修饰生物体	3					20	---Genetics material and Gene modified Organism
2114	3002.9090	---其他	3			0△		20	---Other
	30.03	**两种或两种以上成分混合而成的治病或防病用药品(不包括税号30.02、30.05或30.06的货品),未配定剂量或制成零售包装:**							**Medicaments (excluding goods of heading 30.02,30.05 or 30.06) consisting of two or more constituents which have been mixed together for therapeutic or prophylactic uses, not put up in measured doses or in forms or packings for retail sale:**
		-含有青霉素及具有青霉烷酸结构的青霉素衍生物或链霉素及其衍生物:							-Containing penicillins or derivatives thereof, with a penicillanic acid structure, or streptomycins or their derivatives:
		---青霉素:							---Containing penicillins:
2115	3003.1011	----氨苄青霉素	6					30	----Ampicillin
2116	3003.1012	----羟氨苄青霉素	6					30	----Amoxycillin
2117	3003.1013	----青霉素V	6					30	----Penicillin V
2118	3003.1019	----其他	6					30	----Other
2119	3003.1090	---其他	6					30	---Other
		-含有其他抗菌素:							-Containing other antibiotics:
		---头孢菌素:							---Containing cephamycins:
2120	3003.2011	----头孢噻肟	6					30	----Cefotaxime
2121	3003.2012	----头孢他啶	6					30	----Ceftazidime
2122	3003.2013	----头孢西丁	6					30	----Cefoxitin
2123	3003.2014	----头孢替唑	6					30	----Ceftezole
2124	3003.2015	----头孢克罗	6					30	----Cefaclor
2125	3003.2016	----头孢呋辛	6					30	----Cefuroxime
2126	3003.2017	----头孢三嗪(头孢曲松)	6					30	----Ceftriaxone
2127	3003.2018	----头孢哌酮	6					30	----Cefoperazone
2128	3003.2019	----其他	6					30	----Other
2129	3003.2090	---其他	6					30	---Other

序号 No.	税则号列 Tariff Item	货品名称	最惠国税率(%) M.F.N.	协定税率(%) Agreement 东盟 ASEAN	曼谷协定 BA	香港 H.K.	澳门 Macao	普通税率(%) Gen.	Article Description
		-含有激素或税号 29.37 的其他产品，但不含抗菌素：							-Containing hormones or other products of heading 29. 37 but not containing antibiotics:
2130	3003.3100	--含有胰岛素	5					30	--Containing insulin
2131	3003.3900	--其他	6					30	--Other
		-含有生物碱及其衍生物，但不含抗菌素及税号 29.37 的激素或其他产品：							-Containing alkaloids or derivatives thereof but not containing hormones or other products of heading 29.37 or antibiotics:
2032	3003.4010	---含有奎宁或其盐	5					35	---Containing quinine or its salts
2133	3003.4090	---其他	5					30	---Other
		-其他：							-Other:
2134	3003.9010	---含有磺胺类	6				0	40	---Containing sulfa drugs
2135	3003.9020	---含有青蒿素及其衍生物	5				0	30	---Containing artemisinins and their derivatives
2136	3003.9090	---其他	5				0	30	---Other
	30.04	**由混合或非混合产品构成的治病或防病用药品(不包括税目 30.02、30.05 或 30.06 的货品)，已配定剂量或(包括制成皮肤摄入形式的)制成零售包装：**							**Medicaments (excluding goods of heading 30.02,30.05 or 30.06) consisting of mixed or unmixed products for therapeutic or prophylactic uses, put up in measured doses (including those in the form of transdermal administration systems) or in forms of packings for retail sale:**
		-含有青霉素及具有青霉烷酸结构的青霉素衍生物或链霉素及其衍生物：							-Containing penicillins or derivatives thereof, with a penicillanic acid structure, or streptomycins or their derivatives:
		---青霉素：							---Containing penicillins:
2137	3004.1011	----氨苄青霉素制剂	6		5	0	0	30	----Ampicillin
2138	3004.1012	----羟氨苄青霉素制剂	6		5	0	0	30	----Amoxycillin
2139	3004.1013	----青霉素 V 制剂	6		5	0	0	30	----Penicillin V
2140	3004.1019	----其他	6		5	0	0	30	----Other
2141	3004.1090	---其他	6		5	0	0	30	---Other
		-含有其他抗菌素：							-Containing other antibiotics:
		---头孢菌素：							---Containing cephamycins:
2142	3004.2011	----头孢噻肟制剂	6		5	0		30	----Cefotaxime
2143	3004.2012	----头孢他啶制剂	6		5	0		30	----Ceftazidime
2144	3004.2013	----头孢西丁制剂	6		5	0		30	----Cefoxitin
2145	3004.2014	----头孢替唑制剂	6		5	0		30	----Ceftezole
2146	3004.2015	----头孢克罗制剂	6		5	0		30	----Cefaclor
2147	3004.2016	----头孢呋辛制剂	6		5	0		30	----Cefuroxime
2148	3004.2017	----头孢三嗪(头孢曲松)制剂	6		5	0		30	----Ceftriaxone

序号 No.	税则号列 Tariff Item	货品名称	最惠国税率(%) M.F.N.	协定税率(%)Agreement 东盟 ASEAN	曼谷协定 BA	香港 H.K.	澳门 Macao	普通税率(%) Gen.	Article Description
2149	3004.2018	----头孢哌酮制剂	6		5	0		30	----Cefoperazone
2150	3004.2019	----其他	6		5	0		30	----Other
2151	3004.2090	---其他	6		5	0	0	30	---Other
		-含有激素或税号29.37的其他产品,但不含抗菌素:							-Containing hormones or other products of heading No. 29.37 but not containing antibiotics:
2152	3004.3100	--含有胰岛素	5		4			30	--Containing insulin
2153	3004.3200	--含有皮质甾类激素及其衍生物或结构类似物	5		4	0		30	--Containing corticosteroid hormones, their derivatives and structural analogues
2154	3004.3900	--其他	5		4	0		30	--Other
		-含有生物碱及其衍生物,但不含抗菌素及税号29.37的激素或其他产品:							-Containing alkaloids or derivatives thereof but not containing hormones, other products of heading No. 29.37 or antibiotics:
2155	3004.4010	---含有奎宁或其盐	5		4	0		35	---Containing quinine or its salts
2156	3004.4090	---其他	5		5	0		30	---Other
2157	3004.5000	-含有维生素或税号29.36所列产品的其他药品	6		5			40	-Other medicaments containing vitamins or other products of heading No.29.36
		-其他:							-Other:
2158	3004.9010	---含有磺胺类	6		5			40	---Containing sulfa drugs
2159	3004.9020	---含有联苯双酯	4					30	---Containing biphenyl dicarbxybte
		---中式成药:							---Medicaments of Chinese type:
2160	3004.9051	----中药酒	3		2	0		30	----Medicated liquors or wines
2161	3004.9052	----片仔癀	3		2	0		30	----Pien Tzu Huang
2162	3004.9053	----白药	3		2	0		30	----Bai Yao
2163	3004.9054	----清凉油	3		2	0	0	30	----Essential balm
2164	3004.9059	----其他	3		2	0	0	30	----Other
2165	3004.9060	---含有青蒿素及其衍生物	4		4	0	0	30	---Containing artemisinins and their derivatives
2166	3004.9090	---其他	4		4	0	0	30	---Other
	30.05	**软填料、纱布、绷带及类似物品(例如,敷料、橡皮膏、泥罨剂),经过药物浸涂或制成零售包装供医疗、外科、牙科或兽医用:**							**Wadding, gauze, bandages and similar articles (for example, dressings, adhesive plasters, poultices), impregnated or coated with pharmaceutical substances or put up in forms or packings for retail sale for medical, surgical, dental or veterinary purposes:**
		-胶粘敷料及有胶粘涂层的其他物品:							-Adhesive dressings and other articles having an adhesive layer:
2167	3005.1010	---橡皮膏	5					70	---Adhesive plasters
2168	3005.1090	---其他	5					35	---Other
		-其他:							-Other:
2169	3005.9010	---药棉、纱布、绷带	5		3	0△		70	---Absorbent cotton, gauze, bandages
2170	3005.9090	---其他	5			0△		35	---Other

序号 No.	税则号列 Tariff Item	货品名称	最惠国税率(%) M.F.N.	协定税率(%)Agreement 东盟 ASEAN	曼谷协定 BA	香港 H.K.	澳门 Macao	普通税率(%) Gen.	Article Description
	30.06	**本章注释四所规定的医药用品:**							**Pharmaceutical goods specified in Note 4 to this Chapter:**
2171	3006.1000	-无菌外科肠线、类似的无菌缝合材料及外伤用的无菌粘合胶布;无菌昆布及无菌昆布塞条;外科及牙科用无菌吸收性止血材料	5					30	-Sterile surgical catgut, similar sterile suture materials and sterile tissue adhesives for surgical wound closure; sterile laminaria and sterile laminaria tents; sterile absorbable surgical or dental haemostatics
2172	3006.2000	-血型试剂	3					20	-Blood-grouping reagents
2173	3006.3000	-X 光检查造影剂;用于病人的诊断试剂	4					30	-Opacifying preparations for X-ray examinations; diagnostic reagents designed to be administered to the patient
2174	3006.4000	-牙科粘固剂及其他牙科填料;骨骼粘固剂	5					30	-Dental cements and other dental fillings; bone reconstruction cements
2175	3006.5000	-急救药箱、药包	5					30	-First-aid boxes and kits
		-以激素、税目 29.37 的其他产品或杀精子剂为基本成分的化学避孕药物:							-Chemical contraceptive preparations based on hormones, on other products of heading 29.37 or on spermicides:
2176	3006.6010	---以激素为基本成份的避孕药	0					0	---Contraceptive preparations based on hormones
2177	3006.6090	---其他	0					0	---Other
2178	3006.7000	-专用于人类或作兽药用的凝胶制品,作为外科手术或体检时躯体部位的润滑剂,或者作为躯体和医疗器械之间的偶合剂	6.5			0		30	-Gel preparations designed to be used in human or veterinary medicine as a lubricant for parts of the body for surgical operations or physical examinations or as a coupling agent between the body and medical instruments".
2179	3006.8000	-废药物	5			0		30	-Waste pharmaceuticals

第三十一章
肥　　料

注释：

一、本章不包括：

（一）税号05.11的动物血；

（二）单独的已有化学定义的化合物（符合下列注释二（一）、三（一）、四（一）或五所规定的化合物除外）；

（三）税号38.24的每颗重量不低于2.5克的氯化钾培养晶体（光学元件除外）；氯化钾光学元件（税号90.01）。

二、税号31.02只适用于下列货品，但未制成税号31.05所述形状或包装：

（一）符合下列任何一条规定的货品：

1. 硝酸钠，不论是否纯净；
2. 硝酸铵，不论是否纯净；
3. 硫酸铵及硝酸铵的复盐，不论是否纯净；
4. 硫酸铵，不论是否纯净；
5. 硝酸钙及硝酸铵的复盐（不论是否纯净）或硝酸钙及硝酸铵的混合物；
6. 硝酸钙及硝酸镁的复盐（不论是否纯净）或硝酸钙及硝酸镁的混合物；
7. 氰氨化钙，不论是否纯净或用油处理；
8. 尿素，不论是否纯净。

（二）由上述(一)款任何货品相互混合的肥料。

（三）由氯化铵或上述（一）或（二）款任何货品与白垩、石膏或其他无肥效无机物混合而成的肥料。

（四）由上述（一）2或8项的货品或其混合物溶于水或液氨的液体肥料。

三、税号31.03只适用于下列货品，但未制成税号31.05所述形状或包装：

（一）符合下列任何一条规定的货品：

1. 碱性熔渣；
2. 税号25.10的天然磷酸盐，已焙烧或经过超出清除杂质范围的热处理；

Chapter 31
Fertilizers

Notes:

1. This Chapter does not cover:

(a) Animal blood of heading No.05.11;

(b) Separate chemically defined compounds (other than those answering to the descriptions in Note 2 (A), 3 (A), 4 (A) or 5 below); or

(c) Cultured potassium chloride crystals (other than optical elements) weighing not less than 2.5g each, of heading No.38.24; optical elements of potassium chloride (heading No.90.01).

2. Heading No.31.02 applies only to the following goods, provided that they are not put up in the forms or packages described in heading No.31.05:

(a) Goods which answer to one or other of the descriptions given below:

(1) Sodium nitrate, whether or not pure;

(2) Ammonium nitrate, whether or not pure;

(3) Double salts, whether or not pure, of ammonium sulphate and ammonium nitrate;

(4) Ammonium sulphate, whether or not pure;

(5) Double salts (whether or not pure) or mixtures of calcium nitrate and ammonium nitrate;

(6) Double salts (whether or not pure) or mixtures of calcium nitrate and magnesium nitrate;

(7) Calcium cyanamide, whether or not pure or treated with oil;

(8) Urea, whether or not pure.

(b) Fertilizers consisting of any of the goods described in (A) above mixed together.

(c) Fertilizers consisting of ammonium chloride or of any of the goods described in (A) or (B) above mixed with chalk, gypsum or other inorganic nonfertilizing substances.

(d) Liquid fertilizers consisting of the goods of subparagraph (A) (ii) or (viii) above, or of mixtures of those goods, in an aqueous or ammoniacal solution.

3. Heading No.31.03 applies only to the following goods, provided that they are not put up in the forms or packages described in heading No.31.05:

(a) Goods which answer to one or other of the descriptions given below:

(1) Basic slag;

(2) Natural phosphates of heading No.25.10, calcined or further heat-treated than for the removal of impurities;

3. 过磷酸钙（一过磷酸钙、二过磷酸钙或三过磷酸钙）；

4. 磷酸氢钙，按干燥无水产品重量计含氟量不低于0.2%。

（二）由上述（一）款的任何货品相互混合的肥料，不论含氟量多少。

（三）由上述（一）或（二）款的任何货品与白垩、石膏或其他无肥效无机物混合而成的肥料，不论含氟量多少。

(3) Superphosphates (single, double or triple);

(4) Calcium hydrogenorthophosphate containing not less than 0.2% by weight of flourine calculated on the dry anhydrous product.

(b) Fertilizers consisting of any of the goods described in (A) above mixed together, but with no account being taken of the fluorine content limit.

(c) Fertilizers consisting of any of the goods described in (a) or (b) above, but with no account being taken of the fluorine content limit, mixed with chalk, gypsum or other inorganic non-fertilizing substances.

四、税号31.04只适用于下列货品，但未制成税号31.05所述形状或包装：

（一）符合下列任何一条规定的货品：

1. 天然粗钾盐（例如，光卤石、钾盐镁矾及钾盐）；

2. 氯化钾，不论是否纯净，但上述注释一（三）所述的产品除外；

3. 硫酸钾，不论是否纯净；

4. 硫酸镁钾，不论是否纯净。

（二）由上述(一)款任何货品相互混合的肥料。

4. Heading No.31.04 applies only to the following goods, provided that they are not put up in the forms or packages described in heading No.31.05:

(a) Goods which answer to one or other of the descriptions given below:

(1) Crude natural potassium salts (for example, carnallite, kainite and sylvite);

(2) Potassium chloride, whether or not pure, except as provided in Note 1 (c) above;

(3) Potassium sulphate, whether or not pure;

(4) Magnesium potassium sulphate, whether or not pure.

(b) Fertilizers consisting of any of the goods described in (a) above mixed together.

五、磷酸二氢铵及磷酸氢二铵（不论是否纯净）及其相互之间的混合物应归入税号31.05。

5. Ammonium dihydrogenorthophosphate (monoammonium phosphate) and diammonium hydrogenorthophosphate (diammonium phosphate), whether or not pure, and intermixtures thereof, are to be classified in heading No.31.05.

六、税号31.05所称“其他肥料”，仅适用于其基本成分至少含有氮、磷、钾中一种肥效元素的肥料用产品。

6. For the purposes of heading No.31.05, the term “other fertilizers” applies only to products of a kind used as fertilizers and containing, as an essential constituent, at least one of the fertilizing elements nitrogen, phosphorus or potassium.

序号 No.	税则号列 Tariff Item	货品名称	最惠国税率(%) M.F.N.	协定税率(%) Agreement 东盟 ASEAN	曼谷协定 BA	香港 H.K.	澳门 Macao	普通税率(%) Gen.	Article Description
	31.01	**动物或植物肥料，不论是否相互混合或经化学处理；动植物产品经混合或化学处理制成的肥料：**							**Animal or vegetable fertilizers, whether or not mixed together or chemically treated; fertilizers produced by the mixing or chemical treatment of animal or vegetable products:**
		---未经化学处理：							---Not chemically treated:
2180	3101.0011	----鸟粪	3					11	----Guano

序号 No.	税则号列 Tariff Item	货品名称	最惠国税率(%) M.F.N.	协定税率(%) Agreement 东盟 ASEAN	曼谷协定 BA	香港 H.K.	澳门 Macao	普通税率(%) Gen.	Article Description
2181	3101.0019	----其他	6.5					30	----Other
2182	3101.0090	---其他	4					11	---Other
	31.02	**矿物氮肥及化学氮肥：**							**Mineral or chemical fertilizers, nitrogenous:**
2183	3102.1000	-尿素，不论是否水溶液	见表5		40			见表5	-Urea, whether or not in aqueous solution
		-硫酸铵；硫酸铵和硝酸铵的复盐及混合物：							-Ammonium sulphate; double salts and mixtures of ammonium sulphate and ammonium nitrate:
2184	3102.2100	--硫酸铵	4					11	--Ammonium sulphate
2185	3102.2900	--其他	4					11	--Other
2186	3102.3000	-硝酸铵，不论是否水溶液	4					11	-Ammonium nitrate, whether or not in aqueous solution
2187	3102.4000	-硝酸铵与碳酸钙或其他无肥效无机物的混合物	4					11	-Mixtures of ammonium nitrate with calcium carbonate or other inorganic nonfertilizing substances
2188	3102.5000	-硝酸钠	4					11	-Sodium nitrate
2189	3102.6000	-硝酸钙和硝酸铵的复盐及混合物	4					11	-Double salts and mixtures of calcium nitrate and ammonium nitrate
2190	3102.7000	-氰氨化钙	4					11	-Calcium cyanamide
2191	3102.8000	-尿素及硝酸铵混合物的水溶液或氨水溶液	4					11	-Mixtures of urea and ammonium nitrate in aqueous or ammoniacal solution
2192	3102.9000	-其他，包括上述子目未列名的混合物	4					11	-Other, including mixtures not specified in the foregoing subheadings
	31.03	**矿物磷肥及化学磷肥：**							**Mineral or chemical fertilizers, phosphatic:**
2193	3103.1000	-过磷酸钙	4					11	-Superphosphates
2194	3103.2000	-碱性熔渣	4					11	-Basic slag
2195	3103.9000	-其他	4					11	-Other
	31.04	**矿物钾肥及化学钾肥：**							**Mineral or chemical fertilizers, potassic:**
2196	3104.1000	-光卤石、钾盐及其他天然粗钾盐	3					11	-Carnallite, sylvite and other crude natural potassium salts
		-氯化钾：							-Potassium chloride:
2197	3104.2010	---分析纯的	3					11	---Analytically pure
2198	3104.2090	---其他	3					11	---Other
2199	3104.3000	-硫酸钾	3		3			11	-Potassium sulphate
2200	3104.9000	-其他	3					11	-Other

序号 No.	税则号列 Tariff Item	货品名称	最惠国税率(%) M.F.N.	协定税率(%) Agreement 东盟 ASEAN	曼谷协定 BA	香港 H.K.	澳门 Macao	普通税率(%) Gen.	Article Description
	31.05	**含氮、磷、钾中两种或三种肥效元素的矿物肥料或化学肥料;其他肥料;制成片及类似形状或每包毛重不超过10公斤的本章各项货品:**							**Mineral or chemical fertilizers containing two or three of the fertilizing elements nitrogen, phosphorus and potassium; other fertilizers; goods of this Chapter in tablets or similar forms or in packages of a gross weight not exceeding 10kg:**
2201	3105.1000	-制成片及类似形状或每包毛重不超过10公斤的本章各项货品	4					11	-Goods of this Chapter in tablets or similar forms or in packages of a gross weight not exceeding 10kg
2202	3105.2000	-含氮、磷、钾三种肥效元素的矿物肥料或化学肥料	见表5					见表5	-Mineral or chemical fertilizers containing the three fertilizing elements nitrogen, phosphorus and potassium
2203	3105.3000	-磷酸氢二铵	见表5					见表5	-Diammonium hy drogenorthophosphate (diammonium phosphate)
2204	3105.4000	-磷酸二氢铵及磷酸二氢铵与磷酸氢二铵的混合物	4					11	-Ammonium dihydrogenorthophosphate (monoammonium phosphate) and mixtures thereof with diammonium hydrogenorthophosphate (diammonium phosphate)
		-其他含氮、磷两种肥效元素的矿物肥料或化学肥料:							-Other mineral or chemical fertilizers containing the two fertilizing elements nitrogen and phosphorus:
2205	3105.5100	--含有硝酸盐及磷酸盐	4					11	--Containing nitrates and phosphates
2206	3105.5900	--其他	4					11	--Other
2207	3105.6000	-含磷、钾两种肥效元素的矿物肥料或化学肥料	4					11	-Mineral or chemical fertilizers containing the two fertilizing elements phosphorus and potassium
2208	3105.9000	-其他	4					11	-Other

第三十二章
鞣料浸膏及染料浸膏；鞣酸及其衍生物；染料、颜料及其他着色料；油漆及清漆；油灰及其他类似胶粘剂；墨水、油墨

注释：

一、本章不包括：

（一）单独的已有化学定义的化学元素及化合物（税号 32.03 及 32.04 的货品、税号 32.06 的用作发光体的无机产品、税号 32.07 所述形状的熔融石英或其他熔融硅石制成的玻璃及税号 32.12 的零售形状或零售包装的染料及其他着色料除外）；

（二）税号 29.36 至 29.39、29.41 及 35.01 至 35.04 的鞣酸盐及其他鞣酸衍生物；

（三）沥青胶粘剂（税号 27.15）。

二、税号 32.04 包括生产偶氮染料用的稳定重氮盐与偶合物的混合物。

三、税号 32.03、32.04、32.05 及 32.06 也包括以着色料为基本成分的制品（例如，税号 32.06 包括以税号 25.30 或第二十八章的颜料，金属粉片及金属粉末为基本成分的制品）。该制品是用作原材料着色剂的拼料。但以上税号不包括分散在非水介质中呈液状或浆状的制漆用颜料，例如，税号 32.12 的瓷漆及税号 32.07、32.08、32.09、32.10、32.12、32.13 及 32.15 的其他制品。

四、税号 32.08 包括由税号 39.01 至 39.13 所列产品溶于挥发性有机溶剂的溶液（胶棉除外），但溶剂重量必须超过溶液重量的 50%。

五、本章所称"着色料"，不包括作为油漆填料的产品，不论这些产品能否用于水浆涂料的着色。

六、税号 32.12 所称"压印箔"，只包括用以压印

Chapter 32
Tanning or dyeing extracts; tannins and their derivatives; dyes, pigments and other colouring matter; paints and varnishes; putty and other mastics; inks

Notes:

1. This Chapter does not cover:

(a) Separate chemically defined elements or compounds (except those of heading No.32.03 or 32.04, inorganic products of a kind used as luminophores (heading No.32.06), glass obtained from fused quartz or other fused silica in the forms provided for in heading No.32.07, and also dyes and other colouring matter put up in forms or packings for retail sale, of heading No.32.12);

(b) Tannates or other tannin derivatives of products of headings Nos.29.36 to 29.39, 29.41 or 35.01 to 35.04; or

(c) Mastics of asphalt or other bituminous mastics (heading No.27.15).

2. Heading No.32.04 includes mixtures of stabilized diazonium salts and couplers for the production of azo dyes.

3. Headings Nos.32.03, 32.04, 32.05 and 32.06 apply also to preparations based on colouring matter (including, in the case of heading No.32.06, colouring pigments of heading No.25.30 or Chapter 28, metal flakes and metal powders), of a kind used for colouring any material or used as ingredients in the manufacture of colouring preparations. The headings do not apply, however, to pigments dispersed in non-aqueous media, in liquid or paste form, of a kind used in the manufacture of paints, including enamels (heading No.32.12), or to other preparations of heading No.32.07, 32.08, 32.09, 32.10, 32.12, 32.13 or 32.15.

4. Heading No.32.08 includes solutions (other than collodions) consisting of any of the products specified in headings Nos.39.01 to 39.13 in volatile organic solvents when the weight of the solvent exceeds 50% of the weight of the solution.

5. The expression "colouring matter" in this Chapter does not include products of a kind used as extenders in oil paints, whether or not they are also suitable for colouring distempers.

6. The expression "stamping foils" in heading No.32.12 ap-

诸如书本封面或帽带之类的薄片，这些薄片由以下材料构成：

（一）金属粉（包括贵金属粉）或颜料经胶水、明胶及其他粘合剂凝结而成的；

（二）金属（包括贵金属）或颜料沉积于任何材料衬片上的。

plies only to thin sheets of a kind used for printing, for example, book covers or hat bands, and consisting of:

(a) Metallic powder (including powder of precious metal) or pigment, agglomerated with glue, gelatin or other binder; or

(b) Metal (including precious metal) or pigment, deposited on a supporting sheet of any material.

序号 No.	税则号列 Tariff Item	货品名称	最惠国税率(%) M.F.N.	协定税率(%) Agreement 东盟 ASEAN	曼谷协定 BA	香港 H.K.	澳门 Macao	普通税率(%) Gen.	Article Description
	32.01	**植物鞣料浸膏；鞣酸及其盐、醚、酯和其他衍生物：**							**Tanning extracts of vegetable origin; tannins and their salts, ethers, esters and other derivatives:**
2209	3201.1000	-坚木浸膏	5					35	-Quebracho extract
2210	3201.2000	-荆树皮浸膏	6.5					35	-Wattle extract
		-其他：							-Other:
2211	3201.9010	---其他鞣料浸膏	6.5					40	---Other tanning extracts
2212	3201.9090	---其他	6.5					35	---Other
	32.02	**有机合成鞣料；无机鞣料；鞣料制剂，不论是否含有天然鞣料；预鞣用酶制剂：**							**Synthetic organic tanning substances; inorganic tanning substances; tanning preparations, whether or not containing natural tanning substances; enzymatic preparations for pre-tanning:**
2213	3202.1000	-有机合成鞣料	6.5			0		35	-Synthetic organic tanning substances
2214	3202.9000	-其他	6.5			0		35	-Other
	32.03	**动植物质着色料（包括染料浸膏，但动物炭黑除外），不论是否已有化学定义；本章注释三所述的以动植物质着色料为基本成分的制品：**							**Colouring matter of vegetable or animal origin (including dyeing extracts but excluding animal black), whether or not chemically defined; preparations as specified in Note 3 to this Chapter based on colouring matter of vegetable or animal origin:**
		---植物质着色料及以其为基本成分的制品：							---Colouring matter of vegetable origin and preparations based thereon:
2215	3203.0011	----天然靛蓝及以其为基本成分的制品	6.5					80	----Natural indigo and preparations based thereon
2216	3203.0019	----其他	6.5					45	----Other
2217	3203.0020	---动物质着色料及以其为基本成分的制品	6.5					50	---Colouring matter of animal origin and preparations based thereon
	32.04	**有机合成着色料，不论是否已有化学定义；本章注释三所述的以有机合成着色料为基本成分的制品；用作荧光增白剂或发光体的有机合成产品，不论是否已有化学定义：**							**Synthetic organic colouring matter, whether or not chemically defined; preparations as specified in Note 3 to this Chapter based on synthetic organic colouring matter; synthetic organic products of a kind used as fluorescent brightening agents or as luminophores, whether or not chemically defined:**

序号 No.	税则号列 Tariff Item	货品名称	最惠国税率(%) M.F.N.	协定税率(%) Agreement 东盟 ASEAN	曼谷协定 BA	香港 H.K.	澳门 Macao	普通税率(%) Gen.	Article Description
		-有机合成着色料及本章注释三所述的以有机合成着色料为基本成分的制品：							-Synthetic organic colouring matter and preparations based thereon as specified in Note 3 to this Chapter:
2218	3204.1100	--分散染料及以其为基本成分的制品	8.3		7.7			35	--Disperse dyes and preparations based thereon
2219	3204.1200	--酸性染料(不论是否预金属络合)及以其为基本成分的制品；媒染染料及以其为基本成分的制品	8.3		7.7	0△		35	--Acid dyes, whether or not premetallized, and preparations based thereon; mordant dyes and preparations based thereon
2220	3204.1300	--碱性染料及以其为基本成分的制品	6.5		6			35	--Basic dyes and preparations based thereon
2221	3204.1400	--直接染料及以其为基本成分的制品	6.5		6			35	--Direct dyes and preparations based thereon
		--瓮染料(包括颜料用的)及以其为基本成分的制品：							--Vat dyes (including those usable in that state as pigments) and preparations based thereon:
2222	3204.1510	---合成靛蓝(还原靛蓝)	6.5		6			35	---Synthetic indigo (reductive indigo)
2223	3204.1590	---其他	6.5		6			35	---Other
2224	3204.1600	--活性染料及以其为基本成分的制品	8.3		7.7	0		35	--Reactive dyes and preparations based thereon
2225	3204.1700	--颜料及以其为基本成分的制品	6.5		6	0		35	--Pigments and preparations based thereon
		--其他，包括由子目号3204.11至3204.19中两个或多个子目所列着色料组成的混合物：							--Other, including mixtures of colouring matter of two or more of the subheadings Nos. 3204.11 to 3204.19:
		---硫化染料及以其为基本成分的制品：							---Sulphur dyes and preparations based thereon:
2226	3204.1911	----硫化黑(硫化青)及以其为基本成分的制品	6.5		6			35	----Sulphur black and preparations based thereon
2227	3204.1919	----其他	6.5		6			35	----Other
2228	3204.1990	---其他	6.5		6			35	---Other
2229	3204.2000	-用作荧光增白剂的有机合成产品	6.5		6	0		40	-Synthetic organic products of a kind used as flourescent brightening agents
		-其他：							-Other:
2230	3204.9010	---生物染色剂及染料指示剂	6.5		6			20	---Biological stains and dye indicators
2131	3204.9090	---其他	6.5		6			40	---Other
	32.05	**色淀；本章注释三所述的以色淀为基本成分的制品：**							**Colour lakes; preparations as specified in Note 3 to this Chapter based on colour lakes:**

序号 No.	税则号列 Tariff Item	货品名称	最惠国税率(%) M.F.N.	协定税率(%)Agreement 东盟 ASEAN	曼谷协定 BA	香港 H.K.	澳门 Macao	普通税率(%) Gen.	Article Description
2232	3205.0000	色淀;本章注释三所述的以色淀为基本成分的制品	6.5					35	Colour lakes; preparations as specified in Note 3 to this Chapter based on colour lakes
	32.06	**其他着色料;本章注释三所述的制品,但税号32.03、32.04及32.05的货品除外;用作发光体的无机产品,不论是否已有化学定义:**							**Other colouring matter; preparations as specified in Note 3 to this Chapter, other than those of heading No.32.03, 32.04 or 32.05; inorganic products of a kind used as luminophores, whether or not chemically defined:**
		-以二氧化钛为基本成分的颜料及制品:							-Pigments and preparations based on titanium dioxide:
		--以干物质计二氧化钛含量在80%及以上的:							--Containing 80% or more by weight of titanium dioxide calculated on the dry matter:
2233	3206.1110	---钛白粉	6.5					30	---Titanium White
2234	3206.1190	---其他	6.5					30	---Other
2235	3206.1900	--其他	10			0△		30	--Other
2236	3206.2000	-以铬化合物为基本成分的颜料及制品	6.5					35	-Pigments and preparations based on chromium compounds
2237	3206.3000	-以镉化合物为基本成分的颜料及制品	6.5					35	-Pigments and preparations based on cadmium compounds
		-其他着色料及其他制品:							-Other colouring matter and other preparations:
2238	3206.4100	--群青及以其为基本成分的制品	6.5					35	--Ultramarine and preparations based thereon
		--锌钡白及以硫化锌为基本成分的其他颜料和制品:							--Lithopone and other pigments and preparations based on zinc sulphide:
2239	3206.4210	---锌钡白	6.5					30	---Lithopone
2240	3206.4290	---其他	6.5					30	---Other
2241	3206.4300	--以六氰合高铁酸盐(氰亚铁酸盐及氰铁酸盐)为基本成分的颜料及制品	6.5					35	--Pigments and preparations based on hexacyanoferrates (ferrocyanides and ferricyanides)
2242	3206.4900	--其他	6.5			0		35	--Other
2243	3206.5000	-用作发光体的无机产品	6.5					35	-Inorganic products of a kind used as luminophores
	32.07	**陶瓷、搪瓷及玻璃工业用的调制颜料、遮光剂、着色剂、珐琅和釉料、釉底料(泥釉)、光瓷釉以及类似产品;搪瓷玻璃料及其他玻璃,呈粉、粒或粉片状的:**							**Prepared pigments, prepared opacifiers and prepared colours, vitrifiable enamels and glazes, engobes (slips), liquid lustres and similar preparations, of a kind used in the ceramic, enamelling or glass industry; glass frit and other glass, in the form of powder, granules or flakes:**

序号 No.	税则号列 Tariff Item	货品名称	最惠国税率(%) M.F.N.	协定税率(%)Agreement				普通税率(%) Gen.	Article Description
				东盟 ASEAN	曼谷协定 BA	香港 H.K.	澳门 Macao		
2244	3207.1000	-调制颜料、遮光剂、着色剂及类似制品	5					50	-Prepared pigments, prepared opacifiers, prepared colours and similar preparations
2245	3207.2000	-珐琅和釉料、釉底料(泥釉)及类似制品	5					50	-Vitrifiable enamels and glazes, engobes(slips) and similar preparations
2246	3207.3000	-光瓷釉及类似制品	5					50	-Liquid lustres and similar preparations
2247	3207.4000	-搪瓷玻璃料及其他玻璃,呈粉、粒或粉片状的	5					50	- Glass frit and other glass, in the form of powder, granules or flakes
	32.08	**以合成聚合物或化学改性天然聚合物为基本成分的油漆及清漆(包括瓷漆及大漆),分散于或溶于非水介质的;本章注释四所述的溶液:**							**Paints and varnishes(including enamels and lacquers) based on synthetic polymers or chemically modified natural polymers, dispersed or dissolved in a nonaqueous medium; solutions as defined in Note 4 to this Chapter:**
2248	3208.1000	-以聚酯为基本成分	10		9	0	0	50	-Based on polyesters
		-以丙烯酸聚合物或乙烯聚合物为基本成分:							-Based on acrylic or vinyl polymers:
2249	3208.2010	---以丙烯酸聚合物为基本成分	10		9			50	---Based on acrylic polymers
2250	3208.2020	---以乙烯聚合物为基本成分	10		9			50	---Based on vinyl polymers
		-其他:							-Other:
2251	3208.9010	---以聚胺酯类化合物为基本成分	10		9			50	---Based on polyurethane polymers
2252	3208.9090	---其他	10		9	0		50	---Other
	32.09	**以合成聚合物或化学改性天然聚合物为基本成分的油漆及清漆(包括瓷漆及大漆),分散于或溶于水介质的:**							**Paints and varnishes(including enamels and lacquers) based on synthetic polymers or chemically modified natural polymers, dispersed or dissolved in an aqueous medium:**
2253	3209.1000	-以丙烯酸聚合物或乙烯聚合物为基本成分	10				0	50	-Based on acrylic or vinyl polymers
		-其他:							-Other:
2254	3209.9010	---以环氧树脂为基本成份	10					50	---Based on epoxy resin
2255	3209.9020	---以氟树脂为基本成份	10					50	---Based on fluororesin
2256	3209.9090	---其他	10					50	---Other
	32.10	**其他油漆及清漆(包括瓷漆、大漆及水浆涂料);加工皮革用的水性颜料:**							**Other paints and varnishes (including enamels, lacquers and distempers); prepared water pigments of a kind used for finishing leather:**

序号 No.	税则号列 Tariff Item	货品名称	最惠国税率(%) M.F.N.	协定税率(%) Agreement				普通税率(%) Gen.	Article Description
				东盟 ASEAN	曼谷协定 BA	香港 H.K.	澳门 Macao		
2257	3210.0000	其他油漆及清漆(包括瓷漆、大漆及水浆涂料);加工皮革用的水性颜料	10		9	0	0	50	Other paints and varnishes (including enamels, lacquers and distempers); prepared water pigments of a kind used for finishing leather
	32.11	**配制的催干剂:**							**Prepared driers:**
2258	3211.0000	配制的催干剂	10					50	Prepared driers
	32.12	**制造油漆(含瓷漆)用的颜料(包括金属粉末或金属粉片),分散于非水介质中呈液状或浆状的;压印箔;零售形状及零售包装的染料或其他着色料:**							**Pigments (including metallic powders and flakes) dispersed in non-aqueous media, in liquid or paste form, of a kind used in the manufacture of paints (including enamels); stamping foils; dyes and other colouring matter put up in forms or packings for retail sale:**
2259	3212.1000	-压印箔	15					80	-Stamping foils
2260	3212.9000	-其他	10					50	-Other
	32.13	**艺术家、学生和广告美工用的颜料、调色料、文娱颜料及类似品,片状、管装、罐装、瓶装、扁盒装以及类似形状或包装的:**							**Artists', students' or signboard painters' colours, modifying tints, amusement colours and the like, in tablets, tubes, jars, bottles, pans or in similar forms or packings:**
2261	3213.1000	-成套的颜料	10					70	-Colours in sets
2262	3213.9000	-其他	10		9			70	-Other
	32.14	**安装玻璃用油灰、接缝用油灰、树脂胶泥、嵌缝胶及其他类似胶粘剂;漆工用填料;非耐火涂面制剂,涂门面、内墙、地板、天花板等用:**							**Glaziers' putty, graftig putty, resin cements, caulking compounds and other mastics; painters' fillings; non-refractory surfacing preparations for facades, indoor walls, floors, ceilings or the like:**
2263	3214.1000	-安装玻璃用油灰、接缝用油灰、树脂胶泥、嵌缝胶及其他类似胶粘剂;漆工用填料	9					70	-Glaziers' putty, grafting putty, resin cements, caulking compounds and other mastics; painters' fillings
2264	3214.9000	-其他	9					70	-Other
	32.15	**印刷油墨、书写或绘图墨水及其他墨类,不论是否固体或浓缩:**							**Printing ink, writing or drawing ink and other inks, whether or not concentrated or solid:**
		-印刷油墨:							-Printing ink:
2265	3215.1100	--黑色	6.5				$0^{\triangle}$	45	--Black
2266	3215.1900	--其他	6.5			0	0	45	--Other
		-其他:							-Other:
2267	3215.9010	---书写墨水	6.5				$0^{\triangle}$	70	---Writing or drawing inks
2268	3215.9090	---其他	10				$0^{\triangle}$	70	---Other

第三十三章
精油及香膏；芳香料制品及化妆盥洗品

Chapter 33
Essential oils and resinoids; perfumery, cosmetic or toilet preparations

注释：

一、本章不包括：

（一）税号13.01及13.02的天然油树脂及植物浸膏；

（二）税号34.01的肥皂及其他产品；

（三）税号38.05的脂松节油、木松节油和硫酸盐松节油及其他产品。

二、税号33.02所称“香料”，仅指税号33.01所列的物质、从这些物质离析出来的香料组分以及合成芳香剂。

三、税号33.03至33.07主要包括适合作这些税号所列用途的零售包装产品，不论其是否混合（精油水馏液及水溶液除外）。

四、税号33.07所称“芳香料制品及化妆盥洗品”，主要适用于下列产品：香袋；通过燃烧散发香气的制品；香纸及用化妆品浸渍或涂布的纸；隐形眼镜片或假眼用的溶液；用香水或化妆品浸渍、涂布、包覆的絮胎、毡呢及无纺织物；动物用盥洗品。

Notes:

1. This Chapter does not cover:
   (a) Natural oleoresins or vegetable extracts of heading No.13.01 or 13.02;
   (b) Soap or other products of heading No.34.01; or
   (c) Gum, wood or sulphate turpentine or other products of heading No.38.05.

2. The expression "odoriferous substances" in heading No.33.02 refers only to the substances of heading No.33.01, to odoriferous constituents isolated from those substances or to synthetic aromatics.

3. Headings Nos.33.03 to 33.07 apply, *inter alia*, to products, whether or not mixed (other than aqueous distillates and aqueous solutions of essential oils), suitable for use as goods of these headings and put up in packings of a kind sold by retail for such use.

4. The expression "perfumery, cosmetic or toilet preparations" in heading No.33.07 applies, *inter alia*, to the following products: scented sachets; odoriferous preparations which operate by burning; perfumed papers and papers impregnated or coated with cosmetics; contact lens or artificial eye solutions; wadding, felt and nonwovens, impregnated, coated or covered with perfume or cosmetics; animal toilet preparations.

序号 No.	税则号列 Tariff Item	货品名称	最惠国税率(%) M.F.N.	协定税率(%)Agreement 东盟 ASEAN	曼谷协定 BA	香港 H.K.	澳门 Macao	普通税率(%) Gen.	Article Description
	33.01	**精油（无萜或含萜），包括浸膏及净油；香膏；提取的油树脂；用花香吸取法或浸渍法制成的含浓缩精油的脂肪、固定油、蜡及类似品；精油脱萜时所得的萜烯副产品；精油水馏液及水溶液：**							**Essential oils (terpeneless or not), including concretes and absolutes; resinoids; extracted oleoresins; concentrates of essential oils in fats, in fixed oils, in waxes or the like, obtained by enfleurage or maceration; terpenic byproducts of the deterpenation of essential oils; aqueous distillates and aqueous solutions of essential oils:**
		-柑桔属果实的精油：							-Essential oils of citrus fruit:
2269	3301.1100	--香柠檬油（佛手油）	20					80	--Of bergamot
2270	3301.1200	--橙油	20					80	--Of orange

序号 No.	税则号列 Tariff Item	货品名称	最惠国税率(%) M.F.N.	协定税率(%) Agreement 东盟 ASEAN	曼谷协定 BA	香港 H.K.	澳门 Macao	普通税率(%) Gen.	Article Description
2271	3301.1300	--柠檬油	20					80	--Of lemon
2272	3301.1400	--白柠檬油(酸橙油)	20					80	--Of lime
2273	3301.1900	--其他	20					80	--Other
		-非柑桔属果实的精油:							-Essential oils other than those of citrus fruit:
2274	3301.2100	--老鹳草油(香叶油)	20					80	--Of geranium
2275	3301.2200	--茉莉油	20					80	--Of jasmin
2276	3301.2300	--熏衣草(包括杂种熏衣草)油	20					80	--Of lavender or of lavandin
2277	3301.2400	--胡椒薄荷油	20					90	--Of peppermint(Mentha piperita)
2278	3301.2500	--其他薄荷油	15		14			90	--Of other mints
2279	3301.2600	--岩兰草油	20					80	--Of vetiver
		--其他:							--Other:
2280	3301.2910	---樟脑油	20					90	---Of camphor
2281	3301.2920	---香茅油	15					70	---Of citronella
2282	3301.2930	---茴香油	20					80	---Of aniseed
2283	3301.2940	---桂油	20					80	---Of cassia
2284	3301.2950	---山苍子油	20					80	---Of litsea cubeba
2285	3301.2960	---桉叶油	20					80	---Of eucalyptus
2286	3301.2990	---其他	15					80	---Other
		-香膏:							-Resinoids:
2287	3301.3010	---鸢尾凝脂	20					80	---Balsam of irises
2288	3301.3090	---其他	20					80	---Other
		-其他:							-Other:
2289	3301.9010	---提取的油树脂	20		19		0	80	---Extracted oleoresins
2290	3301.9020	---柑桔属果实的精油脱萜的萜烯副产品	20				0	80	---Terpenic byproducts of the deterpenation of essential oils of citrus fruit
2291	3301.9090	---其他	20		18.1		0	80	---Other
	33.02	**工业原料用的混合香料以及以一种或多种香料为基本成分的混合物(包括酒精溶液);生产饮料用的以香料为基本成分的其他制品:**							**Mixtures of odoriferous substances and mixtures(including alcoholic solutions)with a basis of one or more of these substances,of a kind used as raw materials in industry;other preparations based on odoriferous substances, of a kind used for the manufacture of beverages:**
		-食品或饮料工业用:							-Of a kind used in the food or drink industry:
2292	3302.1010	---生产饮料用的以香料为基本成分的制品,按容量计酒精浓度不超过0.5%的	15		12.8			90	---Preparations based on odoriferous substances, of a kind used for the manufacture of beverages, alcoholic strength by volume not exceeding 0.5% vol.

序号 No.	税则号列 Tariff Item	货品名称	最惠国税率(%) M.F.N.	协定税率(%) Agreement 东盟 ASEAN	曼谷协定 BA	香港 H.K.	澳门 Macao	普通税率(%) Gen.	Article Description
2293	3302.1090	---其他	15					130	---Other
2294	3302.9000	-其他	10			0	0	130	-Other
	33.03	**香水及花露水:**							**Perfumes and toilet waters:**
2295	3303.0000	香水及花露水	10		8.2	0	0	150	Perfumes and toilet waters
	33.04	**美容品或化妆品及护肤品(药品除外),包括防晒油或晒黑油;指(趾)甲化妆品:**							**Beauty or make-up preparations and preparations for the care of the skin (other than medicaments), including sunscreen or sun tan preparations; manicure or pedicure preparations:**
2296	3304.1000	-唇用化妆品	10			0	0	150	-Lip make-up preparations
2297	3304.2000	-眼用化妆品	10			0	0	150	-Eye make-up preparations
2298	3304.3000	-指(趾)甲化妆品	15			0	0	150	-Manicure or pedicure preparations
		-其他:							-Other:
2299	3304.9100	--香粉,不论是否压紧	10					150	--Powdrs, whether or not compressed
2300	3304.9900	--其他	16			0	0	150	--Other
	33.05	**护发品:**							**Preparations for use on the hair:**
2301	3305.1000	-洗发剂(香波)	12.7		10.5			150	-Shampoos
2302	3305.2000	-烫发剂	15					150	-Preparations for permanent waving or straightening
2303	3305.3000	-定型剂	15					150	-Hair lacquers
2304	3305.9000	-其他	10					150	-Other
	33.06	**口腔及牙齿清洁剂,包括假牙模膏及粉;清洁牙缝用的纱线(牙线),单独零售包装的:**							**Preparations for oral or dental hygiene, including denture fixative pastes and powders; yarn used to clean between the teeth (dental floss), in individual retail package:**
		-洁齿品:							-Dentifrices:
2305	3306.1010	---牙膏	10					150	---Toothpastes
2306	3306.1090	---其他	10					150	---Other
2307	3306.2000	-清洁牙缝用的纱线(牙线)	10		9.2			70	-Yarn used to clean between the teeth (dental floss)
2308	3306.9000	-其他	10					70	-Other
	33.07	**剃须用制剂、人体除臭剂、沐浴用制剂、脱毛剂和其他税号未列名的芳香料制品及化妆盥洗品;室内除臭剂,不论是否加香水或消毒剂:**							**Pre-shave, shaving or after-shave preparations, personal deodorants, bath preparations, depilatories and other perfumery, cosmetic or toilet preparations, not elsewhere specified or included; prepared room deodorizers, whether or not perfumed or having disinfectant properties:**

序号 No.	税则号列 Tariff Item	货品名称	最惠国税率(%) M.F.N.	协定税率(%)Agreement 东盟 ASEAN	曼谷协定 BA	香港 H.K.	澳门 Macao	普通税率(%) Gen.	Article Description
2309	3307.1000	-剃须用制剂	10					150	-Pre-shave, shaving or after-shave preparations
2310	3307.2000	-人体除臭剂及止汗剂	10					150	-Personal deodorants and antiperspirants
2311	3307.3000	-香浴盐及其他沐浴用制剂	10					150	-Perfumed bath salts and other bath preparations
		-室内散香或除臭制品，包括宗教仪式用的香：							-Preparations for perfuming or deodorizing rooms, including odoriferous preparations used during religious rites:
2312	3307.4100	--神香及其他通过燃烧散发香气的制品	10			0		150	--Agarbatti and other odoriferous preparations which operate by burning
2313	3307.4900	--其他	10					150	--Other
2314	3307.9000	-其他	9					150	-Other

第三十四章
肥皂、有机表面活性剂、洗涤剂、润滑剂、人造蜡、调制蜡、光洁剂、蜡烛及类似品、塑型用膏、“牙科用蜡”及牙科用熟石膏制剂

注释:

一、本章不包括:

(一) 用作脱模剂的食用动植物油、脂混合物或制品(税号 15.17);

(二) 单独的已有化学定义的化合物;

(三) 含肥皂或其他有机表面活性剂的洗发剂、洁齿品、剃须膏及沐浴用制剂(税号 33.05、33.06 及 33.07)。

二、税号 34.01 所称“肥皂”,只适用于水溶性肥皂。税号 34.01 的肥皂及其他产品可以含有添加料(例如,消毒剂、磨料粉、填料或药料)。含磨料粉的产品,只有条状、块状或模制形状可以归入税号 34.01。其他形状的应作为“去污粉及类似品”归入税号 34.05。

三、税号 34.02 所称“有机表面活性剂”,是指温度在 20℃ 时与水混合配成 0.5% 浓度的水溶液,并在同样温度下搁置一小时后与下列规定相符的产品:

(一) 成为透明或半透明的液体或稳定的乳浊液而未离析出不溶解物质;

(二) 将水的表面张力减低到每厘米 45 达因及以下。

四、税号 34.03 所称“石油及从沥青矿物提取的油类”,适用于第二十七章注释二所规定的产品。

五、税号 34.04 所称“人造蜡及调制蜡”,仅适用于:

(一) 用化学方法生产的具有蜡质特性的有机产品,不论是否为水溶性的;

(二) 各种蜡混合制成的产品;

(三) 以一种或几种蜡为基本原料并含有油脂、

Chapter 34
Soap, organic surface-active agents, washing preparations, lubricating preparations, artificial waxes, prepared waxes, polishing or scouring preparations, candles and similar articles, modelling pastes, "dental waxes" and dental preparations with a basis of plaster

Notes:

1. This Chapter does not cover:
   (a) Edible mixtures or preparations of animal or vegetable fats or oils of a kind used as mould release preparations (heading No.15.17);
   (b) Separate chemically defined compounds; or
   (c) Shampoos, dentifrices, shaving creams and foams, or bath preparations, containing soap or other organic surface-active agents (heading No.33.05, 33.06 or 33.07).

2. For the purposes of heading No.34.01, the expression "soap" applies only to soap soluble in water. Soap and the other products of heading No.34.01 may contain added substances (for example, disinfectants, abrasive powders, fillers or medicaments). Products containing abrasive powders remain classified in heading No.34.01 only if in the form of bars, cakes or moulded pieces or shapes. In other forms they are to be classified in heading No.34.05 as "scouring powders and similar preparations".

3. For the purposes of heading No.34.02, "organic surfaceactive agents" are products which when mixed with water at a concentration of 0.5% at 20℃ and left to stand for one hour at the same temperature:
   (a) give a transparent or translucent liquid or stable emulsion without separation of insoluble matter; and
   (b) reduce the surface tension of water to 4.5×10^{-2} N/m (45 dyen/cm) or less

4. In heading No.34.03 the expression "petroleum oils and oils obtained from bituminous minerals" applies to the products defined in Note 2 to Chapter 27.

5. In heading No.34.04, subject to the exclusions provided below, the expression "artificial waxes and prepared waxes" applies only to:
   (A) Chemically produced organic products of a waxy character, whether or not water-soluble;
   (B) Products obtained by mixing different waxes;
   (C) Products of a waxy character with a basis of one or more

树脂、矿物质或其他原料的具有蜡质特性的产品。

本税号不包括：

（一）税号 15.16、34.02 或 38.23 的产品，不论是否具有蜡质特性；

（二）税号 15.21 的未混合的动物蜡或未混合的植物蜡，不论是否精制或着色；

（三）税号 27.12 的矿物蜡或类似产品，不论是否相互混合或仅经着色；

（四）混合、分散或溶解于液体溶剂的蜡（税号 34.05、38.09 等）。

waxes and containing fats, resins, mineral substances or other materials.

The heading does not apply to:

(a) Products of heading No.15.16, 34.02 or 38.23, even if having a waxy character;

(b) Unmixed animal waxes or unmixed vegetable waxes, whether or not refined or coloured, of heading No.15.21;

(c) Mineral waxes or similar products of heading No.27.12, whether or not intermixed or merely coloured; or

(d) Waxes mixed with, dispersed in or dissolved in a liquid medium (headings Nos.34.05, 38.09, etc.).

序号 No.	税则号列 Tariff Item	货品名称	最惠国税率(%) M.F.N.	协定税率(%) Agreement 东盟 ASEAN	曼谷协定 BA	香港 H.K.	澳门 Macao	普通税率(%) Gen.	Article Description
	34.01	**肥皂；作肥皂用的有机表面活性产品及制品，条状、块状或模制形状的，不论是否含有肥皂；洁肤用的有机表面活性产品及制品，液状或膏状并制成零售包装的，不论是否含有肥皂；用肥皂或洗涤剂浸渍、涂面或包覆的纸、絮胎、毡呢及无纺织物：**							**Soap; organic surface-active products and preparations for use as soap, in the form of bars, cakes, moulded pieces or shapes, whether or not containing soap; organic surfaceactive products and preparations for washing the skin, in the form of liguid or cream and put up for retail sale, whether or not containing soap; paper, wadding, felt and nonwovens, impregnated, coated or covered with soap or detergent:**
		-肥皂及有机表面活性产品及制品，条状、块状或模制形状的，以及用肥皂或洗涤剂浸渍、涂面或包覆的纸、絮胎、毡呢及无纺织物：							-Soap and organic surface-active products and preparations, in the form of bars, cakes, moulded pieces or shapes, and paper, wadding, felt and nonwovens, impregnated, coated or covered with soap or detergent:
2315	3401.1100	--盥洗用（包括含有药物的产品）	10		8.3			130	--For toilet use (including medicated products)
		--其他：							--Other:
2316	3401.1910	---洗衣皂	10					80	---Laundry soap
2317	3401.1990	---其他	15	见表1				130	---Other
2318	3401.2000	-其他形状的肥皂	15	见表1	12.4	0		130	-Soap in other forms
2319	3401.3000	-洁肤用的有机表面活性产品及制品，液状或膏状并制成零售包装的，不论是否含有肥皂	10					130	-Organic surface-active products and preparations for washing the skin, in the form of liquid or cream and put up for retail sale, whether or not containing soap
	34.02	**有机表面活性剂（肥皂除外）；表面活性剂制品、洗涤剂（包括助洗剂）及清洁剂，不论是否含有肥皂，但税号 34.01 的产品除外：**							**Organic surface-active agents (other than soap); surface-active preparations, washing preparations (including auxiliary washing preparations) and cleaning preparations, whether or not containing soap, other than those of heading No.34.01:**

序号 No.	税则号列 Tariff Item	货品名称	最惠国税率(%) M.F.N.	协定税率(%)Agreement 东盟 ASEAN	曼谷协定 BA	香港 H.K.	澳门 Macao	普通税率(%) Gen.	Article Description
		-有机表面活性剂,不论是否零售包装:							-Organic surface-active agents, whether or not put up for retail sale:
2320	3402.1100	--阴离子型	6.5		6	0		30	--Anionic
2321	3402.1200	--阳离子型	6.5		6	0		30	--Cationic
2322	3402.1300	--非离子型	6.5		6	0		30	--Non-ionic
2323	3402.1900	--其他	6.5		6			30	--Other
		-零售包装的制品:							-Preparations put up for retail sale:
2324	3402.2010	---合成洗涤粉	10		9.4			80	---Synthetic detergents in powder form
2325	3402.2090	---其他	10		8.9	0		80	---Other
2326	3402.9000	-其他	9		7.8	0	0	80	-Other
	34.03	**润滑剂(包括以润滑剂为基本成分的切削油制剂、螺栓或螺母松开剂、防锈或防腐蚀制剂及脱模剂)及用于纺织材料、皮革、毛皮或其他材料油脂处理的制剂,但不包括以石油或从沥青矿物提取的油类为基本成分(按重量计不低于70%)的制剂:**							**Lubricating preparations (including cutting-oil preparations, bolt or nut release preparations, anti-rust or anti-corrosion preparations and mould release preparations, based on lubricants) and preparations of a kind used for the oil or grease treatment of textile materials, leather, furskins or other materials, but excluding preparations containing, as basic constituents, 70% or more by weight of petroleum oils or of oils obtained from bituminous minerals:**
		-含有石油或从沥青矿物提取的油类:							-Containing petroleum oils or oils obtained from bituminous minerals:
2327	3403.1100	--处理纺织材料、皮革、毛皮或其他材料的制剂	10			0		50	--Preparations for the treatment of textile materials, leather, furskins or other materials
2328	3403.1900	--其他	10					50	--Other
		-其他:							-Other:
2329	3403.9100	--处理纺织材料、皮革、毛皮或其他材料的制剂	10			0		50	--Preparations for the treatment of textile materials, leather, furskins or other materials
2330	3403.9900	--其他	10				0△	50	--Other
	34.04	**人造蜡及调制蜡:**							**Artificial waxes and prepared waxes:**
2331	3404.1000	-化学改性的褐煤蜡	10					70	-Of chemically modified lignite
2332	3404.2000	-聚氧乙烯(聚乙二醇)蜡	10					70	-Of poly(oxyethylene) (polyethylene glycol)
2333	3404.9000	-其他	10					70	-Other
	34.05	**鞋靴、家具、地板、车身、玻璃或金属用的光洁剂、擦洗膏、去污粉及类似制品(包括用这类制剂浸渍、涂面或包覆的纸、絮胎、毡呢、无纺织物、泡沫塑料或海绵橡胶),但不包括税号34.04的蜡:**							**Polishes and creams, for footwear, furniture, floors, coachwork, glass or metal, scouring pastes and powders and similar preparations (whether or not in the form of paper, wadding, felt, nonwovens, cellular plastics or cellular rubber, impregnated, coated or covered with such preparations), excluding waxes of heading No.34.04:**

序号 No.	税则号列 Tariff Item	货品名称	最惠国税率(%) M.F.N.	协定税率(%) Agreement 东盟 ASEAN	曼谷协定 BA	香港 H.K.	澳门 Macao	普通税率(%) Gen.	Article Description
2334	3405.1000	-鞋靴或皮革用的上光剂及类似制品	10					80	-Polishes, creams and similar preparations for footwear or leather
2335	3405.2000	-保养木制家具、地板或其他木制品用的上光剂及类似制品	10			0		80	-Polishes, creams and similar preparations for the maintenance of wooden furniture, floors or other woodwork
2336	3405.3000	-车身用的上光剂及类似制品,但金属用的光洁剂除外	10					80	-Polishes and similar preparations for coachwork, other than metal polishes
2337	3405.4000	-擦洗膏、去污粉及类似制品	10					80	-Scouring pastes and powders and other scouring preparations
2338	3405.9000	-其他	10			0		80	-Other
	34.06	**各种蜡烛及类似品:**							**Candles, tapers and the like:**
2339	3406.0000	各种蜡烛及类似品	10					130	Candles, tapers and the like
	34.07	**塑型用膏,包括供儿童娱乐用的在内;通称为"牙科用蜡"或"牙科造形膏"的制品,成套、零售包装或制成片状、马蹄形、条状及类似形状的;以熟石膏(煅烧石膏或硫酸钙)为基本成分的牙科用其他制品:**							**Modelling pastes, including those put up for children's amusement; preparations known as "dental wax" or as "dental impression compounds", put up in sets, in packings for retail sale or in plates, horseshoe shapes, sticks or similar forms; other preparations for use in dentistry, with a basis of plaster (of calcined gypsum or calcium sulphate):**
2340	3407.0010	---牙科用蜡及造型膏	6.5					30	---Preparations of a kind known as "dental wax" or as "dental impression compounds"
2341	3407.0020	---以熟石膏为基本成分的牙科用其他制品	6.5					40	---Other preparations for use in dentistry, with a basis of plaster
2342	3407.0090	---其他	10					100	---Other

第三十五章
蛋白类物质；改性淀粉；胶；酶

Chapter 35
Albuminoidal substances; modified starches; glues; enzymes

注释：

一、本章不包括：

（一）酵母（税号 21.02）；

（二）第三十章的血份（非治病、防病用的血清白蛋白除外）、药品及其他产品；

（三）预鞣用酶制剂（税号 32.02）；

（四）第三十四章的加酶的浸透剂、洗涤剂及其他产品；

（五）硬化蛋白（税号 39.13）；

（六）印刷工业用的明胶产品（第四十九章）。

二、税号 35.05 所称"糊精"，是指淀粉的降解产品，其还原糖含量以右旋糖的干重量计不超过 10%。

如果还原糖含量超过 10%，应归入税号 17.02。

Notes:

1. This Chapter does not cover:
   (a) Yeasts (heading No.21.02);
   (b) Blood fractions (other than blood albumin not prepared for therapeutic or prophylactic uses), medicaments or other products of Chapter 30;
   (c) Enzymatic preparations for pre-tanning (heading No.32.02);
   (d) Enzymatic soaking or washing preparations or other products of Chapter 34;
   (e) Hardened proteins (heading No.39.13); or
   (f) Gelatin products of the printing industry (Chapter 49).

2. For the purposes of heading No.35.05, the term "dextrins" means starch degradation products with a reducing sugar content, expressed as dextrose on the dry substance, not exceeding 10%.
   Such products with a reducing sugar content exceeding 10% fall in heading No.17.02.

序号 No.	税则号列 Tariff Item	货品名称	最惠国税率(%) M.F.N.	协定税率(%) Agreement 东盟 ASEAN	曼谷协定 BA	香港 H.K.	澳门 Macao	普通税率(%) Gen.	Article Description
	35.01	**酪蛋白、酪蛋白酸盐及其他酪蛋白衍生物；酪蛋白胶：**							**Casein, caseinates and other casein derivatives; casein glues:**
2343	3501.1000	-酪蛋白	10					35	-Casein
2344	3501.9000	-其他	10					35	-Other
	35.02	**白蛋白(包括按重量计干质成分的乳清蛋白含量超过 80% 的两种或两种以上的乳清蛋白浓缩物)、白蛋白盐及其他白蛋白衍生物：**							**Albumins (including concentrates of two or more whey proteins, containing by weight more than 80% whey proteins, calculated on the dry matter), albuminates and other albumin derivatives:**
		-卵清蛋白：							-Egg albumin:
2345	3502.1100	--干的	10					80	--Dried
2346	3502.1900	--其他	10					80	--Other
2347	3502.2000	-乳白蛋白，包括两种或两种以上的乳清蛋白浓缩物	10					35	-Milk albumin, including concentrates of two or more whey proteins
2348	3502.9000	-其他	10			0△		35	-Other
	35.03	**明胶(包括长方形、正方形明胶薄片，不论是否表面加工或着色)及其衍生物；鱼鳔胶；其他动物胶，但不包括税号35.01的酪蛋白胶：**							**Gelatin (including gelatin in rectangular (including square) sheets, whether or not surface-worked or coloured) and gelatin derivatives; isinglass; other glues of animal origin, excluding casein glues of heading No.35.01:**

序号 No.	税则号列 Tariff Item	货品名称	最惠国税率(%) M.F.N.	协定税率(%)Agreement 东盟 ASEAN	曼谷协定 BA	香港 H.K.	澳门 Macao	普通税率(%) Gen.	Article Description
2349	3503.0010	---明胶及其衍生物	12					35	---Gelatin and gelatin derivatives
2350	3503.0090	---其他	12					50	---Other
	35.04	**蛋白胨及其衍生物;其他税号未列名的蛋白质及其衍生物;皮粉,不论是否加入铬矾:**							**Peptones and their derivatives; other protein substances and their derivatives, not elsewhere specified or included; hide powder, whether or not chromed:**
2351	3504.0010	---蛋白胨	3					11	---Peptones
2352	3504.0090	---其他	8					35	---Other
	35.05	**糊精及其他改性淀粉(例如,预凝化淀粉或酯化淀粉);以淀粉、糊精或其他改性淀粉为基本成分的胶:**							**Dextrins and other modified starches (for example, pregelatinized or esterified starches); glues based on starches, or on dextrins or other modified starches:**
2353	3505.1000	-糊精及其他改性淀粉	12			0		50	-Dextrins and other modified starches
2354	3505.2000	-胶	20					50	-Glues
	35.06	**其他税号未列名的调制胶及其他调制粘合剂;适于作胶或粘合剂用的产品,零售包装每件净重不超过1公斤:**							**Prepared glues and other prepared adhesives, not elsewhere specified or included; products suitable for use as glues or adhesives, put up for retail sale as glues or adhesives, not exceeding a net weight of 1 kg:**
2355	3506.1000	-适于作胶或粘合剂用的产品,零售包装每件净重不超过1公斤	10		9.2	$0^{\triangle}$		90	-Products suitable for use as glues or adhesives, put up for retail sale as glues or adhesives, not exceeding a net weight, of 1 kg
		-其他:							-Other:
		--以橡胶或税目39.01至39.13的聚合物为基本成分的粘合剂:							--Adhesives based on polymers of headings 39.01 to 39.13 or on rubber:
2356	3506.9110	---以聚酰胺为基本成分的	10			0	0	90	---Based on polyamide
2357	3506.9120	---以环氧树脂为基本成分的	10			0	0	90	---Based on epoxy resin
2358	3506.9190	---其他	10			0	0	90	---Other
2359	3506.9900	--其他	10		8.6	0	0	90	--Other
	35.07	**酶;其他税号未列名的酶制品:**							**Enzymes; prepared enzymes not elsewhere specified or included:**
2360	3507.1000	-粗制凝乳酶及其浓缩物	6					30	-Rennet and concentrates thereof
		-其他:							-Other:
2361	3507.9010	---碱性蛋白酶	6					30	---Basic proteinase
2362	3507.9020	---碱性脂肪酶	6					30	---Basic lipase
2363	3507.9090	---其他	6					30	---Other

第三十六章
炸药；烟火制品；火柴；引火合金；易燃材料制品

Chapter 36
Explosives; pyrotechnic products; matches; pyrophoric alloys; certain combustible preparations

注释：

一、本章不包括单独的已有化学定义的化合物，但下列注释二（一）、（二）所述物品除外。

二、税号36.06所称“易燃材料制品”，只适用于：

（一）聚乙醛、六甲撑四胺及类似物质，已制成片、棒或类似形状作燃料用的；以酒精为基本成分的固体或半固体燃料及类似的配制燃料；

（二）直接灌注香烟打火机及类似打火器用的液体燃料或液化气体燃料，其包装容器的容积不超过300立方厘米；

（三）树脂火炬、引火物及类似品。

Notes:

1. This Chapter does not cover separate chemically defined compounds other than those described in Note 2 (a) or (b) below.

2. The expression "articles of combustible materials" in heading No.36.06 applies only to:
   (a) Metaldehyde, hexamethylenetetramine and similar substances, put up in forms (for example, tablets, sticks or similar forms) for use as fuels; fuels with as basis of alcohol, and similar prepared fuels, in solid or semi-solid form;
   (b) Liquid or liquefied-gas fuels in containers of a kind used for filling or refilling cigarette or similar lighters and of a capacity not exceeding 300 cm^3, and
   (c) Resin torches, firelighters and the like.

序号 No.	税则号列 Tariff Item	货品名称	最惠国税率(%) M.F.N.	协定税率(%)Agreement 东盟 ASEAN	曼谷协定 BA	香港 H.K.	澳门 Macao	普通税率(%) Gen.	Article Description
	36.01	**发射药：**							**Propellent powders:**
2364	3601.0000	发射药	9					50	Propellent powders
	36.02	**配制炸药，但发射药除外：**							**Prepared explosives, other than propellent powders:**
2365	3602.0010	---硝铵炸药	9					50	---Based on ammonals nitrate
2366	3602.0090	---其他	9					50	---Other
	36.03	**安全导火索；导爆索；火帽或雷管；引爆器；电雷管：**							**Safety fuses; detonating fuses; percussion or detonating caps; igniters; electric detonators:**
2367	3603.0000	安全导火索；导爆索；火帽或雷管；引爆器；电雷管	9					50	Safety fuses; detonating fuses; percussion or detonating caps; igniters; electric detonators
	36.04	**烟花、爆竹、信号弹、降雨火箭、浓雾信号弹及其他烟火制品：**							**Fireworks, signalling flares, rain rockets, fog signals and other pyrotechnic articles:**
2368	3604.1000	-烟花、爆竹	6					130	-Firoworks
2369	3604.9000	-其他	6					100	-Other
	36.05	**火柴，但税号36.04的烟火制品除外：**							**Matches, other than pyrotechnic articles of heading No.36.04:**
2370	3605.0000	火柴，但税号36.04的烟火制品除外	6					100	Matches, other than pyrotechnic articles of heading No.36.04

序号 No.	税则号列 Tariff Item	货品名称	最惠国税率(%) M.F.N.	协定税率(%) Agreement 东盟 ASEAN	曼谷协定 BA	香港 H.K.	澳门 Macao	普通税率(%) Gen.	Article Description
	36.06	**各种形状的铈铁及其他引火合金;本章注释二所述的易燃材料制品:**							**Ferro-cerium and other pyrophoric alloys in all forms; articles of combustible materials as specified in Note 2 to this Chapter:**
2371	3606.1000	-直接灌注香烟打火机及类似打火器用的液体燃料或液化气体燃料,其包装容器的容积不超过300立方厘米	10			0		80	-Liquid or liquefied-gas fuels in containers of a kind used for filling or refilling cigarette or similar lighters and of a capacity not exceeding 300 cm^3
		--其他:							--Other:
		---铈铁及其他引火合金:							---Ferro-cerium and other pyrophoric alloys:
2372	3606.9011	----已切成形可直接使用	9					80	----Cut to shape, for immediate use
2373	3606.9019	----其他	9					50	----Other
2374	3606.9090	---其他	9					80	---Other

第三十七章
照相及电影用品

Chapter 37
Photographic or cinematographic goods

注释：

一、本章不包括废碎料。

二、本章所称“摄影”，是指光或其他射线作用于感光面上直接或间接形成可见影像的过程。

Notes:

1. This Chapter does not cover waste or scrap.

2. In this Chapter the word “photographic” relates to the process by which visible images are formed, directly or indirectly, by the action of light or other forms of radiation on photosensitive surfaces.

序号 No.	税则号列 Tariff Item	货品名称	最惠国税率(%) M.F.N.	协定税率(%) Agreement 东盟 ASEAN	曼谷协定 BA	香港 H.K.	澳门 Macao	普通税率(%) Gen.	Article Description
	37.01	**未曝光的摄影感光硬片及平面软片，用纸、纸板及纺织物以外任何材料制成；未曝光的一次成像感光平片，不论是否分装：**							**Photographic plates and film in the flat, sensitized, unexposed, of any material other than paper, paperboard or textiles; instant print film in the flat, sensitized, unexposed, whether or not in packs:**
2375	3701.1000	-X 光用	20					40	-For X-ray
2376	3701.2000	-一次成像平片	5					40	-Instant print film
		-其他硬片及软片，任何一边超过 255 毫米：							-Other plates and film, with any side exceeding 255 mm:
		---照相制版用：							---For preparing printing plates or cylinders:
2377	3701.3021	----激光照排片	见表 4				0	见表 4	----Laser phototypesetting film
2378	3701.3022	----PS 版	见表 4				0	见表 4	----Precoated sensitized plate
2379	3701.3023	----CTP 版（计算机直接制版的热敏版材）	见表 4				0	见表 4	----CTP plate
2380	3701.3029	----其他	见表 4				0	见表 4	----Other
2381	3701.3090	---其他	20					70	---Other
		-其他：							-Other:
2382	3701.9100	--彩色摄影用	22					70	--For colour photography(polychrome)
		--其他：							--Other:
2383	3701.9920	---照相制版用	10					40	---For preparing printing plates or cylinders
2384	3701.9990	---其他	25					70	---Other
	37.02	**成卷的未曝光摄影感光胶片，用纸、纸板及纺织物以外任何材料制成；未曝光的一次成像感光卷片：**							**Photographic film in rolls, sensitized, unexposed, of any material other than paper, paperboard or textiles; instant print film in rolls, sensitized, unexposed:**
2385	3702.1000	-X 光用	10					40	-For X-ray
2386	3702.2000	-一次成像卷片	5					40	-Instant print film
		-无齿孔的其他胶片，宽度不超过 105 毫米：							-Other film, without perforations, of a width not exceeding 105 mm:
2387	3702.3100	--彩色摄影用	见表 4					见表 4	--For colour photography(polychrome)

序号 No.	税则号列 Tariff Item	货品名称	最惠国税率(%) M.F.N.	协定税率(%)Agreement 东盟 ASEAN	曼谷协定 BA	香港 H.K.	澳门 Macao	普通税率(%) Gen.	Article Description
		--其他涂卤化银乳液的:							--Other, with silver halide emulsion:
2388	3702.3220	---照相制版用	见表4					见表4	---For preparing printing plates or cylinders
2389	3702.3290	---其他	见表4					见表4	---Other
		--其他:							--Other:
2390	3702.3920	---照相制版用	见表4					见表4	---For preparing printing plates or cylinders
2391	3702.3990	---其他	见表4					见表4	---Other
		-无齿孔的其他胶片,宽度超过105毫米:							-Other film, without perforations, of a width exceeding 105 mm:
2392	3702.4100	--彩色摄影用,宽度超过610毫米,长度超过200米	见表4					见表4	--Of a width exceeding 610 mm and of a length exceeding 200 m, for colour photography(polychrome):
		--非彩色摄影用,宽度超过610毫米,长度超过200米:							--Of a width exceeding 610 mm and of a length exceeding 200 m, other than for colour photography:
		---照相制版用:							---For preparing printing plates or cylinders:
2393	3702.4221	----印刷电路板制造用光致抗蚀干膜	见表4					见表4	----Wide anticorrosive photographic plate for printed circuit processing
2394	3702.4229	----其他	见表4					见表4	----Other
		---其他:							---Other:
2395	3702.4291	----红外或氦氖激光胶片	见表4					见表4	----Infra-red or helium-neon laser film
2396	3702.4299	----其他	见表4					见表4	----Other
		--宽度超过610毫米,长度不超过200米:							--Of a width exceeding 610 mm and of a length not exceeding 200 m:
		---照相制版用:							---For preparing printing plates or cylinders:
2397	3702.4321	----激光照排片	见表4					见表4	----Laser phototypesetting film
2398	3702.4329	----其他	见表4					见表4	----Other
2399	3702.4390	---其他	见表4					见表4	---Other
		--宽度超过105毫米,但不超过610毫米:							--Of a width exceeding 105 mm but not exceeding 610 mm:
		---照相制版用:							---For preparing printing plates or cylinders:
2400	3702.4421	----激光照排片	见表4					见表4	----Laser phototypesetting film
2401	3702.4422	----印刷电路板制造用光致抗蚀干膜	见表4					见表4	----Narrow anticorrosive photographic plate for printed circuit processing
2402	3702.4429	----其他	见表4					见表4	----Other
2403	3702.4490	---其他	见表4					见表4	---Other
		-彩色摄影用的其他胶片:							-Other film, for colour photography (polychrome):
2404	3702.5100	--宽度不超过16毫米,长度不超过14米	见表4					见表4	--Of a width not exceeding 16 mm and of a length not exceeding 14 m

序号 No.	税则号列 Tariff Item	货品名称	最惠国税率(%) M.F.N.	协定税率(%)Agreement 东盟 ASEAN	曼谷协定 BA	香港 H.K.	澳门 Macao	普通税率(%) Gen.	Article Description
2405	3702.5200	--宽度不超过16毫米,长度超过14米	见表4					见表4	--Of a width not exceeding 16 mm and of a length exceeding 14 m
2406	3702.5300	--幻灯片用,宽度超过16毫米,但不超过35毫米,长度不超过30米	见表4					见表4	--Of a width exceeding 16 mm but not exceeding 35 mm and of a length not exceeding 30 m, for slides
		--非幻灯片用,宽度超过16毫米,但不超过35毫米,长度不超过30米:							--Of a width exceeding 16 mm but not exceeding 35 mm and of a length not exceeding 30 m, other than for slides:
2407	3702.5410	---宽度为35毫米,长度不超过2米	见表4					见表4	---Of a width 35 mm and of a length not exceeding 2 m
2408	3702.5490	---其他	见表4					见表4	---Other
		--宽度超过16毫米,但不超过35毫米,长度超过30米:							--Of a width exceeding 16 mm but not exceeding 35 mm and of a length exceeding 30 m:
2409	3702.5520	---电影胶片	见表4					见表4	---Cinematographic film
2410	3702.5590	---其他	见表4					见表4	---Other
		--宽度超过35毫米:							--Of a width exceeding 35 mm:
2411	3702.5620	---电影胶片	见表4					见表4	---Cinematographic film
2412	3702.5690	---其他	见表4					见表4	---Other
		-其他:							-Other:
2413	3702.9100	--宽度不超过16毫米	见表4					见表4	--Of a width not exceeding 16 mm
		--宽度超过16毫米,但不超过35毫米,长度不超过30米:							--Of a width exceeding 16 mm but not exceeding 35 mm and of a length not exceeding 30 m:
2414	3702.9310	---宽度为35毫米,长度不超过2米	见表4					见表4	---Of a width 35 mm and of a length not exceeding 2 m
2415	3702.9390	---其他	见表4					见表4	---Other
		--宽度超过16毫米,但不超过35毫米,长度超过30米:							--Of a width exceeding 16 mm but not exceeding 35 mm and of a length exceeding 30 m:
2416	3702.9420	---电影胶片	见表4					见表4	---Cinematographic film
2417	3702.9490	---其他	见表4					见表4	---Other
		--宽度超过35毫米:							--Of a width exceeding 35 mm:
2418	3702.9520	---电影胶片	见表4					见表4	---Cinematographic film
2419	3702.9590	---其他	见表4					见表4	---Other
	37.03	**未曝光的摄影感光纸、纸板及纺织物:**							**Photographic paper, paperboard and textiles, sensitized, unexposed:**
		-成卷,宽度超过610毫米:							-In rolls of a width exceeding 610 mm:
2420	3703.1010	---感光纸及纸板	22.6					100	---Photographic paper and paperboard
2421	3703.1090	---其他	18					70	---Other
		-其他,彩色摄影用:							-Other, for colour photography (polychrome):

序号 No.	税则号列 Tariff Item	货品名称	最惠国税率(%) M.F.N.	协定税率(%)Agreement 东盟 ASEAN	曼谷协定 BA	香港 H.K.	澳门 Macao	普通税率(%) Gen.	Article Description
2422	3703.2010	---感光纸及纸板	35					100	---Photographic paper and paperboard
2423	3703.2090	---其他	18					70	---Other
		-其他:							-Other:
2424	3703.9010	---感光纸及纸板	35					100	---Photographic paper and paperboard
2425	3703.9090	---其他	18					70	---Other
	37.04	**已曝光未冲洗的摄影硬片、软片、纸、纸板及纺织物:**							**Photographic plates, film, paper, paperboard and textiles, exposed but not developed:**
2426	3704.0010	---电影胶片	8.4					30	---Cinematographic film
2427	3704.0090	---其他	18					70	---Other
	37.05	**已曝光已冲洗的摄影硬片及软片,但电影胶片除外:**							**Photographic plates and film, exposed and developed, other than cinematographic film:**
2428	3705.1000	-供复制胶版用	18					70	-For offset reproduction
		-缩微胶片:							-Microfilms:
2429	3705.2010	---书籍、报刊的	0					0	---For printed books and newspapers
2430	3705.2090	---其他	4					14	---Other
		-其他:							-Other:
2431	3705.9010	---教学专用幻灯片	0					0	---Lantern slides, for educational use only
2432	3705.9090	---其他	18					70	---Other
	37.06	**已曝光已冲洗的电影胶片,不论是否配有声道或仅有声道:**							**Cinematographic film, exposed and developed, whether or not incorporating sound track or consisting only of sound track:**
		-宽度在35毫米及以上:							-Of a width of 35 mm or more:
2433	3706.1010	---教学专用	0					0	---For educational use only
2434	3706.1090	---其他	5			0		14	---Other
		-其他:							-Other:
2435	3706.9010	---教学专用	0					0	---For educational use only
2436	3706.9090	---其他	4			0		14	---Other
	37.07	**摄影用化学制剂(不包括上光漆、胶水、粘合剂及类似制剂);摄影用未混合产品;定量包装或零售包装可立即使用的:**							**Chemical preparations for photographic uses (other than varnishes, glues, adhesives and similar preparations); unmixed products for photographic uses, put up in measured portions or put up for retail sale in a form ready for use:**
2437	3707.1000	-感光乳液	8				0	35	-Sensitizing emulsions
		-其他:							-Other:
2438	3707.9010	---冲洗照相胶卷及相片用	16			0		100	---For use in developing photographic film and photographs
2439	3707.9020	---复印机用	10			0		45	---For use in photo-copying apparatus
2440	3707.9090	---其他	8			0	0	35	---Other

第三十八章
杂项化学产品

注释：

一、本章不包括：

(一) 单独的已有化学定义的元素及化合物，但下列各项除外：

1. 人造石墨（税号 38.01）；
2. 制成税号 38.08 所述的形状或包装的杀虫剂、杀鼠剂、杀菌剂、除草剂、抗萌剂、植物生长调节剂、消毒剂及类似产品；
3. 灭火器的装配药及已装药的灭火弹（税号 38.13）；
4. 下列注释二所规定的检定参照物；
5. 下列注释三（一）及三（三）所规定的产品。

(二) 配制食品用的与食物或其他营养物质混合的化学品（一般归入税号 21.06）；

(三) 符合第二十六章注释三（一）或三（二）的规定，含有金属，砷及其混合物的矿灰和残渣（包括淤渣，但下水道淤泥除外）（税目 26.20）；

(四) 药品（税号 30.03 及 30.04）；

(五) 用于提取贱金属或生产贱金属化合物的废催化剂（税号 26.20），主要用于回收贵金属的废催化剂（税号 71.12），或某种形状（例如，精细粉末或纱网状）的金属或金属合金催化剂（第十四类或第十五类）。

二、

(一) 税目 38.22 所称的“检定参照物”，是指附有证书的参照物，该证书标明了参照物属性的指标、确定这些指标的方法以及与每一指标相关的确定度，这些参照物适用于分析、校准和比较。

(二) 除第二十八章和二十九章的产品外，检定参照物在本目录中应优先归入税目 38.22。

三、税号 38.24 包括不归入本目录其他税号的下列

Chapter 38
Miscellaneous chemical products

Notes:

1. This Chapter does not cover:

(a) Separate chemically defined elements or compounds with the exception of the following:

(1) Artificial graphite (heading No.38.01);

(2) Insecticides, rodenticides, fungicides, herbicides, anti-sprouting products and plant-growth regulators, disinfectants and similar products, put up as described in heading No.38.08;

(3) Products put up as charges for fire-extinguishers or put up in fire-extinguishing grenades (heading No.38.13);

(4) Certified reference materials specified in Note 2 below;

(5) Products specified in Note 3 (a) or 3 (c) below;

(b) Mixtures of chemicals with foodstuffs or other substances with nutritive value, of a kind used in the preparations of human foodstuffs (generally heading No.21.06);

(c) Ash and residues (including sludges, other than sewage sludge), containing metals, arsenic or their mixtures and meeting the requirements of Note 3 (a) or 3 (b) to Chapter 26 (heading 26.20);

(d) Medicaments (heading No.30.03 or 30.04);

(e) Spent catalysts of a kind used for the extraction of base metals or for the manufacture of chemical compounds of base metals (heading No.26.20), spent catalysts of a kind used principally for the recovery of precious metal (heading No.71.12) or catalysts consisting or metals of metal alloys in the form of, for example, finely divided powder or woven gauze (Section XIV or XV).

2.

(a) For the purpose of heading 38.22, the expression "certified reference materials" means reference materials which are accompanied by a certificate which indicates the values of the certified properties, the methods used to determine these values and the degree of certainty associated with each value and which are suitable for analytical, calibrating or referencing purboses.

(b) With the exception of the products of Chapter 28 or 29, for the classification of certified reference materials, heading 38.22 shall take precedence over any other heading in the Nomenclature."

3. Heading No.38.24 includes the following goods which are not

货品：

（一）每颗重量不小于2.5克的氧化镁、碱金属或碱土金属卤化物制成的培养晶体（光学元件除外）；
（二）杂醇油；骨焦油；
（三）零售包装的除墨剂；
（四）零售包装的蜡纸改正液及其他改正液；

（五）可熔性陶瓷测温器（例如，塞格测温锥）。

to be classified in any other heading of the Nomenclature:
(a) Cultured crystals (other than optical elements) weighing not less than 2.5 g each, of magnesium oxide or of the halides of the alkali or alkaline-earth metals;
(b) Fusel oil; Dippel's oil;
(c) Ink removers put up in packings for retail sale;
(d) Stencil correctors and other correcting fluids put up in packings for retail sale; and
(e) Ceramic firing testers, fusible (for example, Segercones).

四、本目录所称"城市垃圾"是指一种从家庭、宾馆、餐馆、医院、商店、办公室等收集来的废物、马路和人行道的垃圾以及建筑垃圾或废墟废物。城市垃圾通常含有大量各种各样的材料，例如，塑料、橡胶、木材、纸张、纺织品、玻璃、金属、食物、破碎家具和其他已损坏或被丢弃的物品，但不包括：

4. Throughout the Nomenclature, "municipal waste" means waste of a kind collected from households, hotels, restaurants, hospitals, shops, offices, etc., road and pavement sweepings, as well as construction and demolition waste. Municipal waste generally contains a large variety of materials such as plastics, rubber, wood, paper, textiles, glass, metals, food materials, broken furniture and other damaged or discarded articles. The term "municipal waste", however, does not cover:

（一）已从垃圾中分拣出来的单独的材料或物品，例如，塑料、橡胶、木材、纸张、纺织品、玻璃、金属及电池的废品，这些材料或物品应归入本目录中适当税目；
（二）工业废物；
（三）在第三十章注释四（十）所规定的废药物；
（四）本章注释六（一）所规定的医疗废物。

(a) Individual materials or articles segregated from the waste, such as wastes of plastics, rubber, wood, paper, textiles, glass or metals and spent batteries which fall in their appropriate headings of the Nomenclature;
(b) Industrial waste;
(c) Waste pharmaceuticals, as defined in Note 4 (k) to Chapter 30; or
(d) Clinical waste, as defined in Note 6 (a) below.

五、税目38.25所称"下水道淤泥"，是指城市污水处理厂产生的淤渣，包括预处理的废物、洗涤污垢和性质不稳定的淤泥。但适合作为肥料用的性质稳定的淤泥除外（第三十一章）。

5. For the purposes of heading 38.25, "sewage sludge" means sludge arising from urban effluent treatment plant and includes pretreatment waste, scourings and unstabilised sludge. Stabilised sludge when suitable for use as fertiliser is excluded (Chapter 31).

六、税目38.25所称"其他废物"适用于：

6. For the purposes of heading 38.25, the expression "other wastes" applies to:

（一）医疗废物，即医学研究、诊断、治疗以及其他内科、外科、牙科或兽医治疗所产生的被污染的废物，通常含有病菌和药物，需作专门的处理（例如，脏的敷料、用过的手套和注射器）；

(a) Clinical waste, that is, contaminated waste arising from medical research, diagnosis, treatment or other medical, surgical, dental or veterinary procedures, which often contain pathogens and pharmaceutical substances and require special disposal procedures (for example, soiled dressings, used gloves and used syringes);

（二）废有机溶剂；
（三）废的金属酸洗液、液压油、制动油和防冻液；
（四）其他化学工业及相关工业的废物。

(b) Waste organic solvents;
(c) Wastes of metal pickling liquors, hydraulic fluids, brake fluids and anti-freezing fluids; and
(d) Other wastes from chemical or allied industries.

但不包括主要含有石油及从沥青矿物提取的油类的废油（税目27.10）。

The expression "other wastes" does not, however, cover wastes which contain mainly petroleum oils or oils obtained from bituminous minerals (heading 27.10)."

子目注释：

子目3825.41和3825.49所称"废有机溶剂"，是指主要含有有机溶剂的废物，不适合再作原产品使用，不论其是否用于回收溶剂。

Subheading Note:

For the purposes of subheadings 3825.41 and 3825.49, "waste organic solvents" are wastes containing mainly organic solvents, not fit for further use as presented as primary products, whether or not intended for recovery of the solvents.

序号 No.	税则号列 Tariff Item	货品名称	最惠国税率(%) M.F.N.	协定税率(%) Agreement 东盟 ASEAN	曼谷协定 BA	香港 H.K.	澳门 Macao	普通税率(%) Gen.	Article Description
	38.01	**人造石墨；胶态或半胶态石墨；以石墨或其他碳为基本成分的糊状、块状、板状制品或半制品：**							**Artificial graphite; colloidal or semi-colloidal graphite; preparations based on graphite or other carbon in the form of pastes, blocks, plates or other semi-manufactures:**
2441	3801.1000	-人造石墨	6.5			0		30	-Artificial graphite
2442	3801.2000	-胶态或半胶态石墨	6.5					30	-Colloidal or semi-colloidal graphite
2443	3801.3000	-电极用碳糊及炉衬用的类似糊	6.5					35	-Carbonaceous pastes for electrodes and similar pastes for furnace linings
2444	3801.9000	-其他	6.5					35	-Other
	38.02	**活性碳；活性天然矿产品；动物炭黑，包括废动物炭黑：**							**Activated carbon; activated natural mineral products; animal black, including spent animal black:**
2445	3802.1000	-活性碳	6.5		6.4			20	-activated carbon
2446	3802.9000	-其他	10					45	-Other
	38.03	**妥尔油，不论是否精炼：**							**Tall oil. whether or not refined:**
2447	3803.0000	妥尔油，不论是否精炼	6.5					35	Tall oil. whether or not refined
	38.04	**木浆残余碱液，不论是否浓缩、脱糖或经化学处理，包括木素磺酸盐，但不包括税号38.03的妥尔油：**							**Residual lyes from the manufacture of wood pulp, whether or not concentrated, desugared or chemically treated, including lignin sulphonates, but excluding tall oil of heading No.38.03:**
2448	3804.0000	木浆残余碱液，不论是否浓缩、脱糖或经化学处理，包括木素磺酸盐，但不包括税号38.03的妥尔油	6.5					35	Residual lyes from the manufacture of wood pulp, whether or not concentrated, desugared or chemically treated, including lignin sulphonates, but excluding tall oil of heading No.38.03
	38.05	**脂松节油、木松节油和硫酸盐松节油及其他萜烯油，用蒸馏或其他方法从针叶木制得；粗制二聚戊烯；亚硫酸盐松节油及其他粗制对异丙基苯甲烷；以α萜品醇为基本成分的松油：**							**Gum, wood or sulphate turpentine and other terpenic oils produced by the distillation or other treatment of coniferous woods; crude dipentene; sulphite turpentine and other crude paracymene; pine oil containing alpha-terpineol as the main constituent:**

序号 No.	税则号列 Tariff Item	货品名称	最惠国税率(%) M.F.N.	协定税率(%) Agreement 东盟 ASEAN	曼谷协定 BA	香港 H.K.	澳门 Macao	普通税率(%) Gen.	Article Description
2449	3805.1000	-脂松节油、木松节油和硫酸盐松节油	6.5					50	-Gum,wood or sulphate turpentine oils
2450	3805.2000	-松油	6.5					50	-Pine oil
2451	3805.9000	-其他	6.5					50	-Other
	38.06	**松香和树脂酸及其衍生物;松香精及松香油;再熔胶:**							**Rosin and resin acids,and derivatives thereof;rosin spirit and rosin oils;run gums:**
		-松香及树脂酸:							-Rosin and resin acids:
2452	3806.1010	---松香	10					70	---Rosin
2453	3806.1020	---树脂酸	10					70	---Resin acides
		-松香盐、树脂酸盐及松香或树脂酸衍生物的盐,但松香加合物的盐除外:							-Salts of rosin, of resin acids or of derivatives of rosin or resin acids,other than salts of rosin adducts:
2454	3806.2010	---松香盐及树脂酸盐	6.5					40	----Salts of rosin,of resin acids
2455	3806.2090	---其他	6.5					40	---Other
2456	3806.3000	-酯胶	6.5					50	-Ester gums
2457	3806.9000	-其他	6.5					40	-Other
	38.07	**木焦油;精制木焦油;木杂酚油;粗木精;植物沥青;以松香、树脂酸或植物沥青为基本成分的啤酒桶沥青及类似制品:**							**Wood tar; wood tar oils; wood creosote;wood naphtha; vegetable pitch; brewers'pitch and similar preparations based on rosin,resin acids or on vegetable pitch:**
2458	3807.0000	木焦油;精制木焦油;木杂酚油;粗木精;植物沥青;以松香、树脂酸或植物沥青为基本成分的啤酒桶沥青及类似制品	6.5					35	Wood tar; wood tar oils; wood creosote; wood naphtha; vegetable pitch; brewers'pitch and similar preparations based on rosin, resin acids or on vegetable pitch
	38.08	**杀虫剂、杀鼠剂、杀菌剂、除草剂、抗萌剂、植物生长调节剂、消毒剂及类似产品,零售形状、零售包装或制成制剂及成品(例如,经硫磺处理的带子、杀虫灯芯、蜡烛及捕蝇纸):**							**Insecticides, rodenticides, fungicides, herbicides, anti-sprouting products and plantgrowth regulators, disinfectants and similar products, put up in forms or packings for retail sale or as preparations or articles(for example, sulphur-treated bands,wicks and candles,and fly-papers):**
		-杀虫剂:							-Insecticides:
		---零售包装:							---Put up for retail sale:
2459	3808.1011	----蚊香	10					80	----Mosquito smudges
2460	3808.1019	----其他	10			0		35	----Other
2461	3808.1090	---其他	6			0		11	---Other
		-杀菌剂:							-Fungicides:
2462	3808.2010	---零售包装	9					35	---Put up for retail sale
2463	3808.2090	---其他	6					11	---Other

序号 No.	税则号列 Tariff Item	货品名称	最惠国税率(%) M.F.N.	协定税率(%) Agreement 东盟 ASEAN	曼谷协定 BA	香港 H.K.	澳门 Macao	普通税率(%) Gen.	Article Description
		-除草剂、抗萌剂及植物生长调节剂:							-Herbicides, anti-sprouting products and plant-growth regulators:
		---除草剂:							---Herbicides:
2464	3808.3011	----零售包装	9					35	----Put up for retai sale
2465	3808.3019	----其他	5					11	----Other
		---其他:							---Other:
2466	3808.3091	----零售包装	9					35	----Put up for retail sale
2467	3808.3099	----其他	6					14	----Other
2468	3808.4000	-消毒剂	9			0		35	-Disinfectants
		-其他:							-Other:
2469	3808.9010	---零售包装	9					35	---Put up for retail sale
2470	3808.9090	---其他	9					14	---Other
	38.09	**纺织、造纸、制革及类似工业用的其他税号未列名的整理剂、染料加速着色或固色助剂及其他产品和制剂(例如,修整剂及媒染剂):**							**Finishing agents, dye carriers to accelerate the dyeing or fixing of dye-stuffs and other products and preparations (for example, dressings and mordants), of a kind used in the textile, paper, leather or like industries, not elsewhere specified or included:**
2471	3809.1000	-以淀粉物质为基本成分	10					35	-With a basis of amylaceous substances
		-其他:							-Other:
2472	3809.9100	--纺织工业及类似工业用	6.5			0		35	--Of a kind used in the textile or like industries
2473	3809.9200	--造纸工业及类似工业用	6.5			0		35	--Of a kind used in the paper or like industries
2474	3809.9300	--制革工业及类似工业用	6.5			0		35	--Of a kind used in the leather or like industries
	38.10	**金属表面酸洗剂;焊接用的焊剂及其他辅助剂;金属及其他材料制成的焊粉或焊膏;作焊条芯子或焊条涂料用的制品:**							**Pickling preparations for metal surfaces; fluxes and other auxiliary preparations for soldering, brazing or welding; soldering, brazing or welding powders and pastes consisting of metal and other materials; preparations of a kind used as cores or coatings for welding electrodes or rods:**
2475	3810.1000	-金属表面酸洗剂;金属及其他材料制成的焊粉或焊膏	6.5			0		35	-Pickling preparations for metal surfaces; soldering, brazing or welding powders and pastes consisting of metal and other materials
2476	3810.9000	-其他	6.5					35	-Other
	38.11	**抗震剂、抗氧剂、防胶剂、粘度改良剂、防腐蚀制剂及其他配制添加剂,用于矿物油(包括汽油)或与矿物油同样用途的其他液体:**							**Anti-knock preparations, oxidation inhibitors, gum inhibitors, viscosity improvers, anti-corrosive preparations and other prepared additives, for mineral oils (including gasoling) or for other liquids used for the same purposes as mineral oils:**

序号 No.	税则号列 Tariff Item	货品名称	最惠国税率(%) M.F.N.	协定税率(%)Agreement 东盟 ASEAN	曼谷协定 BA	香港 H.K.	澳门 Macao	普通税率(%) Gen.	Article Description
		-抗震剂:							-Anti-knock preparations:
2477	3811.1100	--以铅化合物为基本成分	6.5					35	--Based on lead compounds
2478	3811.1900	--其他	6.5					35	--Other
		-润滑油添加剂:							-Additives for lubricating oils:
2479	3811.2100	--含有石油或从沥青矿物提取的油类	6.5					35	--Containing petroleum oils or oils obtained from bituminous minerals
2480	3811.2900	--其他	6.5				0△	35	--Other
2481	3811.9000	-其他	6.5			0	0△	35	-Other
	38.12	**配制的橡胶促进剂;其他税号未列名的橡胶或塑料用复合增塑剂;橡胶或塑料用抗氧制剂及其他复合稳定剂:**							**Prepared rubber accelerators; compounds plasticizers for rubber or plastics, not elsewhere specified or included; anti-oxidizing preparations and other compound stabilizers for rubber or plastics:**
2482	3812.1000	-配制的橡胶促进剂	6					20	-Prepared rubber accelerators
2483	3812.2000	-橡胶或塑料用复合增塑剂	6.5					35	-Compound plasticizers for rubber or plastics
		-橡胶或塑料用抗氧制剂及其他复合稳定剂:							-Anti-oxidizing preparations and other compound stabilizers for rubber or plastics:
2484	3812.3010	---橡胶防老剂	6			0		20	---Rubber antioxidants
2485	3812.3090	---其他	6.5			0		35	---Other
	38.13	**灭火器的装配药;已装药的灭火弹:**							**Preparations and charges for fire-extinguishers; charged fire-extinguishing grenades:**
2486	3813.0010	---灭火器的装配药	6.5					35	---Preparations and charges for fire-extinguishers
2487	3813.0020	---已装药的灭火弹	10					70	---Charged fire-extinguishing grenades
	38.14	**其他税号未列名的有机复合溶剂及稀释剂;除漆剂:**							**Organic composite solvents and thinners, not elsewhere specified or included; prepared paint or varnish removers:**
2488	3814.0000	其他税号未列名的有机复合溶剂及稀释剂;除漆剂	10		9		0	50	Organic composite solvents and thinners, not elsewhere specified or included; prepared paint or varnish removers
	38.15	**其他税号未列名的反应引发剂、反应促进剂、催化剂:**							**Reaction initiators, reaction accelerators and catalytic preparations, not elsewhere specified or included:**
		-载体催化剂:							-Supported catalysts:
2489	3815.1100	--以镍及其化合物为活性物的	6.5					35	With nickel or nickel compounds as the active substance
2490	3815.1200	--以贵金属及其化合物为活性物的	6.5					35	--With precious metal or precious metal compounds as the active substance
2491	3815.1900	--其他	6.5					35	--Other
2492	3815.9000	-其他	6.5				0	35	-Other

序号 No.	税则号列 Tariff Item	货品名称	最惠国税率(%) M.F.N.	协定税率(%) Agreement 东盟 ASEAN	曼谷协定 BA	香港 H.K.	澳门 Macao	普通税率(%) Gen.	Article Description
	38.16	**耐火的水泥、灰泥、混凝土及类似耐火混合制品，但税号38.01的产品除外：**							**Refractory cements, mortars, concretes and similar compositions, other than products of heading No.38.01:**
2493	3816.0000	耐火的水泥、灰泥、混凝土及类似耐火混合制品，但税号38.01的产品除外	6.5					35	Refractory cements, mortars, concretes and similar compositions, other than products of heading No.38.01
	38.17	**混合烷基苯及混合烷基萘，但税号27.07及29.02的货品除外：**							**Mixed alkylbenzenes and mixed alkylnaphthalenes, other than those of heading No.27.07 or 29.02:**
2494	3817.0000	混合烷基苯及混合烷基萘	6.5		6.5			35	Mixed alkylbenzenes and alkylnaphthalenes
	38.18	**经掺杂用于电子工业的化学元素，已切成圆片、薄片或类似形状；经掺杂用于电子工业的化合物：**							**Chemical elements doped for use in electronics, in the form of discs, wafers or similar forms; chemical compounds doped for use in electronics:**
		---直径在7.5厘米及以上的单晶硅切片：							---Monocrystalline sillicon, in the form of discs, wafers or similar form, 7.5 cm or more in diameter:
2495	3818.0011	----直径在15.24厘米及以下的	0					11	----Diameter not exceeding 15.24cm
2496	3818.0019	----其他	0					11	----Other
2497	3818.0090	---其他	0					17	---Other
	38.19	**闸用液压油及其他液压传动用液体，不含石油或从沥青矿物提取的油类，或者按重量计石油或从沥青矿物提取的油类含量低于70%：**							**Hydraulic brake fluids and other prepared liquids for hydraulic transmission, not containing or containing less than 70% by weight of petroleum oils or oils obtained from bituminous minerals:**
2498	3819.0000	闸用液压油及其他液压传动用液体，不含石油或从沥青矿物提取的油类，或者按重量计石油或从沥青矿物提取的油类含量低于70%	6.5				0△	35	Hydraulic brake fluids and other prepared liquids for hydraulic transmission, not containing or containing less than 70% by weight of petroleum oils or oils obtained from bituminous minerals
	38.20	**防冻剂及解冻剂：**							**Anti-freezing preparations and prepared de-icing fluids:**
2499	3820.0000	防冻剂及解冻剂	10				0△	35	Anti-freezing preparations and prepared de-icing fluids
	38.21	**制成的微生物培养基：**							**Prepared culture media for development of micro-organisms:**
2500	3821.0000	制成的微生物培养基	3					11	Prepared culture media for development of micro-organisms

序号 No.	税则号列 Tariff Item	货品名称	最惠国税率(%) M.F.N.	协定税率(%)Agreement 东盟 ASEAN	曼谷协定 BA	香港 H.K.	澳门 Macao	普通税率(%) Gen.	Article Description
	38.22	**附于衬背上的诊断或实验用试剂及不论是否附于衬背上的诊断或实验用配制试剂,但税号30.02及30.06的货品除外;检定参照物:**							**Diagnostic or laboratory reagents on a backing and prepared diagnostic or laboratory reagents whether or not on a backing, other than those of heading No.30.02 or 30.06; certified reference materials:**
2501	3822.0010	---附于衬背上的	4		4			35	---On a backing
2502	3822.0090	---其他	5					35	---Other
	38.23	**工业用单羧脂肪酸;精炼所得的酸性油;工业用脂肪醇:**							**Industrial monocarboxylic fatty acids; acidoils from refining; industrial fatty alcohols:**
		-工业用单羧脂肪酸;精炼所得的酸性油:							-Industrial monocarboxylic fatty acids; acid oils from refining:
2503	3823.1100	--硬脂酸	16	见表1				50	--Stearic acid
2504	3823.1200	--油酸	16					50	--Oleic acid
2505	3823.1300	--妥尔油脂肪酸	16					50	--Tall oil fatty acids
2506	3823.1900	--其他	16			0		50	--Other
2507	3823.7000	-工业用脂肪醇	13					50	-Industrial fatty alcohols
	38.24	**铸模及铸芯用粘合剂;其他税号未列名的化学工业及其相关工业的化学产品及配制品(包括由天然产品混合组成的):**							**Prepared binders for foundry moulds or cores; chemical products and preparations of the chemical or allied industries (including those consisting of mixtures of natural products), not elsewhere spelified or included:**
2508	3824.1000	-铸模及铸芯用粘合剂	6.5					35	-Prepared binders for foundry moulds or cores
		-环烷酸及其水不溶性的盐和酯:							-Naphthenic acids, their water-insoluble salts and their esters:
2509	3824.2010	---环烷酸钴	6.5					35	---Cobalt naphthenate
2510	3824.2090	---其他	6.5					35	---Other
2511	3824.3000	-自身混合或与金属粘合剂混合的未烧结金属碳化物	6.5					35	-Non-agglomerated metal carbides mixed together or with metallic binders
2512	3824.4000	-水泥、灰泥及混凝土用添加剂	6.5			0		35	-Prepared additives for cements, mortars or concretes
2513	3824.5000	-非耐火的灰泥及混凝土	6.5					35	-Non-refractory mortars and concretes
2514	3824.6000	-子目号2905.44以外的山梨醇	14					40	-Sorbitol other than that of subheading No.2905.44
		-含有两种或两种以上不同卤素的无环烃全卤化衍生物的混合物:							-Mixtures containing perhalogenated derivatives of acyclic hydrocarbons containing two or more different halogens:
2515	3824.7100	--仅含氟及氯的无环烃全卤化衍生物的混合物	6.5		6.5			35	--Containing acyclic hydrocarbons perhalogenated only with fluorine and chlorine

序号 No.	税则号列 Tariff Item	货品名称	最惠国税率(%) M.F.N.	协定税率(%)Agreement 东盟 ASEAN	曼谷协定 BA	香港 H.K.	澳门 Macao	普通税率(%) Gen.	Article Description
2516	3824.7900	--其他	6.5		6.5			35	--Other
		-其他：							-Other:
2517	3824.9010	---杂醇油	6.5		5.5			40	---Fusel oil
2518	3824.9020	---除墨剂、蜡纸改正液及类似品	9		8.3			80	---Ink-removers, stencil correctors and the like
2519	3824.9030	---增炭剂	6.5					35	---Carburetant
2520	3824.9040	---多次甲基多苯基异氰酸酯(聚合 MDI)	6.5		6	0		35	---Polymethane polyphenyl Isocyanate Cpolymeric MDI or crude MDI
2521	3824.9090	---其他	6.5		6	0	0	35	---Other
	38.25	**其他税目未列名的化学工业及其相关工业的副产品；城市垃圾；下水道淤泥；本章注释六所规定的其他废物：**							**Residual products of the chemical or allied industries, not elsewhere specified or included; municipal waste; sewage sludge; other wastes specified in Note 6 to this Chapter:**
2522	3825.1000	-城市垃圾	6.5			0		35	-Municipal waste
2523	3825.2000	-下水道淤泥	6.5			0		35	-Sewage sludge
2524	3825.3000	-医疗废物	6.5			0		35	-Clinical waste
		-废有机溶剂：							Waste organic solvents:
2525	3825.4100	--卤化物的	6.5			0		35	--Halogenated
2526	3825.4900	--其他	6.5			0		35	--Other
2527	3825.5000	-废的金属酸液、液压油、制动油及防冻液	10			0		35	-Wastes of metal pickling liguors, hydraulic fluids, brake fluids and anti-freeze fluids
		-其他化学工业及相关工业的废物：							-Other wastes from chemical or allied industries:
2528	3825.6100	--主要含有有机成分的	6.5			0		35	--Mainly containing organic constituents
2529	3825.6900	--其他	6.5			0		35	--Other
2530	3825.9000	-其他	6.5			0		35	-Other

第七类

塑料及其制品；橡胶及其制品

注释：

一、由两种或两种以上单独成分配套的货品，其部分或全部成分属于本类范围以内，混合后则构成第六类或第七类的货品，应按混合后产品归入相应的税号，但其组成成分必须同时符合下列条件：

（一）其包装形式足以表明这些成分不需经过改装就可以一起使用的；

（二）一起进口或出口的；

（三）这些成分的属性及相互比例足以表明是相互配用的。

二、除税号 39.18 或 39.19 的货品外，印有花纹、文字、图画的塑料、橡胶及其制品，如果所印花纹、字画作为其主要用途，应归入第四十九章。

SECTION Ⅶ

PLASTICS AND ARTICLES THEREOF; RUBBER AND ARTICLES THEREOF

Notes:

1. Goods put up in sets consisting of two or more separate constituents, some or all of which fall in this Section and are intended to be mixed together to obtain a product of Section Ⅵ or Ⅶ, are to be classified in the heading appropriate to that product, provided that the constituents are:

(a) having regard to the manner in which they are put up, clearly identifiable as being intended to be used together without first being repacked;

(b) presented together; and

(c) identifiable, whether by their nature or by the relative proportions in which they are present, as being complementary one to another.

2. Except for the goods of heading No.39.18 or 39.19, plastics, rubber, and articles thereof, printed with motifs, characters or pictorial representations, which are not merely incidental to the primary use of the goods, fall in Chapter 49.

第三十九章

塑料及其制品

注释：

一、本目录所称"塑料"，是指税号 39.01 至 39.14 的材料，这些材料能够在聚合时或聚合后在外力（一般是热力和压力，必要时加入溶剂或增塑剂）作用下通过模制、浇铸、挤压、滚轧或其他工序制成一定的形状，成形后除去外力，其形状仍保持不变。

本目录所称"塑料"，还应包括钢纸，但不包括第十一类的纺织材料。

二、本章不包括：

（一）税号 27.12 或 34.04 的蜡；

Chapter 39

Plastics and articles thereof

Notes:

1. Throughout the Nomenclature the expression "plastics" means those materials of headings Nos.39.01 to 39.14 which are or have been capable, either at the moment of polymerization or at some subsequent stage, of being formed under external influence (usually heat and pressure, if necessary with a solvent or plasticizer) by moulding, casting, extruding, rolling or other process into shapes which are retained on the removal of the external influence.

Throughout the Nomenclature any reference to "plastics" also includes vulcanized fibre. The expression, however, does not apply to materials regarded as textile materials of Section Ⅺ.

2. This Chapter does not cover:

(a) Waxes of heading No.27.12 or 34.04;

（二）单独的已有化学定义的有机化合物（第二十九章）；

（三）肝素及其盐（税号 30.01）；

（四）税号 39.01 至 39.13 所列的任何产品溶于挥发性有机溶剂的溶液（胶棉除外），但溶剂的重量必须超过溶液重量的 50%（税号 32.08）；税号 32.12 的压印箔；

（五）有机表面活性剂或税号 34.02 的制剂；

（六）再熔胶及酯胶（税号 38.06）；

（七）附于塑料衬背上的诊断或实验用试剂（税号 38.22）；

（八）第四十章规定的合成橡胶及其制品；

（九）鞍具及挽具（税号 42.01）；税号 42.02 的衣箱、提箱、手提包及其他容器；

（十）第四十六章的缏条、编结品及其他制品；

（十一）税号 48.14 的壁纸；

（十二）第十一类的货品(纺织原料及纺织制品)；

（十三）第十二类的物品（例如，鞋靴、帽类、雨伞、阳伞、手杖、鞭子、马鞭及其零件）；

（十四）税号 71.17 的仿首饰；

（十五）第十六类的物品（机器、机械器具或电气器具）；

（十六）第十七类的航空器零件及车辆零件；

（十七）第九十章的物品（例如，光学元件、眼镜架及绘图仪器）；

（十八）第九十一章的物品（例如，钟壳及表壳）；

（十九）第九十二章的物品（例如，乐器及其零件）；

（二十）第九十四章的物品（例如，家具、灯具、照明装置、灯箱及活动房屋）；

（二十一）第九十五章的物品（例如，玩具、游戏品及运动用品）；

（二十二）第九十六章的物品（例如，刷子、钮扣、拉链、梳子、烟斗的嘴及柄、香烟嘴及类似品、保温瓶的零件及类似品、钢笔、活动铅笔）。

三、税号 39.01 至 39.11 仅适用于化学合成的下列货品：

（一）温度在 300℃ 时，压力转为 1013 毫巴后减压蒸馏出的液体合成聚烯烃以体积计小于 60% 的货品（税号 39.01 及 39.02）；

(b) Separate chemically defined organic compounds (Chapter 29);

(c) Heparin or its salts (heading No.30.01);

(d) Solutions (other than collodions) consisting of any of the products specified in headings Nos.39.01 to 39.13 in volatile organic solvents when the weight of the solvent exceeds 50% of the weight of the solution (heading No.32.08); stamping foils of heading No.32.12;

(e) Organic surface-active agents or preparations of heading No.34.02;

(f) Run gums or ester gums (heading No.38.06);

(g) Diagnostic or laboratory reagents on a backing of plastics (heading No.38.22);

(h) Synthetic rubber, as defined for the purposes of Chapter 40, or articles thereof;

(i) Sadlery or harness (heading No.42.01) or trunks, suitcases, handbags or other containers of heading No.42.02;

(j) Plaits, wickerwork or other articles of Chapter 46;

(k) Wall coverings of heading No.48.14;

(l) Goods of Section XI (textiles and textile articles);

(m) Articles of Section XII (for example, footwear, headgear, umbrellas, sun umbrellas, walking-sticks, whips, riding-crops or parts thereof);

(n) Imitation jewellery of heading No.71.17;

(o) Articles of Section XVI (machines and mechanical or electrical appliances);

(p) Parts of aircraft or vehicles of Section XVII;

(q) Articles of Chapter 90 (for example, optical elements, spectacle frames, drawing instruments);

(r) Articles of Chapter 91 (for example, clock or watch cases);

(s) Articles of Chapter 92 (for example, musical instruments or parts thereof);

(t) Articles of Chapter 94 (for example, furniture, lamps and lighting fittings. illuminated signs, prefabricated buildings);

(u) Articles of Chapter 95 (for example, toys, games, sports requisites); or

(v) Articles of Chapter 96 (for example, brushes, buttons, slide fasteners, combs, mouthpieces or stems for smoking pipes, cigaretteholders or the like, parts of vacuum flasks or the like, pens, propelling pencils).

3. Headings Nos.39.01 to 39.11 apply only to goods of a kind produced by chemical synthesis, falling in the following categories:

(a) Liquid synthetic polyolefins of which less than 60% by volume distils at 300℃, after conversion to 1,013 millibars when a reduced-pressure distillation method is used (headings Nos.39.01 and 39.02);

（二）非高度聚合的苯并呋喃—茚树脂（税号39.11）；

（三）平均至少有五个单体单元的其他合成聚合物；

（四）聚硅氧烷（税号 39.10）；

（五）甲阶酚醛树脂（税号 39.09）及其他预聚物。

(b) Resins, not highly polymerized, of the coumarone-indene type (heading No.39.11);

(c) Other synthetic polymers with an average of at least 5 monomer units;

(d) Silicones (heading No.39.10);

(e) Resols (heading No.39.09) and other prepolymers.

四、所称“共聚物”，包括在整个聚合物中按重量计没有一种单体单元的含量在 95% 及以上的各种聚合物。

在本章中，除条文另有规定的以外，共聚物（包括共缩聚物、共加聚物、嵌段共聚物及接枝共聚物）及聚合物混合体应按聚合物中重量最大的那种共聚单体单元所构成的聚合物归入相应税号。在本注释中，归入同一税号的聚合物的共聚单体单元应作为一种单体单元对待。

如果没有任何一种共聚单体单元重量为最大，共聚物或聚合物混合体应按号列顺序归入其可归入的最末一个税号。

4. The expression “copolymers” covers all polymers in which no single monomer unit contributes 95% or more by weight to the total polymer content.

For the purposes of this Chapter, except where the context otherwise requires, copolymers (including copolycondensates, co-polyaddition products, block copolymers and graft copolymers) and polymer blends are to be classified in the heading covering polymers of that comonomet unit which predominates by weight over every other single comonomer unit. For the purposes of this Note, constituent comonomer units of polymers falling in the same heading shall be taken together.

If no single comonomer unit predominates, copolymens or polymer blends, as the case may be, are to be classified in the heading which occurs last in numerical order among those which equally merit consideration.

五、化学改性聚合物，即聚合物主链上的支链通过化学反应发生了变化的聚合物，应按未改性的聚合物的相应税号归类。本规定不适用于接枝共聚物。

5. Chemically modified polymers, that is those in which only appendages to the main polymer chain have been changed by chemical reaction, are to be classified in the heading appropriate to the unmodified polymer. This provision does not apply to graft copolymers.

六、税号 39.01 至 39.14 所称“初级形状”，只限于下列各种形状：

（一）液状及糊状，包括分散体（乳浊液及悬浮液）及溶液；

（二）不规则形状的块，团、粉（包括压型粉）、颗粒、粉片及类似的散装形状。

6. In headings Nos.39.01 to 39.04, the expression “primary forms” applies only to the following forms:

(a) Liquids and pastes, including dispersions (emulsions and suspensions) and solutions;

(b) Blocks of irregular shape, lumps, powders (including moulding powders), granules, flakes and similar bulk forms.

七、税号 39.15 不适用于已制成初级形状的单一的热塑材料废碎料及下脚料（税号 39.01 至 39.14）。

7. Heading No.39.15 does not apply to waste, parings and scrap of a single thermoplastic material, transformed into primary forms (headings Nos.39.01 to 39.14).

八、税号 39.17 所称“管子”，是指通常用于输送或供给气体或液体的空心制品或半制品（例如，肋纹浇花软管、多孔管），还包括香肠用肠衣及其他扁平管。除肠衣及扁平管外，内截面如果不呈圆形、椭圆形、矩形（其长度不超过宽度的 1.5 倍）或正几何形，则不能视为管子，而应作为异型材。

8. For the purposes of heading No.39.17, the expression “tubes, pipes and hoses” means hollow products, whether semi-manufactures or finished products, of a kind generally used for conveying, conducting or distributing gases or liquids (for example, ribbed garden hose, perforated tubes). This expression also includes sausage casings and other lay-flat tubing. However, except for the last-mentioned, those having an internal cross-section other than round, oval, rect-

angular (in which the length does not exceed 1.5 times the width) or in the shape of a regular polygon are not to be regarded as tubes, pipes and hoses but as profile shapes.

九、税号39.18所称"塑料糊墙品",适用于墙壁或天花板装饰用的宽度不小于45厘米的成卷产品,这类产品是将塑料牢固地附着在除纸张以外任何材料的衬背上,并且在塑料面起纹、压花、着色、印制图案或用其他方法装饰。

9. For the purposes of heading No.39.18, the expression "wall or ceiling coverings of plastics" applies to products in rolls, of a width not less than 45 cm, suitable for wall or ceiling decoration, consisting of plastics fixed permanently on a backing of any material other than paper, the layer of plastics (on the face side) being grained, embossed, coloured, design-printed or otherwise decorated.

十、税号39.20及39.21所称"板、片、膜、箔、扁条",只适用于未切割或仅切割成矩形(包括正方形)(含切割后即可供使用的);但未经进一步加工的板、片、膜、箔、扁条(第五十四章的物品除外)及正几何形块,不论是否经过印制或其他表面加工。

10. In headings Nos.39.20 and 39.21, the expression "plates, sheets, film, foil and strip" applies only to plates, sheets, film, foil and strip (other than those of Chapter 54) and to blocks of regular geometric shape, whether or not printed or otherwise surface-woked, uncut or cut into rectangles (including squares) but not further worked (even if when so cut they become articles ready for use).

十一、税号39.25只适用于第二分章以前各税号未包括的下列物品:

(一)容积超过300升的囤、柜(包括化粪池)、罐、桶及类似容器;
(二)用于地板、墙壁、隔墙、天花板或屋顶等方面的结构件;
(三)槽管及其附件;
(四)门、窗及其框架和门槛;
(五)阳台、栏杆、栅栏、栅门及类似品;
(六)窗板、百叶窗(包括威尼斯式百叶窗)或类似品及其零件、附件;
(七)商店、工棚、仓库等用的拼装式固定大型货架;
(八)建筑用的特色(例如,凹槽、圆顶及鸽棚式)装饰件;
(九)固定装于门窗、楼梯、墙壁或建筑物其他部位的附件及架座,例如,球形把手、拉手、挂钩、托架、毛巾架、开关板及其他护板。

11. Heading No.39.25 applies only to the following articles, not being products covered by any of the earlier headings of sub-Chapter Ⅱ:

(a) Reservoirs, tanks (including septic tanks), vats and similar containers, of a capacity exceeding 300 l;
(b) Structural elements used, for example, in floors, walls or partitions, ceilings or roofs;
(c) Gutters and fittings therefor;
(d) Doors, windows and their frames and thresholds for doors;
(e) Balconies, balustrades, fencing, gates and similar barriers;
(f) Shutters, blinds (including Venetian blinds) and similar articles and parts and fittings thereof;
(g) Large-scale shelving for assembly and permanent installation, for example, in shops, workshops, warehouses;
(h) Ornamental architectural features, for example, flutings, cupolas, dovecotes; and
(i) Fittings and mountings intended for permanent installation in or on doors, windows, staircases, walls or other parts of buildings, for example, knobs, handles, hooks, brackets, towel rails, switch-plates and other protective plates.

子目注释:

属于本章任一税号项下的聚合物(包括共聚物)及化学改性聚合物应按下列规则归类:

一、在同级子目中有一个"其他"子目的:

Subheading Notes:

1. Within any one heading of this Chapter, polymers (including copolymers) and chemically modified polymers are to be classified according to the following provisions:

(a) Where there is a subheading named "Other" in the same series:

（一）子目所列聚合物名称冠有"聚（多）"的（例如，聚乙烯及聚酰胺－6，6），是指列名的该种聚合物单体单元含量在整个聚合物中按重量计必须占95%及以上。

(1) The designation in a subheading of a polymer by the prefix "poly" (e. g., polyethylene and polyamide－6, 6) means that the constituent monomer unit or monomer units of the named polymer taken together must contribute 95% or more by weight of the total polymer content.

（二）子目号3901.30、3903.20、3903.30及3904.30所列的共聚物，如果该种共聚单体单元含量在整个聚合物中按重量计占95%及以上，即应归入上述子目。

(2) The copolymers named in subheadings Nos. 3901.30, 3903.20, 3903.30 and 3904.30 are to be classified in those subheadings, provided that the comonomer units of the named copolymers contribute 95% or more by weight of the total polymer content.

（三）化学改性聚合物如未在其他子目具体列名，应归入列明为"其他"的子目内。

(3) Chemically modified polymers are to be classified in the subheading named "Other", provided that the chemically modified polymers are not more specifically covered by another subheading.

（四）不符合上述（一）、（二）、（三）款规定的聚合物，应按聚合物中重量最大的那种单体单元（与其他各种单一的共聚单体单元相比）所构成的聚合物归入该级其他相应子目。为此，归入同一子目的聚合物单体单元应作为一种单体单元对待。只有在同级子目中的聚合物共聚单体单元才可以进行比较。

(4) Polymers not meeting (1), (2) or (3) above, are to be classified in the subheading, among the remaining subheadings in the series, covering polymers of that monomer unit which predominates by weight over every other single comonomer unit. For this purpose, constituent monomer units of polymers falling in the same subheading shall be taken together. Only the constituent comonomer units of the polymers in the series of subheadings under consideration are to be compared.

二、在同级子目中没有"其他"子目的：

2. Where there is no subheading named "Other" in the same series:

（一）聚合物应按聚合物中重量最大的那种单体单元（与其他各种单一的共聚单体单元相比）所构成的聚合物归入该级相应子目。为此，归入同一子目的聚合物单体单元应作为一种单体单元对待。只有在同级子目中的聚合物共聚单体单元才可以进行比较。

(a) Polymers are to be classified in the subheading covering polymers of that monomer unit which predominates by weight over every other single comonomer unit. For this purpose, constituent monomer units of polymers falling in the same subheading shall be taken together. Only the constituent comonomer units of the polymers in the series under consideration are to be compared.

（二）化学改性聚合物应按相应的未改性聚合物的子目归类。

聚合物混合体应按单体单元比例相等、种类相同的聚合物归入相应子目。

(b) Chemically modified polymers are to be classified in the subheading appropriate to the unmodified polymer.

Polymer blends are to be classified in the same subheading as polymers of the same monomer units in the same proportions.

三、子目3920.43所称"增塑剂"，包括"次级增塑剂"。

3. For the purposes of subheading 3920.43, the term "plasticisers" includes "secondary plasticisers".

序号 No.	税则号列 Tariff Item	货品名称	最惠国税率(%) M.F.N.	协定税率(%) Agreement 东盟 ASEAN	曼谷协定 BA	香港 H.K.	澳门 Macao	普通税率(%) Gen.	Article Description
		第一分章　初级形状							Ⅰ. PRIMARY FORMS
	39.01	**初级形状的乙烯聚合物：**							**Polymers of ethylene, in primary forms:**
2531	3901.1000	-聚乙烯，比重小于0.94	10.3		9.9			45	-Polyethylene having a specific gravity of less than 0.94
2532	3901.2000	-聚乙烯，比重在0.94及以上	10.3		9.9			45	-Polyethylene having a specific gravity of 0.94 or more
2533	3901.3000	-乙烯-乙酸乙烯酯共聚物	6.5		6.2			45	-Ethylene-vinyl acetate copolymers
		-其他：							-Other:
2534	3901.9010	---乙烯-丙烯共聚物（乙丙橡胶）	6.5					45	---Ethylene-propylene copolymers
2535	3901.9020	---线型低密度聚乙烯	6.5					45	---Linearity low density polyethylene
2536	3901.9090	---其他	6.5		6.3			45	---Other
	39.02	**初级形状的丙烯或其他烯烃聚合物：**							**Polymers of propylene or of other olefins, in primary forms:**
2537	3902.1000	-聚丙烯	9.7		9.7	0		45	-Polypropylene
2538	3902.2000	-聚异丁烯	9.7					45	-Polyisobutylene
		-丙烯共聚物：							-Propylene copolymers:
2539	3902.3010	---乙烯-丙烯共聚物（乙丙橡胶）	9.7		9.1			45	---Ethylene-propylene copolymers
2540	3902.3090	---其他	9.7		9.1			45	---Other
2541	3902.9000	-其他	9.7					45	-Other
	39.03	**初级形状的苯乙烯聚合物：**							**Polymers of styrene, in primary forms:**
		-聚苯乙烯：							-Polystyrene:
2542	3903.1100	--可发性的	9.7		9.1			45	--Expansible
2543	3903.1900	--其他	9.7		9.1	0		45	--Other
2544	3903.2000	-苯乙烯-丙烯腈(SAN)共聚物	12					45	-Styrene-acrylonitrile (SAN) copolymers
2545	3903.3000	-丙烯腈-丁二烯-苯乙烯(ABS)共聚物	9.7		9.1	0		45	-Acrylonitrile-butadiene-styrene (ABS) copolymers
2546	3903.9000	-其他	9.7		9.4	0		45	-Other
	39.04	**初级形状的氯乙烯或其他卤化烯烃聚合物：**							**Polymers of vinyl chloride or of other halogenated olefins, in primary forms:**
2547	3904.1000	-聚氯乙烯，未掺其他物质	9.7		9.1			45	-Poly(vinyl chloride), not mixed with any other substances
		-其他聚氯乙烯：							-Other poly(vinyl chloride):
2548	3904.2100	--未塑化	9.7			0		45	--Non-plasticized
2549	3904.2200	--已塑化	9.7			0		45	--Plasticized
2550	3904.3000	-氯乙烯-乙酸乙烯酯共聚物	9					45	-Vinyl chloride-vinyl acetate copolymers
2551	3904.4000	-其他氯乙烯共聚物	12					45	-Other vinyl chloride copolymers
2552	3904.5000	-偏二氯乙烯聚合物	9.7					45	-Vinylidene chloride polymers
		-氟聚合物：							-Fluoro-polymers:

序号 No.	税则号列 Tariff Item	货品名称	最惠国税率(%) M.F.N.	协定税率(%) Agreement				普通税率(%) Gen.	Article Description
				东盟 ASEAN	曼谷协定 BA	香港 H.K.	澳门 Macao		
2553	3904.6100	--聚四氟乙烯	10					45	--Polytetrafluoroethylene
2554	3904.6900	--其他	6.5					45	--Other
2555	3904.9000	-其他	10					45	-Other
	39.05	**初级形状的乙酸乙烯酯或其他乙烯酯聚合物;初级形状的其他乙烯基聚合物:**							**Polymers of vinyl acetate or of other vinyl esters, in primary forms; other vinyl polymers in primary forms:**
		-聚乙酸乙烯酯:							-Poly(vinyl acetate):
2556	3905.1200	--水分散体	10					45	--In aqueous dispersion
2557	3905.1900	--其他	10					45	--Other
		-乙酸乙烯酯共聚物:							-Vinyl acetate copolymers:
2558	3905.2100	--水分散体	10					45	--In aqueous dispersion
2559	3905.2900	--其他	10					45	--Other
2560	3905.3000	-聚乙烯醇,不论是否含有未水解的乙酸酯基	14					45	-Poly (vinyl alcohol), whether or not containing unhydrolyzed acetate groups
		-其他:							-Other:
2561	3905.9100	--共聚物	10					45	--Copolymers
2562	3905.9900	--其他	10					45	--Other
	39.06	**初级形状的丙烯酸聚合物:**							**Acrylic polymers in primary forms:**
2563	3906.1000	-聚甲基丙烯酸甲酯	6.5					45	-Poly(methyl methacrylate)
2564	3906.9000	-其他	6.5			0		45	-Other
	39.07	**初级形状的聚缩醛、其他聚醚及环氧树脂;初级形状的聚碳酸酯、醇酸树脂、聚烯丙基酯及其他聚酯:**							**Polyacetals, other polyethers and epoxide resins, in primary forms; polycarbonates, alkyd resins, polyallyl esters and other polyesters, in primary forms:**
		-聚缩醛:							-Polyacetals:
2565	3907.1010	---聚甲醛	9.7					45	---Polyoxymethylene(POM)
2566	3907.1090	---其他	9.7					45	---Other
2567	3907.2000	-其他聚醚	9.7					45	-Other polyethers
2568	3907.3000	-环氧树脂	9.7			0△		45	-Epoxide resins
2569	3907.4000	-聚碳酸酯	6.5			0		45	-Polycarbonates
2570	3907.5000	-醇酸树脂	10					45	-Alkyd resins
		-聚对苯二甲酸乙二酯:							-Poly(ethylene terephthalate):
		---切片:							---In the form of slices or chips:
2571	3907.6011	----高粘度	9.7					45	----High viscosity
2572	3907.6019	----其他	9.7					45	----Other
2573	3907.6090	---其他	6.5					45	---Other
		-其他聚酯:							-Other polyesters:
2574	3907.9100	--不饱和	9.7					45	--Unsaturated
2575	3907.9900	--其他	6.5		6.2		0	45	--Other
	39.08	**初级形状的聚酰胺:**							**Polyamides in primary forms:**
		-聚酰胺-6、-11、-12、-6,6、-6,9、-6,10或-6,12:							-Polyamide-6, -11, -12, -6,6, -6,9, -6,10 or -6,12:

序号 No.	税则号列 Tariff Item	货品名称	最惠国税率(%) M.F.N.	协定税率(%) Agreement 东盟 ASEAN	曼谷协定 BA	香港 H.K.	澳门 Macao	普通税率(%) Gen.	Article Description
		---切片:							---In the form of slices or chips:
2576	3908.1011	----聚酰胺－6,6 切片	9.7			0		45	----Of polyamide－6,6
2577	3908.1019	----其他	9.7			0		45	----Other
2578	3908.1090	---其他	6.5			0		45	---Other
2579	3908.9000	-其他	10					45	-Other
	39.09	**初级形状的氨基树脂、酚醛树脂及聚氨酯类:**							**Amino-resins, phenolic resins and polyurethanes, in primary forms:**
2580	3909.1000	-尿素树脂;硫尿树脂	6.5					45	-Urea resins; thiourea resins
2581	3909.2000	-蜜胺树脂	6.5					45	-Melamine resins
2582	3909.3000	-其他氨基树脂	6.5					45	-Other amino-resins
2583	3909.4000	-酚醛树脂	6.5					45	-Phenolic resins
2584	3909.5000	-聚氨基甲酸酯	9.7			0	0	45	-Polyurethanes
	39.10	**初级形状的聚硅氧烷:**							**Silicones in primary forms:**
2585	3910.0000	初级形状的聚硅氧烷	6.5			0△		45	Silicones in primary forms
	39.11	**初级形状的石油树脂、苯并呋喃—茚树脂、多萜树脂、多硫化物、聚砜及本章注释三所规定的其他税号未列名产品:**							**Petroleum resins, coumarone-indene resins, polyterpenes, polysulphides, polysulphones and other products specified in Note 3 to this Chapter, not elsewhere specified or included, in primary forms:**
2586	3911.1000	-石油树脂、苯并呋喃树脂、茚树脂、苯并呋喃—茚树脂及多萜树脂	6.5					45	-Petroleum resins, coumarone, indene or coumarone-indene resins and polyterpenes
2587	3911.9000	-其他	6.5					45	-Other
	39.12	**初级形状的其他税号未列名的纤维素及其化学衍生物:**							**Cellulose and its chemical derivatives, not elsewhere specified or included, in primary forms:**
		-乙酸纤维素:							-Cellulose acetates:
2588	3912.1100	--未塑化	6.5					40	--Non-plasticized
2589	3912.1200	--已塑化	6.5					40	--Plasticized
2590	3912.2000	-硝酸纤维素(包括胶棉)	6.5					45	-Cellulose nitrates (including collodions)
		-纤维素醚:							-Cellulose ethers:
2591	3912.3100	--羧甲基纤维素及其盐	6.5					45	--Carboxymethylcellulose and its salts
2592	3912.3900	--其他	6.5					45	--Other
2593	3912.9000	-其他	6.5					45	-Other
	39.13	**初级形状的其他税号未列名的天然聚合物(例如藻酸)及改性天然聚合物(例如,硬化蛋白、天然橡胶的化学衍生物):**							**Natural polymers (for example, alginic acid) and modified natural polymers (for example, hardened proteins, chemical derivatives of natural rubber), not elsewhere specified or included, in primary forms:**
2594	3913.1000	-藻酸及其盐和酯	10					45	-Alginic acid, its salts and esters
2595	3913.9000	-其他	6.5					50	-Other

序号 No.	税则号列 Tariff Item	货品名称	最惠国税率(%) M.F.N.	协定税率(%)Agreement 东盟 ASEAN	曼谷协定 BA	香港 H.K.	澳门 Macao	普通税率(%) Gen.	Article Description
	39.14	**初级形状的离子交换剂,以税号39.01至39.13的聚合物为基本成分的:**							**Ion-exchangers based on polymers of headings Nos.39.01 to 39.13,in primary forms:**
2596	3914.0000	初级形状的离子交换剂,以税号39.01至39.13的聚合物为基本成分的	6.5					45	Ion-exchangers based on polymers of headings Nos.39.01 to 39.13,in primary forms
		第二分章　废碎料及下脚料;半制品;制成品							Ⅱ. WASTE,PARINGS AND SCRAP; SEMI-MANUFACTURES; ARTICLES
	39.15	**塑料的废碎料及下脚料:**							**Waste,parings and scrap,of plastics:**
2597	3915.1000	-乙烯聚合物的	9.7			0		50	-Of polymers of ethylene
2598	3915.2000	-苯乙烯聚合物的	9.7			0		50	-Of polymers of styrene
2599	3915.3000	-氯乙烯聚合物的	9.7			0		50	-Of polymers of vinyl chloride
		-其他塑料的:							-Of other plastics:
2600	3915.9010	---聚对苯二甲酸乙二酯的	9.7			0	0	50	---Of pdyethylene glycol tevephthalate
2601	3915.9090	---其他	9.7			0	0	50	---Other
	39.16	**塑料制的单丝(截面直径超过1毫米)、条、杆、型材及异型材,不论是否经表面加工,但未经其他加工:**							**Monofilament of which any cross-sectional dimension exceeds 1 mm,rods, sticks and profile shapes, whether or not surfaceworked but not otherwise worked,of plastics:**
2602	3916.1000	-乙烯聚合物制	10					45	-Of polymers of ethylene
2603	3916.2000	-氯乙烯聚合物制	10					45	-Of polymers of vinyl chloride
		-其他塑料制:							-Of other plastics:
2604	3916.9010	---聚酰胺制	10					45	---Of polyamides
2605	3916.9090	---其他	10					45	---Other
	39.17	**塑料制的管子及其附件(例如,接头、肘管、法兰):**							**Tubes, pipes and hoses, and fittings therefor(for example,joints,elbows, flanges),of plastics:**
2606	3917.1000	-硬化蛋白或纤维素材料制的人造肠衣(香肠用肠衣)	10					50	-Artificial guts (sausage casings) of hardened protein or of cellulosic materials
		-硬管:							-Tubes,pipes and hoses,rigid:
2607	3917.2100	--乙烯聚合物制	10					45	--Of polymers of ethylene
2608	3917.2200	--丙烯聚合物制	10					45	--Of polymers of propylene
2609	3917.2300	--氯乙烯聚合物制	10			0		45	--Of polymers of vinyl chloride
2610	3917.2900	--其他塑料制	10					45	--Of other plastics
		-其他管:							-Other tubes,pipes and hoses:
2611	3917.3100	--软管,最小爆破压力为27.6兆帕斯卡	10			0		45	--Flexible tubes,pipes and hoses,having a minimum burst pressure of 27.6 MPa

序号 No.	税则号列 Tariff Item	货品名称	最惠国税率(%) M.F.N.	协定税率(%) Agreement 东盟 ASEAN	曼谷协定 BA	香港 H.K.	澳门 Macao	普通税率(%) Gen.	Article Description
2612	3917.3200	--其他未装有附件的管子,未经加强也未与其他材料合制	6.5					45	--Other, not reinforced or otherwise combined with other materials, without fittings
2613	3917.3300	--其他装有附件的管子,未经加强也未与其他材料合制	6.5					45	--Other, not reinforced or otherwise combined with other materials, with fittings
2614	3917.3900	--其他	9.7					45	--Other
2615	3917.4000	-管子附件	10					45	-Fittings
	39.18	**块状或成卷的塑料铺地制品,不论是否胶粘;本章注释九所规定的塑料糊墙品:**							**Floor coverings of plastics, whether or not self-adhesive, in rolls or in the form of tiles; wall or ceiling coverings of plastics, as defined in Note 9 to this Chapter:**
		-氯乙烯聚合物制:							-Of polymers of vinyl chloride:
2616	3918.1010	---糊墙品	10					45	---Wall or ceiling coverings
2617	3918.1090	---其他	10					45	---Other
		-其他塑料制:							-Of other plastics:
2618	3918.9010	---糊墙品	10			0		45	---Wall or ceiling coverings
2619	3918.9090	---其他	10			0		45	---Other
	39.19	**自粘的塑料板、片、膜、箔、带、扁条及其他扁平形状材料,不论是否成卷:**							**Self-adhesive plates, sheets, film, foil, tape, strip and other flat shapes, of plastics, whether or not in rolls:**
		-成卷,宽度不超过20厘米:							-In rolls of a width not exceeding 20 cm:
2620	3919.1010	---丙烯酸树脂类为基本成分	6.5					45	---Based on acrylic resin
		---其他:							---Other:
2621	3919.1091	----胶囊型反光膜	6.5					45	----Encapsulant reflective film
2622	3919.1099	----其他	6.5					45	----Other
		-其他:							-Other:
2623	3919.9010	---胶囊型反光膜	6.5					45	---Encapsulant reflective film
2624	3919.9090	---其他	6.5					45	---Other
	39.20	**其他非泡沫塑料的板、片、膜、箔及扁条,未用其他材料强化、层压、支撑或用类似方法合制:**							**Other plates, sheets, film, foil and strip, of plastics, non-cellular and not reinforced, laminated, supported or similarly combined with other materials:**
		-乙烯聚合物制:							-Of polymers of ethylene:
2625	3920.1010	---乙烯聚合物制电池隔膜	6.5			0		45	---Battery separator, of polymers of ethylene
2626	3920.1090	---其他	6.5			0		45	---Other
		-丙烯聚合物制:							-Of polymers of propylene:
2627	3920.2010	---丙烯聚合物制电池隔膜	6.5			0		45	---Battery separator, of polymers of propylene
2628	3920.2090	---其他	6.5			0		45	---Other

序号 No.	税则号列 Tariff Item	货品名称	最惠国税率(%) M.F.N.	协定税率(%)Agreement 东盟 ASEAN	曼谷协定 BA	香港 H.K.	澳门 Macao	普通税率(%) Gen.	Article Description
2629	3920.3000	-苯乙烯聚合物制	6.5			0		45	-Of polymers of styrene
		-氯乙烯聚合物制:							-Of polymers of vinyl chloride:
2630	3920.4300	--按重量计增塑剂含量不小于6%	9.7			0		45	--Containing by weight not less than 6% of pleasticisers
2631	3920.4900	--其他	9			0		45	--Other
		-丙烯酸聚合物制:							-Of acrylic polymers:
2632	3920.5100	--聚甲基丙烯酸甲酯制	6.5					45	--Of poly(methyl methacrylate)
2633	3920.5900	--其他	6.5					45	--Other
		-聚碳酸酯、醇酸树脂、聚烯丙酯或其他聚酯制:							-Of polycarbonates, alkyd resins, polyallyl esters or other polyesters:
2634	3920.6100	--聚碳酸酯制	6.5					45	--Of polycarbonates
2635	3920.6200	--聚对苯二甲酸乙二酯制	6.5			0		45	--Of poly(ethylene terephthalate)
2636	3920.6300	--不饱和聚酯制	10					45	--Of unsaturated polyesters
2637	3920.6900	--其他聚酯制	10		9			45	--Of other polyesters
		-纤维素及其化学衍生物制:							-Of cellulose or its chemical derivatives:
2638	3920.7100	--再生纤维素制	6.5					45	--Of regenerated cellulose
2639	3920.7200	--钢纸制	10					35	--Of vulcanized fibre
2640	3920.7300	--乙酸纤维素制	6.5			0		45	--Of cellulose acetate
2641	3920.7900	--其他纤维素衍生物制	10					45	--Of other cellulose derivatives
		-其他塑料制:							-Of other plastics:
2642	3920.9100	--聚乙烯醇缩丁醛制	6.5					45	--Of poly(vinyl butyral)
2643	3920.9200	--聚酰胺制	10					45	--Of polyamides
2644	3920.9300	--氨基树脂制	6.5					45	--Of amino-resins
2645	3920.9400	--酚醛树脂制	10					45	--Of phenolic resins
		--其他塑料制:							--Of other plastics:
2646	3920.9910	---聚四氟乙烯制	6.5					45	---Of polytetrafluoroethylene
2647	3920.9990	---其他塑料制	6.5			0		45	---Of other plastics
	39.21	**其他塑料板、片、膜、箔、扁条:**							**Other plates, sheets, film, foil and strip, of plastics:**
		-泡沫塑料的:							-Cellular:
2648	3921.1100	--苯乙烯聚合物制	10		9			45	--Of polymers of styrene
		--氯乙烯聚合物制:							--Of polymers of vinyl chloride:
2649	3921.1210	---人造革及合成革	9			0		70	---Combined with textile fabrics
2650	3921.1290	---其他	6.5					45	---Other
		--氨酯聚合物制:							--Of polyurethanes:
2651	3921.1310	---人造革及合成革	9			0△		70	---Combined with textile fabrics
2652	3921.1390	---其他	6.5			0△		45	---Other
2653	3921.1400	--再生纤维素制	10					45	--Of regenerated cellulose
		--其他塑料制:							--Of other plastics:
2654	3921.1910	---人造革及合成革	9					45	---Combined with textile fabrics
2655	3921.1990	---其他	6.5					45	---Other
		-其他:							-Other:
2656	3921.9020	---聚乙烯嵌有玻璃纤维的板、片	6.5			0		45	---Plates, sheets of polyethylene with glass fibres

序号 No.	税则号列 Tariff Item	货品名称	最惠国税率(%) M.F.N.	协定税率(%)Agreement 东盟 ASEAN	曼谷协定 BA	香港 H.K.	澳门 Macao	普通税率(%) Gen.	Article Description
2657	3921.9030	---聚异丁烯为基本成分的附有人造毛毡的板、片、卷材	6.5			0		45	---Plates, sheets, coils of polyisobutylene with man-made felt
2658	3921.9090	---其他	6.5			0		45	---Other
	39.22	**塑料浴缸、淋浴盘、洗涤槽、盥洗盆、坐浴盆、便盆、马桶座圈及盖、抽水箱及类似卫生洁具:**							**Baths, shower-baths, sinks, wash-basins, bidets, lavatory pans, seats and covers, flushing cisterns and similar sanitary ware, of plastics:**
2659	3922.1000	-浴缸、淋浴盘、洗涤槽及盥洗盆	10					80	-Baths, shower-baths, sinks and wash-basins
2660	3922.2000	-马桶座圈及盖	10					80	-Lavatory seats and covers
2661	3922.9000	-其他	10					80	-Other
	39.23	**供运输或包装货物用的塑料制品;塑料制的塞子、盖子及类似品:**							**Articles for the conveyance or packing of goods, of plastics; stoppers, lids, caps and other closures, of plastics:**
2662	3923.1000	-盒、箱(包括板条箱)及类似品	10			0	0	80	-Boxes, cases, crates and similar articles
		-袋及包(包括锥形的):							-Sacks and bags(including cones):
2663	3923.2100	--乙烯聚合物制	10			0	0	80	--Of polymers of ethylene
2664	3923.2900	--其他塑料制	10			0	0	80	--Of other plastics
2665	3923.3000	-坛、瓶及类似品	8.4			0		80	-Carboys, bottles, flasks and similar articles
2666	3923.4000	-卷轴、纡子、筒管及类似品	10					35	-Spools, cops, bobbins and similar supports
2667	3923.5000	-塞子、盖子及类似品	10			0		80	-Stoppers, lids, caps and other closures
2668	3923.9000	-其他	10			0	0	80	-Other
	39.24	**塑料制的餐具、厨房用具、其他家庭用具及盥洗用具:**							**Tableware, kitchenware, other household articles and toilet articles, of plastics:**
2669	3924.1000	-餐具及厨房用具	10					80	-Tableware and kitchenware
2670	3924.9000	-其他	10					80	-Other
	39.25	**其他税号未列名的建筑用塑料制品:**							**Builders' ware of plastics, not elsewhere specified or included:**
2671	3925.1000	-囤、柜、罐、桶及类似容器,容积超过300升	10					80	-Reservoirs, tanks, vats and similar containers, of a capacity exceeding 300 L
2672	3925.2000	-门、窗及其框架、门槛	10					80	-Doors, windows and their frames and thresholds for doors
2673	3925.3000	-窗板、百叶窗(包括威尼斯式百叶窗)或类似制品及其零件	10					80	-Shutters, blinds (including Venetian blinds) and similar articles and parts thereof
2674	3925.9000	-其他	10			0		80	-Other
	39.26	**其他塑料制品及税号39.01至39.14所列其他材料的制品:**							**Other articles of plastics and articles of other materials of headings Nos. 39.01 to 39.14:**

序号 No.	税则号列 Tariff Item	货品名称	最惠国税率(%) M.F.N.	协定税率(%) Agreement 东盟 ASEAN	曼谷协定 BA	香港 H.K.	澳门 Macao	普通税率(%) Gen.	Article Description
2675	3926.1000	-办公室或学校用品	10					80	-Office or school supplies
2676	3926.2000	-衣服及衣着附件(包括分指手套、连指手套及露指手套)	10			0		90	-Articles of apparel and clothing accessories (including gloves, mittens and mitts)
2677	3926.3000	-家具、车厢或类似品的附件	10					80	-Fittings for furniture, coachwork or the like
2678	3926.4000	-小雕塑品及其他装饰品	10		8.3			100	-Statuettes and other ornamental articles
		-其他:							-Other:
2679	3926.9010	---机器及仪器用零件	10		9	0	0	35	---Of a kind for used in machines or instruments
2680	3926.9090	---其他	10		9.2	0	0	80	---Other

第四十章
橡胶及其制品

注释：

一、除条文另有规定的以外，本目录所称“橡胶”，是指不论是否硫化或硬化的下列产品：天然橡胶、巴拉塔胶、古塔波胶、银胶菊胶、糖胶树胶及类似的天然树胶、合成橡胶、从油类中提取的油膏以及上述物品的再生品。

二、本章不包括：

（一）第十一类的货品（纺织原料及纺织制品）；

（二）第六十四章的鞋靴及其零件；

（三）第六十五章的帽类及其零件（包括游泳帽）；

（四）第十六类的硬质橡胶制的机械器具、电气器具及其零件（包括各种电气用品）；

（五）第九十章、第九十二章、第九十四章或第九十六章的物品；

（六）第九十五章的物品（运动用分指手套、连指手套、露指手套及税号 40.11 至 40.13 的制品除外）。

三、税号 40.01 至 40.03 及 40.05 所称“初级形状”，只限于下列形状：

（一）液状及糊状，包括胶乳（不论是否预硫化）及其他分散体和溶液；

（二）不规则形状的块、团、包、粉、粒、碎屑及类似的散装形状。

四、本章注释一和税号 40.02 所称“合成橡胶”，适用于：

（一）不饱和合成物质，即用硫磺硫化能使其不可逆地变为非热塑物质，这种物质能在温度 18℃ 至 29℃ 之间被拉长到其原长度的三倍而不致断裂，拉长到原长度的两倍时，在五分钟内能回复到不超过原长度的一倍半。为了进行上述试验，可以加入交联所需的硫化活化剂或促进剂；也允许含有注释五（二）2 及 3 所述的物质。但不能加入非交联所需的物质，例如，增量剂、增塑剂及填料；

Chapter
Rubber and articles thereof

Notes:

1. Except where the context otherwise requires, throughout the Nomenclature the expression “rubber” means the following products, whether or not vulcanized or hard: natural rubber, balata, gutta-percha, guayule, chicle and similar natural gums, synthetic rubber, factice derived from oils, and such substances reclaimed.

2. This Chapter does not cover:

(a) Goods of Section XI (texitils and textile articles);

(b) Footwear or parts thereof of Chapter 64;

(c) Headgear or parts thereof (including bathing caps) of Chapter 65;

(d) Mechanical or electrical appliances or parts thereof of Section XVI (including electrical goods of all kinds), of hard rubber;

(e) Articles of Chapter 90, 92, 94 or 96; or

(f) Articles of Chapter 95 (other than sports gloves, mittens and mitts and articles of headings Nos.40.11 to 40.13).

3. In headings Nos.40.01 to 40.03 and 40.05, the expression “primary forms” applies only to the following forms:

(a) Liquids and pastes (including latex, whether or not pre-vulcanized, and other dispersions and solutions);

(b) Blocks of irregular shape, lumps, bales, powders, granules, crumbs and similar bulk forms.

4. In Note I to this Chapter and in heading No.40.02, the expression “synthetic rubber” applies to:

(a) Unsaturated synthetic substances which can be irreversibly transformed by vulcanization with sulphur into non-thermoplastic substances which, at a temperature between 18℃ and 29℃, will not break on being extended to three times their original length and will return, after being extended to twice their original length, within a period of five minutes, to a length not greater than one and a half times their original length. For the purposes of this test, substances necessary for the cross-linking, such as vulcanizing activators or accelerators, may be added; the presence of substances as provided for by Note 5 (b) (ii) and (iii) is also permitted. However, the presence of any substances not necessary for the cross-linking, such as extenders, plasticizers and fillers, is not permitted;

（二）聚硫橡胶（TM）；

（三）与塑料接枝共聚或混合而改性的天然橡胶、解聚天然橡胶以及不饱和合成物质与饱和合成高聚物的混合物，但这些产品必须符合以上（一）款关于硫化、延伸及回复的要求。

(b) Thioplasts (TM); and

(c) Natural rubber modified by grafting or mixing with plastics, depolymerized natural rubber, mixtures of unsaturated synthetic substances with saturated synthetic high polymers provided that all the above-mentioned products comply with the requirements concerning vulcanization, elongation and recovery in (a) above.

五、（一）税号40.01及40.02不适用于任何凝结前或凝结后与下列物质相混合的橡胶或橡胶混合物：

1. 硫化剂、促进剂、防焦剂或活性剂（为制造预硫胶乳所加入的除外）；

2. 颜料或其他着色料，但仅为易于识别而加入的除外；

3. 增塑剂或增量剂（用油增量的橡胶中所加的矿物油除外）、填料、增强剂、有机溶剂或其他物质，但以下（二）款所述的除外；

（二）含有下列物质的橡胶或橡胶混合物，只要仍具有原料的基本特性，应归入税号40.01或40.02：

1. 乳化剂或防粘剂；
2. 少量的乳化剂分解产品；
3. 微量的下列物质：热敏剂（一般为制造热敏胶乳用）、阳离子表面活性剂（一般为制造阳性胶乳用）、抗氧剂、凝固剂、碎裂剂、抗冻剂、胶溶剂、保存剂、稳定剂、粘度控制剂或类似的特殊用途添加剂。

5. (a) Headings Nos. 40.01 and 40.02 do not apply to any rubber or mixture of rubbers which has been compounded, before or after coagulation, with:

(1) vulcanizing agents, accelerators, retarders or activators (other than those added for the preparation of pre-vulcanized rubber latex);

(2) pigments or other colouring matter, other than those added solely for the purpose of identification;

(3) plasticizers or extenders (except mineral oil in the case of oil-extended rubber), fillers, reinforcing agents, organic solvents or any other substances, except those permitted under (b);

(b) The presence of the following substances in any rubber or mixture of rubbers shall not affect its classification in heading No. 40.01 or 40.02, as the case may be, provided that such rubber or mixture of rubbers retains its essential character as a raw material:

(1) emulsifiers or anti-tack agents;

(2) small amounts of breakdown products of emulsifiers;

(3) very small amounts of the following: heat-sensitive agents (generally for obtaining thermosensitive rubber latexes), cationic surface-active agents (generally for obtaining electro-positive rubber latexes), antioxidants, coagulants, crumbling agents, freeze-resisting agents, peptizers, preservatives, stabilizers, viscosity-control agents, or similar special-purpose additives.

六、税号40.04所称“废碎料及下脚料”，是指在橡胶或橡胶制品生产或加工过程中由于切割、磨损或其他原因所造成没有使用价值的废橡胶及下脚料。

6. For the purposes of heading No. 40.04, the expression “waste, parings and scrap” means rubber waste, parings and scrap from the manufacture or working of rubber and rubber goods definitely not usable as such because of cutting-up, wear or other reasons.

七、全部用硫化橡胶制成的线，其任一截面的尺寸超过5毫米的，应作为带、杆或型材及异型材归入税号40.08。

7. Thread wholly of vulcanized rubber, of which any cross-sectional dimension exceeds 5 mm, is to be classified as strip, rods or profile shapes, of heading No. 40.08.

八、税号40.10包括用橡胶浸渍、涂布、包覆或层压的织物制成的或用橡胶浸渍、涂布、包覆或套裹的纱线或绳制成的传动带、输送带。

8. Heading No.40.10 includes conveyor or transmission belts or belting of textile fabric impregnated, coated, covered or laminated with rubber or made from textile yarn or cord impregnated, coated, covered or sheathed with rubber.

九、税号40.01、40.02、40.03、40.05及40.08所称“板”、“片”、“带”，仅指未切割或只简单切割成矩形（包括正方形）的板、片、带及正几何形块，不论是否具有成品的特征，也不论是否经过印制或其他表面加工，但未切割成其他形状或进一步加工。

税号40.08所称“杆”或“型材及异型材”，仅指不论是否切割成一定长度或表面加工，但未经进一步加工的该类产品。

9. In headings Nos.40.01, 40.02, 40.03, 40.05, and 40.08, the expressions “plates”, “sheets” and “strip” apply only to plates, sheets and strip and to blocks of regular geometric shape, uncut or simply cut to rectangular (including square) shape, whether or not having the character of articles and whether or not printed or otherwise surface-worked, but not otherwise cut to shape or further worked.

In heading No.40.08 the expressions “rods” and “profile shapes” apply only to such products, whether or not cut to length or surface-worked but not otherwise worked.

序号 No.	税则号列 Tariff Item	货品名称	最惠国税率(%) M.F.N.	协定税率(%) Agreement 东盟 ASEAN	曼谷协定 BA	香港 H.K.	澳门 Macao	普通税率(%) Gen.	Article Description
	40.01	**天然橡胶、巴拉塔胶、古塔波胶、银胶菊胶、糖胶树胶及类似的天然树胶，初级形状或板、片、带：**							**Natural rubber, balata, gutta-percha, guayule, chicle and similar natural gums, in primary forms or in plates, sheets or strip:**
2681	4001.1000	-天然胶乳，不论是否预硫化	20					40	-Natural rubber latex, whether or not prevulcanized
		-其他形状的天然橡胶：							-Natural rubber in other forms:
2682	4001.2100	--烟胶片	20		17			40	--Smoked sheets
2683	4001.2200	--技术分类天然橡胶(TSNR)	20					40	--Technically specified natural rubber (TSNR)
2684	4001.2900	--其他	20		17			40	--Other
2685	4001.3000	-巴拉塔胶、古塔波胶、银胶菊胶、糖胶树胶及类似的天然树胶	20					40	-Balata, gutta-percha, guayule, chicle and similar natural gums
	40.02	**合成橡胶及从油类提取的油膏，初级形状或板、片、带；税号40.01所列产品与本税号所列产品的混合物，初级形状或板、片、带：**							**Synthetic rubber and factice derived from oils, in primary forms or in plates , sheets or strip; mixtures of any products of heading No.40.01 with any product of this heading, in primary forms or in plates, sheets or strip:**
		-丁苯橡胶(SBR)；羧基丁苯橡胶(XSBR)：							-Styrene-butadiene rubber(SBR); carboxylated styrene-butadiene rubber (XSBR):
		--胶乳：							--Latex:
2686	4002.1110	---羧基丁苯橡胶	7.5			0		14	---Carboxylated styrene-butadiene rubber(XSBR)

序号 No.	税则号列 Tariff Item	货品名称	最惠国税率(%) M.F.N.	协定税率(%)Agreement 东盟 ASEAN	曼谷协定 BA	香港 H.K.	澳门 Macao	普通税率(%) Gen.	Article Description
2687	4002.1190	---其他	7.5			0		14	---Other
		--其他：							--Other：
		---初级形状的：							---In primary forms：
2688	4002.1911	----未经任何加工的丁苯橡胶	7.5			0		14	----SBR，not worked
2689	4002.1912	----充油丁苯橡胶	7.5			0		14	----SBR，oil-filled
2690	4002.1913	----热塑丁苯橡胶	7.5			0		14	----SBR，thermo-plasticated
2691	4002.1914	----充油热塑丁苯橡胶	7.5			0		14	----SBR，oil-filled and thermo-plasticated
2692	4002.1919	----其他	7.5			0		14	----Other
2693	4002.1990	---其他	7.5			0		35	---Other
		-丁二烯橡胶(BR)：							-Butadiene rubber(BR)：
2694	4002.2010	---初级形状的	7.5					14	---In primary forms
2695	4002.2090	---其他	7.5		7			35	---Other
		-异丁烯-异戊二烯(丁基)橡胶(IIR)；卤代丁基橡胶(CIIR或BIIR)：							-Isobutene-isoprene (butyl) rubber (IIR)；halo-isobutene-isoprene rubber (CIIR or BIIR)：
		--异丁烯-异戊二烯(丁基)橡胶(IIR)：							--Isobutene-isoprene (butyl) rubber (IIR)：
2696	4002.3110	---初级形状的	6					14	---In primary forms
2697	4002.3190	---其他	7.5					35	---Other
		--其他：							--Other：
2698	4002.3910	---初级形状的	7.5					14	---In pimary forms
2699	4002.3990	---其他	7.5					35	---Other
		-氯丁二烯(氯丁)橡胶(CR)：							-Chloroprene (chlorobutadiene) rubber (CR)：
2700	4002.4100	--胶乳	7.5					14	--Latex
		--其他：							--Other：
2701	4002.4910	---初级形状的	7.5					14	---In primary forms
2702	4002.4990	---其他	7.5					35	---Other
		-丁腈橡胶(NBR)：							-Acrylonitrile-butadient rubber (NBR)：
2703	4002.5100	--胶乳	7.5					14	---Latex
		--其他：							--Other：
2704	4002.5910	---初级形状的	7.5					14	---In primary forms
2705	4002.5990	---其他	7.5					35	---Other
		-异戊二烯橡胶(IR)：							-Isoprene rubber(IR)：
2706	4002.6010	---初级形状的	3					14	---In primary forms
2707	4002.6090	---其他	5					35	---Other
		-乙丙非共轭二烯橡胶(EPDM)：							-Ethylene-propylene-non-conjugated diene rubber(EPDM)：
2708	4002.7010	---初级形状的	7.5					14	---In primary forms
2709	4002.7090	---其他	7.5					35	---Other
2710	4002.8000	-税号40.01所列产品与本税号所列产品的混合物	7.5					35	-Mixtures of any product of heading No.40.01 with any product of this heading

序号 No.	税则号列 Tariff Item	货品名称	最惠国税率(%) M.F.N.	协定税率(%) Agreement 东盟 ASEAN	曼谷协定 BA	香港 H.K.	澳门 Macao	普通税率(%) Gen.	Article Description
		-其他:							-Other:
2711	4002.9100	--胶乳	7.5					14	--Latex
		--其他:							--Other:
		---其他合成橡胶:							---Other synthetic rubber:
2712	4002.9911	----初级形状的	7.5					14	----In primary forms
2713	4002.9919	----其他	7.5					35	----Other
2714	4002.9990	---其他	4					14	---Other
	40.03	**再生橡胶,初级形状或板、片、带:**							**Reclaimed rubber in primary forms or in plates,sheets or strip:**
2715	4003.0000	再生橡胶,初级形状或板、片、带	8			0		30	Reclaimed rubber in primary forms or in plates,sheets or strip
	40.04	**橡胶(硬质橡胶的除外)的废碎料、下脚料及其粉、粒:**							**Waste, parings and scrap of rubber (other than hard rubber)and powders and granules obtained therefrom:**
2716	4004.0000	橡胶(硬质橡胶的除外)的废碎料、下脚料及其粉、粒	8			0		30	Waste, parings and scrap of rubber (other than hard rubber)and powders and granules obtained therefrom
	40.05	**未硫化的复合橡胶,初级形状或板、片、带:**							**Compounded rubber, unvulcanized, in primary forms or in plates, sheets or strip:**
2717	4005.1000	-与碳黑或硅石混合	8					35	-Compounded with carbon black or silica
2718	4005.2000	-溶液;子目号4005.10以外的分散体	8					35	-Solutions;dispersions other than those of subheading No.4005.10
		-其他:							-Other:
2719	4005.9100	--板、片、带	8					35	--Plates,sheets and strip
2720	4005.9900	--其他	8					35	--Other
	40.06	**其他形状(例如,杆、管或型材及异型材)的未硫化橡胶及未硫化橡胶制品(例如,盘、环):**							**Other forms(for example, rods, tubes and profile shapes) and articles (for example, discs and rings), or unvulcanized rubber:**
2721	4006.1000	-轮胎翻新用胎面补料胎条	8					35	-Camel-back strips for retreading rubber tyres
		-其他:							-Other:
2722	4006.9010	---其他形状的未硫化橡胶	8					35	---Other forms of unvulcanized rubber
2723	4006.9020	---未硫化橡胶制品	14					80	---Articles of unvulcanized rubber
	40.07	**硫化橡胶线及绳:**							**Vulcanized rubber thread and cord:**
2724	4007.0000	硫化橡胶线及绳	14					80	Vulcanized rubber thread and cord
	40.08	**硫化橡胶(硬质橡胶除外)制的板、片、带、杆或型材及异型材:**							**Plates, sheets, strip, rods and profile shapes, of vulcanized rubber other than hard rubber:**
		-海绵橡胶制:							-Of cellular rubber:
2725	4008.1100	--板、片、带	8					35	--Plates,sheets and strip
2726	4008.1900	--其他	8					35	--Other
		-非海绵橡胶制:							-Of non-cellular rubber:

序号 No.	税则号列 Tariff Item	货品名称	最惠国税率(%) M.F.N.	协定税率(%) Agreement 东盟 ASEAN	曼谷协定 BA	香港 H.K.	澳门 Macao	普通税率(%) Gen.	Article Description
2727	4008.2100	--板、片、带	8					35	--Plates, sheets and strip
2728	4008.2900	--其他	8					35	--Other
	40.09	**硫化橡胶(硬质橡胶除外)制的管子,不论是否装有附件(例如,接头、肘管、法兰):**							**Tubes, pipes and hoses, of vulcanized rubber other than hard rubber, with or without their fittings (for example, joints, elbows, flanges):**
		-未经加强或未与其他材料合制:							-Not reinforced or otherwise combined with other materials:
2729	4009.1100	--未装有附件	10.5					40	--Without fittings
2730	4009.1200	--装有附件	10					40	--With fittings
		-用金属加强或只与金属合制:							-Reinforced or otherwise combined only with metal:
2731	4009.2100	--未装有附件	10.5					40	--Without fittings
2732	4009.2200	--未装有附件	10					40	--With fittings
		-用纺织材料加强或只与纺织材料合制:							-Reinforced or otherwise combined only with textile materials:
2733	4009.3100	--未装有附件	10.5					40	--Without fittings
2734	4009.3200	--装有附件	10					40	--With fittings
		-用其他材料加强或与其他材料合制:							-Reinforced or otherwise combined with other materials:
2735	4009.4100	--未装有附件	10.5					40	--Without fittings
2736	4009.4200	--装有附件	10					40	--With fittings
	40.10	**硫化橡胶制的传动带或输送带及带料:**							**Conveyor or transmission belts or belting, of vulcanized rubber:**
		-输送带及带料:							-Conveyor belts or belting:
2737	4010.1100	--仅用金属加强的	10					35	--Reinforced only with metal
2738	4010.1200	--仅用纺织材料加强的	10					35	--Reinforced only with textile materials
2739	4010.1300	--仅用塑料加强的	10					35	--Reinforced only with plastics
2740	4010.1900	--其他	10					35	--Other
		-传动带及带料:							-Transmission belts or belting:
2741	4010.3100	--梯形截面的环形传动带(三角带),V形肋状的,外周长超过60厘米,但不超过180厘米	8					35	--Endless transmission belts of trapezoidal cross-section (V-belts), whether or not grooved, of a circumference exceeding 60 cm but not exceeding 180 cm
2742	4010.3200	--梯形截面的环形传动带(三角带),外周长超过60厘米,但不超过180厘米,V形肋状的除外	8					35	--Endless transmission belts of trapezoidal cross – section (V – belts), V – ribbed, of an outside circumference exceeding 60 cm but not exceeding 180 cm
2743	4010.3300	--梯形截面的环形传动带(三角带),V形肋状的,外周长超过180厘米,但不超过240厘米	8					35	--Endless transmission belts of trapezoidal cross – section (V – belts), other than V – ribbed, of an outside circumference exceeding 180 cm but not exceeding 240 cm

序号 No.	税则号列 Tariff Item	货品名称	最惠国税率(%) M.F.N.	协定税率(%) Agreement 东盟 ASEAN	曼谷协定 BA	香港 H.K.	澳门 Macao	普通税率(%) Gen.	Article Description
2744	4010.3400	--梯形截面的环形传动带(三角带),外周长超过180厘米,但不超过240厘米,V形肋状的除外	8					35	--Endless transmission belts of trapezoidal cross – section (V – belts), V – ribbed, of an outside circumference exceeding 180cm but not exceeding 240 cm
2745	4010.3500	--环形同步带,外周长超过60厘米,但不超过150厘米	10					35	--Endless transmission belts of trapezoidal cross – section(V – belts), other than V – ribbed, of an outside circumference exceeding 60 cm but not exceeding 150 cm
2746	4010.3600	--环形同步带,外周长超过150厘米,但不超过198厘米	10					35	--Endless synchronous belts, of an outside circumference exceeding 150 cm but not exceeding 198 cm
2747	4010.3900	--其他	8					35	--Other
	40.11	**新的充气橡胶轮胎:**							**New pneumatic tyres, of rubber:**
2748	4011.1000	-机动小客车(包括旅行小客车及赛车)用	10		9.5			50	-Of a kind used on motor cars(including station wagons and racing cars)
2749	4011.2000	-客运机动车辆或货运机动车辆用	10		9.5			50	-Of a kind used on buses or lorries
2750	4011.3000	-航空器用	1					11	-Of a kind used on aircraft
2751	4011.4000	-摩托车用	15					80	-Of a kind used on motorcycles
2752	4011.5000	-自行车用	20					80	-Of a kind used on bicycles
		-其他,人字形胎面或类似胎面的:							-Other, Having a "herring-bone" or similar tread:
2753	4011.6100	--农业或林业车辆及机器用	17.5					50	--Of a kind used on agricultural or forestry vehicles and machines
2754	4011.6200	--建筑业或工业搬运车辆及机器用,辋圈尺寸不超过61厘米	17.5					50	--Of a kind used on construction or industrial handling vehicles and machines and having a rim size not exceeding 61 cm
2755	4011.6300	--建筑业或工业搬运车辆及机器用,辋圈尺寸超过61厘米	17.5					50	--Of a kind used on construction or industrial handling vehicles and machines and having a rim size exceeding 61 cm
2756	4011.6900	--其他	17.5					50	--Other
		-其他:							-Other:
2757	4011.9200	--农业或林业车辆及机器用	25					50	--Of a kind used on agricultural or forestry vehicles and machines
2758	4011.9300	--建筑业或工业搬运车辆及机器用,辋圈尺寸不超过61厘米	25					50	--Of a kind used on construction or industrial handling vehicles and machines and having a rim size not exceeding 61 cm
2759	4011.9400	--建筑业或工业搬运车辆及机器用,辋圈尺寸超过61厘米	25					50	--Of a kind used on construction or industrial handling vehicles and machines and having a rim size exceeding 61 cm

序号 No.	税则号列 Tariff Item	货品名称	最惠国税率(%) M.F.N.	协定税率(%) Agreement 东盟 ASEAN	曼谷协定 BA	香港 H.K.	澳门 Macao	普通税率(%) Gen.	Article Description
2760	4011.9900	--其他	25					50	--other
	40.12	**翻新的或旧的充气橡胶轮胎;实心或半实心橡胶轮胎、橡胶胎面及橡胶轮胎衬带:**							**Retreaded or used pneumatic tyres of rubber; solid or cushion tyres, tyre treads and tyre flaps,of rubber:**
		-翻新轮胎:							-Retreaded tyres:
2761	4012.1100	--机动小客车(包括旅行小客车及赛车)用	20					50	--Of a kind used on motor cars,(including station wagons and racing cars)
2762	4012.1200	--机动大客车或货运机动车辆用	20			0		50	--Of a kind used on buses or corries
2763	4012.1300	--航空器用	20			0		50	--Of a kind used on aircratr
2764	4012.1900	--其他	20					50	--Other
		-旧的充气轮胎:							-Used pneumatic tyres:
2765	4012.2010	---汽车用	25			0		50	---Of a kind used on motor cars,buses or lorries
2766	4012.2090	---其他	25			0		80	---Other
		-其他:							-Other:
2767	4012.9010	---航空器用	3					11	---Of a kind used on aircraft
2768	4012.9020	---汽车用	22					50	---Of a kind used on motor cars,buses or lorries
2769	4012.9090	---其他	22					50	---Other
	40.13	**橡胶内胎:**							**Inner tubes,of rubber:**
2770	4013.1000	-机动小客车(包括旅行小客车及赛车)、客运机动车辆或货运机动车辆用	15		13			50	-Of a kind used on motor cars(including station wagons and racing cars), buses or lorries
2771	4013.2000	-自行车用	15					80	-Of a kind used on bicycles
		-其他:							-Other:
2772	4013.9010	---航空器用	3					11	---Of a kind used on aircraft
2773	4013.9090	---其他	15					50	---Other
	40.14	**硫化橡胶(硬质橡胶除外)制的卫生及医疗用品(包括奶嘴),不论是否装有硬质橡胶制的附件:**							**Hygienic or pharmaceutical articles (including teats), of vulcanized rubber other than hard rubber, with or without fittings of hard rubber:**
2774	4014.1000	-避孕套	0					0	-Sheath contraceptives
2775	4014.9000	-其他	17.5					50	-Other
	40.15	**硫化橡胶(硬质橡胶除外)制的衣着用品及附件(包括手套):**							**Articles of apparel and clothing accessories(including gloves), for all purposes,of vulcanized rubber other than hard rubber:**
		-手套:							-Gloves:
2776	4015.1100	--外科用	8					30	--Surgical

序号 No.	税则号列 Tariff Item	货品名称	最惠国税率(%) M.F.N.	协定税率(%) Agreement 东盟 ASEAN	曼谷协定 BA	香港 H.K.	澳门 Macao	普通税率(%) Gen.	Article Description
2777	4015.1900	--其他	18					80	--Other
		-其他:							-Other:
2778	4015.9010	---医疗用	8					30	---For medical purpose
2779	4015.9090	---其他	15					90	---Other
	40.16	**硫化橡胶(硬质橡胶除外)的其他制品:**							**Other articles of vulcanized rubber other than hard rubber:**
		-海绵橡胶制:							-Of cellular rubber:
2780	4016.1010	---机器及仪器用零件	8					30	---Of a kind used in machines or instruments
2781	4016.1090	---其他	15					80	---Other
		-其他:							-Other:
2782	4016.9100	--铺地制品及门垫	18			0		80	--Floor coverings and mats
2783	4016.9200	--橡皮擦	18	见表1				80	--Erasers
		--垫片、垫圈及其他密封垫:							--Gaskets, washers and other seals:
2784	4016.9310	---机器及仪器用	8			0		30	---Of a kind used in machines or instruments
2785	4016.9390	---其他	15			0		80	---Other
2786	4016.9400	--船舶或码头的碰垫,不论是否可充气	18					80	--Boat or dock fenders, whether or not inflatable
2787	4016.9500	--其他可充气制品	18					80	--Other inflatable articles
		--其他:							--Other:
2788	4016.9910	---机器及仪器用零件	8					30	---Of a kind used in machines or instruments
2789	4016.9990	---其他	10					80	---Other
	40.17	**各种形状的硬质橡胶(例如纯硬质胶),包括废碎料;硬质橡胶制品:**							**Hard rubber (for example, ebonite) in all forms, including waste and scrap; articles of hard rubber:**
2790	4017.0010	---各种形状的硬质橡胶,包括废碎料	8					35	---Hard rubber in all forms, including waste and scrap
2791	4017.0020	---硬质橡胶制品	15					90	---Articles of hard rubber

第八类

生皮、皮革、毛皮及其制品；鞍具及挽具；旅行用品、手提包及类似容器；动物肠线（蚕胶丝除外）制品

第四十一章
生皮（毛皮除外）及皮革

注释：

一、本章不包括：

（一）生皮的边角废料（税号05.11）；

（二）税号05.05或67.01的带羽毛或羽绒的整张或部分鸟皮；

（三）带毛生皮或已鞣的带毛皮张（第四十三章）；但下列动物的带毛生皮应归入第四十一章，牛（包括水牛）、马、绵羊及羔羊（不包括阿斯特拉罕、喀拉科尔、波斯羔羊或类似羔羊、印度、中国或蒙古羔羊）、山羊或小山羊（不包括也门或蒙古山羊及小山羊）、猪（包括西猯）、小羚羊、瞪羚、驯鹿、麋、鹿、狍或狗。

二、（一）税目41．04至41．06不包括经逆鞣（包括预鞣）加工的皮（酌情归入税目41．01至41．03）；

（二）税目41．04至41．06所称“坯革”包括在干燥前经复鞣、染色或加油（加脂）的皮。

三、本目录所称“再生皮革”，仅指税号41.15的皮革。

SECTION Ⅷ

RAW HIDES AND SKINS，LEATHER，FURSKINS AND ARTICLES THEREOF；SADDLERY AND HARNESS；TRAVEL GOODS，HANDBAGS AND SIMILAR CONTAINERS；ARTICLES OF ANIMAL GUT（OTHER THAN SILK-WORM GUT）

Chapter 41
Raw hides and skins（other than furskins）and leather

Notes：

1．This Chapter does not cover：

（a）Parings or similar waste，of raw hides or skins（heading No.05.11）；

（b）Birdskins or parts of birdskins，with their feathers or down，of heading No.05.05 or 67.01；or

（c）Hides or skins，with the hair or wool on，raw，tanned or dressed（Chapter 43）；the following are，however，to be classified in Chapter 41，namely，raw hides and skins with the hair or wool on，of bovine animals（including buffalo），of equine animals，of sheep or lambs（except Astrakhan，Broadtail，Caracul，Persian or similar lambs，Indian，Chinese，Mongolian or Tibetan lambs），of goats or kids（except Yemen，Mongolian or Tibetan goats and kids），of swine（including peccary），of chamois，of gazelle，of reindeer，of elk，of deer，of roebucks or of dogs.

2．（a）Headings 41．04 to 41．06 do not cover hides and skins which have undergone a tanning（including pre－tanning）process which is reversible（headings 41．01 to 41．03，as the case may be）.

（b）For the purposes of headings 41．04 to 41．06，the term “crust” includes hides and skins that have been retanned，coloured or fat－liquored（stuffed）prior to drying.”

3．Throughout the Nomenclature the expression “composition leather” means only substances of the kind referred to in heading No.41.15.

序号 No.	税则号列 Tariff Item	货品名称	最惠国税率(%) M.F.N.	协定税率(%) Agreement 东盟 ASEAN	曼谷协定 BA	香港 H.K.	澳门 Macao	普通税率(%) Gen.	Article Description
	41.01	**生牛皮(包括水牛皮)、生马皮(鲜的、盐腌的、干的、石灰浸渍的、浸酸的或以其他方法保藏,但未鞣制、未经羊皮纸化处理或进一步加工的),不论是否去毛或剖层:**							**Raw hides and skins of bovine (including buffalo) or equine animals (fresh, or salted, dried, limed, pickled or otherwise preserved, but not tanned, parchment-dressed or further prepared), whether or not dehaired or split:**
		-整张皮,完全干燥的每张重量不超过8公斤,干盐腌的不超过10公斤,鲜的、湿盐腌的或以其他方法保藏的不超过16公斤:							-Whole hides and skins, of a weight per skin not exceeding 8 kg when simply dried, 10 kg when dry-salted, or 16 kg when fresh, wet-salted or otherwise preserved:
		---牛皮:							---Of bovine animals:
2792	4101.2011	----经退鞣处理的	8		6	0		17	----Have undergone a reversible tanning process
2793	4101.2019	----其他	5			0		17	----Other
2794	4101.2020	---马皮	5			0		30	---Of equine animals
		-整张皮,重量超过16公斤:							-Whole hides and skins, of a weight exceeding 16 kg:
		---牛皮:							---Of bovine animals:
2795	4101.5011	----经退鞣处理的	8.4		7			17	----Have undergone a reversible tanning process
2796	4101.5019	----其他	5					17	----Other
2797	4101.5020	---马皮	5					30	----Of equine animals
		-其他,包括整张或半张的背皮及腹皮:							-Other, including butts, bends and bollies:
		---牛皮:							---Of bovine animals:
2798	4101.9011	----经退鞣处理的	8.4		7			17	----Have undergone a reversible tanning process
2799	4101.9019	----其他	5					17	----Other
2800	4101.9020	---马皮	5					30	----Of equine animals
	41.02	**绵羊或羔羊生皮(鲜的、盐腌的、干的、石灰浸渍的、浸酸的或经其他方法保藏,但未鞣制、未经羊皮纸化处理或进一步加工的),不论是否带毛或剖层,但本章注释一(三)所述不包括的生皮除外:**							**Raw skins of sheep or lambs(fresh, or salted, dried, limed, pickled or otherwise preseved, but not tanned, parchment-dressed or further prepared), whether or not with wool on or split, other than those excluded by Note 1 (c) to this chapter:**
2801	4102.1000	-带毛	7					30	-With wool on
		-不带毛:							-Without wool on:
		--浸酸的:							--Pickled:
2802	4102.2110	---经退鞣处理的	14					30	---Have undergone a reversible tanning process

序号 No.	税则号列 Tariff Item	货品名称	最惠国税率(%) M.F.N.	协定税率(%)Agreement 东盟 ASEAN	曼谷协定 BA	香港 H.K.	澳门 Macao	普通税率(%) Gen.	Article Description
2803	4102.2190	---其他	9		8			30	---Other
		--其他:							--Other:
2804	4102.2910	---经退鞣处理的	14					30	---Have undergone a reversible tanning process
2805	4102.2990	---其他	7		6			30	----Other
	41.03	**其他生皮(鲜的、盐腌的、干的、石灰浸腌的、浸酸的或以其他方法保藏,但未鞣制、未经羊皮纸化处理或进一步加工的),不论是否去毛或剖层,但本章注释一(二)或(三)所述不包括的生皮除外:**							**Other raw hides and skins(fresh, or salted, dried, limed, pickled or otherwise preserved, but not tanned, parchment-dressed or further prepared), whether or not dehaired or split, other than those excluded by Note 1(b) or 1(c) to this Chapter:**
		-山羊或小山羊皮:							-Of goats or kids:
		---山羊板皮:							--Of goats:
2806	4103.1011	----经退鞣处理的	14					35	---Have undergone a reversible tanning process
2807	4103.1019	----其他	9					35	---Other
		---其他:							--Other:
2808	4103.1091	----经退鞣处理的	14					30	---Have undergone a reversible tanning process
2809	4103.1099	----其他	9					30	---Other
2810	4103.2000	-爬行动物皮	9					30	-Of reptiles
2811	4103.3000	-猪皮	9					30	-Of swine
2812	4103.9000	-其他	9					30	-Other
	41.04	**经鞣制的不带毛牛皮(包括水牛皮)、马皮及其坯革,不论是否剖层,但未经进一步加工:**							**Tanned or crust hides and skins of bovine (including buffalo) or equine animals, without hair on, whether or not split, but not further prepared:**
		-湿革(包括蓝湿皮):							-In the wet state (including wet-blue):
		--全粒面未剖层革;粒面剖层革:							--Full grains, unsplit; grain splits:
		---牛皮:							---Of bovine animals:
2813	4104.1111	----蓝湿的	7		6.8			17	----Wet-blue
2814	4104.1119	----其他	8		6	0		35	----Other
2815	4104.1120	---马皮	5		4	0		35	---Of equine animals
		--其他:							--Other:
		---牛皮:							---Of bovine animals:
2816	4104.1911	----蓝湿的	7		6.8			17	----Wet-blue
2817	4104.1919	----其他	7		6	0		35	----Other
2818	4104.1920	---马皮	7		5	0		35	---Of equine animals
		-干革(坯革):							-In the dry state(crust):

序号 No.	税则号列 Tariff Item	货品名称	最惠国税率(%) M.F.N.	协定税率(%) Agreement 东盟 ASEAN	曼谷协定 BA	香港 H.K.	澳门 Macao	普通税率(%) Gen.	Article Description
2819	4104.4100	--全粒面未剖层革粒面剖层革的	5		4	0		35	--Full grain, unsplit; grain splits
		--其他:							--Other:
2820	4104.4910	---机器带用牛、马皮革	5		4			20	---For machinery belting
2821	4104.4990	---其他	7		6	0		35	---Other
	41.05	**经鞣制的不带毛绵羊或者羔羊皮及其坯革,不论是否剖层,但未经进一步加工:**							**Tanned or crust skins of sheep or lambs, without wool on, whether or not split, but not further prepared:**
		-湿革(包括蓝湿皮):							-In the wet state (including wet - blue):
2822	4105.1010	---蓝湿的	14		10			50	---wet - blue
2823	4105.1090	---其他	10		7			50	---Other
2824	4105.3000	-干革(坯革)	8		7			50	---In the dry state(crust)
	41.06	**经鞣制的其他不带毛动物皮及其坯革,不论是否剖层,但未经进一步加工:**							**Tanned or crust hides and skins of other animals, without wool or hair on, whether or not split, but not further prepared:**
		-山羊或小山羊的:							--Of goats or kids:
2825	4106.2100	--湿革(包括蓝湿皮)	14		12			50	--In the wet state (including wet - blue)
2826	4106.2200	--干革(坯革)	14		12			50	--In the dry state (crust)
		-猪的:							-Of swine:
		--湿革(包括蓝湿皮):							--In the wet state (including wet - blue):
2827	4106.3110	---蓝湿的	14					50	---wet - blue
2828	4106.3190	---其他	14					50	---Other
2829	4106.3200	--干革(坯革)	14					50	--In the dry state (crust)
2830	4106.4000	-爬行动物的	14					50	-Of reptiles
		-其他:							-Other:
2831	4106.9100	--湿革(包括蓝湿皮)	14					50	--In the wet state (including wet - blue)
2832	4106.9200	--干革(坯革)	14					50	--In the dry state (crust)
	41.07	**经鞣制或半硝处理后进一步加工的不带毛的牛皮革(包括水牛皮革)及马皮革,包括羊皮纸化处理的皮革,不论是否剖层,但税目41.14的皮革除外:**							**Leather further prepared after tanning or crusting, including parchment - dressed leather, of bovine (including buffalo) or equine animals, without hair on, whether or not split, other than leather of heading 41.14:**
		-整张的:							-Whole hides and skins:
		--全粒面未剖层革:							--Full grains, unsplit:
2833	4107.1110	---牛皮	8			0		50	---Of bovine animals
2834	4107.1120	---马皮	5			0		50	---Of equine animals
		--粒面剖层革:							--Grain splits:
2835	4107.1210	---牛皮	8			0		50	---Of bovine animals

序号 No.	税则号列 Tariff Item	货品名称	最惠国税率(%) M.F.N.	协定税率(%) Agreement 东盟 ASEAN	曼谷协定 BA	香港 H.K.	澳门 Macao	普通税率(%) Gen.	Article Description
2836	4107.1220	---马皮	5			0		50	---Of equine animals
		--其他:							--Other:
2837	4107.1910	---机器带用	5					50	---For machinery belting
2838	4107.1990	---其他	7			0		50	---Other
		-其他,包括半张的:							-Other, including sides:
2839	4107.9100	--全粒面未剖层革	5					50	--Full grains, unsplit
2840	4107.9200	--粒面剖层革	5					50	--Grain splits
		--其他:							--Other:
2841	4107.9910	---机器带用	5					50	---For machinery belting
2842	4107.9990	---其他	7			0		50	---Other
	41.12	**经鞣制或半硝处理后进一步加工的不带毛的绵羊或羔羊皮革,包括羊皮纸化处理的,不论是否剖层,但税目41.14的皮革除外:**							**Leather further prepared after tanning or crusting, including parchment - dressed leather, of sheep or lamb, without wool on, whether or not split, other than leather of heading 41.14:**
2843	4112.0000	经鞣制或半硝处理后进一步加工的不带毛的绵羊或羔羊皮革,包括羊皮纸化处理的,不论是否剖层,但税目41.14的皮革除外:	8		7	0△		50	Leather further prepared after tanning or crusting, including parchment - dressed leather, of sheep or lamb, without wool on, whether or not split, other than leather of heading 41.14:
	41.13	**经鞣制或半硝处理后进一步加工的不带毛的其他动物皮革,包括羊皮纸化处理的,不论是否剖层,但税目41.14的皮革除外:**							**Leather further prepared after tanning or crusting, including parchment - dressed leather, of other animals, without wool or hair on, whether or not split, other than leather of heading 41.14:**
2844	4113.1000	-山羊或小山羊皮的	14		13	0△		50	-Of goats or kids
2845	4113.2000	-猪皮的	14			0△		50	-Of swine
2846	4113.3000	-爬行动物皮的	14					50	-Of reptiles
2847	4113.9000	-其他	14					50	-Other
	41.14	**油鞣皮革(包括结合鞣制的油鞣皮革);漆皮及层压漆皮;镀金属皮革:**							**Chamois (including combination chamois) leather; patent leather and patent laminated leather; metallised leather:**
2848	4114.1000	-油鞣皮革(包括结合鞣制的油鞣皮革)	14					50	-Chamois (including combination chamois) leather
2849	4114.2000	-漆皮及层压漆皮;镀金属皮革	10		9			50	-Patent leather and patent laminated leather; metallised leather
	41.15	**以皮革或皮革纤维为基本成分的再生皮革,成块、成张或成条的,不论是否成卷;皮革或再生皮革的边角废料;不适宜作皮革制品用;皮革粉末:**							**Composition leather with a basis of leather or leather fibre, in slabs, sheets or strip, whether or not in rolls; parings and other waste of leather or of composition leather, not suitable for the manufacture of leather articles; leather dust, powder and flour:**

序号 No.	税则号列 Tariff Item	货品名称	最惠国税率(%) M.F.N.	协定税率(%)Agreement 东盟 ASEAN	曼谷协定 BA	香港 H.K.	澳门 Macao	普通税率(%) Gen.	Article Description
2850	4115.1000	-以皮革或皮革纤维为基本成分的再生皮革，成块、成张或成条，不论是否成卷	14					50	-Composition leather with a basis of leather or leather fibre, in slabs, sheets or strip, whether or not in rolls
2851	4115.2000	-皮革或再生皮革的边角废料，不适宜作皮革制品用；皮革粉末	14					50	-Parings and other waste of leather or of composition leather, not suitable for the manufacture of leather articles; leather dust, powder and flour

第四十二章
皮革制品；鞍具及挽具；旅行用品、手提包及类似容器；动物肠线（蚕胶丝除外）制品

注释：

一、本章不包括：

（一）外科用无菌肠线或类似的无菌缝合材料（税号30.06）；

（二）以毛皮或人造毛皮衬里或作面（仅饰边的除外）的衣服及衣着附件（分指手套、连指手套及露指手套除外）（税目43.03或43.04）；

（三）网线袋及类似品（税号56.08）；

（四）第六十四章的物品；

（五）第六十五章的帽类及其零件；

（六）税号66.02的鞭子、马鞭或其他物品；

（七）袖扣、手镯或其他仿首饰（税号71.17）；

（八）单独进口或出口的挽具附件或装饰物，例如，马镫、马嚼子、马铃铛及类似品、带扣（一般归入第十五类）；

（九）弦线、鼓面皮或类似品及其他乐器零件（税号92.09）；

（十）第九十四章的物品（例如，家具、灯具及照明装置）；

（十一）第九十五章的物品（例如，玩具、游戏品及运动用品）；

（十二）税号96.06的钮扣、揿扣、钮扣芯或这些物品的其他零件、钮扣坯。

二、（一）除上述注释一所规定的以外，税号42.02也不包括：

1. 非供长期使用的带把手塑料薄膜袋，不论是否印制（税号39.23）；

2. 编结材料制品（税号46.02）。

（二）税号42.02及42.03的制品，如果装有用贵金属、包贵金属、天然或养殖珍珠、宝石或半宝石（天然、合成或再造）制的零件，即使这些零件不是仅作为小配件或小饰物的，只要其未构成物品的基本特征，仍应归入上述税号。但如果这些零件已构成物品的基本特征，则应归入第七十一章。

Chapter 42
Articles of leather; saddlery and harness; travel goods, handbags and similar containers; articles of animal gut (other than silk-worm gut)

Notes:

1. This Chapter does not cover:

(a) Sterile surgical catgut or similar sterile suture materials (heading No.30.06);

(b) Articles of apparel or clothing accessories (except gloves, mittens and mitts), lined with furskin or artificial fur or to which furskin or artificial fur is attached on the outside except as mere trimming (heading No.43.03 or 43.04);

(c) Made up articles of netting (heading No.56.08);

(d) Articles of Chapter 64;

(e) Headgear or parts thereor of Chapter 65;

(f) Whips, riding-crops or other articles of heading No.66.02;

(g) Cuff-links, bracelets or other imitation jewellery (heading No.71.17);

(h) Fittings or trimmings for harness, such as stirrups, bits, horse brasses and buckles, separately presented (generally Section XV);

(ij) Strings, skins for drums or the like, or other parts of musical instruments (heading No.92.09);

(k) Articles of Chapter 94 (for example, furniture, lamps and lighting fittings);

(l) Articles of Chapter 95 (for example, toys, games, sports requisites); or

(m) Buttons, press-fasteners, snap-fasteners, pressstuds, button moulds or other parts of these articles, button blanks, of heading No.96.06.

2. (a) In addition to the provisions of Note 1 above, heading No.42.02 does not cover:

(1) Bags made of sheeting of plastics, whether or not printed, with handles, not designed for prolonged use (heading No.39.23);

(2) Articles of plaiting materials (heading No.46.02).

(b) Articles of headings Nos.42.02 and 42.03 which have parts of precious metal or metal clad with precious metal, of natural or cultured pearls, of precious or semi-precious stones (natural, synthetic or reconstructed) remain classified in those headings even if such parts constitute more than minor fittings or minor ornamentation, provided that these parts do not give the articles their essential character. If, on the other hand, the parts give the articles their essential character, the articles are to be classified in Chapter 71.

三、税号 42.03 所称“衣服及衣着附件”，主要包括分指手套、连指手套及露指手套（包括运动及防护手套）、围裙及其他防护用衣着、裤吊带、腰带、子弹带及腕带，但不包括表带（税目 91.13）。

3. For the purpose of heading No.42.03, the expression "ar ticles of apparel and clothing accessories" applies, *interalia*, to gloves, mittens and mitts (including those for sport or for protection), aprons and other protective clothing, braces, belts, bandoliers and wrist straps, but excluding watch straps (heading No.91.13).

序号 No.	税则号列 Tariff Item	货品名称	最惠国税率(%) M.F.N.	协定税率(%) Agreement 东盟 ASEAN	曼谷协定 BA	香港 H.K.	澳门 Macao	普通税率(%) Gen.	Article Description
	42.01	**各种材料制成的鞍具及挽具（包括缰绳、挽绳、护膝垫、口套、鞍褥、马褡裢、狗外套及类似品），适合各种动物用：**							**Saddlery and harness for any animal (including traces, leads, knee pads, muzzles, saddle cloths, saddle bags, dog coats and the like), of any material:**
2852	4201.0000	各种材料制成的鞍具及挽具（包括缰绳、挽绳、护膝垫、口套、鞍褥、马褡裢、狗外套及类似品），适合各种动物用	20		12		0	100	Saddlery and harness for any animal (including traces, leads, knee pads, muzzles, saddle cloths, saddle bags, dog coats and the like), of any material
	42.02	**衣箱、提箱、小手袋、公文箱、公文包、书包、眼镜盒、望远镜盒、照相机套、乐器盒、枪套及类似容器；旅行包、食品或饮料保温包、化妆包、帆布包、手提包、购物袋、钱夹、钱包、地图盒、烟盒、烟袋、工具包、运动包、瓶盒、首饰盒、粉盒、刀叉餐具盒及类似容器，用皮革或再生皮革、塑料片、纺织材料、钢纸或纸板制成，或者全部或主要用上述材料或纸包覆制成：**							**Trunks, suit-cases, vanity-cases, executive-cases, brief-cases, school satchels, spectacle cases, binocular cases, camera cases, musical instrument cases, gun cases, holsters and similar containers; travelling-bags, insulated food or beverages bags, toilet bags, rucksacks, handbags, shopping-bags, wallets, purses, map-cases, cigarett-cases, tobacco-pouches, tool bags, sports bags, bottle-cases, jewellery boxes, powder-boxes, cutlery cases and similar containers, of leather or of composition leather, of sheeting of plastics, of textile materials, of vulcanized fibre or of paperboard, or wholly or mainly covered with such materials or with paper:**
		-衣箱、提箱、小手袋、公文箱、公文包、书包及类似容器：							-Trunks, suit-cases, vanity-cases, executive-cases, brief-cases, school satchels and similar containers:
		--以皮革、再生皮革或漆皮作面：							--With outer surface of leather, of composition leather or of patent leather:
2853	4202.1110	---衣箱	15					100	---Trunks and suit-cases
2854	4202.1190	---其他	10					100	---Other
		--以塑料或纺织材料作面：							--With outer surface of plastics or of textile materials:

序号 No.	税则号列 Tariff Item	货品名称	最惠国税率(%) M.F.N.	协定税率(%)Agreement 东盟 ASEAN	曼谷协定 BA	香港 H.K.	澳门 Macao	普通税率(%) Gen.	Article Description
2855	4202.1210	---衣箱	20					100	---Trunks and suit-cases
2856	4202.1290	---其他	20					100	---Other
2857	4202.1900	--其他	20					100	--Other
		-手提包，不论是否有背带，包括无把手的：							-Handbags, whether or not with shoulder strap, including those without handle:
2858	4202.2100	--以皮革、再生皮革或漆皮作面	10		6.9	0△		100	--With outer surface of leather, of composition leather or of patent leather
2859	4202.2200	--以塑料片或纺织材料作面	10		8.2			100	--With outer surface of plastic sheeting or of textile materials
2860	4202.2900	--其他	20		14			100	---Other
		-通常置于口袋或手提包内的物品：							-Articles of a kind normally carried in the pocket or in the handbag:
2861	4202.3100	--以皮革、再生皮革或漆皮作面	10		6.9			100	--With outer surface of leather, of composition leather or of patent leather
2862	4202.3200	--以塑料片或纺织材料作面	20		14			100	--With outer surface of plastic sheeting or of textile materials
2863	4202.3900	--其他	20		14			100	--Other
		-其他：							-Other:
2864	4202.9100	--以皮革、再生皮革或漆皮作面	10					100	--With outer surface of leather, of composition leather or of patent leather
2865	4202.9200	--以塑料片或纺织材料作面	10					100	--With outer surface of plastic sheeting or of textile materials
2866	4202.9900	--其他	20					100	---Other
	42.03	**皮革或再生皮革制的衣服及衣着附件：**							**Articles of apparel and clothing accessories, of leather or of composition leather:**
2867	4203.1000	-衣服	10			0△	0	100	-Articles of apparel
		-手套，包括连指或露指的：							-Gloves, mittens and mitts:
2868	4203.2100	--专供运动用	20			0△	0△	100	--Specially designed for use in sports
		--其他：							--Other:
2869	4203.2910	---劳保手套	20			0△	0	100	---Working gloves
2870	4203.2990	---其他	20			0△	0	100	---Other
		-腰带及子弹带：							-Belts and bandoliers:
2871	4203.3010	---腰带	0			0△	0	100	---Belts
2872	4203.3020	---子弹带	0			0△		100	---bandoliers
2873	4203.4000	-其他衣着附件	20			0△	0△	100	-Other clothing accessories
	42.04	**机器、机械器具或其他专门技术用途的皮革或再生皮革制品：**							**Articles of leather or of composition leather, of a kind used in machinery or mechanical appliances or for other technical uses:**

序号 No.	税则号列 Tariff Item	货品名称	最惠国税率(%) M.F.N.	协定税率(%) Agreement 东盟 ASEAN	曼谷协定 BA	香港 H.K.	澳门 Macao	普通税率(%) Gen.	Article Description
2874	4204.0000	机器、机械器具或其他专门技术用途的皮革或再生皮革制品	8					35	Articles of leather or of composition leather, of a kind used in machinery or mechanical appliances or for other technical uses
	42.05	**皮革或再生皮革的其他制品:**							**Other articles of leather or of composition leather:**
		-皮革或再生皮革的其他制品:							-Other articles of leather or of composition leather:
2875	4205.0010	---坐具套	0			0		100	---Cover of seat
2876	4205.0090	---其他	0			0		100	---Other
	42.06	**肠线(蚕胶丝除外)、肠膜、膀胱或筋腱制品:**							**Articles of gut (other than silk-worm gut), of goldbeater's skin, of bladders or of tendons:**
2877	4206.1000	-羊肠线	20					90	-Catgut
2878	4206.9000	-其他	20					90	-Other

第四十三章
毛皮、人造毛皮及其制品

Chapter 43
Furskins and artificial fur; manufactures thereof

注释：

一、本目录所称“毛皮”，是指已鞣的各种动物的带毛毛皮，但不包括税号43.01的生毛皮。

二、本章不包括：

（一）带羽毛或羽绒的整张或部分鸟皮（税号05.05或67.01）；

（二）第四十一章的带毛生皮〔见该章注释一（三）〕；

（三）用皮革与毛皮或用皮革与人造毛皮制成的分指手套、连指手套及露指手套（税目42.03）；

（四）第六十四章的物品；

（五）第六十五章的帽类及其零件；

（六）第九十五章的物品（例如，玩具、游戏品及运动用品）。

三、税号43.03包括加有其他材料缝合的毛皮和毛皮部分品，以及缝合成衣服、衣服部分品、衣着附件或其他制品的毛皮和毛皮部分品。

四、以毛皮或人造毛皮衬里或作面（仅饰边的除外）的衣服及衣着附件（不包括注释二所述的货品），应分别归入税号43.03或43.04，但毛皮或人造毛皮仅作为装饰的除外。

五、本目录所称“人造毛皮”，是指以毛、发或其他纤维粘附或缝合于皮革、织物或其他材料之上而构成的仿毛皮，但不包括以机织或针织方法制得的仿毛皮（一般应归入税号58.01或60.01）。

Notes:

1. Throughout the Nomenclature references to “furskins”, other than to raw furskins of heading No.43.01, apply to hides or skins of all animals which have been tanned or dressed with the hair or wool on.

2. This Chapter does not cover:

(a) Birdskins or parts of birdskins, with their feathers or down (heading No.05.05 or 67.01);

(b) Raw hides or skins, with the hair or wool on, of Chapter 41 (see Note 1 (c) to that Chapter);

(c) Gloves, mittens and mitts consisting of leather and furskin or of leather and artificial fur (heading No.42.03);

(d) Articles of Chapter 64;

(e) Headgear or parts thereof of Chapter 65; or

(f) Articles of Chapter 95 (for example, toys, games, sports requisites).

3. Heading No.43.03 includes furskins and parts thereof, assembled with the addition of other materials, and furskins and parts thereof, sewn together in the form of garments or parts or accessories of garments or in the form of other articles.

4. Articles of apparel and clothing accessories (except those excluded by Note 2) lined with furskin or artificial fur or to which furskin or artificial fur is attached on the outside except as mere trimming are to be classified in heading No.43.03 or 43.04 as the case may be.

5. Throughout the Nomenclature the expression “artificial fur” means any imitation of furskin consisting of wool, hair or other fibres gummed or sewn on to leather, woven fabric or other materials, but does not include imitation furskins obtained by weaving or knitting (generally, heading No.58.01 or 60.01).

序号 No.	税则号列 Tariff Item	货品名称	最惠国税率(%) M.F.N.	协定税率(%)Agreement 东盟 ASEAN	曼谷协定 BA	香港 H.K.	澳门 Macao	普通税率(%) Gen.	Article Description
	43.01	**生毛皮(包括适合加工皮货用的头、尾、爪及其他块、片),但税号41.01、41.02或41.03的生皮除外:**							**Raw furskins(including heads, tails, paws and other pieces or cuttings, suitable for furriers'use), other than raw hides and skins of heading No.41.01,41.02 or 41.03:**
2879	4301.1000	-整张水貂皮,不论是否带头、尾或爪	15			0△		100	-Of mink, whole, with or without head, tail or paws
2880	4301.3000	-下列羔羊的整张毛皮,不论是否带头、尾或爪:阿斯特拉罕、喀拉科尔、波斯羔羊及类似羔羊、印度、中国或蒙古羔羊	20			0△		90	-Of lamb, the following: Astrakhan, Broadtail, Caracul, Persian and similar lamb, Indian, Chinese, Mongolian or Tibetan lamb, whole, with or without head, tail or paws
2881	4301.6000	-整张狐皮,不论是否带头、尾或爪	20			0△		100	-Of fox, whole, with or without head, tail or paws
2882	4301.7000	-整张海豹皮,不论是否带头、尾或爪	20			0△		90	-Of seal, whole, with or without head, tail or paws
		-整张的其他毛皮,不论是否带头、尾或爪:							-Other furskins, whole, with or without head, tail or paws:
2883	4301.8010	---整张兔皮,不论是否带头、尾或爪	20			0△		90	---Of rabbit or hare, whole, with or without head, tail or paws
2884	4301.8090	---其他	20			0△		90	---Other
		-适合加工皮货用的头、尾、爪及其他块、片:							-Heads, tails, paws and other pieces or cuttings, suitable for furriers'use:
2885	4301.9010	---黄鼠狼尾	20			0△		50	---Weasel tails
2886	4301.9090	---其他	20			0△		90	---Other
	43.02	**未缝制或已缝制(不加其他材料)的已鞣毛皮(包括头、尾、爪及其他块、片),但税号43.03的货品除外:**							**Tanned or dressed furskins(including heads, tails, paws and other pieces or cuttings), unassembled, or assembled (without the addition of other materials) other than those of heading No.43.03:**
		-未缝制的整张毛皮,不论是否带头、尾或爪:							-Whole skins, with or without head, tail or paws, not assembled:
2887	4302.1100	--水貂皮	12			0△		130	--Of mink
2888	4302.1300	--下列羔羊皮:阿斯特拉罕、喀拉科尔、波斯羔羊及类似羔羊、印度、中国或蒙古羔羊	20			0△		100	--Of lamb, the following: Astrakhan, Broadtail, Caracul, Persian and similar lamb, Indian, Chinese, Mongolian or Tibetan lamb
		--其他:							--Other:
2889	4302.1910	---灰鼠皮、白鼬皮、其他貂皮、狐皮、水獭皮、旱獭皮及猞猁皮	10			0△		130	---Of grey squirrel, ermine, other marten, fox, otter, marmot and lynx
2890	4302.1920	---兔皮	10			0△		100	---Of rabbit or hare
2891	4302.1990	---其他	10			0△		100	---Other
2892	4302.2000	-未缝制的头、尾、爪及其他块、片	20			0△		100	-Heads, tails, paws and other pieces or cuttings, not assembled

序号 No.	税则号列 Tariff Item	货品名称	最惠国税率(%) M.F.N.	协定税率(%) Agreement 东盟 ASEAN	曼谷协定 BA	香港 H.K.	澳门 Macao	普通税率(%) Gen.	Article Description
		-已缝制的整张毛皮及其块、片:							-Whole skins and pieces or cuttings thereof, assembled:
2893	4302.3010	---灰鼠、白鼬、貂、狐、水獭、旱獭及猞猁的整张毛皮及其块、片	20			0△		130	---Of grey squirrel, ermine, other marten, fox, otter, marmot and lynx
2894	4302.3090	---其他	20			0△		100	---Other
	43.03	**毛皮制的衣服、衣着附件及其他物品:**							**Articles of apparel, clothing accessories and other articles of furskin:**
		-衣服及衣着附件:							-Articles of apparel and clothing accessories:
2895	4303.1010	---毛皮衣服	23			0		150	---Articles of appare
2896	4303.1020	---毛皮衣着附件	18			0		150	---Clothing accessories
2897	4303.9000	-其他	18			0		150	---Other
	43.04	**人造毛皮及其制品:**							**Artificial fur and articles thereof:**
2898	4304.0010	---人造毛皮	18			0△		130	---Artificial fur
2899	4304.0020	---人造毛皮制品	18			0△		150	---Articles of artificial fur

第九类

木及木制品；木炭；软木及软木制品；稻草、秸秆、针茅或其他编结材料制品；篮筐及柳条编结品

第四十四章
木及木制品；木炭

注释：

一、本章不包括：

（一）主要作香料、药料、杀虫、杀菌或类似用途的木片、刨花、碎木、木粒或木粉（税号12.11）；

（二）竹或主要作编结用的其他木质材料，未经加工、劈开、纵锯或切段（税号14.01）；

（三）主要作染料或鞣料用的木片、刨花、木粒或木粉（税号14.04）；

（四）活性炭（税号38.02）；

（五）税号42.02的物品；

（六）第四十六章的货品；

（七）第六十四章的鞋靴及其零件；

（八）第六十六章的货品（例如，伞、手杖及其零件）；

（九）税号68.08的货品；

（十）税号71.17的仿首饰；

（十一）第十六类或第十七类的货品（例如，机器零件，机器及器具的箱、罩、壳，车辆部件）；

（十二）第十八类的货品（例如，钟壳、乐器及其零件）；

（十三）火器的零件（税号93.05）；

（十四）第九十四章的物品（例如，家具、灯具及照明器具、活动房屋）；

（十五）第九十五章的物品（例如，玩具、游戏品及运动用品）；

（十六）第九十六章的物品（例如，烟斗及其零件、钮扣、铅笔），但税号96.03所列物品的木身及木柄除外；

SECTION Ⅸ

WOOD AND ARTICLES OF WOOD; WOOD CHARCOAL; CORK AND ARTICLES OF CORK; MANUFACTURES OF STRAW, OF ESPARTO OR OF OTHER PLAITING MATERIALS; BASKETWARE AND WICKERWORK

Chapter 44
Wood and articles of wood; wood charcoal

Notes:

1. This Chapter does not cover:

(a) Wood, in chips, in shavings, crushed, ground or powdered, of a kind used primarily in perfumery, in pharmacy, or for insecticidal, fungicidal or similar purposes (heading No.12.11);

(b) Bamboos or other materials of a woody nature of a kind used primarily for plaiting, in the rough, whether or not split, sawn lengthwise or cut to length (heading No.14.01);

(c) Wood, in chips, in shavings, ground or powdered, of a kind used primarily in dyeing or in tanning (heading No.14.04);

(d) Activated charcoal (heading No.38.02);

(e) Articles of heading No.42.02;

(f) Goods of Chapter 46;

(g) Footwear or parts thereof of Chapter 64;

(h) Goods of Chapter 66 (for example, umbrellas and walking-sticks and parts thereof);

(i) Goods of heading No.68.08;

(j) Imitation jewellery of heading No.71.17;

(k) Goods of Section ⅩⅥ or Section ⅩⅦ (for example, machine parts, cases, covers, cabinets for machines and apparatus and wheel wrights'wares);

(l) Goods of Section ⅩⅧ (for example, clock cases and musical instruments and parts thereof);

(m) Parts of firearms (heading No.93.05);

(n) Articles of Chapter 94 (for example, furniture, lamps and lighting fittings, prefabricated buildings);

(o) Articles of Chapter 95 (for example, toys, games, sports requisites);

(p) Articles of Chapter 96 (for example, smoking pipes and parts thereof, buttons, pencils) excluding bodies and handles, of wood, for articles of heading No.96.03; or

（十七）第九十七章的物品（例如艺术品）。

(q) Articles, of Chapter 97 (for example, works of art).

二、本章所称“强化木”，是指经过化学或物理方法处理（对于多层粘合木材，其处理应超出一般粘合需要），从而增加了密度或硬度并改善了机械强度、抗化学或抗电性能的木材。

2. In this Chapter, the expression "densified wood" means wood which has been subjected to chemical or physical treatment (being, in the case of layers bonded together, treatment in excess of that needed to ensure a good bond), and which has thereby acquired increased density or hardness together with improved mechanical strength or resistance to chemical or electrical agencies.

三、税号 44.14 至 44.21 适用于木质碎料板或类似木质材料板、纤维板、层压板或强化木的制品。

3. Headings Nos. 44.14 to 44.21 apply to articles of the respective descriptions of particle board or similar board, fibreboard, laminated wood or densified wood as they apply to such articles of wood.

四、税号 44.10、44.11 或 44.12 的产品，可以加工成税号 44.09 所述的各种形状，也可以加工成弯曲、瓦楞、多孔或其他形状（正方形或矩形除外），以及经其他任何加工，但未具有其他税号所列制品的特性。

4. Products of heading No. 44.10, 44.11 or 44, 12 may be worked to form the shapes provided for in respect of the goods of heading No. 44.09, curved, corrugated, perforated, cut or formed to shapes other than square or rectangular or submitted to any other operation provided it does not give them the character of articles of other headings.

五、税号 44.17 不包括装有第八十二章注释一所述材料制成的刀片、工作刃、工作面或其他工作部件的工具。

5. Heading No. 44.17 does not apply to tools in which the blade, working edge, working surface or other working part is formed by any of the materials specified in Note 1 to Chapter 82.

六、除上述注释一及其他条文另有规定的以外，本章税目中所称“木”，也包括竹及其他木质材料。

6. Subject to Note 1 above and except where the context otherwise requires, any reference to "wood" in a heading of this Chapter applies also to bamboos and other materials of a woody nature.

子目注释：

子目号 4403.41 至 4403.49、4407.24 至 4407.29、4408.31 至 4408.39 及 4412.13 至 4412.99 所称“热带木”，是指下列木材：

大叶帽柱木、非洲桃花心木、西非红豆木、箭毒木、阿兰木、圭亚那苦油楝木、非洲甘比山榄木、杜楝木、非洲栎柞木、婆罗双木、美洲轻木、白驼峰楝木、黑驼峰楝木、卡蒂沃木、雪松木、西非褐红椴木、深红色红柳桉木，非洲核桃楝木、阿夫苏木、象牙海岸榄仁木、破布木、吉贝木、丝棉木、乔状黄牛木、安哥拉丛花木、巴西胡桃木、皮蚁木、伊罗科木、拟爱神木、夹竹桃木、巴西红木、绒根木、龙脑香木、开姆帕斯木、羯布罗香木、康多非洲楝木、象牙海岸褐红椴木、象牙海岸翼梧桐木、浅红色红柳桉木、非洲榄仁木、南美樟木、圭亚那铁线子木、西印度桃花心木、猴子果木、肖氏夸利亚木、曼孙梧桐木、马来蝴蝶木、巴栲红柳桉

Subheading Note:

For the purposes of subheadings Nos. 4403.41 to 4403.49, 4407.24 to 4407.29, 4408.31 to 4408.39 and 4412.13 to 4412.99, the expression "tropical wood" means one of the following types of wood:

Abura, Acajou d'Afrique, Afrormosia, Ako, Alan, Andiroba, Aningré, Avodiré, Azobé, Balau, Balsa, Bossé clair, Bossé foncé, Cativo, Cedro, Dabema, Dark Red Meranti, Dibétou, Doussié, Framiré, Freijo, Fromager, Fuma, Geronggang, Ilomba, Imbuia, Ipé, Iroko, Jaboty, Jelutong, Jequitiba, Jongkong, Kapur, Kempas, Keruing, Kosipo, Kotibé, Koto, Light Red Meranti, Limba, Louro, Macaranduba, Mahogany, Makoré, Mandioqueira, Mansonia, Mengkulang, Meranti Bakau, Merawan, Merbau, Merpauh, Mersawa, Moabi, Niangon, Nyatoh, Obeche, Okoumé, Onzabili, Orey, Ovengkol, Ozigo, Padauk, Paldao, Palissandre de Guatemala, Palissandre de Para, Palissandre de Rio, Palissandre de Rose,

木、粗轴坡垒木、印茄木、斯温漆木、异翅香木、非洲梨木、非洲银叶木、胶木、非洲白梧桐木、加蓬榄木、蓖麻木、爱里古夷苏木、奥文科尔木、中非蜡烛木、紫檀木、人面子木、危地马拉黑黄檀木、印度黑黄檀木、巴西黑黄檀木、巴西柚、巴西花梨木、白坚木、鸡骨常山木、印马四出香木、大沃契希亚木、东西亚棱柱木、萨撇列木、萌生木棉木、苏帕楠木、西波木、苏古皮拉木、红椿木、柚木、圭亚那考拉玉蕊木、安哥拉香桃花心木、非洲阿勃木、南美肉豆蔻木、白柳桉木、白色红柳桉木、白色柳桉木、黄色红柳桉木。

Pau Amarelo, Pau Marfim, Pulai, Punah, Quaruba, Ramin, Sapelli, Saqui-Saqui, Sepetir, Sipo, Sucupira, Suren, Teak, Tauari, Tiama, Tola, Virola, White Lauan, White Meranti, White Seraya, Yellow Meranti.

序号 No.	税则号列 Tariff Item	货品名称	最惠国税率(%) M.F.N.	协定税率(%) Agreement 东盟 ASEAN	曼谷协定 BA	香港 H.K.	澳门 Macao	普通税率(%) Gen.	Article Description
	44.01	**薪柴(圆木段、块、枝、成捆或类似形状);木片或木粒;锯末、木废料及碎片,不论是否粘结成圆木段、块、片或类似形状:**							**Fuel wood, in logs, in billets, in twigs, in faggots or in similar forms; wood in chips or particles; sawdust and wood waste and scrap, whether or not agglomerated in logs, briquettes, pellets or similar forms:**
2900	4401.1000	-薪柴(圆木段、块、枝、成捆或类似形状)	0					70	-Fuel wood, in logs, in billets, in twigs, in faggots or in similar forms
		-木片或木粒:							-Wood in chips or particles:
2901	4401.2100	--针叶木	0					8	--Coniferous
2902	4401.2200	--非针叶木	0					8	--Non-coniferous
2903	4401.3000	-锯末、木废料及碎片,不论是否粘结成圆木段、块、片或类似形状	0					8	-Sawdust and wood waste and scrap, whether or not agglomerated in logs, briquettes, pellets or similar forms
	44.02	**木炭(包括果壳炭及果核炭),不论是否结块:**							**Wood charcoal (including shell or nut charcoal), whether or not agglomerated:**
2904	4402.0000	木炭(包括果壳炭及果核炭),不论是否结块	10.5					70	Wood charcoal (including shell or nut charcoal), whether or not agglomerated
	44.03	**原木,不论是否去皮、去边材或粗锯成方:**							**Wood in the rough, whether or not stripped of bark or sapwood, or roughly squared:**
2905	4403.1000	-用油漆、着色剂、杂酚油或其他防腐剂处理	0					8	-Treated with paint , stains, creosote or other preservatives
		-其他针叶木:							-Other, coniferous:
2906	4403.2010	---红松和樟子松	0					8	---Korean pine and Mongolian scotch pine
2907	4403.2020	---白松(云杉和冷杉)	0					8	---White pine (spruce and fir)
2908	4403.2030	---辐射松	0					8	---Radiata pine
2909	4403.2040	---落叶松	0					8	---Larch
2910	4403.2090	---其他	0					8	---Other
		-其他本章子目注释一所列的热带木:							-Other, of tropical wood specified in Subheading Note 1 to this Chapter:

序号 No.	税则号列 Tariff Item	货品名称	最惠国税率(%) M.F.N.	协定税率(%) Agreement 东盟 ASEAN	曼谷协定 BA	香港 H.K.	澳门 Macao	普通税率(%) Gen.	Article Description
2911	4403.4100	--深红色红柳安木、浅红色红柳安木及巴栲红柳安木	0					8	--Dark Red Meranti, Light Red Meranti and Meranti Bakau
		--其他(本章子目注释一所列热带非针叶木):							--Other (of tropical non-contiferous wood specified in subheading note 1 to this chapter):
2912	4403.4910	---柚木	0					35	---Teak
2913	4403.4920	---奥克曼(奥克榄)	0					35	---Okoume(Aukoumed Klaineana)
2914	4403.4930	---龙脑香木(克隆)	0					35	---Dipterocarpus spp. Keruing
2915	4403.4940	---山樟(香木)	0					35	---Kapur(Dryobalanops spp.)
2916	4403.4950	---印加木(波罗格)	0					35	---Intsia spp. (Mengaris)
2917	4403.4960	---大干巴豆(门格里斯或康派斯)	0					35	---Koompassia spp. (Mengaris or Kempas)
2918	4403.4970	---异翅香木	0					35	---Anisopter spp.
2919	4403.4990	---其他	0					8	---Other
		-其他:							-Other:
2920	4403.9100	--栎木(橡木)	0					8	--Of oak(*Quercus spp.*)
2921	4403.9200	--山毛榉木	0					8	--Of beech(*Fagus spp.*)
		--其他:							--Other:
2922	4403.9910	---楠木	0					35	---Of nan mu(*Phoebe*)
2923	4403.9920	---樟木	0					35	---Of camphor wood
2924	4403.9930	---红木	0					35	---Of rosewood
2925	4403.9940	---泡桐木	0					8	---Of Kiri(*Paulownia*)
2926	4403.9950	---水曲柳	0					8	---Ash
2927	4403.9960	---北美硬阔叶木(包括樱桃木、黑胡桃木、枫木)	0					8	---North American hard wood (including cherry, walnut, and maple)
2928	4403.9980	---其他未列名的温带非针叶木	0					8	---Other temperate non-coniferous not specified
2929	4403.9990	---其他	0					8	---Other
	44.04	**箍木;木劈条;已削尖但未经纵锯的木桩;粗加修整但未经车圆、弯曲或其他方式加工的木棒,适合制手杖、伞柄、工具把柄及类似品;木片条及类似品:**							**Hoopwood; split poles; piles, pickets and stakes of wood, pointed but not sawn lengthwise; wooden sticks, roughly trimmed but not turned, bent or otherwise worked, suitable for the manufacture of walking-sticks, umbrellas, tool handles or the like; chipwood and the like:**
2930	4404.1000	-针叶木的	8					50	-Coniferous
2931	4404.2000	-非针叶木的	8					50	-Non-coniferous
	44.05	**木丝;木粉:**							**Wood wool; wood flour:**
2932	4405.0000	木丝;木粉	8					40	Wood wool; wood flour
	44.06	**铁道及电车道枕木:**							**Railway or tramway sleepers (cross-ties) of wood:**
2933	4406.1000	-未浸渍	0					14	-Not impregnated
2934	4406.9000	-其他	0					14	-Other

序号 No.	税则号列 Tariff Item	货品名称	最惠国税率(%) M.F.N.	协定税率(%) Agreement 东盟 ASEAN	曼谷协定 BA	香港 H.K.	澳门 Macao	普通税率(%) Gen.	Article Description
	44.07	**经纵锯、纵切、刨切或旋切的木材,不论是否刨平、砂光或指榫接合,厚度超过6毫米:**							**Wood sawn or chipped lengthwise, sliced or peeled, whether or not planed, sanded or finger-jointed, of a thickness exceeding 6 mm:**
		-针叶木:							-Coniferous:
2935	4407.1010	---红松和樟子松	0					14	---Korean pine and Mongolian scotch pine
2936	4407.1020	---白松(云杉和冷杉)	0					14	---White pine(spruce and fir)
2937	4407.1030	---辐射松	0					14	---Rediata pine
2938	4407.1040	---花旗松	0					14	---Douglas fir
2939	4407.1090	---其他	0					14	---Other
		-本章子目注释一所列的热带木:							-Of tropical wood specified in Subheading Note 1 to this Chapter:
2940	4407.2400	--苏里南肉豆蔻木、美洲桃花心木、巴西胡桃木及美洲轻木	0					14	--Virola, Mahogany(*Swietenia spp.*), Imbuia and Balsa
2941	4407.2500	--深红色红柳安木、浅红色红柳安木及巴梣红柳安木	0					14	--Dark Red Meranti, Light Red Meranti and Meranti Bakau
2942	4407.2600	--白柳安木、白色红柳安木、白色柳安木、黄色红柳安木及阿兰木	0					14	--White Lauan, White Meranti, White Seraya, Yellow Meranti and Alan
		--其他:							--Other:
2943	4407.2910	---柚木	0					40	---Teak
2944	4407.2920	---非洲桃花心木、沙比利	0					40	---Acajou, sapele
2945	4407.2930	---波罗格	0					40	---Merban
2946	4407.2990	---其他	0					14	---Other
		-其他:							-Other:
2947	4407.9100	--栎木(橡木)	0					14	--Of oak(*Quercus spp.*)
2948	4407.9200	--山毛榉木	0					14	--Of beech(*Fagus spp.*)
		--其他:							--Other:
2949	4407.9910	---樟木、楠木、红木	0					40	---Of camphor-wood, nanmu or rosewood
2950	4407.9920	---泡桐木	0					14	---Of Paulownia
2951	4407.9930	---北美硬阔叶材(含樱桃木、枫木、黑胡桃木)	0					14	---North American hard wood (including cherry, walnut, and maple)
2952	4407.9980	---其他温带非针叶材	0					14	---Other temperate non-coniferous wood
2953	4407.9990	---其他	0					14	---Other
	44.08	**饰面用单板(包括刨切积层木获得的单板)、制胶合板或类似多层板用单板以及其他经纵锯、刨切或旋切的木材,不论是否刨平、砂光、拼接或端部接合,厚度不超过6毫米:**							**Sheets for veneering (including those obtained by slicing laminated wood), for plywood or for other similar laminated wood and other wood, sawn lengthwise, sliced or peeled, whether or not planed, sanded, spliced or end-jointed, of a thickness not exceeding 6 mm:**

序号 No.	税则号列 Tariff Item	货品名称	最惠国税率(%) M.F.N.	协定税率(%)Agreement 东盟 ASEAN	曼谷协定 BA	香港 H.K.	澳门 Macao	普通税率(%) Gen.	Article Description
		-针叶木:							-Coniferous:
		---饰面用单板:							---veneer sheets:
2954	4408.1011	----用胶合板等多层板制的	8					40	----of laminated plywood
2955	4408.1019	----其他	4					40	----Other
2956	4408.1020	---制胶合板用单板	4					17	---Sheets for plywood
2957	4408.1090	---其他	4					30	---Other
		-本章子目注释一所列的热带木:							-Of tropical wood specified in Subheading Note 1 to this Chapter:
		--深红色红柳安木、浅红色红柳安木及巴栲红柳安木:							--Dark Red Meranti, Light Red Meranti and Meranti Bakau:
		---饰面用单板:							---veneer sheets:
2958	4408.3111	----用胶合板等多层板制的	10					40	----of laminated plywood
2959	4408.3119	----其他	4					40	----Other
2960	4408.3120	---制胶合板用单板	4					17	---Sheets for plywood
2961	4408.3190	---其他	4					30	---Other
		--其他:							--Other:
		---饰面用单板:							---veneer sheets:
2962	4408.3911	----用胶合板等多层板制的	10					40	----of laminated plywood
2963	4408.3919	----其他	4					40	----Other
2964	4408.3920	---制胶合板用单板	4					17	---Sheets for plywood
2965	4408.3990	---其他	4					30	---Other
		-其他:							-Other:
		---饰面用单板:							---veneer sheets:
2966	4408.9011	----用胶合板等多层板制的	4					40	----of laminated plywood
2967	4408.9012	----温带非针叶木制	3					40	----Of temperate non-coniferous wood
2968	4408.9019	----其他	3					40	----Other
		---胶合板用单板:							---Sheets for plywood:
2969	4408.9021	----温带非针叶木制	3					17	----Of temperate non-coniferous wood
2970	4408.9029	----其他	3					17	----Other
		---其他:							---Other:
2971	4408.9091	----温带非针叶木制	3					30	----Of temperate non-coniferous wood
2972	4408.9099	----其他	3					30	----Other
	44.09	**任何一边端或面制成连续形状(舌榫、槽榫、半槽榫、斜角、V形接头、珠榫、缘饰、刨圆及类似形状)的木材(包括未装拼的拼花地板用板条及缘板),不论是否刨平、砂光或端部接合:**							**Wood(including strips and friezes for parquet flooring, not assembled) continuously shaped(tongues, grooved, rebated, chamfered, v-jointed, beaded, moulded, rounded or the like) along any of its "edges, ends or faces, whether or not planed, sanded or end-jointed":**
		-针叶木:							-Coniferous:
2973	4409.1010	---地板条(块)	7.5					50	---Floov board stvips

序号 No.	税则号列 Tariff Item	货品名称	最惠国税率(%) M.F.N.	协定税率(%) Agreement 东盟 ASEAN	曼谷协定 BA	香港 H.K.	澳门 Macao	普通税率(%) Gen.	Article Description
2974	4409.1090	---其他	7.5					50	---Other
		-非针叶木:							-Non-coniferous:
		---地板条(块):							---Floov boavd stvips:
2975	4409.2011	----竹的	4					50	----Of bamboo
2976	4409.2019	----其他	4					50	----Other
2977	4409.2090	---其他	4					50	---Other
	44.10	**木质碎料板及其他类似木质材料板,(例如,定向板及华夫板),不论是否用树脂或其他有机粘合剂粘合:**							**Particle board and similar board of wood or other ligneous materials (for example, oriented strand board and waferboabd), whether or not agglomerated with resins or other organic binding substances:**
		-木质的定向板及华夫板							-Oriented strand board and waferboard, of wood:
2978	4410.2100	--未加工或除砂光外未经进一步加工的	4					40	--Unworked or not further worked than sanded
2979	4410.2900	--其他	4					40	--Other
		-其他木质板							-Other, of wood
2980	4410.3100	--未加工或除砂光外未经进一步加工的	4					40	--Unworked or not further worked than sanded
2981	4410.3200	--用蜜胺浸纸覆面的	4					40	--Surface-covered with melamine-impregnated paper
2982	4410.3300	--用塑料装饰薄片覆面的	4					40	--Surface-covered with decorative laminates of plastics
2983	4410.3900	--其他	4					40	--Other
2984	4410.9000	-其他	7.5					40	-Other
	44.11	**木纤维板或其他木质材料纤维板,不论是否用树脂或其他有机粘合剂粘合:**							**Fibreboard of wood or other ligneous materials, whether or not bonded with resins or other organic substances:**
		-密度超过每立方厘米0.8克:							-Fibreboard of a density exceeding 0.8 g/cm^3:
2985	4411.1100	--未经机械加工或盖面的	4					40	--Not mechanically worked or surface covered
2986	4411.1900	--其他	7.5					40	--Other
		-密度超过每立方厘米0.5克,但未超过每立方厘米0.8克:							-Fibreboard of a density exceeding 0.5 g/cm^3 but not exceeding 0.8g/cm^3:
2987	4411.2100	--未经机械加工或盖面的	4		4			40	--Not mechanically worked or surface covered
2988	4411.2900	--其他	4		4			40	--Other
		-密度超过每立方厘米0.35克,但未超过每立方厘米0.5克:							-Fibreboard of a density exceeding 0.35 g/cm^3 but not exceeding 0.5/cm^3:
2989	4411.3100	--未经机械加工或盖面的	7.5		7.5			40	--Not mechanically worked or surface covered

序号 No.	税则号列 Tariff Item	货品名称	最惠国税率(%) M.F.N.	协定税率(%)Agreement 东盟 ASEAN	曼谷协定 BA	香港 H.K.	澳门 Macao	普通税率(%) Gen.	Article Description
2990	4411.3900	--其他	7.5		7.5			40	--Other
		-其他：							-Other：
2991	4411.9100	--未经机械加工或盖面的	7.5		7.5			40	--Not mechanically worked or surface covered
2992	4411.9900	--其他	4		4			40	--Other
	44.12	**胶合板、单板饰面板及类似的多层板：**							**Plywood, veneered panels and similar laminated wood：**
		-仅由薄木板制的胶合板，每层厚度不超过6毫米：							-Plywood consisting solely of sheets of wood, each ply not exceeding 6 mm thickness：
2993	4412.1300	--至少有一表层是本章子目注释所列的热带木	12		12			30	--With at least one outer ply of tropical wood specified in Subheading Note 1 to this Chapter
		--其他，至少有一表层是非针叶木：							--Other, with at least one outer ply of nonconiferous wood：
2994	4412.1410	---至少有一表层是温带非针叶木	4		4			30	---With at least one outer ply of temperate non-coniferous wood
2995	4412.1420	---由两层或两层以上竹板层叠胶合而成的其他多层板	4		4			30	---Multilayer board with two or move plies of bamboo
2996	4412.1490	---其他	4		4			30	---Other
2997	4412.1900	--其他	4		4			30	--Other
		-其他，至少有一表层是非针叶木：							-Other, with at least one outer ply of nonconiferous wood：
2998	4412.2200	--至少有一层是本章子目注释一所列的热带木	10		10			30	--With at least one ply of tropical wood specified in Subheading Note 1 to this Chapter
2999	4412.2300	--其他，至少含有一层木碎料板	10		10			30	--Other, containing at least one layer of particle board
		--其他：							--Other：
3000	4412.2910	---至少有一层是温带非针叶木	10		10			30	---With at least one ply of temperate non-coniferous wood
3001	4412.2990	---其他	10		10			30	---Other
		-其他：							-Other：
3002	4412.9200	--至少有一层是本章子目注释一所列的热带木	8		8			30	--With at least one ply of tropical wood specified in Subheading Note 1 to this Chapter
3003	4412.9300	--其他，至少含有一层木碎料板	10		10			30	--Other, containing at least one layer of particle board
		--其他：							--Other：
3004	4412.9910	---至少有一层是温带非针叶木	4		4			30	---With at least one ply of temperate non-coniferous wood
3005	4412.9990	---其他	4		4			30	---Other
	44.13	**强化木，成块、板、条或异型的：**							**Densified wood, in blocks, plates, strips or profile shapes：**

序号 No.	税则号列 Tariff Item	货品名称	最惠国税率(%) M.F.N.	协定税率(%) Agreement 东盟 ASEAN	曼谷协定 BA	香港 H.K.	澳门 Macao	普通税率(%) Gen.	Article Description
3006	4413.0000	强化木,成块、板、条或异型的	6					20	Densified wood, in blocks, plates, strips or profile shapes
	44.14	**木制的画框、相框、镜框及类似品:**							**Wooden frames for paintings, photographs, mirrors or similar objects:**
3007	4414.0000	木制的画框、相框、镜框及类似品	20					100	Wooden frames for paintings, photographs, mirrors or similar objects
	44.15	**包装木箱、木盒、板条箱、圆桶及类似的包装容器;木制电缆卷筒;木托板、箱形托盘及其他装载用木板;木制的托盘护框:**							**Packing cases, boxes, crates, drums and similar packings, of wood; cable-drums of wood; pallets, box pallets and other load boards, of wood; pallet collars of wood:**
3008	4415.1000	-箱、盒、板条箱、圆桶及类似的包装容器;电缆卷筒	7.5					80	-Cases, boxes, crates, drums and similar packing; cable-drums
3009	4415.2000	-木托板、箱形托盘及其他装载用木板;木制的托盘护框	7.5					80	-Pallets, box pallets and other load boards; pallet collars
	44.16	**木制大桶、琵琶桶、盆和其他木制箍桶及其零件,包括桶板:**							**Casks, barrels, vats, tubs and other coopers' products and parts thereof, of wood, including staves:**
3010	4416.0000	木制大桶、琵琶桶、盆和其他木制箍桶及其零件,包括桶板	16					80	Casks, barrels, vats, tubs and other coopers' products and parts thereof, of wood, including staves:
	44.17	**木制的工具、工具支架、工具柄、扫帚及刷子的身及柄;木制鞋靴楦及楦头:**							**Tools, tool bodies, tool handles, broom or brush bodies and handles, of wood; boot or shoe lasts and trees, of wood:**
3011	4417.0000	木制的工具、工具支架、工具柄、扫帚及刷子的身及柄;木制鞋靴楦及楦头	16					80	Tools, tool bodies, tool handles, broom or brush bodies and handles, of wood; boot or shoe lasts and trees, of wood
	44.18	**建筑用木工制品,包括蜂窝结构木镶板、已装拼的拼花地板、木瓦及盖屋板:**							**Builders' joinery and carpentry of wood, including cellular wood panels, assembled parquet panels, shingles and shakes:**
3012	4418.1000	-窗、法兰西式(落地)窗及其框架	4					70	-Windows, French-windows and their frames
3013	4418.2000	-门及其框架和门槛	4					70	-Doors and their frames and thresholds
3014	4418.3000	-拼花地板	4					70	-Parquet panels
3015	4418.4000	-水泥构件的模板	4					70	-Shuttering for concrete constructional work
3016	4418.5000	-木瓦及盖屋板	7.5					70	-Shingles and shakes
3017	4418.9000	-其他	4					70	-Other

序号 No.	税则号列 Tariff Item	货品名称	最惠国税率(%) M.F.N.	协定税率(%) Agreement 东盟 ASEAN	曼谷协定 BA	香港 H.K.	澳门 Macao	普通税率(%) Gen.	Article Description
	44.19	**木制餐具及厨房用具:**							**Tableware and kitchenware, of wood:**
		木制餐具及厨房用具:							Tableware and kitchenware, of wood:
3018	4419.0010	---木制一次性筷子	0					100	---One-time chopsticks, of wood
3019	4419.0020	---竹制一次性筷子	0					100	---One-time chopsticks, of bamboos
3020	4419.0090	---其他	0					100	---Other
	44.20	**镶嵌木(包括细工镶嵌木);装珠宝或刀具用的木制盒子和小匣子及类似品;木制小雕像及其他装饰品;第九十四章以外的木制家具:**							**Wood marquetry and inlaid wood; caskets and cases for jewellery or cutlery, and similar articles, of wood; statuettes and other ornaments, of wood; wooden articles or furniture not falling in Chapter 94:**
		-木制小雕像及其他装饰品:							-Statuettes and other ornaments, of wood:
3021	4420.1010	---木刻及竹刻	0					100	---Wood or bamboo carvings
3022	4420.1020	---木扇	0					100	---Wooden fans
3023	4420.1090	---其他	0					100	---Other
		-其他:							-Other:
3024	4420.9010	---镶嵌木	0					45	---Wood marquetry and inlaid wood
3025	4420.9090	---其他	0					100	---Other
	44.21	**其他木制品:**							**Other articles of wood:**
3026	4421.1000	-衣架	0					90	-Clothes hangers
		-其他:							-Other:
3027	4421.9010	---卷轴、纡子、筒管、缝纫用线轴及类似品	0					35	---Spools, cops, bobbins, sewing thread reels and the like
3028	4421.9090	---其他	0					90	---Other

第四十五章
软木及软木制品

Chapter 45
Cork and articles of cork

注释：

本章不包括：

一、第六十四章的鞋靴及其零件；

二、第六十五章的帽类及其零件；

三、第九十五章的物品（例如，玩具、游戏品及运动用品）。

Notes:

This Chapter does not cover:

(a) Footwear or parts of footwear of Chapter 64;

(b) Headgear or parts of headgear of Chapter 65; or

(c) Articles of Chapter 95 (for example, toys, games, sports requisites).

序号 No.	税则号列 Tariff Item	货品名称	最惠国税率(%) M.F.N.	协定税率(%)Agreement 东盟 ASEAN	曼谷协定 BA	香港 H.K.	澳门 Macao	普通税率(%) Gen.	Article Description
	45.01	**未加工或简单加工的天然软木；软木废料；碎的、粒状的或粉状的软木：**							**Natural cork, raw or simply prepared; waste cork; crushed, granulated or ground cork:**
3029	4501.1000	-未加工或简单加工的天然软木	6					17	-Natrual cork, raw or simply prepared
3030	4501.9000	-其他	0					17	-Other
	45.02	**天然软木，除去表皮或粗切成方形，或成长方块、正方块、板、片或条状（包括作塞子用的方块坯料）：**							**Natural cork, debarked or roughly squared, or in rectangular (including square) blocks, plates, sheets or strip, (including sharp-edged blanks for corks or stoppers):**
3031	4502.0000	天然软木，除去表皮或粗切成方形，或成长方块、正方块、板、片或条状（包括作塞子用的方块坯料）	8					30	Natural cork, debarked or roughly squared, or in rectangular (including square) blocks, plates, sheets or strip, (including sharp-edged blanks for corks or stoppers)
	45.03	**天然软木制品：**							**Articles of natural cork:**
3032	4503.1000	-塞子	8					50	-Corks and stoppers
3033	4503.9000	-其他	10.5					50	-Other
	45.04	**压制软木（不论是否使用粘合剂压成）及其制品：**							**Agglomerated cork (with or without a binding substance) and articles of agglomerated cork:**
3034	4504.1000	-块、板、片及条；任何形状的砖、瓦；实心圆柱体，包括圆片	8.4					30	-Blocks, plates, sheets and strip; tiles of any shape; solid cylinders, including discs
3035	4504.9000	-其他	0					50	-Other

第四十六章
稻草、秸杆、针茅或其他编结材料制品；篮筐及柳条编结品

Chapter 46
Manufactures of straw, of esparto or of other plaiting materials; basketware and wickerwork

注释：

一、本章所称“编结材料”，是指其状态或形状适于编结、交织或类似加工的材料，包括稻草、秸杆、柳条、竹、灯芯草、芦苇、木片条、其他植物材料扁条（例如，树皮条、狭叶、酒椰叶纤维或其他从阔叶获取的条）、未纺的天然纺织纤维、塑料单丝及扁条、纸带，但不包括皮革、再生皮革、毡呢或无纺织物的扁条、人发、马毛、纺织粗纱或纱线以及第五十四章的单丝和扁条。

二、本章不包括：

（一）税号 48.14 的壁纸；

（二）不论是否编结而成的线、绳、索、缆（税号 56.07）；

（三）第六十四章和第六十五章的鞋靴、帽类及其零件；

（四）编结而成的车辆或车身（第八十七章）；

（五）第九十四章的物品（例如，家具、灯具及照明装置）。

三、税号 46.01 所称“平行连结的成片编结材料、缏条或类似的编结材料产品”，是指编结材料、缏条及类似的编结材料产品平行排列连结成片的制品，其连结材料不论是否为纺制的纺织材料。

Notes:

1. In this Chapter the expression “plaiting materials” means materials in a state or form suitable for plaiting, interlacing or similar processes; it includes straw, osier or willow, bamboos, rushes, reeds, strips of wood, strips of other vegetable material (for example, strips of bark, narrow leaves and raffia or other strips obtained from broad leaves), unspun natural textile fibres, monofilament and strip and the like of plastics and strips of paper, but not strips of leather or composition leather or of felt or nonwovens, human hair, horsehair, textile rovings or yarns, or monofilament and strip and the like of Chapter 54.

2. This Chapter does not cover:

(a) Wall coverings of heading No.48.14;

(b) Twine, cordage, ropes or cables, plaited or not (heading No.56.07);

(c) Footwear or headgear or parts thereof of Chapter 64 or 65;

(d) Vehicles or bodies for vehicles of basketware (Chapter 87); or

(e) Articles of Chapter 94 (for example, furniture, lamps and lighting fittings).

3. For the purposes of heading No.46.01, the expression “plaiting materials, plaits and similar products of plaiting materials, bound together in parallel strands” means plaiting materials, plaits and similar products of plaiting materials, placed side by side and bound together, in the form of sheets, whether or not the binding materials are to spun textile materials.

序号 No.	税则号列 Tariff Item	货品名称	最惠国税率(%) M.F.N.	协定税率(%) Agreement 东盟 ASEAN	曼谷协定 BA	香港 H.K.	澳门 Macao	普通税率(%) Gen.	Article Description
	46.01	用编结材料编成的缏条及类似产品，不论是否缝合成宽条；平行连结或编结的成片材料、缏条或类似的编结材料产品，不论是否制成品（例如，席子、席料、帘子）：							**Plaits and similar products of plaiting materials, whether or not assembled into strips; plaiting materials, plaits and similar products of plaiting materials, bound together in parallel strands or woven, in sheet form, whether or not being finished articles (for example, mats, matting, screens):**

序号 No.	税则号列 Tariff Item	货品名称	最惠国税率(%) M.F.N.	协定税率(%) Agreement 东盟 ASEAN	曼谷协定 BA	香港 H.K.	澳门 Macao	普通税率(%) Gen.	Article Description
		-植物材料制的席子、席料及帘子:							-Mats, matting and screens of vegetable materials:
3036	4601.2010	---藤制的	9					100	---Of rattans
		---草制的:							---Of grass or straw:
3037	4601.2021	----灯心草属材料制的	9					90	----Of rushes
3038	4601.2029	----其他	9					90	----Other
		---芦苇制的:							---Of reeds:
3039	4601.2031	----苇帘	9					90	----Screens of reeds
3040	4601.2039	----其他	9					90	----Other
3041	4601.2040	----竹制的	9					90	---of bamboos
3042	4601.2090	---其他	9					90	---Other
		-其他:							-Other:
		--植物材料制:							-Of vegetable materials:
		---藤制的							---Of rattans
3043	4601.9111	----鞭条及类似产品,不论是否缝合成宽条	9					100	----Plaits and similar products of plaiting materials, whether or not assembled into strips
3044	4601.9119	----其他	9					90	----Other
		---稻草制的:							---Of straw:
3045	4601.9121	----缏条(绳)	10					90	----Plaits
3046	4601.9129	----其他	10					90	----Other
		---其他:							---Other:
3047	4601.9191	----鞭条及类似产品,不论是否缝合成宽条	9					100	----Plaits and similar products of plaiting materials, whether or not assembled into strips
3048	4601.9199	----其他	9					90	----Other
		--其他:							--Other:
3049	4601.9910	---鞭条及类似产品,不论是否缝合成宽条	9					90	---Plaits and similar products of plaiting materials,whether or not assembled into strips
3050	4601.9990	---其他	9					90	---Other
	46.02	**用编结材料直接编成或用税号46.01所列货品制成的篮筐、柳条编结品及其他制品;丝瓜络制品:**							**Basketwork, wickerwork and other articles, made directly to shape from plaiting materials or made up from goods of heading No.46.01; articles of loofah:**
		-植物材料制:							-Of vegetable materials:
3051	4602.1010	---藤制的	9					100	---Of rattans
3052	4602.1020	---草制的	9					100	---Of grass or straw
3053	4602.1030	---竹制的	9					100	---Of bamboos
3054	4602.1040	---玉米皮制的	9					100	---Of maize-shuck
3055	4602.1050	---柳条制的	9					100	---Of osier
3056	4602.1090	---其他	9					100	---Other
3057	4602.9000	-其他	9		8			100	-Other

第十类

SECTION X

木浆及其他纤维状纤维素浆、回收（废碎）纸或纸板；纸、纸板及其制品

PULP OF WOOD OR OF OTHER FIBROUS CELLULOSIC MATERIAL; ERCOVERED（WASTE AND SCRAP）PAPER OR PAPERBOARD; PAPER AND PAPERBOARD AND ARTICLES THEREOF

第四十七章
木浆及其他纤维状纤维素浆；回收（废碎）纸或纸板

Chapter 47
Pulp of wood or of other fibrous cellulosic material; recovered（waste and scrap）paper or paperboard

注释：

税号47.02所称“化学木浆，溶解级”，是指温度在20℃时浸入含18%氢氧化钠的苛性碱溶液内，一小时后，按重量计含有92%及以上的不溶级分的碱木浆或硫酸盐木浆，或者含有88%及以上的不溶级分的亚硫酸盐木浆。对于亚硫酸盐木浆，按重量计灰分含量不得超过0.15%。

Notes:

For the purposes of heading No.47.02, the expression “chemical wood pulp, dissolving grades” means chemical wood pulp having by weight an insoluble fraction of 92% or more for soda or sulphate wood pulp of of 88% or more for sulphite wood pulp after one hour in a caustic soda solution containing 18% sodium bydroxide（NaOH）at 20℃, and for sulphite wood pulp an ash content that does not exceed 0.15% by weight.

序号 No.	税则号列 Tariff Item	货品名称	最惠国税率(%) M.F.N.	协定税率(%)Agreement				普通税率(%) Gen.	Article Description
				东盟 ASEAN	曼谷协定 BA	香港 H.K.	澳门 Macao		
	47.01	**机械木浆：**							**Mechanical wood pulp:**
3058	4701.0000	机械木浆	0					8	Mechanical wood pulp
	47.02	**化学木浆，溶解级：**							**Chemical wood pulp, dissolving grades:**
3059	4702.0000	化学木浆，溶解级	0					8	Chemical wood pulp, dissolving grades
	47.03	**碱木浆或硫酸盐木浆，但溶解级的除外：**							**Chemical wood pulp, soda or sulphate, other than dissolving grades:**
		-未漂白：							-Unbleached:
3060	4703.1100	--针叶木的	0					8	--Coniferous
3061	4703.1900	--非针叶木的	0					8	--Non-coniferous
		-半漂白或漂白：							-Semi-bleached or bleached:
3062	4703.2100	--针叶木的	0					8	--Coniferous
3063	4703.2900	--非针叶木的	0					8	--Non-coniferous
	47.04	**亚硫酸盐木浆，但溶解级的除外：**							**Chemical wood pulp, sulphite, other than dissolving grades:**
		-未漂白：							-Unbleached:
3064	4704.1100	--针叶木的	0					8	--Coniferous
3065	4704.1900	--非针叶木的	0					8	--Non-coniferous
		-半漂白或漂白：							-Semi-bleached or bleached:
3066	4704.2100	--针叶木的	0					8	--Coniferous

序号 No.	税则号列 Tariff Item	货品名称	最惠国税率(%) M.F.N.	协定税率(%)Agreement 东盟 ASEAN	曼谷协定 BA	香港 H.K.	澳门 Macao	普通税率(%) Gen.	Article Description
3067	4704.2900	--非针叶木的	0					8	--Non-coniferous
	47.05	**用机械与化学联合制浆法制得的木浆:**							**Wood pulp obtained by a combination of mechanical and chemical pulping processes:**
3068	4705.0000	用机械与化学联合制浆法制得的木浆	0					8	Wood pulp obtained by a combination of mechanical and chemical pulping processes
	47.06	**从回收(废碎)纸或纸板提取的纤维浆或其他纤维状纤维素浆:**							**Pulps of fibres derived from recovered (waste and scrap) paper or paperboard or of other fibrous cellulosic material:**
3069	4706.1000	-棉短绒纸浆	0					8	-Cotton linters pulp
3070	4706.2000	-从回收(废碎)纸或纸板提取的纤维浆	0					8	-Pulps of fibres derived from recovered (waste and scrap)paper or paperboard
		-其他:							-Other:
3071	4706.9100	--机械浆	0					8	--Mechanical
3072	4706.9200	--化学浆	0					8	--Chemical
3073	4706.9300	--半化学浆	0					8	--Semi-chemical
	47.07	**回收(废碎)纸或纸板:**							**Recovered(waste and scrap) paper or paperboard:**
3074	4707.1000	-未漂白的牛皮纸或纸板及瓦楞纸或纸板	0					8	-Unbleached kraft paper or paperboard or of corrugated paper or paperboard
3075	4707.2000	-主要由漂白化学木浆制成未经本体染色的其他纸和纸板	0					8	-Other paper or paperboard made mainly of bleached chemical pulp, not coloured in the mass
3076	4707.3000	-主要由机械浆制成的纸或纸板(例如,报纸、杂志及类似印刷品)	0					8	-Paper or paperboard made mainly of mechanical pulp (for example, newspapers, journals and similar printed matter)
3077	4707.9000	-其他,包括未分选的废碎品	0					8	-Other, including unsorted waste and scrap

第四十八章
纸及纸板；纸浆、纸或纸板制品

注释：

一、除条文另有规定的以外，本章所称“纸”包括“纸板”（不论其厚度或每平方米重量如何）。

二、本章不包括：

（一）第三十章的物品；

（二）税号32.12的压印箔；

（三）香纸及用化妆品浸渍或涂布的纸（第三十三章）；

（四）用肥皂或洗涤剂浸渍、覆盖或涂布的纸或纤维素絮纸（税号34.01）和用光洁剂、擦光膏及类似制剂浸渍、覆盖或涂布的纸或纤维素絮纸（税号34.05）；

（五）税号37.01至37.04的感光纸或感光纸板；

（六）用诊断或实验用试剂浸渍的纸（税号38.22）；

（七）第三十九章的用纸强化的层压塑料板，用塑料覆盖或涂布的单层纸或纸板（塑料部分占总厚度的一半以上），以及上述材料的制品，但税号48.14的壁纸除外；

（八）税号42.02的物品（例如旅行用品）；

（九）第四十六章的物品（编结材料制品）；

（十）纸纱线或纸纱线纺织物（第十一类）；

（十一）第六十四章或第六十五章的物品；

（十二）税号68.05的砂纸或税号68.14的用纸或纸板衬底的云母（但涂布云母粉的纸及纸板归入本章）；

（十三）用纸或纸板衬底的金属箔（第十五类）；

（十四）税号92.09的制品；

（十五）第九十五章的物品（例如，玩具、游戏品及运动用品）或第九十六章的物品（例如钮扣）。

三、除注释七另有规定的以外，税号48.01至48.05包括经研光、高度研光、釉光或类似处理、仿水印、表面施胶的纸及纸板；同时还包括用各种方法本体着色或染成斑纹的纸、

Chapter 48
Paper and paperboard; articles of paper pulp, of paper or of paperboard

Notes:

1. For the purposes of this Chapter, except where the context otherwise requires, a reference to “paper” includes references to paperboard (irrespective of thickness or weight per m^2).

2. This Chapter does not cover:

(a) Articles of Chapter 30;

(b) Stamping foils of heading No.32.12;

(c) Perfumed papers or papers impregnated or coated with cosmetics (Chapter 33);

(d) Paper or cellulose wadding impregnated, coated or covered with soap or detergent (heading No.34.01), or with polishes, creams or similar preparations (heading No.34.05);

(e) Sensitized paper or paperboard of headings Nos.37.01 to 37.04;

(f) Paper impregnated with diagnostic or laboratory reagents (heading No.38.22);

(g) Paper-reinforced stratified sheeting of plastics, or one layer of paper or paperboard coated or covered with a layer of plastics, the latter constituting more than half the total thickness, or articles of such materials, other than wall coverings of heading No.48.14 (Chapter 39);

(h) Articles of heading No.42.02 (for example, travel goods);

(i) Articles of Chapter 46 (manufactures of plaiting material);

(j) Paper yarn or textile articles of paper yarn (Section Ⅺ);

(k) Articles of Chapter 64 or Chapter 65;

(l) Abrasive paper or paperboard (heading No.68.05) or paper-or paperboard-backed mica (heading No.68.14) (paper and paperboard coated with mica powder are, however, to be classified in this Chapter);

(m) Metal foil backed with paper or paperboard (Section ⅩⅤ);

(n) Articles of heading No.92.09; or

(o) Articles of Chapter 95 (for example, toys, games, sports requisites) or Chapter 96 (for example, buttons).

3. Subject to the provisions of Note 7, headings Nos.48.01 to 48.05 include paper and paperboard which have been subjected to calendering, super-calendering, glazing or similar finishing, false watermarking or surface sizing, and also paper,

纸板、纤维素絮纸及纤维素纤维网纸。除税号48.03另有规定的以外，上述税号不适用于经过其他方法加工的纸、纸板、纤维素絮纸或纤维素纤维网纸。

paperboard, cellulose wadding and webs of cellulose fibres, coloured or marbled throughout the mass by any method. Except where heading No.48.03 otherwise requires, these headings do not apply to paper, paperboard, cellulose wadding or webs of cellulose fibres which have been otherwise processed.

四、本章所称“新闻纸”，是指所含用机械或化学－机械方法制得的木纤维不少于全部纤维重量的65%的未经涂布的报刊用纸，未施胶或微施胶，每面的粗糙度〔帕克印刷面粗糙度（1兆帕）〕超过2.5微米，每平方米重量不小于40克，但不超过65克。

4. In this Chapter the expression “newsprint” means uncoated paper of a kind used for the printing of newspapers, of which not less than 65% by weight of the total fibre content consists of wood fibres obtained by a mechanical or chemi-mechanical process, unsized or very lightly sized, having a surface roughness Parker Print Surf (1 MPa) on each side exceeding 2.5 micrometres (microns), weighing not less than 40 g/m^2 and not more than 65g/m^2.

五、税目48.02所称“书写、印刷或类似用途的纸及纸板”及“未打孔的穿孔卡片纸及穿孔纸带纸”，是指主要用漂白纸浆或用机械或化学—机械方法制得的纸浆制成的纸及纸板，并且符合下列任一标准：

5. For the purposes of heading 48.02, the expressions “paper and paperboard, of a kind used for writing, printing or other graphic purposes” and “non perforated punch-cards and punch tape paper” mean paper and paperboard made mainly from bleached pulp or from pulp obtained by a mechanical or chemi-mechanical process and satisfying any of the following criteria:

每平方米重量不超过150克的纸或纸板：

(一) 用机械或化学—机械方法制得的纤维含量在10%及以上，并且
   1. 每平方米重量不超过80克；或
   2. 本体着色；

(二) 灰分含量在8%以上，并且
   1. 每平方米重量不超过80克；或
   2. 本体着色；

(三) 灰分含量在3%以上，亮度在60%及以上；

(四) 灰分含量在3%以上，但不超过8%，亮度低于60%，耐破指数等于或小于2.5千帕斯卡·平方米/克；

(五) 灰分含量在3%及以下，亮度在60%及以上，耐破指数等于或小于2.5千帕斯卡·平方米/克。

每平方米重量超过150克的纸或纸板：

(一) 本体着色；或

(二) 亮度在60%及以上，并且
   1. 厚度在225微米及以下；或
   2. 厚度在225微米以上，但不超过508微米，灰分含量在3%以上；

(三) 亮度低于60%，厚度不超过254微米，灰分含量在8%以上。

For paper or paperboard weighing not more than 150 g/m^2:

(a) containing 10% or more of fibres obtained by a mechanical or chemi-mechanical process, and
   ①weighing not more than 80 g/m^2, or
   ②coloured throughout the mass; or

(b) containing more than 8% ash, and
   ①weighing not more than 80 g/m^2, or
   ②coloured throughout the mass; or

(c) containing more than 3% ash and having a brightness of 60% or more; or

(d) containing more than 3% but not more than 8% ash, having a brightness less than 60%, and a burst index equal to or less than 2.5 $kPa/m^2/g$; or

(e) containing 3% ash or less, having a brightness of 60% or more and a burst index equal to or less than 2.5 $kPa/m^2/g$.

For paper or paperboard weighing more than 150g/m^2:

(a) coloured throughout the mass; or

(b) having a brightness of 60% or more, and
   ①a caliper of 225 micrometres (microns) or less, or
   ②a caliper more than 225 micrometres (microns) but not more than 508 micrometres (microns) and an ash content more than 3%; or

(c) having a brightness of less than 60% a caliper of 254 micrometres (microns) or less and an ash content more than 8%.

税目48.02不包括滤纸及纸板（含茶袋纸）或

Heading No.48.02 does not, however, cover filter paper

毡纸及纸板。

and paperboard (including tea-bag paper) or felt paper or paperboard.

六、本章所称“牛皮纸及纸板”，是指所含用硫酸盐法或烧碱法制得的纤维不少于全部纤维重量的80%的纸及纸板。

6. In this Chapter “kraft paper and paperboard” means paper and paperhoard of which not less than 80% by weight of the total fibre content consists of fibres obtained by the chemical sulphate or soda processes.

七、除税目条文另有规定的以外，符合税目48.01至48.11中两个或两个以上税号所规定的纸、纸板、纤维素絮纸及纤维素纤维网纸，应按号列顺序归入有关税号中的最末一个税号。

7. Except where the terms of the headings otherwise require, paper, paperboard, cellulose wadding and webs of cellulose fibres answering to a description in two or more of the headings Nos.48.01 to 48.11 are to be classified under that one of such headings which occurs last in numerical order in the Nomenclature.

八、税号48.01及48.03至48.09仅适用于下列规格的纸、纸板、纤维素絮纸及纤维素纤维网纸：

(一) 成条或成卷，宽度超过36厘米；

(二) 成张矩形（包括正方形），一边超过36厘米，另一边超过15厘米（以未折叠计）。

8. Headings Nos.48.01 and 48.03 to 48.09 apply only to paper, paperboard, cellulose wadding and webs of cellulose fibres:

(A) in strips or rolls of s width exceeding 36 cm; or

(B) in rectangular (including square) sheets with one side exceeding 36 cm and the other side exceeding 15 cm in the unfolded state.

九、税号48.14所称“壁纸及类似品”，仅限于：

(一) 适合作墙壁或天花板装饰用的成卷纸张，宽度不小于45厘米，但不超过160厘米：

1. 起纹、压花、染面、印有图案或经其他装饰的（例如起绒），不论是否用透明的防护塑料涂布或覆盖；

2. 表面饰有木粒或草粒而凹凸不平的；

3. 表面用塑料涂布或覆盖并起纹、压花、染面、印有图案或经其他装饰的；

4. 表面用不论是否平行连结或编织的编结材料覆盖的；

(二) 适于装饰墙壁或天花板用的经上述加工的纸边及纸条，不论是否成卷；

(三) 由几幅拼成的壁纸，成卷或成张，贴到墙上可组成印刷的风景或图案。

既可作铺地制品，也可作壁纸的以纸或纸板为底的产品，应归入税号48.15。

9. For the purposes of heading No.48.14, the expression “wallpaper and similar wall coverings” applies only to:

(a) Paper in rolls, of a width of not less than 45 cm and not more than 160 cm, suitable for wall or ceiling decoration:

①Grained, embossed, surface-coloured, designprinted or otherwise surface-decorated (e.g., with textile flock), whether or not coated or covered with transparent protective plastics;

②With an uneven surface resulting from the incorporation of particles of wood, straw, etc.;

③Coated or covered on the face side with plastics, the layer of plastics being grained, embossed, coloured, design-printed or otherwise decorated; or

④Covered on the face side with plaiting material, whether or not bound together in parallel strands or woven;

(b) Borders and friezes, of paper, treated as above, whether or not in rolls, suitable for wall or ceiling decoration;

(c) Wall coverings of paper made up of several panels, in rolls or sheets, printed so as to make up a scene, design or motif when applied to a wall.

Products on a base of paper or paperboard, suitable for use both as floor coverings and as wall coverings, are to be classified in heading No.48.15.

十、税号48.20不包括切成一定尺寸的活页纸张或

10. Heading No.48.20 does not cover loose sheets or cards, cut

卡片，不论是否印制、压花、打孔。

十一、税号48.23主要适用于提花机或类似机器用的穿孔纸或卡片，以及纸花边。

十二、除税号48.14及48.21的货品外，印有图案、文字或图画的纸、纸板、纤维素絮纸及其制品，如果所印图案、文字或图画作为其主要用途，应归入第四十九章。

子目注释：

一、子目号4804.11及4804.19所称“牛皮挂面纸”，是指所含用硫酸盐法或烧碱法制得的木纤维不少于全部纤维重量的80%的成卷机器上光或研光纸及纸板，每平方米重量超过115克，并且最低缪伦耐破度符合下表所示（其他重量的耐破度可参照下表换算）：

重　　量 克/平方米	最低耐破度 千帕斯卡
115	393
125	417
637	200
824	300
961	400

二、子目号4804.21及4804.29所称“袋用牛皮纸”，是指所含用硫酸盐法或烧碱法制得的木纤维不少于全部纤维重量的80%的成卷机器上光纸，每平方米重量不小于60克，但不超过115克，并且符合下列一种规格：

（一）缪伦耐破指数不小于3.7千帕斯卡·平方米/克，并且横向伸长率大于4.5%，纵向伸长率大于2%；

（二）至少能达到下表所示的最小撕裂度和抗张强度（其他重量的可参照下表换算）：

to size, whether or not printed, embossed or perforated.

11. Heading No.48.23 applies, *inter alia* , to perforated paper or paperboard cards for Jacquard or similar machines and paper lace.

12. Except for the goods of heading No.48.14 or 48.21, paper, paperboard, cellulose wadding and articles thereof, printed with motifs, characters or pictorial representations, which are not merely incidental to the primary use of the goods, fall in Chapter 49.

Subheading Notes:

1. For the purposes of subheadings Nos.4804.11 and 4804.19. "kraftliner" means machine-finished or machine-glazed paper and paperboard, of which not less than 80% by weight of the total fibre content consists of wood fibres obtained by the chemical sulphate or soda processes, in rolls, weighing more than 115 g/m^2 and having a minimum Mullen bursting strength as indicated in the following table or the linearly interpolated or extrapolated equivalent for any other weight.

Weight g/m^2	Minimum Mullen bursting strength kPa
115	393
125	417
637	200
824	300
961	400

2. For the puposes of subheadings Nos.4804.21 and 4804.29, "sack kraft paper" means machine-finished paper, of which not less than 80% by weight of the total fibre content consists of fibres obtained by the chemical sulphate or soda processes, in rolls, weighing not less than 60 g/m^2 but not more than 115 g/m^2 and meeting one of the following sets of specifications:

(a) Having a Mullen burst index of not less than 3.7 kPa·m^2/g and a stretch factor of more than 4.5% in the cross direction and of more than 2% in the machine direction.

(b) Having minima for tear and tensile as indicated in the following table or the linearly interpolated equivalent for any other weight:

重　量 克/平方米	最小撕裂度 毫牛顿		最小抗张强度 千牛顿/米	
	纵　向	纵向加横向	横　向	纵向加横向
60	700	1510	1.9	6
70	830	1790	2.3	7.2
8.3	80	965	2070	2.8
10.6	100	1230	2635	3.7
12.3	115	1425	3060	4.4

Weight g/m^2	Minimum tear mN		Minimum tensile kN/m	
	Machine direction	Machine direction plus cross direction	Cross direction	Machine direction plus cross direction
60	700	1510	1.9	6
70	830	1790	2.3	7.2
8.3	80	965	2070	2.8
10.6	100	1230	2635	3.7
12.3	115	1425	3060	4.4

三、子目号 4805.11 所称“半化学的瓦楞纸”，是指所含用半化学制浆法制得的未漂白硬木纤维不少于全部纤维重量的 65% 的成卷纸张，并且在温度为 23℃ 和相对湿度为 50% 时，经过 30 分钟的瓦楞芯纸平压强度测定（CMT30），抗压强度超过 1.8 牛顿/克/平方米。

3. For the purposes of subheading No. 4805.11, “semi-chemical fluting paper” means paper, in rolls, of which not less than 65% by weight of the total fibre content consists of unbleached hardwood fibres obtained by a semi-chemical pulping process, and having a CMT 30 (Corrugated Medium Test with 30 minutes of conditioning) crush resistance exceeding 1.8 newtons/g/m^2 at 50% relative humidity, at 23℃.

四、子目 4805.12 包括主要用半化学法制得草浆制成的成卷纸张，每平方米重量在 130 克及以上，并且在温度为 23°C 和相对湿度为 50% 时，经过 30 分钟瓦楞芯纸平压强度测定（CMT30），抗压强度超过 1.4 牛顿/克/平方米。

4. Subheading 4805.12 covers paper, in rolls, made mainly of straw pulp obtained by a semi-chemical process, weighing 130g/m^2 or more, and having a CMT 30 (Corrugated Medium Test with 30 minutes of conditioning) crush resistance exceeding 1.4 newtons/g/m^2 at 50% relative humidity, at 23°C.

五、子目 4805.24 及 4805.25 包括全部或主要用回收（废碎）纸及纸板制得的纸浆制成的纸及纸板，强韧箱纸板也可以有一面用染色纸或由漂白或未漂白的非再生浆制得的纸做表层，这些产品缪伦耐破指数不小于 2 千帕斯卡·平方米/克。

5. Subheadings 4805.24 and 4805.25 cover paper and paperboard made wholly or mainly of pulp of recovered (waste and scrap) paper or paperboard. Testliner may also have a surface layer of dyed paper or of paper made of bleached or unbleached non-recovered pulp. These products have a Mullen burst index of not less than 2 kPa·m^2/g.

六、子目号 4805.30 所称“亚硫酸盐包装纸”，是指所含用亚硫酸盐法制得的木纤维超过全部纤维重量的 40% 的机器研光纸，灰分含量不超过 8%，并且缪伦耐破指数不小于 1.47 千帕斯卡·平方米/克。

6. For the purposes of subheading No. 4805.30, “sulphite wrapping paper” means machine-glazed paper, of which more than 40% by weight of the total fibre content consists of wood fibres obtained by the chemical sulpohite process, having an ash content not exceeding 8% and having a Mullen burst index of not less than 1.47 kPa·m^2/g.

七、子目号 4810.22 所称“轻质涂布纸”，是指双面涂布纸，其每平方米总重量不超过 72 克，每面每平方米的涂层重量不超过 15 克，原纸中所含用机械方法制得的木纤维不少于全部纤维重量的 50%。

7. For the purposes of subheading No. 4810.22, “lightweight coated paper” means paper, coated on both sides, of a total weight not exceeding 72 g/m^2, with a coating weight not exceeding 15g/m^2 per side, on a base of which not less than 50% by weight of the total fibre content consists of wood fibres obtained by a mechanical process.

序号 No.	税则号列 Tariff Item	货品名称	最惠国税率(%) M.F.N.	协定税率(%) Agreement 东盟 ASEAN	曼谷协定 BA	香港 H.K.	澳门 Macao	普通税率(%) Gen.	Article Description
	48.01	**成卷或成张的新闻纸:**							**Newsprint, in rolls or sheets:**
3078	4801.0000	成卷或成张的新闻纸	7.5					30	Newsprint, in rolls or sheets
	48.02	**书写、印刷或类似用途的未经涂布的纸及纸板、未打孔的穿孔卡片纸及穿孔纸带纸、成卷或成张矩形(包括正方形),任何尺寸,但税目48.01或48.03的纸除外;手工制纸及纸板:**							**Uncoated paper and paperboard, of a kind used for writing, printing or other graphic purposes, and non perforated punch-cards and punch tape paper, in rolls or rectangular (including square) sheets, of any size, other than paper of heading No.48.01 or 48.03; hand-made paper and paperboard:**
3079	4802.1000	-手工制纸及纸板	7.5					70	-Hand-made paper and paperboard
		-光敏、热敏、电敏纸及纸板的原纸和原纸板:							-Paper and paperboard of a kind used as a base for photo-sensitive, heat-sensitive or electro-sensitive paper or paperboard:
3080	4802.2010	---照相原纸	7.5			0		40	---photo poper bese
3081	4802.2090	---其他	7.5			0		40	---Other
3082	4802.3000	-碳化原纸	7.5			0		40	-Carbonizing base paper
3083	4802.4000	-壁纸原纸	7.5			0		40	-Wallpaper base
		-其他纸及纸板,不含用机械或化学—机械方法制得的纤维或所含前述纤维不超过全部纤维重量的10%:							-Other paper and paperboard, not containing fibres obtained by a mechanical or chemi-mechanical process or of which not more than 10% by weight of the total fibre content consists of such fibres:
3084	4802.5400	--每平方米重量小于40克	7.5			0		30	--Weighing less than 40 g/m^2
3085	4802.5500	--每平方米重量在40克及以上,但不超过150克,成卷	5			0		30	--Weighing 40 g/m^2 or more but not more than 150 g/m^2, in rolls
3086	4802.5600	--每平方米重量在40克及以上,但不超过150克,一边不越过435毫米,另一边不超过297毫米(以未折叠计)成张的	5			0		30	--Weighing 40 g/m^2 or more but not more than 150 g/m^2, in sheets with one side not exceeding 435 mm and the other side not exceeding 297 mm in the unfolded state
3087	4802.5700	--其他,每平方米重量在40克及以上,但不超过150克	5			0		30	--Other, weighing 40 g/m^2 or more but not more than 150 g/m^2
3088	4802.5800	--每平方米重量超过150克	5			0		30	--Weighing more than 150 g/m^2
		-其他纸及纸板,所含用机械或化学—机械方法制得的纤维超过全部纤维重量的10%:							-Other paper and paperboard, of which more than 10% by weight of the total fibre content consists of fibres obtained by a mechanical or chemi-mechanical process:

序号 No.	税则号列 Tariff Item	货品名称	最惠国税率(%) M.F.N.	协定税率(%) Agreement 东盟 ASEAN	曼谷协定 BA	香港 H.K.	澳门 Macao	普通税率(%) Gen.	Article Description
		--成卷的:							--In rolls:
3089	4802.6110	---新闻纸	7.5					30	---Newsprint
3090	4802.6190	---其他	5			0		30	---Other
3091	4802.6200	--成张的,一边不超过435毫米,另一边不超过297毫米(以未折叠计)	5			0		30	--In sheets with one side not exceeding 435 mm and the other side not exceeding 297 mm in the unfolded state
		--其他:							---Other:
3092	4802.6910	---新闻纸	7.5					30	---Newsprint
3093	4802.6990	---其他	5			0		30	---Other
	48.03	**卫生纸、面巾纸、餐巾纸以及家庭或卫生用的类似纸、纤维素絮纸和纤维素纤维网纸,不论是否起纹、压花、打孔、染面、饰面或印花,成卷或成张的:**							**Toilet or facial tissue stock, towel or napkin stock and similar paper of a kind used for household or sanitary purposes, cellulose wadding and webs of cellulose fibres, whether or not creped, crinkled, embossed, perforated, surface-coloured, surface-decorated or printed, in rolls or sheets:**
3094	4803.0000	卫生纸、面巾纸、餐巾纸以及家庭或卫生用的类似纸、纤维素絮纸和纤维素纤维网纸,不论是否起纹、压花、打孔、染面、饰面或印花,成卷或成张的	7.5					40	Toilet or facial tissue stock, towel or napkin stock and similar paper of a kind used for household or sanitary purposes, cellulose wadding and webs of cellulose fibres, whether or not creped, crinkled, embossed, perforated, surface-coloured, surface-decorated or printed, in rolls or sheets
	48.04	**成卷或成张的未经涂布的牛皮纸及纸板,但不包括税号48.02或48.03的货品:**							**Uncoated kraft paper and paperboard, in rolls or sheets, other than that of heading No.48.02 or 48.03:**
		-牛皮挂面纸:							-Kraftliner:
3095	4804.1100	--未漂白	5		5			30	--Unbleached
3096	4804.1900	--其他	5		5			30	--Other
		-袋用牛皮纸:							-Sack kraft paper:
3097	4804.2100	--未漂白	5		5			30	--Unbleached
3098	4804.2900	--其他	5		5			30	--Other
		-其他牛皮纸及纸板,每平方米重量不超过150克:							-Other kraft paper and paperboard weighing 150 g/m^2 or less:
3099	4804.3100	--未漂白	2		2			30	--Unbleached
3100	4804.3900	--其他	2		2			30	--Other
		-其他牛皮纸及纸板,每平方米重量超过150克,但小于225克:							-Other kraft paper and paperboard weighing more than 150 g/m^2 but less than 225 g/m^2:
3101	4804.4100	--未漂白	2		2			30	--Unbleached

序号 No.	税则号列 Tariff Item	货品名称	最惠国税率(%) M.F.N.	协定税率(%) Agreement 东盟 ASEAN	曼谷协定 BA	香港 H.K.	澳门 Macao	普通税率(%) Gen.	Article Description
3102	4804.4200	--本体均匀漂白,所含用化学方法制得的木纤维超过全部纤维重量的95%	5		5			30	--Bleached uniformly throughout the mass and of which more than 95% by weight of the total fibre content consists of wood fibres obtained by a chemical process
3103	4804.4900	--其他	2		2			30	--Other
		-其他牛皮纸及纸板,每平方米重量在225克及以上:							-Other kraft paper and paperboard weighing 225 g/m² or more:
3104	4804.5100	--未漂白	2		2			30	--Unbleached
3105	4804.5200	--本体均匀漂白,所含用化学方法制得的木纤维超过全部纤维重量的95%	5		5			30	--Bleached uniformly throughout the mass and of which more than 95% by weight of the total fibre content consists of wood fibres obtained by a chemical process
3106	4804.5900	--其他	2		2			30	--Other
	48.05	**成卷或成张的其他未经涂布的纸及纸板,加工程度不超过本章注释三所列范围:**							**Other uncoated paper and paperboard, in rolls or sheets, not further worked or processed than as specified in Note 3 to this Chapter:**
		-瓦楞原纸:							-fluting paper:
3107	4805.1100	--半化学的瓦楞原纸	7.5			0		30	--Semi-chemical fluting paper
3108	4805.1200	--草浆瓦楞原纸	7.5			0		30	--Straw fluting paper
3109	4805.1900	--其他	7.5			0		30	--Other
		-强韧箱纸板(再生挂面纸板):							-Testliner(recycled liner board):
3110	4805.2400	--每平方米重量在150克及以下	7.5			0		30	--Weighing 150g/m² or less
3111	4805.2500	--每平方米重量超过150克	7.5			0		30	--Weighing more than 150g/m²
3112	4805.3000	-亚硫酸盐包装纸	7.5					30	-Sulphite wrapping paper
3113	4805.4000	-滤纸及纸板	7.5					30	-Filter paper and paperboard
3114	4805.5000	-毡纸及纸板	7.5					30	-Felt paper and paperboard
		-其他:							-Other:
		--每平方米重量在150克及以下:							--Weighing 150 g/m² or less:
3115	4805.9110	---电解电容器原纸	7.5			0		30	---Paper for electrolytic capacitor
3116	4805.9190	---其他	7.5			0		30	---Other
3117	4805.9200	--每平方米重量超过150克,但小于225克	7.5					30	--Weighing more than 150 g/m² but less than 225 g/m²
3118	4805.9300	--每平方米重量在225克及以上	7.5			0		30	--Weighing 225 g/m² or more.
	48.06	**成卷或成张的植物羊皮纸、防油纸、描图纸、半透明纸及其他高光泽透明或半透明纸:**							**Vegetable parchment, greaseproof papers, tracing papers and glassine and other glazed transparent or translucent papers, in rolls or sheets:**

序号 No.	税则号列 Tariff Item	货品名称	最惠国税率(%) M.F.N.	协定税率(%) Agreement 东盟 ASEAN	曼谷协定 BA	香港 H.K.	澳门 Macao	普通税率(%) Gen.	Article Description
3119	4806.1000	-植物羊皮纸	7.5					40	-Vegetable parchment
3120	4806.2000	-防油纸	7.5					40	-Greaseproof papers
3121	4806.3000	-描图纸	7.5					30	-Tracing papers
3122	4806.4000	-高光泽透明或半透明纸	7.5					40	-Glassine and other glazed transparent or translucent papers
	48.07	**成卷或成张的复合纸及纸板(用粘合剂粘合各层纸或纸板制成),未经表面涂布或未浸渍,不论内层是否有加强材料:**							**Composite paper and paperboard (made by sticking flat layers of paper or paperboard together with an adhesive), not surface-coated or impregnated, whether or not internally reinforced, in rolls or sheets:**
3123	4807.0000	成卷或成张的复合纸及纸板(用粘合剂粘合各层纸或纸板制成),未经表面涂布或未浸渍,不论内层是否有加强材料	7.5					40	Composite paper and paperboard (made by sticking flat layers of paper or paperboard together with an adhesive), not surface-coated or impregnated, whether or not internally reinforced, in rolls or sheets.
	48.08	**成卷或成张的瓦楞纸及纸板(不论是否与平面纸胶合)、皱纹纸及纸板、压纹纸及纸板、穿孔纸及纸板,但税号48.03的纸除外:**							**Paper and paperboard, corrugated (with or without glued flat surface sheets), creped, crinkled, embossed or perforated, in rolls or sheets, other than paper of the kind described in heading No.48.03:**
3124	4808.1000	-瓦楞纸及纸板,不论是否穿孔	7.5					30	-Corrugated paper and paperboard, whether or not perforated
3125	4808.2000	-袋用皱纹牛皮纸,不论是否压花或穿孔	7.5					40	-Sack kraft paper, creped or crinkled, whether or not embossed or perforated
3126	4808.3000	-其他皱纹牛皮纸,不论是否压花或穿孔	7.5					40	-Other kraft paper, creped or crinkled, whether or not embossed or perforated
3127	4808.9000	-其他	7.5					40	-Other
	48.09	**复写纸、自印复写纸及其他拷贝或转印纸(包括涂布或浸渍的油印蜡纸或胶印版纸),不论是否印制,成卷或成张的:**							**Carbon paper, self-copy paper and other copying or transfer papers (including coated or impregnated paper for duplicator stencils or offset plates), whether or not printed, in rolls or sheets:**
3128	4809.1000	-复写纸及类似拷贝纸	7.5					40	-Carbon or similar copying papers
3129	4809.2000	-自印复写纸	7.5					40	-Self-copy paper
3130	4809.9000	-其他	7.5					40	-Other
	48.10	**成卷或成张矩形(包括正方形)的仟何尺寸的单面或双面涂布高岭土或其他无机物质(不论是否加粘合剂)的纸及纸板,未涂布其他涂料,不论是否染面、饰面或印花:**							**Paper and paperboard, coated on one or both sides with kaolin (China clay) or other inorganic substances, with or without a binder, and with no other coating, whether or not surface-coloured, surface-decorated or printed, in rolls or rectangular (including square) sheets, of any size:**

序号 No.	税则号列 Tariff Item	货品名称	最惠国税率(%) M.F.N.	协定税率(%) Agreement 东盟 ASEAN	曼谷协定 BA	香港 H.K.	澳门 Macao	普通税率(%) Gen.	Article Description
		-书写、印刷或类似用途的纸及纸板,不含用机械或化学—机械方法制得的纤维或所含前述纤维不超过全部纤维重量的10%:							-Paper and paperboard of a kind used for writing, printing or other graphic purposes, not containing fibres obtained by a mechanical or chemi-mechanical process or of which not more than 10% by weight of the total fibre content consists of such fibres:
3131	4810.1300	--成卷的	5		5	0		40	--In rolls
3132	4810.1400	--成张的,一边不超过435毫米,另一边不超过297毫米(以未折叠计)	5		5	0		40	--In sheets with one side not exceeding 435 mm and the other side not exceeding 297 mm in the unfolded state
3133	4810.1900	--其他	5		5	0		40	--Other
		-书写、印刷或类似用途的纸及纸板,所含用机械或化学—机械方法制得的纤维超过全部纤维重量的10%:							-Paper and paperboard of a kind used for writing, printing or other graphic purposes, of which more than 10% by weight of the total fibre content consists of fibres obtained by a mechanical or chemi-mechanical process:
3134	4810.2200	--轻质涂布纸	5					40	--Light-weight coated paper
3135	4810.2900	--其他	5			0		40	--Other
		-牛皮纸及纸板,但书写、印刷或类似用途的除外:							-Kraft paper and paperboard, other than that of a kind used for writing, printing or other graphic purposes:
3136	4810.3100	--本体均匀漂白,所含用化学方法制得的木纤维超过全部纤维重量的95%,每平方米重量不超过150克	5			0	0	40	--Bleached uniformly throughout the mass and of which more than 95% by weight of the total fibre content consists of wood fibres obtained by a chemical process, and weighing 150 g/m^2 or less
3137	4810.3200	--本体均匀漂白,所含用化学方法制得的木纤维超过全部纤维重量的95%,每平方米重量超过150克	5			0	0	40	--Bleached uniformly throughout the mass and of which more than 95% by weight of the total fibre content consists of wood fibres obtained by a chemical process, and weighing more than 150 g/m^2
3138	4810.3900	--其他	5			0	0	40	--Other
		-其他纸及纸板:							-Other paper and paperboard:
3139	4810.9200	--多层的	5			0	0	40	--Multi-ply
3140	4810.9900	--其他	7.5			0	0	40	--Other
	48.11	**成卷或成张矩形(包括正方形)的任何尺寸的经涂布、浸渍、覆盖、染面、饰面或印花的纸、纸板、纤维素絮纸及纤维素纤维网纸,但税目48.03、48.09或48.10的货品除外:**							**Paper, paperboard, cellulose wadding and webs of cellulose fibres, coated, impregnated, covered, surface coloured, surface-decorated or printed, in rolls or rectangular (including square) sheets, of any size, other than goods of the kind described in heading No.48.03, 48.09 or 48.10:**

序号 No.	税则号列 Tariff Item	货品名称	最惠国税率(%) M.F.N.	协定税率(%) Agreement 东盟 ASEAN	曼谷协定 BA	香港 H.K.	澳门 Macao	普通税率(%) Gen.	Article Description
3141	4811.1000	-焦油纸及纸板、沥青纸及纸板	7.5			0	0	40	-Tarred, bituminised or asphalted paper and paperboard
		-胶粘纸及纸板:							-Gummed or adhesive paper and paperboard:
3142	4811.4100	--自粘的	7.5			0	0	40	--Self-adhesive
3143	4811.4900	--其他	7.5			0	0	40	--Other
		-用塑料(不包括粘合剂)涂布、浸渍或覆盖的纸及纸板:							-Paper and paperboard coated, impregnated or covered with plastics(excluding adhesives):
		--漂白的,每平方米重量超过150克:							--Bleached, weighing more than 150 g/m^2:
3144	4811.5110	---彩色相纸用双面涂塑纸	7.5			0	0	40	---Paper coated on both sides with plastics for colour photography
3145	4811.5190	---其他	7.5			0	0	40	---Other
		--其他:							--Other:
3146	4811.5910	---绝缘纸及纸板	7.5			0	0	30	---Insulating paper and paperboard
3147	4811.5990	---其他	7.5			0	0	40	---Other
		-用蜡、石蜡、硬脂精、油或甘油涂布、浸渍、覆盖的纸及纸板:							-Paper and paperboard, coated, impregnated or covered with wax, paraffin wax, stearin, oil or glyecrol:
3148	4811.6010	---绝缘纸及纸板	7.5			0	0	30	---Insulating paper and paperboard
3149	4811.6090	---其他	7.5			0	0	40	---Other
3150	4811.9000	-其他纸、纸板、纤维素絮纸及纤维素纤维网纸	7.5			0	0	40	-Other paper, paperboard, cellulose wadding and webs of cellulose fibres
	48.12	**纸浆制的滤块、滤板及滤片:**							**Filter blocks, slabs and plates, of paper pulp:**
3151	4812.0000	纸浆制的滤块、滤板及滤片	7.5					40	Filter blocks, slabs and plates, of paper pulp
	48.13	**烟卷纸,不论是否切成一定尺寸、成小本或管状:**							**Cigarette paper, whether or not cut to size or in the form of booklets or tubes:**
3152	4813.1000	-成小本或管状	7.5					100	-In the form of booklets or tubes
3153	4813.2000	-宽度不超过5厘米成卷的	7.5					100	-In rolls of a width not exceeding 5 cm
3154	4813.9000	-其他	7.5					100	-Other
	48.14	**壁纸及类似品;窗用透明纸:**							**Wallpaper and similar wall coverings; window transparencies of paper:**
3155	4814.1000	-用木粒或草粒等饰面的壁纸	7.5					50	-"Ingrain"paper
3156	4814.2000	-用塑料涂面或盖面的壁纸及类似品,起纹、压花、着色、印刷图案或经其他装饰	7.5					50	-Wallpaper and similar wall coverings, consisting of paper coated or covered, on the face side, with a grained, embossed, coloured, design-printed or otherwise decorated layer of plastics

序号 No.	税则号列 Tariff Item	货品名称	最惠国税率(%) M.F.N.	协定税率(%) Agreement 东盟 ASEAN	曼谷协定 BA	香港 H.K.	澳门 Macao	普通税率(%) Gen.	Article Description
3157	4814.3000	-用编结材料(不论是否平行连结或编织)盖面的壁纸及类似品	7.5					50	-Wallpaper and similar wall coverings, consisting of paper covered, on the face side, with plaiting material, whether or not bound together in parallel strands or woven
3158	4814.9000	-其他	7.5					50	-Other
	48.15	**以纸或纸板为底制成的铺地制品,不论是否切成一定尺寸:**							**Floor coverings on a base of paper or of paperboard, whether or not cut to size:**
3159	4815.0000	以纸或纸板为底制成的铺地制品,不论是否切成一定尺寸	7.5					90	Floor coverings on a base of paper or of paperboard, whether or not cut to size
	48.16	**复写纸、自印复写纸及其他拷贝或转印纸(不包括税号 48.09 的纸)、油印蜡纸或胶印版纸,不论是否盒装:**							**Carbon paper, self-copy paper and other copying or transfer papers (other than those of heading No.48.09), duplicator stencils and offset plates, of paper, whether or not put up in boxes:**
3160	4816.1000	-复写纸或类似拷贝纸	7.5					70	-Carbon or similar copying papers
3161	4816.2000	-自印复写纸	7.5					70	-Self-copy paper
3162	4816.3000	-油印蜡纸	7.5					70	-Duplicator stencils
		-其他:							-Other:
3163	4816.9010	---热敏转印纸	7.5					40	---Heat transfer paper
3164	4816.9090	---其他	7.5					70	---Other
	48.17	**纸或纸板制的信封、封缄信片、素色明信片及通信卡片;纸或纸板制的盒子、袋子及夹子,内装各种纸制文具:**							**Envelopes, letter cards, plain postcards and correspondence cards, of paper or paperboard; boxes, pouches, wallets and writing compendiums, of paper or paperboard, containing an assortment of paper stationery:**
3165	4817.1000	-信封	7.5					80	-Envelopes
3166	4817.2000	-封缄信片、素色明信片及通信卡片	7.5					80	-Letter cards, plain postcards and correspondence cards
3167	4817.3000	-纸或纸板制的盒子、袋子及夹子,内装各种纸制文具	7.5					80	-Boxes, pouches, wallets and writing compendiums, of paper or paperboard, containing an assortment of paper stationery
	48.18	**卫生纸及类似纸,家庭或卫生用纤维素絮纸及纤维素纤维网纸,成卷宽度不超过 36 厘米或切成一定尺寸或形状的;纸浆、纸、纤维素絮纸或纤维素纤维网纸制的手帕、面巾、台布、餐巾、尿布、止血塞、床单及类似的家庭、卫生或医院用品、衣服及衣着附件:**							**Toilet paper and similar paper, cellulose wadding or webs of cellulose fibres, of a kind used for household or sanitary purposes, in rolls of a width not exceeding 36 cm, or cut to size or shape; handkerchiefs, cleansing tissues, towels, tablecloths, serviettes, napkins for babies, tampons, bed sheets and similar household, sanitary or hospital articles, articles of apparel and clothing accessories, of paper pulp, paper, cellulose wadding or webs of cellulose fibres:**

序号 No.	税则号列 Tariff Item	货品名称	最惠国税率(%) M.F.N.	协定税率(%) Agreement 东盟 ASEAN	曼谷协定 BA	香港 H.K.	澳门 Macao	普通税率(%) Gen.	Article Description
3168	4818.1000	-卫生纸	7.5					80	-Toilet paper
3169	4818.2000	-纸手帕及纸面巾	7.5					90	-Handkerchiefs, cleansing or facial tissues and towels
3170	4818.3000	-纸台布及纸餐巾	7.5					90	-Tablecloths and serviettes
3171	4818.4000	-纸卫生巾及止血塞、婴儿纸尿布、尿布衬里及类似的卫生用品	7.5					80	-Sanitary towels and tampons, napkins and napkin liners for babies and similar sanitary articles
3172	4818.5000	-衣服及衣着附件	7.5					90	-Articles of apparel and clothing accessories
3173	4818.9000	-其他	7.5					90	-Other
	48.19	**纸、纸板、纤维素絮纸或纤维素纤维网纸制的箱、盒、匣、袋及其他包装容器;纸或纸板制的卷宗盒、信件盘及类似品,供办公室、商店及类似场所使用的:**							**Cartons, boxes, cases, bags and other packing containers, of paper, paperboard, cellulose wadding or webs of cellulose fibres; box files, letter trays and similar articles, of paper or paperboard of a kind used in offices, shops or the like:**
3174	4819.1000	-瓦楞纸或纸板制的箱、盒、匣	5			0	0	80	-Cartons, boxes and cases, of corrugated paper or paperboard
3175	4819.2000	-非瓦楞纸或纸板制的可折叠箱、盒、匣	5			0	0	80	-Folding cartons, boxes and cases, of noncorrugated paper or paperboard
3176	4819.3000	-底宽40厘米及以上的纸袋	7.5					80	-Sacks and bags, having a base of a width of 40 cm or more
3177	4819.4000	-其他纸袋,包括锥形袋	7.5					80	-Other sacks and bags, including cones
3178	4819.5000	-其他包装容器,包括唱片套	7.5					80	-Other packing containers, including record sleeves
3179	4819.6000	-办公室、商店及类似场所使用的卷宗盒、信件盘、存储盒及类似品	7.5					80	-Box files, letter trays, storage boxes and similar articles, of a kind used in offices, shops or the like
	48.20	**纸或纸板制的登记本、账本、笔记本、定货本、收据本、信笺本、记事本、日记本及类似品、练习本、吸墨纸本、活动封面(活页及非活页)、文件夹、卷宗皮、多联商业表格纸、页间夹有复写纸的本及其他文具用品;纸或纸板制的样品簿、粘贴簿及书籍封面:**							**Registers, account books, note books, order books, receipt books, letter pads, memorandum pads, diaries and similar articles, exercise books, blotting-pads, binders (loose-leaf or other), folders, file covers, manifold business forms, interleaved carbon sets and other articles of stationery, of paper or paperboard; albums for samples or for collections and book covers, of paper or paperboard:**
3180	4820.1000	-登记本、账本、笔记本、定货本、收据本、信笺本、记事本、日记本及类似品	7.5					80	-Registers, account books, note nooks, order books, receipt books, letter pads, memorandum pads, diaries and similar articles
3181	4820.2000	-练习本	7.5					80	-Exercise books

序号 No.	税则号列 Tariff Item	货品名称	最惠国税率(%) M.F.N.	协定税率(%) Agreement 东盟 ASEAN	曼谷协定 BA	香港 H.K.	澳门 Macao	普通税率(%) Gen.	Article Description
3182	4820.3000	-活动封面(书籍封面除外)、文件夹及卷宗皮	7.5					80	-Binders (other than book covers), folders and file covers
3183	4820.4000	-多联商业表格纸、页间夹有复写纸的本	7.5					80	-Manifold business forms and interleaved carbon sets
3184	4820.5000	-样品簿及粘贴簿	7.5					80	-Albums for samples or for collections
3185	4820.9000	-其他	7.5					80	-Other
	48.21	**纸或纸板制的各种标签,不论是否印制:**							**Paper or paperboard labels of all kinds, whether or not printed:**
3186	4821.1000	-印制	7.5			0	0	50	-Printed
3187	4821.9000	-其他	7.5					50	-Other
	48.22	**纸浆、纸或纸板(不论是否穿孔或硬化)制的筒管、卷轴、纡子及类似品:**							**Bobbins, spools, cops and similar supports of paper pulp, paper or paperboard (whether or not perforated or hardened):**
3188	4822.1000	-纺织纱线用	7.5					35	-Of a kind used for winding textile yarn
3189	4822.9000	-其他	7.5					70	-Other
	48.23	**切成一定尺寸或形状的其他纸、纸板、纤维素絮纸及纤维素纤维网纸;纸浆、纸、纸板、纤维素絮纸及纤维素纤维网纸制的其他物品:**							**Other paper, paperboard, cellulose wadding and webs of cellulose fibres, cut to size or shape; other articles of paper pulp, paper, paperboard, cellulose wadding or webs of cellulose fibres:**
		-成条或成卷的胶粘纸:							-Gummed or adhesive paper, in strips or rolls:
3190	4823.1200	--自粘的	7.5					50	--Self-adhesive
3191	4823.1900	--其他	7.5					50	--Other
3192	4823.2000	-滤纸及纸板	7.5					30	-Filter paper and paperboard
3193	4823.4000	-已印制的自动记录器用打印纸卷、纸张及纸盘	7.5					30	-Rolls, sheets and dials, printed for self-recording apparatus
3194	4823.6000	-纸或纸板制的盘、碟、盆、杯及类似品	7.5					90	-Trays, dishes, plates, cups and the like, of paper or paperboard
3195	4823.7000	-压制或模制纸浆制品	7.5					90	-Moulded or pressed articles of paper pulp
		-其他:							-Other:
3196	4823.9020	---神纸及类似用品	7.5		7.5			180	---Joss paper and the like
3197	4823.9030	---纸扇	7.5		7.5			90	---Paper fans
3198	4823.9090	---其他	7.5		7.5	0	0	90	---Other

第四十九章

书籍、报纸、印刷图画及其他印刷品；手稿、打字稿及设计图纸

注释：

一、本章不包括：

(一) 透明基的照相负片或正片（第三十七章）；

(二) 立体地图、设计图表或地球仪、天体仪，不论是否印刷（税号 90.23）；

(三) 第九十五章的扑克牌或其他物品；

(四) 雕版画、印刷画、石印画的原本（税号 97.02），税号 97.04 的邮票、印花税票、纪念封、首日封、邮政信笺及类似品，以及第九十七章的超过一百年的古物或其他物品。

二、第四十九章所称"印刷"，也包括用胶版复印机、油印机印制，在自动数据处理设备控制下打印绘制，压印、冲印、感光复印、热敏复印或打字。

三、用纸以外材料装订成册的报纸、杂志和期刊，以及一期以上装订在同一封面里的成套报纸、杂志和期刊，应归入税号 49.01，不论是否有广告材料。

四、税号 49.01 还包括：

(一) 附有说明文字，每页编有号数以便装订成一册或几册的整集印刷复制品，例如，美术作品、绘画；

(二) 随同成册书籍的图画附刊；

(三) 供装订书籍或小册子用的散页、集页或书帖形式的印刷品，已构成一部作品的全部或部分。

但没有说明文字的印刷图画或图解，不论是否散页或书帖形式，应归入税号 49.11。

五、除本章注释三另有规定的以外，税号 49.01 不包括主要作广告用的出版物（例如，小册子，散页印刷品、商业目录、同业公会出版的年鉴、旅游宣传品），这类出版物应归入税号

Chapter 49

Printed books, newspapers, pictures and other products of the printing industry; manuscripts, typescripts and plans

Notes:

1. This Chapter does not cover:

(a) Photographic negatives or positives on transparent bases (Chapter 37);

(b) Maps, plans or globes, in relief, whether or not printed (heading No.90.23);

(c) Playing cards or other goods of Chapter 95; or

(d) Original engravings, prints or lithographs (heading No.97.02), postage or revenue stamps, stamp-postmarks, first-day covers, postal stationery or the like of heading No.97.04, antiques of an age exceeding one hundred years or other articles of Chapter 97.

2. For the purposes of Chapter 49, the term "printed" also means reproduced by means of a duplicating machine, produced under the control of an automatic data processing machine, embossed, photographed, photocopied, thermocopied or typewritten.

3. Newspapers, journals and periodicals which are bound otherwise than in paper, and sets of newspapers, journals or periodicals comprising more than one number under a single cover are to be classified in heading No.49.01, whether or not containing advertising material.

4. Heading No.49.01 also covers:

(a) A collection of printed reproductions of, for example, works of art or drawings, with a relative text, put up with numbered pages in a form suitable for binding into one or more volumes;

(b) A pictorial supplement accompanying, and subsidiary to, a bound volume; and

(c) Printed parts of books or booklets, in the form of assembled or separate sheets or signatures, constituting the whole or a part of a complete work and designed for binding.

However, printed pictures or illustrations not bearing a text, whether in the form of signatures or separate sheets, fall in heading No.49.11.

5. Subject to Note 3 to this Chapter, heading No.49.01 does not cover publications which are essentially devoted to advertising (for example, brochures, pamphlets, leaflets, trade catalogues, year books published by trade associations,

49.11。

tourist propaganda). Such publications are to be classified in heading No.49.11.

六、税号 49.03 所称“儿童图画书”，是指以图画为主、文字为辅，供儿童阅览的书籍。

6. For the purposes of heading No.49.03, the expression “children's picture books” means books for children in which the pictures form the principal interest and the text is subsidiary

序号 No.	税则号列 Tariff Item	货品名称	最惠国税率(%) M.F.N.	协定税率(%) Agreement 东盟 ASEAN	曼谷协定 BA	香港 H.K.	澳门 Macao	普通税率(%) Gen.	Article Description
	49.01	**书籍、小册子、散页印刷品及类似印刷品，不论是否单张：**							**Printed books, brochures, leaflets and similar printed matter, whether or not in single sheets:**
3199	4901.1000	-单张的，不论是否折叠	0					0	-In single sheets, whether or not folded
		-其他：							-Other:
3200	4901.9100	--字典或百科全书及其连续出版的分册	0					0	--Dictionaries and encyclopaedias, and serial instalments thereof
3201	4901.9900	--其他	0					0	--Other
	49.02	**报纸、杂志及期刊，不论有无插图或广告材料：**							**Newspapers, journals and periodicals, whether or not illustrated or containing advertising material:**
3202	4902.1000	-每周至少出版四次	0					0	-Appearing at least four times a week
3203	4902.9000	-其他	0					0	-Other
	49.03	**儿童图画书、绘画或涂色书：**							**Children's picture, drawing or colouring books:**
3204	4903.0000	儿童图画书、绘画或涂色书	0					0	Children's picture, drawing or colouring books
	49.04	**乐谱原稿或印本，不论是否装订或印有插图：**							**Music, printed or in manuscript, whether or not bound or illustrated:**
3205	4904.0000	乐谱原稿或印本，不论是否装订或印有插图	0					0	Music, printed or in manuscript, whether or not bound or illustrated
	49.05	**各种印刷的地图、水道图及类似图表，包括地图册、挂图、地形图及地球仪、天体仪：**							**Maps and hydrographic or similar charts of all kinds, including atlases, wall maps, topographical plans and globes, printed:**
3206	4905.1000	-地球仪、天体仪	0					0	-Globes
		-其他：							-Other:
3207	4905.9100	--成册的	0					0	--In book form
3208	4905.9900	--其他	0					0	--Other
	49.06	**手绘的建筑、工程、工业、商业、地形或类似用途的设计图纸原稿；手稿；用感光纸照相复印或用复写纸誊写的上述物品复制件：**							**Plans and drawings for architectural, engineering, industrial, commercial, topographical or similar purposes, being originals drawn by hand; hand-written text; photographic reproductions on sensitized paper and carbon copies of the foregoing:**

序号 No.	税则号列 Tariff Item	货品名称	最惠国税率(%) M.F.N.	协定税率(%)Agreement 东盟 ASEAN	曼谷协定 BA	香港 H.K.	澳门 Macao	普通税率(%) Gen.	Article Description
3209	4906.0000	手绘的建筑、工程、工业、商业、地形或类似用途的设计图纸原稿;手稿;用感兴纸照相复印或用复写纸誊写的上述物品复制件	0					0	Plans and drawings for architectural, engineering, industrial, commercial topographical or similar purposes, being originals drawn by hand; hand-written text; photographic reqroductions on sensitized paper and carbon copies of the foregoing
	49.07	**在承认或将承认其面值的国家流通或新发行并且未经使用的邮票、印花税票及类似票证;印有邮票或印花税票的纸品;钞票;空白支票;股票、债券及类似所有权凭证:**							**Unused postage, revenue or similar stamps of current or new issue in the country in which they have, or will have, a recognised face value; stamp-impressed paper; banknotes; cheque forms; stock, share or bond certificates and similar documents of title:**
3210	4907.0010	---邮票	7.5					50	---Postage
3211	4907.0020	---钞票	0					50	---Banknotes
3212	4907.0030	---证券凭证	0					50	---Documents of title
3213	4907.0090	---其他	7.5					50	---Other
	49.08	**转印贴花纸(移画印花法用图案纸):**							**Transfers(decalcomanias):**
3214	4908.1000	-釉转印贴花纸(移画印花法用图案纸)	7.5					50	-Transfers(decalcomanias), vitrifiable
3215	4908.9000	-其他	7.5					50	-Other
	49.09	**印刷或有图画的明信片;印有个人问候、祝贺、通告的卡片,不论是否有图画、带信封或饰边:**							**Printed or illustrated postcards; printed cards bearing personal greetings, messages or announcements, whether or not illustrated, with or without envelopes or trimmings:**
3216	4909.0010	---印刷或有图画的明信片	7.5					50	---Printed or illustrated postcards
3217	4909.0090	---其他	7.5					50	---Other
	49.10	**印刷的各种日历,包括日历芯:**							**Calendars of any kind, printed, including calendar blocks:**
3218	4910.0000	印刷的各种日历,包括日历芯	7.5					50	Calendars of any kind, printed, including calendar blocks
	49.11	**其他印刷品,包括印刷的图片及照片:**							**Other printed matter, including printed pictures and photographs:**
		-商业广告品、商品目录及类似印刷品:							-Trade advertising material, commercial catalogues and the like:
3219	4911.1010	---无商业价值的	0					0	---No commercial value
3220	4911.1090	---其他	7.5			0		50	---Other
		-其他:							-Other:
3221	4911.9100	--图片、设计图样及照片	7.5					50	--Pictures, designs and photographs
3222	4911.9900	--其他	7.5			0		50	--Other

第十一类

纺织原料及纺织制品

注释：

一、本类不包括：

（一）制刷用的动物鬃、毛（税号 05.02）；马毛及废马毛（税号 05.03）；

（二）人发及人发制品（税号 05.01、67.03 或 67.04），但通常用于榨油机或类似机器的滤布除外（税号 59.11）；

（三）第十四章的棉短绒或其他植物材料；

（四）税号 25.24 的石棉、税号 68.12 或 68.13 的石棉制品或其他产品；

（五）税号 30.05 或 30.06 的物品（例如，医疗、外科、牙科或兽医用的软填料、纱布、绷带及类似品、外科用无菌缝合材料）；税号 33.06 的用于清洁牙缝的纱线（牙线），单独零售包装的；

（六）税号 37.01 至 37.04 的感光布；

（七）截面尺寸超过 1 毫米的塑料单丝和表面宽度超过 5 毫米的塑料扁条及类似品（例如人造草）（第三十九章），以及上述单丝或扁条的缏条、织物、篮筐或柳条编结品（第四十六章）；

（八）第三十九章的用塑料浸渍、涂布、包覆或层压的机织物、针织物或钩编织物、毡呢或无纺织物及其制品；

（九）第四十章的用橡胶浸渍、涂布、包覆或层压的机织物、针织物或钩编织物、毡呢或无纺织物及其制品；

（十）带毛皮张（第四十一章或第四十三章）、税号 43.03 或 43.04 的毛皮制品、人造毛皮及其制品；

（十一）税号 42.01 或 42.02 的用纺织材料制成的物品；

（十二）第四十八章的产品或物品（例如纤维素絮纸）；

（十三）第六十四章的鞋靴及其零件、护腿、裹腿及类似品；

（十四）第六十五章的发网、其他帽类及其零件；

（十五）第六十七章的货品；

（十六）涂有研磨料的纺织材料（税号 68.05）

SECTION Ⅺ

TEXTILES AND TEXTILE ARTICLES

Notes：

1. This Section does not cover：

(a) Animal brush making bristles or hair (heading No. 05.02); horsehair or horsehair waste (heading No. 05.03);

(b) Human hair or articles of human hair (heading No. 05.01, 67.03 or 67.04), except straining cloth of a kind commonly used in oil presses or the like (heading No.59.11);

(c) Cotton linters or other vegetable materials of Chapter 14;

(d) Asbestos of heading No.25.24 or articles of asbestos or other products of heading No.68.12 or 68.13;

(e) Articles of heading No.30.05 or 30.06 (for example, wadding, gauze, bandages and similar articles for medical, surgical, dental or veterinary purposes, sterile surgical suture materials); yarn used to clean between the teeth, in individual retail packages (dental floss), of heading No.33.06;

(f) Sensitized textiles of headings No.37.01 to 37.04;

(g) Monofilament of which any cross-sectional dimension exceeds 1mm or strip or the like (for example, artificial straw) of an apparent width exceeding 5mm, of plastics (Chapter 39), or plaits or fabrics or other basketware or wickerwork of such monofilament or strip (Chapter 46);

(h) Woven, knitted or crocheted fabrics, felt or nonwovens, impregnated, coated, covered or laminated with plastics, or articles thereof, of Chapter 39;

(i) Woven, knitted or crocheted fabrics, felt or nonwovens, impregnated, coated, covered or laminated with rubber, or articles thereof, of Chapter 40;

(j) Hides or skins with their hair or wool on (Chapter 41 or 43) or artcles of furskin, artificial fur or articles thereof, of heading No.43.03 or 43.04;

(k) Articles of textile materials of heading No.42.01 or 42.02;

(l) Products or articles of Chapter 48 (for example, cellulose wadding);

(m) Footwear or parts of footwear, gaiters or leggings or similar articles of Chapter 64;

(n) Hair-nets or other headgear or parts thereof of Chapter 65;

(o) Goods of Chapter 67;

(p) Abrasive-coated textile material (Heading No.68.05)

以及税号68.15的碳纤维及其制品；

（十七）玻璃纤维及其制品，但可见底布的玻璃线刺绣品除外（第七十章）；

（十八）第九十四章的物品（例如，家具、寝具、灯具及照明装置）；

（十九）第九十五章的物品（例如，玩具、游戏品、运动用品及网具）；

（二十）第九十六章的物品（例如，刷子、旅行用成套缝纫用具、拉链及打字机色带）；

（二十一）第九十七章的物品。

二、（一）可归入第五十章至第五十五章及税号58.09或59.02的由两种或两种以上纺织材料混合制成的货品，应按其中重量最大的那种纺织材料归类。

当没有一种纺织材料重量较大时，应按可归入的有关税号中最后一个税号所列的纺织材料归类。

（二）应用上述规定时：

1. 马毛粗松螺旋花线（税号51.10）和含金属纱线（税号56.05）均应作为一种单一的纺织材料，其重量应为它们在纱线中的合计重量；在机织物的归类中，金属线应作为一种纺织材料；

2. 在选择合适的税号时，应首先确定章，然后再确定该章的有关税号，至于不归入该章的其他材料可不予考虑；

3. 当归入第五十四章及第五十五章的货品与其他章的货品进行比较时，应将这两章作为一个单一的章对待；

4. 同一章或同一税号所列各种不同的纺织材料应作为单一的纺织材料对待。

（三）上述（一）、（二）两款规定亦适用于以下注释三、四、五或六所述纱线。

三、（一）本类的纱线(单纱、多股纱线或缆线)除下列(二)款另有规定的以外，凡符合以下规格的应作为“线、绳、索、缆”：

and also carbon fibres or articles of carbon fibres of heading No. 68.15;

(q) Glass fibres or articles of glass fibres, other than embroidery with glass thread on a visible ground of fabric (Chapter 70);

(r) Articles of Chapter 94 (for example, furniture, bedding, lamps and lighting fittings);

(s) Articles of Chapter 95 (for example, toys games, sports requisites and nets);

(t) Articles of Chapter 96 (for example, brushes, travel sets for sewing, slide fasteners and typewriter ribbons); or

(u) Articles of Chapter 97.

2. (A) Goods classifiable in Chapters 50 to 55 or in heading No. 58.09 or 59.02 and of a mixture of two or more textile materials are to be classified as if consisting wholly of that one textile material which predominates by weight over any other single textile material.

When no one textile material predominates by weight, the goods are to be classified as if consisting wholly of that one textile material which is covered by the heading which occurs last in numerical order among those which equally merit consideration.

(B) For the purposes of the above rule:

(a) Gimped horsehair yarn (heading No. 51.10) and metallized yarn (heading No. 56.05) are to be treated as a single textile material the weight of which is to be taken as the aggregate of the weights of its components; for the classification of woven fabrics, metal thread is to be regarded as a textile material;

(b) The choice of appropriate heading shall be effected by determining first the Chapter and then the applicable heading within that Chapter, disregarding any materials not classified in that Chapter;

(c) When both Chapters 54 and 55 are involved with any other Chapter, Chapters 54 and 55 are to be treated as a single Chapter;

(d) Where a Chapter or a heading refers to goods of different textile materials, such materials are to be treated as a single textile material.

(C) The provisions of paragraphs (A) and (B) above apply also to the yarns referred to in Notes 3, 4, 5 or 6 below.

3. (A) For the purposes of this Section, and subject to the exceptions in paragraph (B) below, yarns (single, multiple (folded) or cabled) of the following descriptions are to be treated as “twine, cordage, ropes and cables”:

1. 丝或绢丝纱线，细度在 20000 分特以上；
2. 化学纤维纱线（包括第五十四章的用两根及以上单丝纺成的纱线），细度在 10000 分特以上；
3. 大麻或亚麻纱线：
(1) 加光或上光的，细度在 1429 分特及以上；
(2) 未加光或上光的，细度在 20000 分特以上；
4. 三股或三股以上的椰壳纤维纱线；
5. 其他植物纤维纱线，细度在 20000 分特以上；
6. 用金属线加强的纱线。

(二) 下列各项不按上述（一）款规定办理：
1. 羊毛或其他动物毛纱线及纸纱线，但用金属线加强的纱线除外；
2. 第五十五章的化学纤维长丝丝束以及第五十四章的未加捻或捻度每米少于 5 转的复丝纱线；
3. 税号 50.06 的蚕胶丝及第五十四章的单丝；
4. 税号 56.05 的含金属纱线；但用金属线加强的纱线按上述（一）款 6 项规定办理；
5. 税号 56.06 的绳绒线、粗松螺旋花线及纵行起圈纱线。

四、(一) 除下列（二）款另有规定的以外，第五十章、第五十一章、第五十二章、第五十四章和第五十五章所称"供零售用"纱线，是指以下列方式包装的纱线（单纱、股纱线或缆线）：
1. 绕于纸板、线轴、纱管或类似芯子上，其重量(含线芯)符合下列规定：
(1) 丝、绢丝或化学纤维长丝纱线，不超过 85 克；
(2) 其他纱线，不超过 125 克。
2. 绕成团、绞或束，其重量符合下列规定：
(1) 细度在 3000 分特以下的化学纤维长丝纱线，丝或绢丝纱线，不超过 85 克；
(2) 细度在 2000 分特以下的任何其他纱线，不超过 125 克；
(3) 其他纱线，不超过 500 克。
3. 绕成绞或束，每绞或每束中有若干用线分开的小绞或小束，每小绞或小束的重量相等，并且符合下列规定：
(1) 丝、绢丝或化学纤维长丝纱线，

(a) Of silk or waste silk, measuring more than 20000 decitex;
(b) Of man-made fibres (including yarn of two or more monofilaments of Chapter 54), measuring more than 10000 decitex;
(c) Of true hemp or flax:
(1) Polished or glazed, measuring 1429 decitex or more; or
(2) Not polished or glzed, measuring more than 20000 decitex;
(d) Of coir, consisting of three or more plies;
(e) Of other vegetable fibres, measuring more than 20000 decitex, or
(f) Reinforced with metal thread.

(B) Exceptions:
(a) Yarn of wool or other animal hair and paper yarn, other than yarn, reinforced with metal thread;
(b) Man-made filament tow of Chapter 55 and multifilament yarn without twist or with a twist of less than 5 turns per metre of Chapter 54;
(c) Silk worm gut of heading No.50.06, and monofilaments of Chapter 54;
(d) Metallized yarn of Heading No.56.05; yarn reinforced with metal thread is subject to paragraph (A) (f) above; and
(e) Chenille, yarn, gimped yarn and loop wale-yarn of heading No.56.06.

4. (A) For the purposes of Chapters 50, 51, 52, 54 and 55, the expression "put up for retail sale" in relation to yarn means, subject to the exceptions in paragraph (B) below, yarn (single, multiple (folded) or cabled) put up:
(a) On cards, reels, tubes or similar supports, of a weight (including support) not exceeding:
(1) 85g in the case of silk, waste silk or manmade filament yarn; or
(2) 125g in other cases;
(b) In balls, hanks or skeins of a weight not exceeding:
(1) 85g in the case of man-made filament yarn of less than 3000 decitex, silk or silk waste;
(2) 125g in the case of all other yarns of less than 2000 decitex; or
(3) 500g in other cases;
(c) In hanks or skeins comprising several smaller hanks or skeins separated by dividing threads which render them independent one of the other, each of uniform weight not exceeding:
(1) 85g in the case of silk, waste silk or manmade

不超过85克；

(2) 其他纱线，不超过125克。

(二) 下列各项不按上述(一)款规定办理：

1. 各种纺织材料制的单纱，但下列两种除外：

(1) 未漂白的羊毛或动物细毛单纱；

(2) 漂白、染色或印色的羊毛或动物细毛单纱，细度在5000分特以上。

2. 未漂白的多股纱线或缆线：

(1) 丝或绢丝制的，不论何种包装；

(2) 除羊毛或动物细毛外其他纺织材料制，成绞或成束的。

3. 漂白、染色或印色丝或绢丝制的多股纱线或缆线，细度在133分特及以下；

4. 任何纺织材料制的单纱、多股纱线或缆线：

(1) 交叉绕成绞或束的；

(2) 绕于纱芯上或以其他方式卷绕，明显用于纺织工业的(例如，绕于纱管、加捻管、纬纱管、锥形筒管或锭子上的或者绕成蚕茧状以供绣花机使用的纱线)。

五、税号52.04、54.01及55.08所称“缝纫线”，是指下列多股纱线或缆线：

(一) 绕于芯子(例如，线轴、纱管)上，重量(包括纱芯)不超过1000克；

(二) 作为缝纫线上过浆的；

(三) 终捻为反手(Z)捻的。

六、本类所称“高强力纱”，是指断裂强度大于下列标准的纱线：

尼龙、其他聚酰胺或聚酯制的单纱——60厘牛顿/特克斯；

尼龙、其他聚酰胺或聚酯制的多股纱线或缆线——53厘牛顿/特克斯；

粘胶纤维制的单纱、多股纱线或缆线——27厘牛顿/特克斯。

七、本类所称“制成的”，是指：

(一) 裁剪成除正方形或长方形以外的其他形状的；

(二) 呈制成状态，无需缝纫或其他进一步加

filament yarn; or

(2) 125 g in other cases.

(B) Exceptions:

(a) Single yarn of any textile material, except:

(1) Single yarn of wool or fine animal hair, unbleached; and;

(2) Single yarn of wool or fine animal hair, bleached, dyed or printed, measuring more than 5000 decitex;

(b) Multiple (folded) or cabled yarn, unbleached:

(1) Of silk or waste silk, however put up; or

(2) Of other textile material except wool or fine animal hair, in hanks or skeins;

(c) Multiple (folded) or cabled yarn of silk or waste silk, bleached, dyed or printed, measuring 133 decitex or less; and

(d) Single, multiple (folded) or cabled yarn of any textile material:

(1) In cross-reeled hanks or skeins; or

(2) Put up on supports or in some other manner indicating its use in the textile industry (for example, on cops, twisting mill tubes, pirns, conical bobbins or spindles, or reeled in the form of cocoons for embroidery looms).

5. For the purposes of headings Nos.52.04, 54.01 and 55.08 the expression “sewing thread” means multiple (folded) or cabled yarn:

(a) Put up on supports (for example, reels, tubes) of a weight (including support) not exceeding 1000g;

(b) Dressed for use as sewing thread; and

(c) With a final “Z” twist.

6. For the purposes of this Section, the expression “high tenacity yarn” means yarn having a tenacity, expressed in cN/tex (centinewtons per tex), greater than the following:

Single yarn of nylon or other polyamides, or of polyesters 60 cN/tex

Multiple (folded) or cabled yarn of nylon or other polyamides, or of polyesters 53 cN/tex

Single, multiple (folded) or cabled yarn of viscose rayon 27 cN/tex

7. For the purposes of this Section, the expression “made up” means:

(a) Cut otherwise than into squares or rectangles;

(b) Produced in the finished state, ready for use (or merely

工（或仅需剪断分隔联线）即可使用的（例如，某些抹布、毛巾、台布、方披巾、毯子）；

needing separation by cutting dividing threads) without sewing or other working (for example, certain dusters, towels, table cloths, scarf squares, blankets);

（三）已缝边或滚边，或者在任一边带有结制的流苏，但不包括为防止剪边脱纱而锁边或用其他简单方法处理的织物；

（四）裁剪成一定尺寸并经抽纱加工的；

（五）缝合、胶合或用其他方法拼合而成的（将两段或两段以上同样料子的织物首尾连接而成的匹头，以及由两层或两层以上的织物，不论中间有无胎料，层迭而成的匹头除外）；

（六）针织或钩编成一定形状，不论进口或出口时是单件还是以若干件相连成幅的。

(c) Hemmed or with rolled edges, or with a knotted fringe at any of the edges, but excluding fabrics the cut deges of which have been prevented from unravelling by whipping or by other simple means;

(d) Cut to size and having undergone a process of drawn thread work;

(e) Assembled by sewing, gumming or otherwise (other than piece goods consisting of two or more lengths of identical material joined end to end and piece goods composed of two or more textiles assembled in layers, whether or not padded);

(f) Knitted or crocheted to shape, whether presented as separate items or in the form of a number of items in the length.

八、对于第五十章至第六十章：

（一）第五十章至第五十五章和第六十章，以及除条文另有规定以外的第五十六章至第五十九章，不适用于上述注释七所规定的制成货品；

（二）第五十章至第五十五章及第六十章不包括第五十六章至第五十九章的货品。

8. For the purposes of Chapters 50 to 60:

(a) Chapters 50 to 55 and 60 and, except where the context otherwise requires, Chapters 56 to 59 do not apply to goods made up within the meaning of Note 7 above; and;

(b) Chapters 50 to 55 and 60 do not apply to goods of Chapters 56 to 59.

九、第五十章至第五十五章的机织物包括由若干层平行纱线以锐角或直角相互层迭，在纱线交叉点用粘合剂或以热粘合法粘合而成的织物。

9. The woven fabrics of Chapters 50 to 55 include fabrics consisting of layers of parallel textile yarns superimposed on each other at acute or right angles. These layers are bonded at the intersections of the yarns by an adhesive or by thermal bonding;

十、用纺织材料和橡胶线制成的弹性产品归入本类。

10. Elastic products consisting of textile materials combined with rubber threads are classified in this Section.

十一、本类所称"浸渍"，包括"浸泡"。

11. For the purposes of this Section, the expression "impregnated" includes "dipped".

十二、本类所称"聚酰胺"，包括"芳族聚酰胺"。

12. For the purposes of this Section, the expression "polyamides" includes "aramides".

十三、除条文另有规定的以外，各种服装即使成套包装供零售用，也应按各自税号分别归类。本注释所称"纺织服装"，是指税号61.01至61.14及税号62.01至62.11所列的各种服装。

13. Unless the context otherwise requires, textile garments of different headings are to be classified in their own headings even if put up in sets for retail sale. For the purposes of this Note, the expression "textile garments" means garments of headings Nos.61.01 to 61.14 and heading Nos.62.01 to 62.11.

子目注释：

一、本类及本目录所用有关名词解释如下：

Subheading Notes:

1. In the Section and , where applicable , throughsout the

（一）弹性纱线

合成纤维纺织材料制成的长丝纱线，包括单丝（变形丝除外）。这些丝可拉伸至原长的三倍而不断裂，并可在拉伸至原长两倍后五分钟内回复到不超过原长度一倍半。

（二）未漂白纱线

1. 带有纤维自然色泽并且未经漂染（不论是否整体染色）或印色的纱线；

2. 从回收纤维制得，色泽未定的纱线（本色纱）。

这种纱线可用无色浆料或易褪色染料（可轻易地用肥皂洗去）处理，如果是化学纤维纱线，则整体用消光剂（例如二氧化钛）进行处理。

（三）漂白纱线

1. 经漂白加工、用漂白纤维制得或经染白（除条文另有规定的以外）（不论是否整体染色）及用白浆料处理的纱线；
2. 用未漂白纤维和漂白纤维混纺制得的纱线；
3. 用未漂白纱和漂白纱纺成多股纱线或缆线。

（四）着色（染色或印色）纱线

1. 染成彩色（不论是否整体染色，但白色或易褪色除外）或印色的纱线，以及用染色或印色纤维纺制的纱线；
2. 用各色染色纤维混合纺制或用未漂白或漂白纤维与着色纤维混合制得的纱线（夹色纱或混色纱），以及用一种或几种颜色间隔印色而获得点纹印迹的纱线；
3. 用已经印色的纱条或粗纱纺制的纱线；

4. 用未漂白纱和漂白纱与着色纱纺成的多股纱线或缆线。

上述定义作相应调整后适用于第五十四章的单丝、扁条或类似产品。

Nomenclature, the following expressions have the meanings hereby assigned to them:

(a) Elastomeric yarn

Filament yarn, including monofilament, of synthetic textile material, other than textured yarn, which does not break on being extended to three times its original length and which returns, after being extended to twice its original length, within a period of five minutes, to a length not greater than one and a half times its original length.

(b) Unbleached yarn

Yarn which:

(1) has the natural colour of its constituent fibres and has not been bleached, dyed (whether or not in the mass) or printed; or

(2) is of indeterminate colour ("grey yarn"), manufactured from garnetted stock.

Such yarn may have been treated with a colourless dressing or fugitive dye (which disappears after simple washing with soap) and, in the case of man-made fibres, treated in the mass with delustring agent (for example, titanium dioxide).

(c) Bleached yarn

Yarn which:

(1) has undergone a bleaching process, is made of bleached fibres or, unless the context otherwise requires, has been dyed white (whether or not in the mass) or treated with a white dressing;

(2) consists of a mixture of unbleached and bleached fibres; or

(3) is multiple (folded) or cabled and consists of unbleached and bleached yarns.

(d) Coloured (dyed or printed) yarn

Yarn which:

(1) is dyed (whether or not in the mass) other than white or in a fugitive colour, or printed, or made from dyed or printed fibres;

(2) consists of a mixture of dyed fibres of different colours or of a mixture of unbleached or bleached fibres with coloured fibres (marl or mixture yarns), or is printed in one or more colours at intervals to give the impression of dots;

(3) is obtained from slivers or rovings which have been printed; or

(4) is multiple (folded) or cabled and consists of unbleached or bleached yarn and coloured yarn.

The above definitions also apply, *mutatis mutandis*, to monofilament and to strip or the like of Chapter 54.

（五）未漂白机织物

用未漂白纱线织成后未经漂白、染色或印花的机织物。这类织物可用无色浆料或易褪色染料处理。

（六）漂白机织物

1. 经漂白、染白或用白浆料处理（除条文另有规定的以外）的成匹机织物；

2. 用漂白纱线织成的机织物；

3. 用未漂白纱线和漂白纱线织成的机织物；

（七）染色机织物

1. 除条文另有规定的以外，染成白色以外的其他单一颜色或用白色以外的其他有色整理剂处理的成匹机织物；

2. 用单一颜色的着色纱线织成的机织物。

（八）色织机织物

除印花机织物以外的下列机织物：

1. 用各种不同颜色纱线或同一颜色不同深浅（纤维的自然色彩除外）纱线织成的机织物；

2. 用未漂白或漂白与着色纱线织成的机织物；

3. 用夹色纱线或混色纱线织成的机织物。

不论何种情况，布边或布头的纱线均可忽略不计。

（九）印花机织物

成匹印花的机织物，不论是否用各色纱线织成。

用刷子或喷枪，经转印纸转印、植绒或蜡防印花等方法印成花纹图案的机织物亦可视为印花机织物。

上述各类纱线或织物如经丝光工艺处理并不影响其归类。

上述（五）至（九）的定义在作必要修改后适用于针织或钩编织物。

（十）平纹组织

每根纬纱在并排的经纱间上下交错而过，而每根经纱也在并排的纬纱间上下交错而过的织物组织。

二、（一）含有两种或两种以上纺织材料的第五十六章至第六十三章的产品，应根据本类注释二对第五十章至第五十五章或税目58.09的此类纺织材料产品归类的规定

(e) Unbleached woven fabric

Woven fabric made from unbleached yarn and which has not been bleached, dyed or printed. Such fabric may have treated with a colourless dressing or a fugitive dye.

(f) Bleached woven fabric

Woven fabric which:

(1) has been bleached or, unless the context otherwise requires, dyed white or treated with a white dressing, in the piece;

(2) consists of bleached yarn; or

(3) consists of unbleached and bleached yarn.

(g) Dyed woven fabric

Woven fabric which:

(1) is dyed a single uniform colour other than white (unless the context otherwise requires) or has been treated with a coloured finish other than withe (unless the context otherwise requires), in the piece; or

(2) consists of coloured yarn of a single uniform colour.

(h) Woven fabric of yarns of different colours

Woven fabric (other than printed woven fabric) which:

(1) consists of yarns of different colours or yarns of different shades of the same colour (other than the natural colour of the constituent fibres);

(2) consists of unbleached or bleached yarn and coloured yarn; or

(3) consists of marl or mixture yarns.

(In all cases, the yarn used in selvedges and piece ends is not taken into consideration.)

(i) Printed woven fabric

Woven fabric which has been printed in the piece, whether or not made from yarns of differenent colours. (The following are also regarded as printed woven fabrics: woven fabrics bearing designs made, for example, with a brush or spray gun, by means of transfer paper, by flocking or by the batik process.) The process of mercerization does not affect the classification of yarns or fabrics within the above categories.

The definitions at (e) to (ij) above apply, mutatis mutandis, or to knitted crocheted fabrics.

(j) Plain weave

A fabric construction in which each yarn of the weft passes alternately over and under successive yarns of the warp and each yarn of the warp passes alternately over and under successive yarns of the weft.

2. (a) Products of Chapters 56 to 63 containing two or more textile materials are to be regarded as consisting wholly of that textile material which would be selected under Note 2 to this Section for the classification of the product

来确定归类。

of Chapters 50 to 55 or of heading 58. 09 consisting of the same textile materials.

（二）运用本条规定时：

1. 应酌情考虑按归类总规则第三条来确定归类；

2. 对由底布和绒面或毛圈面构成的纺织品，在归类时可不考虑底布的属性；

3. 对税号58.10的刺绣品及其制品，归类时应只考虑底布的属性，但不见底布的刺绣品及其制品应根据绣线的属性确定归类。

(b) For the application of this rule:

(1) where appropriate, only the part which determines the classification under interpretative Rule 3 shall be taken into account;

(2) in the case of textile products consisting of a ground fabric and a pile or looped surface, no account shall be taken of the ground fabric;

(3) in the case of embroidery of heading No.58.10 and goods thereof, only the ground fabric shall be taken into account. However, embroidery without visible ground, and goods thereof, shall be classified with reference to the embroidering threads alone.

第五十章 蚕丝

Chapter 50 Silk

序号 No.	税则号列 Tariff Item	货品名称	最惠国税率(%) M.F.N.	协定税率（%）Agreement 东盟 ASEAN	曼谷协定 BA	香港 H.K.	澳门 Macao	普通税率(%) Gen.	Article Description
	50.01	**适于缫丝的蚕茧：**							**Silk-worm cocoons suitable for reeling:**
3223	5001.0010	---适于缫丝的桑蚕茧	6					70	---Bombyx mori cocoons (Mulberry feeding silk-worm cocoons)
3224	5001.0090	---其他	6					70	---Other
	50.02	**生丝(未加捻)：**							**Raw silk(not thrown):**
		---桑蚕丝：							---Steam filature silk:
3225	5002.0011	----厂丝	9					80	----Plant reeled(filature silk)
3226	5002.0012	----土丝	9					80	----Home reeled
3227	5002.0013	----双宫丝	9					80	----Doupion
3228	5002.0019	----其他	9					80	----Other
3229	5002.0020	---柞蚕丝	9					80	---Tussah silk
3230	5002.0090	---其他	9					80	---Other
	50.03	**废丝(包括不适于缫丝的蚕茧、废纱及回收纤维)：**							**Silk waste(including cocoons unsuitable for reeling, yarn waste and garnetted stock):**
3231	5003.1000	-未梳	9					70	-Not carded or combed
3232	5003.9000	-其他	9					70	-Other
	50.04	**丝纱线(绢纺纱线除外)，非供零售用：**							**Silk yarn(other than yarn spun from silk waste)not put up for retail sale:**
3233	5004.0000	丝纱线(绢纺纱线除外)，非供零售用	6				0	90	Silk yarn(other than yarn spun from silk waste)not put up for retail sale
	50.05	**绢纺纱线，非供零售用：**							**Yarn spun from silk waste, not put up for retail sale:**
3234	5005.0010	---细丝纱线	6					90	---Spun from noil

序号 No.	税则号列 Tariff Item	货品名称	最惠国税率(%) M.F.N.	协定税率(%) Agreement 东盟 ASEAN	曼谷协定 BA	香港 H.K.	澳门 Macao	普通税率(%) Gen.	Article Description
3235	5005.0090	---其他	6					90	---Other
	50.06	**丝纱线及绢纺纱线,供零售用;蚕胶丝:**							**Silk yarn and yarn spun from silk waste, put up for retail sale; silk-worm gut:**
3236	5006.0000	丝纱线及绢纺纱线,供零售用;蚕胶丝	6					100	Silk yarn and yarn spun from silk waste, put up for retail sale; silk-worm gut
	50.07	**丝或绢丝机织物:**							**Woven fabrics of silk or of silk waste:**
		-细丝机织物:							-Fabrics of noil silk:
3237	5007.1010	---未漂白(包括未练白或练白)或漂白	10					130	---Unbleaded(unscoured or scoured) or bleached
3238	5007.1090	---其他	10					130	---Other
		-其他机织物,按重量计丝或绢丝(细丝除外)含量在85%及以上:							-Other fabrics, containing 85% or more by weight of silk or of silk waste other than noil silk:
		---桑蚕丝机织物:							---Of Bombyx mori silk:
3239	5007.2011	----未漂白(包括未练白或练白)或漂白	10		9			130	----Unbleached(unscoured or scoured) or bleached
3240	5007.2019	----其他	10		9	0		130	----Other
		---柞蚕丝机织物:							---Of tussah silk:
3241	5007.2021	----未漂白(包括未练白或练白)或漂白	10		9			130	----Unbleaded(unscoured or scoured) or bleached
3242	5007.2029	----其他	10		9			130	----Other
		---绢丝机织物:							---Of silk waste other than noil silk:
3243	5007.2031	----未漂白(包括未练白或练白)或漂白	10		9			130	----Unbleached(unscoured or scoured) or bleached
3244	5007.2039	----其他	10		9			130	----Other
3245	5007.2090	---其他	10		9			130	---Of other silk
		-其他机织物:							-Other fabrics:
3246	5007.9010	---未漂白(包括未练白或练白)或漂白	10					130	---Unbleached(unscoured or scoured) or bleached
3247	5007.9090	---其他	10					130	---Other

第五十一章
羊毛、动物细毛或粗毛；马毛纱线及其机织物

Chapter 51
Wool, fine or coarse animal hair; horsehair yarn and woven fabric

注释：

本目录所称：

一、"羊毛"，是指绵羊或羔羊身上长的天然纤维；

二、"动物细毛"，是指下列动物的毛：羊驼、美洲驼、驼马、骆驼、牦牛、安哥拉山羊、西藏山羊、喀什米尔山羊及类似山羊（普通山羊除外）、家兔（包括安哥拉兔）、野兔、海狸、河狸鼠或麝鼠；

三、"动物粗毛"，是指以上未提及的其他动物的毛，但不包括制刷用鬃、毛（税号05.02）以及马毛（税号05.03）。

Note:

Throughout the Nomenclature:

(a) "Wool" means the natural fibre grown by sheep or lambs;

(b) "Fine animal hair" means the hair of alpaca, llama, vicuna, camel, yak, Angora, Tibetan, Kashmir or similar goats (but not common goats), rabbit (including Angora rabbit), hare, beaver, nutria or muskrat;

(c) "Coarse animal hair" means the hair of animals not mentioned above, excluding brush-making hair and bristles (heading No.05.02) and horsehair (heading No.05.03).

序号 No.	税则号列 Tariff Item	货品名称	最惠国税率(%) M.F.N.	协定税率(%) Agreement 东盟 ASEAN	曼谷协定 BA	香港 H.K.	澳门 Macao	普通税率(%) Gen.	Article Description
	51.01	**未梳的羊毛：**							**Wool, not carded or combed:**
		-含脂羊毛，包括剪前水洗毛：							-Greasy, including fleece-washed wool:
3248	5101.1100	--剪羊毛	见表5					见表5	--Shorn wool
3249	5101.1900	--其他	见表5 Table5					见表5	1--Other
		-脱脂羊毛，未碳化：							-Degreased, not carbonized:
3250	5101.2100	--剪羊毛	见表5					见表5	--Shorn wool
3251	5101.2900	--其他	见表5					见表5	--Other
3252	5101.3000	-碳化羊毛	见表5				0△	见表5	-Carbonized
	51.02	**未梳的动物细毛或粗毛：**							**Fine or coarse animal hair, not carded or combed:**
		-细毛：							-Fine animal hair:
3253	5102.1100	--喀什米尔山羊的	9					45	--Of kashmir(cashmere)goats
		--其他							--Other
3254	5102.1910	---兔毛	9					50	---Of rabbit and hare
3255	5102.1920	---其他山羊绒	9					45	---Of other goats
3256	5102.1930	---骆驼毛、骆驼绒	9					45	---Of camel
3257	5102.1990	---其他	9					45	---Other
3258	5102.2000	-粗毛	9					50	-Coarse animal hair
	51.03	**羊毛或动物细毛或粗毛的废料，包括废纱线，但不包括回收纤维：**							**Waste of wool or of fine or coarse animal hair, including yarn waste but excluding garnetted stock:**
		-羊毛或动物细毛的落毛：							-Noils of wool or of fine animal hair:
3259	5103.1010	---羊毛落毛	见表5					见表5	---Of wool

序号 No.	税则号列 Tariff Item	货品名称	最惠国税率(%) M.F.N.	协定税率(%) Agreement 东盟 ASEAN	曼谷协定 BA	香港 H.K.	澳门 Macao	普通税率(%) Gen.	Article Description
3260	5103.1090	---其他	9					50	---Other
		-羊毛或动物细毛的其他废料:							-Other waste of wool or of fine animal hair:
3261	5103.2010	---羊毛废料	13.5					20	---Ofwool
3262	5103.2090	---其他	9					50	---Other
3263	5103.3000	-动物粗毛废料	9					50	-Waste of coarse animal hair
	51.04	**羊毛或动物细毛或粗毛的回收纤维:**							**Garnetted stock of wool or of fine or coarse animal hair:**
3264	5104.0010	---羊毛的回收纤维	15					20	---Of wool
3265	5104.0090	---其他	5					50	---Other
	51.05	**已梳的羊毛及动物细毛或粗毛(包括精梳片毛):**							**Wool and fine or coarse animal hair, carded or combed (including combed wool in fragments):**
3266	5105.1000	-粗梳羊毛	见表 5					见表 5	-Carded wool
		-羊毛条及其他精梳羊毛:							-Wool tops and other combed wool:
3267	5105.2100	--精梳片毛	见表 5					见表 5 Table5	--Combed wool in fragments
3268	5105.2900	--其他	见表 5					见表 5	--Other
		-已梳动物细毛:							-Fine animal hair, carded or combed:
3269	5105.3100	--喀什米尔山羊的	5					50	--Of kashmir (cashmere) goats
		--其他:							--Other:
3270	5105.3910	---兔毛	5					70	---Of rabbit or hare
		---其他山羊绒:							---Of other goats:
3271	5105.3921	----无毛山羊绒	5					50	----Dehaired goats wool
3272	5105.3929	----其他	5					50	----Other
3273	5105.3990	---其他	5					50	---Other
3274	5105.4000	-已梳动物粗毛	5					50	-Coarse animal hair, carded or combed
	51.06	**粗梳羊毛纱线,非供零售用:**							**Yarn of carded wool, not put up for retailsale:**
3275	5106.1000	-按重量计羊毛含量在85%及以上	5			0△	0	70	-Containing 85% or more by weight of wool
3276	5106.2000	-按重量计羊毛含量在85%以下	5			0	0	70	-Containing less than 85% by weight of wool
	51.07	**精梳羊毛纱线,非供零售用:**							**Yarn of combed wool, not put up for retail sale:**
3277	5107.1000	-按重量计羊毛含量在85%及以上	5			0		70	-Containing 85% or more by weight of wool
3278	5107.2000	-按重量计羊毛含量在85%以下	5					70	-Containing less than 85% by weight of wool
	51.08	**动物细毛(粗梳或精梳)纱线,非供零售用:**							**Yarn of fine animal hair (carded or combed), not put up for retail sale:**
3279	5108.1000	-粗梳	5					70	-Carded
3280	5108.2000	-精梳	5					70	-Combed
	51.09	**羊毛或动物细毛的纱线,供零售用:**							**Yarn of wool or of fine animal hair, put up for retail sale:**

序号 No.	税则号列 Tariff Item	货品名称	最惠国税率(%) M.F.N.	协定税率(%) Agreement 东盟 ASEAN	曼谷协定 BA	香港 H.K.	澳门 Macao	普通税率(%) Gen.	Article Description
3281	5109.1000	-按重量计羊毛或动物细毛含量在85%及以上	6					80	-Containing 85% or more by weight of wool or of fine animal hair
3282	5109.9000	-其他	6					80	-Other
	51.10	**动物粗毛或马毛的纱线(包括马毛粗松螺旋花线),不论是否供零售用:**							**Yarn of coarse animal hair or of horsehair(including gimped horsehair yarn), whether or not put up for retail sale:**
3283	5110.0000	动物粗毛或马毛的纱线(包括马毛粗松螺旋花线),不论是否供零售用	6					70	Yarn of coarse animal hair or of horsehair (including gimped horsehair yarn), whether or not put up for retail sale
	51.11	**粗梳羊毛或粗梳动物细毛的机织物:**							**Woven fabrics of carded wool or of carded fine animal hair:**
		-按重量计羊毛或动物细毛含量在85%及以上:							-Containing 85% or more by weight of wool or of fine animal hair:
3284	5111.1100	--每平方米重量不超过300克	10					130	--Of a weight not exceeding 300 g/m²
3285	5111.1900	--其他	10					130	--Other
3286	5111.2000	-其他,主要或仅与化学纤维长丝混纺	10					130	-Other, mixed mainly or solely with manmade filaments
3287	5111.3000	-其他,主要或仅与化学纤维短纤混纺	10					130	-Other, mixed mainly or solely with manmade staple fibres
3288	5111.9000	-其他	10					130	-Other
	51.12	**精梳羊毛或精梳动物细毛的机织物:**							**Woven fabrics of combed wool or of combed fine animal hair:**
		-按重量计羊毛或动物细毛含量在85%及以上:							-Containing 85% or more by weight of wool or of fine animal hair:
3289	5112.1100	--每平方米重量不超过200克	10		8.9	0△		130	--Of a weight not exceeding 200 g/m²
3290	5112.1900	--其他	10			0		130	--Other
3291	5112.2000	-其他,主要或仅与化学纤维长丝混纺	10					130	-Other, mixed mainly or solely with manmade filaments
3292	5112.3000	-其他,主要或仅与化学纤维短纤混纺	10					130	-Other, mixed mainly or solely with manmade staple fibres
3293	5112.9000	-其他	10					130	-Other
	51.13	**动物粗毛或马毛的机织物:**							**Woven fabrics of coarse animal hair or of horsehair:**
3294	5113.0000	动物粗毛或马毛的机织物	10					130	Woven fabrics of coarse animal hair or of horsehair

第五十二章 棉 花

Chapter 52 Cotton

子目注释：

子目号5209.42及5211.42所称"粗斜纹布（劳动布）"，是指用不同颜色的纱线织成的三线或四线斜纹织物，包括破斜纹组织的织物，这种织物以经纱为面，经纱染成一种相同的颜色，纬纱未漂白或经漂白、染成灰色或比经纱稍浅的颜色。

Subheading Note:

For the purposes of subheadings Nos.5209.42 and 5211.42, the expression "denim" means fabrics of yarns of different colours, of 3-thread or 4-thread twill, including broken twill, warp faced, the warp yarns of which are of one and the same colour and the weft yarns of which are unbleached, bleached, dyed grey or coloured a lighter shade of the colour of the warp yarns.

序号 No.	税则号列 Tariff Item	货品名称	最惠国税率(%) M.F.N.	协定税率(%) Agreement 东盟 ASEAN	曼谷协定 BA	香港 H.K.	澳门 Macao	普通税率(%) Gen.	Article Description
	52.01	**未梳的棉花：**							**Cotton, not carded or combed:**
3295	5201.0000	未梳的棉花	见表5 Table5					见表5	Cotton, not carded or combed
	52.02	**废棉(包括废棉纱线及回收纤维)：**							**Cotton waste (including yarn waste and garnetted stock):**
3296	5202.1000	-废棉纱线(包括废棉线)	10					30	-Yarn waste(including thread waste)
		-其他：							-Other:
3297	5202.9100	--回收纤维	10					30	--Garnetted stock
3298	5202.9900	--其他	10					30	--Other
	52.03	**已梳的棉花：**							**Cotton, carded or combed:**
3299	5203.0000	已梳的棉花	见表5					见表5	Cotton, carded or combed
	52.04	**棉制缝纫线，不论是否供零售用：**							**Cotton sewing thread, whether or not put up for retail sale:**
		-非供零售用：							-Not put up for retail sale:
3300	5204.1100	--按重量计含棉量在85%及以上	5					40	--Containing 85% or more by weight of cotton
3301	5204.1900	--其他	5					40	--Other
3302	5204.2000	-供零售用	5					50	-Put up for retail sale
	52.05	**棉纱线(缝纫线除外)，按重量计含棉量在85%及以上，非供零售用：**							**Cotton yarn (other than sewing thread), containing 85% or more by weight of cotton, not put up for retail sale:**
		-未精梳纤维纺制的单纱：							-Single yarn, of uncombed fibres:
3303	5205.1100	--细度在714.29分特及以上(不超过14公支)	5			0	0	40	--Measuring 714.29 decitex or more (not exceeding 14 metric number)
3304	5205.1200	--细度在714.29分特以下，但不细于232.56分特(超过14公支，但不超过43公支)	5			0		40	--Measuring less than 714.29 decitex but not less than 232.56 decitex(exceeding 14 metric number but not exceeding 43 metric number)

序号 No.	税则号列 Tariff Item	货品名称	最惠国税率(%) M.F.N.	协定税率(%) Agreement 东盟 ASEAN	曼谷协定 BA	香港 H.K.	澳门 Macao	普通税率(%) Gen.	Article Description
3305	5205.1300	--细度在232.56分特以下,但不细于192.31分特(超过43公支,但不超过52公支)	5			0		40	--Measuring less than 232.56 decitex but not less than 192.31 decitex(exceeding 43 metric number but not exceeding 52 metric number)
3306	5205.1400	--细度在192.31分特以下,但不细于125分特(超过52公支,但不超过80公支)	5			0		40	--Measuring less than 192.31 decitex but not less than 125 decitex (exceeding 52 metric number but not exceeding 80 metric number)
3307	5205.1500	--细度在125分特以下(超过80公支)	5					40	--Measuring less than 125 decitex(exceeding 80 metric number)
		-精梳纤维纺制的单纱:							-Single yarn,of combed fibres:
3308	5205.2100	--细度在714.29分特及以上(不超过14公支)	5			0		40	--Measuring 714.29 decitex or more (not exceeding 14 metric number)
3309	5205.2200	--细度在714.29分特以下,但不细于232.56分特(超过14公支,但不超过43公支)	5			0		40	--Measuring less than 714.29 decitex but not less than 232.56 decitex(exceeding 14 metric number but not exceeding 43 metric number)
3310	5205.2300	--细度在232.56分特以下,但不细于192.31分特(超过43公支,但不超过52公支)	5			0		40	--Measuring less than 232.56 decitex but not less than 192.31 decitex(exceeding 43 metric number but not exceeding 52 metric number)
3311	5205.2400	--细度在192.31分特以下,但不细于125分特(超过52公支,但不超过80公支)	5			0		40	--Measuring less than 192.31 decitex but not less than 125 decitex (exceeding 52 metric number but not exceeding 80 metric number)
3312	5205.2600	--细度在125分特以下,但不细于106.38分特(超过80公支,但不超过94公支)	5			0		40	--Measuring less than 125 decitex but not less than 106.38 decitex(exceeding 80 metric number but not exceeding 94 metric number)
3313	5205.2700	--细度在106.38分特以下,但不细于83.33分特(超过94公支,但不超过120公支)	5			0		40	--Measuring less than 106.38 decitex but not less than 83.33 decitex(exceeding 94 metric number but not exceeding 120 metric number)
3314	5205.2800	--细度在83.33分特以下(超过120公支)	5			0△		40	--Measuring less than 83.33 decitex (exceeding 120 metric number)
		-未精梳纤维纺制的多股纱线或缆线:							-Multiple (folded) or cabled yarn, of uncombed fibres:
3315	5205.3100	--每根单纱细度在714.29分特及以上(每根单纱不超过14公支)	5		4.5	0		40	--Measuring per single yarn 714.29 decitex or more (not exceeding 14 metric number per single yarn)

序号 No.	税则号列 Tariff Item	货品名称	最惠国税率(%) M.F.N.	协定税率(%) Agreement 东盟 ASEAN	曼谷协定 BA	香港 H.K.	澳门 Macao	普通税率(%) Gen.	Article Description
3316	5205.3200	--每根单纱细度在714.29分特以下,但不细于232.56分特(每根单纱超过14公支,但不超过43公支)	5			0		40	--Measuring per single yarn less than 714.29 decitex but not less than 232.56 decitex(exceeding 14 metric number but not exceeding 43 metric number per single yarn)
3317	5205.3300	--每根单纱细度在232.56分特以下,但不细于192.31分特(每根单纱超过43公支,但不超过52公支)	5			0△		40	--Measuring per single yarn less than 232.56 decitex but not less than 192.31 decitex(exceeding 43 metric number but not exceeding 52 metric number per single yarn)
3318	5205.3400	--每根单纱细度在192.31分特以下,但不细于125分特(每根单纱超过52公支,但不超过80公支)	5			0		40	--Measuring per single yarn less than 192.31 decitex but not less than 125 decitex(exceeding 52 metric number but not exceeding 80 metric number per single yarn)
3319	5205.3500	--每根单纱细度在125分特以下(每根单纱超过80公支)	5					40	--Measuring per single yarn less than 125 decitex (exceeding 80 metric number per single yarn)
		-精梳纤维纺制的多股纱线或缆线:							-Multiple (folded) or cabled yarn, of combed fibres:
3320	5205.4100	--每根单纱细度在714.29分特及以上(每根单纱不超过14公支)	5		4.5	0△		40	--Measuring per single yarn 714. 29 decitex or more (not exceeding 14 metric number per single yarn)
3321	5205.4200	--每根单纱细度在714.29分特以下,但不细于232.56分特(每根单纱超过14公支,但不超过43公支)	5		4.5	0		40	--Measuring per single yarn less than 714.29 decitex but not less than 232.56 decitex(exceeding 14 metric number but not exceeding 43 metric number per single yarn)
3322	5205.4300	--每根单纱细度在232.56分特以下,但不细于192.31分特(每根单纱超过43公支,但不超过52公支)	5			0		40	--Measuring per singleyarn less than 232.56 decitex but not less than 192.31 decitex(exceeding 43 metric number but not exceeding 52 metric number per single yarn)
3323	5205.4400	--每根单纱细度在192.31分特以下,但不细于125分特(每根单纱超过52公支,但不超过80公支)	5			0		40	--Measuring per single yarn less than 192.31 decitex but not less than 125 decitex(exceeding 52 metric number but not exceeding 80 metric number per single yarn)
3324	5205.4600	--每根单纱细度在125分特以下,但不细于106.38分特(每根单纱超过80公支,但不超过94公支)	5		4.5	0△		40	--Measuring per single yarn less than 125 decitex but not less than 106.38 decitex(exceeding 80 metric number but not exceeding 94 metric number per single yarn)

序号 No.	税则号列 Tariff Item	货品名称	最惠国税率(%) M.F.N.	协定税率(%) Agreement 东盟 ASEAN	曼谷协定 BA	香港 H.K.	澳门 Macao	普通税率(%) Gen.	Article Description
3325	5205.4700	--每根单纱细度在106.38分特以下,但不细于83.33分特(每根单纱超过94公支,但不超过120公支)	5		4.5	0△		40	--Measuring per singleyarn less than 106.38 decitex but not less than 83.33 decitex (exceeding 94 metric number but not exceeding 120 metric number per single yarn)
3326	5205.4800	--每根单纱细度在83.33分特以下(每根单纱超过120公支)	5		4.5	0△		40	--Measuring per single yarn less than 83.33 decitex(exceeding 120 metric number per single yarn)
	52.06	**棉纱线(缝纫线除外),按重量计含棉量在85%以下,非供零售用:**							**Cotton yarn (other than sewing thread),containing less than 85% by weight of cotton,not put up for retail sale:**
		-未精梳纤维纺制的单纱:							-Single yarn,of uncombed fibres:
3327	5206.1100	--细度在714.29分特及以上(不超过14公支)	5			0	0	40	--Measuring 714.29 decitex or more (not exceeding 14 metric number)
3328	5206.1200	--细度在714.29分特以下,但不细于232.56分特(超过14公支,但不超过43公支)	5			0		40	--Measuring less than 714.29 decitex but not less than 232.56 decitex(exceeding 14 metric number but not exceeding 43 metric number)
3329	5206.1300	--细度在232.56分特以下,但不细于192.31分特(超过43公支,但不超过52公支)	5			0△		40	--Measuring less than 232.56 decitex but not less than 192.31 decitex(exceeding 43 metric number but not exceeding 52 metric number)
3330	5206.1400	--细度在192.31分特以下,但不细于125分特(超过52公支,但不超过80公支)	5			0		40	--Measuring less than 192.31 decitex but not less than 125 decitex (exceeding 52 metric number but not exceeding 80 metric number)
3331	5206.1500	--细度在125分特以下(超过80公支)	5			0△		40	--Measuring less than 125 decitex(exceeding 80 metric number)
		-精梳纤维纺制的单纱:							-Single yarn,of combed fibres:
3332	5206.2100	--细度在714.29分特及以上(不超过14公支)	5		4.5	0		40	--Measuring 714.29 decitex or more (not exceeding 14 metric number)
3333	5206.2200	--细度在714.29分特以下,但不细于232.56分特(超过14公支,但不超过43公支)	5			0		40	--Measuring less than 714.29 decitex but not less than 232.56 decitex(exceeding 14 metric number but not exceeding 43 metric number)
3334	5206.2300	--细度在232.56分特以下,但不细于192.31分特(超过43公支,但不超过52公支)	5			0		40	--Measuring less than 232.56 decitex but not less than 192.31 decitex(exceeding 43 metric number but not exceeding 52 metric number)

序号 No.	税则号列 Tariff Item	货品名称	最惠国税率(%) M.F.N.	协定税率(%) Agreement 东盟 ASEAN	曼谷协定 BA	香港 H.K.	澳门 Macao	普通税率(%) Gen.	Article Description
3335	5206.2400	--细度在192.31分特以下,但不细于125分特(超过52公支,但不超过80公支)	5			0		40	-Measuring less than 192.31 decitex but not less than 125 decitex(exceeding 52 metric number but not exceeding 80 metric number)
3336	5206.2500	--细度在125分特以下(超过80公支)	5			0		40	--Measuring less than 125 decitex(exceeding 80 metric number)
		-未精梳纤维纺制的多股纱线或缆线:							-Multiple (folded) or cabled yarn, of uncombed fibres:
3337	5206.3100	--每根单纱细度在714.29分特及以上(每根单纱不超过14公支)	5			0		40	--Measuring per single yarn 714.29 decitex or more (not exceeding 14 metric number per single yarn)
3338	5206.3200	--每根单纱细度在714.29分特以下,但不细于232.56分特(每根单纱超过14公支,但不超过43公支)	5			0		40	--Measuring per single yarn less than 714.29 decitex but not less than 232.56 decitex(exceeding 14 metric number but not exceeding 43 metric number per single yarn)
3339	5206.3300	--每根单纱细度在232.56分特以下,但不细于192.31分特(每根单纱超过43公支,但不超过52公支)	5			0△		40	--Measuring per single yarn less than 232.56 decitex but not less than 192.31 decitex(exceeding 43 metric number but not exceeding 52 metric number per single yarn)
3340	5206.3400	--每根单纱细度在192.31分特以下,但不细于125分特(每根单纱超过52公支,但不超过80公支)	5			0		40	--Measuring per single yarn less than 192.31 decitex but not less than 125 decitex(exceeding 52 metric number but not exceeding 80 metric number per single yarn)
3341	5206.3500	--每根单纱细度在125分特以下(每根单纱超过80公支)	5			0△		40	--Measuring per single yarn less than 125 decitex (exceeding 80 metric number per single yarn)
		-精梳纤维纺制的多股纱线或缆线:							-Multiple (folded) or cabled yarn, of combed fibres:
3342	5206.4100	--每根单纱细度在714.29分特及以上(每根单纱不超过14公支)	5			0△		40	--Measuring per single yarn 714.29 decitex or more (not exceeding 14 metric number per single yarn)
3343	5206.4200	--每根单纱细度在714.29分特以下,但不细于232.56分特(每根单纱超过14公支,但不超过43公支)	5			0		40	--Measuring per single yarn less than 714.29 decitex but not less than 232.56 decitex(exceeding 14 metric number but not exceeding 43 metric number per single yarn)

序号 No.	税则号列 Tariff Item	货品名称	最惠国税率(%) M.F.N.	协定税率(%)Agreement 东盟 ASEAN	曼谷协定 BA	香港 H.K.	澳门 Macao	普通税率(%) Gen.	Article Description
3344	5206.4300	--每根单纱细度在232.56分特以下,但不细于192.31分特(每根单纱超过43公支,但不超过52公支)	5			0		40	--Measuring per single yarn less than 232.56 decitex but not less then 192.31 decitex(exceeding 43 metric number but not exceeding 52 metric number per single yarn)
3345	5206.4400	--每根单纱细度在192.31分特以下,但不细于125分特(每根单纱超过52公支,但不超过80公支)	5			0		40	--Measuring per single yarn less than 192.31 decitex but not less than 125 decitex(exceeding 52 metric number but not exceeding 80 metric number per single yarn)
3346	5206.4500	--每根单纱细度在125分特以下(每根单纱超过80公支)	5			0△		40	--Measuring per single yarn less than 125 decitex (exceeding 80 metric number per single yarn)
	52.07	**棉纱线(缝纫线除外),供零售用:**							**Cotton yarn (other than sewing thread) put up for retail sale:**
3347	5207.1000	-按重量计含棉量在85%及以上	6		5			50	-Containing 85% or more by weight of cotton
3348	5207.9000	-其他	6					50	-Other
	52.08	**棉机织物,按重量计含棉量在85%及以上,每平方米重量不超过200克:**							**Woven fabrics of cotton, containing 85% or more by weight of cotton, weighing not more than 200 g/m^2:**
		-未漂白:							-Unbleached:
3349	5208.1100	--平纹机织物,每平方米重量不超过100克	10			0△		70	--Plain weave, weighing not more than 100 g/m^2
3350	5208.1200	--平纹机织物,每平方米重量超过100克	10			0△		70	--Plain weave, weighing more than 10 g/m^2
3351	5208.1300	--三线或四线斜纹机织物,包括双面斜纹机织物	10			0△		70	--3-thread or 4-thread twill, including cross twill
3352	5208.1900	--其他机织物	10					70	--Other fabrics
		-漂白:							-Bleached:
3353	5208.2100	--平纹机织物,每平方米重量不超过100克	10					70	--Plain weave, weighing not more than 100 g/m^2
3354	5208.2200	--平纹机织物,每平方米重量超过100克	10					70	--Plain weave, weighing more than 100 g/m^2
3355	5208.2300	--三线或四线斜纹机织物,包括双面斜纹机织物	12					70	--3-thread or 4-thread twill, icluding cross twill
3356	5208.2900	--其他机织物	10					70	--Other fabrics
		-染色:							-Dyed:
3357	5208.3100	--平纹机织物,每平方米重量不超过100克	10					70	--Plain weave, weighing not more than 100 g/m^2

序号 No.	税则号列 Tariff Item	货品名称	最惠国税率(%) M.F.N.	协定税率(%)Agreement 东盟 ASEAN	曼谷协定 BA	香港 H.K.	澳门 Macao	普通税率(%) Gen.	Article Description
3358	5208.3200	--平纹机织物,每平方米重量超过100克	10			0		70	--Plain weave, weighing more than 100 g/m^2
3359	5208.3300	--三线或四线斜纹机织物,包括双面斜纹机织物	10			0		70	--3-thread or 4-thread twill, including cross twill
3360	5208.3900	--其他机织物	10		9	0		70	--Other fabrics
		-色织:							-Of yarns of different colours:
3361	5208.4100	--平纹机织物,每平方米重量不超过100克	10					70	--Plain weave, weighing not more than 100 g/m^2
3362	5208.4200	--平纹机织物,每平方米重量超过100克	10			0		70	--Plain weave, weighing more than 100 g/m^2
3363	5208.4300	--三线或四线斜纹机织物,包括双面斜纹机织物	10					70	--3-thread or 4-thread twill, including cross twill
3364	5208.4900	--其他机织物	10		9	0		70	--Other fabrics
		-印花:							-Printed.
3365	5208.5100	--平纹机织物,每平方米重量不超过100克	10					70	--Plain weave, weighing not more than 100 g/m^2
3366	5208.5200	--平纹机织物,每平方米重量超过100克	10					70	--Plain weave, weighing more than 100 g/m^2
3367	5208.5300	--三线或四线斜纹机织物,包括双面斜纹机织物	10					70	--3-thread or 4-thread twill, including cross twill
3368	5208.5900	--其他机织物	10					70	--Other fabrics
	52.09	**棉机织物,按重量计含棉量在85%及以上,每平方米重量超过200克:**							**Woven fabrics of cotton, containing 85% or more by weight of cotton, weighing more than 200 g/m^2:**
		-未漂白							-Unbleached:
3369	5209.1100	--平纹机织物	10			0		70	--Plain weave
3370	5209.1200	--三线或四线斜纹机织物,包括双面斜纹机织物	10			0		70	--3-thread or 4-thread twill, including cross twill
3371	5209.1900	--其他机织物	10			0		70	--Other fabrics
		-漂白:							-Bleached:
3372	5209.2100	--平纹机织物	12					70	--Plain weave
3373	5209.2200	--三线或四线斜纹机织物,包括双面斜纹机织物	12					70	--3-thread or 4-thread twill, including cross twill
3374	5209.2900	--其他机织物	12					70	--Other fabrics
		-染色:							-Dyed:
3375	5209.3100	--平纹机织物	10		9.3	0		70	--Plain weave
3376	5209.3200	--三线或四线斜纹机织物,包括双面斜纹机织物	10			0		70	--3-thread or 4-thread twill, including cross twill
3377	5209.3900	--其他机织物	10			0		70	--Other fabrics

序号 No.	税则号列 Tariff Item	货品名称	最惠国税率(%) M.F.N.	协定税率(%) Agreement 东盟 ASEAN	曼谷协定 BA	香港 H.K.	澳门 Macao	普通税率(%) Gen.	Article Description
		-色织：							-Of yarns of different colours:
3378	5209.4100	--平纹机织物	10			0		70	--Plain weave
3379	5209.4200	--粗斜纹布(劳动布)	10			0	0	70	--Denim
3380	5209.4300	--三线或四线斜纹机织物，包括双面斜纹机织物	10		9.3	0		70	--Other fabrics of 3-thread or 4-thread twill, including cross twill
3381	5209.4900	--其他机织物	10					70	--Other fabrics
		-印花：							-Printed:
3382	5209.5100	--平纹机织物	10		9.3			70	--Plain weave
3383	5209.5200	--三线或四线斜纹机织物，包括双面斜纹机织物	10					70	--3-thread or 4-thread twill, including cross twill
3384	5209.5900	--其他机织物	10					70	--Other fabrics
	52.10	**棉机织物，按重量计含棉量在85%以下，主要或仅与化学纤维混纺，每平方米重量不超过200克：**							**Woven fabrics of cotton, containing less than 85% by weight of cotton, mixed mainly or solely with man-made fibres, weighing not more than 200 g/m^2:**
		-未漂白：							-Unbleached:
3385	5210.1100	--平纹机织物	12		11	0		90	--Plain weave
3386	5210.1200	--三线或四线斜纹机织物，包括双面斜纹机织物	12			0		90	--3-thread or 4-thread twill, including cross twill
3387	5210.1900	--其他机织物	12					90	--Other fabrics
		-漂白：							-Bleached:
3388	5210.2100	--平纹机织物	14					90	--Plain weave
3389	5210.2200	--三线或四线斜纹机织物，包括双面斜纹机织物	14					90	--3-thread or 4-thread twill, including cross twill
3390	5210.2900	--其他机织物	14					90	--Other fabrics
		-染色：							-Dyed:
3391	5210.3100	--平纹机织物	10			0		90	--Plain weave
3392	5210.3200	--三线或四线斜纹机织物，包括双面斜纹机织物	10			0		90	--3-thread or 4-thread twill, including cross twill
3393	5210.3900	--其他机织物	10					90	--Other fabrics
		-色织：							-Of yarns of different colours:
3394	5210.4100	--平纹机织物	10					90	--Plain weave
3395	5210.4200	--三线或四线斜纹机织物，包括双面斜纹机织物	10			0		90	--3-thread or 4-thread twill, including cross twill
3396	5210.4900	--其他机织物	10					90	--Other fabrics
		-印花：							-Printed:
3397	5210.5100	--平纹机织物	10					90	--Plain weave
3398	5210.5200	--三线或四线斜纹机织物，包括双面斜纹机织物	10					90	--3-thread or 4-thread twill, including cross twill

序号 No.	税则号列 Tariff Item	货品名称	最惠国税率(%) M.F.N.	协定税率(%) Agreement 东盟 ASEAN	曼谷协定 BA	香港 H.K.	澳门 Macao	普通税率(%) Gen.	Article Description
3399	5210.5900	--其他机织物	10					90	--Other fabrics
	52.11	**棉机织物,按重量计含棉量在85%以下,主要或仅与化学纤维混纺,每平方米重量超过200克:**							**Woven fabrics of cotton, containing less than 85% by weight of cotton, mixed mainly or solely with man-made fibres, weighing more than 200 g/m²:**
		-未漂白:							-Unbleached:
3400	5211.1100	--平纹机织物	12			0		90	--Plain weave
3401	5211.1200	--三线或四线斜纹机织物,包括双面斜纹机织物	12			0		90	--3-thread or 4-thread twill, including cross twill
3402	5211.1900	--其他机织物	12			0		90	--Other fabrics
		-漂白:							-Bleached:
3403	5211.2100	--平纹机织物	14					90	--Plain weave
3404	5211.2200	--三线或四线斜纹机织物,包括双面斜纹机织物	14					90	--3-thread or 4-thread twill, including cross twill
3405	5211.2900	--其他机织物	14					90	--Other fabrics
		-染色:							-Dyed:
3406	5211.3100	--平纹机织物	10			0		90	--Plain weave
3407	5211.3200	--三线或四线斜纹机织物,包括双面斜纹机织物	10			0		90	--3-thread or 4-thread twill, including cross twill
3408	5211.3900	--其他机织物	10			0		90	--Other fabrics
		-色织:							-Of yarns of different colours:
3409	5211.4100	--平纹机织物	10					90	--Plainweave
3410	5211.4200	--粗斜纹布(劳动布)	10			0		90	--Denim
3411	5211.4300	--三线或四线斜纹机织物,包括双面斜纹机织物	10					90	--Other fabrics of 3-thread or 4-thread twill, including cross twill
3412	5211.4900	--其他机织物	10					90	--Other fabrics
		-印花:							-Printed:
3413	5211.5100	--平纹机织物	10					90	--Plain weave
3414	5211.5200	--三线或四线斜纹机织物,包括双面斜纹机织物	10					90	--3-thread or 4-thread twill, including cross twill
3415	5211.5900	--其他机织物	10					90	--Other fabrics
	52.12	**其他棉机织物:**							**Other woven fabrics of cotton:**
		-每平方米重量不超过200克:							-Weighing not more than 200 g/m²:
3416	5212.1100	--未漂白	12			0△		80	--Unbleached
3417	5212.1200	--漂白	14			0△		80	--Bleached
3418	5212.1300	--染色	10			0△		80	--Dyed
3419	5212.1400	--色织	10					80	--Of yarns of different colours
3420	5212.1500	--印花	10					80	--Printed
		-每平方米重量超过200克:							-Weighing more than 200 g/m²:

序号 No.	税则号列 Tariff Item	货品名称	最惠国税率(%) M.F.N.	协定税率(%)Agreement 东盟 ASEAN	曼谷协定 BA	香港 H.K.	澳门 Macao	普通税率(%) Gen.	Article Description
3421	5212.2100	--未漂白	12			0△		80	--Unbleached
3422	5212.2200	--漂白	14			0△		80	--Bleached
3423	5212.2300	--染色	10			0△		80	--Dyed
3424	5212.2400	--色织	10					80	--Of yarns of different colours
3425	5212.2500	--印花	10					80	--Printed

第五十三章
其他植物纺织纤维；纸纱线及其机织物

Chapter 53
Other vegetable textile fibres; paper yarn and woven fabrics of paper yarn

序号 No.	税则号列 Tariff Item	货品名称	最惠国税率(%) M.F.N.	协定税率(%) Agreement 东盟 ASEAN	曼谷协定 BA	香港 H.K.	澳门 Macao	普通税率(%) Gen.	Article Description
	53.01	**亚麻，生的或经加工但未纺制的；亚麻短纤及废麻（包括废麻纱线及回收纤维）：**							**Flax, raw or processed but not spun; flax tow and waste (including yarn waste and garnetted stock):**
3426	5301.1000	-生的或经沤制的亚麻	6					30	-Flax, raw or retted
		-破开、打成、栉梳或经其他加工但未纺制的亚麻：							-Flax, broken, scutched, hackled or otherwise processed, but not spun:
3427	5301.2100	--破开的或打成的	6					30	--Broken or scutched
3428	5301.2900	--其他	6					30	--Other
3429	5301.3000	-亚麻短纤及废麻	6					30	-Flax tow and waste
	53.02	**大麻，生的或经加工但未纺制的；大麻短纤及废麻（包括废麻纱线及回收纤维）：**							**True hemp (*Cannabis sativa L*), raw or processed but not spun; tow and waste of true hemp (including yarn waste and garnetted stock):**
3430	5302.1000	-生的或经沤制的大麻	6					30	-True hemp, raw or retted
3431	5302.9000	-其他	6					30	-Other
	53.03	**黄麻及其他纺织用韧皮纤维（不包括亚麻、大麻及苎麻），生的或经加工但未纺制的；上述纤维的短纤及废麻（包括废纱线及回收纤维）：**							**Jute and other textile bast fibres (excluding flax, true hemp and ramie), raw or processed but not spun; tow and waste of these fibres (including yarn waste and garnetted stock):**
3432	5303.1000	-生的或经沤制的黄麻及其他纺织用韧皮纤维	5					20	-Jute and other textile bast fibres, raw or retted
3433	5303.9000	-其他	5					30	-Other
	53.04	**西沙尔麻及其他纺织用龙舌兰类纤维，生的或经加工但未纺制的；上述纤维的短纤及废麻（包括废纱线及回收纤维）：**							**Sisal and other textile fibres of the genus *Agave*, raw or processed but not spun; tow and waste of these fibres (including yarn waste and garnetted stock):**
3434	5304.1000	-生的西沙尔麻及其他纺织用龙舌兰类纤维	5					20	-Sisal and other textile fibres of the genus *Agave*, raw
3435	5304.9000	-其他	5					30	-Other

序号 No.	税则号列 Tariff Item	货品名称	最惠国税率(%) M.F.N.	协定税率(%) Agreement 东盟 ASEAN	曼谷协定 BA	香港 H.K.	澳门 Macao	普通税率(%) Gen.	Article Description
	53.05	**椰壳纤维、蕉麻(马尼拉麻)、苎麻及其他税号未列名的纺织用植物纤维,生的或经加工但未纺制的;上述纤维的短纤、落麻及废料(包括废纱线及回收纤维):**							**Coconut, abaca (Manila hemp or *Musatextilis Nee*), ramie and other vegetable textile fibres, not elsewhere specified or included, raw or processed but not spun; tow, noils and waste of these fibres (including yarn waste and garnetted stock):**
		-椰壳纤维:							-Of coconut(coir):
3436	5305.1100	--生的	5		4			30	--Raw
3437	5305.1900	--其他	5					30	--Other
		-蕉麻:							-Of abaca:
3438	5305.2100	--生的	3					20	--Raw
3439	5305.2900	--其他	3					20	--Other
		-其他:							-Other:
		---生的:							---Raw:
3440	5305.9011	----苎麻	5					30	----Ramie
3441	5305.9019	----其他	5					30	----Other
		---其他:							---Other:
3442	5305.9091	----经加工但未纺制的苎麻	5					30	----Ramie, processed but not spun
3443	5305.9092	----苎麻的短纤及废麻	5					30	----Ramie, tow and waste
3444	5305.9099	----其他	5					30	---- Other
	53.06	**亚麻纱线:**							**Flax yarn:**
3445	5306.1000	-单纱	6			0	0	50	-Single
3446	5306.2000	-多股纱线或缆线	10			0△	0	50	-Multiple(folded)or cabled
	53.07	**黄麻纱线或税号53.03的其他纺织用韧皮纤维纱线:**							**Yarn of jute or of other textile bast fibres of heading No.53.03:**
3447	5307.1000	-单纱	6					35	-Single
3448	5307.2000	-多股纱线或缆线	6					35	-Multiple(folded)or cabled
	53.08	**其他植物纺织纤维纱线;纸纱线:**							**Yarn of other vegetable textile fibres; paper yarn:**
3449	5308.1000	-椰壳纤维纱线	6					45	-Coir yarn
3450	5308.2000	-大麻纱线	6					45	-True hemp yarn
		-其他:							-Other:
		---苎麻纱线:							---Ramie yarn:
3451	5308.9011	----按重量计苎麻含量在85%及以上的未漂白或漂白纱线	6		6		0	50	----Containing 85% or more by weight of ramie, unbleached or bleached yarn
3452	5308.9012	----按重量计苎麻含量在85%及以上的色纱线	6		6		0	50	----Containing 85% or more by weight of ramie, coloured yarn
3453	5308.9013	----按重量计苎麻含量在85%以下的未漂白或漂白纱线	6		6		0	50	----Containing less than 85% by weight of ramie, unbleached or bleached yarn

序号 No.	税则号列 Tariff Item	货品名称	最惠国税率(%) M.F.N.	协定税率(%)Agreement 东盟 ASEAN	曼谷协定 BA	香港 H.K.	澳门 Macao	普通税率(%) Gen.	Article Description
3454	5308.9014	----按重量计苎麻含量在85%以下的色纱线	6		6		0	50	----Containing less than 85% by weight of coloured yarn
		---其他:							---Other:
3455	5308.9091	----纸纱线	6		6			70	----Paper yarn
3456	5308.9099	----其他	6		6			45	----Other
	53.09	**亚麻机织物:**							**Woven fabrics of flax:**
		-按重量计亚麻含量在85%及以上:							-Containing 85% or more by weight of flax:
		--未漂白或漂白:							--Unbleached or bleached:
3457	5309.1110	---未漂白	10					80	---Unbleached
3458	5309.1120	---漂白	10					80	---Bleached
3459	5309.1900	--其他	10		9.3	0		80	--Other
		-按重量计亚麻含量在85%以下:							-Containing less than 85% by weight of flax:
		--未漂白或漂白:							--Unbleached or bleached:
3460	5309.2110	---未漂白	10					80	---Unbleached
3461	5309.2120	---漂白	10					80	---Bleached
3462	5309.2900	--其他	10			0		80	--Other
	53.10	**黄麻或税号53.03的其他纺织用韧皮纤维机织物:**							**Woven fabrics of jute or of other textilebast fibres of heading No.53.03:**
3463	5310.1000	-未漂白	10					40	-Unbleached
3464	5310.9000	-其他	10					40	-Other
	53.11	**其他纺织用植物纤维机织物;纸纱线机织物:**							**Woven fabrics of other vegetable textile fibres; woven fabrics of paper yarn:**
		---苎麻机织物:							---Woven fabrics of ramie:
3465	5311.0012	----按重量计苎麻含量在85%及以上的未漂白机织物	10					80	----Containing 85% or more by weight of ramie, unbleached woven fabrics
3466	5311.0013	----按重量计苎麻含量在85%及以上的其他机织物	12					80	----Containing 85% or more by weight of ramie, other woven fabrics
3467	5311.0014	----按重量计苎麻含量在85%以下的未漂白机织物	10					80	----Containing less than 85% by weight of ramie, unbleached woven fabrics
3468	5311.0015	----按重量计苎麻含量在85%以下的其他机织物	12					80	----Containing less than 85% by weight of ramie, other woven fabrics
3469	5311.0020	---纸纱线机织物	10					90	---Woven fabrics of paper yarn
3470	5311.0090	---其他	10		9.3			50	---Other

第五十四章
化学纤维长丝

Chapter 54
Man-made filaments

注释：

一、本目录所称"化学纤维"，是指通过下列任一方法加工制得的有机聚合物的短纤或长丝：

(一) 将有机单体物质加以聚合而制得，例如，聚酰胺、聚酯、聚氨基甲酸酯或聚乙烯衍生物；

(二) 将天然有机聚合物（例如，纤维素、酪素、蛋白质或海藻）经化学变化而制得，例如，粘胶纤维、醋酸纤维素纤维、铜铵纤维或藻酸盐纤维。

对于化学纤维，所称"合成"，是指(一)款所述的纤维；所称"人造"，是指(二)款所述的纤维。

对于纺织材料，所称"化学纤维"、"合成纤维"及"人造纤维"，其含义应与上述解释相同。

二、税号54.02及54.03不适用于第五十五章的合成纤维或人造纤维的长丝丝束。

Notes:

1. Throughout the Nomenclature, the term "man-made fibres" means staple fibres and filaments of organic polymers produced by manufacturing processes, either:

   (a) By polymerization of organic monomers, such as polyamides, polyesters, polyurethanes or polyvinyl derivatives; or

   (b) By chemical transformation of natural organic polymers (for example, cellulose, casein, proteins or algae), such as viscose rayon, cellulose acetate, cupro or alginates.

   The terms "synthetic" and "artificial", used in relation to fibres, mean: synthetic: fibres as defined at (a); artificial: fibres as defined at (b).

   The terms "man-made", "synthetic" and "artificial" shall have the same meanings when used in relation to "textile materials".

2. Headings Nos.54.02 and 54.03 do not apply to synthetic or artificial filament tow of Chapter 55.

序号 No.	税则号列 Tariff Item	货品名称	最惠国税率(%) M.F.N.	协定税率(%)Agreement				普通税率(%) Gen.	Article Description
				东盟 ASEAN	曼谷协定 BA	香港 H.K.	澳门 Macao		
	54.01	**化学纤维长丝纺制的缝纫线，不论是否供零售用：**							**Sewing thread of man-made filaments, whether or not put up for retail sale:**
		-合成纤维长丝纺制：							-Of synthetic filaments:
3471	5401.1010	---非供零售用	5			0	0	70	---Not put up for retail sale
3472	5401.1020	---供零售用	5					90	---Put up for retail sale
		-人造纤维长丝纺制：							-Of artificial filaments:
3473	5401.2010	---非供零售用	5					35	---Not put up for retail sale
3474	5401.2020	---供零售用	5					90	---Put up for retail sale
	54.02	**合成纤维长丝纱线（缝纫线除外），非供零售用，包括细度在67分特以下的合成纤维单丝：**							**Synthetic filament yarn (other than sewing thread), not put up for retail sale, including synthetic monofilament of less than 67 decitex:**
		-尼龙或其他聚酰胺纺制的高强力纱：							-High tenacity yarn of nylon or other polyamides:
3475	5402.1010	---聚酰胺-6(尼龙-6)纺制	5				0	70	---Of nylon-6
3476	5402.1020	---聚酰胺-6,6(尼龙-6,6)纺制	5				0	70	---Of nylon-6,6
3477	5402.1030	---芳香族聚酰胺纺制	5				0	70	---Of aramides

序号 No.	税则号列 Tariff Item	货品名称	最惠国税率(%) M.F.N.	协定税率(%)Agreement 东盟 ASEAN	曼谷协定 BA	香港 H.K.	澳门 Macao	普通税率(%) Gen.	Article Description
3478	5402.1090	---其他	5				0	70	---Other
3479	5402.2000	-聚酯高强力纱	5				0	70	-High tenacity yarn of polyesters
		-变形纱线：							-Textured yarn:
		--尼龙或其他聚酰胺纺制，每根单纱细度不超过50特：							--Of nylon or other polyamides, measuring per single yarn not more than 50 tex:
		---弹力丝：							---Elastic filament:
3480	5402.3111	----聚酰胺-6(尼龙-6)纺制	5			0		80	----Of nylon-6
3481	5402.3112	----聚酰胺-6,6(尼龙-6,6)纺制	5			0		80	----Of nylon-6,6
3482	5402.3113	----芳香族聚酰胺纺制	5			0		80	---Of aramides
3483	5402.3119	----其他	5			0		80	----Other
3484	5402.3190	---其他	5			0		70	---Other
		--尼龙或其他聚酰胺纺制，每根单纱细度超过50特：							--Of nylon or other polyamides, measuring per single yarn more than 50 tex:
		---弹力丝：							---Elastic filament:
3485	5402.3211	----聚酰胺-6(尼龙-6)纺制	5			0		80	----Of nylon-6
3486	5402.3212	----聚酰胺-6,6(尼龙-6,6)纺制	5			0		80	----Of nylon-6,6
3487	5402.3213	----芳香族聚酰胺纺制	5			0		80	----Of aramides
3488	5402.3219	----其他	5			0		80	----Other
3489	5402.3290	---其他	5		4.5	0		70	---Other
		--聚酯纺制：							--Of polyesters:
3490	5402.3310	---弹力丝	5					90	---Elastic filament
3491	5402.3390	---其他	5					70	---Other
		--其他：							--Other:
3492	5402.3910	---聚丙烯纱线	5					70	---Of polypropylene
3493	5402.3990	---其他	5				0	70	---Other
		-其他单纱，未加捻或捻度每米不超过50转：							Other yarn, single, untwisted or with a twist not exceeding 50 turns per metre:
		--尼龙或其他聚酰胺纱线：							--Of nylon or other polyamides:
3494	5402.4110	---聚酰胺-6(尼龙-6)制	5		4.7		0	70	---Of nylon-6
3495	5402.4120	---聚酰胺-6,6(尼龙-6,6)制	5		4.7		0	70	---Of nylon-6,6
3496	5402.4130	---芳香族聚酰胺制	5		4.7		0	70	---Of aramides
3497	5402.4190	---其他	5		4.7		0	70	---Other
3498	5402.4200	--部分定向聚酯纱线	5					70	--Of polyesters, partially oriented
3499	5402.4300	--其他聚酯纱线	5		4.9			70	--Of polyesters, other
		--其他：							--Other:

序号 No.	税则号列 Tariff Item	货品名称	最惠国税率(%) M.F.N.	协定税率(%) Agreement 东盟 ASEAN	曼谷协定 BA	香港 H.K.	澳门 Macao	普通税率(%) Gen.	Article Description
3500	5402.4910	---聚丙烯纱线	5				0	70	---Of polypropylene
3501	5402.4920	---氨纶纱线	5			0	0	70	---Of polyurethane
3502	5402.4990	---其他	5				0	70	---Other
		-其他单纱,捻度每米超过 50 转:							-Other, yarn, single, with a twist exceeding 50 turns per metre:
		--尼龙或其他聚酰胺纱线:							--Of nylon or other polyamides:
3503	5402.5110	---聚酰胺-6(尼龙-6)制	5				0	70	---Of nylon-6
3504	5402.5120	---聚酰胺-6,6(尼龙-6,6)制	5				0	70	---Of nylon-6,6
3505	5402.5130	---芳香族聚酰胺制	5				0	70	---Of aramides
3506	5402.5190	---其他	5				0	70	---Other
3507	5402.5200	--聚酯纱线	5		4.3		0	70	--Of polyesters
		--其他:							--Other:
3508	5402.5910	---聚丙烯纱线	5					70	---Of polypropylene
3509	5402.5990	---其他	5					70	---Other
		-其他纱线(多股纱线或缆线):							-Other yarn, multiple (folded) or cabled:
		--尼龙或其他聚酰胺纺制:							--Of nylon or other polyamides:
3510	5402.6110	---聚酰胺-6(尼龙-6)制	5				0	70	---Of nylon-6
3511	5402.6120	---聚酰胺-6,6(尼龙-6,6)制	5				0	70	---Of nylon-6,6
3512	5402.6130	---芳香族聚酰胺制	5				0	70	---Of aramides
3513	5402.6190	---其他	5				0	70	---Other
3514	5402.6200	--聚酯纺制	5					70	--Of polyesters
		--其他:							--Other:
3515	5402.6910	---聚丙烯纱线	5				0	70	---Of polypropylene
3516	5402.6920	---氨纶纱线	5				0	70	---Of polyurethane
3517	5402.6990	---其他	5				0	70	---Other
	54.03	**人造纤维长丝纱线(缝纫线除外),非供零售用,包括细度在 67 分特以下的人造纤维单丝:**							**Artificial filament yarn (other than sewing thread), not put up for retail sale, including artificial monofilament of less than 67 decitex:**
3518	5403.1000	-粘胶纤维纺制的高强力纱	5					35	-High tenacity yarn of viscose rayon
3519	5403.2000	-变形纱线	5					35	-Textured yarn
		-其他单纱:							-Other yarn, single:
3520	5403.3100	--粘胶纤维纺制,未加捻或捻度每米不超过 120 转	5					35	--Of viscose rayon, untwisted or with a twist not exceeding 120 turns per metre
3521	5403.3200	--粘胶纤维纺制,捻度每米超过 120 转	5				0	35	--Of viscose rayon, with a twist exceeding 120 turns per metre
		--醋酸纤维纺制:							--Of cellulose acetate:

序号 No.	税则号列 Tariff Item	货品名称	最惠国税率(%) M.F.N.	协定税率(%) Agreement 东盟 ASEAN	曼谷协定 BA	香港 H.K.	澳门 Macao	普通税率(%) Gen.	Article Description
3522	5403.3310	---二醋酸纤维纺制	5					40	---Of cellulose diacetate
3523	5403.3390	---其他	5					35	---Other
3524	5403.3900	--其他	5					35	--Other
		-其他纱线(多股纱线或缆线):							-Other, yarn, multiple (folded) or cabled:
3525	5403.4100	--粘胶纤维纺制	5					35	--Of viscose rayon
3526	5403.4200	--醋酸纤维纺制	5					35	--Of cellulose acetate
3527	5403.4900	--其他	5					35	--Other
	54.04	**截面尺寸不超过1毫米,细度在67分特及以上的合成纤维单丝;表观宽度不超过5毫米的合成纤维纺织材料制扁条及类似品(例如人造草):**							**Synthetic monofilament of 67 decitex or more and of which no cross-sectional dimension exceeds 1 mm; strip and the like (for example, artificial straw) of synthetic textile materials of an apparent width not exceeing 5 mm:**
3528	5404.1000	-单丝	5			0		80	-Monofilament
3529	5404.9000	-其他	5					80	-Other
	54.05	**截面尺寸不超过1毫米,细度在67分特及以上的人造纤维单丝;表观宽度不超过5毫米的人造纤维纺织材料制扁条及类似品(例如人造草):**							**Artificial monofilament of 67 decitex or more and of which no cross-sectional dimension exceeds 1 mm; strip and the like (for example, artificial straw) of artificial textile materials of an apparent width not exceeding 5 mm:**
3530	5405.0000	截面尺寸不超过1毫米,细度在67分特及以上的人造纤维单丝;表观宽度不超过5毫米的人造纤维纺织材料制扁条及类似品(例如人造草)	5					80	Artificial monofilament of 67 decitex or more and of which no cross-sectional dimension exceeds 1 mm; strip and the like (for example, artificial straw) of artificial textile materials of an apparent width not exceeding 5 mm
	54.06	**化学纤维长丝纱线(缝纫线除外),供零售用:**							**Man-made filament yarn (other than sewing thread), put up for retail sale:**
3531	5406.1000	-合成纤维长丝纱线	5					90	-Synthetic filament yarn
3532	5406.2000	-人造纤维长丝纱线	5					90	-Artificial filament yarn
	54.07	**合成纤维长丝纱线的机织物,包括税号54.04所列材料的机织物:**							**Woven fabrics of synthetic filament yarn, including woven fabrics obtained from materials of heading No. 54.04:**
		-尼龙或其他聚酰胺高强力纱、聚酯高强力纱纺制的机织物:							-Woven fabrics obtained from high tenacity yarn of nylon or other polyamides or of polyesters:
3533	5407.1010	---尼龙或其他聚酰胺高强力纱纺制	10					130	---Of nylon or other polyamides

序号 No.	税则号列 Tariff Item	货品名称	最惠国税率(%) M.F.N.	协定税率(%)Agreement 东盟 ASEAN	曼谷协定 BA	香港 H.K.	澳门 Macao	普通税率(%) Gen.	Article Description
3534	5407.1020	---聚酯高强力纱纺制	10					130	---Of polyesters
3535	5407.2000	-扁条及类似品的机织物	10					130	-Woven fabrics obtained from strip or the like
3536	5407.3000	-第十一类注释九所列的机织物	10					130	-Fabrics specified in Note 9 to Section XI
		-其他机织物,按重量计尼龙或其他聚酰胺长丝含量在85%及以上:							-Other woven fabrics, containing 85% or more by weight of filaments of nylon or other polyamides:
3537	5407.4100	--未漂白或漂白	10					130	--Unbleached or bleached
3538	5407.4200	--染色	10		9.7	0	0	130	--Dyed
3539	5407.4300	--色织	10			0		130	--Of yarns of different colours
3540	5407.4400	--印花	10					130	--Printed
		-其他机织物,按重量计聚酯变形长丝含量在85%及以上:							-Other woven fabrics, containing 85% or more by weight of textured polyester filaments:
3541	5407.5100	--未漂白或漂白	10					130	--Unbleached or bleached
3542	5407.5200	--染色	10		9.7	0		130	--Dyed
3543	5407.5300	--色织	10					130	--Of yarns of different colours
3544	5407.5400	--印花	10					130	--Printed
		-其他机织物,按重量计聚酯长丝含量在85%及以上:							-Other woven fabrics, containing 85% or more by weight of polyester filaments:
3545	5407.6100	--按重量计聚酯非变形长丝含量在85%及以上	10		9.7	0	0	130	--Containing 85% or more by weight of non-textured polyester filaments
3546	5407.6900	--其他	10		9.7			130	--Other
		-其他机织物,按重量计其他合成纤维长丝含量在85%及以上:							-Other woven fabrics, containing 85% or more by weight of synthetic filaments:
3547	5407.7100	--未漂白或漂白	10		8.8			130	--Unbleached or bleached
3548	5407.7200	--染色	10		9.7	0		130	--Dyed
3549	5407.7300	--色织	10					130	--Of yarns of different colours
3550	5407.7400	--印花	10		9.7			130	--Printed
		-其他机织物,按重量计其他合成纤维长丝含量在85%以下,主要或仅与棉混纺:							-Other woven fabrics, containing less than 85% by weight of synthetic filaments, mixed mainly or solely with cotton:
3551	5407.8100	--未漂白或漂白	10					130	--Unbleached or bleached
3552	5407.8200	--染色	10					130	--Dyed
3553	5407.8300	--色织	10					130	--Of yarns of different colours
3554	5407.8400	--印花	10					130	--Printed
		-其他机织物:							-Other woven fabrics:
3555	5407.9100	--未漂白或漂白	10					130	--Unbleached or bleached
3556	5407.9200	--染色	10			0		130	--Dyed
3557	5407.9300	--色织	10					130	--Of yarns of different colours
3558	5407.9400	--印花	10					130	--Printed

序号 No.	税则号列 Tariff Item	货品名称	最惠国税率(%) M.F.N.	协定税率(%) Agreement 东盟 ASEAN	曼谷协定 BA	香港 H.K.	澳门 Macao	普通税率(%) Gen.	Article Description
	54.08	**人造纤维长丝纱线的机织物,包括税号54.05所列材料的机织物:**							**Woven fabrics of artificial filament yarn, including woven fabrics obtained from materials of heading No.54.05:**
3559	5408.1000	-粘胶纤维高强力纱的机织物	10	130				130	-Woven fabrics obtained from high tenacity yarnof viscose rayon
		-其他机织物,按重量计人造纤维长丝、扁条或类似品含量在85%及以上:							-Other woven fabrics, containing 85% or more by weight of artificial filament or strip or the like:
		--未漂白或漂白:							--Unbleached or bleached:
3560	5408.2110	---粘胶纤维制	12					130	---Of yarns of viscose rayon
3561	5408.2120	---醋纤纤维制	12					130	---Of yarns of cellulose acetate
3562	5408.2190	---其他	12					130	---Other
		--染色:							--Dyed:
3563	5408.2210	---粘胶纤维制	10					130	---Of yarns of viscose rayon
3564	5408.2220	---醋纤纤维制	10					130	---Of yarns of cellulose acetate
3565	5408.2290	---其他	10					130	---Other
		--色织:							--Of yarns of different colours:
3566	5408.2310	---粘胶纤维制	10					130	---Of yarns of viscose rayon
3567	5408.2320	---醋纤纤维制	10					130	---Of yarns of cellulose acetate
3568	5408.2390	---其他	10					130	---Other
		--印花:							--Printed:
3569	5408.2410	---粘胶纤维制	10					130	---Of yarns of viscose rayon
3570	5408.2420	---醋纤纤维制	10					130	---Of yarns of cellulose acetate
3571	5408.2490	---其他	10					130	---Other
		-其他机织物:							-Other woven fabrics:
3572	5408.3100	--未漂白或漂白	10					130	--Unbleached or bleached
3573	5408.3200	--染色	10		9.9			130	--Dyed
3574	5408.3300	--色织	10					130	--Of yarns of different colours
3575	5408.3400	--印花	10					130	--Printed

第五十五章
化学纤维短纤

Chapter 55
Man-made staple fibres

注释:

税号 55.01 和 55.02 仅适用于每根与丝束长度相等的平行化学纤维长丝丝束。前述丝束应同时符合下列规格:

一、丝束长度超过 2 米;
二、捻度每米少于 5 转;
三、每根长丝细度在 67 分特以下;
四、合成纤维长丝丝束,须经拉伸处理,即本身不能被拉伸至超过本身长度的一倍;
五、丝束总细度大于 20000 分特。

丝束长度不超过 2 米的归入税号 55.03 或 55.04。

Notes:

Headings Nos. 55.01 and 55.02 apply only to man-made filament tow, consisting of parallel filaments of a uniform length equal to the length of the tow, meeting the following specifications:

(1) Length of tow exceeding 2 m;
(2) Twist less than 5 turns per metre;
(3) Measuring per filament less than 67 decitex;
(4) Synthetic filament tow only: the tow must be drawn, that is to say, be incapable of being stretched by more than 100% of its length;
(5) Total measurement of tow more than 20000 decitex.

Tow of a length not exceeding 2 m is to be classified in heading No. 55.03 or 55.04.

序号 No.	税则号列 Tariff Item	货品名称	最惠国税率(%) M.F.N.	协定税率(%) Agreement 东盟 ASEAN	曼谷协定 BA	香港 H.K.	澳门 Macao	普通税率(%) Gen.	Article Description
	55.01	**合成纤维长丝丝束:**							**Synthetic filament tow:**
3576	5501.1000	-尼龙或其他聚酰胺制	5					70	-Of nylon or other polyamides
3577	5501.2000	-聚酯制	5					70	-Of polyesters
3578	5501.3000	-聚丙烯腈或变性聚丙烯腈制	5		4.5			35	-Acrylic or modacrylic
3579	5501.9000	-其他	5			0		70	-Other
	55.02	**人造纤维长丝丝束:**							**Artificial filament tow:**
3580	5502.0010	---二醋酸纤维丝束	3					40	---Cellulose diacetate filament tow
3581	5502.0090	---其他	5					35	---Other
	55.03	**合成纤维短纤,未梳或未经其他纺前加工:**							**Synthetic staple fibres, not carded, combed or otherwise processed for spinning:**
3582	5503.1000	-尼龙或其他聚酰胺制	5					70	-Of nylon or other polyamides
3583	5503.2000	-聚酯制	5		4.5			70	-Of polyesters
3584	5503.3000	-聚丙烯腈或变性聚丙烯腈制	5		4.5			35	-Acrylic or modacrylic
3585	5503.4000	-聚丙烯制	5					70	-Of polypropylene
3586	5503.9000	-其他	5					70	-Other
	55.04	**人造纤维短纤,未梳或未经其他纺前加工:**							**Artificial staple fibres, not carded, combed or otherwise processed for spinning:**
3587	5504.1000	-粘胶纤维制	5					35	-Of viscose rayon
3588	5504.9000	-其他	5					35	-Other
	55.05	**化学纤维废料(包括落绵、废纱及回收纤维):**							**Waste(including noils, yarn waste and garnetted stock) of man-made fibres:**
3589	5505.1000	-合成纤维的	5					70	-Of synthetic fibres
3590	5505.2000	-人造纤维的	5			0		70	-Of artificial fibres

序号 No.	税则号列 Tariff Item	货品名称	最惠国税率(%) M.F.N.	协定税率(%)Agreement 东盟 ASEAN	曼谷协定 BA	香港 H.K.	澳门 Macao	普通税率(%) Gen.	Article Description
	55.06	**合成纤维短纤,已梳或经其他纺前加工:**							**Synthetic staple fibres, carded, combed or otherwise processed for spinning:**
3591	5506.1000	-尼龙或其他聚酰胺制	5					70	-Of nylon or other polyamides
3592	5506.2000	-聚酯制	5		4.5			70	-Of polyesters
3593	5506.3000	-聚丙烯腈或变性聚丙烯腈制	5		4.5			35	-Acrylic or modacrylic
3594	5506.9000	-其他	5					70	-Other
	55.07	**人造纤维短纤,已梳或经其他纺前加工:**							**Artificial staple fibres, carded, combed or otherwise processed for spinning:**
3595	5507.0000	人造纤维短纤,已梳或经其他纺前加工	5					35	Artificial staple fibres, carded, combed or otherwise processed for spinning
	55.08	**化学纤维短纤纺制的缝纫线,不论是否供零售用:**							**Sewing thread of man-made staple fibres, whether or not put up for retail sale:**
3596	5508.1000	-合成纤维短纤纺制	5		4.6	0	0	90	-Of synthetic staple fibres
3597	5508.2000	-人造纤维短纤纺制	5					70	-Of artificial staple fibres
	55.09	**合成纤维短纤纺制的纱线(缝纫线除外),非供零售用:**							**Yarn (other than sewing thread) of synthetic staple fibres, not put up for retail sale:**
		-按重量计尼龙或其他聚酰胺短纤含量在85%及以上:							-Containing 85% or more by weight of staple fibres of nylon or other polyamides:
3598	5509.1100	--单纱	5			0		90	--Single yarn
3599	5509.1200	--多股纱线或缆线	5			0		90	--Multiple (folded) or cabled yarn
		-按重量计聚酯短纤含量在85%及以上:							-Containing 85% or more by weight of polyester staple fibres:
3600	5509.2100	--单纱	5			0		90	--Single yarn
3601	5509.2200	--多股纱线或缆线	5			0		90	--Multiple (folded) or cabled yarn
		-按重量计聚丙烯腈或变性聚丙烯腈短纤含量在85%及以上:							-Containing 85% or more by weight of acrylic or modacrylic staple fibres:
3602	5509.3100	--单纱	5			0△		90	--Single yarn
3603	5509.3200	--多股纱线或缆线	5			0	0	90	--Multiple (folded) or cabled yarn
		-其他纱线,按重量计合成纤维短纤含量在85%及以上:							-Other yarn, containing 85% or more by weight of synthetic staple fibres:
3604	5509.4100	--单纱	5			0△		90	--Single yarn
3605	5509.4200	--多股纱线或缆线	5			0△		90	--Multiple (folded) or cabled yarn
		-其他聚酯短纤纺制的纱线:							-Other yarn, of polyester staple fibres:
3606	5509.5100	--主要或仅与人造纤维短纤混纺	5			0△		90	--Mixed mainly or solely with artificial staple fibres
3607	5509.5200	--主要或仅与羊毛或动物细毛混纺	5			0△		90	--Mixed mainly or solely with wool or fine animal hair
3608	5509.5300	--主要或仅与棉混纺	5		4.3	0		90	--Mixed maninly or solely with cotton

序号 No.	税则号列 Tariff Item	货品名称	最惠国税率(%) M.F.N.	协定税率(%) Agreement 东盟 ASEAN	曼谷协定 BA	香港 H.K.	澳门 Macao	普通税率(%) Gen.	Article Description
3609	5509.5900	--其他	5			0		90	--Other
		-其他聚丙烯腈或变性聚丙烯腈短纤纺制的纱线:							-Other yarn, of acrylic or modacrylic staple fibres:
3610	5509.6100	--主要或仅与羊毛或动物细毛混纺	5			$0^{\triangle}$		90	--Mixed mainly or solely with wool or fine animal hair
3611	5509.6200	--主要或仅与棉混纺	5			0		90	--Mixed mainly or solely with cotton
3612	5509.6900	--其他	5			$0^{\triangle}$		90	--Other
		-其他纱线:							-Other yarn:
3613	5509.9100	--主要或仅与羊毛或动物细毛混纺	5			$0^{\triangle}$		90	--Mixed nainly or solely with wool or fine animal hair
3614	5509.9200	--主要或仅与棉混纺	5			0		90	--Mixed mainly or solely with cotton
3615	5509.9900	--其他	5			$0^{\triangle}$		90	--Other
	55.10	**人造纤维短纤纺制的纱线(缝纫线除外),非供零售用:**							**Yarn(other than sewing thread)of artificial staple fibres, not put up for retail sale:**
		-按重量计人造纤维短纤含量在85%及以上:							-Containing 85% or more by weight of artificial staple fibres:
3616	5510.1100	--单纱	5		4.5	0		70	--Single yarn
3617	5510.1200	--多股纱线或缆线	5			0		70	--Multiple(folded)or cabled yarn
3618	5510.2000	-其他纱线,主要或仅与羊毛或动物细毛混纺	5			$0^{\triangle}$		70	-Other yarn, mixed mainly or solely with wool or fine animal hair
3619	5510.3000	-其他纱线,主要或仅与棉混纺	5			0		70	-Other yarn, mixed mainly or solely with cotton
3620	5510.9000	-其他	5		4.5	0		70	-Other yarn
	55.11	**化学纤维短纤纺制的纱线(缝纫线除外),供零售用:**							**Yarn (other than sewing thread) of man-made staple fibres, put up for retail sale:**
3621	5511.1000	-按重量计合成纤维短纤含量在85%及以上	5					90	-Of synthetic staple fibres, containing 85% or more by weight of such fibres
3622	5511.2000	-按重量计合成纤维短纤含量在85%以下	5					90	-Of synthetic staple fibres, containing less than 85% by weight of such fibres
3623	5511.3000	-人造纤维短纤纺制	5					90	-Of artificial staple fibres
	55.12	**合成纤维短纤纺制的机织物,按重量计合成纤维短纤含量在85%及以上:**							**Woven fabrics of synthetic staple fibres, containing 85% or more by weight of synthetic staple fibres:**
		-按重量计聚酯短纤含量在85%及以上:							-Containing 85% or more by weight of polyester staple fibres:
3624	5512.1100	--未漂白或漂白	18					130	--Unbleached or bleached
3625	5512.1900	--其他	10			0		130	--Other
		-按重量计聚丙烯腈或变性聚丙烯腈短纤含量在85%及以上:							-Containing 85% or more by weight of acrylic or modacrylic staple fibres:

序号 No.	税则号列 Tariff Item	货品名称	最惠国税率(%) M.F.N.	协定税率(%) Agreement 东盟 ASEAN	曼谷协定 BA	香港 H.K.	澳门 Macao	普通税率(%) Gen.	Article Description
3626	5512.2100	--未漂白或漂白	13					130	--Unbleached or bleached
3627	5512.2900	--其他	10		9.3			130	--Other
		-其他:							-Other:
3628	5512.9100	--未漂白或漂白	18					130	--Unbleached or bleached
3629	5512.9900	--其他	10			0	0	130	--Other
	55.13	**合成纤维短纤纺制的机织物,按重量计合成纤维短纤含量在85%以下,主要或仅与棉混纺,每平方米重量不超过170克:**							**Woven fabrics of synthetic staple fibres, containing less than 85% by weight of such fibres, mixed mainly or solely with cotton, of a weight not exceeding 170 g/m^2:**
		-未漂白或漂白:							-Unbleached or bleached:
		--聚酯短纤纺制的平纹机织物:							--Of polyester staple fibres, plain weave:
3630	5513.1110	---未漂白	16			0		130	---Unbleached
3631	5513.1120	---漂白	15			0		130	---Bleached
		--聚酯短纤纺制的三线或四线斜纹机织物,包括双面斜纹机织物:							--3-thread or 4-thread twill, including cross twill, of polyester staple fibres:
3632	5513.1210	---未漂白	16			0		130	---Unbleached
3633	5513.1220	---漂白	18			0		130	---Bleached
		--其他聚酯短纤纺制的机织物:							--Other woven fabrics of polyester staple fibres:
3634	5513.1310	---未漂白	16					130	---Unbleached
3635	5513.1320	---漂白	18					130	---Bleached
3636	5513.1900	--其他机织物	18					130	--Other woven fabrics
		-染色:							-Dyed:
3637	5513.2100	--聚酯短纤纺制的平纹机织物	10		9.1	0	0	130	--Of polyester staple fibres, plain weave
3638	5513.2200	--聚酯短纤纺制的三线或四线斜纹机织物,包括双面斜纹机织物	10			0		130	--3-thread or 4-thread twil, including cross twill, of polyester staple fibres
3639	5513.2300	--其他聚酯短纤纺制的机织物	10					130	--Other woven fabrics of polyester staple fibres
3640	5513.2900	--其他机织物	10					130	--Other woven fabrics
		-色织:							-Of yarns of different colours:
3641	5513.3100	--聚酯短纤纺制的平纹机织物	10					130	--Of polyester staple fibres, plain weave
3642	5513.3200	--聚酯短纤纺制的三线或四线斜纹机织物,包括双面斜纹机织物	10		8.8			130	--3-thread or 4-thread twill, including cross twill, of polyester staple fibres
3643	5513.3300	--其他聚酯短纤纺制的机织物	10			0		130	--Other woven fabrics of polyester staple fibres
3644	5513.3900	--其他机织物	10					130	--Other woven fabrics
		-印花:							-Printed:

序号 No.	税则号列 Tariff Item	货品名称	最惠国税率(%) M.F.N.	协定税率(%)Agreement 东盟 ASEAN	曼谷协定 BA	香港 H.K.	澳门 Macao	普通税率(%) Gen.	Article Description
3645	5513.4100	--聚酯短纤纺制的平纹机织物	10					130	--Of polyester staple fibres, plain weave
3646	5512.4200	--聚酯短纤纺制的三线或四线斜纹机织物，包括双面斜纹机织物	10					130	--3-thread or 4-thread twill, including cross twill, of polyester staple fibres
3647	5513.4300	--其他聚酯短纤纺制的机织物	10					130	--Other woven fabrics of polyester staple fibres
3648	5513.4900	--其他机织物	10					130	--Other woven fabrics
	55.14	**合成纤维短纤纺制的机织物，按重量计合成纤维短纤含量在85%以下，主要或仅与棉混纺，每平方米重量超过170克：**							**Woven fabrics of synthetic staple fibres, containing less than 85% by weight of such fibres, mixed mainly or solely with cotton, of a weight exceeding 170g/m^2:**
		-未漂白或漂白：							-Unbleached or bleached:
		--聚酯短纤纺制的平纹机织物：							--Of polyester staple fibres, plain weave:
3649	5514.1110	---未漂白	16		14.4			130	---Unbleached
3650	5514.1120	---漂白	18					130	---Bleached
		--聚酯短纤纺制的三线或四线斜纹机织物，包括双面斜纹机织物：							--3-thread or 4-thread twill, including cross twill, of polyester staple fibres:
3651	5514.1210	---未漂白	16			0		130	---Unbleached
3652	5514.1220	---漂白	18			0		130	---Bleached
		--其他聚酯短纤纺制的机织物：							--Other woven fabrics of polyester staple fibres:
3653	5514.1310	---未漂白	16					130	---Unbleached
3654	5514.1320	---漂白	18					130	---Bleached
3655	5514.1900	--其他机织物	16					130	--Other woven fabrics
		-染色：							-Dyed:
3656	5514.2100	--聚酯短纤纺制的平纹机织物	10					130	--Of polyester staple fibres, plain weave
3657	5514.2200	--聚酯短纤纺制的三线或四线斜纹机织物，包括双面斜纹机织物	10			0		130	--3-thread or 4-thread twill, including cross twill, of polyester staple fibres
3658	5514.2300	--其他聚酯短纤纺制的机织物	10					130	--Other woven fabrics of polyester staple fibres
3659	5514.2900	--其他机织物	10					130	--Other woven fabrics
		-色织：							-Of yarns of different colours:
3660	5514.3100	--聚酯短纤纺制的平纹机织物	10					130	--Of polyester staple fibres, plain weave
3661	5514.3200	--聚酯短纤纺制的三线或四线斜纹机织物，包括双面斜纹机织物	10					130	--3-thread or 4-thread twill, including cross twill, of polyester staple fibres

序号 No.	税则号列 Tariff Item	货品名称	最惠国税率(%) M.F.N.	协定税率(%)Agreement 东盟 ASEAN	曼谷协定 BA	香港 H.K.	澳门 Macao	普通税率(%) Gen.	Article Description
3662	5514.3300	--其他聚酯短纤纺制的机织物	10					130	--Other woven fabrics of polyester staple fibres
3663	5514.3900	--其他机织物	10					130	--Other woven fabrics
		-印花:							-Printed:
3664	5514.4100	--聚酯短纤纺制的平纹机织物	10					130	--Of polyester staple fibres, plain weave
3665	5514.4200	--聚酯短纤纺制的三线或四线斜纹机织物,包括双面斜纹机织物	10					130	--3-thread or 4-thread twill, including cross twill, of polyester staple fibres
3666	5514.4300	--其他聚酯短纤纺制的机织物	10					130	--Other woven fabrics of polyester staple fibres
3667	5514.4900	--其他机织物	10					130	--Other woven fabrics
	55.15	**合成纤维短纤纺制的其他机织物:**							**Other woven fabrics of synthetic staple fibres:**
		-聚酯短纤纺制:							-Of polyester staple fibres:
3668	5515.1100	--主要或仅与粘胶纤维短纤混纺	10		8.8			130	--Mixed mainly or solely with viscose rayon staple fibres
3669	5515.1200	--主要或仅与化学纤维长丝混纺	10					130	--Mixed mainly or solely with man-made filaments
3670	5515.1300	--主要或仅与羊毛或动物细毛混纺	10					130	--Mixed mainly or solely with wool or fine animal hair
3671	5515.1900	--其他	10					130	--Other
		-聚丙烯腈或变性聚丙烯腈短纤纺制:							-Of acrylic or modacrylic staple fibres:
3672	5515.2100	--主要或仅与化学纤维长丝混纺	10					130	--Mixed mainly or solely with man-made filaments
3673	5515.2200	--主要或仅与羊毛或动物细毛混纺	12					130	--Mixed mainly or solely with wool or fine animal hair
3674	5515.2900	--其他	10					130	--Other
		-其他机织物:							-Other woven fabrics:
3675	5515.9100	--主要或仅与化学纤维长丝混纺	10					130	--Mixed mainly or solely with man-made filaments
3676	5515.9200	--主要或仅与羊毛或动物细毛混纺	10					130	--Mixed mainly or solely with wool or fine animal hair
3677	5515.9900	--其他	10					130	--Other
	55.16	**人造纤维短纤纺制的机织物:**							**Woven fabrics of artificial staple fibres:**
		-按重量计人造纤维短纤含量在85%及以上:							-Containing 85% or more by weight of artificial staple fibres:
3678	5516.1100	--未漂白或漂白	12					130	--Unbleached or bleached
3679	5516.1200	--染色	10			0		130	--Dyed
3680	5516.1300	--色织	10					130	--Of yarns of different colours
3681	5516.1400	--印花	10					130	--Printed

序号 No.	税则号列 Tariff Item	货品名称	最惠国税率(%) M.F.N.	协定税率(%)Agreement 东盟 ASEAN	曼谷协定 BA	香港 H.K.	澳门 Macao	普通税率(%) Gen.	Article Description
		-按重量计人造纤维短纤含量在85%以下,主要或仅与化学纤维长丝混纺:							-Containing less than 85% by weight of artificial staple fibres,mixed mainly or solely with man-made filaments:
3682	5516.2100	--未漂白或漂白	12					130	--Unbleached or bleached
3683	5516.2200	--染色	10		9.9			130	--Dyed
3684	5516.2300	--色织	10					130	--Of yarns of different colours
3685	5516.2400	--印花	10					130	--Printed
		-按重量计人造纤维短纤含量在85%以下,主要或仅与羊毛或动物细毛混纺:							-Containing less than 85% by weight of artificial staple fibres,mixed mainly or solely with wool or fine animal hair:
3686	5516.3100	--未漂白或漂白	12					130	--Unbleached or bleached
3687	5516.3200	--染色	10					130	--Dyed
3688	5516.3300	--色织	10					130	--Of yarns of different colours
3689	5516.3400	--印花	10					130	--Printed
		-按重量计人造纤维短纤含量在85%以下,主要或仅与棉混纺:							--Containing less than 85% by weight of artificial staple fibres,mixed mainly or solely with cotton:
3690	5516.4100	--未漂白或漂白	12					130	--Unbleached or bleached
3691	5516.4200	--染色	12					130	--Dyed
3692	5516.4300	--色织	10					130	--Of yarns of different colours
3693	5516.4400	--印花	10					130	--Printed
		-其他:							-Other:
3694	5516.9100	--未漂白或漂白	12					130	--Unbleached or bleached
3695	5516.9200	--染色	10					130	--Dyed
3696	5516.9300	--色织	10					130	--Of yarns of different colours
3697	5516.9400	--印花	10		9.4			130	--Printed

第五十六章
絮胎、毡呢及无纺织物；特种纱线；线、绳、索、缆及其制品

注释：

一、本章不包括：

（一）用各种物质或制剂（例如，第三十三章的香水或化妆品、税号 34.01 的肥皂或洗涤剂、税号 34.05 的光洁剂及类似制剂、税号 38.09 的织物柔软剂）浸渍、涂布、包覆的絮胎、毡呢或无纺织物，其中的纺织材料仅作为承载介质；

（二）税号 58.11 的纺织产品；

（三）以毡呢或无纺织物为底的砂布及类似品（税号 68.05）；

（四）以毡呢或无纺织物为底的粘聚或复制云母（税号 68.14）；

（五）以毡呢或无纺织物为底的金属箔（第十五类）。

二、所称“毡呢”，包括针刺机制毡呢以及纤维本身通过缝编工序增强了抱合力的纺织纤维网状织物。

三、税号 56.02 及 56.03 分别包括用各种性质（紧密结构或泡沫状）的塑料或橡胶浸渍、涂布、包覆或层压的毡呢及无纺织物。

税号 56.03 还包括用塑料或橡胶作粘合材料的无纺织物。

但税号 56.02 及 56.03 不包括：

（一）用塑料或橡胶浸渍、涂布、包覆或层压，按重量计纺织材料含量在 50% 及以下的毡呢或者完全嵌入塑料或橡胶之内的毡呢（第三十九章或第四十章）；

（二）完全嵌入塑料或橡胶之内的无纺织物，以及用肉眼可辨别出两面都用塑料或橡胶涂布、包覆的无纺织物，涂布或包覆所引起的颜色变化可不予考虑（第三十九章或第四十章）；

（三）与毡呢或无纺织物混制的泡沫塑料或海绵橡胶板、片或扁条，纺织材料仅在其中起增强作用（第三十九章或第四十章）。

Chapter 56
Wadding, felt and nonwovens; special yarns; twine, cordage, ropes and cables and articles thereof

Notes:

1. This Chapter does not cover:

(a) Wadding, felt or nonwovens, impregnated, coated or covered with substances or preparations (for example, pefumes or cosmetics of Chapter 33, soaps or detergents of heading No.34.01, polishes, creams or similar preparations of heading No.34.05, fabric softeners of heading No.38.09) where the textile material is present merely as a carrying medium;

(b) Textile products of heading No.58.11;

(c) Natural or artificial abrasive powder or grain, on a backing of felt or nonwovens (heading No.68.05);

(d) Agglomerated or reconstituted mica, on a backing of felt or nonwovens (heading No.68.14); or

(e) Metal foil on a backing of felt or nonwovens (Section XV).

2. The term “felt” includes needleloom felt and fabrics consisting of a web of textile fibres the cohesion of which has been enhanced by a stitch-bonding process using fibres from the web itself.

3. Headings Nos.56.02 and 56.03 cover respectively felt and nonwovens, impregnated, coated, covered or laminated with plastics or rubber whatever the nature of these materials (compact or cellular).

Heading No.56.03 also includes nonwovens in which plastics or rubber forms the bonding substance.

Headings Nos.56.02 and 56.03 do not, however, cover:

(a) Felt impregnated, coated, covered or laminated with plastics or rubber, containing 50% or less by weight of textile material or felt completely embedded in plastics or rubber (Chapter 39 or 40);

(b) Nonwovens, either completely embedded in plastics or rubber, or entirely coated or covered on both sides with such materials, provided that such coating or covering can be seen with the naked eye with no account being taken of any resulting change of colour (Chapter 39 or 40); or

(c) Plates, sheets or strips of cellular plastics or cellular rubber combined with felt or nonwovens, where the textile material is present merely for reinforcing purposes (Chapter 39 or 40).

四、税号56.04不包括用肉眼无法辨别出是否经过浸渍，涂布或包覆的纺织纱线或税号54.04或54.05的扁条及类似品（通常归入第五十章至第五十五章）；运用本条规定，可不考虑浸渍、涂布或包覆所引起的颜色变化。

4. Heading No. 56.04 does not cover textile yarn, or strip or the like of heading No. 54.04 or 54.05, in which the impregnation, coating or covering cannot be seen with the naked eye (usually Chapters 50 to 55); for the purpose of this provision, no account should be taken of any resulting change of colour.

序号 No.	税则号列 Tariff Item	货品名称	最惠国税率(%) M.F.N.	协定税率(%) Agreement 东盟 ASEAN	曼谷协定 BA	香港 H.K.	澳门 Macao	普通税率(%) Gen.	Article Description
	56.01	**纺织材料絮胎及其制品；长度不超过5毫米的纺织纤维（纤维屑）、纤维粉末及球结：**							**Wadding of textile materials and articles thereof; textile fibres, not exceeding 5 mm in length (flock), textile dust and mill neps:**
3698	5601.1000	-絮胎制的卫生巾及止血塞、婴儿尿布或尿布垫及类似卫生用品	10					80	-Sanitary towels and tampons, napkins and napkin liners for babies and similar sanitary articles, of wadding
		-絮胎；其他絮胎制品：							-Wadding; other articles of wadding:
3699	5601.2100	--棉制	10					50	--Of cotton
		--化学纤维制：							--Of man-made fibres:
3700	5601.2210	---卷烟滤嘴	12					100	---Cigarette filter tips
3701	5601.2290	---其他	12					100	---Other
3702	5601.2900	--其他	10					90	--Other
3703	5601.3000	-纤维屑、纤维粉末及球结	10					100	-Textile flock and dust and mill neps
	56.02	**毡呢，不论是否浸渍、涂布、包覆或层压：**							**Felt, whether or not impregnated, coated, covered or laminated:**
3704	5602.1000	-针刺机制毡呢及纤维缝编织物	10					100	-Needleloom felt and stitch-bonded fibre fabrics
		-其他毡呢，未浸渍、涂布、包覆或层压：							-Other felt, not impregnated, coated, covered or laminated:
3705	5602.2100	--羊毛或动物细毛制	10					100	--Of wool or fine animal hair
3706	5602.2900	--其他纺织材料制	10					100	--Of other textile materials
3707	5602.9000	-其他	10					100	-Other
	56.03	**无纺织物，不论是否浸渍、涂布、包覆或层压：**							**Nonwovens, whether or not impregnated, coated, covered or laminated:**
		-化学纤维长丝制：							-Of man-made filaments:
		--每平方米重量不超过25克：							--Weighing not more than 25 g/m^2
3708	5603.1110	---经浸渍、涂布、包覆或层压	10		9.7			70	---Impregnated, coated, covered or laminated
3709	5603.1190	---其他	10					130	---Other
		--每平方米重量超过25克，但不超过70克：							--Weighing more than 25 g/m^2 but not more than 70 g/m^2
3710	5603.1210	---经浸渍、涂布、包覆或层压	10		9.7			70	---Impregnated, coated, covered or laminated
3711	5603.1290	---其他	10					130	---Other
		--每平方米重量超过70克，但不超过150克：							--Weighing more than 70 g/m^2 but not more than 150 g/m^2

序号 No.	税则号列 Tariff Item	货品名称	最惠国税率(%) M.F.N.	协定税率(%) Agreement 东盟 ASEAN	曼谷协定 BA	香港 H.K.	澳门 Macao	普通税率(%) Gen.	Article Description
3712	5603.1310	---经浸渍、涂布、包覆或层压	10		9.7			70	---Impregnated, coated, covered or laminated
3713	5603.1390	---其他	10					130	---Other
		--每平方米重量超过150克:							--Weighing more than 150 g/m²
3714	5603.1410	---经浸渍、涂布、包覆或层压	10		9.7			70	---Impregnated, coated, covered or laminated
3715	5603.1490	---其他	10					130	---Other
		-其他:							-Other:
		--每平方米重量不超过25克:							--Weighing not more than 25 g/m²:
3716	5603.9110	---经浸渍、涂布、包覆或层压	10		9.7			70	---Impregnated, coated, covered or laminated
3717	5603.9190	---其他	10		9.3			85	---Other
		--每平方米重量超过25克,但不超过70克:							--Weighing more than 25 g/m² but not more than 70 g/m²:
3718	5603.9210	---经浸渍、涂布、包覆或层压	10		9.7	0		70	---Impregnated, coated, covered or laminated
3719	5603.9290	---其他	10		9.3	0		85	---Other
		--每平方米重量超过70克,但不超过150克:							--Weighing! more than 70 g/m² but not more than 150 g/m²:
3720	5603.9310	---经浸渍、涂布、包覆或层压	10		9.7	0		70	---Impregnated, coated, covered or laminated
3721	5603.9390	---其他	10		9.3	0		85	---Other
		--每平方米重量超过150克:							--Weighing more than 150 g/m²:
3722	5603.9410	---经浸渍、涂布、包覆或层压	10		9.7	0		70	---Impregnated, coated, covered or laminated
3723	5603.9490	---其他	10		9.3	0		85	---Other
	56.04	**用纺织材料包覆的橡胶线及绳;用橡胶或塑料浸渍、涂布、包覆或套裹的纺织纱线及税号54.04或54.05的扁条及类似品:**							**Rubber thread and cord, textile covered; textile yarn, and strip and the like of heading No. 54.04 or 54.05, impregnated, coated, covered or sheathed with rubber or plastics:**
3724	5604.1000	-用纺织材料包覆的橡胶线及绳	5					80	-Rubber thread and cord, textile covered
3725	5604.2000	-浸渍或涂布的聚酯、尼龙、其他聚酰胺或粘胶纤维制的高强力纱	5					40	-High tenacity yarn of polyesters, of nylon or other polyamides or of viscose rayon, impregnated or coated
3726	5604.9000	-其他	5					80	-Other
	56.05	**含金属纱线,不论是否螺旋花线,由纺织纱线或税号54.04或54.05的扁条及类似品与金属线、扁条或粉末混合制得或用金属包覆制得:**							**Metallized yarn, whether or not gimped, being textile yarn, or strip or the like of heading No. 54.04 or 54.05, combined with metal in the form of thread, strip or powder or covered with metal:**

序号 No.	税则号列 Tariff Item	货品名称	最惠国税率(%) M.F.N.	协定税率(%) Agreement 东盟 ASEAN	曼谷协定 BA	香港 H.K.	澳门 Macao	普通税率(%) Gen.	Article Description
3727	5605.0000	含金属纱线,不论是否螺旋花线,由纺织纱线或税号 54.04 或 54.05 的扁条及类似品与金属线、扁条或粉末混合制得或用金属包覆制得	5					70	Metallzied yarn, whether or not gimped, being textile yarn, or strip or the like of heading No. 54. 04 or 54. 05, combined with metal in the form of thread, strip or powder or covered with metal
	56.06	**粗松螺旋花线,税号 54.04 或 54.05 的扁条及类似品制的螺旋花线(税号 56.05 的货品及马毛粗松螺旋花线除外);绳绒线(包括植绒绳绒线);纵行起圈纱线:**							**Gimped yarn, and strip and the like of heading No. 54. 04 or 54. 05, gimped (other than those of heading No.56.05 and gimped horsehair yarn); chenille yarn (including flock chenille yarn); loop waleyarn:**
3728	5606.0000	粗松螺旋花线,税号 54.04 或 54.05 的扁条及类似品制的螺旋花线(税号 56.05 的货品及马毛粗松螺旋花线除外);绳绒线(包括植绒绳绒线);纵行起圈纱线	5					70	Gimped yarn, and strip and the like of heading No. 54. 04 or 54. 05, gimped (other than those of heading No.56.05 and gimped horsehair yarn); chenille yarn (including flock chenille yarn); loopwale-yarn
	56.07	**线、绳、索、缆,不论是否编织或编结而成,也不论是否用橡胶或塑料浸渍,涂布、包覆或套裹:**							**Twine, cordage, ropes and cables, whether or not plaited or braided and whether or not impregnated, coated, covered or sheathed with rubber or plastics:**
3729	5607.1000	-黄麻或税号 53.03 的其他韧皮纺织纤维纺制	5					50	-Of jute or other textile bast fibres of heading No. 53.03
		-西沙尔麻或其他纺织用龙舌兰类纤维纺制:							-Of sisal or other textile fibres of the genus Agave:
3730	5607.2100	--包扎用绳	5					50	--Binder or baler twine
3731	5607.2900	--其他	5					50	--Other
		-聚乙烯或聚丙烯纺制:							-Of polyethylene or polypropylene:
3732	5607.4100	--包扎用绳	5					100	--Binder or baler twine
3733	5607.4900	--其他	5					100	--Other
3734	5607.5000	-其他合成纤维纺制	5					100	-Of other synthetic fibres
3735	5607.9010	---蕉麻(马尼拉麻)或其他硬质(叶)纤维纺制	5					50	---Of abaca (Manila hemp or *Musa textilis Nee*) or other hard (leaf) fibres
3736	5607.9090	---其他	5					100	---Other
	56.08	**线、绳或索结制的网料;纺织材料制成的渔网及其他网:**							**Knotted netting of twine, cordage or rope; made up fishing nets and other made up nets, of textile materials:**

序号 No.	税则号列 Tariff Item	货品名称	最惠国税率(%) M.F.N.	协定税率(%)Agreement				普通税率(%) Gen.	Article Description
				东盟 ASEAN	曼谷协定 BA	香港 H.K.	澳门 Macao		
		-化学纤维材料制:							-Of man-made textile materials:
3737	5608.1100	--制成的渔网	10					50	--Made up fishing nets
3738	5608.1900	--其他	12					100	--Other
3739	5608.9000	-其他	10			0		100	-Other
	56.09	**用纱线、税号54.04或54.05的扁条及类似品或线、绳、索、缆制成的其他税号未列名物品:**							**Articles of yarn, strip or the like of heading No. 54.04 or 54.05, twine, cordage, rope or cables, not elsewhere specified or included:**
3740	5609.0000	用纱线、税号54.04或54.05的扁条及类似品或线、绳、索、缆制成的其他税号未列名物品	10					100	Articles of yarn, strip or the like of heading No. 54.04 or 54.05, twine, cordage, rope or cables, not elsewhere specified or included

第五十七章
地毯及纺织材料的其他铺地制品

Chapter 57
Carpets and other textile floor coverings

注释：

一、本章所称"地毯及纺织材料的其他铺地制品"，是指使用时以纺织材料作面的铺地制品，也包括具有纺织材料铺地制品特征但作其他用途的物品。

二、本章不包括铺地制品衬垫。

Notes:

1. For the purposes of this Chapter, the term "carpets and other textilefloor coverings" means floor coverings in which textile materials serve as the exposed surface of the article when in use and includes articles having the characteristics of textile floor coverings but intended for use for other purposes.

2. This Chapter does not cover floor covering underlays.

序号 No.	税则号列 Tariff Item	货品名称	最惠国税率(%) M.F.N.	协定税率(%)Agreement 东盟 ASEAN	曼谷协定 BA	香港 H.K.	澳门 Macao	普通税率(%) Gen.	Article Description
	57.01	**结织栽绒地毯及纺织材料的其他结织栽绒铺地制品，不论是否制成的：**							**Carpets and other textile floor coverings knotted, whether or not made up:**
3741	5701.1000	-羊毛或动物细毛制	14					130	-Of wool or fine animal hair
		-其他纺织材料制：							-Of other textile materials:
3742	5701.9010	---化学纤维制	16					130	---Of man-made textile materials
3743	5701.9020	---丝制	14					100	---Of silk
3744	5701.9090	---其他	14					100	---Other
	57.02	**机织地毯及纺织材料的其他机织铺地制品，未簇绒或未植绒，不论是否制成的，包括"开来姆"、"苏麦克"、"卡拉马尼"及类似的手织地毯：**							**Carpets and other textile floor coverings, woven, not tufted or flocked, whether or not made up, including "Kelem", "Schumacks", "Karamanie" and similar handwoven rugs:**
3745	5702.1000	-"开来姆"、"苏麦克"、"卡拉马尼"及类似的手织地毯	14					130	-"Kelem", "Schumacks", "Karamanie" and similar hand-woven rugs
3746	5702.2000	-椰壳纤维制的铺地制品	14					100	-Floor coverings of coconut fibres (coir)
		-其他起绒结构的铺地制品，未制成的：							-Other, of pile construction, not made up:
3747	5702.3100	--羊毛或动物细毛制	10					130	--Of wool or fine animal hair
3748	5702.3200	--化学纤维制	16					130	--Of man-made textile materials
3749	5702.3900	--其他纺织材料制	14					100	--Of other textile materials
		-其他起绒结构的铺地制品，制成的：							-Ohter, of pile construction, made up:
3750	5702.4100	--羊毛或动物细毛制	10					130	--Of wool or fine animal hair
3751	5702.4200	--化学纤维制	10					130	--Of man-made textile materials
3752	5702.4900	--其他纺织材料制	14		9.3			100	--Of other textile materials

序号 No.	税则号列 Tariff Item	货品名称	最惠国税率(%) M.F.N.	协定税率(%) Agreement 东盟 ASEAN	曼谷协定 BA	香港 H.K.	澳门 Macao	普通税率(%) Gen.	Article Description
		-其他非起绒结构的铺地制品,未制成的:							-Ohter, not of pile construction, not made up:
3753	5702.5100	--羊毛或动物细毛制	14					130	--Of wool or fine animal hair
3754	5702.5200	--化学纤维制	16					130	--Of man-made textile materials
3755	5702.5900	--其他纺织材料制	14					100	--Of other textile materials
		-其他非起绒结构的铺地制品,制成的:							-Other, not of pile construction, made up:
3756	5702.9100	--羊毛或动物细毛制	14					130	--Of wool or fine animal hair
3757	5702.9200	--化学纤维制	16					130	--Of man-made textile materials
3758	5702.9900	--其他纺织材料制	14					100	--Of other textile materials
	57.03	**簇绒地毯及纺织材料的其他簇绒铺地制品,不论是否制成的:**							**Carpets and other textile floor coverings, tufted, whether or not made up:**
3759	5703.1000	-羊毛或动物细毛制	14					130	-Of wool or fine animal hair
3760	5703.2000	-尼龙或其他聚酰胺制	10					130	-Of nylon or other polyamides
3761	5703.3000	-其他化学纤维制	10					130	-Of other man-made textile materials
3762	5703.9000	-其他纺织材料制	14					100	-Of other textile materials
	57.04	**毡呢地毯及纺织材料的其他毡呢铺地制品,未簇绒或未植绒,不论是否制成的:**							**Carpets and other textile floor coverings, of felt, not tufted or flocked, whether or not made up:**
3763	5704.1000	-最大表面面积不超过0.3平方米	14					130	-Tiles, having a maximum surface area of 0.3 m^2
3764	5704.9000	-其他	10					130	-Other
	57.05	**其他地毯及纺织材料的其他铺地制品,不论是否制成的:**							**Other carpets and other textile floor coverings, whether or not made up:**
3765	5705.0010	---羊毛或动物细毛制	14					130	---Of wool or fine animal hair
3766	5705.0020	---化学纤维制	10					130	---Of man-made textile materials
3767	5705.0090	---其他纺织材料制	14					100	---Other

第五十八章
特种机织物；簇绒织物；花边；装饰毯；装饰带；刺绣品

Chapter 58
Special woven fabrics；tufted textile fabrics；lace；tapestries；trimmings；embroidery

注释：

一、本章不适用于经浸渍、涂布、包覆或层压的第五十九章注释一所述的纺织物或第五十九章的其他货品。

二、税号 58.01 也包括因未将浮纱割断而使表面无竖绒的纬起绒织物。

三、税号 58.03 所称“纱罗”，是指经线全部或部分由地经纱和绞经纱构成的织物，其中绞经纱绕地经纱半圈、一圈或几圈而形成圈状，纬纱从圈中穿过。

四、税号 58.04 不适用于税号 56.08 的线、绳、索结制的网状织物。

五、税号 58.06 所称“狭幅机织物”，是指：

（一）幅宽不超过 30 厘米的机织物，不论是否织成或从宽幅料剪成，但两侧必须有织成的、胶粘的或用其他方法制成的布边；

（二）压平宽度不超过 30 厘米的圆筒机织物；

（三）折边的斜裁滚条布，其未折边时的宽度不超过 30 厘米。

流苏状的狭幅织物归入税号 58.08。

六、税号 58.10 所称“刺绣品”，除了一般纺织材料绣线绣制的刺绣品外，还包括在可见底布上用金属线或玻璃线刺绣的刺绣品，也包括用珠片、饰珠、纺织材料或其他材料制的装饰用花纹图案所缝绣的贴花织物。但不包括手工针绣嵌花装饰毯（税号 58.05）。

七、除税号 58.09 的产品外，本章还包括金属线制的用于衣着、装饰及类似用途的物品。

Notes:

1. This Chapter does not apply to textile fabrics referred to in Note 1 to Chapter 59, impregnated, coated, covered or laminated, or to other goods of Chapter 59.

2. Heading No.58.01 also includes woven weft pile fabrics which have not yet had the floats cut, at which stage they have no pile standing up.

3. For the purpose of heading No.58.03, “gauze” means a fabric with a warp composed wholly or in part of standing or ground threads and crossing or doup threads which cross the standing or ground threads making a half turn, a complete turn or more to form loops through which weft threads pass.

4. Heading No.58.04 does not apply to knotted net fabrics of twine, cordage or rope, of heading No.56.08.

5. For the purposes of heading No.58.06, the expression “narrow woven fabrics” means:
   (a) Woven fabrics of a width not exceeding 30 cm, whether woven as such or cut from wider pieces, provided with selvedges (woven, gummed or otherwise made) on both edges;
   (b) Tubular woven fabrics of a flattened width not exceeding 30 cm; and
   (c) Bias binding with folded edges, of a width when unfolded not exceeding 30 cm.

   Narrow woven fabrics with woven fringes are to be classified in heading No.58.08.

6. In heading No.58.10, the expression “embroidery” means, *inter alia*, embroidery with metal or glass thread on a visible ground of textile fabric, and sewn applique work of sequins, beads or ornamental motifs of textile or other materials. The heading does not apply to meedlework tapestry (heading No.58.05).

7. In addition to the products of heading No.58.09, this Chapter also includes articles made of metal thread and of a kind used in apparel, as furnishing fabrics or for similar purposes.

序号 No.	税则号列 Tariff Item	货品名称	最惠国税率(%) M.F.N.	协定税率(%)Agreement 东盟 ASEAN	曼谷协定 BA	香港 H.K.	澳门 Macao	普通税率(%) Gen.	Article Description
	58.01	**起绒机织物及绳绒织物，但税号58.02或58.06的织物除外：**							**Woven pile fabrics and chenille fabrics, other than fabrics of heading No. 58.02 or 58.06:**
3768	5801.1000	-羊毛或动物细毛制	10					130	-Of wool or fine animal hair
		-棉制：							-Of cotton:
3769	5801.2100	--不割绒的纬起绒织物	12					70	--Uncut weft pile fabrics
3770	5801.2200	--割绒的灯芯绒	10			0		70	--Cut corduroy
3771	5801.2300	--其他纬起绒织物	10					70	--Other weft pile fabrics
3772	5801.2400	--不割绒的经起绒织物（棱纹绸）	10					70	--Warp pile fabrics, èpinglè(uncut)
3773	5801.2500	--割绒的经起绒织物	10					70	--Warp pile fabrics, cut
3774	5801.2600	--绳绒织物	10					70	--Chenille fabrics
		-化学纤维制：							-Of man-made fibres:
3775	5801.3100	--不割绒的纬起绒织物	10					130	--Uncut weft pile fabrics
3776	5801.3200	--割绒的灯芯绒	10					130	--Cut corduroy
3777	5801.3300	--其他纬起绒织物	10					130	--Other weft pile fabrics
3778	5801.3400	--不割绒的经起绒织物（棱纹绸）	10					130	--Warp pile fabrics, èpinglè(uncut)
3779	5801.3500	--割绒的经起绒织物	10					130	--Warp pile fabrics, cut
3780	5801.3600	--绳绒织物	10					130	--Chenille fabrics
		-其他纺织材料制：							-Of other textile materials:
3781	5801.9010	---丝及绢丝制	10					130	---Of silk or silk waste
3782	5801.9090	---其他	10					80	---Other
	58.02	**毛巾织物及类似的毛圈机织物，但税号58.06的狭幅织物除外；簇绒织物，但税号57.03的产品除外：**							**Terry towelling and similar woven terry fabrics, other than narrow fabrics of heading No. 58.06; tufted textile fabrics, other than products of heading No. 57.03:**
		-棉制毛巾织物及类似毛圈机织物：							-Terry towelling and similar woven terry fabrics, of cotton:
3783	5802.1100	--未漂白	12					70	--Unbleached
3784	5802.1900	--其他	10					70	--Other
		-其他纺织材料制的毛巾织物及类似的毛圈机织物：							-Terry towelling and similar woven terry fabrics, of other textile materials:
3785	5802.2010	---丝及绢丝制	12					130	---Of silk or silk waste
3786	5802.2020	---羊毛或动物细毛制	12					130	---Of wool or fine animal hair
3787	5802.2030	---化学纤维制	14					130	---Of man-made fibres
3788	5802.2090	---其他	12					80	---Other
		-簇绒织物：							-Tufted textile fabrics:
3789	5802.3010	---丝及绢丝制	10					130	---Of silk or silk waste
3790	5802.3020	---羊毛或动物细毛制	10					130	---Of wool or fine animal hair
3791	5802.3030	---棉或麻制	10					70	---Of cotton or bast fibre
3792	5802.3040	---化学纤维制	10					130	---Of man-made fibres
3793	5802.3090	---其他纺织材料制	10					80	---Of other textile materials

序号 No.	税则号列 Tariff Item	货品名称	最惠国税率(%) M.F.N.	协定税率(%) Agreement 东盟 ASEAN	曼谷协定 BA	香港 H.K.	澳门 Macao	普通税率(%) Gen.	Article Description
	58.03	**纱罗，但税号58.06的狭幅织物除外：**							**Gauze, other than narrow fabrics of heading No.58.06:**
3794	5803.1000	-棉制	10					70	-Of cotton
		-其他纺织材料制：							--Of other textile materials:
3795	5803.9010	---丝及绢丝制	10					130	---Of silk or silk waste
3796	5803.9020	---化学纤维制	12					130	---Of man-made fibres
3797	5803.9090	---其他	10					80	---Other
	58.04	**网眼薄纱及其他网眼织物，但不包括机织物、针织物或钩编织物；成卷、成条或成小块图案的花边，但税号60.02的织物除外：**							**Tulles and other net fabrics, not including woven, knitted or crocheted fabrics; lace in the piece, in strips or in motifs, other than fabrics of heading No.60.02:**
		-网眼薄纱及其他网眼织物：							-Tulles and other net fabrics:
3798	5804.1010	---丝及绢丝制	10					130	---Of silk or silk waste
3799	5804.1020	---棉制	10					70	---Of cotton
3800	5804.1030	---化学纤维制	12					130	---Of man-made fibres
3801	5804.1090	---其他纺织材料制	10					90	---Of other textile materials
		-机制花边：							-Mechanically made lace:
3802	5804.2100	--化学纤维制	10			0	0	130	--Of man-made fibres
		--其他纺织材料制：							---Of other textile materials:
3803	5804.2910	---丝及绢丝制	10					130	---Of silk or silk waste
3804	5804.2920	---棉制	10					70	---Of cotton
3805	5804.2990	---其他	10					90	---Other
3806	5804.3000	-手工制花边	10					100	-Hand-made lace
	58.05	**"哥白林"、"弗朗德"、"奥步生"、"波威"及类似式样的手织装饰毯，以及手工针绣嵌花装饰毯（例如，小针脚或十字绣），不论是否制成的：**							**Hand-woven tapestries of the type Gobelins, Flanders, Aubusson, Beauvais and the like, and meedle-worked tapestries (for example, petit point, cross stitch), whether or not made up:**
3807	5805.0010	---手工针绣嵌花装饰毯	12					130	---Needle-worked tapestries
3808	5805.0090	---其他	12					130	---Other
	58.06	**狭幅机织物，但税号58.07的货品除外；用粘合剂粘合制成的有经纱而无纬纱的狭幅织物（包扎匹头用带）：**							**Narrow woven fabrics, other than goods of heading No.58.07; narrow fabrics consisting of warp without weft assembled by means of an adhesive (bolducs):**
		-起绒机织物（包括毛巾织物及类似的毛圈织物）及绳绒织物：							-Woven pile fabrics (including terry towelling and similar terry fabris) and chenille fabrics:

序号 No.	税则号列 Tariff Item	货品名称	最惠国税率(%) M.F.N.	协定税率(%) Agreement 东盟 ASEAN	曼谷协定 BA	香港 H.K.	澳门 Macao	普通税率(%) Gen.	Article Description
3809	5806.1010	---棉或麻制	10					70	---Of cotton or bast fibres
3810	5806.1090	---其他纺织材料制	10		9.1			80	---Of other textile materials
3811	5806.2000	-按重量计弹性纱线或橡胶线含量在5%及以上的其他机织物	10			0	0	100	-Other woven fabrics, containing by weight 5% or more of elastomeric yarn or rubber thread
		-其他机织物:							-Other woven fabrics:
3812	5806.3100	--棉制	10					70	--Of cotton
3813	5806.3200	--化学纤维制	10				0	130	--Of man-made fibres
		--其他纺织材料制:							--Of other textile materials:
3814	5806.3910	---丝及绢丝制	10					130	---Of silk or silk waste
3815	5806.3920	---羊毛或动物细毛制	10					130	---Of wool or fine animal hair
3816	5806.3990	---其他	10					80	---Other
		-用粘合剂粘合制成的有经纱而无纬纱的织物(包扎匹头用带):							-Fabrics consisting of warp without weft assembled by means of an adhesive(bolducs):
3817	5806.4010	---棉或麻制	10					70	---Of cotton or bast fibres
3818	5806.4090	---其他纺织材料制	10					80	---Of other textile materials
	58.07	**非绣制的纺织材料制标签、徽章及类似品,成匹、成条或裁成一定形状或尺寸:**							**Labels, badges and similar articles of textile materials, in the piece, in strips or cut to shape or size, not embroidered:**
3819	5807.1000	-机织	10		9.5	0	0	100	-Woven
3820	5807.9000	-其他	10					100	-Other
	58.08	**成匹的编带;非绣制的成匹装饰带,但针织或钩编的除外;流苏、绒球及类似品:**							**Braids in the piece; ornamental trimmings in the piece, without embroidery, other than knitted or crocheted; tassels, pompons and similar articles:**
3821	5808.1000	-成匹的编带	10					100	-Braids in the piece
3822	5808.9000	-其他	10					100	-Other
	58.09	**其他税号未列名的金属线机织物及税号56.05所列含金属纱线的机织物,用于衣着、装饰及类似用途:**							**Woven fabrics of metal thread and woven fabrics of metallized yarn of heading No. 56.05, of a kind used in apparel, as furnishing fabrics or for similar purposes, not elsewhere specified or included:**
3823	5809.0010	---与棉混制	10					90	---Combined with cotton
3824	5809.0020	---与化学纤维混制	10					130	---Combined with man-made fibres
3825	5809.0090	---其他	10					100	---Other
	58.10	**成匹、成条或成小块图案的刺绣品:**							**Embroidery in the piece, in strips or in motifs:**
3826	5810.1000	-不见底布的刺绣品	10					130	-Embroidery without visible ground
		-其他刺绣品:							-Other embroidery:
3827	5810.9100	--棉制	10					130	--Of cotton
3828	5810.9200	--化学纤维制	10					130	--Of man-made fibres

序号 No.	税则号列 Tariff Item	货品名称	最惠国税率(%) M.F.N.	协定税率(%) Agreement 东盟 ASEAN	曼谷协定 BA	香港 H.K.	澳门 Macao	普通税率(%) Gen.	Article Description
3829	5810.9900	--其他纺织材料制	10					130	--Of other textile materials
	58.11	**用一层或几层纺织材料与胎料经绗缝或其他方法组合制成的被褥状纺织品,但税号58.10的刺绣品除外:**							**Quilted textile products in the piece, composed of one or more layers of textile materials assembled with padding by stitching or otherwise, other than embroidery of heading No. 58.10:**
3830	5811.0010	---丝及绢丝制	10					130	---Of silk or silk waste
3831	5811.0020	---羊毛或动物细毛制	10					130	---Of wool or fine animal hair
3832	5811.0030	---棉制	10					80	---Of cotton
3833	5811.0040	---化学纤维制	12					130	---Of man-made fibres
3834	5811.0090	---其他纺织材料制	10					90	---Of other textile materials

第五十九章
浸渍、涂布、包覆或层压的纺织物；工业用纺织制品

注释：

一、除条文另有规定的以外，本章所称“纺织物”，仅适用于第五十章至第五十五章、税号58.03及58.06的机织物、税号58.08的成匹编带和装饰带及税号60.02至60.06的针织物或钩编织物。

二、税号59.03适用于：

（一）用塑料浸渍、涂布、包覆或层压的纺织物，不论每平方米重量多少以及塑料的性质如何（紧密结构或泡沫状的），但下列各项除外：

1. 用肉眼无法辨别出是否经过浸渍、涂布、包覆或层压的织物（通常归入第五十章至第五十五章、第五十八章或第六十章），但由于浸渍、涂布、包覆或层压所引起的颜色变化可不予考虑；

2. 温度在15℃至30℃时，用手工将其绕于直径7毫米的圆柱体上会发生断裂的产品（通常归入第三十九章）；

3. 纺织物完全嵌入塑料内或在其两面均用塑料完全包覆或涂布，而这种包覆或涂布用肉眼是能够辨别出的产品（但由于包覆或涂布所引起的颜色变化可不予考虑）（第三十九章）；

4. 用塑料部分涂布或包覆并由此而形成图案的织物（通常归入第五十章至第五十五章、第五十八章或第六十章）；

5. 与纺织物混制而其中纺织物仅起增强作用的泡沫塑料板、片或带（第三十九章）；

6. 税号58.11的纺织品。

（二）由税号56.04的用塑料浸渍、涂布、包覆或套裹的纱线、扁条或类似品制成的织物。

三、税号59.05所称“糊墙织物”，是指以纺织材料作面，固定在一衬背上或在背面进行处理（浸渍或涂布以便于裱糊），适于装饰墙壁或天花板，且宽度不小于45厘米的成卷产品。

Chapter 59
Impregnated, coated, covered or laminated textile fabrics; textile articles of a kind suitable for industrial use

Notes:

1. Except where the context otherwise requires, for the purposes of this Chapter the expression "textile fabrics" applies only to the woven fabrics of Chapters 50 to 55 and headings Nos.58.03 and 58.06, the braids and ornamental trimmings in the piece of heading No.58.08 and the knitted or crocheted fabrics of headings 60.02 to 60.06.

2. Heading No. 59.03 applies to:

(a) Textile fabrics, impregnated, coated, covered or laminated with plastics, whatever the weight per square metre and whatever the nature of the plastic material (compact or cellular), other than:

(1) Fabrics in which the impregnation, coating, or covering cannot be seen with the naked eye (usually Chapters 50 to 55,58 or 60); for the purpose of this provision, no account should be taken of any resulting change of colour;

(2) Products which cannot, without fracturing, be bent manually around a cylinder of a diameter of 7 mm, at a temperature between 15℃ and 30℃ (usually Chapter 39);

(3) Products in which the textile fabric is either completely embedded in plastics or entirely coated or covered on both sides with such materal, provided that such coating or covering can be seen with the naked eye with no account being taken of any resulting change of colour (Chapter 39);

(4) Fabrics partially coated or partially covered with plastics and bearing designs resulting from these treatments (usually Chapters 50 to 55,58 or 60);

(5) Plates, sheets of strip of cellular plastics, combined with textile fabric, where the textile fabric is present merely for reinforcing purposes (Chapter 39); or

(6) Textile products of heading No. 58.11;

(b) Fabrics made from yarn, strip or the like, impregnated,coated, covered or sheathed with plastics, of heading No. 56.04.

3. For the purposes of heading No. 59.05, the expression "textile wall coverings" applies to products in rolls, of a width of not less than 45 cm, suitable for wall or ceiling decoration, consisting of a textile surface which has beenfixed

但本税号不适用于以纺织纤维屑或粉末直接粘于纸上（税号48.14）或布底上（通常归入税号59.07）的糊墙物品。

on a backing or has been treated on the back (impregnated or coated to permit pasting).

This heading does not, however, apply to wall coverings consisting of textile flock or dust fixed directly on backing of paper (heading No. 48.14) or on a textile backing (generally heading No.59.07).

四、税号59.06所称“用橡胶处理的纺织物”是指：

（一）用橡胶浸渍、涂布、包覆或层压的纺织物：

1．每平方米重量不超过1500克；

2．每平方米重量超过1500克，按重量计纺织材料含量在50%以上；

（二）由税号56.04的用橡胶浸渍、涂布、包覆或套裹的纱线、扁条或类似品制成的织物；

（三）平行纺织纱线经橡胶粘合的织物，不论每平方米重量多少。

但本税号不包括与纺织物混制而其中纺织物仅起增强作用的海绵橡胶板、片或带（第四十章），也不包括税号58.11的纺织品。

4. For the purposes of heading No. 59.06, the expression “rubberized textile fabrics” means:

(a) Textile fabrics impregnated, coated, covered or laminated with rubber:

(Ⅰ) Weighing not more than 1,500g/m^2; or

(Ⅱ) Weighing more than 1,500g/m^2 and containing more than 50% by weight of textile material;

(b) Fabrics made from yarn, strip or the like, impregnated, coated, covered or sheathed with rubber, of heading 56.04; and

(c) Fabrics composed of parallel textile yarns agglomerated with rubber, irrespective of their weight per square metre.

This heading does not, however, apply to plates, sheets or strip of cellular rubber, combined with textile fabric, where the textile fabric is present merely for reinforcing purposes (Chapter 40), or textile products of heading No. 58.11.

五、税号59.07不适用于：

（一）用肉眼无法辨别出是否经过浸渍、涂布或包覆的织物（通常归入第五十章至第五十五章、第五十八章或第六十章），但由于浸渍、涂布或包覆所引起的颜色变化可不予考虑；

（二）绘有图画的织物（作为舞台、摄影布景或类似品的已绘制的画布除外）；

（三）用短绒、粉末、软木粉或类似品部分覆面并由此而形成图案的织物，但仿绒织物仍归入本税号；

（四）以淀粉或类似物质为基本成分的普通浆料上浆整理的织物；

（五）以纺织物为底的木饰面板（税号44.08）；

（六）以纺织物为底的砂布及类似品（税号68.05）；

（七）以纺织物为底的粘聚或复制云母片（税号68.14）；

（八）以纺织物为底的金属箔（第十五类）。

5. Heading No. 59.07 does not apply to:

(a) Fabrics in which the impregnation, coating or covering cannot be seen with the naked eye (usually Chapters 50 to 55,58 or 60); for the purpose of this provision, no account should be taken of any resulting change of colour;

(b) Fabrcs painted with designs (other than painted canvas being theatrical scenery, studio back-cloths or the like);

(c) Fabrics partially covered with flock, dust, powdered cork or the like and bearing designs resulting from these treatments; however, imitation pile fabrics remain classified in this heading;

(d) Fabrics finished with normal dressings having a basis of amylaceous or similar substances;

(e) Wood veneered on a backing of textile fabrics (heading No. 44.08);

(f) Natural or artificial abrasive powder or grain, on a backing of textile fabrics (heading No. 68.05);

(g) Agglomerated or reconstituted mica, on a backing of fextile fabrics (heading No. 68.14); or

(h) Metal foil on a backing of textile fabrics(Section XV).

六、税号59.10不适用于：

（一）厚度小于3毫米的纺织材料制传动带或

6. Heading No.59.10 does not apply to:

(a) Transmission or conveyor belting, of textile material, of

输送带；

（二）用橡胶浸渍、涂布、包覆或层压的织物制成的或用橡胶浸渍、涂布、包覆或套裹的纱线或绳制成的传动带及运输带（税号40.10）。

七、税号59.11适用于下列不能归入第十一类其他税号的货品：

（一）下列成匹的、裁成一定长度或仅裁成矩形（包括正方形）的纺织产品（具有税号59.08至59.10所列产品特征的产品除外）：

1．用橡胶、皮革或其他材料涂布、包覆或层压的作针布用的纺织物、毡呢及毡呢衬里机织物，以及其他专门技术用途的类似织物，包括用橡胶浸渍的用于包覆纺锤（织轴）的狭幅丝绒织物；

2．筛布；

3．用于榨油机器或类似机器的纺织材料制或人发制滤布；

4．用多股经纱或纬纱平织而成的纺织物，不论是否毡化、浸渍或涂布，通常用于机械或其他专门技术用途；

5．专门技术用途的增强纺织物；

6．工业上作填塞或润滑材料的线绳、编带及类似品，不论是否涂布、浸渍或用金属加强。

（二）专门技术用途的纺织制品（税号59.08至59.10的货品除外），例如，造纸机器或类似机器（如制浆机或制石棉水泥的机器）用的环状或装有联接装置的纺织物或毡呢、密封垫、垫圈、抛光盘及其他机器零件。

a thickness of less than 3 mm; or

(b) Transmission or conveyor belts or belting of textile fabric impregnated, coated, covered or laminated with rubber or made from textile yarn or cord impregnated, coated, covered or sheathed with rubber (heading No. 40.10).

7. Heading No. 59.11 applies to the following goods, which do not fall in any other heading of Section XI:

(a) Textile products in the piece, cut to length or simply cut to rectangular (including square) shape (other than those having the character of the products of headings Nos, 59.08 to 59.10), the following only:

(1) Textile fabrics, felt and felt-lined woven fabrics, coated, covered or laminated with rubber, leather or other material, of a kind used for card clothing, and similar fabrics of a kind used for other technical purposes, including narrow fabrics made of velvet impregnated with rubber, for covering weaving spindles (weaving beams);

(2) Bolting cloth;

(3) Straining cloth of a kind used in oil presses or the like, of textile material or of human hair;

(4) Flat woven textile fabrics with multiple warp or weft, whether or not felted, impregnated or coated, of a kind used in machinery or for other technical purposes;

(5) Textile fabrics reinforced with metal, of a kind used for technical purposes;

(6) Cords, braids and the like, whether or not coated, impregnated or reinforced with metal, of a kind used in industry as packing or lubricating materials;

(b) Textile articles (other than those of headings Nos.59.08 to 59.10) of a kind used for technical purposes, for example, textile fabrics and felts, endless or fitted with linking devicse, of a kind used in paper-making or similar machines (for example, for pulp or asbestos-cement), gaskets, washers, polish-ing discs and other machinery parts.

序号 No.	税则号列 Tariff Item	货品名称	最惠国税率(%) M.F.N.	协定税率(%)Agreement 东盟 ASEAN	曼谷协定 BA	香港 H.K.	澳门 Macao	普通税率(%) Gen.	Article Description
	59.01	**用胶或淀粉物质涂布的纺织物,作书籍封面及类似用途的;描图布;制成的油画布;作帽里的硬衬布及类似硬挺纺织物:**							**Textile fabrics coated with gum or amylaceous substances, of a kind used for the outer covers of books or the like; tracing cloth; prepared painting canvas; buckram and similar stiffened textile fabrics of a kind used for hat foundations:**
		-用胶或淀粉物质涂布的纺织物,作书籍封面及类似用途的:							-Textile fabrics coated with gum or amylaceous substances, of a kind used for the outer covers of books or the like:
3835	5901.1010	---棉或麻制	10					80	---Of cotton or bast fibres
3836	5901.1020	---化学纤维制	10					130	---Of man-made fibres
3837	5901.1090	---其他	10					100	---Other
		-其他:							-Other:
3838	5901.9010	---制成的油画布	10					50	---Prepared painting canvas
		---其他:							---Other:
3839	5901.9091	----棉或麻制	10					80	----Of cotton or bast fibres
3840	5901.9092	----化学纤维制	10					130	----Of man-made fibres
3841	5901.9099	----其他	10					100	----Other
	59.02	**尼龙或其他聚酰胺,聚酯或粘胶纤维高强力纱制的帘子布:**							**Tyre cord fabric of high tenacity yarn of nylon or other polyamides, polyesters or viscose rayon:**
3842	5902.1000	-尼龙或其他聚酰胺制	10					40	-Of nylon or other polyamides
3843	5902.2000	-聚酯制	10					40	-Of polyesters
3844	5902.9000	-其他	10					40	-Other
	59.03	**用塑料浸渍、涂布、包覆或层压的纺织物,但税号59.02的货品除外:**							**Textile fabrics impregnated, coated, covered or laminated with plastics, other than those of heading No. 59.02:**
		-用聚氯乙烯浸渍、涂布、包覆或层压的:							-With poly(vinyl chloride):
3845	5903.1010	---绝缘布或带	10		9			40	---Insulating cloth or tape
3846	5903.1020	---人造革	10		9.2	0		70	---Imitation leather
3847	5903.1090	---其他	10		9.1	0		90	---Other
		-用聚氨基甲酸酯浸渍、涂布、包覆或层压的:							-With polyurethane:
3848	5903.2010	---绝缘布或带	10		9			40	---Insulating cloth or tape
3849	5903.2020	---人造革	10		9.2			70	---Imitation leather
3850	5903.2090	---其他	10		9.1	0		90	---Other
		-其他:							-Other:
3851	5903.9010	---绝缘布或带	10		9			40	---Insulating cloth or tape
3852	5903.9020	---人造革	10		9.2			70	---Imitation leather
3853	5903.9090	---其他	10		9.1	0		90	---Other

序号 No.	税则号列 Tariff Item	货品名称	最惠国税率(%) M.F.N.	协定税率(%) Agreement 东盟 ASEAN	曼谷协定 BA	香港 H.K.	澳门 Macao	普通税率(%) Gen.	Article Description
	59.04	**列诺伦(亚麻油地毡),不论是否剪切成形;以织物为底布经涂布或覆面的铺地制品,不论是否剪切成形:**							**Linoleum, whether or not cut to shape; floor coverings consisting of a coating or covering applied on a textile backing, whether or not cut to shape:**
3854	5904.1000	-列诺伦(亚麻油地毡)	14					90	-Linoleum
3855	5904.9000	-其他	14					90	-Other
	59.05	**糊墙织物:**							**Textile wall coverings:**
3856	5905.0000	糊墙织物	10					80	Textile wall coverings
	59.06	**用橡胶处理的纺织物,但税号59.02的货品除外:**							**Rubberized teztile fabrics, other than those of heading No.59.02:**
		-宽度不超过20厘米的胶粘带:							-Adhesive tape of a width not exceeding 20 cm:
3857	5906.1010	---绝缘带	10					40	---Insulating tape
3858	5906.1090	---其他	10					100	---Other
		其他:							-Other:
3859	5906.9100	--针织或钩编的	10					130	--Knitted or crocheted
		--其他:							--Other:
3860	5906.9910	---绝缘布或带	10					40	---Insulating cloth or tape
3861	5906.9990	---其他	10					100	---Other
	59.07	**用其他材料浸渍、涂布或包覆的纺织物;作舞台、摄影布景或类似用途的已绘制画布:**							**Textile fabrics otherwise impregnated, coated or covered; painted canvas being theatrical scenery, studio backcloths or the like:**
3862	5907.0010	---绝缘布或带	10					40	---Insulating cloth or tape
3863	5907.0020	---已绘制画布	10					50	---Painted canvas
3864	5907.0090	---其他	10					100	---Other
	59.08	**用纺织材料机织、编结或针织而成的灯芯、炉芯、打火机芯、烛芯或类似品;煤气灯纱筒及纱罩,不论是否浸渍:**							**Textile wicks, woven, platied or knitted, for lamps, stoves, lighters, candles or the like; incandescent gas mantles and tubular knitted gas mantle fabric therefor, whether or not impregnated:**
3865	5908.0000	用纺织材料机织、编结或针织而成的灯芯、炉芯、打火机芯、烛芯或类似品;煤气灯纱筒及纱罩,不论是否浸渍	10					70	Textile wicks, woven, plaited or knitted, for lamps, stoves, lighters, candles or the like; incandescent gas mantles and tubular knitted gas mantle fabric therefor, whether or not impregnated
	59.09	**纺织材料制的水龙软管及类似的管子,不论有无其他材料作衬里、护套或附件:**							**Textile hosepiping and similar textile tubing, with or without lining, armour or accessories of other materials:**
3866	5909.0000	纺织材料制的水龙软管及类似的管子,不论有无其他材料作衬里、护套或附件	8					35	Textile hosepiping and similar textile tubing, with or without lining, armour or accessories of other materials

序号 No.	税则号列 Tariff Item	货品名称	最惠国税率(%) M.F.N.	协定税率(%)Agreement 东盟 ASEAN	曼谷协定 BA	香港 H.K.	澳门 Macao	普通税率(%) Gen.	Article Description
	59.10	**纺织材料制的传动带或输送带及带料,不论是否用塑料浸渍、涂布、包覆或压层,也不论是否用金属或其他材料加强:**							**Transmission or conveyor belts or belting, of textile material, whether or not impregnated, coated, covered or laminated with plastics, or reinforced with metal or other material:**
3867	5910.0000	纺织材料制的传动带或输送带及带料,不论是否用塑料浸渍、涂布、包覆或压层,也不论是否用金属或其他材料加强	8					35	Transmission or conveyor belts or belting, of textile material, whether or not impregnated, coated, covered or laminated with plastics, or reinforced with metal or other material
	59.11	**本章注释七所规定的作专门技术用途的纺织产品及制品:**							**Textile products and articles, for technical uses, specified in Note 7 to this Chapter:**
		-用橡胶、皮革或其他材料涂布、包覆或层压的作针布用的纺织物、毡呢及毡呢衬里机织物,以及作专门技术用途的类似织物,包括用橡胶浸渍的、用于包覆纺绖(织轴)的狭幅丝绒织物:							-Textile fabrics, felt and felt-lined woven fabrics, coated, covered or laminated with rubber, leather or other material, of a kind used for card clothing, and similar fabrics of a kind used for other technical purposes, including narrow fabrics made of velvet impregnated with rubber, for covering weaving spindles(weaving beams):
3868	5911.1010	---用橡胶浸渍的、用于包覆纺绖(织轴)的狭幅丝绒织物	8		7.2			75	---Narrow fabrics made of velvet impregnated with rubber, for covering weaving spindles(weaving beams)
3869	5911.1090	---其他	8					35	---Other
3870	5911.2000	-筛布,不论是否制成的	8					35	-Bolting cloth, whether or not made up
		-环状或装有联接装置的纺织物及毡呢,用于造纸机器或类似机器(例如,制浆机或制石棉水泥的机器):							-Textile fabrics and felts, endless or fitted with linking devices, of a kind used in paper-making or similar machines(for example, for pulp or asbestos-cement):
3871	5911.3100	--每平方米重量在650克以下	8					35	--Weighing less than $650g/m^2$
3872	5911.3200	--每平方米重量在650克及以上	8					35	--Weighing $650g/m^2$ or more
3873	5911.4000	-用于榨油机器或类似机器的滤布,包括人发制滤布	8					35	-Straining cloth of a kind used in oil presses or the like, including that of human hair
3874	5911.9000	-其他	8					35	-Other

第六十章
针织物及钩编织物

Chapter 60
Knitted or crocheted fabrics

注释：

一、本章不包括：

（一）税号58.04的钩编花边；

（二）税号58.07的针织或钩编的标签、徽章及类似品；

（三）第五十九章的经浸渍、涂布、包覆或层压的针织物及钩编织物。但经浸渍、涂布、包覆或层压的起绒针织物及起绒钩编织物仍归入税号60.01。

二、本章还包括用金属线制的用于衣着、装饰或类似用途的织物。

三、本目录所称"针织物"，包括由纺织纱线用链式针法构成的缝编织物。

Notes:

1. This Chapter does not cover:
   (a) Crochet lace of heading No.58.04;
   (b) Labels, badges or similar articles, knitted or crocheted, of heading No. 58.07; or
   (c) Knitted or crocheted fabrics, impregnated, coated, covered or laminated, of Chapter 59. However, knitted or crocheted plie fabrics, impregnated, coated, covered or laminated, remain classified in heading No. 60.01.

2. This Chapter also includes fabrics made of metal thread and of a kind used in apparel, as furnishing fabrics or for similar purposes.

3. Throughout the Nomenclature any reference to "knitted" goods includes a reference to stitch-bonded goods in which the chain stitches are formed of textile yarn.

序号 No.	税则号列 Tariff Item	货品名称	最惠国税率(%) M.F.N.	协定税率(%) Agreement 东盟 ASEAN	曼谷协定 BA	香港 H.K.	澳门 Macao	普通税率(%) Gen.	Article Description
	60.01	**针织或钩编的起绒织物，包括"长毛绒"织物及毛圈织物：**							**Pile fabrics, including "long pile" fabrics and terry fabrics, knitted or crocheted:**
3875	6001.1000	-"长毛绒"织物	10		9.4			130	-"Long pile" fabrics
		-毛圈绒头织物：							-Looped pile fabrics:
3876	6001.2100	--棉制	10					70	--Of cotton
3877	6001.2200	--化学纤维制	10					130	--Of man-made fibres
3878	6001.2900	--其他纺织材料制	12					130	--Of other textile materials
		-其他：							-Other:
3879	6001.9100	--棉制	10					70	--Of cotton
3880	6001.9200	--化学纤维制	13		9.4	0	0	130	--Of man-made fibres
3881	6001.9900	--其他纺织材料制	12					130	--Of other textile materials
	60.02	**宽度不超过30厘米，按重量计弹性纱线或橡胶线含量在5%及以上的针织物或钩编织物，但税目60.01的货品除外：**							**Knitted or crocheted fabrics of a width not exceeding 30cm, containing by weight 5% or more of elastomeric yarn or rubber thread, other than those of heading 60.01:**
		-按重量计弹性纱线含量在5%及以上，但不含橡胶线：							-Containing by weight 5% or more of elastomeric yarn but not containing rubber thread:
3882	6002.4010	---棉制	10					70	---Of cotton
3883	6002.4020	---丝及绢丝制	10					130	---Of silk or silk waste

序号 No.	税则号列 Tariff Item	货品名称	最惠国税率(%) M.F.N.	协定税率(%)Agreement 东盟 ASEAN	曼谷协定 BA	香港 H.K.	澳门 Macao	普通税率(%) Gen.	Article Description
3884	6002.4030	---合成纤维制	10					130	---Of synthetic fibres
3885	6002.4040	---人造纤维制	10					130	---Of artificial fibres
3886	6002.4090	---其他	10					130	---Other
		-其他:							-Other:
3887	6002.9010	---棉制	10					70	---Of cotton
3888	6002.9020	---丝及绢丝制	10					130	---Of silk or silk waste
3889	6002.9030	---合成纤维制	10					130	---Of synthetic fibres
3890	6002.9040	---人造纤维制	10					130	---Of artificial fibres
3891	6002.9090	---其他	10					130	---Other
	60.03	**宽度不超过30厘米的针织或钩编织物,但税目60.01或60.02的货品除外:**							**Knitted or crocheted fabrics of a width not exceeding 30cm, other than those of heading 60.01 or 60.02:**
3892	6003.1000	-羊毛或动物细毛制	10					130	-Of wool or fine animal hair
3893	6003.2000	-棉制	10					70	-Of cotton
3894	6003.3000	-合成纤维制	10		9.4			130	-Of synthetic fibres
3895	6003.4000	-人造纤维制	10		9.4			130	-Of artificial fibres
3896	6003.9000	-其他	10					130	-Other
	60.04	**宽度超过30厘米,按重量计弹性纱线或橡胶线含量在5%及以上的针织物或钩编织物,但税目60.01的货品除外:**							**Knitted or crocheted fabrics of a width exceeding 30cm, containing by weight 5% or more elastomeric yarn or rubber thread, other than those of heading 60.01:**
		-按重量计弹性纱线含量在5%及以上,但不含橡胶线:							-Containing by weight 5% or more of elastomeric yarn but not containing rubber thread:
3897	6004.1010	---棉制	10			0		70	---Of cotton
3898	6004.1020	---丝及绢丝制	10					130	---Of silk or silk waste
3899	6004.1030	---合成纤维制	10			0	0	130	---Of synthetic fibres
3900	6004.1040	---人造纤维制	10			0	0	130	---Of artificial fibres
3901	6004.1090	---其他	10			0	0	130	---Other
		-其他:							-Other:
3902	6004.9010	---棉制	10			0		70	---Of cotton
3903	6004.9020	---丝及绢丝制	10					130	---Of silk or silk waste
3904	6004.9030	---合成纤维制	10			0	0	130	---Of synthetic fibres
3905	6004.9040	---人造纤维制	10			0	0	130	---Of artificial fibres
3906	6004.9090	---其他	10			0	0	130	---Other
	60.05	**经编织物(包括由花边针织机织成的),但税目60.01至60.04的货品除外:**							**Warp knit fabrics (including those made on galloon knitting machines), other than those of headings 60.01 to 60.04:**
3907	6005.1000	-羊毛或动物细毛制	12					130	-Of wool or fine animal hair
		-棉制:							-Of cotton:
3908	6005.2100	--未漂白或漂白	10			0		70	--Unbleached or bleached
3909	6005.2200	--染色	10			0		70	--Dyed

序号 No.	税则号列 Tariff Item	货品名称	最惠国税率(%) M.F.N.	协定税率(%)Agreement 东盟 ASEAN	曼谷协定 BA	香港 H.K.	澳门 Macao	普通税率(%) Gen.	Article Description
3910	6005.2300	--色织	10			0		70	--Of yarns of different colours
3911	6005.2400	--印花	10			0		70	--Printed
		-合成纤维制							--Of synthetic fibres:
3912	6005.3100	--未漂白或漂白	10		9.4			130	--Unbleached or bleached
3913	6005.3200	--染色	10		9.4			130	--Dyed
3914	6005.3300	--色织	10		9.4			130	--Of yarns of different colours
3915	6005.3400	--印花	10		9.4			130	--Printed
		-人造纤维制:							--Of artificial fibres:
3916	6005.4100	--未漂白或漂白	10		9.4			130	--Unbleached or bleached
3917	6005.4200	--染色	10		9.4			130	--Dyed
3918	6005.4300	--色织	10		9.4			130	--Of yarns of different colours
3919	6005.4400	--印花	10		9.4			130	--Printed
3920	6005.9000	--其他	12					130	--Other
	60.06	**其他针织或钩编织物:**							**Other knitted or crocheted fabrics:**
3921	6006.1000	-羊毛或动物细毛制	12					130	-Of wool or fine animal hair
		-棉制:							-Of cotton:
3922	6006.2100	--未漂白或漂白	10			0	0	70	--Unbleached or bleached
3923	6006.2200	--染色	10			0	0	70	--Dyed
3924	6006.2300	--色织	10			0	0	70	--Of yarns of different colours
3925	6006.2400	--印花	10			0	0	70	--Printed
		-合成纤维制:							-Of synthetic fibres:
3926	6006.3100	--未漂白或漂白	10		9.4	0	0	130	--Unbleached or bleached
3927	6006.3200	--染色	10		9.4	0	0	130	--Dyed
3928	6006.3300	--色织	10		9.4	0	0	130	--Of yarns of different colours
3929	6006.3400	--印花	10		9.4	0	0	130	--Printed
		-人造纤维制:							-Of artificial fibres:
3930	6006.4100	--未漂白或漂白	10		9.4	0	0	130	--Unbleached or bleached
3931	6006.4200	--染色	10		9.4	0	0	130	--Dyed
3932	6006.4300	--色织	10		9.4	0	0	130	--Of yarns of different colours
3933	6006.4400	--印花	10		9.4	0	0	130	--Printed
3934	6006.9000	-其他	12		10.9			130	-Other

第六十一章
针织或钩编的服装及衣着附件

注释：

一、本章仅适用于制成的针织品或钩编织品。

二、本章不包括：

（一）税号 62.12 的货品；

（二）税号 63.09 的旧衣着或其他旧物品；

（三）矫形器具、外科手术带、疝气带及类似品（税号 90.21）。

三、税号 61.03 及 61.04 所称：

（一）“西服套装”，是指面料用相同的织物制成的两件套或三件套的下列成套服装：

一件人体上半身穿着的外套或短上衣，除袖子外，其面料数为四片或四片以上；也可附带一件西服背心，这件背心的前片面料应与套装其他各件的面料相同，后片面料则应与外套或短上衣的衬里料相同；以及一件人体下半身穿着的服装，即不带背带或护胸的长裤、马裤、短裤（游泳裤除外）、裙子或裙裤。

西服套装各件面料质地、颜色及构成必须相同，其款式也必须相同，尺寸大小还须相互般配，但可以用不同织物滚边（缝口上缝入长条织物）。

如果数件人体下半身穿着的服装同时进口或出口（例如，两条长裤、长裤与短裤、裙子或裙裤与长裤），构成西服套装下装的应是一条长裤，而对于女式西服套装，则应是一条裙子或裙裤，其他服装应分别归类。

所称“西服套装”，包括不论是否完全符合上述条件的下列配套服装：

1. 常礼服，由一件后襟下垂并下端开圆弧形叉的素色短上衣和一条条纹长裤组成；

Chapter 61
Articles of apparel and clothing accessories, knitted or crocheted

Notes:

1. This Chapter applies only to made up knitted or crocheted articles.

2. This Chapter does not cover:

(a) Goods of heading No.62.12;

(b) Worn clothing or other worn articles of heading No.63.09; or

(c) Orthopaedic appliances, surgical belts, trusses or the like (heaing No.90.21).

3. For the purposes of headings Nos.61.03 and 61.04:

(a) The term “suit” means a set of garments composed of two or three pieces made up, in respect of their outer surface, in identical fabric and comprising:

- One suit coat or jacket the outer shell of which, exclusive of sleeves, consists of four or more panels, designed to cover the upper part of the body, possibly with a tailored waistcoat in addition whose front panel is made from the same fabric as the outer shell of the other components of the set and whose rear panel is made from the same fabric as the lining of the suit coat or jacket, and
- One garment designed to cover the lower part of the body and consisting of trousers, breeches or shorts (other than swimwear), a skirt or a divided skirt, having neither braces nor bibs.

All of the components of a suit must be of the same fabric construction, style, colour and composition; they must also be of corresponding or compatible size. However, these components may have bindings of different fabric (the narrow fabric stitched to finish raw edges). If several separate components to cover the lower part of the body are presented together (for example, trousers and shorts, or a skirt or divided skirt and trousers), the constituent lower part shall be one pair of trousers, or, in the case of women's or girls'suits, the skirt or divided skirt, the other garments being considered separately.

The term “suit” includes the following sets of garments, whether or not they fulfill all the above conditions:

- morning dress, comprising a plain jacket (cutaway) with rounded tails hanging well down at the back and striped trousers;

2. 晚礼服（燕尾服），一般用黑色织物制成，上衣前襟较短且不闭合，背后有燕尾；

3. 无燕尾套装夜礼服，其中上衣款式与普通上衣相似（可以更为显露衬衣前胸），但有光滑丝质或仿丝质的翻领。

（二）“便服套装”，是指面料相同并作零售包装的下列成套服装（西服套装及税号61.07、61.08或61.09的物品除外）：

一件人体上半身穿着的服装，但套衫及背心除外，因为套衫可在两件套服装中作为内衣，背心也可作为内衣；以及一件或两件不同的人体下半身穿着的服装，即长裤、护胸背带工装裤、马裤、短裤（游泳裤除外）、裙子或裙裤。

便服套装各件面料质地、款式、颜色及构成必须相同；尺寸大小也须相互般配。所称“便服套装”，不包括税号61.12的运动服及滑雪服。

- evening dress (tailcoat), generally made of black fabric, the jacket of which is relatively short at the front, does not close and has narrow skirts cut in at the hips and hanging down behind;
- dinner jacket suits, in which the jacket is similar in style to an ordinary jacket (though perhaps revealing more of the shirt front), but has shiny silk or imitation silk lapels.

(b) The term "ensemble" means a set of garments (other than suits and articles of heading No.61.07, 61.08 or 61.09), composed of several pieces made up in identical fabric, put up for retail sale, and comprising:

- One garment designed to cover the upper part of the body, with the exception of pullovers which may form a sceond upper garment in the sole context of twin sets, and of waistcoats which may also form a second upper garment, and
- One or two different garments, designed to cover the lower part of the body and consisting of trousers, bib and brace overalls, breeches, shorts (other than swimwear), a skirt or a divided skirt.

All of the components of an ensemble must be of the same fabric construction, style, colour and composition; they also must be of corresponding or compatible size. The term "ensemble" does not apply to track suits or ski suits, of heading No.61.12.

四、税号61.05及61.06不包括在腰围以下有口袋的服装、带有罗纹腰带及以其他方式收紧下摆的服装或其织物至少在10厘米×10厘米的面积内沿各方向的直线长度上平均每厘米少于10针的服装。税号61.05不包括无袖服装。

4. Headings Nos.61.05 and 61.06 do not cover garments with pockets below the waist, with a ribbed waistband or other means of tightening at the bottom of the garment, or garments having an average of less than 10 stitches per linear centimetre in each direction counted on an area measuring at least 10cm × 10cm. Heading No.61.05 does not cover sleeveless garments.

五、税号61.09不包括带有束带、罗纹腰带或其他方式收紧下摆的服装。

5. Heading No. 61.09 does not cover garments with a drawstring, a ribbed waistband or other means of tightening at the bottom of the garment.

六、对于税号61.11：

（一）所称“婴儿服装及衣着附件”，是指用于身高不超过86厘米幼儿的服装；也包括婴儿尿布；

（二）既可归入税号61.11，也可归入本章其他税号的物品，应归入税号61.11。

6. For the purposes of heading No.61.11:

(a) The expression "babies'garments and clothing accessories" means articles for young children of a body height not exceeding 86cm; it also covers babies'napkins;

(b) Articles which are, *prima facie*, classifiable both in heading No. 61.11and in other headings of this Chapter are to be classified in heading No.61.11.

七、税号61.12所称“滑雪服”，是指从整个外观和织物质地来看，主要在滑雪（速度滑雪或高山滑雪）时穿着的下列服装或成套服装：

7. For the purposes of heading No.61.12, "ski suits" means garments or sets of garments which, by their general appearance and texture, are identifiable as intended to be worn

（一）“滑雪连身服”，即上下身连在一起的单件服装；除袖子和领子外，滑雪连身服可有口袋或脚带；或

（二）“滑雪套装”即由两件或三件构成一套并作零售包装的下列服装：

一件用一条拉链扣合的带风帽的厚夹克、防风衣、防风短上衣或类似的服装，可以附带一件背心；以及一条不论是否过腰的长裤、一条马裤或一条护胸背带工装裤。

“滑雪套装”也可由一件类似以上（一）款所述的连身服和一件可套在连身服外面的有胎料背心组成。

“滑雪套装”各件颜色可以不同，但面料质地、款式及构成必须相同；尺寸大小也须相互般配。

principally for sking (cross-country or alpine). They consist either of :

(a) a "ski overall", that is, a one-piece garment designed to cover the upper and the lower parts of the body; in addition to sleeves and a collar the ski overall may have pockets or footstraps; or

(b) a "ski ensemble", that is, a set of garments composed of two or three pieces, put up for retail sale and comprising:

- one garment such as an anorak, wind-cheater, wind-jacket or similar article, closed by a slide fastener (zipper), possibly with a waistcoat in addition, and
- one pair of trousers whether or not extending above waist-level, one pair of breeches or one bib and brace overall.

The "ski ensemble" may also consist of an overall similar to the one mentioned in paragraph (a) above and a type of padded, sleeveless jacket worn over the overall.

All the components of a "ski ensemble" must be made up in a fabric of the same texture, style and composition whether or not of the same colour; they also must be of corresponding or compatible size.

八、既可归入税号61.13，也可归入本章其他税号的服装，除税号61.11所列的仍归入该税号外，其余的应一律归入税号61.13。

8. Garments which are, *prima facie*, classifiable both in heading No.61.13 and in other headings of this Chapter, excluding heading No.61.11, are to be classified in heading No.61.13.

九、本章的服装，凡门襟为左压右的，应视为男式；右压左的，应视为女式。但本规定不适用于其式样已明显为男式或女式的服装。无法区别是男式还是女式的服装，应按女式服装归入有关税号。

9. Garments of this Chapter desinged for left over right closure at the front shall be regarded as men's or boys' garments, and those designed for right over left closure at the front as women's or girls'garments. These provisions do not apply where the cut of the garment clearly indicates that it is designed for one or other of the sexes. Garments which cannot be identified as either men's or boys'garments or as women's or girls'garments are to be classified in the headings covering women's or girls'garments.

十、本章物品可用金属线制成。

10. Articles of this Chapter may be made of metal thread.

序号 No.	税则号列 Tariff Item	货品名称	最惠国税率(%) M.F.N.	协定税率(%) Agreement 东盟 ASEAN	曼谷协定 BA	香港 H.K.	澳门 Macao	普通税率(%) Gen.	Article Description
	61.01	**针织或钩编的男式大衣、短大衣、斗篷、短斗篷、带风帽的防寒短上衣(包括滑雪短上衣)、防风衣、防风短上衣及类似品,但税号61.03的货品除外:**							**Men's or boys'overcoats, car-coats, capes, cloaks, anoraks (including ski-jackets), wind-cheaters, wind-jackets and similar articles, knitted or crocheted, other than those of heading No.61.03:**
3935	6101.1000	-羊毛或动物细毛制	25		18	0	0	130	-Of wool or fine animal hair
3936	6101.2000	-棉制	17.5		14	0	0	90	-Of cotton
3937	6101.3000	-化学纤维制	17.5		12.4	0	0	130	-Of man-made fibres
3938	6101.9000	-其他纺织材料制	17.5		12.4	0	0	130	-Of other textile materials
	61.02	**针织或钩编的女式大衣、短大衣、斗篷、短斗篷、带风帽的防寒短上衣(包括滑雪短上衣)、防风衣、防风短上衣及类似品,但税号61.04的货品除外:**							**Women's or girls' overcoats, car-coats, capes, cloaks, anoraks (including ski-jack-ets), wind-cheaters, wind-jackets and similar articles, knitted or crocheted, other than those of heading No.61.04:**
3939	6102.1000	-羊毛或动物细毛制	25		18	0	0	130	-Of wool or fine animal hair
3940	6102.2000	-棉制	17.5		14	0	0	90	-Of cotton
3941	6102.3000	-化学纤维制	17.5		12.4	0	0	130	-Of man-made fibres
3942	6102.9000	-其他纺织材料制	20		13.8	0	0	130	-Of other textile materials
	61.03	**针织或钩编的男式西服套装、便服套装、上衣、长裤、护胸背带工装裤、马裤及短裤(游泳裤除外):**							**Men's or boys' suits, ensembles, jackets, blazers, trousers, bib and brace overalls, breeches and shorts (other than swimwear), knitted or crocheted:**
		-西服套装:							-Suits:
3943	6103.1100	--羊毛或动物细毛制	25		18		0△	130	--Of wool or fine animal hair
3944	6103.1200	--合成纤维制	25		18		0	130	--Of synthetic fibres
3945	6103.1900	--其他纺织材料制	17.5		12.4		0	130	--Of other textile materials
		-便服套装:							-Ensembles:
3946	6103.2100	--羊毛或动物细毛制	25		18		0△	130	--Of wool or fine animal hair
3947	6103.2200	--棉制	20		14.8		0	90	--Of cotton
3948	6103.2300	--合成纤维制	25		18		0	130	--Of synthetic fibres
3949	6103.2900	--其他纺织材料制	25		18		0△	130	--Of other textile materials
		-上衣:							-Jackets and blazers:
3950	6103.3100	--羊毛或动物细毛制	16		11.7	0	0△	130	--Of wool or fine animal hair
3951	6103.3200	--棉制	16		12	0	0	90	--Of cotton
3952	6103.3300	--合成纤维制	19		13.6	0	0	130	--Of synthetic fibres
3953	6103.3900	--其他纺织材料制	16		11.7	0	0△	130	--Of other textile materials
		-长裤、护胸背带工装裤、马裤及短裤:							-Trousers, bib and brace overalls, breeches and shorts:

序号 No.	税则号列 Tariff Item	货品名称	最惠国税率(%) M.F.N.	协定税率(%)Agreement				普通税率(%) Gen.	Article Description
				东盟 ASEAN	曼谷协定 BA	香港 H.K.	澳门 Macao		
3954	6103.4100	--羊毛或动物细毛制	16		11.7		0△	130	--Of wool or fine animal hair
3955	6103.4200	--棉制	16		13.6		0	90	--Of cotton
3956	6103.4300	--合成纤维制	17.5		12.4		0	130	--Of synthetic fibres
3957	6103.4900	--其他纺织材料制	16		11.7		0	130	--Of other textile materials
	61.04	**针织或钩编的女式西服套装、便服套装、上衣、连衣裙、裙子、裙裤、长裤、护胸背带工装裤、马裤及短裤(游泳服除外):**							**Women's or girls' suits, ensembles, jackets, blazers, dresses, skirts, divided skirts, trousers, bib and brace overalls, breeches and shorts (other than swimwear), knitted or crocheted:**
		-西服套装:							-Suits:
3958	6104.1100	--羊毛或动物细毛制	17.5		12.4		0△	130	--Of wool or fine animal hair
3959	6104.1200	--棉制	17.5		14		0	90	--Of cotton
3960	6104.1300	--合成纤维制	25		18		0	130	--Of synthetic fibres
3961	6104.1900	--其他纺织材料制	17.5		12.4		0	130	--Of other textile materials
		-便服套装:							-Ensembles:
3962	6104.2100	--羊毛或动物细毛制	17.5		12.4		0△	130	--Of wool or fine animal hair
3963	6104.2200	--棉制	17.5		14		0	90	--Of cotton
3964	6104.2300	--合成纤维制	25		18		0	130	--Of synthetic fibres
3965	6104.2900	--其他纺织材料制	15		10.4		0△	130	--Of other textile materials
		-上衣:							-Jackets and blazers:
3966	6104.3100	--羊毛或动物细毛制	16		11.7	0	0	130	--Of wool or fine animal hair
3967	6104.3200	--棉制	16		13.6	0	0	90	--Of cotton
3968	6104.3300	--合成纤维制	19		13.6	0	0	130	--Of synthetic fibres
3969	6104.3900	--其他纺织材料制	16		11.7	0	0	130	--Of other textile materials
		-连衣裙:							-Dresses:
3970	6104.4100	--羊毛或动物细毛制	16		11.7	0	0	130	--Of wool or fine animal hair
3971	6104.4200	--棉制	16		13.6	0	0	90	--Of cotton
3972	6104.4300	--合成纤维制	17.5		12.4	0	0	130	--Of synthetic fibres
3973	6104.4400	--人造纤维制	16		11.7		0	130	--Of artificial fibres
3974	6104.4900	--其他纺织材料制	16		11.7	0	0	130	--Of other textile materials
		-裙子及裙裤:							-Skirts and divided skirts:
3975	6104.5100	--羊毛或动物细毛制	14		10.8	0	0	130	--Of wool or fine animal hair
3976	6104.5200	--棉制	14		10.3	0	0	90	--Of cotton
3977	6104.5300	--合成纤维制	16		11.7	0	0	130	--Of synthetic fibres
3978	6104.5900	--其他纺织材料制	14		10.8	0	0	130	--Of other textile materials
		-长裤、护胸背带工装裤、马裤及短裤:							-Trousers, bib and brace overalls, breeches and shorts:
3979	6104.6100	--羊毛或动物细毛制	16		11.7		0	130	--Of wool or fine animal hair
3980	6104.6200	--棉制	16		13.6	0	0	90	--Of cotton
3981	6104.6300	--合成纤维制	17.5		12.4		0	130	--Of synthetic fibres
3982	6104.6900	--其他纺织材料制	16		12		0	130	--Of other textile materials
	61.05	**针织或钩编的男衬衫:**							**Men's or boys' shirts, knitted or crocheted:**
3983	6105.1000	-棉制	16		13.6	0	0	90	-Of cotton
3984	6105.2000	-化学纤维制	17.5		12.4	0	0	130	-Of man-made fibres

序号 No.	税则号列 Tariff Item	货品名称	最惠国税率(%) M.F.N.	协定税率(%) Agreement				普通税率(%) Gen.	Article Description
				东盟 ASEAN	曼谷协定 BA	香港 H.K.	澳门 Macao		
3985	6105.9000	-其他纺织材料制	16		11.7	0	0	130	-Of other textile materials
	61.06	**针织或钩编的女衬衫:**							**Women's or girls' blouses, shirts and shirtblouses, knitted or crocheted:**
3986	6106.1000	-棉制	16		13.6	0	0	90	-Of cotton
3987	6106.2000	-化学纤维制	17.5		12.4	0	0	130	-Of man-made fibres
3988	6106.9000	-其他纺织材料制	16		11.7	0	0	130	-Of other textile materials
	61.07	**针织或钩编的男式内裤、三角裤、长睡衣、睡衣裤、浴衣、晨衣及类似品:**							**Men's or boys' underpants, briefe, night-shirts, pyjamas, bathrobes, dressing gowns and similar articles, knitted or crocheted:**
		-内裤及三角裤:							-Underpants and briefs:
3989	6107.1100	--棉制	14			0	0	90	--Of cotton
3990	6107.1200	--化学纤维制	16		10.9			130	--Of man-made fibres
		--其他纺织材料制:							--Of other textile materials
3991	6107.1910	---丝及绢丝制	14		9.3			130	---Of silk or silk waste
3992	6107.1990	---其他	14		9.3			130	---Other
		-长睡衣及睡衣裤:							-Nightshirts and pyjamas:
3993	6107.2100	--棉制	14					90	--Of cotton
3994	6107.2200	--化学纤维制	16		10.9		0	130	--Of man-made fibres
		--其他纺织材料制:							--Of other textile materials
3995	6107.2910	---丝及绢丝制	14		9.3			130	---Of silk or silk waste
3996	6107.2990	---其他	14		9.3			130	---Other
		-其他:							-Other:
3997	6107.9100	--棉制	14					90	--Of cotton
3998	6107.9200	--化学纤维制	16		10.9			130	--Of man-made fibres
3999	6107.9900	--其他纺织材料制	14		9.3			130	--Of other textile materials
	61.08	**针织或钩编的女式长衬裙、衬裙、三角裤、短衬裤、睡衣、睡衣裤、浴衣、晨衣及类似品:**							**Women's or girls'slips, petticoats, briefs, panties, nightdresses, pyjamas, nègligès, bathrobes, dressing gowns and similar articles, knitted or crocheted:**
		-长衬裙及衬裙:							-Slips and petticoats:
4000	6108.1100	--化学纤维制	16		10.9			130	--Of man-made fibres
		--其他纺织材料制:							--Of other textile materials:
4001	6108.1910	---棉制	14					90	---Of cotton
4002	6108.1920	---丝及绢丝制	14					130	---Of silk or silk waste
4003	6108.1990	---其他	14		9.3			130	---Other
		-三角裤及短衬裤:							-Briefs and panties:
4004	6108.2100	--棉制	14			0	0	90	--Of cotton
4005	6108.2200	--化学纤维制	16		10.9	0	0	130	--Of man-made fibres
		--其他纺织材料制:							--Of other textile materials:
4006	6108.2910	---丝及绢丝制	14		9.3			130	---Of silk or silk waste
4007	6108.2990	---其他	14		9.3			130	---Other
		-睡衣及睡衣裤:							-Nightdresses and pyjamas:
4008	6108.3100	--棉制	14			0	0	90	--Of cotton
4009	6108.3200	--化学纤维制	16		10.9	0	0	130	--Of man-made fibres

序号 No.	税则号列 Tariff Item	货品名称	最惠国税率(%) M.F.N.	协定税率(%) Agreement 东盟 ASEAN	曼谷协定 BA	香港 H.K.	澳门 Macao	普通税率(%) Gen.	Article Description
		--其他纺织材料制:							--Of other textile materials
4010	6108.3910	---丝及绢丝制	14		9.3			130	---Of silk or silk waste
4011	6108.3990	---其他	14		9.3			130	---Other
		-其他:							-Other:
4012	6108.9100	--棉制	14				0	90	--Of cotton
4013	6108.9200	--化学纤维制	16		10.9			130	--Of man-made fibres
4014	6108.9900	--其他纺织材料制	14		9.3			130	--Of other textile materials
	61.09	**针织或钩编的T恤衫、汗衫及其他背心:**							**T-shirts, singlets and other vests, knitted or crocheted:**
4015	6109.1000	-棉制	14		9.4	0	0	90	-Of cotton
		-其他纺织材料制:							-Of other textile materials
4016	6109.9010	---丝及绢丝制	14		9.3		0	130	---Of silk or silk waste
4017	6109.9090	---其他	14		9.3	0	0	130	---Other
	61.10	**针织或钩编的套头衫、开襟衫、背心及类似品:**							**Jerseys, pullovers, cardigans, waistcoats and similar articles, knitted or crocheted:**
		-羊毛或动物细毛制:							-Of wool or fine animal hair:
4018	6110.1100	--羊毛制	14		9.3	0	0	130	--Of wool
4019	6110.1200	--喀什米尔山羊细毛制	14		9.3	0	0	130	--Of kashmir(cashmere)goats
		--其他:							--Other:
4020	6110.1910	---其他山羊细毛制	14		9.3	0	0	130	---Of other goats
4021	6110.1920	---兔毛制	14		9.3	0	0	130	---Of rabbit and hare
4022	6110.1990	---其他	14		9.3	0	0	130	---Other
4023	6110.2000	-棉制	14		11	0	0	90	-Of cotton
4024	6110.3000	-化学纤维制	16		10.9	0	0	130	-Of man-made fibres
		-其他纺织材料制							-Of other textile materials
4025	6110.9010	---丝及绢丝制	14		9.3	0	0	130	---Of silk or silk waste
4026	6110.9090	---其他	14		9.3	0	0	130	---Other
	61.11	**针织或钩编的婴儿服装及衣着附件:**							**Babies' garments and clothing accessories, knitted or crocheted:**
4027	6111.1000	-羊毛或动物细毛制	14		9.3			130	-Of wool or fine animal hair
4028	6111.2000	-棉制	14			0	0	90	-Of cotton
4029	6111.3000	-合成纤维制	16		10.9		0	130	-Of synthetic fibres
4030	6111.9000	-其他纺织材料制	14		9.3			130	-Of other textile materials
	61.12	**针织或钩编的运动服、滑雪服及游泳服:**							**Track suits, ski suits and swimwear, knitted or crocheted:**
		-运动服:							-track suits:
4031	6112.1100	--棉制	16		13.6			90	--Of cotton
4032	6112.1200	--合成纤维制	17.5		12.4		0	130	--Of synthetic fibres
4033	6112.1900	--其他纺织材料制	16		11.7			130	--Of other textile materials
		-滑雪服:							-Ski suits:
4034	6112.2010	---棉制	16		13.6			90	---Of cotton
4035	6112.2090	---其他	19		13.6			130	---Other
		-男式游泳服:							-Men's or boys' swimwear:
4036	6112.3100	--合成纤维制	17.5		12.4	0		130	--Of synthetic fibres

序号 No.	税则号列 Tariff Item	货品名称	最惠国税率(%) M.F.N.	协定税率(%)Agreement 东盟 ASEAN	曼谷协定 BA	香港 H.K.	澳门 Macao	普通税率(%) Gen.	Article Description
4037	6112.3900	--其他纺织材料制	16		11.7			130	--Of other textile materials
		-女式游泳服：							-Women's or girls' swimwear:
4038	6112.4100	--合成纤维制	16		12.4	0		130	--Of synthetic fibres
4039	6112.4900	--其他纺织材料制	16		11.7			130	--Of other textile materials
	61.13	**用税号59.03、59.06或59.07的针织物或钩编织物制成的服装：**							**Garments, made up of knitted or crocheted fabrics of heading No. 59.03, 59.06 or 59.07:**
4040	6113.0000	用税号59.03、59.06或59.07的针织物或钩编织物制成的服装	16		11.7			130	Garments, made up of knitted or crocheted fabrics of heading No. 59.03, 59.06 or 59.07
	61.14	**针织或钩编的其他服装：**							**Other garments, knitted or crocheted:**
4041	6114.1000	-羊毛或动物细毛制	16					130	-Of wool or fine animal hair
4042	6114.2000	-棉制	16				0	90	-Of cotton
4043	6114.3000	-化学纤维制	17.5				0	130	-Of man-made fibres
4044	6114.9000	-其他纺织材料制	16					130	-Of other textile materials
	61.15	**针织或钩编的连裤袜、紧身裤袜、长统袜、短袜及其他袜类，包括用以治疗静脉曲张的长统袜和无外绱鞋底的鞋类：**							**Panty hose, tights, stockings, socks and other hosiery, including stockings for varicose veins and footwear without applied soles, knitted or crocheted:**
		-连裤袜及紧身裤袜：							-Panty hose and tights:
4045	6115.1100	--每根单丝细度在67分特以下的合成纤维制	16					130	--Of synthetic fibres, measuring per single yarn less than 67 decitex
4046	6115.1200	--每根单丝细度在67分特及以上的合成纤维制	16		10.9			130	--Of synthetic fibres, measuring per single yarn 67 decitex or more
		--其他纺织材料制：							--Of other textile materials:
4047	6115.1910	---棉制	14					90	---Of cotton
4048	6115.1990	---其他	14		9.3			130	---Other
4049	6115.2000	-女式长统袜及中统袜，每根单丝细度在67分特以下	14		9.3			130	-Women's full-length or knee-length hosiery, measuring per single yarn less than 67 decitex
		-其他：							-Other:
4050	6115.9100	--羊毛或动物细毛制	14					130	--Of wool or fine animal hair
4051	6115.9200	--棉制	14					90	--Of cotton
4052	6115.9300	--合成纤维制	16		10.9			130	--Of synthetic fibres
4053	6115.9900	--其他纺织材料制	14		9.3			130	--Of other textile materials
	61.16	**针织或钩编的分指手套、连指手套及露指手套：**							**Gloves, mittens and mitts, knitted or crocheted:**

序号 No.	税则号列 Tariff Item	货品名称	最惠国税率(%) M.F.N.	协定税率(%)Agreement 东盟 ASEAN	曼谷协定 BA	香港 H.K.	澳门 Macao	普通税率(%) Gen.	Article Description
4054	6116.1000	-用塑料或橡胶浸渍、涂布或包覆的	14					130	-Gloves impregnated, coated or covered with plastics or rubber
		-其他:							-Other:
4055	6116.9100	--羊毛或动物细毛制	14					130	--Of wool of fine animal hair
4056	6116.9200	--棉制	14				0	90	--Of cotton
4057	6116.9300	--合成纤维制	16		10.9		0	130	--Of synthetic fibres
4058	6116.9900	--其他纺织材料制	14		9.3			130	--Of other textile materials
	61.17	**其他制成的针织或钩编的衣着附件;服装或衣着附件的针织或钩编的零件:**							**Other made up clothing accessories, knitted or crocheted; knitted or crocheted parts of garments or of clothing accessories:**
4059	6117.1000	-披巾、头巾、围巾、披纱、面纱及类似品	14		9.3			130	-Shawls, scarves, mufflers, mantillas, veils and the like
4060	6117.2000	-领带及领结	14		9.3			130	-Ties, bow ties and cravats
4061	6117.8000	-其他附件	14		9.3			130	-Other accessories
4062	6117.9000	-零件	14		9.3	0	0	130	-Parts

第六十二章
非针织或非钩编的服装及衣着附件

注释：

一、本章仅适用于除絮胎以外任何纺织物的制成品，但不适用于针织品或钩编织品（税号62.12的除外）。

二、本章不包括：

（一）税号63.09的旧衣着或其他旧物品；

（二）矫形器具、外科手术带、疝气带及类似品（税号90.21）。

三、税号62.03及62.04所称：

（一）“西服套装”，是指面料用相同的织物制成的两件套或三件套的下列成套服装：

一件人体上半身穿着的外套或短上衣，除袖子外，其面料数为四片或四片以上；也可附带一件西服背心，这件背心的前片面料应与套装其他各件的面料相同，后片料则应与外套或短上衣的衬里料相同；以及一件人体下半身穿着的服装，即不带背带或护胸的长裤、马裤、短裤（游泳裤除外）、裙子或裙裤。

西服套装各件面料质地、颜色及构成必须相同，其款式也必须相同，尺寸大小还须相互般配，但可以用不同织物滚边（缝口上缝入长条织物）。

如果数件人体下半身穿着的服装同时进口或出口（例如，两条长裤、长裤与短裤、裙子或裙裤与长裤），构成西服套装下装的应是一条长裤，而对于女式西服套装，则应是一条裙子或裙裤，其他服装应分别归类。

所称“西服套装”，包括不论是否完全符合上述条件的下列配套服装：

1. 常礼服，由一件后襟下垂并下端开圆弧形叉的素色短上衣和一条条纹长裤组成；

Chapter 62
Articles of apparel and clothing accessories, not knitted or crocheted

Notes:

1. This Chapter applies only to made up articles of any textile fabric other than wadding, excluding knitted or crocheted articles (other than those of heading No.62.12).

2. This Chapter does not cover:

(a) Worn clothing or other worn articles of heading No.63.09; or

(b) Orthopaedic appliances, surgical belts, trusses or the like (heading No.90.21).

3. For the purposes of headings Nos.62.03 and 62.04:

(a) The term "suit" means a set of garments composed of two or three pieces made up in identical fabric and comprising:

-one suit coat or jacket the outer shell of which, exclusive of sleeves, consists of four or more panels, designed to cover the upper part of the body, possibly with a tailored waistcoat in addition whose front panel is made from the same fabric as the outer shell of the other components of the set and whose rear panel is made from the same fabric as the lining of the suit coat or jacket, and

- One garment designed to cover the lower part of the body and consisting of trousers, breeches or shorts (other than swimwear), a skirt or a divided skirt, having neither braces nor bibs.

All of the components of a suit must be of the same fabric construction, style, colour and composition; they must also be of corresponding or compatible size. However, these components may have bindings of different fabric (the narrow fabric stitched to finish raw edges). If several separate components to cover the lower part of the body are presented together (for example, trousers and shorts, or a skirt or divided skirt and trousers), the constituent lower part shall be the trousers, or, in the case of women's or girls' suits, the skirt or divided skirt, the other garments being considered separately.

The term "suit" includes the following sets of garments, whether or not they fulfil all the above conditions:

-morning dress, comprising a plain jacket (cutaway) with rounded tails hanging well down at the back and striped trousers;

2. 晚礼服（燕尾服），一般用黑色织物制成，上衣前襟较短且不闭合，背后有燕尾；

3. 无燕尾套装夜礼服，其中上衣款式与普通上衣相似（可以更为显露衬衣前胸），但有光滑丝质或仿丝质的翻领。

（二）“便服套装”，是指面料相同并作零售包装的下列成套服装（西服套装及税号62.07或62.08的物品除外）：

一件人体上半身穿着的服装，但背心除外，因为背心可作为内衣；以及一件或两件不同的人体下半身穿着的服装，即长裤、护胸背带工装裤、马裤、短裤（泳裤除外）、裙子或裙裤。

便服套装各件面料质地、款式、颜色及构成必须相同；尺寸大小也须相互般配。所称“便服套装”，不包括税号62.11的运动服及滑雪服。

四、对于税号62.09：

（一）所称“婴儿服装及衣着附件”，是指用于身高不超过86厘米幼儿的服装；也包括婴儿尿布；

（二）既可归入税号62.09，也可归入本章其他税号的物品，应归入税号62.09；

五、既可归入税号62.10，也可归入本章其他税号的服装，除税号62.09所列的仍归入该税号外，其余的应一律归入税号62.10。

六、税号62.11所称“滑雪服”，是指从整个外观和织物质地来看，主要在滑雪（速度滑雪和高山滑雪）时穿着的下列服装或成套服装：

（一）“滑雪连身服”，即上下身连在一起的单件服装；除袖子和领子外，滑雪连身服可有口袋或脚带；或

（二）“滑雪套装”，即由两件或三件构成一套并作零售包装的下列服装：

- evening dress (tailcoat), generally made of black cloth, the jacket of which is relatively short at the front, does not close and has narrow skirts cut in at the hips and hanging down behind;
- dinner jacket suits, in which the jacket is similar in style to an ordinary jacket (though perhaps revealing more of the shirt front), but has shiny silk or imitation silk lapels.

(b) The term "ensemble" means a set of garments (other than suits and articles of heading No.62.07 or 62.08) composed of several pieces made up in identical fabric, put up for retail sale, and comprising:

- one garment designed to cover the upper part of the body, with the exception of waistcoats which may also form a second upper garment, and
- one or two different garments, designed to cover the lower part of the body and consisting of trousers, bib and brace overalls, breeches, shorts (other than swimwear), a skirt or a divided skirt.

All of the components of an ensemble must be of the same fabric construction, style, colour and composition; they also must be of corresponding or compatible size. The term "ensemble" does not apply to track suits or ski suits, of heading No.62.11.

4. For the purposes of heading No.62.09:

(a) The expression "babies' garments and clothing accessories" means articles for young children of a body height not exceeding 86cm; it also covers babies' napkins;

(b) Articles which are, *prima facie*, classifiable both in heading No.62.09 and in other headings of this Chapter are to be classified in heading No.62.09.

5. Garments which are, *prima facie*, classifiable both in heading No.62.10 and in other headings of this Chapter, excluding heading No.62.09, are to be classified in heading No.62.10.

6. For the purposes of heading No.62.11, "ski suits" means garments or sets of garments which, by their general appearance and texture, are identifiable as intended to be worn principally for skiing (cross-country or alpine). They consist either of:

(a) a "ski overall", that is, a one-piece garment designed to cover the upper and the lower parts of the body; in addition to sleeves and a collar the ski overall may have pockets or footstraps; or

(b) a "ski ensemble", that is, a set of garments composed of two or three pieces, put up for retail sale and comprising:

一件用一条拉链扣合的带风帽的厚夹克、防风衣、防风短上衣或类似的服装，可以附带一件背心；以及一条不论是否过腰的长裤、一条马裤或一条护胸背带工装裤。

“滑雪套装”也可由一件类似以上（一）款所述的连身服和一件可套在连身服外面的有胎料背心组成。

“滑雪套装”各件颜色可以不同，但面料质地、款式及构成必须相同；尺寸大小也须相互般配。

- one garment such as an anorak, wind-cheater, wind-jacket or similar article, closed by a slide fastener (zipper), possibly with a waistcoat in addition, and
- one pair of trousers whether or not extending above waist - level, one pair of breeches or one bib and brace overall.

The "ski ensemble" may also consist of an overall similar to the one mentioned in paragraph (a) above and a type of padded, sleeveless jacket worn over the overall.

All the components of a "ski ensemble" must be made up in a fabric of the same texture, style and composition whether or not of the same colour; they also must be of corresponding or compatible size.

七、正方形或近似正方形的围巾及围巾式样的物品，如果每边均不超过60厘米，应作为手帕归类（税号62.13）。任何一边超过60厘米的手帕，应归入税号62.14。

7. Scarves and articles of the scarf type, square or approximately square, of which no side exceeds 60cm, are to be classified as handkerchiefs (heading No.62.13). Handkerchiefs of which any side exceeds 60cm are to be classified in heading No.62.14.

八、本章的服装，凡门襟为左压右的，应视为男式；右压左的，应视为女式。但本规定不适用于其式样已明显为男式或女式的服装。无法区别是男式还是女式的服装，应按女式服装归入有关税号。

8. Garments of this Chapter designed for left over right closure at the front shall be regarded as men's or boys' garments, and those designed for right over left closure at the front as women's or girls' garments. These provisions do not aplly where the cut of the garment clearly indicates that it is designed for other of sexes. Garments which cannot be identified as either men's or boys' garments or as women's or girls' garments are to be classified in the headings concerning women's or girls' garments.

九、本章物品可用金属线制成。

9. Articles of this chapter may be made of metal thread.

序号 No.	税则号列 Tariff Item	货品名称	最惠国税率(%) M.F.N.	协定税率(%) Agreement 东盟 ASEAN	曼谷协定 BA	香港 H.K.	澳门 Macao	普通税率(%) Gen.	Article Description
	62.01	**男式大衣、短大衣、斗篷、短斗篷、带风帽的防寒短上衣(包括滑雪短上衣)、防风衣、防风短上衣及类似品，但税号62.03的货品除外：**							**Men's or boys' overcoats, car-coats, capes, cloaks, anoraks (including ski-jackets), wind-cheaters, wind-jackets and similar articles, other than those of heading No.62.03:**
		-大衣、雨衣、短大衣、斗篷、短斗篷及类似品：							-Overcoats, raincoats, car-coats, capes, cloaks and similar articles:
4063	6201.1100	--羊毛或动物细毛制	16		11.7	0	0	130	--Of wool or fine animal hair
		--棉制：							--Of cotton:
4064	6201.1210	---羽绒服	16			0	0	90	---Padded with feathers or down
4065	6201.1290	---其他	16			0	0	90	---Other
		--化学纤维制：							--Of man-made fibres:
4066	6201.1310	---羽绒服	17.5		12.4	0△	0	130	---Padded with feathers or down
4067	6201.1390	---其他	17.5		12.4	0△	0	130	---Other

序号 No.	税则号列 Tariff Item	货品名称	最惠国税率(%) M.F.N.	协定税率(%)Agreement				普通税率(%) Gen.	Article Description
				东盟 ASEAN	曼谷协定 BA	香港 H.K.	澳门 Macao		
4068	6201.1900	--其他纺织材料制	16		11.7	0		100	--Of other textile materials
		-其他:							-Other:
4069	6201.9100	--羊毛或动物细毛制	16		11.7	0	0	130	--Of wool or fine animal hair
		--棉制:							--Of cotton:
4070	6201.9210	---羽绒服	16			0	0	90	---Padded with feathers or down
4071	6201.9290	---其他	16			0	0	90	---Other
		--化学纤维制:							--Of man-made fibres:
4072	6201.9310	---羽绒服	17.5		12.4	0	0	130	---Padded with feathers or down
4073	6201.9390	---其他	17.5		12.4	0	0	130	---Other
4074	6201.9900	--其他纺织材料制	16		11.7	0	0	100	--Of other textile materials
	62.02	**女式大衣、短大衣、斗篷、短斗篷、带风帽的防寒短上衣(包括滑雪短上衣)、防风衣、防风短上衣及类似品,但税号62.04的货品除外:**							**Women's or girls' overcoats, car-coats, capes, cloaks, anoraks (including ski-jackets), wind-cheaters, wind-jackets and similar articles, other than those of heading No.62.04:**
		-大衣、雨衣、短大衣、斗篷、短斗篷及类似品:							-Overcoats, raincoats, car-coats, capes, cloaks and similar articles:
4075	6202.1100	--羊毛或动物细毛制	16		11.7	0	0	130	--Of wool or fine animal hair
		--棉制:							--Of cotton:
4076	6202.1210	---羽绒服	16				0	90	---Padded with feathers or down
4077	6202.1290	---其他	16				0	90	---Other
		--化学纤维制:							--Of man-made fibres:
4078	6202.1310	---羽绒服	19		13.6	0△	0	130	---Padded with feathers or down
4079	6202.1390	---其他	19		13.6	0△	0	130	---Other
4080	6202.1900	--其他纺织材料制	16		11.7		0	100	--Of other textile materials
		-其他:							-Other:
4081	6202.9100	--羊毛或动物细毛制	16		11.7	0△	0	130	--Of wool or fine animal hair
		--棉制:							--Of cotton:
4082	6202.9210	---羽绒服	16			0△	0	90	---Padded with feathers or down
4083	6202.9290	---其他	16			0△	0	90	---Other
		--化学纤维制:							--Of man-made fibres:
4084	6202.9310	---羽绒服	17.5		12.4	0△	0	130	---Padded with feathers or down
4085	6202.9390	---其他	17.5		12.4	0△	0	130	---Other
4086	6202.9900	--其他纺织材料制	16		11.7	0△	0	100	--Of other textile materials
	62.03	**男式西服套装、便服套装、上衣、长裤、护胸背带工装裤、马裤及短裤(游泳裤除外):**							**Men's or boys' suits, ensembles, jackets, blazers, trousers, bib and brace overalls, breeches and shorts (other than swim wear):**
		-西服套装:							-Suits:
4087	6203.1100	--羊毛或动物细毛制	17.5		12.4	0	0	130	--Of wool or fine animal hair
4088	6203.1200	--合成纤维制	17.5		12.4	0	0△	130	--Of synthetic fibres
		--其他纺织材料制:							--Of other textile materials:
4089	6203.1910	---丝及绢丝制	17.5		12.4	0	0	100	---Of silk or silk waste
4090	6203.1990	---其他	17.5		12.4	0	0	100	---Other
		-便服套装:							-Ensembles:

序号 No.	税则号列 Tariff Item	货品名称	最惠国税率(%) M.F.N.	协定税率(%) Agreement 东盟 ASEAN	曼谷协定 BA	香港 H.K.	澳门 Macao	普通税率(%) Gen.	Article Description
4091	6203.2100	--羊毛或动物细毛制	17.5		12.4		0△	130	--Of wool or fine animal hair
4092	6203.2200	--棉制	17.5				0	90	--Of cotton
4093	6203.2300	--合成纤维制	17.5		12.4		0	130	--Of synthetic fibres
		--其他纺织材料制:							--Of other textile materials:
4094	6203.2910	---丝及绢丝制	17.5		12.4		0△	130	---Of silk or silk waste
4095	6203.2990	---其他	17.5		12.4		0△	100	---Other
		-上衣:							-Jackets and blazers:
4096	6203.3100	--羊毛或动物细毛制	16		11.7	0	0	130	--Of wool or fine animal hair
4097	6203.3200	--棉制	16			0	0	90	--Of cotton
4098	6203.3300	--合成纤维制	17.5		12.4	0	0	130	--Of synthetic fibres
		--其他纺织材料制:							--Of other textile materials:
4099	6203.3910	---丝及绢丝制	16		11.7	0	0	130	---Of silk or silk waste
4100	6203.3990	---其他	16		11.7	0	0	100	---Other
		-长裤、护胸背带工装裤、马裤及短裤:							-Trousers, bib and brace overalls, breeches and shorts:
4101	6203.4100	--羊毛或动物细毛制	16		11.7	0	0	130	--Of wool or fine animal hair
		--棉制:							--Of cotton:
4102	6203.4210	---阿拉伯裤	16		12.9	0	0	90	---Arabian trousers
4103	6203.4290	---其他	16		12.9	0	0	90	---Other
		--合成纤维制:							--Of synthetic fibres:
4104	6203.4310	---阿拉伯裤	17.5		13.1	0	0	130	---Arabian trousers
4105	6203.4390	---其他	17.5		13.1	0	0	130	---Other
		--其他纺织材料制:							--Of other textile materials:
4106	6203.4910	---阿拉伯裤	16		11.7	0	0	100	---Arabian trousers
4107	6203.4990	---其他	16		11.7	0	0	100	---Other
	62.04	**女式西服套装、便服套装、上衣、连衣裙、裙子、裙裤、长裤、护胸背带工装裤、马裤及短裤(游泳服除外):**							**Women's or girls' suits, ensembles, jackets, blazers dresses, skirts, divided skirts, trousers, bib and brace overalls, breeches and shorts (other than swimwear):**
		-西服套装:							-Suits:
4108	6204.1100	--羊毛或动物细毛制	17.5		12.4		0△	130	--Of wool or fine animal hair
4109	6204.1200	--棉制	17.5				0	90	--Of cotton
4110	6204.1300	--合成纤维制	17.5		12.4		0	130	--Of synthetic fibres
		--其他纺织材料制:							--Of other textile materials:
4111	6204.1910	---丝及绢丝制	17.5		12.4		0△	100	---Of silk or silk waste
4112	6204.1990	---其他	17.5		12.4		0△	100	---Other
		-便服套装:							-Ensembles:
4113	6204.2100	--羊毛或动物细毛制	17.5		12.4		0△	130	--Of wool or fine animal hair
4114	6204.2200	--棉制	17.5				0	90	--Of cotton
4115	6204.2300	--合成纤维制	20		13.8		0	130	--Of synthetic fibres
		--其他纺织材料制:							--Of other textile materials:
4116	6204.2910	---丝及绢丝制	20		13.8		0	130	---Of silk or silk waste
4117	6204.2990	---其他	14		9.9		0△	100	---Other
		-上衣:							-Jackets and blazers:
4118	6204.3100	--羊毛或动物细毛制	16		11.7	0	0	130	--Of wool or fine animal hair

序号 No.	税则号列 Tariff Item	货品名称	最惠国税率(%) M.F.N.	协定税率(%)Agreement 东盟 ASEAN	曼谷协定 BA	香港 H.K.	澳门 Macao	普通税率(%) Gen.	Article Description
4119	6204.3200	--棉制	16			0	0	90	--Of cotton
4120	6204.3300	--合成纤维制	17.5		12.4	0	0	130	--Of synthetic fibres
		--其他纺织材料制:							--Of other textile materials:
4121	6204.3910	---丝及绢丝制	16		11.7	$0^{\triangle}$	0	130	---Of silk or silk waste
4122	6204.3990	---其他	16		11.7	0	0	100	---Other
		-连衣裙:							-Dresses:
4123	6204.4100	--羊毛或动物细毛制	16		11.7		0	130	--Of wool or fine animal hair
4124	6204.4200	--棉制	16				0	90	--Of cotton
4125	6204.4300	--合成纤维制	17.5		12.4	0	0	130	--Of synthetic fibres
4126	6204.4400	--人造纤维制	16		11.7		0	130	--Of artificial fibres
		--其他纺织材料制:							--Of other textile materials:
4127	6204.4910	---丝及绢丝制	16		11.7		0	130	---Of silk or silk waste
4128	6204.4990	---其他	16		11.7	0	0	100	---Other
		-裙子及裙裤:							-Skirts and divided skirts:
4129	6204.5100	--羊毛或动物细毛制	14		9.3	$0^{\triangle}$	0	130	--Of wool or fine animal hair
4130	6204.5200	--棉制	14			0	0	90	--Of cotton
4131	6204.5300	--合成纤维制	16		11.7	0	0	130	--Of synthetic fibres
		--其他纺织材料制:							--Of other textile materials
4132	6204.5910	---丝及绢丝制	14		9.3		0	130	---Of silk or silk waste
4133	6204.5990	---其他	14		9.3	0	0	100	---Other
		-长裤、护胸背带工装裤、马裤及短裤:							-Trousers, bib and brace overalls, breeches and shorts:
4134	6204.6100	--羊毛或动物细毛制	16		11.7	0	0	130	--Of wool or fine animal hair
4135	6204.6200	--棉制	16			0	0	90	--Of cotton
4136	6204.6300	--合成纤维制	17.5		12.4	0	0	130	--Of synthetic fibres
4137	6204.6900	--其他纺织材料制	16		12	0	0	100	--Of other textile materials
	62.05	**男衬衫:**							**Men's or boys'shirts:**
4138	6205.1000	-羊毛或动物细毛制	16		11.7		$0^{\triangle}$	100	-Of wool or fine animal hair
4139	6205.2000	-棉制	16		13.6	0	0	90	-Of cotton
4140	6205.3000	-化学纤维制	16		11.7	0	0	130	-Of man-made fibres
		-其他纺织材料制:							-Of other textile materials
4141	6205.9010	---丝及绢丝制	16		11.7		0	130	---Of silk or silk waste
4142	6205.9090	---其他	16		11.7	0	0	100	---Other
	62.06	**女衬衫:**							**Women's or girls'blouses, shirts and shirtblouses:**
4143	6206.1000	-丝及绢丝制	16		11.7		0	130	-Of silk or silk waste
4144	6206.2000	-羊毛或动物细毛制	16		11.7		$0^{\triangle}$	130	-Of wool or fine animal hair
4145	6206.3000	-棉制	16		13.6	0	0	90	-Of cotton
4146	6206.4000	-化学纤维制	17.5		12.4	0	0	130	-Of man-made fibres
4147	6206.9000	其他纺织材料制	16		11.7	0	0	100	-Of other textile materials
	62.07	**男式背心及其他内衣、内裤、三角裤、长睡衣、睡衣裤、浴衣、晨衣及类似品:**							**Men's or boys' singlets and other vests, underpants, briefs, nightshirts, pyjamas, bathrobes, dressing gowns and similar articles:**
		-内裤及三角裤:							-Underpants and briefs:

序号 No.	税则号列 Tariff Item	货 品 名 称	最惠国税率(%) M.F.N.	协定税率(%) Agreement				普通税率(%) Gen.	Article Description
				东盟 ASEAN	曼谷协定 BA	香港 H.K.	澳门 Macao		
4148	6207.1100	--棉制	14		10.3		0	90	--Of cotton
		--其他纺织材料制:							--Of other textile materials:
4149	6207.1910	---丝及绢丝制	14					130	---Of silk or silk waste
4150	6207.1920	---化学纤维制	16					130	---Of man-made fibres
4151	6207.1990	---其他	14					100	---Other
		-长睡衣及睡衣裤:							-Nightshirts and pyjamas:
4152	6207.2100	--棉制	14					90	--Of cotton
4153	6207.2200	--化学纤维制	16					130	--Of man-made fibres
		--其他纺织材料制:							--Of other textile materials:
4154	6207.2910	---丝及绢丝制	14					130	---Of silk or silk waste
4155	6207.2990	---其他	14					100	---Other
		-其他:							-Other:
4156	6207.9100	--棉制	14			0		90	--Of cotton
4157	6207.9200	--化学纤维制	16					130	--Of man-made fibres
		-其他纺织材料制:							--Of other textile materials:
4158	6207.9910	---丝及绢丝制	14			0		130	---Of silk or silk waste
4159	6207.9990	---其他	14			0		100	---Other
	62.08	**女式背心及其他内衣、长衬裙、衬裙、三角裤、短衬裤、睡衣、睡衣裤、浴衣、晨衣及类似品:**							**Women's or girls' singlets and other vests, slips, petticoats, briefs, panties, nightdresses, pyjamas, nègligès, bathrobes, dressing gowns and similar articles:**
		-长衬裙及衬裙:							-Slips and petticoats:
4160	6208.1100	--化学纤维制	16					130	--Of man-made fibres
		--其他纺织材料制:							--Of other textile materials:
4161	6208.1910	---丝及绢丝制	14					130	---Of silk or silk waste
4162	6208.1920	---棉制	14					90	---Of cotton
4163	6208.1990	---其他	14					100	---Other
		-睡衣及睡衣裤:							-Nightdresses and pyjamas:
4164	6208.2100	--棉制	14		10.3		0	90	--Of cotton
4165	6208.2200	--化学纤维制	16			0	0	130	--Of man-made fibres
		--其他纺织材料制:							--Of other textile materials:
4166	6208.2910	---丝及绢丝制	14					130	---Of silk or silk waste
4167	6208.2990	---其他	14					100	---Other
		-其他:							-Other:
4168	6208.9100	--棉制	14				0	90	--Of cotton
4169	6208.9200	--化学纤维制	16					130	--Of man-made fibres
		--其他纺织材料制:							--Of other textile materials:
4170	6208.9910	---丝及绢丝制	14					130	---Of silk or silk waste
4171	6208.9990	---其他	14					100	---Other
	62.09	**婴儿服装及衣着附件:**							**Babies' garments and clothing accessories:**
4172	6209.1000	-羊毛或动物细毛制	14					130	-Of wool or fine animal hair
		-棉制:							-Of cotton:
4173	6209.2010	---尿布	14					80	---Babies' napkins(diapers)
4174	6209.2090	---其他	14				0	90	---Other

序号 No.	税则号列 Tariff Item	货品名称	最惠国税率(%) M.F.N.	协定税率(%)Agreement 东盟 ASEAN	曼谷协定 BA	香港 H.K.	澳门 Macao	普通税率(%) Gen.	Article Description
4175	6209.3000	-合成纤维制	16					130	-Of synthetic fibres
4176	6209.9000	-其他纺织材料制	14					100	-Of other textile materials
	62.10	**用税号56.02、56.03、59.03、59.06或59.07的织物制成的服装：**							**Garments, made up of fabrics of heading No.56.02, 56.03, 59.03, 59.06 or 59.07:**
		-用税号56.02或56.03的织物制成的服装：							-Of fabrics of heading No.56.02 or 56.03:
4177	6210.1010	---羊毛或动物细毛制	16		11.7			130	---Of wool or fine animal hair
4178	6210.1020	---棉或麻制	16					90	---Of cotton or bast fibres
4179	6210.1030	---化学纤维制	17.5		12.4			130	---Of man-made fibres
4180	6210.1090	---其他纺织材料制	16					100	---Of other textile materials
4181	6210.2000	-子目号6201.11至6201.19所列类型的其他服装	16		11.7			100	-Other garments, of the type described in subheadings 6201.11 to 6201.19
4182	6210.3000	-子目号6202.11至6202.19所列类型的其他服装	16		11.7			100	-Other garments, of the type described in subheadings 6202.11 to 6202.19
4183	6210.4000	-其他男式服装	16		11.7			100	-Other men's or boys' garments
4184	6210.5000	-其他女式服装	16		11.7			100	-Other women's or girls garments
	62.11	**运动服、滑雪服及游泳服；其他服装：**							**Track suits, ski suits and swimwear; other garments:**
		-游泳服：							-Swimwear:
4185	6211.1100	--男式	16		11.7		0	130	--Men's or boys'
4186	6211.1200	--女式	16		11.7			130	--Women's or girls'
		-滑雪服：							-Ski suits:
4187	6211.2010	---棉制	16					90	---Of cotton
4188	6211.2090	---其他纺织材料制	19		13.6			130	---Of other textile materials
		-其他男式服装：							-Other garments, men's or boys':
4189	6211.3100	--羊毛或动物细毛制	16		11.7			130	--Of wool or fine animal hair
		--棉制：							--Of cotton:
4190	6211.3210	---阿拉伯袍	16			0△		90	---Arabian robes
4191	6211.3290	---其他	16			0△		90	---Other
		--化学纤维制：							--Of man-made fibres:
4192	6211.3310	---阿拉伯袍	17.5		12.4	0△		130	---Arabian robes
4193	6211.3390	---其他	17.5		12.4	0△	0	130	---Other
		--其他纺织材料制：							--Of other textile materials:
4194	6211.3910	---丝及绢丝制	16		11.7			130	---Of silk or silk waste
4195	6211.3990	---其他	16		11.7			100	---Other
		-其他女式服装：							-Other garments, women's or girls':
4196	6211.4100	--羊毛或动物细毛制	16		11.7			130	--Of wool or fine animal hair
4197	6211.4200	--棉制	16				0	90	--Of cotton
4198	6211.4300	--化学纤维制	17.5		12.4		0	130	--Of man-made fibres
		--其他纺织材料制：							--Of other textile materials:
4199	6211.4910	---丝及绢丝制	16		11.7			130	---Of silk or silk waste
4200	6211.4990	---其他	16		11.7			100	---Other

序号 No.	税则号列 Tariff Item	货品名称	最惠国税率(%) M.F.N.	协定税率(%) Agreement 东盟 ASEAN	曼谷协定 BA	香港 H.K.	澳门 Macao	普通税率(%) Gen.	Article Description
	62.12	**胸罩、束腰带、紧身胸衣、吊裤带、吊袜带、束袜带和类似品及其零件,不论是否针织或钩编的:**							**Brassières, girdles, corsets, braces, suspenders, garters and similar articles and parts thereof, whether or not knitted or crocheted:**
		-胸罩:							-Brassières:
4201	6212.1010	---化学纤维制	16			0		130	---Of man-made fibres
4202	6212.1090	---其他纺织材料制	14			0	0	100	---Of other textile materials
		-束腰带及腹带:							-Girdles and panty-girdles:
4203	6212.2010	---化学纤维制	16			0		130	---Of man-made fibres
4204	6212.2090	---其他纺织材料制	14			0		100	---Of other textile materials
		-束腰胸衣:							-Corselettes:
4205	6212.3010	---化学纤维制	16			0		130	---Of man-made fibres
4206	6212.3090	---其他纺织材料制	14			0		100	---Of other textile materials
		-其他:							-Other:
4207	6212.9010	---化学纤维制	16			0	0	130	---Of man-made fibres
4208	6212.9090	---其他纺织材料制	14			0	0	100	---Of other textile materials
	62.13	**手帕:**							**Handkerchiefs:**
		-丝及绢丝制:							-Of silk or silk waste:
4209	6213.1010	---刺绣的	14					130	---Embroidered
4210	6213.1090	---其他	14					130	---Other
		-棉制:							-Of cotton:
4211	6213.2010	---刺绣的	14					90	---Embroidered
4212	6213.2090	---其他	14					90	---Other
		-其他纺织材料制:							-Of other textile materials:
4213	6213.9010	---刺绣的	14					100	---Embroidered
4214	6213.9090	---其他	14					100	---Other
	62.14	**披巾、领巾、围巾、披纱、面纱及类似品:**							**Shawls, scarves, mufflers, mantillas, veils and the like:**
4215	6214.1000	-丝或绢丝制	14					130	-Of silk or silk waste
4216	6214.2000	-羊毛或动物细毛制	14					130	-Of wool or fine animal hair
4217	6214.3000	-合成纤维制	16					130	-Of synthetic fibres
4218	6214.4000	-人造纤维制	14					130	-Of artificial fibres
4219	6214.9000	-其他纺织材料制	14					100	-Of other textile materials
	62.15	**领带及领结:**							**Ties, bow ties and cravats:**
4220	6215.1000	-丝或绢丝制	14					130	-Of silk or silk waste
4221	6215.2000	-化学纤维制	16					130	-Of man-made fibres
4222	6215.9000	-其他纺织材料制	14					100	-Of other textile materials
	62.16	**分指手套、连指手套及露指手套:**							**Gloves, mittens and mitts:**
4223	6216.0000	分指手套、连指手套及露指手套	14				0	100	Gloves, mittens and mitts
	62.17	**其他制成的衣着附件;服装或衣着附件的零件,但税号62.12的货品除外:**							**Other made up clothing accessories; parts of garments or of clothing accessories, other than those of heading No.62.12:**
		-附件:							-Accessories:

序号 No.	税则号列 Tariff Item	货品名称	最惠国税率(%) M.F.N.	协定税率(%)Agreement				普通税率(%) Gen.	Article Description
				东盟 ASEAN	曼谷协定 BA	香港 H.K.	澳门 Macao		
4224	6217.1010	---袜子及袜套	14		12.2			130	---Stocking, socks and sockettes
4225	6217.1020	---和服腰带	14		12.2			100	---Kimono belts
4226	6217.1090	---其他	14		12.2			100	---Other
4227	6217.9000	-零件	14			0	0	100	-Parts

第六十三章
其他纺织制成品；成套物品；旧衣着及旧纺织品；碎织物

Chapter 63
Other made up textile articles; sets; worn clothing and worn textile articles; rags

注释：

一、第一分章仅适用于各种纺织物制成的物品。

二、第一分章不包括：

（一）第五十六章至第六十二章的货品；

（二）税号63.09的旧衣着或其他旧物品。

三、税号63.09仅适用于下列货品：

（一）纺织材料制品：

1. 衣着和衣着附件及其零件；

2. 毯子及旅行毯；

3. 床上、餐桌、盥洗及厨房用的织物制品；

4. 装饰用织物制品，但税号57.01至57.05的地毯及税号58.05的装饰毯除外。

（二）用石棉以外其他任何材料制成的鞋帽类。

上述物品只有同时符合下列两个条件才能归入本税号：

1. 必须明显看得出穿用过；

2. 必须以散装、捆装、袋装或类似的大包装形式进口或出口。

Notes:

1. Sub-Chapter 1 applies only to made up articles, of any textile fabric.

2. Sub-Chapter 1 does not cover:

(a) Goods of Chapters 56 to 62; or

(b) Worn clothing or other worn articles of heading No.63.09.

3. Heading No.63.09 applies only to the following goods:

(a) Articles of textile materials:

(1) Clothing and clothing accessories, and parts thereof;

(2) Blankets and travelling rugs;

(3) Bed linen, table linen, toilet lined and kitchen linen;

(4) Furnishing articles, other than carpets of headings Nos.57.01 to 57.05 and tapestries of heading No.58.05;

(b) Footwear and headgear of any material other than asbestos.

In order to be classified in this heading, the articles mentioned above must comply with both of the following requirements:

(1) they must show signs of appreciable wear, and

(2) they must be presented in bulk or in bales, sacks or similar packings.

序号 No.	税则号列 Tariff Item	货品名称	最惠国税率(%) M.F.N.	协定税率(%) Agreement 东盟 ASEAN	曼谷协定 BA	香港 H.K.	澳门 Macao	普通税率(%) Gen.	Article Description
		第一分章　其他纺织制成品							Ⅰ. -OTHER MADE UP TEXTILE ARTICLES
	63.01	**毯子及旅行毯：**							**Blankets and travelling rugs:**
4228	6301.1000	-电暖毯	16					100	-Electric blankets
4229	6301.2000	-羊毛或动物细毛制的毯子(电暖毯除外)及旅行毯	16					130	-Blankets(other than electric blankets) and travelling rugs, of wool or of fine animal hair
4230	6301.3000	-棉制的毯子(电暖毯除外)及旅行毯	16					90	-Blankets(other than electric blankets) and travelling rugs, of cotton
4231	6301.4000	-合成纤维制的毯子(电暖毯除外)旅行毯	17.5					130	-Blankets(other than electric blankets) and travelling rugs, of synthetic fibres
4232	6301.9000	-其他毯子及旅行毯	16					90	-Other blankets and travelling rugs

序号 No.	税则号列 Tariff Item	货品名称	最惠国税率(%) M.F.N.	协定税率(%) Agreement 东盟 ASEAN	曼谷协定 BA	香港 H.K.	澳门 Macao	普通税率(%) Gen.	Article Description
	63.02	**床上、餐桌、盥洗及厨房用的织物制品:**							**Bed linen, table linen, toilet linen and kitchen linen:**
		-针织或钩编的床上用织物制品:							-Bed linen, knitted or crocheted:
4233	6302.1010	---棉制	14					90	---Of cotton
4234	6302.1090	---其他纺织材料制	14					130	---Other
		-其他印花的床上用织物制品:							-Other bed linen, printed:
		--棉制:							--Of cotton:
4235	6302.2110	---床单	14					90	---Bed sheets
4236	6302.2190	---其他	14					90	---Other
		--化学纤维制:							--Of man-made fibres:
4237	6302.2210	---床单	16					130	---Bed sheets
4238	6302.2290	---其他	16					130	---Other
		--其他纺织材料制:							--Of other textile materials:
4239	6302.2910	---丝及绢丝制	14					130	---Of silk or silk waste
4240	6302.2920	---麻制	14					90	---Of bast fibres
4241	6302.2990	---其他	14					100	---Other
		-其他床上用织物制品:							-Other bed linen:
		--棉制:							--Of cotton:
4242	6302.3110	---刺绣的	14					90	---Embroidered
		---其他:							---Other:
4243	6302.3191	----床单	14					90	----Bed sheets
4244	6302.3192	----毛巾被	14					90	----Towelling coverlets
4245	6302.3199	----其他	14					90	----Other
		--化学纤维制:							--Of man-made fibres:
4246	6302.3210	---刺绣的	16					130	---Embroidered
4247	6302.3290	---其他	16					130	---Other
		--其他纺织材料制:							--Of other textile materials:
4248	6302.3910	---丝及绢丝制	14					130	---Of silk or silk waste
		---麻制:							---Of bast fibres:
4249	6302.3921	----刺绣的	14					90	----Embroidered
4250	6302.3929	----其他	14					90	----Other
		---其他:							---Other:
4251	6302.3991	----刺绣的	14					100	----Embroidered
4252	6302.3999	----其他	14					100	----Other
		-针织或钩编的餐桌用织物制品:							-Table linen, knitted or crocheted:
4253	6302.4010	---手工制	14					100	---Hand-worked
4254	6302.4090	---其他	14					100	---Other
		-其他餐桌用织物制品:							-Other table linen:
		--棉制:							--Of cotton:
4255	6302.5110	---刺绣的	14			0		90	---Embroidered
4256	6302.5190	---其他	14			0	0	90	---Other
		--亚麻制:							--Of flax:
4257	6302.5210	---刺绣的	14					90	---Embroidered
4258	6302.5290	---其他	14					90	---Other

序号 No.	税则号列 Tariff Item	货品名称	最惠国税率(%) M.F.N.	协定税率(%) Agreement 东盟 ASEAN	曼谷协定 BA	香港 H.K.	澳门 Macao	普通税率(%) Gen.	Article Description
		--化学纤维制:							--Of man-made fibres:
4259	6302.5310	---刺绣的	14					130	---Embroidered
4260	6302.5390	---其他	16					130	---Other
4261	6302.5900	--其他纺织材料制	14					100	--Of other textile materials
		-盥洗及厨房用棉制毛巾织物或类似的毛圈织物的制品:							-Toilet linen and kitchen linen, of terry towelling or similar terry fabrics, of cotton:
4262	6302.6010	---浴巾	14					90	---Bath towels
4263	6302.6090	---其他	14					90	---Other
		-其他:							-Other:
4264	6302.9100	--棉制	14					90	--Of cotton
4265	6302.9200	--亚麻制	14					90	--Of flax
4266	6302.9300	--化学纤维制	16			0		130	--Of man-made fibres
4267	6302.9900	--其他纺织材料制	14					100	--Of other textile materials
	63.03	**窗帘(包括帷帘)及帐幔;帘帷或床帷:**							**Curtains(including drapes)and interior blinds; curtain or bed valances:**
		-针织或钩编的:							-Knitted or crocheted:
		--棉制:							--Of cotton:
4268	6303.1110	---针织的	14					90	---Knitted
4269	6303.1120	---钩编的	14					90	---Crocheted
		--合成纤维制:							--Of synthetic fibres:
4270	6303.1210	---针织的	16					130	---Knitted
4271	6303.1220	---钩编的	16					130	---Crocheted
		--其他纺织材料制:							--Of other textile materials:
4272	6303.1910	---针织的	14					130	---Knitted
4273	6303.1920	---钩编的	14					130	---Crocheted
		-其他:							-Other:
4274	6303.9100	--棉制	14					90	--Of cotton
4275	6303.9200	-合成纤维制	16			0		130	--Of synthetic fibres
4276	6303.9900	-其他纺织材料制	14					100	--Of other textile materials
	63.04	**其他装饰用织物制品,但税号94.04的货品除外:**							**Other furnishing articles, excluding those of heading No.94.04:**
		-床罩:							-Bedspreads:
		-针织或钩编的:							--Knitted or crocheted:
		---针织的:							---Knitted:
4277	6304.1121	----手工制	14					100	----Hand-worked
4278	6304.1129	----其他	14					100	----Other
		---钩编的:							---Crocheted:
4279	6304.1131	----手工制	14					100	----Hand-worked
4280	6304.1139	----其他	14					100	----Other
		--其他:							--Other:
4281	6304.1910	---丝及绢丝制	14					130	---Of silk or silk waste
		---棉或麻制:							---Of cotton or bast fibres:
4282	6304.1921	----刺绣的	14					90	----Embroidered
4283	6304.1929	----其他	14					90	----Other

序号 No.	税则号列 Tariff Item	货品名称	最惠国税率(%) M.F.N.	协定税率(%) Agreement 东盟 ASEAN	曼谷协定 BA	香港 H.K.	澳门 Macao	普通税率(%) Gen.	Article Description
		---化学纤维制:							---Of man-made fibres:
4284	6304.1931	----刺绣的	16					130	----Embroidered
4285	6304.1939	----其他	16					130	----Other
		---其他纺织材料制:							---Of other textile materials:
4286	6304.1991	----刺绣的	14					100	----Embroidered
4287	6304.1999	----其他	14					100	----Other
		-其他:							-Other:
		--针织或钩编的:							--Knitted or crocheted:
		---针织的:							---Knitted:
4288	6304.9121	----手工制	14					100	----Hand-worked
4289	6304.9129	----其他	14					100	----Other
		---钩编的:							---Crocheted:
4290	6304.9131	----手工制	14					100	----Hand-worked
4291	6304.9139	----其他	14					100	----Other
		--非针织或非钩编的,棉制:							--Not knitted or crocheted,of cotton:
4292	6304.9210	---刺绣的	14					90	---Embroidered
4293	6304.9290	---其他	14					90	---Other
		--非针织或非钩编的,合成纤维制:							--Not knitted or crocheted,of synthetic fibres:
4294	6304.9310	---刺绣的	16					130	---Embroidered
4295	6304.9390	---其他	16					130	---Other
		--非针织或非钩编的,其他纺织材料制:							--Not knitted or crocheted, of other textile materials:
4296	6304.9910	---丝及绢丝制	14					130	---Of silk or silk waste
		---麻制:							---Of bast fibres:
4297	6304.9921	----刺绣的	14					90	----Embroidered
4298	6304.9929	----其他	14					90	----Other
4299	6304.9990	---其他	14					100	---Other
	63.05	**货物包装用袋:**							**Sacks and bags,of a kind used for the packing of goods:**
4300	6305.1000	-黄麻或税号53.03的其他韧皮纺织纤维制	10					40	-Of jute or of other textile bast fibres of heading No.53.03
4301	6305.2000	-棉制	16					90	-Of cotton
		-化学纤维材料制:							-Of man-made textile materials:
4302	6305.3200	--散装货物储运软袋	16					100	--Flexible intermediate bulk containers
4303	6305.3300	--其他,聚乙烯、聚丙烯扁条或类似材料制	16					100	--Other,of polyethylene or polypropylene strip or the like
4304	6305.3900	--其他	16					100	--Other
4305	6305.9000	-其他纺织材料制	14					90	-Of other textile materials
	63.06	**油苫布、天篷及遮阳篷;帐篷;风帆;野营用品:**							**Tarpaulins, awnings and sunblinds; tents; sails for boats, sailboards or landcraft;camping goods:**
		-油苫布、天篷及遮阳篷:							-Tarpaulins,awnings and sunblinds:

序号 No.	税则号列 Tariff Item	货 品 名 称	最惠国税率(%) M.F.N.	协定税率(%) Agreement 东盟 ASEAN	曼谷协定 BA	香港 H.K.	澳门 Macao	普通税率(%) Gen.	Article Description
4306	6306.1100	--棉制	14					80	--Of cotton
4307	6306.1200	--合成纤维制	16					130	--Of synthetic fibres
		--其他纺织材料制:							--Of other textile materials:
4308	6306.1910	---麻制	14					80	---Of bast fibres
4309	6306.1990	---其他	14					100	---Other
		-帐篷:							-Tents:
4310	6306.2100	--棉制	14					80	--Of cotton
4311	6306.2200	--合成纤维制	16					130	--Of synthetic fibres
4312	6306.2900	--其他纺织材料制	14					100	--Of other textile materials
		-风帆:							-Sails:
4313	6306.3100	--合成纤维制	16					130	--Of synthetic fibres
4314	6306.3900	--其他纺织材料制	14					100	--Of other textile materials
		-充气褥垫:							-Pneumatic mattresses:
4315	6306.4100	--棉制	14					80	--Of cotton
		--其他纺织材料制:							--Of other textile materials:
4316	6306.4910	---化学纤维制	16					130	---Of man-made fibres
4317	6306.4990	---其他	14					100	---Other
		-其他:							-Other:
4318	6306.9100	--棉制	14					80	--Of cotton
		--其他纺织材料制:							--Of other textile materials:
4319	6306.9910	---麻制	14					80	---Of bast fibres
4320	6306.9920	---化学纤维制	16					130	---Of man-made fibres
4321	6306.9990	---其他	14					100	---Other
	63.07	**其他制成品,包括服装裁剪样:**							**Other made up articles, including dress patterns:**
4322	6307.1000	-擦地布、擦碗布、抹布及类似擦拭用布	14					130	-Floor-cloths, dish-cloths, dusters and similar cleaning cloths
4323	6307.2000	-救生衣及安全带	14					70	-Life-jackets and life-belts
4324	6307.9000	-其他	14					100	-Other
		第二分章 成套物品							Ⅱ.-SETS
	63.08	**由机织物及纱线构成的零售包装成套物品,不论是否带附件,用以制作小地毯、装饰毯、绣花台布、餐巾或类似的纺织物品:**							**Sets consisting of woven fabric and yarn, whether or not with accessories, for making up into urgs, tapestries, embroidered table cloths or serviettes, or similar textile articles, put up in packings for retail sale:**
4325	6308.0000	由机织物及纱线构成的零售包装成套物品,不论是否带附件,用以制作小地毯、装饰毯、绣花台布、餐巾或类似的纺织物品	14					130	Sets consisting of woven fabric and yarn, whether or not with accessories, for making up into rugs, tapestries, embroidered table cloths or serviettes, or similar textile articles, put up in packings for retail sale
		第三分章 旧衣着及旧纺织品;碎织物							Ⅲ. -WORN CLOTHING AND WORN TEXTILE ARTICLES; RAGS
	63.09	**旧衣物:**							**Worn clothing and other worn articles:**

序号 No.	税则号列 Tariff Item	货品名称	最惠国税率(%) M.F.N.	协定税率(%) Agreement 东盟 ASEAN	曼谷协定 BA	香港 H.K.	澳门 Macao	普通税率(%) Gen.	Article Description
4326	6309.0000	旧衣物	14					130	Worn clothing and other worn articles
	63.10	**纺织材料的新的或旧的碎织物及废线、绳、索、缆及其制品:**							**Used or new rags, scrap twine, cordage, rope and cables and worn out articles of twine, cordage, rope or cables, of textile materials:**
4327	6310.1000	-经分拣的	14					50	-Sorted
4328	6310.9000	-其他	14					50	-Other

第十二类

鞋、帽、伞、杖、鞭及其零件；已加工的羽毛及其制品；人造花；人发制品

SECTION Ⅻ

FOOTWEAR, HEADGEAR, UMBRELLAS, SUN UMBRELLAS, WALKING-STICKS, SEAT-STICKS, WHIPS, RIDING-CROPS AND PARTS THEREOF; PREPARED FEATHERS AND ARTICLES MADE THEREWITH; ARTIFICIAL FLOWERS; ARTICLES OF HUMAN HAIR

第六十四章

鞋靴、护腿和类似品及其零件

Chapter 64

Footwear, gaiters and the like; parts of such articles

注释：

一、本章不包括：

（一）易损材料（例如，纸、塑料薄膜）制的无外绱鞋底的一次性鞋靴罩或套，这些产品应按其构成材料归类；

（二）纺织材料制的鞋靴，没有用粘、缝或其他方法将外底固定或安装在鞋面上的（第十一类）；

（三）税号 63.09 的旧鞋靴；

（四）石棉制品（税号 68.12）；

（五）矫形鞋靴或其他矫形器具及其零件（税号 90.21）；

（六）玩具鞋及装有冰刀或轮子的滑冰鞋；护胫或类似的运动防护服装（第九十五章）。

二、税号 64.06 所称“零件”，不包括鞋钉、护鞋铁掌、鞋眼、鞋钩、鞋扣、饰物、编带、鞋带、绒球或其他装饰带（应分别归入相应税号）及税号 96.06 的钮扣或其他货品。

三、本章所称：

（一）“橡胶”及“塑料”，包括能用肉眼辨出其外表有一层橡胶或塑料的机织物或其他纺织产品，运用本款时，橡胶或塑料仅引起颜色变化的不计在内。

（二）“皮革”，是指税号 41.07 及 41.12 至 41.14 的货品。

Notes:

1. This Chapter does not cover:

(a) Disposable foot or shoe coverings of flimsy material (for example, paper, sheeting of plastics) without applied soles. These products are classified according to their constituent material;

(b) Footwear of textile material, without an outer sole glued, sewn or otherwise affixed or applied to the upper (Section Ⅺ);

(c) Worn footwear of heading No.63.09;

(d) Articles of asbestos (heading No.68.12);

(e) Orthopaedic footwear or other orthopaedic appliances, or parts thereof (heading No.90.21); or

(f) Toy footwear or skating boots with ice or roller skates attached; shin-guards or similar protective sportswear (Chapter 95).

2. For the purposes of heading No.64.06, the term “parts” does not include pegs, protectors, eyelets, hooks, buckles, ornaments, braid, laces, pompons or other trimmings (which are to be classified in their appropriate headings) or buttons or other goods of heading No.96.06.

3. For the purposes of this Chapter:

(a) the terms “rubber” and “plastics” include woven fabrics or other textile products with an external layer of rubber or plastics being visible to the naked eye; for the purpose of this provision, no account should be taken of any resulting change of colour; and

(b) the term “leather” refers to the goods of headings 41.07and 41.12 to 41.14.

四、除本章注释三另有规定的以外：

（一）鞋面的材料应以占表面面积最大的那种材料为准，计算表面面积可不考虑附件及加固件，例如，护踝、裹边、饰物、扣子、拉襻、鞋眼或类似附属件；

（二）外底的主要材料应以与地面接触最广的那种材料为准，计算接触面时可不考虑鞋底钉、铁掌或类似附属件。

4. Subject to Note 3 to this Chapter:

(a) the material of the upper shall be taken to be the constituent material having the greatest external surface area, no account being taken of accessories or reinforcements such as ankle patches, edging, ornamentation, buckles, tabs, eyelet stays or similar attachments;

(b) the constituent material of the outer sole shall be taken to be the material having the greatest surface area in contact with the ground, no account being taken of accessories or reinforcements such as spikes, bars, nails, protectors or similar attachments.

子目注释：

子目号 6402.12、6402.19、6403.12、6403.19 及 6404.11 所称"运动鞋靴"，仅适用于：

一、带有或可装鞋底钉、止滑柱、夹钳、马蹄掌或类似品的体育专用鞋靴；

二、滑冰靴、滑雪靴及越野滑雪用鞋靴、滑雪板靴、角力靴、拳击靴及赛车鞋。

Subheading Notes:

For the purposes of subheadings Nos. 6402.12, 6402.19, 6403.12, 6403.19 and 6404.11, the expression "sports footwear" applies only to:

1. footwear which is designed for a sporting activity and has, or has provsion for the attachment of, spikes, sprigs, stops, clips, bars or the like;

2. skating boots, ski-boots and cross-country ski footwear, snowboard boots, wrestling boots, boxing boots and cycling shoes.

序号 No.	税则号列 Tariff Item	货品名称	最惠国税率 (%) M.F.N.	协定税率 (%) Agreement 东盟 ASEAN	曼谷协定 BA	香港 H.K.	澳门 Macao	普通税率 (%) Gen.	Article Description
	64.01	**橡胶或塑料制外底及鞋面的防水鞋靴，其鞋面不是用缝、铆、钉、旋、塞或类似方法固定在鞋底上的：**							**Waterproof footwear with outer soles and uppers of rubber or of plastics, the uppers of which are neither fixed to the sole nor assembled by stitching, riveting, nailing, screwing, plugging or similar processes:**
4329	6401.1000	-装有金属防护鞋头的鞋靴	24					100	-Footwear incorporating a protective metal toe-cap
		-其他鞋靴：							-Other footwear:
4330	6401.9100	--高统靴(过膝)	24					100	--Covering the knee
4331	6401.9200	--中、短统靴(过踝但未到膝)	24					100	--Covering the ankle but not covering the knee
4332	6401.9900	--其他	24					100	--Other
	64.02	**橡胶或塑料制外底及鞋面的其他鞋靴：**							**Other footwear with outer soles and uppers of rubber or plastics:**
		-运动鞋靴：							-Sports footwear:
4333	6402.1200	--滑雪靴、越野滑雪鞋靴及滑雪板靴	10					100	--Ski-boots, cross-country ski footwear and snowboard boots
4334	6402.1900	--其他	24					100	--Other
4335	6402.2000	-用栓塞方法将鞋面条带装配在鞋底上的鞋	24					100	-Footwear with upper straps or thongs assembled to the sole by means of plugs
4336	6402.3000	-装有金属防护鞋头的其他鞋靴	24					100	-Other footwear, incorporating a protective metal toe-cap
		-其他鞋靴：							-Other footwear:

序号 No.	税则号列 Tariff Item	货品名称	最惠国税率(%) M.F.N.	协定税率(%) Agreement 东盟 ASEAN	曼谷协定 BA	香港 H.K.	澳门 Macao	普通税率(%) Gen.	Article Description
4337	6402.9100	--短统靴(过踝)	24				0	100	--Covering the ankle
4338	6402.9900	--其他	24				0	100	--Other
	64.03	**橡胶、塑料、皮革或再生皮革制外底,皮革制鞋面的鞋靴:**							**Footwear with outer soles of rubber, plastics, leather or composition leather and uppers of leather:**
		-运动鞋靴:							-Sports footwear:
4339	6403.1200	--滑雪靴、越野滑雪鞋靴及滑雪板靴	24					100	--Ski-boots, cross-country ski footwear and snowboard boots
4340	6403.1900	--其他	15				0	100	--Other
4341	6403.2000	-皮革制外底,由交叉于脚背并绕大脚趾的皮革条带构成鞋面的鞋	24				0	100	-Footwear with outer soles of leather, and uppers which consist of leather straps across the instep and around the big toe
4342	6403.3000	-不带内底或金属防护鞋头的木底鞋	24					100	-Footwear made on a base or platform of wood, not having an inner sole or a protective metal toe-cap
4343	6403.4000	-装有金属防护鞋头的其他鞋靴	24					100	-Other footwear, incorporating a protective metal toe-cap
		-皮革制外底的其他鞋靴:							-Other footwear with outer soles of leather:
4344	6403.5100	--短统靴(过踝)	10					100	--Covering the ankle
4345	6403.5900	--其他	10					100	--Other
		-其他鞋靴:							-Other footwear:
4346	6403.9100	--短统靴(过踝)	10				0	100	--Covering the ankle
4347	6403.9900	--其他	10				0	100	--Other
	64.04	**橡胶、塑料、皮革或再生皮革制外底,用纺织材料制鞋面的鞋靴:**							**Footwear with outer soles of rubber, plastics, leather or composition leather and uppers of textile materials:**
		-橡胶或塑料制外底的鞋靴:							-Footwear with outer soles of rubber or plastics:
4348	6404.1100	--运动鞋靴;网球鞋、篮球鞋、体操鞋、训练鞋及类似鞋	24				0	100	--Sports footwear, tennis shoes, basketball shoes, gym shoes, training shoes and the like
4349	6404.1900	--其他	24				0	100	--Other
4350	6404.2000	-皮革或再生皮革制外底的鞋靴	24					100	-Footwear with outer soles of leather or composition leather
	64.05	**其他鞋靴:**							**Other footwear:**
4351	6405.1000	-皮革或再生皮革制鞋面的	24					100	-With uppers of leather or composition leather
4352	6405.2000	-纺织材料制鞋面的	22					100	-With uppers of textile materials
4353	6405.9000	-其他	15					100	-Other
	64.06	**鞋靴零件(包括鞋面,不论是否带有除外底以外的其他鞋底);活动式鞋内底、跟垫及类似品;护腿、裹腿和类似品及其零件:**							**Parts of footwear (including uppers whether or not attached to soles other than outer soles); removable in-soles, heel cushions and similar articles; gaiters, leggings and similar articles, and parts thereof:**

序号 No.	税则号列 Tariff Item	货品名称	最惠国税率(%) M.F.N.	协定税率(%) Agreement 东盟 ASEAN	曼谷协定 BA	香港 H.K.	澳门 Macao	普通税率(%) Gen.	Article Description
4354	6406.1000	-鞋面及其零件,但硬衬除外	15		13.2			90	-Uppers and parts thereof, other than stiffeners
4355	6406.2000	-橡胶或塑料制的外底及鞋跟	15					90	-Orter soles and heels, of rubber or plastics
		-其他:							-Other:
4356	6406.9100	--木制	15					90	--Of wood
4357	6406.9900	--其他材料制	15		13.2	0	0	90	--Of other materials

第六十五章
帽类及其零件

Chapter 65
Headgear and parts thereof

注释：

一、本章不包括：

（一）税号63.09的旧帽类；

（二）石棉制帽类（税号68.12）；

（三）第九十五章的玩偶帽、其他玩具帽或狂欢节用品。

二、税号65.02不包括缝制的帽坯，但仅将条带缝成螺旋形的除外。

Notes:

1. This Chapter does not cover:

(a) Worn headgear of heading No.63.09;

(b) Asbestos headgear (Heading No.68.12); or

(c) Dolls' hats, other toy hats or carnival articles of Chapter 95.

2. Heading No.65.02 does not cover hat-shapes made by sewing, other than those obtained simply by sewing strips in spirals.

序号 No.	税则号列 Tariff Item	货品名称	最惠国税率(%) M.F.N.	协定税率(%) Agreement 东盟 ASEAN	曼谷协定 BA	香港 H.K.	澳门 Macao	普通税率(%) General	Article Description
	65.01	**毡呢制的帽坯、帽身及帽兜，未楦制成形，也未加帽边；毡呢制的圆帽片及制帽用的毡呢筒(包括裁开的毡呢筒)：**							**Hat-forms, hat bodies and hoods of felt, neither blocked to shape nor with made brims; plateaux and manchons (including slit manchons), of felt:**
4358	6501.0000	毡呢制的帽坯、帽身及帽兜，未楦制成形，也未加帽边；毡呢制的圆帽片及制帽用的毡呢筒(包括裁开的毡呢筒)	22					100	Hat-forms, hat bodies and hoods of felt, neither blocked to shape nor with made brims; plateaux and manchons(including slit manchons), of felt
	65.02	**编结的帽坯或用任何材料的条带拼制而成的帽坯，未楦制成形，也未加帽边、衬里或装饰物：**							**Hat-shapes, plaited or made by assembling strips of any material, neither blocked to shape, nor with made brims, nor lined, nor trimmed:**
4359	6502.0000	编结的帽坯或用任何材料的条带拼制而成的帽坯，未楦制成形，也未加帽边、衬里或装饰物	20					100	Hat-shapes, plaited or made by assembling strips of any material, neither blocked to shape, nor with made brims, nor lined, nor trimmed
	65.03	**用税号65.01的帽身、帽兜或圆帽片制成的毡呢帽类，不论有无衬里或装饰物：**							**Felt hats and other felt headgear, made from the hat bodies, hoods or plateaux of heading No.65.01, whether or not lined or trimmed:**
4360	6503.0000	用税号65.01的帽身、帽兜或圆帽片制成的毡呢帽类，不论有无衬里或装饰物	22					130	Felt hats and other felt headgear, made from the hat bodies, hoods or plateaux of heading No.65.01, whether or not lined or trimmed

序号 No.	税则号列 Tariff Item	货品名称	最惠国税率(%) M.F.N.	协定税率(%)Agreement				普通税率(%) General	Article Description
				东盟 ASEAN	曼谷协定 BA	香港 H.K.	澳门 Macao		
	65.04	**编结帽或用任何材料的条带拼制而成的帽类,不论有无衬里或装饰物:**							**Hats and other headgear, plaited or made by assembling strips of any material, whether or not lined or trimmed:**
4361	6504.0000	编结帽或用任何材料的条带拼制而成的帽类,不论有无衬里或装饰物	20					130	Hats and other headgear, plaited or made by assembling strips of any material, whether or not lined or trimmed
	65.05	**针织或钩编的帽类,用成匹的花边、毡呢或其他纺织物(条带除外)制成的帽类,不论有无衬里或装饰物;任何材料制的发网,不论有无衬里或装饰物:**							**Hats and other headgear, knitted or crocheted, or made up from lace, felt or other textile fabric, in the piece (by not in strips), whether or not lined or trimmed; hair-nets of any material, whether or not lined or trimmed:**
4362	6505.1000	-发网	10					130	-Hair-nets
		-其他:							-Other:
4363	6505.9010	---钩编的帽类	20					130	---Hats and other headgear, knitted or crocheted
4364	6505.9090	---其他	20				0	130	---Other
	65.06	**其他帽类,不论有无衬里或装饰物:**							**Other headgear, whether or not lined or trimmed:**
4365	6506.1000	-安全帽	10					100	-Safety headgear
		-其他:							-Other:
4366	6506.9100	--橡胶或塑料制	10					100	--Of rubber or of plastics
4367	6506.9200	--毛皮制	10			0		130	--Of furskin
		--其他材料制:							--Of other materials:
4368	6506.9910	---皮革制	10					130	---Of leather
4369	6506.9990	---其他	24					100	---Other
	65.07	**帽圈、帽衬、帽套、帽帮、帽骨架、帽舌及帽颏带:**							**Head-bands, linings, covers, hat foundations, hat frames, peaks and chinstraps, for headgear:**
4370	6507.0000	帽圈、帽衬、帽套、帽帮、帽骨架、帽舌及帽颏带	24				0	100	Head-bands, linings, covers, hat foundations, hat frames, peaks and chinstraps, for headgear

第六十六章
雨伞、阳伞、手杖、鞭子、马鞭及其零件

Chapter 66
Umbrellas, sun umbrellas, walking-sticks, seat-sticks, whips, riding-crops and parts thereof

注释：

一、本章不包括：

（一）丈量用杖及类似品（税号 90.17）；

（二）火器手杖、刀剑手杖、灌铅手杖及类似品（第九十三章）；

（三）第九十五章的货品（例如，玩具雨伞、玩具阳伞）。

二、税号 66.03 不包括纺织材料制的零件、附件及装饰品或者任何材料制的罩套、流苏、鞭梢、伞套及类似品。此类货品即使与税号 66.01 或 66.02 的物品一同进口或出口，只要未装配在一起，则不应视为上述税号所列物品的组成零件，而应分别归入各有关税号。

Notes:

1. This Chapter does not cover:
   (a) Measure walking-sticks or the like (heading No.90.17);
   (b) Firearm-sticks, sword-sticks, loaded walking-sticks or the like (Chapter 93); or
   (c) Goods of Chapter 95 (for example, toy umbrellas, toy sun umbrellas).

2. Heading No.66.03 does not cover parts, trimmings or accessories of textile material, or covers, tassels, thongs, umbrella cases or the like, of any material. Such goods presented with, but not fitted to, articles of heading No.66.01 or 66.02 are to be classified separately and not to be treated as forming part of those articles.

序号 No.	税则号列 Tariff Item	货品名称	最惠国税率(%) M.F.N.	协定税率(%) Agreement 东盟 ASEAN	曼谷协定 BA	香港 H.K.	澳门 Macao	普通税率(%) Gen.	Article Description
	66.01	**雨伞及阳伞(包括手杖伞、庭园用伞及类似伞)：**							**Umbrellas and sun umbrellas (including walking-stick umbrellas, garden umbrellas and similar umbrellas):**
4371	6601.1000	-庭园用伞及类似伞	14					130	-Garden or similar umbrellas
		-其他：							-Other:
4372	6601.9100	--折叠伞	10					130	--Having a telescopic shaft
4373	6601.9900	--其他	10					130	--Other
	66.02	**手杖、带座手杖、鞭子、马鞭及类似品：**							**Walking-sticks, seat-sticks, whips, ridingcrops and the like:**
4374	6602.0000	手杖、带座手杖、鞭子、马鞭及类似品	10					130	Walking-sticks, seat-sticks, whips, ridingcrops and the like
	66.03	**税号 66.01 或 66.02 所列物品的零件及装饰品：**							**Parts, trimmings and accessories of articles of heading No.66.01 or 66.02:**
4375	6603.1000	-把柄	14					130	-Handles and knobs
4376	6603.2000	-伞骨，包括装在伞柄上的伞骨	14					130	-Umbrella frames, including frames mounted on shafts (sticks)
4377	6603.9000	-其他	14					130	-Other

第六十七章
已加工羽毛、羽绒及其制品；人造花；人发制品

Chapter 67
Prepared feathers and down and articles made of feathers or of down; artificial flowers; articles of human hair

注释：

一、本章不包括：

(一) 人发制滤布（税号59.11）；

(二) 花边、刺绣品或其他纺织物制成的花卉图案（第十一类）；

(三) 鞋靴（第六十四章）；

(四) 帽类及发网（第六十五章）；

(五) 玩具、运动用品或狂欢节用品（第九十五章）；

(六) 羽毛掸帚、粉扑及人发制的筛子（第九十六章）。

二、税号67.01不包括：

(一) 羽毛或羽绒仅在其中作为填充料的物品(例如税号94.04的寝具)；

(二) 羽毛或羽绒仅作为饰物或填充料的衣服或衣着附件；

(三) 税号67.02的人造花、叶及其部分品，以及它们的制成品。

三、税号67.02不包括：

(一) 玻璃制品（第七十章）；

(二) 用陶器、石料、金属、木料或其他材料经模铸、锻造、雕刻、冲压或其他方法整件制成形的人造花、叶或果实；用捆扎、胶粘及类似方法以外的其他方法将部分品组合而成的上述制品。

Notes:

1. This Chapter does not cover:
   (a) Straining cloth of human hair (heading No.59.11);
   (b) Floral motifs of lace, of embroidery or other textile fabric (Section XI);
   (c) Footwear (Chapter 64);
   (d) Headgear or hair-nets (Chapter 65);
   (e) Toys, sports requisites or carnival articles (Chapter 95); or
   (f) Feather dusters, powder-puffs or hair sieves (Chapter 96).

2. Heading No.67.01 does not cover:
   (a) Articles in which feathers or down constitute only filling or padding (for example, bedding of heading No.94.04);
   (b) Articles of apparel or clothing accessories in which feathers or down constitute no more than mere trimming or padding; or
   (c) Artificial flowers or foliage or parts thereof or madeup articles of heading No.67.02.

3. Heading No.67.02 does not cover:
   (a) Articles of glass (Chapter 70); or
   (b) Artificial flowers, foliage or fruit of pottery, stone, metal, wood or other materials, obtained in one piece by moulding, forging, carving, stamping or other process, or consisting of parts assembled otherwise than by binding, glueing, fitting into one another or similar methods.

序号 No.	税则号列 Tariff Item	货品名称	最惠国税率(%) M.F.N.	协定税率(%) Agreement 东盟 ASEAN	曼谷协定 BA	香港 H.K.	澳门 Macao	普通税率(%) Gen.	Article Description
	67.01	带羽毛或羽绒的鸟皮及鸟体其他部分、羽毛、部分羽毛、羽绒及其制品(税号05.05的货品和经加工的羽管及羽轴除外)：							Skins and other parts of birds with their feathers or down, feathers, parts of feathers, down and articles thereof (other than goods of heading No.05.05 and worked quills and scapes):

序号 No.	税则号列 Tariff Item	货品名称	最惠国税率(%) M.F.N.	协定税率(%)Agreement 东盟 ASEAN	曼谷协定 BA	香港 H.K.	澳门 Macao	普通税率(%) Gen.	Article Description
4378	6701.0000	带羽毛或羽绒的鸟皮及鸟体其他部分、羽毛、部分羽毛、羽绒及其制品(税号05.05的货品和经加工的羽管及羽轴除外)	20				0	130	Skins and other parts of birds with their feathers or down, feathers, parts of feathers, down and articles thereof (other than goods of heading No.05.05 and worked quills and scapes)
	67.02	**人造花、叶、果实及其零件;用人造花、叶或果实制成的物品:**							**Artificial flowers, foliage and fruit and parts thereof; articles made of artificial flowers, foliage or fruit:**
4379	6702.1000	-塑料制	20					130	-Of plastics
		-其他材料制:							-Of other materials:
4380	6702.9010	---羽毛制	20					130	---Of feathers or down
4381	6702.9020	---丝及绢丝制	24					130	---Of silk or silk waste
4382	6702.9030	---化学纤维制	24					130	---Of man-made fibres
4383	6702.9090	---其他	20					130	---Other
	67.03	**经梳理、稀疏、脱色或其他方法加工的人发;作假发及类似品用的羊毛、其他动物毛或其他纺织材料:**							**Human hair, dressed, thinned, bleached or otherwise worked; wool or other animal hair or other textile materials, prepared for use in making wigs or the like:**
4384	6703.0000	经梳理、稀疏、脱色或其他方法加工的人发;作假发及类似品用的羊毛、其他动物毛或其他纺织材料	20		18			100	Human hair, dressed, thinned, bleached or otherwise worked; wool or other animal hair or other textile materials, prepared for use in making wigs or the like
	67.04	**人发、动物毛或纺织材料制的假发、假胡须、假眉毛、假睫毛及类似品;其他税号未列名的人发制品:**							**Wigs, false beards, eyebrows and eyelashes, switches and the like, of human or animal hair or of textile materials; articles of human hair not elsewhere specified or included:**
		-合成纤维纺织材料制:							-Of synthetic textile materials:
4385	6704.1100	--整头假发	25					130	--Complete wigs
4386	6704.1900	--其他	25					130	--Other
4387	6704.2000	-人发制	15					130	-Of human hair
4388	6704.9000	-其他材料制	25					130	-Of other materials

第十三类

石料、石膏、水泥、石棉、云母及类似材料的制品；陶瓷产品；玻璃及其制品

SECTION XIII

ARTICLES OF STONE, PLASTER, CEMENT, ASBESTOS, MICA OR SIMILAR MATERIALS; CERAMIC PRODUCTS; GLASS AND GLASSWARE

第六十八章
石料、石膏、水泥、石棉、云母及类似材料的制品

Chapter 68
Articles of stone, plaster, cement, asbestos, mica or similar materials

注释：

一、本章不包括：

（一）第二十五章的货品；

（二）税号 48.10 或 48.11 的经涂布、浸渍或覆盖的纸及纸板（例如，用云母粉或石墨涂布的纸及纸板、沥青纸及纸板）；

（三）第五十六章或第五十九章的经涂布、浸渍或包覆的纺织物（例如，用云母粉、沥青涂布或包覆的织物）；

（四）第七十一章的物品；

（五）第八十二章的工具及其零件；

（六）税号 84.42 的印刷用石板；

（七）绝缘子（税号 85.46）或绝缘材料制的零件（税号 85.47）；

（八）牙科用磨锉（税号 90.18）；

（九）第九十一章的物品（例如，钟及钟壳）；

（十）第九十四章的物品（例如，家具、灯具及照明装置、活动房屋）；

（十一）第九十五章的物品（例如，玩具、游戏品及运动用品）；

（十二）用第九十六章注释二（二）所述材料制成的税号 96.02 的物品或税号 96.06 的物品（例如钮扣）、税号 96.09 的物品（例如石笔）或税号 96.10 的物品（例如绘画石板）；

（十三）第九十七章的物品（例如艺术品）。

二、税号 68.02 所称"已加工的碑石或建筑用石"，不仅适用于已加工的税号 25.15、25.16 的各种石料，也适用于所有经类似加工的其他天然石料（例如，石英岩、燧石、白云石及冻石），但不适用于板岩。

Notes:

1. This Chapter does not cover:

(a) Goods of Chapter 25;

(b) Coated, impregnated or covered paper and paperboard of heading No.48.10 or 48.11 (for example, paper and paperboard coated with mica powder or graphite, bituminized or asphalted paper and paperboard);

(c) Coated, impregnated or covered textile fabric of Chapter 56 or 59 (for example, fabric coated or covered with mica powder, bituminized or asphalted fabric);

(d) Articles of Chapter 71;

(e) Tools or parts of tools, of Chapter 82;

(f) Lithographic stones of heading No.84.42;

(g) Electrical insulators (heading No.85.46) or fittings of insulating material of heading No.85.47;

(h) Dental burrs (heading No.90.18);

(i) Articles of Chapter 91 (for example, clocks and clock cases);

(j) Articles of Chapter 94 (for example, furniture, lamps and lighting fittings, prefabricated buildings);

(k) Articles of Chapter 95 (for example, toys, games and sports requisites);

(l) Articles of heading No.96.02, if made of materials specified in Note 2 (b) to Chapter 96, or of heading No.96.06 (for example, buttons), No.96.09 (for example, slate pencils) or No.96.10 (for example, drawing slates); or

(m) Articles of Chapter 97 (for example, works of art).

2. In heading No.68.02, the expression "worked monumental or building stone" applies not only to the varieties of stone referred to in heading No.25.15 or 25.16, but also to all other natural stone (for example, quartzite, flint, dolomite and steatite) similarly worked; it does not, however, apply to slate.

序号 No.	税则号列 Tariff Item	货品名称	最惠国税率(%) M.F.N.	协定税率(%) Agreement 东盟 ASEAN	曼谷协定 BA	香港 H.K.	澳门 Macao	普通税率(%) Gen.	Article Description
	68.01	**天然石料(不包括板岩)制的长方砌石、路缘石、扁平石:**							**Setts, curbstones and flagstones, of natural stone(except slate):**
4389	6801.0000	天然石料(不包括板岩)制的长方砌石、路缘石、扁平石	12					70	Setts, curbstones and flagstones, of natural stone(except slate)
	68.02	**已加工的碑石或建筑用石(不包括板岩)及其制品,但税号68.01的货品除外;天然石料(包括板岩)制的镶嵌石(马赛克)及类似品,不论是否有衬背;天然石料(包括板岩)制的人工染色石粒、石片及石粉:**							**Worked monumental or building stone (except slate) and articles thereof, other than goods of heading No.68.01; mosaic cubes and the like, of natural stone (including slate), whether or not on a backing; artificially coloured granules, chippings and powder, of natural stone (including slate):**
		-砖、瓦、方块及类似品,不论是否为矩形(包括正方形),其最大表面积以可置入边长小于7厘米的方格为限;人工染色的石粒、石片及石粉:							-Tiles, cubes and similar articles, whether or not rectangular(including square), the largest surface area of which is capable of being enclosed in a square the side of which is less than 7cm; artificially coloured granules, chippings and powder:
4390	6802.1010	---大理石	24					90	---Marble
4391	6802.1090	---其他	20					90	---Other
		-简单切削或锯开并具有一个平面的其他碑石或建筑用石及其制品:							-Other monumental or building stone and articles thereof, simply cut or sawn, with a flat or even surface:
		--大理石、石灰华及蜡石:							--Marble, travertine and alabaster:
4392	6802.2110	---大理石	10				0	90	---Marble
4393	6802.2120	---石灰华	24				0	90	---Travertine
4394	6802.2190	---其他	24					90	---Other
4395	6802.2200	--其他石灰石	24					90	--Other calcareous stone
4396	6802.2300	--花岗岩	10		9.2		0	90	--Granite
4397	6802.2900	--其他石	15				0	90	--Other stone
		-其他:							-Other:
		--大理石、石灰华及蜡石:							--Marble, travertine and alabaster:
4398	6802.9110	---石刻	24					90	---Carvings
4399	6802.9190	---其他	10					90	---Other
		--其他石灰石:							--Other calcareous stone:
4400	6802.9210	---石刻	24					90	---Carvings
4401	6802.9290	---其他	10					90	---Other
		--花岗岩:							--Granite:

序号 No.	税则号列 Tariff Item	货 品 名 称	最惠国税率(%) M.F.N.	协定税率(%)Agreement				普通税率(%) Gen.	Article Description
				东盟 ASEAN	曼谷协定 BA	香港 H.K.	澳门 Macao		
4402	6802.9310	---石刻	24					90	---Carvings
4403	6802.9390	---其他	10		9.2			90	---Other
		--其他石：							--Other stone:
4404	6802.9910	---石刻	24					90	---Carvings
4405	6802.9990	---其他	24					90	---Other
	68.03	**已加工的板岩及板岩或粘聚板岩的制品：**							**Worked slate and articles of slate or of agglomerated slate:**
		已加工的板岩及板岩或粘聚板岩的制品：							Worked slate and articles of slate or of agglomerated slate:
4406	6803.0010	---板岩制	20				0	80	---Of slate
4407	6803.0090	---其他	20					80	---Other
	68.04	**未装支架的石磨、石碾、砂轮和类似品及其零件，用于研磨、磨刃、抛光、整形或切割，以及手用磨石、抛光石及其零件，用天然石料、粘聚的天然磨料、人造磨料或陶瓷制成，不论是否装有由其他材料制成的零件：**							**Millstones, grindstones, grinding wheels and the like, without frameworks, for grinding, sharpening, polishing, trueing or cutting , hand sharpening or polishing stones, and parts thereof, of natural stone, of agglomerated natural or artificial abrasives, or of ceramics, with or without parts of other materials:**
4408	6804.1000	-碾磨或磨浆用石磨、石碾	8					40	-Millstones and grindstones for milling, grinding or pulping
		-其他石磨、石碾、砂轮及类似品：							-Other millstones, grindstones, grinding wheels and the like:
4409	6804.2100	--粘聚合成或天然金刚石制	8					17	--Of agglomerated synthetic or natural diamond
		--其他粘聚磨料制或陶瓷制：							--Of other agglomerated abrasives or of ceramics:
4410	6804.2210	---砂轮	8					17	---Grinding wheels
4411	6804.2290	---其他	8					40	---Other
		--天然石料制：							--Of natural stone:
4412	6804.2310	---砂轮	8					17	---Grinding wheels
4413	6804.2390	---其他	8					40	---Other
		-手用磨石及抛光石：							-Hand sharpening or polishing stones:
4414	6804.3010	---琢磨油石	8					17	---Oilstones
4415	6804.3090	---其他	8					40	---Other
	68.05	**砂布、砂纸及以其他材料为底的类似品，不论是否裁切、缝合或用其他方法加工成形：**							**Natural or artificial abrasive powder or grain, on a base of textile material, of paper, of paperboard or of other materials, whether or not cut to shape or sewn or otherwise made up:**
4416	6805.1000	-砂布	8					40	-On a base of woven textile fabric only
4417	6805.2000	-砂纸	8					40	-On a base of paper or paperboard only
4418	6805.3000	-其他	8					40	-On a base of other materials

序号 No.	税则号列 Tariff Item	货品名称	最惠国税率(%) M.F.N.	协定税率(%)Agreement 东盟 ASEAN	曼谷协定 BA	香港 H.K.	澳门 Macao	普通税率(%) Gen.	Article Description
	68.06	**矿渣棉、岩石棉及类似的矿质棉;页状蛭石、膨胀粘土、泡沫矿渣及类似的膨胀矿物材料;具有隔热、隔音或吸音性能的矿物材料的混合物及制品,但税号68.11、68.12或第六十九章的货品除外:**							**Slag wool, rock wool and similar mineral wools; exfoliated vermiculite, expanded clays, foamed slag and similar expanded mineral materials; mixtures and articles of heat-insulating, sound-insulating or soundabsorbing mineral materials, other than those of heading No.68.11or 68.12 or of Chapter 69:**
4419	6806.1000	-矿渣棉、岩石棉及类似的矿质棉(包括其相互混合物),块状、成片或成卷	10.5					40	-Slag wool, rock wool and similar mineral wools (including intermixtures thereof), in bulk, sheets or rolls
4420	6806.2000	-页状蛭石、膨胀粘土、泡沫矿渣及类似的膨胀矿物材料(包括其相互混合物)	10.5					40	-Exfoliated vermiculite, expanded clays, foamed slag and similar expanded mineral materials (including intermixtures thereof)
4421	6806.9000	-其他	10					50	-Other
	68.07	**沥青或类似原料(例如,石油沥青或煤焦油沥青)的制品:**							**Articles of asphalt or of similar material (for example, petroleum bitumen or coal tar pitch):**
4422	6807.1000	-成卷	12					50	-In rolls
4423	6807.9000	-其他	12					50	-Other
	68.08	**镶板、平板、瓦、砖及类似品,用水泥、石膏及其他矿物粘合材料粘合植物纤维、稻草、刨花、木片屑、木粉、锯末或木废料制成:**							**Panels, boards, tiles, blocks and similar articles of vegetable fibre, of straw or of shavings, chips, particles, sawdust or other waste, of wood, agglomerated with cement, plaster or other mineral binders:**
4424	6808.0000	镶板、平板、瓦、砖及类似品,用水泥、石膏及其他矿物粘合材料粘合植物纤维、稻草、刨花、木片屑、木粉、锯末或木废料制成	10.5					40	Panels, boards, tiles, blocks and similar articles of vegetable fibre, of straw or of shavings, chips, particles, sawdust or other waste, of wood, agglomerated with cement, plaster or other mineral binders
	68.09	**石膏制品及以石膏为基本成分的混合材料制品:**							**Articles of plaster or of compositions based on plaster:**
		-未经装饰的板、片、砖、瓦及类似品:							-Boards, sheets, panels, tiles and similar articles, not ornamented:
4425	6809.1100	--仅用纸、纸板贴面或加强的	28					100	--Faced or reinforced with paper or paper board only
4426	6809.1900	--其他	25					100	--Other
4427	6809.9000	-其他制品	25					100	-Other articles
	68.10	**水泥、混凝土或人造石制品,不论是否加强:**							**Articles of cement, of concrete or of artificial stone, whether or not reinforced:**

序号 No.	税则号列 Tariff Item	货品名称	最惠国税率(%) M.F.N.	协定税率(%) Agreement 东盟 ASEAN	曼谷协定 BA	香港 H.K.	澳门 Macao	普通税率(%) Gen.	Article Description
		-砖、瓦、扁平石及类似品:							-Tiles, flagstones, bricks and similar articles:
4428	6810.1100	--建筑用砖及石砌块	10.5			0		40	--Building blocks and bricks
		--其他:							--Other:
4429	6810.1910	---人造石制	10.5				0	70	---Of artificial stone
4430	6810.1990	---其他	10.5					70	---Other
		-其他制品:							-Other articles:
		--建筑或土木工程用的预制结构件:							--Prefabricated structural components for building or civil engineering:
4431	6810.9110	---钢筋混凝土和预应力混凝土管、杆、板、桩等	10.5			0		40	---Reinforced concrete and prestressed concrete tubes, pipes, rods, plates, piles and similar articles
4432	6810.9190	---其他	10.5			0		40	---Other
		--其他:							--Other:
4433	6810.9910	---铁道用水泥枕	8					14	---Railway sleepers of concrete
4434	6810.9990	---其他	10.5					70	---Other
	68.11	**石棉水泥、纤维素水泥或类似材料的制品:**							**Articles of asbestos-cement, of cellulose fibre-cement or the like:**
4435	6811.1000	-瓦楞板	5					40	-Corrugated sheets
4436	6811.2000	-其他片、板、砖、瓦及类似制品	10.5					40	-Other sheets, panels, tiles and similar articles
4437	6811.3000	-管子及管子附件	8					40	-Tubes, pipes and tube or pipe fittings
4438	6811.9000	-其他制品	8.4					40	-Other articles
	68.12	**已加工的石棉纤维;以石棉为基本成分或以石棉和碳酸镁为基本成分的混合物;上述混合物或石棉的制品(例如,纱线、机织物、服装、帽类、鞋靴、衬垫),不论是否加强,但税号 68.11 或 68.13 的货品除外:**							**Fabricated asbestos fibres; mixtures with a basis of asbestos or with a basis of asbestos and magnesium carbonate; articles of such mixtures or of asbestos (for example, thread, woven fabric, clothing, headgear, footwear, gaskets), whether or not reinforced, other than goods of heading No.68.11 or 68.13:**
4439	6812.5000	-服装、衣着附件、帽类及鞋靴	10.5					40	-Clothing, clothing accessories, footwear and headgear
4440	6812.6000	-纸、麻丝板及毡子	10.5					40	-Paper, millboard and felt
4441	6812.7000	-成片或成卷的压缩石棉纤维接合材料	10.5					40	-Compressed asbestos fibre jointing, in sheets or rolls
4442	6812.9000	-其他	10					40	-Other
	68.13	**以石棉、其他矿物质或纤维素为基本成分的未装配摩擦材料及其制品(例如,片、卷、带、盘、圈、垫及扇形),适于作制动器、离合器及类似品,不论是否与织物或其他材料结合而成:**							**Friction material and articles thereof (for example, sheets, rolls, strips, segments, discs, washers, pads), not mpunted, for brakes, for clutches or the like, with a basis of asbestos, of other mineral substances or of cellulose, whether or not combined with textile or other materials:**

序号 No.	税则号列 Tariff Item	货品名称	最惠国税率(%) M.F.N.	协定税率(%) Agreement 东盟 ASEAN	曼谷协定 BA	香港 H.K.	澳门 Macao	普通税率(%) Gen.	Article Description
4443	6813.1000	-闸衬、闸垫	10					40	-Brake linings and pads
4444	6813.9000	-其他	12					40	-Other
	68.14	**已加工的云母及其制品,包括粘聚或复制的云母,不论是否附于纸、纸板或其他材料上:**							**Worked mica and articles of mica, including agglomerated or reconstituted mica, whether or not on a support of paper, paperboard or other meterials:**
4445	6814.1000	-粘聚或复制云母制的板、片、带,不论是否附于其他材料上	10.5					35	-Plates, sheets and strips of agglomerated or reconstituted mica, whether or not on a support
4446	6814.9000	-其他	10.5					35	-Other
	68.15	**其他税号未列名的石制品及其他矿物制品(包括碳纤维及其制品和泥煤制品):**							**Articles of stone or of other mineral substances (including carbon fibres, articles of carbon fibres and articles of peat), not elsewhere specified or included:**
4447	6815.1000	-非电器用的石墨或其他碳精制品	15					70	-Non-electrical articles of graphite or other carbon
4448	6815.2000	-泥煤制品	15					70	-Articles of peat
		-其他制品:							-Other articles:
4449	6815.9100	--含有菱镁矿、白云石或铬铁矿的	15					70	--Containing magnesite, dolomite or chromite
4450	6815.9900	--其他	17.5					70	--Other

第六十九章
陶瓷产品

Chapter 69
Ceramic products

注释：

一、本章仅适用于成形后经过烧制的陶瓷产品。税号69.04至69.14仅适用于不能归入税号69.01至69.03的产品。

二、本章不包括：

（一）税号28.44的产品；
（二）税号68.04的物品；
（三）第七十一章的物品（例如，仿首饰）；
（四）税号81.13的金属陶瓷；
（五）第八十二章的物品；
（六）绝缘子（税号85.46）或绝缘材料制的零件（税号85.47）；
（七）假牙（税号90.21）；
（八）第九十一章的物品（例如，钟及钟壳）；
（九）第九十四章的物品（例如，家具、灯具及照明装置、活动房屋）；
（十）第九十五章的物品（例如，玩具、游戏品及运动用品）；
（十一）税号96.06的物品（例如，钮扣）或税号96.14的物品（例如，烟斗）；
（十二）第九十七章的物品（例如，艺术品）。

Notes:

1. This Chapter applies only to ceramic products which have been fired after shaping. Headings Nos.69.04 to 69.14 apply only to such products other than those classifiable in headings Nos.69.01 to 69.03.

2. This Chapter does not cover:
(a) Products of heading No.28.44;
(b) Articles of heading No.68.04;
(c) Articles of Chapter 71 (for example, imitation jewellery);
(d) Cermets of heading No.81.13;
(e) Articles of Chapter 82;
(f) Electrical insulators (heading No.85.46) or fittings of insulating material of heading No.85.47;
(g) Artificial teeth (heading No.90.21);
(h) Articles of Chapter 91 (for example, clocks and clock cases);
(i) Articles of Chapter 94 (for example, furniture, lamps and lighting fittings, prefabricated buildings);
(j) Articles of Chapter 95 (for example, toys, games and sports requisites);
(k) Articles of heading No.96.06 (for example, buttons) or of heading No.96.14 (for example, smoking pipes); or
(l) Articles of Chapter 97 (for example, works of art).

序号 No.	税则号列 Tariff Item	货品名称	最惠国税率(%) M.F.N.	协定税率(%)Agreement 东盟 ASEAN	曼谷协定 BA	香港 H.K.	澳门 Macao	普通税率(%) Gen.	Article Description
		第一分章 硅化石粉或类似硅土及耐火材料制品							1. -GOODS OF SILICEOUS FOSSIL MEALS OR OF SIMILAR SILICEOUS EARTHS, AND REFRACTORY GOODS
	69.01	**硅质化石粉(例如各种硅藻土)或类似硅土制的砖、块、瓦及其他陶瓷制品:**							**Bricks, blocks, tiles and other ceramic goods of siliceous fossil meals (for example, kieselguhr, tripolite or diatimite) or of similar siliceous earths:**
4451	6901.0000	硅质化石粉(例如各种硅藻土)或类似硅土制的砖、块、瓦及其他陶瓷制品	8					50	Bricks, blocks, tiles and other ceramic goods of siliceous fossil meals (for example, kieselguhr, tripolite or diatomite) or of similar siliceous earths
	69.02	**耐火砖、块、瓦及类似耐火陶瓷建材制品,但硅质化石粉及类似硅土制的除外:**							**Refractory bricks, blocks, tiles and similar refractory ceramic constructional goods, other than those of siliceous fossil meals or similar siliceous earths:**

序号 No.	税则号列 Tariff Item	货品名称	最惠国税率(%) M.F.N.	协定税率(%) Agreement 东盟 ASEAN	曼谷协定 BA	香港 H.K.	澳门 Macao	普通税率(%) Gen.	Article Description
4452	6902.1000	-单独或同时含有按重量计超过50%的镁、钙或铬(分别以氧化镁、氧化钙及三氧化二铬的含量计)	8					30	C-Containing by weight, singly or together, more than 50% of the elements Mg, Ca or Cr, expressed as MgO, CaO or Cr_2O_3
4453	6902.2000	-含有按重量计超过50%的三氧化二铝、二氧化硅或其混合物或化合物	8					30	-Containing by weight more than 50% of alumina(Al_2O_3), of silica(SiO_2) or of a mixture or compound of these products
4454	6902.9000	-其他	8					30	-Other
	69.03	**其他耐火陶瓷制品(例如,甑、坩埚、马弗罩、喷管、栓塞、支架、烤钵、管子、护套及棒条),但硅质化石粉及类似硅土制的除外:**							**Other refractory ceramic goods (for example, retorts, crucibles, muffles, nozzles, plugs, supports, cupels, tubes, pipes, sheaths and rods), other than those of siliceous fossil meals or of similar siliceous earthes:**
4455	6903.1000	-含有按重量计超过50%的石墨、其他碳或其混合物	8					20	-Containing by weight more than 50% of graphite or other carbon or of a mixture of these products
4456	6903.2000	-含有按重量计超过50%的三氧化二铝或三氧化二铝和二氧化硅的混合物或化合物	8					20	-Containing by weight more than 50% of alumina(Al_2O_3) or of a mixture of compound of alumina and of silica (SiO_2)
4457	6903.9000	-其他	8					20	-Other
		第二分章 其他陶瓷产品							Ⅱ. -OTHER CERAMIC PRODUCTS
	69.04	**陶瓷制建筑用砖、铺地砖、支撑或填充用砖及类似品:**							**Ceramic building bricks, flooring blocks, support or filler tiles and the like:**
4458	6904.1000	-建筑用砖	15					90	-Building bricks
4459	6904.9000	-其他	24.5					90	-Other
	69.05	**屋顶瓦、烟囱罩、通风帽、烟囱衬壁、建筑装饰物及其他建筑用陶瓷制品:**							**Roofing tiles, chimney-pots, cowls, chimney liners, architectural ornaments and other ceramic constructional goods:**
4460	6905.1000	-屋顶瓦	24.5					90	-Roofing tiles
4461	6905.9000	-其他	24.5					90	-Other
	69.06	**陶瓷套管、导管、槽管及管子附件:**							**Ceramic pipes, conduits, guttering and pipe fittings:**
4462	6906.0000	陶瓷套管、导管、槽管及管子附件	15					90	Ceramic pipes, conduits, guttering and pipe fittings
	69.07	**未上釉的陶瓷贴面砖、铺面砖,包括炉面砖及墙面砖;未上釉的陶瓷镶嵌砖(马赛克)及类似品,不论是否有衬背:**							**Unglazed ceramic flags and paving, hearth or wall tiles; unglazed ceramic mosaic cubes and the like, whether or not on a backing:**

序号 No.	税则号列 Tariff Item	货品名称	最惠国税率(%) M.F.N.	协定税率(%) Agreement 东盟 ASEAN	曼谷协定 BA	香港 H.K.	澳门 Macao	普通税率(%) Gen.	Article Description
4463	6907.1000	-砖、瓦、块及类似品,不论是否矩形,其最大表面积以可置入边长小于7厘米的方格为限	24.5					90	-Tiles, cubes and similar articles, whether or not rectangular, the largest surface area of which is capable of being enclosed in a square the side of which is less than 7cm
4464	6907.9000	-其他	12					90	-Other
	69.08	**上釉的陶瓷贴面砖、铺面砖,包括炉面砖及墙面砖;上釉的陶瓷镶嵌砖(马赛克)及类似品,不论是否有衬背:**							**Glazed ceramic flags and paving, hearth or wall tiles; glazed ceramic mosaic cubes and the like, whether or not on a backing:**
4465	6908.1000	-砖、瓦、块及类似品,不论是否矩形,其最大表面积以可置入边长小于7厘米的方格为限	12		10.4			100	-Tiles, cubes and similar articles, whether or not rectangular, the largest surface area of which is capable of being enclosed in a square the side of which is less than 7cm
4466	6908.9000	-其他	12					100	-Other
	69.09	**实验室、化学或其他专门技术用途的陶瓷器;农业用陶瓷槽、缸及类似容器;通常供运输及盛装货物用的陶瓷罐、坛及类似品;**							**Ceramic wares for laboratory, chemical or other technical uses; ceramic troughs, tubs and similar receptacles of a kind used in agriculture; ceramic pots, jars and similar articles of a kind used for the conveyance or packing of goods:**
		-实验室、化学或其他专门技术用途的陶瓷器:							-Ceramic wares for laboratory, chemical or other technical uses:
4467	6909.1100	--瓷制	8					30	--Of porcelain or china
4468	6909.1200	--莫氏硬度为9或以上的物品	8					30	--Articles having a hardness equivalent to 9 or more on the Mohs scale
4469	6909.1900	--其他	8					30	--Other
4470	6909.9000	-其他	21					90	-Other
	69.10	**陶瓷洗涤槽、脸盆、脸盆座、浴缸、坐浴盆、抽水马桶、水箱、小便池及类似的固定卫生设备:**							**Ceramic sinks, wash basins, wash basin pedestals, baths, bidets, water closet pans, flushing cisterns, urinals and similar sanitary fixtures:**
4471	6910.1000	-瓷制	10					100	-Of porcelain or china
4472	6910.9000	-其他	10					100	-Other
	69.11	**瓷餐具、厨房器具及其他家用或盥洗用瓷器:**							**Tableware, kitchenware, other household articles and toilet articles, of porcelain or china:**
		-餐具及厨房器具:							-Tableware and kitchenware:
4473	6911.1010	---餐具	12		10			100	---Tableware
4474	6911.1020	---厨房器具	15		12.5			100	---Kitchenware
4475	6911.9000	-其他	24.5		20			100	-Other
	69.12	**陶餐具、厨房器具及其他家用或盥洗用陶器:**							**Ceramic tableware, kitchenware, other household articles and toilet articles, other than of procelain or china:**

序号 No.	税则号列 Tariff Item	货品名称	最惠国税率(%) M.F.N.	协定税率(%)Agreement 东盟 ASEAN	曼谷协定 BA	香港 H.K.	澳门 Macao	普通税率(%) Gen.	Article Description
4476	6912.0010	---餐具	15					100	---Tableware
4477	6912.0090	---其他	15					100	---Other
	69.13	**塑像及其他装饰用陶瓷制品:**							**Statuettesa and other ornamental ceramic articles:**
4478	6913.1000	-瓷制	15					100	-Of porcelain or china
4479	6913.9000	-其他	15					100	-Other
	69.14	**其他陶瓷制品:**							**Other ceramic articles:**
4480	6914.1000	-瓷制	24.5					100	-Of porcelain or china
4481	6914.9000	-其他	10					100	-Other

第七十章
玻璃及其制品

注释：

一、本章不包括：

（一）税号32.07的货品（例如，珐琅和釉料、搪瓷玻璃料及其他玻璃粉、粒或粉片）；

（二）第七十一章的物品（例如，仿首饰）；

（三）税号85.44的光缆、税号85.46的绝缘子或税号85.47所列绝缘材料制的零件；

（四）光导纤维、经光学加工的光学元件、注射用针管、假眼、温度计、气压计、液体比重计或第九十章的其他物品；

（五）有永久固定电光源的灯具及照明装置、灯箱标志或名牌和类似品及其零件（税号94.05）；

（六）玩具、游戏品、运动用品、圣诞树装饰品及第九十五章的其他物品（供玩偶或第九十五章其他物品用的无机械装置的玻璃假眼除外）；

（七）钮扣、保温瓶、香水喷雾器和类似的喷雾器及第九十六章的其他物品。

二、对于税号70.03、70.04及70.05：

（一）玻璃在退火前的各种处理都不视为“已加工”；

（二）玻璃切割成一定形状并不影响其作为板片归类；

（三）所称“吸收、反射或非反射层”，是指极薄的金属或化合物（例如，金属氧化物）镀层，该镀层可以吸收红外线等光线或可以提高玻璃的反射性能，同时仍然使玻璃具有一定程度的透明性或半透明性；或者该镀层可以防止光线在玻璃表面的反射。

三、税号70.06所述产品，不论是否具有制成品的特性仍归入该税号。

Chapter 70
Glass and glassware

Notes:

1. This Chapter does not cover:
   (a) Goods of heading No.32.07 (for example, vitrifiable enamels and glazes, glass frit, other glass in the form of powder, granules or flakes);
   (b) Articles of Chapter 71 (for example, imitation jewellery);
   (c) Optical fibre cables (heading No.85.44), electrical insulators (heading No.85.46) or fitting of insulating material of heading No.85.47;
   (d) Optical fibres, optically worked optical elements, hypodermic syringes, artificial eyes, thermometers, barometers, hydrometers or other articles of Chapter 90;
   (e) Lamps or lighting fittings, illuminated signs, illuminated nameplates or the like, having a permanently fixed light source, or parts thereof of heading No.94.05;
   (f) Toys, games, sports requisites, Christmas tree ornaments or other articles of Chapter 95 (excluding glass eyes without mechanisms for dolls or for other articles of Chapter 95); or
   (g) Buttons, fitted vacuum flasks, scent or similar sprays or other articles of Chapter 96.

2. For the purposes of headings Nos.70.03, 70.04 and 70.05:
   (a) glass is not regarded as “worked” by reason of any process it has undergone before annealing;
   (b) cutting to shape does not affect the classification of glass in sheets;
   (c) the expression “absorbent, reflecting or non-reflecting layer” means a microscopically thin coating of metal or of a chemical compound (for example, metal oxide) which absorbs, for example, infra-red light or improves the reflecting qualities of the glass while still allowing it to retain a degree of transparency or translucency; or which prevents light from being reflected on the surface of the glass.

3. The products referred to in heading No.70.06 remain classified in that heading whether or not they have the character of articles.

四、税号70.19所称"玻璃棉"，是指：

（一）按重量计二氧化硅的含量在60%及以上的矿质棉；

（二）按重量计二氧化硅的含量在60%以下，但碱性氧化物（氧化钾或氧化钠）的含量在5%以上或氧化硼的含量在2%以上的矿质棉。

不符合上述规定的矿质棉归入税号68.06。

4. For the purposes of heading No.70.19, the expression "glass wool" means:

(a) Mineral wools with a silica (SiO_2) content not less than 60% by weight;

(b) Mineral wools with a silica (SiO_2) content less than 60% but with an alkaline oxide (K_2O or Na_2O) content exceeding 5% by weight or a boric oxide (B_2O_3) content exceeding 2% by weight.

Mineral wools which do not comply with the above specifications fall in heading No.68.06.

五、本目录所称"玻璃"，包括熔融石英及其他熔融硅石。

5. Throughout the Nomenclature, the expression "glass" includes fused quartz and other fused silica.

子目注释：

子目号7013.21、7013.31及7013.91所称"铅晶质玻璃"，仅指按重量计氧化铅含量不低于24%的玻璃。

Subheading Note:

For the purposes of subheadings Nos.7013.21, 7013.31 and 7013.91, the expression "lead crystal" means only glass having a minimum lead monoxide (PbO) content by weight of 24%.

序号 No.	税则号列 Tariff Item	货品名称	最惠国税率(%) M.F.N.	协定税率(%) Agreement 东盟 ASEAN	曼谷协定 BA	香港 H.K.	澳门 Macao	普通税率(%) Gen.	Article Description
	70.01	**碎玻璃及废玻璃；玻璃块料：**							**Cullet and other waste and scrap of glass; glass in the mass:**
4482	7001.0000	碎玻璃及废玻璃；玻璃块料	12					50	Cullet and other waste and scrap of glass; glass in the mass
	70.02	**未加工的玻璃球、棒及管（税号70.18的微型玻璃球除外）：**							**Glass in balls (other than microspheres of heading No. 70.18), roads or tubes, unworked:**
4483	7002.1000	-玻璃球	12					50	-Balls
		-玻璃棒：							-Rods:
4484	7002.2010	---光导纤维预制棒	6					50	---Preformed bars for drawing optical fibre
4485	7002.2090	---其他	12					50	---Other
		-玻璃管：							-Tubes:
		--熔融石英或其他熔融硅石制：							--Of fused quartz or other fused silica:
4486	7002.3110	---光导纤维用波导级石英玻璃管	5					17	---Waveguide quartz tubes for optical fibres use
4487	7002.3190	---其他	14					50	---Other
4488	7002.3200	--温度在0℃至300℃时线膨胀系数不超过5×10^{-6}/开尔文的其他玻璃制	12					50	--Of other glass having a linear coefficient of expansion not exceeding 5×10^{-6} per Kelvin within a temperature range of 0℃ to 300℃
4489	7002.3900	--其他	12					50	--Other
	70.03	**铸制或轧制玻璃板、片或型材及异型材，不论是否有吸收、反射或非反射层，但未经其他加工：**							**Cast glass and rolled glass, in sheets or profiles, whether or not having an absorbent, reflecting or non-reflecting layer, but not otherwise worked:**

序号 No.	税则号列 Tariff Item	货品名称	最惠国税率(%) M.F.N.	协定税率(%)Agreement 东盟 ASEAN	曼谷协定 BA	香港 H.K.	澳门 Macao	普通税率(%) Gen.	Article Description
		-非夹丝玻璃板、片:							-Non-wired sheets:
4490	7003.1200	--整块着色、不透明、镶色或具有吸收、反射或非反射层的	15					50	--Coloured throughout the mass(body tinted), opacified, flashed or having an absorbent, or non-reflecting reflecting layer, but not otherwise worked
4491	7003.1900	--其他	17.5					50	--Other
4492	7003.2000	-夹丝玻璃板、片	15					50	-Wired sheets
4493	7003.3000	-型材及异型材	15					50	-Profiles
	70.04	**拉制或吹制玻璃板、片,不论是否有吸收、反射或非反射层,但未经其他加工:**							**Drawn glass and blown glass, in sheets, whether or not having an absorbent, reflecting or non-reflecting layer, but not otherwise worked:**
4494	7004.2000	-整块着色、不透明、镶色或具有吸收、反射或非反射层的	17.5					50	-Glass, coloured throughout the mass (body tinted), opacified, flashed or having an absorbent, reflecting or non-reflecting layer
4495	7004.9000	-其他玻璃	17.5					50	-Other glass
	70.05	**浮法玻璃板、片及表面研磨或抛光玻璃板、片,不论是否有吸收、反射或非反射层,但未经其他加工:**							**Float glass and surface ground or polished glass, in sheets, whether or not having an absorbent, reflecting or non-reflecting layer, but not otherwise worked:**
4496	7005.1000	-具有吸收、反射或非反射层的非夹丝玻璃	15					50	-Non-wired glass, having an absorbent, reflecting or noon-reflecting layer
		-其他非夹丝玻璃:							-Other non-wired glass:
4497	7005.2100	--整块着色、不透明、镶色或仅表面研磨的	15					50	--Coloured throughout the mass(body tinted), opacified, flashed or merely surface ground
4498	7005.2900	--其他	15					50	--Other
4499	7005.3000	-夹丝玻璃	17.5					50	-Wired glass
	70.06	**经弯曲、磨边、镂刻、钻孔、涂珐琅或其他加工的税号 70.03、70.04 或 70.05 的玻璃,但未用其他材料镶框或装配:**							**Glass of heading No.70.03,70.04 or 70.05, bent, edge-worked, engraved, drilled, enamelled or otherwise worked, but not framed or fitted with other materials:**
4500	7006.0000	经弯曲、磨边、镂刻、钻孔、涂珐琅或其他加工的税号 70.03、70.04 或 70.05 的玻璃,但未用其他材料镶框或装配	15					50	Glass of heading No. 70.03,70.04 or 70. 05, bent, edge-worked, engraved, drilled, enamelled or otherwise worked, but not framed or fitted with other materials
	70.07	**钢化或层压玻璃制的安全玻璃:**							**Safety glass, consisting of toughened (tempered) or laminated glass:**
		-钢化安全玻璃:							-Toughened(tempered)safety glass:
		--规格及形状适于安装在车辆、航空器、航天器及船舶上:							--Of size and shape suitable for incorporation in vehicles, aircraft, spacecraft or vessels:

序号 No.	税则号列 Tariff Item	货品名称	最惠国税率(%) M.F.N.	协定税率(%) Agreement 东盟 ASEAN	曼谷协定 BA	香港 H.K.	澳门 Macao	普通税率(%) Gen.	Article Description
4501	7007.1110	---航空器、航天器及船舶用	2			0△		11	---For aircraft, spacecraft or vessels
4502	7007.1190	---其他	10			0△		50	---Other
4503	7007.1900	--其他	14					50	--Other
		-层压安全玻璃:							-Laminated safety glass:
		--规格及形状适于安装在车辆、航空器、航天器及船舶上:							--Of size and shape suitable for incorporation in vehicles, aircraft, spacecraft or vessels:
4504	7007.2110	---航空器、航天器及船舶用	2					11	---for aircraft, spacecraft or vessels
4505	7007.2190	---其他	20					50	---Other
4506	7007.2900	--其他	14					50	--Other
	70.08	**多层隔温、隔音玻璃组件:**							**Multiple-walled insulating units of glass:**
		多层隔温、隔音玻璃组件:							Multiple-walled insulating units of glass:
4507	7008.0010	---中空或真空隔温、隔音玻璃	14					50	---Sealed or vacuum insulating glass
4508	7008.0090	---其他	14					50	---Other
	70.09	**玻璃镜(包括后视镜),不论是否镶框:**							**Glass mirrors, whether or not framed, including rear-view mirrors:**
4509	7009.1000	-车辆后视镜	10					100	-Rear-view mirrors for vehicles
		-其他:							-Other:
4510	7009.9100	--未镶框	21					70	--Unframed
4511	7009.9200	--已镶框	12					100	--Framed
	70.10	**玻璃制的坛、瓶、缸、罐、安瓿及其他容器,用于运输或盛装货物;玻璃制保藏罐;玻璃塞、盖及类似的封口器:**							**Carboys, bottles, flasks, jars, pots, phials, ampoules and other containers, of glass, of a kind used for the conveyance or packing of goods; preserving jars of glass; stoppers, lids and other closures, of glass:**
4512	7010.1000	-安瓿	14					50	-Ampoules
4513	7010.2000	-塞、盖及类似的封口器	14					50	-Stoppers, lids and other closures
		-其他:							-Other, of a capacity:
4514	7010.9010	---超过1升	14					50	---Exceeding 1L
4515	7010.9020	---超过0.33升,但不超过一升	14					50	---Exceeding 0.33 L but not exceeding 1L
4516	7010.9030	---超过0.15升,但不超过0.33升	14					50	---Exceeding 0.15 L but not exceeding 0.33L
4517	7010.9090	---不超过0.15升	14					50	---Not exceeding 0.15L
	70.11	**制灯泡、阴极射线管及类似品用的未封口玻璃外壳(包括玻璃泡及管)及其玻璃零件,但未装有配件:**							**Glass envelopes (including bulbs and tubes), open, and glass parts thereof, without fittings, for electric lamps, cathode-ray tubes or the like:**
4518	7011.1000	-电灯用	21					80	-For electric lighting

序号 No.	税则号列 Tariff Item	货品名称	最惠国税率(%) M.F.N.	协定税率(%) Agreement 东盟 ASEAN	曼谷协定 BA	香港 H.K.	澳门 Macao	普通税率(%) Gen.	Article Description
		-阴极射线管用							-For cathode-ray tubes:
4519	7011.2010	---显像管玻壳及其零件	10	见表1 Table 1				35	---Glass envelopes for kinescope and glass parts thereof
4520	7011.2090	---其他	10	见表1 Table 1				35	---Other
		-其他:							-Other:
4521	7011.9010	---电子管用(阴极射线管用的除外)	8					35	---For electronic tubes and valves(other than cathode-ray tubes)
4522	7011.9090	---其他	21					80	---Other
	70.12	**保温瓶或其他保温容器用的玻璃胆:**							**Glass inners for vacuum flasks or for other vacuum vessels:**
4523	7012.0000	保温瓶或其他保温容器用的玻璃胆	21			0		100	Glass inners for vacuum flasks or for other vacuum vessels
	70.13	**玻璃器,供餐桌、厨房、盥洗室、办公室、室内装饰或类似用途(税号70.10或70.18的货品除外):**							**Glassware of a kind used for table, kitchen, toilet, office, indoor decoration or similar purposes (other than that of heading No.70.10 or 70.18):**
4524	7013.1000	-玻璃陶瓷制	24.5					100	-Of glass-ceramics
		-杯子,但玻璃陶瓷制的除外:							-Drinking glasses other than of glass-ceramics:
4525	7013.2100	--铅晶质玻璃制	24.5					100	--Of lead crystal
4526	7013.2900	--其他	8					100	--Other
		-餐桌或厨房用玻璃器皿(不包括杯子),但玻璃陶瓷制的除外:							-Glassware of a kind used for table (other than drinking glasses) or kitchen purposes other than of glass-ceramics:
4527	7013.3100	--铅晶质玻璃制	24.5					100	--Of lead crystal
4528	7013.3200	--温度在0℃至300℃时线膨胀系数不超过5×10^{-6}/开尔文的其他玻璃制	13.3					100	--Of glass having a linear coefficient of expansion not exceeding 5×10^{-6} per Kelvin within a temperature range of 0℃ to 300℃
4529	7013.3900	--其他	10					100	--Other
		-其他玻璃器:							-Other glassware:
4530	7013.9100	--铅晶质玻璃制	10					100	--Of lead crystal
4531	7013.9900	--其他	10					100	--Other
	70.14	**未经光学加工的信号玻璃器及玻璃制光学元件(税号70.15的货品除外):**							**Signalling glassware and optical elements of glass (other than those of heading No. 70.15), not optically worked:**
4532	7014.0010	---光学仪器用光学元件毛坯	10				0	40	---Blanks of optical elements, for optical instruments
4533	7014.0090	---其他	17.5					80	---Other
	70.15	**钟表玻璃及类似玻璃、视力矫正或非视力矫正眼镜用玻璃,呈弧面、弯曲、凹形或类似形状但未经光学加工的;制造上述玻璃用的凹面圆形及扇形玻璃:**							**Clock or watch glasses and similar glasses, glasses for non-corrective or corrective spectacles, curved, bent, hollowed or the like, not optically worked; hollow glass spheres and their segments, for the manufacture of such glasses:**

序号 No.	税则号列 Tariff Item	货品名称	最惠国税率(%) M.F.N.	东盟 ASEAN	曼谷协定 BA	香港 H.K.	澳门 Macao	普通税率(%) Gen.	Article Description
		-视力矫正眼镜用玻璃:							-Glasses for corrective spectacles:
4534	7015.1010	---变色镜片坯件	21					80	---Blanks for photochromic spectacles
4535	7015.1090	---其他	17.5					70	---Other
		-其他:							-Other:
4536	7015.9010	---钟表玻璃	17.5					70	---Clock and watch glasses
4537	7015.9020	---平光变色镜片坯件	18					80	---Blanks for plane photochromic spectacles
4538	7015.9090	---其他	12					80	---Other
	70.16	**建筑用压制或模制的铺面用玻璃块、砖、片、瓦及其他制品,不论是否夹丝;供镶嵌或类似装饰用的玻璃马赛克及其他小件玻璃品,不论是否有衬背;花饰铅条窗玻璃及类似品;多孔或泡沫玻璃块、板、片及类似品:**							**Paving blocks, slabs, bricks, squares, tiles and other articles of pressed or moulded glass, whether or not wired, of a kind used for building or construction purposes; glass cubes and other glass smallwares, whether or not on a backing, for mosaics or similar decorative purposes; leaded lights and the like; multicellular or foam glass in blocks, panels, plates, shells or similar forms:**
4539	7016.1000	-供镶嵌或类似装饰用的玻璃马赛克及其他小件玻璃品,不论是否有衬背	22					100	-Glass cubes and other glass smallwares, whether or not on a backing, for mosaics or similar decorative purposes
		-其他:							-Other:
4540	7016.9010	---花饰铅条窗玻璃及类似品	24					90	---Leaded lights and the like
4541	7016.9090	---其他	18					90	---Other
	70.17	**实验室、卫生及配药用的玻璃器,不论有无刻度或标量:**							**Laboratory, hygienic or pharmaceutical glassware, whether or not graduated or calibrated:**
4542	7017.1000	-熔融石英或其他熔融硅石制	0					30	-Of fused quartz or other fused silica
4543	7017.2000	-温度在0℃至300℃时线膨胀系数不超过5×10^{-6}/开尔文的其他玻璃制	8					30	-Of other glass having a linear coefficient of expansion not exceeding 5×10^{-6} per Kelvin within a temperature range of 0℃ to 300℃
4544	7017.9000	-其他	8					30	-Other
	70.18	**玻璃珠、仿珍珠、仿宝石或仿半宝石和类似小件玻璃品及其制品,但仿首饰除外;玻璃假眼,但医用假眼除外;灯工方法制作的玻璃塑像及其他玻璃装饰品,但仿首饰除外;直径不超过1毫米的微型玻璃球:**							**Glass beads, imitation pearls, imitation precious or semi-precious stones and similar glass smallwares, and articles thereof other than imitation jewellery; glass eyes other than prosthetic articles; statuettes and other ornaments of lampworked gla ss, other than imitation jewellery; glass microspheres not exceeding 1mm in diameter:**

序号 No.	税则号列 Tariff Item	货品名称	最惠国税率(%) M.F.N.	协定税率(%) Agreement 东盟 ASEAN	曼谷协定 BA	香港 H.K.	澳门 Macao	普通税率(%) Gen.	Article Description
4545	7018.1000	-玻璃珠、仿珍珠、仿宝石或仿半宝石及类似小件玻璃品	10					100	-Glass beads, imitation pearls, imitation precious or semi-precious stones and similar glass smallwares
4546	7018.2000	-直径不超过1毫米的微型玻璃球	20					100	-Glass microspheres not exceeding 1mm in diameter
4547	7018.9000	-其他	20					100	-Other
	70.19	**玻璃纤维(包括玻璃棉)及其制品(例如,玻璃纤维纱线及其织物):**							**Glass fibres (including glass wool) and articles thereof (for example, yarn, woven fabrics):**
		-梳条、粗纱、纱线及短切纤维:							-Slivers, rovings, yarn and chopped strands:
4548	7019.1100	--长度不超过50毫米的短切纤维	12					50	--Chopped strands, of a length of not more than 50mm
4549	7019.1200	--粗纱	12					50	--Rovings
4550	7019.1900	--其他	10				0	50	--Other
		-薄片(巴厘纱)、纤维网、席、垫、板及类似无纺产品:							-Thin sheets (voiles), webs, mats, mattresses, boards and similar nonwoven products:
4551	7019.3100	--席	5					40	--Mats
4552	7019.3200	--薄片(巴厘纱)	14					40	--Thin sheets(voiles)
4553	7019.3900	--其他	10.5					40	--Other
4554	7019.4000	-粗纱机织物	12					40	-Woven fabrics of rovings
		-其他机织物:							-Other woven fabrics:
4555	7019.5100	--宽度不超过30厘米的	12					40	--Of a width not exceeding 30cm
4556	7019.5200	--宽度超过30厘米的长丝平纹织物,每平方米重量不超过250克,单根纱线细度不超过136特克斯	12					40	--Of a width exceeding 30cm, plain weave, weighing less than 250g/m^2, of filaments measuring per single yarn not more than 136 tex
4557	7019.5900	--其他	12			0	0	40	--Other
4558	7019.9000	-其他	7				0	40	-Other
	70.20	**其他玻璃制品:**							**Other articles of glass:**
		---工业用:							---For technical use:
4559	7020.0011	----导电玻璃	10.5			0		40	----Conductivity glass
4560	7020.0012	----绝缘子用玻璃伞盘	10.5					40	----Glass umbrella for insulator
4561	7020.0019	----其他	10.5					40	----Other
4562	7020.0090	---其他	15					100	---Other

第十四类

天然或养殖珍珠、宝石或半宝石、贵金属、包贵金属及其制品；仿首饰；硬币

SECTION XIV

NATURAL OR CULTURED PEARLS, PRECIOUS OR SEMI-PRECIOUS STONES, PRECIOUS METALS, METALS CLAD WITH PRECIOUS METAL, AND ARTICLES THEREOF; IMITATION JEWELLERY; COIN

第七十一章

天然或养殖珍珠、宝石或半宝石、贵金属、包贵金属及其制品；仿首饰；硬币

Chapter 71

Natural or cultured pearls, precious or semi-precious stones, precious metals, metals clad with precious metal, and articles thereof; imitation jewellery; coin

注释：

一、除第六类注释一（一）及下列各款另有规定的以外，凡制品的全部或部分由下列物品构成，均应归入本章：

（一）天然或养殖珍珠、宝石或半宝石（天然、合成或再造）；

（二）贵金属或包贵金属。

二、（一）税号 71.13、71.14 及 71.15 不包括带有贵金属或包贵金属制的小零件或小装饰品（例如，交织字母、套、圈、套环）的制品，上述注释一（二）也不适用于这类制品；

（二）税号 71.16 不包括含有贵金属或包贵金属（仅作为小零件或小装饰品的除外）的制品。

三、本章不包括：

（一）贵金属汞齐及胶态贵金属（税号 28.43）；

（二）第三十章的外科用无菌缝合材料、牙科填料或其他货品；

（三）第三十二章的货品（例如光瓷釉）；

（四）载体催化剂（税号 38.15）；

（五）第四十二章注释二（二）所述的税号 42.02 或 42.03 的物品；

（六）税号 43.03 或 43.04 的物品；

Notes:

1. Subject to Note 1 (a) to Section VI and except as provided below, all articles consisting wholly or partly:

(a) Of natural or cultured pearls or of precious or semi-precious stones (natural, synthetic or reconstructed), or

(b) Of precious metal or of metal clad with precious metal, are to be classified in this Chapter.

2. (a) Headings Nos.71.13, 71.14 and 71.15 do not cover articles in which precious metal or metal clad with precious metal is present as minor constituents only, such as minor fittings or minor ornamentation (for example, monograms, ferrules and rims), and paragraph (b) of the foregoing Note does not apply to such articles;

(b) Heading No.71.16 does not cover articles containing precious metal or metal clad with precious metal (other than as minor constituents).

3. This Chapter does not cover:

(a) Amalgams of precious metal, or colloidal precious metal (heading No.28.43);

(b) Sterile surgical suture materials, dental fillings or other goods of Chapter 30;

(c) Goods of Chapter 32 (for example, lustres);

(d) Supported catalysts (heading No.38.15);

(e) Articles of heading No.42.02 or 42.03 referred to in Note 2 (B) to Chapter 42;

(f) Articles of heading No.43.03 or 43.04;

（七）第十一类的货品（纺织原料及纺织制品）；

（八）第六十四章或第六十五章的鞋靴、帽类及其他物品；

（九）第六十六章的伞、手杖及其他物品；

（十）税号 68.04 或 68.05 及第八十二章含有宝石或半宝石（天然或合成）粉末的研磨材料制品；第八十二章装有宝石或半宝石（天然、合成或再造）工作部件的器具；第十六类的机器、机械器具、电气设备及其零件。然而，完全以宝石或半宝石（天然、合成或再造）制成的物品及其零件，除未安装的唱针用已加工蓝宝石或钻石外（税号 85.22），其余仍应归入本章；

（十一）第九十章、第九十一章或第九十二章的物品（科学仪器、钟表及乐器）；

（十二）武器及其零件（第九十三章）；

（十三）第九十五章注释二所述物品；

（十四）根据第九十六章注释四应归入该章的物品；

（十五）雕塑品原件（税号 98.03）、收藏品（税号 98.05）或超过一百年的古物（税号 98.06）、但天然或养殖珍珠、宝石及半宝石除外。

四、（一）所称"贵金属"，是指银、金及铂。

（二）所称"铂"，是指铂、铱、锇、钯、铑及钌。

（三）所称"宝石或半宝石"，不包括第九十六章注释二（二）所述任何物质。

五、含有贵金属的合金（包括烧结及化合的），只要其中任何一种贵金属的含量达到合金重量的 2%，即应视为本章的贵金属合金。贵金属合金应按下列规则归类：

（一）按重量计含铂量在 2% 及以上的合金，应视为铂合金；

（二）按重量计含金量在 2% 及以上，但不含铂或按重量计含铂量在 2% 以下的合金，应视为金合金；

（三）按重量计含银量在 2% 及以上的其他合金，应视为银合金。

(g) Goods of Section Ⅺ (textiles and textile articles);

(h) Footwear, headgear or other articles of Chapter 64 or 65;

(i) Umbrellas, walking-sticks or other articles of Chapter 66;

(j) Abrasive goods of heading No.68.04 or 68.05 or Chapter 82, containing dust or powder of precious or semi-precious stones (natural or synthetic); articles of Chapter 82 with a working part of precious or semi-precious stones (natural, synthetic or reconsturcted); machinery, mechanical appliances or electrical goods, or parts thereof, of Section ⅩⅥ. However, articles and parts thereof, wholly of precious or semiprecious stones (natural, synthetic or reconstructed) remain classified in this Chapter, except unmounted worked sapphires and diamonds for styli (heading No.85.22);

(k) Articles of Chapter 90, 91 or 92 (scientific instruments, clocks and watches, musical instruments);

(l) Arms or parts thereof (Chapter 93);

(m) Articles covered by Note 2 to Chapter 95;

(n) Articles classified in Chapter 96 by virtue of Note 4 to that Chapter;

(o) Original sculptures or statuary (heading No.97.03), collectors'pieces (heading No.97.05) or antiques of an age exceeding one hundred years (heading No.97.06), other than natural or cultured pearls or precious or semi-precious stones.

4. (a) The expression "precious metal" means silver, gold and platinum.

(b) The expression "platinum" means platinum, iridium, osmium, palladium, rhodium and ruthenium.

(c) The expression "precious or semi-precious stones" does not include any of the substances specified in Note 2 (b) to Chapter 96.

5. For the purposes of this Chapter, any alloy (including a sintered mixture and an inter-metallic compound) containing precious metal is to be treated as an alloy of precious metal if any one precious metal constitutes as much as 2%, by weight, of the alloy. Alloys of precious metal are to be classified according to the following rules:

(a) An alloy containing 2% or more, by weight, of platinum is to be treated as an alloy of platinum;

(b) An alloy containing 2% or more, by weight, of gold but no platinum, or less than 2%, by weight, of platinum, is to be treated as analloy of gold;

(c) Other alloys containing 2% or more, by weight, of silver are to be treated as alloys of silver.

六、除条文另有规定的以外，本目录所称贵金属应包括上述注释五所规定的贵金属合金，但不包括包贵金属或表面镀以贵金属的贱金属及非金属。

6. Except where the context otherwise requires, any reference in the Nomenclature to precious metal or to any particular precious metal includes a reference to alloys treated as alloys of precious metal or of the particular metal in accordance with the rules in Note 5 above, but not to metal clad with precious metal or to base metal or nonmetals plated with precious metal.

七、本目录所称"包贵金属"，是指以贱金属为底料，在其一面或多面用焊接、熔接、热轧或类似机械方法覆盖一层贵金属的材料。除条文另有规定的以外，也包括镶嵌贵金属的贱金属。

7. Throughout the Nomenclature the expression "metal clad with precious metal" means material made with a base of metal upon one or more surfaces of which there is affixed by soldering, brazing, welding, hot-rolling or similar mechanical means a covering of precious metal. Except where the context otherwise requires, the expression also covers base metal inlaid with precious metal.

八、除第六类注释一（一）另有规定的以外，凡符合税号 71.12 规定的货品，应归入该税号而不归入本目录的其他税号。

8. Subject to Note 1 (a) to Section Ⅵ, goods answering to a description in heading No.71.12 are to be classified in that heading and in no other heading of the Nomenclature.

九、税号 71.13 所称"首饰"，是指：

（一）个人用小饰物（不论是否镶嵌宝石）（例如，戒指、手镯、项圈、饰针、耳环表链、表链饰物、垂饰、领带别针、袖扣、饰扣、宗教性或其他勋章及徽章）；

（二）通常放置在衣袋、手提包或佩戴在身上的个人用品（例如，烟盒、粉盒、链袋、口香丸盒）。

9. For the purposes of heading No.71.13, the expression "articles of jewellery" means:

(a) Any small objects of personal adornment (gem-set or not) (for example, rings, bracelets, necklaces, brooches, ear-rings, watch-chains, fobs, pendants, tie-pins, cuff-links, dress-studs, religious or other medals and insignia); and

(b) Articles of personal use of a kind normally carried in the pocket, in the handbag or on the person (such as cigarette cases, powder boxes, chain purses, cachou boxes).

十、税号 71.14 所称"金银器"，包括装饰品、餐具、梳妆用具、吸烟用具及类似的家庭、办公室或宗教用的其他物品。

10. For the purposes of heading No.71.14, the expression "articles of goldsmiths'or silversmiths'wares" includes such articles as ornaments, tableware, toilet-ware, smokers'requisites and other articles of household, office or religious use.

十一、税号 71.17 所称"仿首饰"，是指不含天然或养殖珍珠、宝石或半宝石（天然、合成或再造）及贵金属或包贵金属（仅作为镀层或小零件、小装饰品的除外）的上述注释九（一）所述的首饰（不包括税号 96.06 的钮扣及其他物品或税号 96.15 的梳子、发夹及类似品）。

11. For the purposes of heading No.71.17, the expression "imitation jewellery" means articles of jewellery within the meaning of paragraph (a) of Note 9 above (but not including buttons or other articles of heading No.96.06, or dress-combs, hair-slides or the like, or hairpins, of heading No.96.15), not incorporating natural or cultured pearls, precious or semi-precious stones (natural, anthetic or reconstructed) nor (except as plating or as minor constituents) precious metal or metal clad with precious metal.

子目注释：

一、子目号 7106.10、7108.11、7110.11、7110.21、7110.31 及 7110.41 所称“粉末”，是指按重量计 90％及以上可从网眼孔径为 0.5 毫米的筛子通过的产品。

二、子目号 7110.11 及 7110.19 所称“铂”，可不受本章注释四（二）的规定约束，不包括铱、锇、钯、铑及钌。

三、对于税号 71.10 项下的子目所列合金的归类，按其所含铂、钯、铑、铱、锇或钌中重量最大的一种金属归类。

Subheading Notes:

1. For the purposes of subheadings Nos. 7106.10, 7108.11, 7110.11, 7110.21, 7110.31 and 7110.41, the expressions “powder” and “in powder form” mean products of which 90% or more by weight passes through a sieve having a mesh aperture of 0.5mm.

2. Notwithstanding the provisions of Chapter Note 4 (b), for the purposes of subheadings Nos. 7110.11 and 7110.19, the expression “platinum” does not include indium, osmium, palladium, rhodium or ruthenium.

3. For the classification of alloys in the subheadings of heading No. 71.10, eash alloy is to be classified with that metal, platinum, palladium, rhodium, iridium, osmium or ruthenium which predominates by weight over each other of these metals.

序号 No.	税则号列 Tariff Item	货品名称	最惠国税率(%) M.F.N.	协定税率(%) Agreement 东盟 ASEAN	曼谷协定 BA	香港 H.K.	澳门 Macao	普通税率(%) Gen.	Article Description
		第一分章 天然或养殖珍珠、宝石或半宝石							Ⅰ.-NATURAL OR CULTURED PEARLS AND PRECIOUS OR SEMI-PRECIOUS STONES
	71.01	**天然或养殖珍珠，不论是否加工或分级，但未成串或镶嵌；天然或养殖珍珠，为便于运输而暂穿成串：**							**Pearls, natural or cultured, whether or not worked or graded but not strung, mounted or set; ungraded pearls, natural or cultured, temporarily strung for convenience of transport:**
		-天然珍珠：							-Natural pearls:
		---未分级：							---Ungraded:
4563	7101.1011	----黑珍珠	21					100	----Tahitian pearls
4564	7101.1019	----其他	21					100	----Other
		---其他：							---Other:
4565	7101.1091	----黑珍珠	21					130	----Tahitian pearls
4566	7101.1099	----其他	21					130	----Other
		-养殖珍珠：							-Cultured pearls:
		--未加工：							--Unworked:
4567	7101.2110	---未分级	21					100	---Ungraded
4568	7101.2190	---其他	21					130	---Other
		--已加工：							--Worked:
4569	7101.2210	---未分级	21					100	---Ungraded
4570	7101.2290	---其他	21					130	---Other
	71.02	**钻石，不论是否加工，但未镶嵌：**							**Diamonds, whether or not worked, but not mounted or set:**
4571	7102.1000	-未分级	3					14	-Unsorted
		-工业用：							-Industrial:
4572	7102.2100	--未加工或经简单锯开、劈开或粗磨	0					14	--Unworked or simply sawn, cleaved or bruted
4573	7102.2900	--其他	0					14	--Other

序号 No.	税则号列 Tariff Item	货品名称	最惠国税率(%) M.F.N.	协定税率(%)Agreement 东盟 ASEAN	曼谷协定 BA	香港 H.K.	澳门 Macao	普通税率(%) Gen.	Article Description
		-非工业用:							-Non-industrial:
4574	7102.3100	--未加工或经简单锯开、劈开或粗磨	3			0		14	--Unworked or simply sawn, cleaved or bruted
4575	7102.3900	--其他	8		5			35	--Other
	71.03	**宝石(钻石除外)或半宝石,不论是否加工或分级,但未成串或镶嵌;未分级的宝石(钻石除外)或半宝石,为便于运输而暂穿成串:**							**Precious stones (other than diamonds) and semi-precious stones, whether or not worked or graded but not strung, mounted or set; ungraded precious stones (other than diamonds) and semi-precious stones, temporarily strung for covenience of transport:**
4576	7103.1000	-未加工或经简单锯开或粗制成形	3		2.8			14	-Unworked or simply sawn or roughly shaped
		-经其他加工:							-Otherwise worked:
4577	7103.9100	--红宝石、蓝宝石、绿宝石	8		7			35	--Rubies, sapphires and emeralds
		--其他:							--Other:
4578	7103.9910	---翡翠	8		7			35	---Jadeite
4579	7103.9990	---其他	8		7			35	---Other
	71.04	**合成或再造的宝石或半宝石,不论是否加工或分级,但未成串或镶嵌的;未分级的合成或再造的宝石或半宝石,为便于运输而暂穿成串:**							**Synthetic or reconstructed precious or semiprecious stones, whether or not worked or graded but not strung, mounted or set; ungraded synthetic or reconstructed precious or semi-precious stones, temporarily strung for convenience of transport:**
4580	7104.1000	-压电石英	6					14	-Piezo-electric quartz
		-其他,未加工或经简单锯开或粗制成形:							-Other, unworked or simply sawn or roughly shaped:
4581	7104.2010	---钻石	0					14	---diamonds
4582	7104.2090	---其他	0					14	---other
		-其他:							-Other:
		---工业用:							---For technical use:
4583	7104.9011	----钻石	6					14	----diamonds
4584	7104.9019	----其他	6					14	----other
		---其他:							---Other:
4585	7104.9091	----钻石	8					35	----diamonds
4586	7104.9099	----其他	8					35	----other
	71.05	**天然或合成的宝石或半宝石的粉末:**							**Dust and powder of natural or synthetic precious or semi-precious stones:**
4587	7105.1000	-钻石的	0					17	-Of diamonds
4588	7105.9000	-其他	0					17	-Other
		第二分章 贵金属及包贵金属							Ⅱ. -PRECIOUS METALS AND METALS CLAD WITH PRECIOUS METAL

序号 No.	税则号列 Tariff Item	货品名称	最惠国税率(%) M.F.N.	协定税率(%) Agreement 东盟 ASEAN	曼谷协定 BA	香港 H.K.	澳门 Macao	普通税率(%) Gen.	Article Description
	71.06	**银(包括镀金、镀铂的银),未锻造、半制成或粉末状:**							**Silver (including silver plated with gold or platinum), unwrought or in semi-manufactured forms, or in powder form:**
		-银粉							-Powder:
4589	7106.1011	---非片状粉末:	0					0	---Not Flake:
4590	7106.1019	----平均粒径小于3微米	0					0	----Average diameter less than 3 micron
		----其他							----Other
		---片状粉末:							---Flake:
4591	7106.1021	----平均粒径小于10微米	0					0	----Average diameter less than 10 micron
4592	7106.1029	----其他	0					0	----Other
		-其他:							-Other:
		--未锻造:							--Unwrought:
4593	7106.9110	---纯度达99.99%及其上	0					0	---Of a purity of 99.99 percent or more
4594	7106.9190	---其他	0					0	---Other
		--半制成:							--Semi-manufactured:
4595	7106.9210	---纯度达99.99%及以上	0					50	---Of a purity of 99.99 percent or more
4596	7106.9290	---其他	0					50	---Other
	71.07	**以贱金属为底的包银材料:**							**Base metals clad with silver, not further worked than semi-manufactured:**
4597	7107.0000	以贱金属为底的包银材料	10.5					50	Base metals clad with silver, not further worked than semi-manufactured
	71.08	**金(包括镀铂的金),未锻造、半制成或粉末状:**							**Gold (including gold plated with platinum) unwrought or in semi-manufactured forms, or in powder form:**
		-非货币用:							-Non-monetary:
4598	7108.1100	--金粉	0					0	--Powder
4599	7108.1200	--其他未锻造形状	0					0	--Other unwrought forms
4600	7108.1300	--其他半制成形状	0					50	--Other semi-manufactured forms
4601	7108.2000	-货币用	0					0	-Monetary
	71.09	**以贱金属或银为底的包金材料:**							**Base metals or silver, clad with gold, not further worked than semi-manufactured:**
4602	7109.0000	以贱金属或银为底的包金材料	10.5					50	Base metals or silver, clad with gold, not further worked than semi-manufactured
	71.10	**铂,未锻造、半制成或粉末状:**							**Platinum, unwrought or in semi-manufac-tured forms, or in powder form:**
		-铂:							-Platinum:
4603	7110.1100	--未锻造或粉末状	0					0	--Unwrought or in powder form
		--其他:							--Other:
4604	7110.1910	---板、片	0					0	---Plates and sheets
4605	7110.1990	---其他	3			0		11	---Other

序号 No.	税则号列 Tariff Item	货品名称	最惠国税率(%) M.F.N.	协定税率(%)Agreement 东盟 ASEAN	曼谷协定 BA	香港 H.K.	澳门 Macao	普通税率(%) Gen.	Article Description
		-钯:							-Palladium:
4606	7110.2100	--未锻造或粉末状	0					10	--Unwrought or in powder form
		--其他:							--Other:
4607	7110.2910	---板、片	0					0	---Plates and sheets
4608	7110.2990	---其他	3			0		11	---Other
		-铑:							-Rhodium:
4609	7110.3100	--未锻造或粉末状	0					0	--Unwrought or in powder form
		--其他:							--Other:
4610	7110.3910	---板、片	0					0	---Plates and sheets
4611	7110.3990	---其他	3					11	---Other
		-铱、锇及钌:							-Iridium,osmium and ruthenium:
4612	7110.4100	--未锻造或粉末状	0					0	--Unwrought or in powder form
		--其他:							--Other:
4613	7110.4910	---板、片	0					0	---Plates and sheets
4614	7110.4990	---其他	3					11	---Other
	71.11	**以贱金属、银或金为底的包铂材料:**							**Base metals, silver or gold, clad with platinum, not further worked than semi-manu-factured:**
4615	7111.0000	以贱金属、银或金为底的包铂材料	3					11	Base metals, silver or gold, clad with platinum, not further worked than semi-manu-factured
	71.12	**贵金属或包贵金属的废碎料;含有贵金属或贵金属化合物的其他废碎料,主要用于回收贵金属:**							**Waste and scrap of precious metal or of metal clad with precious metal; other waste and scrap containing precious metal or precious metal compounds, of a kind used principally for the recovery of precious metal:**
		-含有贵金属或金属化合物的灰:							-Ash containning precious metal or precious metal compounds
4616	7112.3010	---含有银或银化合物的	8					50	---Of silver or silver compounds
4617	7112.3090	---其他	6					50	---Other
		-其他:							-Other:
		--金及包金的废碎料,但含有其他贵金属的除外:							--Of gold, including metal clad with gold but excluding sweepings contianing other precious metals:
4618	7112.9110	---金及包金的废碎料	0					0	---Of gold or gold compounds
4619	7112.9120	---含有金或金化合物的废碎料	6					35	---Waste and scrap with gold or gold compounds
		--铂及包铂的废碎料,但含有其他贵金属的除外:							--Of platinum, including metal clad with platinum but excluding sweepings containing other precious metals
4620	7112.9210	---铂及包铂的废碎料	0					0	---Of platinum
4621	7112.9220	---含有铂或铂化合物的废碎料	6					35	---Wasted and scrap with plutinum
		--其他:							--Other:

序号 No.	税则号列 Tariff Item	货品名称	最惠国税率(%) M.F.N.	协定税率(%)Agreement 东盟 ASEAN	曼谷协定 BA	香港 H.K.	澳门 Macao	普通税率(%) Gen.	Article Description
4622	7112.9910	---含有银或银化合物的废碎料	8					35	---Waste and scrap with silver or silver compounds
4623	7112.9920	---含有其他贵金属或贵金属化合物的废碎料	6					35	---Waste and scrap with other precious moutals
4624	7112.9990	---其他	0					50	---Other
		第三分章 珠宝首饰、金银器及其他制品							Ⅲ. JEWELLERY, GOLDSMITHS' AND SILVERSMITHS' WARES AND OTHER ARTICLES
	71.13	**贵金属或包贵金属制的首饰及其零件:**							**Articles of jewellery and parts thereof, of precious metal or of metal clad with precious metal:**
		-贵金属制,不论是否包、镀贵金属:							-Of precious metal whether or not plated or clad with precious metal:
		--银制,不论是否包、镀其他贵金属							--Of silver, whether or not plated or clad with other precious metal
4625	7113.1110	---镶嵌钻石的	20		16.5	0	0	130	---Diomond mounted or set
4626	7113.1190	---其他	20		16.5	0	0	130	---Other
		--其他贵金属制,不论是否包、镀贵金属:							--Of other precious metal, whether or not plated or clad with predious metal:
		---黄金制:							---Of gold:
4627	7113.1911	----镶嵌钻石的	20		16.5	0	0	130	----Diomond mounted
4628	7113.1919	----其他	20		16.5	0	0	130	----Other
		---其他:							---Other:
4629	7113.1991	----镶嵌钻石的	35		30	0	0	130	----Diomond mounted or set
4630	7113.1999	----其他	35		30	0	0	130	----Other
		-以贱金属为底的包贵金属制							-Of base metal clad with precious metal
4631	7113.2010	---镶嵌钻石的	35		30	0	0	130	---Diomond mounted or set
4632	7113.2090	---其他	35		30	0	0	130	---Other
	71.14	**贵金属或包贵金属制的金银器及其零件:**							**Articles of goldsmiths' or silversmiths' wares and parts thereof, of precious metal or of metal clad with precious metal:**
		-贵金属制,不论是否包、镀贵金属:							-Of precious metal whether or not plated or clad with precious metal:
4633	7114.1100	--银制,不论是否包、镀其他贵金属	35			0	0	100	--Of silver, whether or not plated or clad with other precious metal
4634	7114.1900	--其他贵金属制,不论是否包、镀贵金属	35			0	0	100	--Of other precious metal, whether or not plated or clad with precious metal
4635	7114.2000	-以贱金属为底的包贵金属制	35			0	0	100	-Of base metal clad with precious metal
	71.15	**贵金属或包贵金属的其他制品:**							**Other articles of precious metal or of metal clad with precious metal:**
4636	7115.1000	-金属丝布或格栅形状的铂催化剂	3			0		11	-Catalysts in the form of wire cloth or grill, of platinum

序号 No.	税则号列 Tariff Item	货品名称	最惠国税率(%) M.F.N.	协定税率(%)Agreement 东盟 ASEAN	曼谷协定 BA	香港 H.K.	澳门 Macao	普通税率(%) Gen.	Article Description
		-其他:							-Other:
4637	7115.9010	---工业或实验室用	3			0	0	11	---For technical or laboratory use
4638	7115.9090	---其他	35			0	0	100	---Other
	71.16	**用天然或养殖珍珠、宝石或半宝石(天然、合成或再造)制成的物品:**							**Articles of natural or cultured pearls, precious or semi-precious stones (natural, synthetic or reconstructed):**
4639	7116.1000	-天然或养殖珍珠制	35			0	0	130	-Of natural or cultured pearls
4640	7116.2000	-宝石或半宝石(天然、合成或再造)制	35			0	0	130	-Of precious or semi-precious stones (natural, synthetic or reconstructed)
	71.17	**仿首饰:**							**Imitation jewellery:**
		-贱金属制,不论是否镀贵金属:							-Of base metal, whether or not plated with precious metal:
4641	7117.1100	--袖扣、饰扣	35			0	0	130	--Cuff-links and studs
4642	7117.1900	--其他	17			0	0	130	--Other
4643	7117.9000	-其他	35		30	0	0	130	-Other
	71.18	**硬币:**							**Coin:**
4644	7118.1000	-非法定货币的硬币(金币除外)	0					0	-Coin (other than gold coin), not being legal tender
4645	7118.9000	-其他	0					0	-Other

第十五类

贱金属及其制品

注释：

一、本类不包括：

（一）以金属粉末为基本成分的调制油漆、油墨或其他产品（税号 32.07 至 32.10、32.12、32.13 或 32.15）；

（二）铈铁或其他引火合金（税号 36.06）；

（三）税号 65.06 或 65.07 的帽类及其零件；

（四）税号 66.03 的伞骨及其他物品；

（五）第七十一章的货品（例如，贵金属合金、以贱金属为底的包贵金属、仿首饰）；

（六）第十六类的物品（机器、机械器具及电气设备）；

（七）已装配的铁路或电车轨道（税号 86.08）或第十七类的其他物品（车辆、船舶、航空器）；

（八）第十八类的仪器及器具，包括钟表发条；

（九）做弹药用的铅弹（税号 93.06）或第十九类的其他物品（武器、弹药）；

（十）第九十四章的物品（例如，家具、弹簧床垫、灯具及照明装置、发光标志、活动房屋）；

（十一）第九十五章的物品（例如，玩具、游戏品及运动用品）；

（十二）手用筛子、钮扣、钢笔、铅笔套、钢笔尖或第九十六章的其他物品（杂项制品）；

（十三）第九十七章的物品（例如，艺术品）。

二、本目录所称“通用零件”，是指：

（一）税号 73.07、73.12、73.15、73.17 或 73.18 的物品及其他贱金属制的类似品；

（二）贱金属制的弹簧及弹簧片，但钟表发条（税号 91.14）除外；

（三）税号 83.01、83.02、83.08、83.10 的物品及税号 83.06 的贱金属制的框架及镜子。

第七十三章至第七十六章（税号 73.15 除外）及第七十八章至第八十二章所列货品的零件，

SECTION XV

BASE METALS AND ARTICLES OF BASE METAL

Notes:

1. This Section does not cover:

(a) Prepared paints, inks or other products with a basis of metallic flakes or powder (headings Nos.32.07 to 32.10, 32.12, 32.13 or 32.15);

(b) Ferro-cerium or other pyrophoric alloys (heading No.36.06);

(c) Headgear or parts thereof of heading No.65.06 or 65.07;

(d) Umbrella frames or other articles of heading No.66.03;

(e) Goods of Chapter 71 (for example, precious metal alloys, base metal clad with precious metal, imitation jewellery);

(f) Articles of Section XVI (machinery, mechanical appliances and electrical goods);

(g) Assembled railway or tramway track (heading No.86.08) or other articles of Section XVII (vehicles, ships and boats, aircraft);

(h) Instruments or apparatus of Section XVIII, including clock or watch springs;

(ij) Lead shot prepared for anmmunition (heading No.93.06) or other articles of Section XIX (arms and ammunition);

(k) Articles of Chapter 94 (for example, furniture, mattress supports, lamps and lighting fittings, illuminated signs, prefabricated buildings);

(l) Articles of Chapter 95 (for example, toys, games, sports requisites);

(m) Hand sieves, buttons, pens, pencil-holders, pen nibs or other articles of Chapter 96 (miscellaneous manufactured articles); or

(n) Articles of Chapter 97 (for example, works of art).

2. Throughout the Nomenclature, the expression "parts of general use" means:

(a) Articles of heading No.73.07, 73.12, 73.15, 73.17 or 73.18 and similar articles of other base metal;

(b) Springs and leaves for springs, of base metal, other than clock or watch springs (heading No.91.14); and

(c) Articles of headings Nos.83.01, 83.02, 83.08, 83.10 and frames and mirrors, of base metal, of heading No.83.06.

In Chapters 73 to 76 and 78 to 82 (but not in heading No.73.15) references to parts of goods do not include refer-

不包括上述的通用零件。

除上段及第八十三章注释一另有规定的以外，第七十二章至第七十六章及第七十八章至第八十一章不包括第八十二章、第八十三章的物品。

三、本目录所称“贱金属”是指：铁及钢、铜、镍、铝、铅、锌、锡、钨、钼、钽、镁、钴、铋、镉、钛、锆、锑、锰、铍、铬、锗、钒、镓、铪、铟、铌（钶）、铼及铊。

四、本目录所称“金属陶瓷”是指金属与陶瓷成分以极细微粒不均匀结合而成的产品。“金属陶瓷”包括硬质合金（金属碳化物与金属烧结而成）。

五、合金的归类规则（第七十二章、第七十四章所规定的铁合金及母合金除外）：

（一）贱金属的合金按其所含重量最大的金属归类；

（二）由本类的贱金属和非本类的元素构成的合金，如果所含贱金属的总重量等于或超过所含其他元素的总重量，应作为本类贱金属合金归类；

（三）本类所称“合金”，包括金属粉末的烧结混合物、熔化而得的不均匀紧密混合物（金属陶瓷除外）及金属间化合物。

六、除条文另有规定的以外，本目录所称的贱金属包括贱金属合金，这类合金应按上述注释五的规则进行归类。

七、复合材料制品的归类规则：

除各税号另有规定的以外，贱金属制品（包括根据“归类总规则”作为贱金属制品的混合材料制品）如果含有两种或两种以上贱金属的，按其所含重量最大的贱金属的制品归类。为此：

（一）钢、铁或不同种类的钢铁，均视为一种金属；

（二）按照注释五的规定作为某一种金属归类的合金，应视为一种金属；

ences to parts of general use as defined above.

subject to the preceding paragraph and to Note 1 to Chapter 83, the articles of Chapter 82 or 83 are excluded from Chapters 72 to 76 and 78 to 81.

3. Throughout the Nomenclature, the expression “base metals” means: iron and steel, copper, nickel, aluminium, lead, zinc, tin, tungsten (wolfram), molybdenum, tantalum, magnesium, cobalt, bismuth, cadmium, titanium, zirconium, antimony, manganese, beryllium, chromium, germanium, vanadium, gallium, hafnium, indium, niobium (columbium), rhenium and thallium.

4. Throughout the Nomenclature, the term “cermets” means products containing a microscopic heterogeneous combination of a metallic component and a ceramic component. The term “cermets” includes sintered metal carbides (metal carbides sintered with a metal).

5. Classification of alloys (other than ferro-alloys and master alloys as defined in Chapters 72 and 74):

(a) An alloy of base metals is to be classified as an alloy of the metal which predominates by weight over each of the other metals;

(b) An alloy composed of base metals of this Section and of elements not falling within this Section is to be treated as an alloy of base metals of this Section if the total weight of such metals equals or exceeds the total weight of the other elements present;

(c) In this Section the term “alloys” includes sintered mixtures of metal powders, heterogeneous intimate mixtures obtained by melting (other than cermets) and intermetallic compounds.

6. Unless the context otherwise requires, any reference in the Nomenclature to base metal includes a reference to alloys which, by virtue of Note 5 above, are to be classified as alloys of that metal.

7. Classification of composite articles:

Except where the headings otherwise require, articles of base metal (including articles of mixed materials treated as articles of base metal under the Interpretative Rules) containing two or more base metals are to be treated as articles of the base metal predominating by weight over each of the other metals. For this purpose:

(a) Iron and steel, or different kinds of iron or steel, are regarded as one and the same metal;

(b) An alloy is regarded as being entirely composed of that metal as an alloy of which, by virtue of Note 5, it is classified; and

（三）税号 81.13 的金属陶瓷，应视为一种贱金属。

（c） A cermet of heading No.81.13 is regarded as a single base metal.

八、本类所用有关名词解释如下：

8. In this Section, the following expressions have the meanings hereby assigned to them:

（一）废碎料

在金属生产或机械加工中产生的废料及碎屑以及因破裂、切断、磨损及其他原因而明显不能作为原物使用的金属货品。

（a） Waste and scrap

Metal waste and scrap from the manufacture of mechanical working of metals, and metal goods definitely not usable as such because of breakage, cutting-up, wear or other reasons.

（二）粉末

按重量计 90%及以上可从网眼孔径为 1 毫米的筛子通过的产品。

（b） Powders

Products of which 90% or more by weight passes through a sieve having a mesh aperture of 1mm.

第七十二章
钢　铁

Chapter 72
Iron and steel

注释：

一、本章所述有关名词解释如下（本条注释（四）、（五）、（六）适用于本目录其他各章）：

Notes:

1. In this Chapter and, in the case of Notes (d), (e) and (f) throughout the Nomenclature, the following expressions have the meanings hereby assigned to them:

（一）生铁

无实用可锻性的铁碳合金，按重量计含碳量在 2%以上并可含有一种或几种下列含量范围的其他元素：

铬不超过 10%；
锰不超过 6%；
磷不超过 3%；
硅不超过 8%；
其他元素合计不超过 10%。

（a） Pig iron

Iron-carbon alloys not usefully malleable, containing more than 2% by weight of carbon and which may contain by weight one or more other elements within the following limits:

-not more than 10% of chromium
-not more than 6% of manganese
-not more than 3% of phosphorus
-not more than 8% of silicon
-a total of not more than 10% of other elements.

（二）镜铁

按重量计含锰量在 6%以上，但不超过 30%的铁碳合金，其他方面符合上述（一）款所列标准。

（b） Spiegeleisen

Iron-carbon alloys containing by weight more than 6% but not more than 30% of manganese and otherwise conforming to the specification at (a) above.

（三）铁合金

锭、块、团或类似初级形状、连续铸造而形成的各种形状及颗粒、粉末状的合金，不论是否烧结，通常用于其他合金生产过程中的添加剂或在黑色金属冶炼中作除氧剂、脱硫剂及类似用途，一般无实用可锻性，按重量计铁元素含量在 4%及以上并含有下列一种或几种元素：

铬超过 10%；
锰超过 30%；
磷超过 3%；
硅超过 8%；

（c） Ferro-alloys

Alloys in pigs, blocks, lumps or similar primary forms, in forms obtained by continuous casting and also in granular or powder forms, whether or not agglomerated, commonly used as an additive in the manufacture of other alloys or as de-oxidants, de-sulphurizing agents or for similar uses in ferrous metal-lurgy and generally not usefully malleable, containing by weight 4% or more of the element iron and one or more of the following:

-more than 10% of chromium
-more than 30% of manganese
-more than 3% of phosphorus
-more than 8% of silicon

除碳以外的其他元素，合计超过10%，但最高含铜量不得超过10%。

-a total of more than 10% of other elements, excluding carbon, subject to a maximum content of 10% in the case of copper.

（四）钢

除税号72.03以外的黑色金属材料（某些铸造而成的种类除外），具有实用可锻性，按重量计含碳量在2%及以下，但铬钢可具有较高的含碳量。

(d) Steel

Ferrous materials other than those of heading No.72.03 which (with the exception of certain types produced in the form of castings) are usefully malleable and which contain by weight 2% or less of carbon. However, chromium steels may contain higher proportions of carbon.

（五）不锈钢

按重量计含碳量在1.2%及以下，含铬量在10.5%及以上的合金钢，不论是否含有其他元素。

(e) Stainless steel

Alloy steels containing, by weight, 1.2% or less of carbon and 10.5% or more of chromium, with or without other elements.

（六）其他合金钢

不符合以上不锈钢定义的钢，含有一种或几种按重量计符合下列含量比例的元素：

铝0.3%及以上；
硼0.0008%及以上；
铬0.3%及以上；
钴0.3%及以上；
铜0.4%及以上；
铅0.4%及以上；
锰1.65%及以上；
钼0.08%及以上；
镍0.3%及以上；
铌0.06%及以上；
硅0.6%及以上；
钛0.05%及以上；
钨0.3%及以上；
钒0.1%及以上；
锆0.05%及以上；
其他元素（硫、磷、碳及氮除外）单项含量在0.1%及以上。

(f) Other alloy steel

Steels not complying with the definition of stainless steel and containing by weight one or more of the following elements in the proportion shown:

-0.3% or more of aluminium
-0.0008% or more of boron
-0.3% or more of chromium
-0.3% or more of cobalt
-0.4% or more of copper
-0.4% or more of lead
-1.65% or more of manganese
-0.08% or more of molybdenum
-0.3% or more of nickel
-0.06% or more of niobium
-0.6% or more of silicon
-0.05% or more of titanium
-0.3% or more of tungsten (wolfram)
-0.1% or more of vanadium
-0.05% or more of zirconium
-0.1% or more of other elements (except sulphur, phosphorus, carbon and nitrogen), taken separately.

（七）供再熔的碎料钢铁锭

粗铸成形无缩孔或冒口的锭块产品，表面有明显瑕疵，化学成分不同于生铁、镜铁及铁合金。

(g) Remelting scrap ingots of iron or steel

Products roughly cast in the form of ingots without feeder-heads or hot tops, or of pigs, having obvious surface faults and not complying with the chemical composition of pig iron, spiegeleisen or ferro-alloys.

（八）颗粒

按重量计不到90%可从网眼孔径为1毫米的筛子通过，而90%及以上可从网眼孔径为5毫米的筛子通过的产品。

(h) Granules

Products of which less than 90% by weight passes through a sieve with a mesh aperture of 1mm and of which 90% or more by weight passes through a sieve with a mesh aperture of 5mm.

（九）半制成品

连续铸造的实心产品，不论是否初步热轧；其他实心产品，除经初步热轧或锻造粗制成形以外未经进一步加工，包括角材、型材及异型材的坯件。

本类产品不包括成卷的产品。

(ij) Semi-finished products

Continuous cast products of solid section, whether or not subjected to primary hot-rolling; and

Other products of solid section, which have not been further worked than subjected to primary hot-rolling or roughly shaped by forging, including blanks for angles, shapes or sections.

These products are not presented in coils.

（十）平板轧材

截面为矩形（正方形除外）并且不符合以上第（九）款所述定义的下列形状实心轧制产品：

1. 层叠的卷材；

2. 平直形状，其厚度如果在4.75毫米以下，则宽度至少是厚度的十倍；其厚度如果在4.75毫米及以上，其宽度应超过150毫米，并且至少应为厚度的两倍。

平板轧材包括直接轧制而成并有凸起式样（例如，凹槽、肋条形、格槽、珠粒、菱形）的产品以及穿孔、抛光或制成瓦楞形的产品，但不具有其他税号所列制品或产品的特征。

各种规格的平板轧材（矩形或正方形除外），但不具有其他税号所列制品或产品的特征，都应作为宽度为600毫米及以上的产品归类。

(k) Flat-rolled products

Rolled products of solid rectangular (other than square) cross-section, which do not conform to the definition at (ij) above in the form of:

-coils of successively superimposed layers, or

-straight lenghths, which if of a thickness less than 4.75mm are of a width measuring at least ten times the thickness or if of a thickness of 4.75mm or more are of a width which exceeds 150mm and measures at least twice the thickness.

Flat-rolled products include those with patterns in relief derived directly from rolling (for example, grooves, ribs, chequers, tears, buttons, lozenges) and those which have been perforated, corrugated or polished, provided that they do not thereby assume the character of articles or products of other headings.

Flat-rolled products of a shape other than rectangular or square, of any size, are to be classified as products of a width of 600mm or more, provided that they do not assume the character of articles or products of other headings.

（十一）不规则盘绕的热轧条、杆

经热轧不规则盘绕的实心产品，其截面为圆形、扇形、椭圆形、矩形（包括正方形）、三角形或其他外凸多边形（包括"扁圆形"及"变形矩形"，即相对两边为弧拱形，另外两边为等长平行直线形）。这类产品可带有在轧制过程中产生的凹痕、凸缘、槽沟或其他变形（钢筋）。

(l) Bars and rods, hot-rolled, in irregularly wound coils

Hot-rolled products in irregularly wound coils, which have a solid cross-section in the shape of circles, segments of circles, ovals, rectangles (including squares), triangles or other convex polygons (including "flattened circles" and modified rectangles of which two opposite sides are convex arcs, the other two sides being straight, of equal length and paralled). These products may have indentations, ribs, grooves or other deformations produced during the rolling process (reinforcing bars and rods).

（十二）其他条、杆

不符合上述（九）、（十）、（十一）款或"丝"定义的实心产品，其全长截面均为圆形、扇形、椭圆形、矩形（包括正方形）、三角形或其他外凸多边形（包括"扁圆形"及"变形矩形"，即相对两边为弧拱形，另两边为等长平行直线形）。这些产品可以：

1. 带有在轧制过程中产生的凹痕、凸缘、槽沟或其他变形（钢筋）；

2. 轧制后扭曲的。

(m) Other bars and rods

Products which do not conform to any of the definitions at (ij), (k) or (l) above or to the definition of wire, which have a uniform solid cross-section along their whole length in the shape of circles, segments of circles, ovals, rectangles (including squares), triangles or other convex polygons (including "flattened circles" and "modified rectangles", of which two opposite sides are convex arcs, the other two sides being straight, of equal length and parallel). These products may:

-have indentations, ribs, grooves or other deformations produced during the rolling process (reinforcing bars and rods);

-be twisted after rolling.

（十三）角材、型材及异型材

不符合上述（九）、（十）、（十一）、

(n) Angles, shapes and sections

Products having a uniform solid cross-section along

（十二）款或“丝”定义，但其全长截面均为同样形状的实心产品。

第七十二章不包括税号73.01或73.02的产品。

（十四）丝

不符合平板轧材定义但全长截面均为同样形状的盘卷冷成形实心产品。

（十五）空心钻钢

适合钻探用的各种截面的空心条、杆，其最大外形尺寸超过15毫米但不超过52毫米，最大内孔尺寸不超过最大外形尺寸的二分之一。不符合本定义的钢铁空心条、杆应归入税号73.04。

their whole length which do not conform to any of the definitions at (ij), (k), (l) or (m) above or to the definition of wire.

Chapter 72 does not include products of heading No.73.01 or 73.02.

(o) Wire

Cold－formed products in coils, of any uniform solid cross－section along their whole length, which do not conform to the definition of flat-rolled products.

(p) Hollow drill bars and rods

Hollow bars and rods of any cross-section, suitable for drills, of which the greatest external dimension of the cross-section exceeds 15mm but does not exceed 52mm, and of which the greatest internal dimension does not exceed one half of the greatest external dimension. Hollow bars and rods of iron or steel not conforming to this definition are to be classified in heading No.73.04.

二、用一种黑色金属包覆不同种类的黑色金属，应按其中重量最大的材料归类。

2. Ferrous metals clad with another ferrous metal are to be classified as products of the ferrous metal predominating by weight.

三、用电解沉积法、压铸法或烧结法所得的钢铁产品，应按其形状、成分及外观归入本章类似热轧产品的相应税号。

3. Iron or steel products obtained by electrolytic deposition, by pressure casting or by sintering are to be classified, according to their form, their composition and their appearance, in the headings of this Chapter appropriate to similar hot-rolled products.

子目注释：

一、本章所用有关名词解释如下：

（一）合金生铁

按重量计含有一种或几种下列比例的元素的生铁：

铬0.2%以上；

铜0.3%以上；

镍0.3%以上；

0.1%以上的任何下列元素：

铝、钼、钛、钨、钒。

（二）非合金易切削钢

按重量计含有一种或几种下列比例的元素的非合金钢：

硫0.08%及以上；

铅0.1%及以上；

硒0.05%以上；

碲0.01%以上；

铋0.05%以上。

（三）硅电钢

按重量计含硅量至少为0.6%但不超过

Subheading Notes:

1. In this Chapter the following expressions have the meanings hereby assigned to them:

(a) Alloy pig iron

Pig iron containing, by weight, one or more of the following elements in the specified proportions:

-more than 0.2% of chromium

-more than 0.3% of copper

-more than 0.3% of nickel

-more than 0.1% of any of the following elements: aluminium, molybdenum, titanium, tungsten (wolfram), vanadium.

(b) Non-alloy free-cutting steel

Non-alloy steel containing, by weight, one or more of the following elements in the specified proportions:

-0.08% or more of sulphur

-0.1% or more of lead

-more than 0.05% of selenium

-more than 0.01% of tellurium

-more than 0.05% of bismuth.

(c) Silicon-electrical steel

Alloy steels containing by weight at least 0.6% but not

6%，含碳量不超过0.08%的合金钢。这类钢还可含有按重量计不超过1%的铝，但所含其他元素的比例并不使其具有其他合金钢的特性。

more than 6% of silicon and not more than 0.08% of carbon. They may also contain by weight not more than 1% of aluminium but no other element in a proportion that would give the steel the characteristics of another alloy steel.

（四）高速钢

不论是否含有其他元素，但至少含有按重量计合计含量在7%及以上的钼、钨、钒中两种元素的合金钢，按重量计其含碳量在0.6%及以上，含铬量在3%～6%。

(d) High speed steel

Alloy steels containing, with or without other elements, at least two of the three elements molybdenum, tungsten and vanadium with a combined content by weight of 7% or more, 0.6% or more of carbon and 3 to 6% of chromium.

（五）硅锰钢

按重量计同时含有下列元素的合金钢：

碳不超过0.7%；

锰0.5%及以上，但不超过1.9%；

硅0.6%及以上，但不超过2.3%；

所含其他元素的比例并不使其具有其他合金钢的特性。

(e) Silico-manganese steel

Alloy steels containing by weight:

- not more than 0.7% of carbon,

- 0.5% or more but not more than 1.9% of manganese, and

- 0.6% or more but not more than 2.3% of silicon, but no other element in a proportion that would give the steel the characteristics of another alloy steel.

二、税号72.02项下的子目所列铁合金，应按照下列规则归类：

对于只有一种元素超出本章注释一（三）规定的最低百分比的铁合金，应作为二元合金归入相应的子目号。以此类推，如果有两种或三种合金元素超出了最低百分比的，则可分别作为三元或四元合金。

在运用本规定时，本章注释一（三）所述的未列名的“其他元素”，按重量计单项含量必须超过10%。

2. For the classification of ferro-alloys in the subheadings of heading No. 72.02 the following rule should be observed: A ferro-alloy is considered as binary and classified under the relevant subheading (if it exists) if only one of the alloy elements exceeds the minimum percentage laid down in Chapter Note 1 (c); by analogy, it is considered respectively as ternary or quaternary if two or three alloy elements exceed the minimum percentage.

For the application of this rule the unspecified “other elements” referred to in Chapter Note 1 (c) must each exceed 10% by weight.

序号 No.	税则号列 Tariff Item	货品名称	最惠国税率(%) M.F.N.	协定税率(%) Agreement 东盟 ASEAN	曼谷协定 BA	香港 H.K.	澳门 Macao	普通税率(%) Gen.	Article Description
		第一分章 原料；粒状及粉状产品							I .-PRIMARY MATERIALS; PRODUCTS IN GRANULAR OR POWDER FORM
	72.01	**生铁及镜铁，锭、块或其他初级形状：**							**Pig iron and spiegeleisen in pigs, blocks or other primary forms:**
4646	7201.1000	-非合金生铁，按重量计含磷量在0.5%及以下	1					8	-Non-alloy pig iron containing by weight 0.5% or less of phosphorus
4647	7201.2000	-非合金生铁，按重量计含磷量在0.5%以上	1					8	-Non-alloy pig iron containing by weight more than 0.5% of phosphorus
4648	7201.5000	-合金生铁；镜铁	1					8	-Alloy pig iron; spiegeleisen
	72.02	**铁合金：**							**Ferro-alloys:**
		-锰铁：							-Ferro-manganese:
4649	7202.1100	--按重量计含碳量在2%以上	2					11	--Containing by weight more than 2% of carbon

序号 No.	税则号列 Tariff Item	货品名称	最惠国税率(%) M.F.N.	协定税率(%) Agreement 东盟 ASEAN	曼谷协定 BA	香港 H.K.	澳门 Macao	普通税率(%) Gen.	Article Description
4650	7202.1900	--其他	2					11	--Other
		-硅铁:							-Ferro-silicon:
4651	7202.2100	--按重量计含硅量在55%以上	2					11	--Containing by weight more than 55% of silicon
4652	7202.2900	--其他	2					11	--Other
4653	7202.3000	-硅锰铁	2					11	-Ferro-silico-manganese
		-铬铁:							-Ferro-chromium:
4654	7202.4100	--按重量计含碳量在4%以上	2					8	--Containing by weight more than 4% of carbon
4655	7202.4900	--其他	2					8	--Other
4656	7202.5000	-硅铬铁	2					11	-Ferro-silico-chromium
4657	7202.6000	-镍铁	2					11	-Ferro-nickel
4658	7202.7000	-钼铁	2					11	-Ferro-molybdenum
		-钨铁及硅钨铁:							-Ferro-tungsten and ferro-silico-tungsten:
4659	7202.8010	---钨铁	2					11	---Ferro-tungsten
4660	7202.8020	---硅钨铁	2					11	---Ferro-silico-tungsten
		-其他:							-Other:
4661	7202.9100	--钛铁及硅钛铁	2					11	--Ferro-titanium and ferro-silico-titanium
		--钒铁:							--Ferro-vanadium:
4662	7202.9210	---按重量计含钒量在75%及以上	9					30	---Containing by weight more than 75% of ranadium
4663	7202.9290	---其他	9					30	---Other
4664	7202.9300	--铌铁	2					11	--Ferro-niobium
4665	7202.9900	--其他	2					11	--Other
	72.03	**直接从铁矿还原所得的铁产品及其他海绵铁产品,块、团、团粒及类似形状;按重量计纯度在99.94%及以上的铁,块、团、团粒及类似形状:**							**Ferrous products obtained by direct reduction of iron ore and other spongy ferrous products, in lumps, pellets or similar forms; iron having a minimum purity by weight of 99.94%, in lumps, pellets or similar forms:**
4666	7203.1000	-直接从铁矿还原所得的铁产品	2					8	-Ferrous products obtained by direct reduction of iron ore
4667	7203.9000	-其他	2					8	-Other
	72.04	**钢铁废碎料;供再熔的碎料钢铁锭:**							**Ferrous waste and scrap; remelting scrap ingots of iron steel:**
4668	7204.1000	-铸铁废碎料	2					8	-Waste and scrap of cast iron
		-合金钢废碎料:							-Waste and scrap of alloy steel:
4669	7204.2100	--不锈钢废碎料	0					8	--Of stainless steel
4670	7204.2900	--其他	0					8	--Other
4671	7204.3000	-镀锡钢铁废碎料	2					8	-Waste and scrap of tinned iron or steel
		-其他废碎料:							-Other waste and scrap:

序号 No.	税则号列 Tariff Item	货品名称	最惠国税率(%) M.F.N.	协定税率(%) Agreement 东盟 ASEAN	曼谷协定 BA	香港 H.K.	澳门 Macao	普通税率(%) Gen.	Article Description
4672	7204.4100	--车、刨、铣、磨、锯、锉、剪、冲加工过程中产生的废料,不论是否成捆	2					8	--Turnings, shavings, chips, milling waste, sawdust, filings, trimmings and stampings, whether or not in bundles
4673	7204.4900	--其他	0					8	--Other
4674	7204.5000	-供再熔的碎料钢铁锭	0					8	-Remelting scrap ingots
	72.05	**生铁、镜铁及钢铁的颗粒和粉末:**							**Granules and powders, of pig iron, spiegeleisen, iron or steel:**
4675	7205.1000	-颗粒	2					30	-Granules
		-粉末:							-Powders:
4676	7205.2100	--合金钢的	2					17	--Of alloy steel
4677	7205.2900	--其他	2					17	--Other
		第二分章　铁及非合金钢							Ⅱ. -IRON AND NON-ALLOY STEEL
	72.06	**铁及非合金钢,锭状或其他初级形状(税号72.03的铁除外):**							**Iron and non-alloy steel in ingots or other primary forms (excluding iron of heading No.72.03):**
4678	7206.1000	-锭状	2					11	-Ingots
4679	7206.9000	-其他	2					11	-Other
	72.07	**铁及非合金钢的半制成品:**							**Semi-finished products of iron or non-alloy steel:**
		-按重量计含碳量在0.25%以下:							-Containing by weight less than 0.25% of carbon:
4680	7207.1100	--矩形(包括正方形)截面,宽度小于厚度的两倍	2					11	--Of rectangular (including square) cross-section, the width measuring less than twice the thickness
4681	7207.1200	--其他矩形(正方形除外)截面的	2					11	--Other, of rectangular (other than square) cross-section
4682	7207.1900	--其他	2					11	--Other
4683	7207.2000	-按重量计含碳量在0.25%及以上	2					11	-Containing by weight 0.25% or more of carbon
	72.08	**宽度在600毫米及以上的铁或非合金钢平板轧材,经热轧,但未经包覆、镀层或涂层:**							**Flat-rolled products of iron or non-alloy steel of a width of 600mm or more, hot-rolled, not clad, plated or coated:**
4684	7208.1000	-除热轧外未经进一步加工的卷材,已轧压花纹	5					14	-In coils, not further worked than hotrolled, with patterns in relief
		-其他经酸洗的卷材,除热轧外未经进一步加工:							-Other, in coils, not further worked than hot-rolled, pickled:
4685	7208.2500	--厚度在4.75毫米及以上	5					14	--Of a thickness of 4.75mm or more
		--厚度在3毫米及以上,但小于4.75毫米:							--Of a thickness of 3mm or more but less than 4.75mm:

序号 No.	税则号列 Tariff Item	货 品 名 称	最惠国税率(%) M.F.N.	协定税率(%) Agreement 东盟 ASEAN	曼谷协定 BA	香港 H.K.	澳门 Macao	普通税率(%) Gen.	Article Description
4686	7208.2610	---屈服强度大于355牛顿/平方毫米	5					14	---Of a yield strength exceeding 355N/mm^2
4687	7208.2690	---其他	5					14	---Other
		--厚度小于3毫米:							--Of a thickness of less than 3mm:
4688	7208.2710	---厚度小于1.5毫米	5					14	---Of a thickness of less than 1.5mm
4689	7208.2790	---其他	5					14	---Other
		-其他卷材,除热轧外未经进一步加工:							-Other, in coils, not further worked than hot-rolled:
4690	7208.3600	--厚度超过10毫米	6					14	--Of a thickness exceeding 10mm
4691	7208.3700	--厚度在4.75毫米及以上,但不超过10毫米	5					14	--Of a thickness of 4.75mm or more but not exceeding 10mm
		--厚度在3毫米及以上,但小于4.75毫米:							--Of a thickness of 3mm or more but less than 4.75mm:
4692	7208.3810	---屈服强度大于355牛顿/平方毫米	5					14	---Of a yield strength exceeding 355N/mm^2
4693	7208.3890	---其他	5					14	---Other
		--厚度小于3毫米:							--Of a thickness of less than 3mm:
4694	7208.3910	---厚度小于1.5毫米	3					14	---Of a thickness of less than 1.5mm
4695	7208.3990	---其他	3					14	---Other
4696	7208.4000	-已轧压花纹的非卷材,除热轧外未经进一步加工	6					17	-Not in coils, not further worked than hotrolled, with patterns in relief
		-其他非卷材,除热轧外未经进一步加工:							-Other, not in coils, not further worked than hot-rolled:
		--厚度超过10毫米:	6					17	--Of a thickness exceeding 10mm:
4697	7208.5110	---厚度超过50毫米	6					17	---Of a thickness exceeding 50mm
4698	7208.5120	---厚度在20毫米以上,但不超过50毫米	6					17	---Of a thickness exceeding 20mm but not exceeding 50mm
4699	7208.5190	---其他	6					17	---Other
4700	7208.5200	--厚度在4.75毫米及以上,但不超过10毫米	6					17	--Of a thickness of 4.75mm or more but not exceeding 10mm
		--厚度在3毫米及以上,但小于4.75毫米:							--Of a thickness of 3mm or more but less than 4.75mm:
4701	7208.5310	---屈服强度大于355牛顿/平方毫米	6					17	---Of a yield strength exceeding 355N/mm^2
4702	7208.5390	---其他	6					17	---Other
		--厚度小于3毫米:							--Of a thickness of less than 3mm:
4703	7208.5410	---厚度小于1.5毫米	6					17	---Of a thickness of less than 1.5mm
4704	7208.5490	---其他	6					17	---Other
4705	7208.9000	-其他	6					17	-Other
	72.09	**宽度在600毫米及以上的铁或非合金钢平板轧材,经冷轧,但未经包覆、镀层或涂层:**							**Flat-rolled products of iron or non-alloy steel, of a width of 600mm or more, coldrolled (cold-reduced), not clad, plated or coated:**
		-卷材,除冷轧外未经进一步加工:							-In coils, not further worked than cold-rolled(cold-reduced):

序号 No.	税则号列 Tariff Item	货品名称	最惠国税率(%) M.F.N.	协定税率(%) Agreement 东盟 ASEAN	曼谷协定 BA	香港 H.K.	澳门 Macao	普通税率(%) Gen.	Article Description
		--厚度在3毫米及以上:							--Of a thickness of 3mm or more:
4706	7209.1510	---屈服强度大于355牛顿/平方毫米	6					17	---Of a yield strength exceeding $355N/mm^2$
4707	7209.1590	---其他	6					17	---Other
		--厚度超过1毫米,但小于3毫米:							--Of a thickness exceeding 1mm but less than 3mm:
4708	7209.1610	---屈服强度大于275牛顿/平方毫米	6					17	---Of a yield strength exceeding $275N/mm^2$
4709	7209.1690	---其他	6					17	---Other
		--厚度在0.5毫米及以上,但不超过1毫米:							--Of a thickness of 0.5mm or more but not exceeding 1mm:
4710	7209.1710	---屈服强度大于275牛顿/平方毫米	3					17	---Of a yield strength exceeding $275N/mm^2$
4711	7209.1790	---其他	3					17	---Other
		--厚度小于0.5毫米:							--Of a thickness of less than 0.5mm:
4712	7209.1810	---厚度小于0.3毫米	6		6			17	---Of a thickness less than 0.3mm
4713	7209.1890	---其他	6		6			17	---Other
		-非卷材,除冷轧外未经进一步加工:							-Not in coils, not further worked than cold-rolled(cold-reduced):
4714	7209.2500	--厚度在3毫米及以上	6					17	--Of a thickness of 3mm or more
4715	7209.2600	--厚度超过1毫米,但小于3毫米	6					17	--Of a thickness exceeding 1mm but less than 3mm
4716	7209.2700	--厚度在0.5毫米及以上,但不超过1毫米	6					17	--Of a thickness of 0.5mm or more but not exceeding 1mm
4717	7209.2800	--厚度小于0.5毫米	6		6			17	--Of a thickness of less than 0.5mm
4718	7209.9000	-其他	6					17	-Other
	72.10	**宽度在600毫米及以上的铁或非合金钢平板轧材,经包覆、镀层或涂层:**							**Flat-rolled products of iron or non-alloy steel, of a width of 600mm of more, clad, plated or coated:**
		-镀或涂锡的:							-Plated or coated with tin:
4719	7210.1100	--厚度在0.5毫米及以上	10		10			20	--Of a thickness of 0.5mm or more
4720	7210.1200	--厚度小于0.5毫米	5		5			20	--Of a thickness of less than 0.5mm
4721	7210.2000	-镀或涂铅的,包括镀铅锡钢板	4		4			20	-Plated or coated with lead, including terneplate
4722	7210.3000	-电镀锌的	8		8			20	-Electrolytically plated or coated with zinc
		-用其他方法镀或涂锌的:							-Otherwise plated or coated with zinc:
4723	7210.4100	--瓦楞形	8		8			20	--Corrugated
4724	7210.4900	--其他	4		4			20	--Other
4725	7210.5000	-镀或涂氧化铬或铬及氧化铬的	8		8			20	-Plated or coated with chromium oxides or with chromium and chromium oxides
		-镀或涂铝的:							-Plated or coated with aluminium:

序号 No.	税则号列 Tariff Item	货品名称	最惠国税率(%) M.F.N.	协定税率(%) Agreement 东盟 ASEAN	曼谷协定 BA	香港 H.K.	澳门 Macao	普通税率(%) Gen.	Article Description
4726	7210.6100	--镀或涂铝锌合金的	8		8			20	--Plated or coated with aluminium-zinc alloys
4727	7210.6900	--其他	8		8			20	--Other
4728	7210.7000	-涂漆或涂塑的	4		4	0		20	-Painted, varnished or coated with plastics
4729	7210.9000	-其他	8		8			20	-Other
	72.11	**宽度小于600毫米的铁或非合金钢平板轧材,但未经包覆、镀层或涂层:**							**Flat-rolled products of iron or non-alloy steel, of a width of less than 600mm, not clad, plated or coated:**
		-除热轧外未经进一步加工:							-Not further worked than hot-rolled:
4630	7211.1300	--经四面轧制或在闭合匣内轧制的非卷材,宽度超过150毫米,厚度不小于4毫米,未轧压花纹	6					30	--Rolled on four faces or in a closed box pass, of a width exceeding 150mm and a thickness of not less than 4mm, not in coils and without patterns in relief
4731	7211.1400	--其他,厚度在4.75毫米及以上	6					30	--Other, of a thickness of 4.75mm or more
4732	7211.1900	--其他	6					30	--Other
		-除冷轧外未经进一步加工:							-Not further worked than cold-rolled (cold-reduced):
4733	7211.2300	--按重量计含碳量低于0.25%	6					30	--Containing by weight less than 0.25% of carbon
4734	7211.2900	--其他	6					30	--Other
4735	7211.9000	-其他	6					30	-Other
	72.12	**宽度小于600毫米的铁或非合金钢平板轧材,经包覆、镀层或涂层:**							**Flat-rolled products of iron or non-alloy steel, of a width of less than 600mm, clad, plated or coated:**
4736	7212.1000	-镀或涂锡的	5					20	-Plated or coated with tin
4737	7212.2000	-电镀锌的	8					20	-Electrolytically plated or coated with zinc
4738	7212.3000	-用其他方法镀或涂锌的	8					20	-Otherwise plated or coated with zinc
4739	7212.4000	-涂漆或涂塑的	4			0△		20	-Painted, varnished or coated with plastics
4740	7212.5000	-镀或涂其他材料的	8					20	-Otherwise plated or coated
4741	7212.6000	-经包覆的	8					20	-Clad
	72.13	**不规则盘卷的铁及非合金钢的热轧条、杆:**							**Bars and rods, hot-rolled, in irregularly wound coils, of iron or non-alloy steel:**
4742	7213.1000	-带有轧制过程中产生的凹痕、凸缘、槽沟及其他变形的	3					20	-Containing indentations, ribs, grooves or other deformations produced during the rolling process
4743	7213.2000	-其他,易切削钢制	3					20	-Other, of free-cutting steel
		-其他:							-Other:

序号 No.	税则号列 Tariff Item	货品名称	最惠国税率(%) M.F.N.	协定税率(%)Agreement 东盟 ASEAN	曼谷协定 BA	香港 H.K.	澳门 Macao	普通税率(%) Gen.	Article Description
4744	7213.9100	--直径小于14毫米圆形截面的	5					20	--Of circular cross-section measuring less than 14mm in diameter
4745	7213.9900	--其他	5					20	--Other
	72.14	**铁或非合金钢的其他条、杆,除锻造、热轧、热拉拔或热挤压外未经进一步加工,包括轧制后扭曲的:**							**Other bars and rods of iron or non-alloy steel, not further worked than forged, hot rolled, hot-drawn or hot-extruded, but including those twisted after rolling:**
4746	7214.1000	-锻造的	7					10	-Forged
4747	7214.2000	-带有轧制过程中产生的凹痕、凸缘、槽沟或其他变形以及轧制后扭曲的	3		3	0		20	-Containing indentations, ribs, grooves or other deformations produced during the rolling process or twisted after rolling
4748	7214.3000	-其他,易切削钢制	7					20	-Other, of free-cutting steel
		-其他:							-Other:
4749	7214.9100	--矩形(正方形除外)截面的	3					20	--Of rectangular cross section (other than square)
4750	7214.9900	--其他	3					20	--Other
	72.15	**铁及非合金钢的其他条、杆:**							**Other bars and rods of iron or non-alloy steel:**
4751	7215.1000	-易切削钢制,除冷成形或冷加工外未经进一步加工	7					20	-Of free-cutting steel, not further worked than cold-formed or cold-finished
4752	7215.5000	-其他,除冷成形或冷加工外未经进一步加工	7					20	-Other, not further worked than cold-formed or cold-finished
4753	7215.9000	-其他	3					20	-Other
	72.16	**铁或非合金钢的角材、型材及异型材:**							**Angles, shapes and sections of iron or nonalloy steel:**
		-槽钢、工字钢及H型钢,除热轧、热拉拔或热挤压外未经进一步加工,截面高度低于80毫米:							-U, I or H sections, not further worked than hot-rolled, hot-drawn or extruded, of a height of less than 80mm:
4754	7216.1010	---H型钢	3					14	---H sections
4755	7216.1020	---工字钢	3					14	---I sections
4756	7216.1090	---其他	3					14	---Other
		-角钢及丁字钢,除热轧、热拉拔或热挤压外未经进一步加工,截面高度低于80毫米:							-L or T sections, not further worked than hot-rolled, hot-drawn or extruded, of a height of less than 80mm:
4757	7216.2100	--角钢	6					17	--L sections
4758	7216.2200	--丁字钢	6					14	--T sections
		-槽钢、工字钢及H型钢,除热轧、热拉拔或热挤压外未经进一步加工,截面高度在80毫米及以上:							-U, I or H sections, not further worked than hot-rolled, hot-drawn or extruded of a height of 80mm or more:

序号 No.	税则号列 Tariff Item	货品名称	最惠国税率(%) M.F.N.	协定税率(%)Agreement 东盟 ASEAN	曼谷协定 BA	香港 H.K.	澳门 Macao	普通税率(%) Gen.	Article Description
4759	7216.3100	--槽钢	6					14	--U sections
		--工字钢:							--I sections:
4760	7216.3210	---截面高度在200毫米以上	6					14	---Of a height exceeding 200mm
4761	7216.3290	---其他	6					14	---Other
		--H型钢:							--H sections:
		---截面高度在200毫米以上:							---Of a height exceeding 200mm:
4762	7216.3311	----截面高度在800毫米以上	6					14	----Of a height exceeding 800mm
4763	7216.3319	----其他	6					14	----Other
4764	7216.3390	---其他	6					14	---Other
		-角钢及丁字钢,除热轧、热拉拔或热挤压外未经进一步加工,截面高度在80毫米及以上:							-L or T sections, not further worked than hot-rolled, hot-drawn or extruded, of a height of 80mm or more:
4765	7216.4010	---角钢	3					17	---L sections
4766	7216.4020	---丁字钢	3					14	---T sections
		-其他角材、型材及异型材,除热轧、热拉拔或热挤压外未经进一步加工:							-Other angles, shapes and sections, not further worked than hot-rolled, hot-drawn or extruded:
4767	7216.5010	---乙字钢	6					14	---Z sections
4768	7216.5090	---其他	3					20	---Other
		-角材、型材及异型材,除冷成形或冷加工外未经进一步加工:							-Angles, shapes and sections, not further worked than cold-formed or cold-finished:
4769	7216.6100	--平板轧材制的	3					20	--Obtained from flat-rolled products
4770	7216.6900	--其他	3					20	--Other
		-其他:							-Other:
4771	7216.9100	--平板轧材经冷成形或冷加工制的	3					20	--Cold-formed or cold-finished from flatrolled products
4772	7216.9900	--其他	3					20	--Other
	72.17	**铁丝或非合金钢丝:**							**Wire of iron or non-alloy steel:**
4773	7217.1000	-未经镀或涂层,不论是否抛光	8					40	-Not plated or coated, whether or not polished
4774	7217.2000	-镀或涂锌的	8					40	-Plated or coated with zinc
4775	7217.3000	-镀或涂其他贱金属的	8					40	-Plated or coated with other base metals
4776	7217.9000	-其他	8			0△		40	-Other
		第三分章 不锈钢							Ⅲ. -STAINLESS STEEL
	72.18	**不锈钢,锭状或其他初级形状;不锈钢半制成品:**							**Stainless steel in ingots or other primary forms; semi-finished products of stainless steel:**
4777	7218.1000	-锭状及其他初级形状	2					11	-Ingots and other primary forms
		-其他:							-Other:
4778	7218.9100	--矩形(正方形除外)截面的	2					11	--Of rectangular (other than square) crosssection

序号 No.	税则号列 Tariff Item	货品名称	最惠国税率(%) M.F.N.	协定税率(%) Agreement 东盟 ASEAN	曼谷协定 BA	香港 H.K.	澳门 Macao	普通税率(%) Gen.	Article Description
4779	7218.9900	--其他	2					11	--Other
	72.19	**不锈钢平板轧材,宽度在600毫米及以上:**							**Flat-rolled products of stainless steel, of a width of 600mm or more:**
		-除热轧外未经进一步加工的卷材:							-Not further worked than hot-rolled, in coils:
4780	7219.1100	--厚度超过10毫米	4		4			14	--Of a thickness exceeding 10mm
4781	7219.1200	--厚度在4.75毫米及以上,但不超过10毫米	4		4			14	--Of a thickness of 4.75mm or more but not exceeding 10mm
		--厚度在3毫米及以上,但小于4.75毫米:							--Of a thickness of 3mm or more but less than 4.75mm:
4782	7219.1310	---未经酸洗的	4		4			14	---Not acid Pickled
4783	7219.1320	---经酸洗的	4		4			14	---Acid pickled
		--厚度小于3毫米:							--Of a thickness of less than 3mm:
4784	7219.1410	---未经酸洗的	4		4			14	---Not acid Pickled
4785	7219.1420	---经酸洗的	4		4			14	---Acid pickled
		-除热轧外未经进一步加工的非卷材:							-Not further worked than hot-rolled, not in coils:
4786	7219.2100	--厚度超过10毫米	10		9.3			40	--Of a thickness exceeding 10mm
4787	7219.2200	--厚度在4.75毫米及以上,但不超过10毫米	10		9.3			40	--Of a thickness of 4.75mm or more but not exceeding 10mm
4788	7219.2300	--厚度在3毫米及以上,但小于4.75毫米	10		9.3			40	--Of a thickness of 3mm or more but less than 4.75mm
		--厚度小于3毫米:							--Of a thickness of less than 3mm:
4789	7219.2410	---厚度超过1毫米但小于3毫米	10		9.3			40	---Of a thickness exceeding 1mm but less than 3mm
4790	7219.2420	---厚度在0.5毫米及以上,但不超过1毫米	10		9.3			40	---Of a thickness of 0.5mm or more but not exceeding 1mm
4791	7219.2430	---厚度小于0.5毫米	10		9.3			40	---Of a thickness of less than 0.5mm
		-除冷轧外未经进一步加工:							-Not further worked than cold-rolled (cold-reduced):
4792	7219.3100	--厚度在4.75毫米及以上	10		10			40	--Of a thickness of 4.75mm or more
4793	7219.3200	--厚度在3毫米及以上,但小于4.75毫米	10		10			40	--Of a thickness of 3mm or more but less than 4.75mm
4794	7219.3300	--厚度超过1毫米,但小于3毫米	10		10	0△		40	--Of a thickness exceeding 1mm but less than 3mm
4795	7219.3400	--厚度在0.5毫米及以上,但不超过1毫米	10		10	0△		40	--Of a thickness of 0. 5mm or more but not exceeding 1mm
4796	7219.3500	--厚度小于0.5毫米	10		10	0△		40	--Of a thickness of less than 0.5mm
4797	7219.9000	-其他	10		10			40	-Other
	72.20	**不锈钢平板轧材,宽度小于600毫米:**							**Flat-rolled products stainless steel, of a width of less than 600mm:**
		-除热轧外未经进一步加工:							-Not further worked than hot-rolled:
4798	7220.1100	--厚度在4.75毫米及以上	10					20	--Of a thickness of 4.75mm or more
4799	7220.1200	--厚度小于4.75毫米	10					20	--Of a thickness of less than 4.75mm

序号 No.	税则号列 Tariff Item	货品名称	最惠国税率(%) M.F.N.	协定税率(%) Agreement 东盟 ASEAN	曼谷协定 BA	香港 H.K.	澳门 Macao	普通税率(%) Gen.	Article Description
		-除冷轧外未经进一步加工:							-Not further worked than cold-rolled (cold-reduced):
4800	7220.2010	---宽度小于300毫米	10					20	---Of a width of less than 300mm
4801	7220.2090	---其他	10					20	---Other
4802	7220.9000	-其他	10					20	-Other
	72.21	**不规则盘卷的不锈钢热轧条、杆:**							**Bars and rods, hot-rolled, in irregularly wound coils, of stainless steel:**
4803	7221.0000	不规则盘卷的不锈钢热轧条、杆	10		8			20	Bars and rods, hot-rolled, in irregularly wound coils, of stainless steel
	72.22	**不锈钢其他条、杆;不锈钢角材、型材及异型材:**							**Other bars and rods of stainless steel: angles, shapes and sections of stainless steel:**
		-条、杆,除热轧、热拉拔或热挤压外未经进一步加工:							-Bars and rods, not further worked than hot-rolled, hot-drawn or extruded:
4804	7222.1100	--圆形截面的	10		9			40	--Of circular cross-section
4805	7222.1900	--其他	10		9			40	--Other
4806	7222.2000	-条、杆,除冷成形或冷加工外未经进一步加工	10					40	-Bars and rods, not further worked than cold-formed or cold-finished
4807	7222.3000	-其他条、杆	10		8.9			40	-Other bars and rods
4808	7222.4000	-角材、型材及异型材	10					17	-Angles, shapes and sections
	72.23	**不锈钢丝:**							**Wire of stainless steel:**
4809	7223.0000	不锈钢丝	10			0		20	Wire of stainless steel
		第四分章 其他合金钢;合金钢或非合金钢制的空心钻钢							Ⅳ. -OTHER ALLOY STEEL; HOLLOW DRILL BARS AND RODS, OF ALLOY OR NON-ALLOY STEEL
	72.24	**其他合金钢,锭状或其他初级形状;其他合金钢制的半制成品:**							**Other alloy steel in ingots or other primary forms; semi-finished products of other alloy steel:**
4810	7224.1000	-锭状及其他初级形状	2					11	-Ingots and other primary forms
		-其他:							-Other:
4811	7224.9010	---单件重量在10吨及以上的粗铸锻件坯	2					11	---Raw casting forging stocks, individual piece weight of 10T or more
4812	7224.9090	---其他	2					11	---Other
	72.25	**其他合金钢平板轧材,宽度在600毫米及以上:**							**Flat-rolled products of other alloy steel, of a width of 600mm or more:**
		-硅电钢制:							-Of silicon-electrical steel:
4813	7225.1100	--取向性硅电钢:	3					20	--Grain-oriented
4814	7225.1900	--其他	6					20	--Other
4815	7225.2000	-高速钢制	3					17	-Of high speed steel
4816	7225.3000	-其他卷材,除热轧外未经进一步加工	3					14	-Other, not further worked than hot-rolled, in coils

序号 No.	税则号列 Tariff Item	货品名称	最惠国税率(%) M.F.N.	协定税率(%)Agreement 东盟 ASEAN	曼谷协定 BA	香港 H.K.	澳门 Macao	普通税率(%) Gen.	Article Description
4817	7225.4000	-其他非卷材,除热轧外未经进一步加工	3					17	-Other, not further worked than hot-rolled, not in coils
4818	7225.5000	-其他,除冷轧外未经进一步加工	3					17	-Other, not further worked than cold-rolled(cold-reduced)
		-其他:							-Other:
4819	7225.9100	--电镀或涂锌的	7					17	--Electrolytically plated or coated with zinc
4820	7225.9200	--用其他方法镀或涂锌的	7					17	--Otherwise plated or coated with zinc
4821	7225.9900	--其他	7					17	--Other
	72.26	**其他合金钢平板轧材,宽度小于600毫米:**							**Flat-rolled products of other alloy steel, of a width of less than 600mm:**
		-硅电钢制:							-Of silicon-electrical steel:
4822	7226.1100	--取向性硅电钢:	3					20	--Grain-oriented
4823	7226.1900	--其他:	3					20	--Other
4824	7226.2000	-高速钢制	3					20	-Of high speed steel
		-其他:							-Other:
4825	7226.9100	--除热轧外未经进一步加工	3					20	--Not further worked than hot-rolled
4826	7226.9200	--除冷轧外未经进一步加工	3					20	--Not further worked than cold-rolled (cold-reduced)
4827	7226.9300	--电镀或涂锌的	7					20	--Electrolytically plated or coated with zinc
4828	7226.9400	--用其他方法镀或涂锌的	7					20	--Otherwise plated or coated with zinc
4829	7226.9900	--其他	7					20	--Other
	72.27	**不规则盘卷的其他合金钢热轧条、杆:**							**Bars and rods, hot-rolled, in irregularly wound coils, of other alloy steel:**
4830	7227.1000	-高速钢制	3					20	-Of high speed steel
4831	7227.2000	-硅锰钢制	6					20	-Of silico-manganese steel
4832	7227.9000	-其他	3					20	-Other
	72.28	**其他合金钢条、杆;其他合金钢角材、型材及异型材;合金钢或非合金钢制的空心钻钢:**							**Other bars and rods of other alloy steel; angles, shapes and sections, of other alloy steel; hollow drill bars and rods, of alloy or non-alloy steel:**
4833	7228.1000	-高速钢条、杆	3					20	-Bars and rods, of high speed steel
4834	7228.2000	-硅锰钢条、杆	6					20	-Bars and rods, of silico-manganese steel
4835	7228.3000	-其他条、杆,除热轧、热拉拔或热挤压外未经进一步加工	3					20	-Other bars and rods, not further worked than hot-crolled, hot-drawn or extruded
4836	7228.4000	-其他条、杆,除锻造外未经进一步加工	3					20	-Other bars and rods, not further worked than forged
4837	7228.5000	-其他条、杆,除冷成形或冷加工外未经进一步加工	3					20	-Other bars and rods, not further worked than cold-formed of cold-finished

序号 No.	税则号列 Tariff Item	货品名称	最惠国税率(%) M.F.N.	协定税率(%)Agreement 东盟 ASEAN	曼谷协定 BA	香港 H.K.	澳门 Macao	普通税率(%) Gen.	Article Description
4838	7228.6000	-其他条、杆	3					20	-Other bars and rods
		-角材、型材及异型材:							-Angles, shapes and sections:
4839	7228.7010	---履带板型钢	6					17	---Shapes of crawler tread
4840	7228.7090	---其他	6					17	---Other
4841	7228.8000	-空心钻钢	7					35	-Hollow drill bars and rods
	72.29	**其他合金钢丝:**							**Wire of other alloy steel:**
4842	7229.1000	-高速钢制	3					20	-Of high speed steel
4843	7229.2000	-硅锰钢制	7					20	-Of silico-manganese steel
4844	7229.9000	-其他	7					20	-Other

第七十三章
钢 铁 制 品

Chapter 73
Articles of iron or steel

注释：

一、本章所称"铸铁"，适用于经铸造而得的产品，按重量计其铁元素含量超过其他元素单项含量并与第七十二章注释一（四）所述的钢的化学成分不同。

二、本章所称"丝"，是指热或冷成形的任何截面形状的产品，但其截面尺寸均不超过16毫米。

Notes:

1. In this Chapter the expression "cast iron" applies to products obtained by casting in which iron predominates by weight over each of the other elements and which do not comply with the chemical composition of steel as defined in Note 1 (d) to Chapter 72.

2. In this Chapter the word "wire" means hot or cold-formed products of any cross-sectional shape, of which no cross-sectional dimension exceeds 16mm.

序号 No.	税则号列 Tariff Item	货品名称	最惠国税率(%) M.F.N.	协定税率(%) Agreement 东盟 ASEAN	曼谷协定 BA	香港 H.K.	澳门 Macao	普通税率(%) Gen.	Article Description
	73.01	**钢铁板桩，不论是否钻孔、打眼或组装；焊接的钢铁角材、型材及异型材：**							**Sheet piling of iron or steel, whether or not drilled, punched or made from assembled elements; welded angles, shapes and sections, of iron or steel:**
4845	7301.1000	-钢铁板桩	7					20	-Sheet piling
4846	7301.2000	-角材、型材及异型材	7					30	-Angles, shapes and sections
	73.02	**铁道及电车道铺轨用钢铁材料(钢轨、护轨、齿轨、道岔尖轨、辙叉、尖轨拉杆及其他叉道段体、轨枕、鱼尾板、轨座、轨座楔、钢轨垫板、钢轨夹、底板、固定板及其他专门用于连接或加固路轨的材料)：**							**Railway or tramway track construction material of iron or steel, the following: rails, check-rails and rack rails, switch blades, crossing frogs, point rods and other crossing pieces, sleepers (cross-ties), fishplates, chairs, chair wedges, sole plates (base plates), rail clips, bedplates, ties and other material specialized for jointing or fixing rails:**
4847	7302.1000	-钢轨	6					14	-Rails
4848	7302.3000	-道岔尖轨、辙叉、尖轨拉杆及其他叉道段体	8					17	-Switch blades, crossing frogs, point rods and other crossing pieces
4849	7302.4000	-鱼尾板及钢轨垫板	7					17	-Fish-plates and sole plates
		-其他							-Other:
4850	7302.9010	---轨枕	6					14	---Sleepers (cross-ties)
4851	7302.9090	---其他	7					17	---Other
	73.03	**铸铁管及空心异型材：**							**Tubes, pipes and hollow profiles, of cast iron:**
4852	7303.0010	---内径在500毫米及以上的圆形截面管	4					40	---Tubes and pipes of circular cross-section, of the internal diameter of 500mm or more
4853	7303.0090	---其他	4					40	---Other

序号 No.	税则号列 Tariff Item	货品名称	最惠国税率(%) M.F.N.	协定税率(%) Agreement 东盟 ASEAN	曼谷协定 BA	香港 H.K.	澳门 Macao	普通税率(%) Gen.	Article Description
	73.04	**无缝钢铁管及空心异型材(铸铁的除外):**							**Tubes, pipes and hollow profiles, seamless, of iron (other than cast iron) or steel:**
		-石油或天然气管道管:							-Line pipe of a kind used for oil or gas pipelines:
4854	7304.1010	---外径大于等于215.9毫米,但不超过406.4毫米	5					17	---Having an outside diameter of 215.9mm or more but not exceeding 406.4mm
4855	7304.1020	---外径超过114.3毫米,但小于215.9毫米	5					17	---Having an outside diameter exceeding 114.3mm but less than 215.9mm
4856	7304.1030	---外径不超过114.3毫米	5					17	---Having an outside diameter not exceeding 114.3mm
4857	7304.1090	---其他	5					17	---Other
		-钻探石油及天然气用的套管、导管及钻管:							-Casing, tubing and drill pipe, of a kind used in drilling for oil or gas:
		--钻管:							--Drill pipe:
4858	7304.2110	---外径不超过168.3毫米	4					17	---Having an outside diameter not exceeding 168.3mm
4859	7304.2190	---其他	4					17	---Other
4860	7304.2900	--其他	4					17	--Other
		-铁或非合金钢的其他圆形截面管:							-Other, of circular cross-section, of iron or non-alloysteel:
		--冷拔或冷轧的:							--Cold-drawn or cold-rolled (cold-reduced):
4861	7304.3110	---锅炉管	4					17	---Boiler tubes and pipes
4862	7304.3120	---地质钻管、套管	8					17	---Geological casing and drill pipes
4863	7304.3190	---其他	4					17	---Other
		--其他:							--Other:
4864	7304.3910	---锅炉管	4			0		17	---Boiler tubes and pipes
4865	7304.3920	---地质钻管、套管	5			0		17	---Geological casing and drill pipes
4866	7304.3990	---其他	4			0		17	---Other
		-不锈钢的其他圆形截面管:							-Other, of circular cross-section, of stainless steel:
		--冷拔或冷轧的:							--Cold-drawn or cold-rolled (cold-reduced):
4867	7304.4110	---锅炉管	10					17	---Boiler tubes and pipes
4868	7304.4190	---其他	10					40	---Other
		--其他:							--Other:
4869	7304.4910	---锅炉管	10					17	---Boiler tubes and pipes
4870	7304.4990	---其他	10					40	---Other
		-其他合金钢的其他圆形截面管:							-Other, of circular cross-section, of other alloy steel:
		--冷拔或冷轧的:							--Cold-drawn or cold-rolled (cold-reduced):

序号 No.	税则号列 Tariff Item	货品名称	最惠国税率(%) M.F.N.	协定税率(%) Agreement 东盟 ASEAN	曼谷协定 BA	香港 H.K.	澳门 Macao	普通税率(%) Gen.	Article Description
4871	7304.5110	---锅炉管	4					17	---Boiler tubes and pipes
4872	7304.5120	---地质钻管、套管	4					17	---Geological casing and drill pipes
4873	7304.5190	---其他	4					17	---Other
		--其他:							--Other:
4874	7304.5910	---锅炉管	4					17	---Boiler tubes and pipes
4875	7304.5920	---地质钻管、套管	4					17	---Geological casing and drill pipes
4876	7304.5990	---其他	4					17	---Other
4877	7304.9000	-其他	4					17	-Other
	73.05	**其他圆形截面钢铁管(例如,焊、铆及用类似方法接合的管),外径超过406.4毫米:**							**Other tubes and pipes (for example, welded, riveted or similarly closed), having circular cross-sections, the external diameter of which exceeds 406.4mm, of iron or steel:**
		-石油或天然气管道管:							-Line pipe of a kind used for oil or gas pipelines:
4878	7305.1100	--纵向埋弧焊接的	7					17	--Longitudinally submerged arc welded
4879	7305.1200	--其他纵向焊接的	3					17	--Other, longitudinally welded
4880	7305.1900	--其他	7					17	--Other
4881	7305.2000	-钻探石油或天然气用套管	7					17	-Casing of a kind used in drilling for oil or gas
		-其他焊接的:							-Other, welded:
4882	7305.3100	--纵向焊接的	6					30	--Longitudinally welded
4883	7305.3900	--其他	6					30	--Other
4884	7305.9000	-其他	6					30	-Other
	73.06	**其他钢铁管及空心异型材(例如,辊缝、焊、铆及类似方法接合的):**							**Other tubes, pipes and hollow profiles (for example, open seam or welded, riveted or similarly closed), of iron or steel:**
4885	7306.1000	-石油及天然气管道管	7					17	Line pipe of a kind used for oil or gas pipelines
4886	7306.2000	-钻探石油及天然气用的套管及导管	3					17	-Casing and tubing of a kind used in drilling for oil or gas
4887	7306.3000	-铁或非合金钢的其他圆形截面焊缝管	3					30	-Other, welded, of circular cross-section, of iron or non-alloy steel
4888	7306.4000	-不锈钢的其他圆形截面焊缝管	6					30	-Other, welded, of circular cross-section, of stainless steel
4889	7306.5000	-其他合金钢的圆形截面焊缝管	3					30	-Other, welded, of circular cross-section, of other alloy steel
4890	7306.6000	-非圆形截面的其他焊缝管	3					30	-Other, welded, of non-circular cross-section
4891	7306.9000	-其他	6					30	-Other
	73.07	**钢铁管子附件(例如,接头、肘管、管套):**							**Tube or pipe fittings (for example, couplings, elbows, sleeves), of iron or steel:**

序号 No.	税则号列 Tariff Item	货品名称	最惠国税率(%) M.F.N.	协定税率(%) Agreement 东盟 ASEAN	曼谷协定 BA	香港 H.K.	澳门 Macao	普通税率(%) Gen.	Article Description
		-铸件:							-Cast fittings:
4892	7307.1100	--无可锻性铸铁制	5					20	--Of non-malleable cast iron
4893	7307.1900	--其他	8					20	--Other
		-其他,不锈钢制:							-Other,of stainless steel:
4894	7307.2100	--法兰	8.4					20	--Flanges
4895	7307.2200	--螺纹肘管、弯管及管套	8.4					20	--Threaded elbows,bends and sleeves
4896	7307.2300	--对焊件	8.4					20	--Butt welding fittings
4897	7307.2900	--其他	8.4					20	--Other
		-其他:							-Other:
4898	7307.9100	--法兰	7					20	--Flanges
4899	7307.9200	--螺纹肘管、弯管及管套	4					20	--Threaded elbows,bends and sleeves
4900	7307.9300	--对焊件	7					20	--Butt welding fittings
4901	7307.9900	--其他	4					20	--Other
	73.08	**钢铁结构体(税号94.06的活动房屋除外)及其部件(例如,桥梁及桥梁体段、闸门、塔楼、格构杆、屋顶、屋顶框架、门窗及其框架、门槛、百叶窗、栏杆、支柱及立柱);上述结构体用的已加工钢铁板、杆、角材、型材、异型材、管子及类似品:**							**Structures (excluding prefabricated buildings of heading No.94.06) and parts of structures (for example, bridges and bridge-sections, lock-gates, towers, lattice masts, roofs, roofing frameworks, doors and windows and their frames and thresholds for doors, shutters, balustrades, pillars and columns), of iron or steel; plates, rods, angles, shapes, sections, tubes and the like, prepared for use in structures, of iron or steel:**
4902	7308.1000	-桥梁及桥梁体段	8					30	-Bridges and bridge-sections
4903	7308.2000	-塔楼及格构杆	8.4					30	-Towers and lattice masts
4904	7308.3000	-门窗及其框架、门槛	10			0		50	-Doors, windows and their frames and thresholds for doors
4905	7308.4000	-脚手架、模板或坑道支撑用的支柱及类似设备	8.4					30	-Equipment for scaffolding, shuttering, propping or pit-propping
4906	7308.9000	-其他	4			0		30	-Other
	73.09	**盛装物料用的钢铁囤、柜、罐、桶及类似容器(装压缩气体或液化气体的除外),容积超过300升,不论是否衬里或隔热,但无机械或热力装置:**							**Reservoirs, tanks, vats and similar containers for any material (other than compressed or liquefied gas), of iron or steel, of a capacity exceeding 300L, whether or not lined or heat-insulated, but not fitted with mechanical or thermal equipment:**

序号 No.	税则号列 Tariff Item	货品名称	最惠国税率(%) M.F.N.	协定税率(%) Agreement 东盟 ASEAN	曼谷协定 BA	香港 H.K.	澳门 Macao	普通税率(%) Gen.	Article Description
4907	7309.0000	盛装物料用的钢铁囤、柜、罐、桶及类似容器(装压缩气体或液化气体的除外),容积超过300升,不论是否衬里或隔热,但无机械或热力装置	10.5					35	Reservoirs, tanks, vats and similar containers for any material (other than compressed or liquefied gas), of iron or steel, of a capacity exceeding 300 l, whether or not lined or heat-insulated, but not fitted with mechanical or thermal equipment
	73.10	**盛装物料用的钢铁柜、桶、罐、听、盒及类似容器(装压缩气体或液化气体的除外),容积不超过300升,不论是否衬里或隔热,但无机械或热力装置:**							**Tanks, casks, drums, cans, boxes and similar containers, for any material (other than compressed or liquefied gas), of iron or steel, of a capacity not exceeding 300 l, whether or not lined or heat-insulated, but not fitted with mechanical or thermal equipment:**
4908	7310.1000	-容积在50升及以上	10.5					40	-Of a capacity of 50 l or more
		-容积在50升以下:							-Of a capacity of less than 50 l:
4909	7310.2100	--焊边或卷边接合的罐	17.5			0		70	--Cans which are to be closed by soldering or crimping
4910	7310.2900	--其他	17.5					70	--Other
	73.11	**装压缩气体或液化气体用的钢铁容器:**							**Containers for compressed or liquefied gas, of iron or steel:**
4911	7311.0010	---零售包装用	17.5					70	---For retail packing
4912	7311.0090	---其他	8					17	---Other
	73.12	**非绝缘的钢铁绞股线、绳、缆、编带、吊索及类似品:**							**Stranded wire, ropes, cables, plaited bands, slings and the like, of iron or steel, not electrically insulated:**
4913	7312.1000	-绞股线、绳、缆	4					20	-Stranded wire, ropes and cables
4914	7312.9000	-其他	4					20	-Other
	73.13	**带刺钢铁丝;围篱用的钢铁绞带或单股扁丝(不论是否带刺)及松绞的双股丝:**							**Barbed wire of iron or steel; twisted hoop or single flat wire, barbed or not, and loosely twisted double wire, of a kind used for fencing, of iron or steel:**
4915	7313.0000	带刺钢铁丝;围篱用的钢铁绞带或单股扁丝(不论是否带刺)及松绞的双股丝	7					70	Barbed wire of iron or steel; twisted hoop or single flat wire, barbed or not, and loosely twisted double wire, of a kind used for fencing, of iron or steel
	73.14	**钢铁丝制的布(包括环形带)、网、篱、格栅;网眼钢铁板:**							**Cloth (including endless bands), grill, netting and fencing, of iron or steel wire; expanded metal of iron or steel:**
		-机织品:							-Woven cloth:
		--不锈钢制的机器用环形带:							--Endless bands for machinery, of stainless steel:
4916	7314.1210	---工业用	12					20	---For technical use
4917	7314.1290	---其他	12					70	---Other

序号 No.	税则号列 Tariff Item	货品名称	最惠国税率(%) M.F.N.	协定税率(%) Agreement 东盟 ASEAN	曼谷协定 BA	香港 H.K.	澳门 Macao	普通税率(%) Gen.	Article Description
		--其他机器用环形带:							--Other endless bands for machinery:
4918	7314.1310	---工业用	8					20	---For technical use
4919	7314.1390	---其他	12					70	---Other
		--不锈钢制的其他机织品:							--Other woven cloth, of stainless steel:
4920	7314.1410	---工业用	12					20	---For technical use
4921	7314.1490	---其他	12					70	---Other
		--其他:							--Other:
4922	7314.1910	---工业用	7					20	---For technical use
4923	7314.1990	---其他	12					70	---Other
4924	7314.2000	-交点焊接的网、篱及格栅,其丝的最大截面尺寸在3毫米及以上,网眼尺寸在100平方厘米及以上	7			0		70	-Grill, netting and fencing, welded at the intersection, of wire with a maximum cross-sectional dimension of 3mm or more and having a mesh size of 100cm^2 or more
		-其他交点焊接的网、篱及格栅:							-Other grill, netting and fencing, welded at the intersection:
4925	7314.3100	--镀或涂锌的	7			0		70	--Plated or coated with zinc
4926	7314.3900	--其他	7			0		70	--Other
		-其他网、篱及格栅:							-Other grill, netting and fencing:
		--镀或涂锌的:							--Plated or coated with zinc:
4927	7314.4110	---工业用	8					20	---For technical use
4928	7314.4190	---其他	10					70	---Other
		--涂塑的:							--Coated whith plastics;
4929	7314.4210	---工业用	8					20	---For technical use
4930	7314.4290	---其他	10					70	---Other
		--其他:							--Other:
4931	7314.4910	---工业用	8					20	---For technical use
4932	7314.4990	---其他	10					70	---Other
4933	7314.5000	-网眼钢铁板	8			0		70	-Expanded metal
	73.15	**钢铁链及其零件:**							**Chain and parts thereof, of iron or steel:**
		-铰接链及其零件:							-Articulated link chain and parts thereof:
		--滚子链:							--Roller chain:
4934	7315.1110	---自行车用	12			0		80	---For bicycles
4935	7315.1120	---摩托车用	12			0		80	---For motorcycles
4936	7315.1190	---其他	12			0		80	---Other
4937	7315.1200	--其他链	12					80	--Other chain
4938	7315.1900	--零件	12					80	--Parts
4939	7315.2000	-防滑链	12					80	-Skid chain
		-其他链:							-Other chain:
4940	7315.8100	--日字环节链	12					80	--Stud-link
4941	7315.8200	--其他焊接链	12					80	--Other, welded link
4942	7315.8900	--其他	12					80	--Other
4943	7315.9000	-其他零件	10					80	-Other parts

序号 No.	税则号列 Tariff Item	货品名称	最惠国税率(%) M.F.N.	协定税率(%) Agreement 东盟 ASEAN	曼谷协定 BA	香港 H.K.	澳门 Macao	普通税率(%) Gen.	Article Description
	73.16	**钢铁锚、多爪锚及其零件:**							**Anchors, grapnels and parts thereof, of iron or steel:**
4944	7316.0000	钢铁锚、多爪锚及其零件	10					40	Anchors, grapnels and parts thereof, of iron or steel
	73.17	**钢铁制的钉、平头钉、图钉、波纹钉、U形钉(税号83.05的货品除外)及类似品,不论钉头是否用其他材料制成,但不包括铜头钉:**							**Nails, tacks, drawing pins, corrugated nails, staples (other than those of heading No.83.05) and similar articles, of iron or steel, whether or not with heads of other material, but excluding such articles with heads of copper:**
4945	7317.0000	钢铁制的钉、平头钉、图钉、波纹钉、U形钉(税号83.05的货品除外)及类似品,不论钉头是否用其他材料制成,但不包括铜头钉	10					80	Nails, tacks, drawing pins, corrugated nails, staples (other than those of heading No.83.05) and similar articles, of iron or steel, whether or not with heads of other material, but excluding such articles with heads of copper
	73.18	**钢铁制的螺钉、螺栓、螺母、方头螺钉、钩头螺钉、铆钉、销、开尾销、垫圈(包括弹簧垫圈)及类似品:**							**Screws, bolts, nuts, coach screws, screw hooks, rivets, cotters, cotter-pins, washers (including spring washers) and similar articles, of iron of steel:**
		-螺纹制品:							-Threaded articles:
4946	7318.1100	--方头螺钉	10					80	--Coach screws
4947	7318.1200	--其他木螺钉	10					80	--Other wood screws
4948	7318.1300	--钩头螺钉及环头螺钉	10					80	--Screw hooks and screw rings
4949	7318.1400	--自攻螺钉	10					80	--Self-tapping screws
4950	7318.1500	--其他螺钉及螺栓,不论是否带有螺母或垫圈	8			0		80	--Other screws and bolts, whether or not with their nuts or washers
4951	7318.1600	--螺母	8					80	--Nuts
4952	7318.1900	--其他	5					80	--Other
		-无螺纹制品:							-Non-threaded articles:
4953	7318.2100	--弹簧垫圈及其他防松垫圈	10			0		80	--Spring washers and other lock washers
4954	7318.2200	--其他垫圈	10					80	--Other washers
4955	7318.2300	--铆钉	10			0		80	--Rivets
4956	7318.2400	--销及开尾销	10					80	--Cotters and cotter-pins
4957	7318.2900	--其他	10					80	--Other
	73.19	**钢铁制的手工缝针、编织针、引针、钩针、刺绣穿孔锥及类似制品;其他税号未列名的钢铁制安全别针及其他别针:**							**Sewing needles, knitting needles, bodkins, crochet hooks, embroidery stilettos and similar articles, for use in the hand, of iron or steel; safety pins and other pins of iron or steel, not elsewhere specified or included:**
4958	7319.1000	-缝针、织补针或刺绣针	10					80	-Sewing, darning or embroidery needles

序号 No.	税则号列 Tariff Item	货品名称	最惠国税率(%) M.F.N.	协定税率(%) Agreement 东盟 ASEAN	曼谷协定 BA	香港 H.K.	澳门 Macao	普通税率(%) Gen.	Article Description
4959	7319.2000	-安全别针	10					90	-Safety pins
4960	7319.3000	-其他别针	10					90	-Other pins
4961	7319.9000	-其他	10					80	-Other
	73.20	**钢铁制弹簧及弹簧片:**							**Springs and leaves for springs, of iron or steel:**
		-片簧及簧片:							-Leaf-springs and leaves thereof:
4962	7320.1010	---铁道车辆用	6					14	---For railway locomotives and rolling-stock
4963	7320.1020	---汽车用	10					14	---For motor vehicles
4964	7320.1090	---其他	10					50	---Other
		-螺旋弹簧:							-Helical springs:
4965	7320.2010	---铁道车辆用	6					14	---For railway locomotives and rolling-stock
4966	7320.2090	---其他	10					50	---Other
		-其他:							-Other:
4967	7320.9010	---铁道车辆用	6					14	---For railway locomotives and rolling-stock
4968	7320.9090	---其他	12					50	---Other
	73.21	**非电热的钢铁制家用炉、灶(包括附有集中供暖用的热水锅的炉)、烤肉架、烤炉、煤气灶、加热板和类似非电热的家用器具及其零件:**							**Stoves, ranges, grates, cookers (including those with subsidiary boilers for central heating), barbecues, braziers, gas-rings, plate warmers and similar non-electric domestic appliances, and parts thereof, of iron or steel:**
		-炊事器具及加热板:							-Cooking appliances and plate warmers:
4969	7321.1100	--使用气体燃料或可使用气体燃料及其他燃料的	15					80	--For gas fuel or for both gas and other fuels
		--使用液体燃料的:							--For liquid fuel:
4970	7321.1210	---煤油炉	21					80	---Kerosene cooking stoves
4971	7321.1290	---其他	21					80	---Other
4972	7321.1300	--使用固体燃料的	21					80	--For solid fuel
		-其他器具:							-Other appliances:
4973	7321.8100	--使用气体燃料或可使用气体燃料及其他燃料的	23					80	--For gas fuel or for both gas and other fuels
4974	7321.8200	--使用液体燃料的	21					80	--For liquid fuel
4975	7321.8300	--使用固体燃料的	21					80	--For solid fuel
4976	7321.9000	-零件	12					80	-Parts
	73.22	**非电热的钢铁制集中供暖用散热器及其零件;非电热的钢铁制空气加热器、暖气分布器(包括可分布新鲜空气或调节空气的)及其零件,装有电动风扇或鼓风机:**							**Radiators for central heating, not electrically heated, and parts thereof, of iron or steel; air heaters and hot air distributors (including distributors which can also distribute fresh or conditioned air), not electrically heated, incorporating a motor-dirven fan or blower, and parts thereof, of iron or steel:**
		-散热器及其零件:							-Radiators and parts thereof:

序号 No.	税则号列 Tariff Item	货品名称	最惠国税率(%) M.F.N.	协定税率(%) Agreement 东盟 ASEAN	曼谷协定 BA	香港 H.K.	澳门 Macao	普通税率(%) Gen.	Article Description
4977	7322.1100	--铸铁制	21					80	--Of cast iron
4978	7322.1900	--其他	21					80	--Other
4979	7322.9000	-其他	20					80	-Other
	73.23	**餐桌、厨房或其他家用钢铁器具及其零件;钢铁丝绒;钢铁制擦锅器、洗刷擦光用的块垫、手套及类似品:**							**Table, kitchen or other household articles and parts thereof, of iron or steel; iron or steel wool; pot scourers and scouring or polishing pads, gloves and the like, of iron or steel:**
4980	7323.1000	-钢铁丝绒;擦锅器及洗刷擦光用的块垫、手套及类似品	14					80	-Iron or steel wool; pot scourers and scouring or polishing pads, gloves and the like
		-其他:							-Other:
4981	7323.9100	--铸铁制,未搪瓷	20					80	--Of cast iron, not enamelled
4982	7323.9200	--铸铁制,已搪瓷	20					100	--Of cast iron, enamelled
4983	7323.9300	--不锈钢制	12			0	0	80	--Of stainless steel
		--钢铁(铸铁除外)制,已搪瓷:							--Of iron (other than cast iron) or steel, enamelled:
4984	7323.9410	---面盆	20					100	---Basin
4985	7323.9420	---烧锅	20					100	---Casserole
4986	7323.9430	---烧烤炉	20					100	---Barbecue
4987	7323.9490	---其他	20					100	---Other
4988	7323.9900	--其他	20					80	--Other
	73.24	**钢铁制卫生器具及其零件:**							**Sanitary ware and parts thereof, of iron or steel:**
4989	7324.1000	-不锈钢制洗涤槽及脸盆	18			0△		80	-Sinks and wash basins, of stainless steel
		-浴缸:							-Baths:
4990	7324.2100	--铸铁制,不论是否搪瓷	10					100	--Of cast iron, whether or not enamelled
4991	7324.2900	--其他	30					100	--Other
4992	7324.9000	-其他,包括零件	25					100	-Other, including parts
	73.25	**其他钢铁铸造制品:**							**Other cast articles of iron or steel:**
		-无可锻性铸铁制:							-Of non-malleable cast iron:
4993	7325.1010	---工业用	7					40	---For technical use
4994	7325.1090	---其他	20					90	---Other
		-其他:							-Other:
4995	7325.9100	--研磨机用的研磨球及类似品	10.5					40	--Grinding balls and similar articles for mills
		--其他:							--Other:
4996	7325.9910	---工业用	10.5					40	---For technical use
4997	7325.9990	---其他	20					90	---Other
	73.26	**其他钢铁制品:**							**Other articles of iron or steel:**
		-经锻造或冲压,但未经进一步加工:							-Forged or stamped, but not further worked:
4998	7326.1100	--研磨机用的研磨球及类似品	10.5					40	--Grinding balls and similar articles for mills
		--其他:							--Other:

序号 No.	税则号列 Tariff Item	货 品 名 称	最惠国税 率 (%) M.F.N.	协定税率(%)Agreement				普通税率 (%) Gen.	Article Description
				东盟 ASEAN	曼谷协定 BA	香港 H.K.	澳门 Macao		
4999	7326.1910	---工业用	10.5					40	---For technical use
5000	7326.1990	---其他	20					90	---Other
		-钢铁丝制品:							-Articles of iron or steel wire:
5001	7326.2010	---工业用	10					40	---For technical use
5002	7326.2090	---其他	18					90	---Other
		-其他:							-Other:
5003	7326.9010	---工业用	10.5			0		40	---For technical use
5004	7326.9090	---其他	8			0		90	---Other

第七十四章
铜及其制品

注释：

本章所用有关名词解释如下：

一、精炼铜

按重量计含铜量至少为 99.85% 的金属；或按重量计含铜量至少为 97.5%，但其他各种元素的含量不超过下表中规定的限量的金属：

其他元素表

元	素	所含重量百分比
Ag	银	0.25
As	砷	0.5
Cd	镉	1.3
Cr	铬	1.4
Mg	镁	0.8
Pb	铅	1.5
S	硫	0.7
Sn	锡	0.8
Te	碲	0.8
Zn	锌	1
Zr	锆	0.3
其他元素*，每种		0.3

*其他元素，例如，铝、铍、钴、铁、锰、镍、硅。

二、铜合金

除未精炼铜以外的金属物质，按重量计含铜量大于其他元素单项含量，但：

1. 按重量计至少有一种其他元素的含量超过上表中规定的限量；

2. 按重量计其他元素的总含量超过 2.5%。

三、铜母合金

含有其他元素，但按重量计含铜量超过 10% 的合金，该合金无实用可锻性，通常用作生产其他合金的添加剂或用作冶炼有色金属的脱氧剂、脱硫剂及类似用途。但按重量计含磷量超过 15% 的磷化铜归入税号 28.48。

Chapter 74
Copper and articles thereof

Notes:

In this Chapter the following expressions have the meanings hereby assigned to them:

1. Refined copper

Metal containing at least 99.85% by weight of copper; or metal containing at least 97.5% by weight of copper, provided that the content by weight of any other element does not exceed the limit specified in the following table:

TABLE-Other elements

Element		Limiting content % by weight
Ag	Silver	0.25
As	Arsenic	0.5
Cd	Cadmium	1.3
Cr	Chromium	1.4
Mg	Magnesium	0.8
Pb	Lead	1.5
S	Sulphur	0.7
Sn	Tin	0.8
Te	Tellurium	0.8
Zn	Zinc	1
Zr	Zirconium	0.3
Other elements*, each		0.3

* Other elements are, for example, Al, Be, Co, Fe, Mn, Ni, Si.

2. Copper alloys

Metallic substances other than unrefined copper in which copper predominates by weight over each of the other elements, provided that:

(1) the content by weight of at least one of the other elements is greater than the limit specified in the foregoing table; or

(2) the total content by weight of such other elements exceeds 2.5%.

3. Master alloys

Alloys containing with other elements more than 10% by weight of copper, not usefully malleable and commonly used as an additive in the manufacture of other alloys or as de-oxidants, de-sulphurizing agents or for similar uses in the metallurgy of non-ferrous metals. However, copper phosphide (phosphor copper) con-taining more than 15% by weight of phosphorus falls in heading No.28.48.

四、条、杆

轧、挤、拔或锻制的实心产品，非成卷的，其全长截面均为圆形、椭圆形、矩形（包括正方形）、等边三角形或规则外凸多边形（包括相对两边为弧拱形，另外两边为等长平行直线的“扁圆形”及“变形矩形”）。对于矩形（包括正方形）、三角形或多边形截面的产品，其全长边角可经磨圆。矩形（包括“变形矩形”）截面的产品，其厚度应大于宽度的十分之一。所述条、杆也包括同样形状及尺寸的铸造或烧结产品。该产品在铸造或烧结后再经加工（简单剪修或去氧化皮的除外），但不具有其他税号所列制品或产品的特征。

线锭及坯段，已具锥形尾端或经其他简单加工以便送入机器制成盘条或管子等的，仍应作为未锻轧铜归入税号74.03。

五、型材及异型材

轧、挤、拔、锻制的产品或其他成型产品，不论是否成卷，其全长截面相同，但与条、杆、丝、板、片、带、箔、管的定义不相符合。同时也包括同样形状的铸造或烧结产品。该产品在铸造或烧结后再经加工（简单剪修或去氧化皮的除外），但不具有其他税号所列制品或产品的特征。

六、丝

盘卷的轧、挤或拔制实心产品，其全长截面均为圆形、椭圆形、矩形（包括正方形）、等边三角形或规则外凸多边形（包括相对两边为弧拱形，另外两边为等长平行直线的“扁圆形”及“变形矩形”）。对于矩形（包括正方形）、三角形或多边形截面的产品，其全长边角可经磨圆。矩形（包括“变形矩形”）截面的产品，其厚度应大于宽度的十分之一。

4. Bars and rods

Rolled, extruded, drawn or forged products, not in coils, which have a uniform solid cross-section along their whole length in the shape of circles, ovals, rectangles (including squares), equilateral triangles or regular convex polygons (including “flattened circles” and “modified rectangles”, of which two opposite sides are convex arcs, the other two sides being straight, of equal length and parallel). Products, with a rectangular (including square), triangular or polygonal cross-section may have corners rounded along their whole length. The thickness of such products which have a rectangular (including “modified rectangular”) cross-section exceeds one-tenth of the width. The expression also covers cast or sintered products, of the same forms and dimensions, which have been subsequently worked after production (otherwise than by simple trimming or de-scaling), provided that they have not thereby assumed the character of articles or products of other headings.

Wire-bars and billets with their ends tapered or otherwise worked simply to facilitate their entry into machines for converting them into, for example, drawing stock (wire-rod) or tubes, are however to be taken to be unwrought copper of heading No. 74.03.

5. Profiles

Rolled, extruded, drawn, forged or formed pruducts, coiled or not, of a uniform cross-section along their whole length, which do not conform to any of the definitions of bars, rods, wire, plates, sheets, strip, foil, tubes or pipes. The expression also covers cast or sintered products, of the same forms, which have been subsequently worked after production (otherwise than by simple trimming or de-scaling), provided that they have not thereby assumed the character of articles or products of other headings.

6. Wire

Rolled, extruded or drawn products, in coils, which have a uniform solid cross-section along their whole length in the shape of circles, ovals, rectangles (including squares), equilateral triangles or regular convex polygons (including “flattened circles” and “modified rectangles”, of which two opposite sides are convex arcs, the other two sides being straight, of equal length and parallel). Products with a rectangular (including square), triangular or polygonal cross-section may have corners rounded along their whole length. The thickness of such products which have a rectangular (including “modified rectangular”) cross-section exceeds one-tenth of the width.

但税号74.14所称"丝"，仅适用于截面尺寸不超过6毫米的各种截面形状的产品（不论是否盘卷）。

In the case of heading No.74.14, however, the term "wire" applies only to products, whether or not in coils, of any cross-sectional shape, of which no cross-sectional dimension exceeds 6mm.

七、板、片、带、箔

成卷或非成卷的平面产品（税号74.03的未锻轧产品除外），截面均为厚度相同的实心矩形（不包括正方形），不论边角是否磨圆（包括相对两边为弧拱形，另外两边为等长平行直线的"变形矩形"），并且符合以下规格：

1. 矩形（包括正方形）的，厚度不超过宽度的十分之一；
2. 矩形或正方形以外形状的，任何尺寸，但不具有其他税号所列制品或产品的特征。

税号74.09及74.10还适用于具有花样（例如，凹槽、肋条形、格槽、珠粒及菱形）的板、片、带、箔以及穿孔、抛光、涂层或制成瓦楞形的这类产品，但不具有其他税号所列制品或产品的特征。

7. Plates, sheets, strip and foil

Flat-surfaced products (other than the unwrought products of heading No.74.03), coiled or not, of soild rectangular (other than square) cross-section with or without rounded corners (including "modified rectangles" of which two opposite sides are convex arcs, the other two sides being straight, of equal length and parallel) of a uniform thickness, which are:

-of rectanglular (including square) shape with a thickness not exceeding one-tenth of the width,

-of a shape other than rectangular or square, of any size, provided that they do not assume the character of articles or products of other headings.

Headings Nos.74.09 and 74.10 apply, *inter alia*, to plates, sheets, strip and foil with patterns (for example, grooves, ribs, chequers, tears, buttons, lozenges) and to such products which have been perforated, corrugated, polished or coated, provided that they do not thereby assume the character of articles or products of other headings.

八、管

全长截面及管壁厚度相同并只有一个闭合空间的空心产品，成卷或非成卷的，其截面为圆形、椭圆形、矩形（包括正方形）、等边三角形或规则外凸多边形。对于截面为矩形（包括正方形）、等边三角形或规则外凸多边形的产品，不论全长边角是否磨圆，只要其内外截面为同一圆心并为同样形状及同一轴向，也可视为管子。上述截面的管子可经抛光、涂层、弯曲、攻丝、钻孔、缩腰、胀口、成锥形或装法兰、颈圈或套环。

8. Tubes and pipes

Hollow products, coiled or not, which have a uniform cross-section with only one enclosed void along their whole length in the shape of circles, ovals, rectangles (including squares), equilateral triangles or regular convex polygons, and which have a uniform wall thickness. Products with a rectangular (including square), equilateral triangular or regular convex polygonal cross-section, which may have corners rounded along their whole length, are also to be taken to be tubes and pipes provided the inner and outer cross-sections are concentric and have the same form and orientation. Tubes and pipes of the foregoing cross-sections may be polished, coated, bent, threaded, drilled, waisted, expanded, cone-shaped or fitted with flanges, collars or rings.

子目注释：

本章所用有关名词解释如下：

一、铜锌合金（黄铜）

铜与锌的合金，不论是否含有其他元素。含有其他元素时：

按重量计含锌量应大于其他各种元素的单项含量；

按重量计含镍量应低于5%［参见铜镍锌合金（德银）］；

按重量计含锡量应低于3%［参见铜锡合金

Subheading Notes:

In this Chapter the following expressions have the meanings hereby assigned to them:

1. Copper-zinc base alloys (brasses)

Alloys of copper and zinc, with or without other elements. When other elments are present:

-zinc predominates by weight over each of such other elements;

-any nickel content by weight is less than 5% (see copper-nickel-zinc alloys (nickel silvers)); and

-any tin content by weight is less than 3% (see copper-tin

(青铜)]。

alloys (bronzes)).

二、铜锡合金(青铜)

铜与锡的合金,不论是否含有其他元素。含有其他元素时,按重量计含锡量应大于其他各种元素的单项含量。当按重量计含锡量在3%及以上时,锌的含量可大于锡的含量,但必须小于10%。

2. Copper-tin base alloys (bronzes)

Alloys of copper and tin, with or without other elements. When other elements are present, tin predominates by weight over each of such other elements, except that when the tin content is 3% or more the zinc content by weight may exceed that of tin but must be less than 10%.

三、铜镍锌合金(德银)

铜、镍、锌的合金,不论是否含有其他元素,按重量计含镍量在5%及以上[参见铜锌合金(黄铜)]。

3. Copper-nickel-zinc base alloys (nickel silvers)

Alloys of copper, nickel and zinc, with or without other elements. The nickel content is 5% or more by weight (see copper-zinc alloys (brasses)).

四、铜镍合金

铜与镍的合金,不论是否含有其他元素,但按重量计含锌量不得大于1%。含有其他元素时,按重量计含镍量应大于其他各种元素的单项含量。

4. Copper-nickel base alloys

Alloys of copper and nickel, with or without other elements but in any case containing by weight not more than 1% of zinc. When other elements are present, nickel predominates by weight over each of such other elements.

序号 No.	税则号列 Tariff Item	货品名称	最惠国税率(%) M.F.N.	协定税率(%) Agreement				普通税率(%) Gen.	Article Description
				东盟 ASEAN	曼谷协定 BA	香港 H.K.	澳门 Macao		
	74.01	**铜锍;沉积铜(泥铜):**							**Copper mattes; cement copper (precipitated copper):**
5005	7401.1000	-铜锍	2					11	-Copper mattes
5006	7401.2000	-沉积铜(泥铜)	2					11	-Cement copper (precipitated copped)
	74.02	**未精炼铜;电解精炼用的铜阳极:**							**Unrefined copper; copper anodes for electrolytic refining:**
5007	7402.0000	未精炼铜;电解精炼用的铜阳极	2					11	Unrefined copper; copper anodes for electrolytic refining
	74.03	**未锻轧的精炼铜及铜合金:**							**Refined copper and copper alloys, unwrought:**
		-精炼铜:							-Refined copper:
5008	7403.1100	--阴极及阴极型材	2					11	--Cathodes and sections of cathodes
5009	7403.1200	--线锭	2					11	--Wire-bars
5010	7403.1300	--坯段	2					11	--Billets
5011	7403.1900	--其他	2					11	--Other
		-铜合金:							-Copper alloys:
5012	7403.2100	--铜锌合金(黄铜)	1				0△	14	--Copper-zinc base alloys (brass)
5013	7403.2200	--铜锡合金(青铜)	1					17	--Copper-tin base alloys (bronze)
5014	7403.2300	--铜镍合金(白铜)或铜镍锌合金(德银)	1					35	--Copper-nickel base alloys (cupronickel) or copper-nickel-zinc base alloys (nickel silver)
5015	7403.2900	--其他铜合金(税号74.05的铜母合金除外)	1					17	--Other copper alloys (other than master alloys of heading No.74.05)

序号 No.	税则号列 Tariff Item	货品名称	最惠国税率(%) M.F.N.	协定税率(%)Agreement 东盟 ASEAN	曼谷协定 BA	香港 H.K.	澳门 Macao	普通税率(%) Gen.	Article Description
	74.04	**铜废碎料:**							**Copper waste and scrap:**
5016	7404.0000	铜废碎料	1.5			0		11	Copper waste and scrap
	74.05	**铜母合金:**							**Master alloys of copper:**
5017	7405.0000	铜母合金	4			0		17	Master alloys of copper
	74.06	**铜粉及片状粉末:**							**Copper powders and flakes:**
		-非片状粉末:							-Powders of non-lamellar structure:
5018	7406.1010	---精炼铜制	3					14	---Of refined copper
5019	7406.1020	---铜镍合金(白铜)或铜镍锌合金(德银)制	6					40	---Of copper-nickel base alloys (cupronickel) or copper-nickel-zinc base alloys (nickel silver)
5020	7406.1030	---铜锌合金(黄铜)制	6					30	---Of copper-zinc base alloys (brass)
5021	7406.1040	---铜锡合金(青铜)制	6					30	---Of copper-tin base alloys (bronze)
5022	7406.1090	---其他铜合金制	6					30	---Other
		-片状粉末:							-Powders of lamellar structure;flakes:
5023	7406.2010	---精炼铜制	4					14	---Of refined copper
5024	7406.2020	---铜镍合金(白铜)或铜镍锌合金(德银)制	6					40	---Of copper-nickel base alloys (cupronickel) or copper-nickel-zinc base alloys (nickel silver)
5025	7406.2090	---其他铜合金制	6					30	---Other
	74.07	**铜条、杆、型材及异型材:**							**Copper bars,rods and profiles:**
5026	7407.1000	-精炼铜制	4					14	-Of refined copper
		-铜合金制:							-Of copper alloys:
5027	7407.2100	--铜锌合金(黄铜)	7				0△	20	--Of copper-zinc base alloys (brass)
5028	7407.2200	--铜镍合金(白铜)或铜镍锌合金(德银)	7					40	--Of copper-nickel base alloys (cupronickel) or copper-nickel-zinc base alloys (nickel silver)
5029	7407.2900	--其他	7					20	--Other
	74.08	**铜丝:**							**Copper wire:**
		-精炼铜制:							-Of refined copper:
5030	7408.1100	--最大截面尺寸超过6毫米	4					14	--Of which the maximum cross-sectional dimension exceeds 6mm
5031	7408.1900	--其他	4			0		14	--Other
		-铜合金制:							-Of copper alloys:
5032	7408.2100	--铜锌合金(黄铜)	7					20	--Of copper-zinc base alloys (brass)
5033	7408.2200	--铜镍合金(白铜)或铜镍锌合金(德银)	8					40	--Of copper-nickel base alloys (cupronickel) or copper-nickel-zinc base alloys (nickel silver)
5034	7408.2900	--其他	7					20	--Other
	74.09	**铜板、片及带,厚度超过0.15毫米:**							**Copper plates, sheets and strip, of a thickness exceeding 0.15mm:**
		-精炼铜制:							-Of refined copper:
5035	7409.1100	--盘卷的	4					14	--In coils

序号 No.	税则号列 Tariff Item	货品名称	最惠国税率(%) M.F.N.	协定税率(%) Agreement 东盟 ASEAN	曼谷协定 BA	香港 H.K.	澳门 Macao	普通税率(%) Gen.	Article Description
5036	7409.1900	--其他	4					14	--Other
		-铜锌合金(黄铜)制:							-Of copper-zinc base alloys (brass):
5037	7409.2100	--盘卷的	7			0		20	--In coils
5038	7409.2900	--其他	7			0		20	--Other
		-铜锡合金(青铜)制:							-Of copper-tin base alloys (bronze):
5039	7409.3100	--盘卷的	7					20	--In coils
5040	7409.3900	--其他	7					20	--Other
5041	7409.4000	-铜镍合金(白铜)或铜镍锌合金(德银)制	7					40	-Of copper-nickel base alloys (cupronickel) or copper-nickel-zinc base alloys (nickel silver)
5042	7409.9000	-其他铜合金制	7			0		20	-Of other copper alloys
	74.10	**铜箔(不论是否印花或用纸、纸板、塑料或类似材料衬背),厚度(衬背除外)不超过0.15毫米:**							**Copper foil (whether or not printed or backed with paper, paperboard, plastics or similar backing materials) of a thickness (excluding any backing) not exceeding 0.15mm:**
		-无衬背:							-Not backed:
5043	7410.1100	--精炼铜制	4			0		14	--Of refined copper
		--铜合金制:							--Of copper alloys:
5044	7410.1210	---铜镍合金(白铜)或铜镍锌合金(德银)	7					40	---Of copper-nickel base alloys (cupronickel) or copper-nickel-zinc base alloys (nickel silver)
5045	7410.1290	---其他	7					20	---Other
		-有衬背:							-Backed:
5046	7410.2100	--精炼铜制	4			0	0	14	--Of refined copper
		--铜合金制:							--Of copper alloys:
5047	7410.2210	---铜镍合金(白铜)或铜镍锌合金(德银)	7					40	---Of copper-nickel base alloys (cupronickel) or copper-nickel-zinc base alloys (nickel silver)
5048	7410.2290	---其他	7					20	---Other
	74.11	**铜管:**							**Copper tubes and pipes:**
5049	7411.1000	-精炼铜制	4					14	-Of refined copper
		-铜合金制:							-Of copper alloys:
5050	7411.2100	--铜锌合金(黄铜)	7			0		20	--Of copper-zinc base alloys(brass)
5051	7411.2200	--铜镍合金(白铜)或铜镍锌合金(德银)	7					40	--Of copper-nickel base alloys (cupronickel) or copper-nickel-zinc base alloys (nickel silver)
5052	7411.2900	--其他	7					20	--Other
	74.12	**铜制管子附件(例如,接头、肘管、管套):**							**Copper tube or pipe fittings (for example, couplings, elbows, sleeves):**
5053	7412.1000	-精炼铜制	4					14	-Of refined copper
		-铜合金制:							-Of copper alloys:
5054	7412.2010	---铜镍合金(白铜)或铜镍锌合金(德银)	7					40	---Of copper-nickel base alloys (cupronickel) or copper-nickel-zinc base alloys (nickel silver)

序号 No.	税则号列 Tariff Item	货品名称	最惠国税率(%) M.F.N.	协定税率(%) Agreement 东盟 ASEAN	曼谷协定 BA	香港 H.K.	澳门 Macao	普通税率(%) Gen.	Article Description
5055	7412.2090	---其他	7					20	---Other
	74.13	**非绝缘的铜丝绞股线、缆、编带及类似品:**							**Stranded wire, cables, plaited bands and the like, of copper, not electrically insulated:**
5056	7413.0000	非绝缘的铜丝绞股线、缆、编带及类似品	5					14	Stranded wire, cables plaited bands and the like, of copper, not electrically insulated
	74.14	**铜丝制的布(包括环形带)、网、格栅、网眼铜板:**							**Cloth (including endless bands), grill and netting, of copper wire; expanded metal of copper:**
		-布:							-Cloth:
5057	7414.2010	---工业用	7					20	---For technical use
5058	7414.2090	---其他	15					70	---Other
		-其他:							-Other:
5059	7414.9010	---工业用	8					20	---For technical use
5060	7414.9090	---其他	15					70	---Other
	74.15	**铜制或钢铁制带铜头的钉、平头钉、图钉、U形钉(税号83.05的货品除外)及类似品;铜制螺钉、螺栓、螺母、钩头螺钉、铆钉、销、开尾销、垫圈(包括弹簧垫圈)及类似品:**							**Nails, tacks, drawing pins, staples (other than those of heading No. 83.05) and similar articles, of copper or of iron or steel with heads of copper; screws, bolts, nuts, screw hooks, rivets, cotters, cotter-pins, washers (including spring washers) and similar articles, of copper:**
5061	7415.1000	-钉、平头钉、图钉、U形钉及类似品	8					80	-Nails and tacks, drawing pins, staples and similar articles
		-其他无螺纹制品:							-Other articles, not threaded:
5062	7415.2100	--垫圈(包括弹簧垫圈)	10			0		80	--Washers (including spring washers)
5063	7415.2900	--其他	10			0		80	--Other
		-其他螺纹制品:							-Other threaded articles:
		--螺钉;螺栓及螺母:							--Screws; bolts and nuts
5064	7415.3310	---木螺钉	8			0		80	---Screws for wood
5065	7415.3390	---其他	8			0		80	---Other
5066	7415.3900	--其他	10			0		80	--Other
	74.16	**铜弹簧:**							**Copper springs:**
5067	7416.0000	铜弹簧	10					40	Copper springs
	74.17	**非电热的铜制家用烹饪或供暖器具及其零件:**							**Cooking or heating apparatus of a kind used for domestic purposes, non-electric, and parts thereof, of copper:**
5068	7417.0000	非电热的铜制家用烹饪或供暖器具及其零件	20					80	Cooking or heating apparatus of a kind used for domestic purposes, non-electric, and parts thereof, of copper

序号 No.	税则号列 Tariff Item	货品名称	最惠国税率 (%) M.F.N.	协定税率(%)Agreement				普通税率 (%) Gen.	Article Description
				东盟 ASEAN	曼谷协定 BA	香港 H.K.	澳门 Macao		
	74.18	**餐桌、厨房或其他家用铜制器具及其零件;铜制刷锅器、洗刷擦光用的块垫、手套及类似品;铜制卫生器具及其零件:**							**Table, kitchen or other household articles and parts thereof, of copper; pot scourers and scouring or polishing pads, gloves and the like, of copper; sanitary ware and parts thereof, of copper:**
		-餐桌、厨房或其他家用器具及其零件;擦锅器及洗刷擦光用的块垫、手套及类似品:							-Table, kitchen or other household articles and parts thereof; pot scourers and scouring or polishing pads, gloves and the like:
5069	7418.1100	--擦锅器及洗刷、擦光用的块垫、手套及类似品	18					80	--Pot scourers and scouring or polishing pads, gloves and the like
5070	7418.1900	--其他	18					80	--Other
5071	7418.2000	-卫生器具及其零件	18					80	-Sanitary ware and parts thereof
	74.19	**其他铜制品:**							**Other articles of copper:**
5072	7419.1000	-链条及其零件	14			0		80	-Chain and parts thereof
		-其他:							-Other:
		--铸造、模压、冲压或锻造,但未经进一步加工的:							--Cast, moulded, stamped or forged, but not further worked:
5073	7419.9110	---工业用	10					40	---For technical use
5074	7419.9190	---其他	20					80	---Other
		--其他:							--Other:
5075	7419.9910	---工业用	10					40	---For technical use
5076	7419.9990	---其他	20					80	---Other

第七十五章
镍及其制品

注释：

本章所用有关名词解释如下：

一、条、杆

轧、挤、拔或锻制的实心产品，非成卷的，其全长截面均为圆形、椭圆形、矩形（包括正方形）、等边三角形或规则外凸多边形（包括相对两边为弧拱形，另外两边为等长平行直线的“扁圆形”及“变形矩形”）。对于矩形（包括正方形）、三角形或多边形截面的产品，其全长边角可经磨圆。矩形（包括“变形矩形”）截面的产品，其厚度应大于宽度的十分之一。所述条、杆也包括同样形状及尺寸的铸造或烧结产品。该产品在铸造或烧结后再经加工（简单剪修或去氧化皮的除外），但不具有其他税号所列制品或产品的特征。

二、型材及异型材

轧、挤、拔、锻制的产品或其他成型产品，不论是否成卷，其全长截面相同，但与条、杆、丝、板、片、带、箔、管的定义不相符合。同时也包括同样形状的铸造或烧结产品。该产品在铸造或烧结后再经加工（简单剪修或去氧化皮的除外），但不具有其他税号所列制品或产品的特征。

三、丝

盘卷的轧、挤或拔制实心产品，其全长截面均为圆形、椭圆形、矩形（包括正方形）、等边三角形或规则外凸多边形（包括相对两边为弧拱形，另外两边为等长平行直线的“扁圆形”及“变形矩形”）。对于矩形（包括正方形）、三角形或多边形截面的产品，其全长边角可经磨圆。矩形（包括“变形矩形”）截面的产品，其厚度应大于宽度的十分之一。

Chapter 75
Nickel and articles thereof

Notes:

In this Chapter the following expressions have the meanings hereby assigned to them:

1. Bars and rods

   Rolled, extruded, drawn or forged products, not in coils, which have a uniform soild cross-section along their whole length in the shape of circles, ovals, rectangles (including squares), equilateral triangles or regular convex polygons (including “flattened circles” and “modified rectangles”, of which two opposite sides are convex arcs, the other two sides being straight, of equal length and parallel). Products with a rectangular (including square), triangular or polygonal cross-section may have corners rounded along their whole length. The thickness of such products which have a rectangular (including “modified rectangular”) cross-section exceeds one-tenth of the width. The expression also covers cast or sintered products, of the same forms and dimensions, which have been subsequently worked after production (otherwise than by simple trimming or de-scaling), provided that they have not thereby assumed the character of articles or products of other headings.

2. Profiles

   Rolled, extruded, drawn, forged or formed products, coiled or not, of a uniform cross-section along their whole length, which do not conform to any of the definitions of bars, rods, wire, plates, sheets, strip, foil, tubes or pipes. The expression also covers cast or sintered products, of the same forms, which have been subsequently worked after production (otherwise than by simple trimming or de-scaling), provided that they have not thereby assumed the character of articles or products of other headings.

3. Wire

   Rolled, extruded or drawn products, in coils, which have a uniform soild cross-section along their whole length in the shape of circles, ovals, rectangles (including squares), equilateral triangles or regular convex polygons (including “flattened circles” and “modified rectangles”, of which two opposite sides are convex arcs, the other two sides being straight, of equal length and parallel). Products with a rectangular (including square), triangular or polygonal cross-section may have corners rounded along their whole length. The thickness of such products which have a rectangular (including “modified rectangular") cross-section exceeds one-tenth of the width.

四、板、片、带、箔

成卷或非成卷的平面产品（税号75.02的未锻轧产品除外），截面均为厚度相同的实心矩形（不包括正方形），不论边角是否磨圆（包括相对两边为弧拱形，另外两边为等长平行直线的"变形矩形"），并且符合以下规格：

1. 矩形（包括正方形）的，厚度不超过宽度的十分之一；
2. 矩形或正方形以外形状的，任何尺寸，但不具有其他税号所列制品或产品的特征。

税号75.06还适用于具有花样（例如，凹槽、肋条形、格槽、珠粒及菱形）的板、片、带、箔以及穿孔、抛光、涂层或制成瓦楞形的这类产品，但不具有其他税号所列制品或产品的特征。

4. Plates, sheets, strip and foil

Flat-surfaced products (other than the unwrought products of heading No. 75.02), coiled or not, of soild rectangular (other than square) cross-section with or without rounded corners (including "modified rectangles" of which two opposite sides are convex arcs, the other two sides being straight, of equal length and parallel) of a uniform thickness, which are:

-of rectanglular (including square) shape with a thickness not exceeding one-tenth of the width.

-of a shape other than rectangular or square, of any size, provided that they do not assume the character of articles or products of other headings.

Heading No. 75.06 applies, *inter alia*, to plates, sheets, strip and foil with patterns (for example, grooves, ribs, chequers, tears, buttons, lozenges) and to such products which have been perforated, corrugated, polished or coated, provided that they do not thereby assume the character of articles or products of other headings.

五、管

全长截面及管壁厚度相同并只有一个闭合空间的空心产品，成卷或非成卷的，其截面为圆形、椭圆形、矩形（包括正方形）、等边三角形或规则外凸多边形。对于截面为矩形（包括正方形）、等边三角形或规则外凸多边形的产品，不论全长边角是否磨圆，只要其内外截面为同一圆心并为同样形状及同一轴向，也可视为管子。上述截面的管子可经抛光、涂层、弯曲、攻丝、钻孔、缩腰、胀口、成锥形或装法兰、颈圈或套环。

5. Tubes and pipes

Hollow products, coiled or not, which have a uniform cross-section with only one enclosed void along their whole length in the shape of circles, ovals, rectangles (including squares), equilateral triangles of regular convex polygons, and which have a uniform wall thickness. Products with a rectangular (including square), equilateral triangular, or regular convex polygonal cross-section, which may have corners rounded along their whole length, are also to be considered as tubes and pipes provided the inner and outer cross-sections are concentric and have the same form and orientation. Tubes and pipes of the foregoing cross-sections may be polished, coated, bent, threaded, drilled, waisted, expanded, cone-shaped or fitted with flanges, collars or rings.

子目注释：

一、本章所用有关名词解释如下：

（一）非合金镍

按重量计镍及钴的含量至少为99%的金属，但：

1. 按重量计含钴量不超过1.5%；
2. 按重量计其他各种元素的含量不超过下表中规定的限量：

Subheading Notes:

1. In this Chapter the following expressions have the meanings hereby assigned to them:

(a) Nickel, not alloyed

Metal containing by weight at least 99% of nickel plus cobalt, provided that:

(1) the cobalt by weight does not exceed 1.5%, and

(2) the content by weight of any other element does not exceed the limit specified in the following table:

其他元素表

元　　素		所含重量百分比
Fe	铁	0.5
O	氧	0.4
其他元素，每种		0.3

TABLE-Other elements

Element		Limiting content % by weight
Fe	Iron	0.5
O	Oxygen	0.4
Other elements, each		0.3

（二）镍合金

按重量计含镍量大于其他元素单项含量的金属物质，但：

1. 按重量计含钴量超过1.5%；
2. 按重量计至少有一种其他元素的含量超过上表中规定的限量；或
3. 除镍及钴以外，按重量计其他元素的总含量超过1%。

(b) Nickel alloys

Metallic substances in which nickel predominates by weight over each of the other elements provided that:

(1) the content by weight of cobalt exceeds 1.5%,
(2) the content by weight of at least one of the other elements is greater than the limit specified in the foregoing table, or
(3) the total content by weight of elements other than nickel plus cobalt exceeds 1%.

二、子目号7508.10所称"丝"，不受本章注释三的限制，仅适用于截面尺寸不超过6毫米的任何截面形状的产品，不论是否盘卷。

2. Notwithstanding the provisions of Chapter Note 1 (c), for the purposes of subheading No.7508.10 the term "wire" applies only to products, whether or not in coils, of any cross-sectional shape, of which no cross-sectional dimension exceeds 6mm.

序号 No.	税则号列 Tariff Item	货品名称	最惠国税率(%) M.F.N.	协定税率(%)Agreement 东盟 ASEAN	曼谷协定 BA	香港 H.K.	澳门 Macao	普通税率(%) Gen.	Article Description
	75.01	**镍锍、氧化镍烧结物及镍冶炼的其他中间产品：**							**Nickel mattes, nickel oxide sinters and other intermediate products of nickel metallurgy:**
5077	7501.1000	-镍锍	3					11	-Nickel mattes
5078	7501.2000	-氧化镍烧结物及镍冶炼的其他中间产品	3					11	-Nickel oxide sinters and other intermediate products of nickel metallurgy
	75.02	**未锻轧镍：**							**Unwrought nickel:**
5079	7502.1000	-非合金镍	3					11	-Nickel, not alloyed
5080	7502.2000	-镍合金	3					11	-Nickel, alloys
	75.03	**镍废碎料：**							**Nickel waste and scrap:**
5081	7503.0000	镍废碎料	1.5					11	Nickel waste and scrap
	75.04	**镍粉及片状粉末：**							**Nickel powders and flakes:**
5082	7504.0010	---非合金镍粉及片状粉末	4					17	---Nickel powders and flakes, not alloyed
5083	7504.0020	---合金镍粉及片状粉末	4					17	---Nickel powders and flakes, alloys
	75.05	**镍条、杆、型材及异型材或丝：**							**Nickel bars, rods, profiles and wire:**
		-条、杆、型材及异型材：							-Bars, rods and profiles:
5084	7505.1100	--非合金镍制	6					14	--Of nickel, not alloyed
5085	7505.1200	--镍合金制	6					14	--Of nickel alloys
		-丝：							-Wire:
5086	7505.2100	--非合金镍制	6					17	--Of nickel, not alloyed

序号 No.	税则号列 Tariff Item	货品名称	最惠国税率(%) M.F.N.	协定税率(%)Agreement 东盟 ASEAN	曼谷协定 BA	香港 H.K.	澳门 Macao	普通税率(%) Gen.	Article Description
5087	7505.2200	--镍合金制	6					17	--Of nickel alloys
	75.06	**镍板、片、带、箔:**							**Nickel plates, sheets, strip and foil:**
5088	7506.1000	-非合金镍制	6					14	-Of nickel, not alloyed
5089	7506.2000	-镍合金制	6					14	-Of nickel alloys
	75.07	**镍管及管子附件(例如,接头、肘管、管套):**							**Nickel tubes, pipes and tube or pipe fittings (for example, couplings, elbows, sleeves):**
		-镍管:							-Tubes and pipes:
5090	7507.1100	--非合金镍制	6					17	--Of nickel, not alloyed
5091	7507.1200	--镍合金制	6					17	--Of nickel alloys
5092	7507.2000	-管子附件	6					17	-Tube or pipe fittings
	75.08	**其他镍制品:**							**Other articles of nickel:**
		-镍丝布、网及格栅:							-Cloth, grill and netting, of nickel wire:
5093	7508.1010	---镍丝布	6					20	---Wire cloth
5094	7508.1080	---其他工业用镍制品	6					40	---Other articles of nickel, for technical use
5095	7508.1090	---其他	6					70	---Other
		-其他:							-Other:
5096	7508.9010	---电镀用镍阳极	4					14	---Electroplating anodes
5097	7508.9080	---其他工业用镍制品	6					40	---Other articles of nickel, for technical use
5098	7508.9090	---其他	6					70	---Other

第七十六章
铝及其制品

Chapter 76
Aluminium and articles thereof

注释：

本章所用有关名词解释如下：

一、条、杆

轧、挤、拔或锻制的实心产品，非成卷的，其全长截面均为圆形、椭圆形、矩形（包括正方形）、等边三角形或规则外凸多边形（包括相对两边为弧拱形，另外两边为等长平行直线的“扁圆形”及“变形矩形”）。对于矩形（包括正方形）、三角形或多边形截面的产品，其全长边角可经磨圆。矩形（包括“变形矩形”）截面的产品其厚度应大于宽度的十分之一。所述条、杆也包括同样形状及尺寸的铸造或烧结产品。该产品在铸造或烧结后再经加工（简单剪修或去氧化皮的除外），但不具有其他税号所列制品或产品的特征。

二、型材及异型材

轧、挤、拔、锻制的产品或其他成型产品，不论是否成卷，其全长截面相同，但与条、杆、丝、板、片、带、箔、管的定义不相符合。同时也包括同样形状的铸造或烧结产品。该产品在铸造或烧结后再经加工（简单剪修或去氧化皮的除外），但不具有其他税号所列制品或产品的特征。

三、丝

盘卷的轧、挤或拔制实心产品，其全长截面均为圆形、椭圆形、矩形（包括正方形）等边三角形或规则外凸多边形（包括相对两边为弧拱形，另外两边为等长平行直线的“扁圆形”及“变形矩形”）。对于矩形（包括正方形）、三角形或多边形截面的产品，其全长边角可经磨圆。矩形（包括“变形矩形”）截面的产品，其厚度应大于宽度的十分之一。

Notes:

In this Chapter the following expressions have the meanings hereby assigned to them:

1. Bars and rods

Rolled, extruded, drawn or forged products, not in coils, which have a uniform solid cross-section along their whole length in the shape of circles, ovals, rectangles (including squares), equilateral triangles or regular convex polygons (including "flattened circles" and "modified rectangles", of which two opposite sides are convex arcs, the other two sides being straight, of equal length and parallel). Products with a rectangular (including square), triangular or polygonal cross-section may have corners rounded along their whole length. The thickness of such products which have a rectangular (including "modified rectangular") cross-section exceeds one-tenth of the width. The expression also covers cast or sintered products, of the same forms and dimensions, which have been subsequently worked after production (otherwise than by simple trimming or de-scaling), provided that they have not thereby assumed the character of articles or products of other headings.

2. Profiles

Rolled, extruded, drawn, forged or formed products, coiled or not, of a uniform cross-section along their whole length, which do not conform to any of the definitions of bars, rods, wire, plates, sheets, strip, foil, tubes or pipes. The expression also covers cast or sintered products, of the same forms, which have been subsequently worked after production (otherwise than by simple trimming or de-scaling), provided that they have not thereby assumed the character of articles or products of other headings.

3. Wire

Rolled, extruded or drawn products, in coils, which、 have a uniform soild cross-section along their whole length in the shape of circles, ovals, rectangles (including squares), equilateral triangles or regular convex polygons (including "flattened circles" and "modified rectangles", of which two opposite sides are convex arcs, the other two sides being straight, of equal length and parallel). Products with a rectangular (including square), triangular or polygonal cross-section may have corners rounded along their whole length. The thickness of such products which have a rectangular (including "modified rectangular") cross-section exceeds one-tenth of the width.

四、板、片、带、箔

成卷或非成卷的平面产品（税号 76.01 的未锻轧产品除外），截面均为厚度相同的实心矩形（不包括正方形），不论边角是否磨圆（包括相对两边为弧拱形，另外两边为等长平行直线的"变形矩形"），并且符合以下规格：

1. 矩形（包括正方形）的，厚度不超过宽度的十分之一；
2. 矩形或正方形以外形状的，任何尺寸，但不具有其他税号所列制品或产品的特征。

税号 76.06 和 76.07 还适用于具有花样（例如，凹槽、肋条形、格槽、珠粒及菱形）的板、片、带、箔以及穿孔、抛光、涂层或制成瓦楞形的这类产品，但不具有其他税号所列制品或产品的特征。

4. Plates, sheets, strip and foil

Flat-surfaced products (other than the unwrought products of heading No. 76.01), coiled or not, of solid rectangular (other than square) cross-section with or without rounded corners (including "modified rectangles" of which two opposite sides are convex arcs, the other two sides being straight, of equal length and parallel) of a uniform thickness, which are:

-of rectangular (including square) shape with a thickness not exceeding one-tenth of the width,

-of a shape other than rectangular or square, of any size, provided that they do not assume the character of articles or products of other headings.

Headings No. 76.06 and 76.07 apply, *inter alia*, to plates, sheets, strip and foil with patterns (for example, grooves, ribs, chequers, tears, buttons, lozenges) and to such products which have been perforated, corrugated, polished or coated, provided that they do not thereby assume the character of articles or products of other headings.

五、管

全长截面及管壁厚度相同并只有一个闭合空间的空心产品，成卷或非成卷的，其截面为圆形、椭圆形、矩形（包括正方形）、等边三角形或规则外凸多边形。对于截面为矩形（包括正方形）、等边三角形或规则外凸多边形的产品，不论全长边角是否磨圆，只要其内外截面为同一圆心并为同样形状及同一轴向，也可视为管子。上述截面的管子可经抛光、涂层、弯曲、攻丝、钻孔、缩腰、胀口、成锥形或装法兰、颈圈或套环。

5. Tubes and pipes

Hollow products, coiled or not, which have a uniform cross-section with only one enclosed void along their whole length in the shape of circles, ovals, rectangles (including squares), equilateral triangles or regular convex polygons, and which have a uniform wall thickness. Products with a rectangular (including square), equilateral triangular or regular convex polygonal cross-section, which may have corners rounded along their whole length, are also to be considered as tubes and pipes provided the inner and outer cross-sections are concentric and have the same form and orientation. Tubes and pipes of the foregoing cross-sections may be polished, coated, bent, threaded, drilled, waisted, expanded, cone-shaped or fitted with flanges, collars or rings.

子目注释：

一、本章所用有关名词解释如下：

（一）非合金铝

按重量计含铝量至少为 99% 的金属，但其他各种元素的含量不超过下表中规定的限量：

Subheading Notes:

1. In this Chapter the following expressions have the meanings hereby assigned to them:

(a) Aluminium, not alloyed

Metal containing by weight at least 99% of aluminium, provided that the content by weight of any other element does not exceed the limit specified in he following table:

其他元素表

元　　素	所含重量百分比
Fe+Si（铁+硅）	1
其他元素（1），每种	0.1（2）

（1）其他元素，例如，铬、铜、镁、锰、镍、锌。

（2）含铜成分可大于 0.1%，但不得大于 0.2%，且铬和锰的含量均不得超过 0.05%。

TABLE-Other elements

Element	Limiting content % by weight
Fe+Si（iron plus silicon）	1
Other elements（1），each	0.1（2）

（1）Other elements are，for example，Cr，Cu，Mg，Mn，Ni，Zn.

（2）Copper is permitted in a proportion greater than 0.1% but not more than 0.2%，provided that neither the chromium nor manganese content exceeds 0.05%.

（二）铝合金

按重量计含铝量大于其他元素单项含量的金属物质，但：

1. 按重量计至少有一种其他元素或铁加硅的含量大于上表中规定的限量；或

2. 按重量计其他元素的总含量超过 1%。

（b）Aluminium alloys

Metallic substances in which aluminium predominates by weight over each of the other elements，provided that：

（1）the content by weight of at least one of the other elements or of iron plus silicon taken together is greater than the limit specified in the foregoing table；or

（2）the total content by weight of such other elements exceeds 1%.

二、子目号 7616.91 所称"丝"，不受本章注释三的限制，仅适用于截面尺寸不超过 6 毫米的任何截面形状的产品，不论是否盘卷。

2. Notwithstanding the provisions of Chapter Note 1（c），for the purposes of subheading No.7616.91 the term "wire" applies only to products，whether or not in coils，of any cross-sectional shape，of which no cross-sectional dimension exceeds 6mm.

序号 No.	税则号列 Tariff Item	货品名称	最惠国税率（%）M.F.N.	协定税率（%）Agreement 东盟 ASEAN	曼谷协定 BA	香港 H.K.	澳门 Macao	普通税率（%）Gen.	Article Description
	76.01	未锻轧铝：							**Unwrought aluminium：**
5099	7601.1000	-非合金铝	5			0		14	-Aluminium，not alloyed
5100	7601.2000	-铝合金	7				0△	14	-Aluminium alloys
	76.02	铝废碎料：							**Aluminium waste and scrap：**
5101	7602.0000	铝废碎料	1.5			0		14	Aluminium waste and scrap
	76.03	铝粉及片状粉末：							**Aluminium powders and flakes：**
5102	7603.1000	-非片状粉末	6					30	-Powders of non-lamellar structure
5103	7603.2000	-片状粉末	7					30	-Powders of lamellar structure；flakes
	76.04	铝条、杆、型材及异型材：							**Aluminium bars，rods and profiles：**
5104	7604.1000	-非合金铝制	5					30	-Of aluminium，not alloyed
		-铝合金制：							-Of aluminium alloys：
5105	7604.2100	--空心异型材	5					30	--Hollow profiles
5106	7604.2900	--其他	5					30	--Other
	76.05	铝丝：							**Aluminium wire：**

序号 No.	税则号列 Tariff Item	货品名称	最惠国税率(%) M.F.N.	协定税率(%) Agreement 东盟 ASEAN	曼谷协定 BA	香港 H.K.	澳门 Macao	普通税率(%) Gen.	Article Description
		-非合金铝制:							-Of aluminium, not alloyed:
5107	7605.1100	--最大截面尺寸超过7毫米	8					17	--Of which the maximum cross-sectional dimension exceeding 7mm
5108	7605.1900	--其他	8			0△		17	--Other
		-铝合金制:							-Of aluminium alloys:
5109	7605.2100	--最大截面尺寸超过7毫米	8					17	--Of which the maximum cross-sectional dimension exceeding 7mm
5110	7605.2900	--其他	8					17	--Other
	76.06	**铝板、片及带,厚度超过0.2毫米:**							**Aluminium plates, sheets and strip, of a thickness exceeding 0.2mm:**
		-矩形(包括正方形):							-Rectangular(including square):
		--非合金铝制:							--Of aluminium, not alloyed:
5111	7606.1120	---厚度在0.30毫米及以上,但不超过0.36毫米	6					50	---Of a thickness of 0.30mm or more but not exceeding 0.36mm
5112	7606.1190	---其他	6			0		30	---Other
		--铝合金制:							--Of aluminium alloys:
5113	7606.1220	---厚度小于0.28毫米	6		5	0		30	---Of a thickness less than 0.28mm
5114	7606.1230	---厚度在0.28毫米及以上,但不超过0.35毫米	6		5	0		30	---Of a thickness of 0.28mm or more but not exceeding 0.35mm
5115	7606.1240	---厚度超过0.35毫米	6		5.5	0		50	---Of a thickness exceeding 0.35mm
		-其他:							-Other:
5116	7606.9100	--非合金铝制	6			0		30	--Of aluminium, not alloyed
5117	7606.9200	--铝合金制	10			0		30	--Of aluminium alloys
	76.07	**铝箔(不论是否印花或用纸、纸板、塑料或类似材料衬背),厚度(衬背除外)不超过0.2毫米:**							**Aluminium foil(whether or not printed or backed with paper, paperboard, plastics or similar backing materlals) of a thickness (excluding any backing) not exceeding 0.2mm:**
		-无衬背:							-Not backed:
		--轧制后未经进一步加工的:							--Rolled but not further worked:
5118	7607.1110	---厚度不超过0.007毫米	6		5.7			35	---Of a thickness not exceeding 0.007mm
5119	7607.1190	---其他	6		5.7			35	---Other
5120	7607.1900	--其他	6					35	--Other
5121	7607.2000	-有衬背	6			0		35	-Backed
	76.08	**铝管:**							**Aluminium tubes and pipes:**
5122	7608.1000	-非合金铝制	8					30	-Of aluminium, not alloyed
5123	7608.2000	-铝合金制	8					30	-Of aluminium alloys
	76.09	**铝制管子附件(例如,接头、肘管、管套):**							**Aluminium tube or pipe fittings(for example, couplings, elbows, sleeves):**
5124	7609.0000	铝制管子附件(例如,接头、肘管、管套)	8					35	Aluminium tube or pipe fittings(for example, couplings, elbows, sleeves)

序号 No.	税则号列 Tariff Item	货品名称	最惠国税率(%) M.F.N.	协定税率(%) Agreement 东盟 ASEAN	曼谷协定 BA	香港 H.K.	澳门 Macao	普通税率(%) Gen.	Article Description
	76.10	**铝制结构体(税号94.06的活动房屋除外)及其部件(例如,桥梁及桥梁体段、塔、格构杆、屋顶、屋顶框架、门窗及其框架、门槛、栏杆、支柱及立柱);上述结构体用的已加工铝板、杆、型材、异型材、管子及类似品:**							**Aluminium structures(excluding pre-fabricated buildings of heading No. 94.06)and parts of structures(for example, bridges and bridge-sections, towers, lattice masts, roofs, roofing frameworks, doors and windows and their frames and thresholds for doors, balustrades, pillars and columns); aluminium plates, rods, profiles, tubes and the like, prepared for use in structures:**
5125	7610.1000	-门窗及其框架、门槛	25					80	-Doors, windows and their frames and thresholds for doors
5126	7610.9000	-其他	6					50	-Other
	76.11	**盛装物料用的铝制囤、柜、罐、桶及类似容器(装压缩气体或液化气体的除外),容积超过300升,不论是否衬里或隔热,但无机械或热力装置:**							**Aluminium reservoirs, tanks, vats and similar containers, for any material (other than compressed or liquefied gas), of a capacity exceeding 300 l, whether or not lined or heat-insulated, but not fitted with mechanical or thermal equipment:**
5127	7611.0000	盛装物料用的铝制囤、柜、罐、桶及类似容器(装压缩气体或液化气体的除外),容积超过300升,不论是否衬里或隔热,但无机械或热力装置	12					35	Aluminium reservoirs, tanks, vats and similar containers, for any material (other than compressed or liquefied gas), of a capacity exceeding 300 l, whether or not lined or heat-insulated, but not fitted with mechanical or thermal equipment
	76.12	**盛装物料用的铝制桶、罐、听、盒及类似容器,包括软管容器及硬管容器(装压缩气体或液化气体的除外),容积不超过300升,不论是否衬里或隔热,但无机械或热力装置:**							**Aluminium casks, drums, cans, boxes and similar containers(including rigid or collapsible tubular containers), for any material (other than compressed or liquefied gas), of a capacity not exceeding 300 l, whether or not lined or heat-insulated, but not fitted with mechanical or thermal equipment:**
5128	7612.1000	-软管容器	12					50	-Collapsible tubular containers
		-其他:							-Other:
5129	7612.9010	---易拉罐及罐体	30					100	---Tear tab ends and bodies thereof
5130	7612.9090	---其他	12					70	---Other
	76.13	**装压缩气体或液化气体用的铝制容器:**							**Aluminium containers for compressed or liquefied gas:**
5131	7613.0010	---零售包装用	12					70	---For retail packing
5132	7613.0090	---其他	6					17	---Other

序号 No.	税则号列 Tariff Item	货品名称	最惠国税率(%) M.F.N.	协定税率(%)Agreement 东盟 ASEAN	曼谷协定 BA	香港 H.K.	澳门 Macao	普通税率(%) Gen.	Article Description
	76.14	**非绝缘的铝制绞股线、缆、编带及类似品:**							**Stranded wire, cables, plaited bands and the like, of aluminium, not electrically insulated:**
5133	7614.1000	-带钢芯的	6					20	-With steel core
5134	7614.9000	-其他	6					20	-Other
	76.15	**餐桌、厨房或其他家用铝制器具及其零件;铝制擦锅器、洗刷擦光用的块垫、手套及类似品;铝制卫生器具及其零件:**							**Table, kitchen or other household articles and parts thereof, of aluminium; pot scourers and scouring or polishing pads, gloves and the like, of aluminium; sanitary ware and parts thereof, of aluminium:**
		-餐桌、厨房或其他家用器具及其零件;擦锅器及洗刷擦光用的块垫、手套及类似品:							-Table, kitchen or other household articles and parts thereof; pot scourers and scouring or polishing pads, gloves and the like:
5135	7615.1100	--擦锅器、洗刷、擦光用的块垫、手套及类似品	18					90	--Pot scourers and scouring or polishing pads, gloves and the like
5136	7615.1900	--其他	15					90	--Other
5137	7615.2000	-卫生器具及其零件	18					90	-Sanitary ware and parts thereof
	76.16	**其他铝制品:**							**Other articles of aluminium:**
5138	7616.1000	-钉、平头钉、U形钉(税号83.05的货品除外)、螺钉、螺栓、螺母、钩头螺钉、铆钉、销、开尾销、垫圈及类似品	10			0		80	-Nails, tacks, staples (other than those of heading No. 83.05), screws, bolts, nuts, screw hooks, rivets, cotters, cotter-pins, washers and similar articles
		其他:							-Other:
		--铝丝制的布、网、篱及格栅:							--Cloth, grill, netting and fencing, of aluminium wire:
5139	7616.9110	---工业用	10					40	---For technical use
5140	7616.9190	---其他	18					80	---Other
		--其他:							--Other:
5141	7616.9910	---工业用	10					40	---For technical use
5142	7616.9990	---其他	15					80	---Other

第七十八章
铅及其制品

注释：

本章所用有关名词解释如下：

一、条、杆

轧、挤、拔或锻制的实心产品，非成卷的，其全长截面均为圆形、椭圆形、矩形（包括正方形）、等边三角形或规则外凸多边形（包括相对两边为弧拱形，另外两边为等长平行直线的“扁圆形”及“变形矩形”）。对于矩形（包括正方形）、三角形或多边形截面的产品，其全长边角可经磨圆。矩形（包括“变形矩形”）截面的产品，其厚度应大于宽度的十分之一。所述条、杆也包括同样形状及尺寸的铸造或烧结产品。该产品在铸造或烧结后再经加工（简单剪修或去氧化皮的除外），但不具有其他税号所列制品或产品的特征。

二、型材及异型材

轧、挤、拔、锻制的产品或其他成型产品，不论是否成卷，其全长截面相同，但与条、杆、丝、板、片、带、箔、管的定义不相符合。同时也包括同样形状的铸造或烧结产品。该产品在铸造或烧结后再经加工（简单剪修或去氧化皮的除外），但不具有其他税号所列制品或产品的特征。

三、丝

盘卷的轧、挤或拔制实心产品，其全长截面均为圆形、椭圆形、矩形（包括正方形）、等边三角形或规则外凸多边形（包括相对两边为弧拱形，另外两边为等长平行直线的“扁圆形”及“变形矩形”）。对于矩形（包括正方形）、三角形或多边形截面的产品，其全长边角可经磨圆。矩形（包括“变形矩形”）截面的产品，其厚度应大于宽度的十分之一。

Chapter 78
Lead and articles thereof

Notes:

In this Chapter the following expressions have the meanings hereby assigned to them:

1. Bars and rods

Rolled, extruded, drawn or forged products, not in coils, which have a uniform solid cross-section along their whole length in the shape of circles, ovals, rectangles (including squares), equilateral triangles or regular convex polygons (including “flattened circles” and “modified rectangles”, of which two opposite sides are convex arcs, the other two sides being straight, of equal length and parallel). Products with a rectangular (including square), triangular or polygonal cross-section may have corners rounded along their whole length. The thickness of such products which have a rectangular (including “modified rectangular”) cross-section exceeds one-tenth of the width. The expression also covers cast or sintered products, of the same forms and dimensions, which have been subsequently worked after production (otherwise than by simple trimming or de-scaling), provided that they have not thereby assumed the character of articles or products of other headings.

2. Profiles

Rolled, extruded, drawn, forged or formed products, coiled or not, of a uniform cross-section along their whole length, which do not conform to any of the definitions of bars, rods, wire, plates, sheets, strip, foil, tubes or pipes. The expression also covers cast or sintered products, of the same forms, which have been subsequently worked after production (otherwise than by simple trimming of de-scaling), provided that they have not thereby assumed the character of articles or products of other headings.

3. Wire

Rolled, extruded or drawn products, in coils, which have a uniform soild cross-section along their whole length in the shape of circles, ovals, rectangles (including squares), equilateral triangles or regular convex polygons (including “flattened circles” and “modified rectangles”, of which two opposite sides are convex arcs, the other two sides being straight, of equal length and parallel). Products with a rectangular (including square), triangular or polygonal cross-section may have corners rounded along their whole length. The thickness of such products which have a rectangular (including “modified rectangular”) cross-section

exceeds one-tenth of the width.

四、板、片、带、箔

成卷或非成卷的平面产品（税号 78.01 的未锻轧产品除外），截面均为厚度相同的实心矩形（不包括正方形），不论边角是否磨圆（包括相对两边为弧拱形，另外两边为等长平行直线的“变形矩形”），并且符合以下规格：

1. 矩形（包括正方形）的，厚度不超过宽度的十分之一；
2. 矩形或正方形以外形状的，任何尺寸，但不具有其他税号所列制品或产品的特征。

税号 78.04 还适用于具有花样（例如，凹槽、肋条形、格槽、珠粒及菱形）的板、片、带、箔以及穿孔、抛光、涂层或制成瓦楞形的这类产品，但不具有其他税号所列制品或产品的特征。

4. Plates, sheets, strip and foil

Flat-surfaced products (other than the unwrought products of heading No. 78.01), coiled or not, of solid rectangular (other than square) cross-section with or without rounded corners (including "modified rectanges" of which two opposite sides are convex arcs, the other two sides being straight, of equal length and parallel) of a uniform thickness, which are:

-of rectangular (including square) shape with a thickness not exceeding, one-tenth of the width,

-of a shape other than rectangular or square, of any size, proviede that they do not assume the character of articles or products of other headings.

Heading No. 78.04 applies, *inter alia*, to plates, sheets, strip and foil with patterns (for example, grooves, ribs, chequers, tears, buttons, lozenges) and to such products which have been perforated, corrugated, polished or coated, provided that they do not thereby assume the character of articles or products of other headings.

五、管

全长截面及管壁厚度相同并只有一个闭合空间的实心产品，成卷或非成卷的，其截面为圆形、椭圆形、矩形（包括正方形）、等边三角形或规则外凸多边形。对于截面为矩形（包括正方形）、等边三角形或规则外凸多边形的产品，不论全长边角是否磨圆，只要其内外截面为同一圆心并为同样形状及同一轴向，也可视为管子。上述截面的管子可经抛光、涂层、弯曲、攻丝、钻孔、缩腰、胀口、成锥形或装法兰、颈圈或套环。

5. Tubes and pipes

Hollow products, coiled or not, which have a uniform cross-section with only one enclosed void along their whole length in the shape of circles, ovals, rectangles (including squares), equilateral triangles or regular convex polygons, and which have a uniform wall thickness. Products with a rectangular (including square), equilateral triangular or regular convex polygonal cross-section, which may have corners rounded along their whole length, are also to be considered as tubes and pipes provided the inner and outer cross-sections are concentric and have the same form and orientation. Tubes and pipes of the foregoing cross-sections may be polished, coated, bent, threaded, drilled, waisted, expanded, cone-shaped or fitted with flanges, collars or rings.

子目注释：

本章所称“精炼铅”，是指：

按重量计含铅量至少为 99.9% 的金属，但其他各种元素的含量不超过下表中规定的限量：

Subheading Note:

In this Chapter the expression "refined lead" means:

Metal containing by weight at least 99.9% of lead, provided that the content by weight of any other element does not exceed the limit specified in the following table:

其他元素表

元　　素	所含重量百分比
Ag　银	0.02
As　砷	0.005
Bi　铋	0.05
Ca　钙	0.002
Cd　镉	0.002
Cu　铜	0.08
Fe　铁	0.002
S　硫	0.002
Sb　锑	0.005
Sn　锡	0.005
Zn　锌	0.002
其他(例如碲),每种	0.001

TABLE—Other elements

Element	Limiting conten % by weight
Ag　Silver	0.02
As　Arsenic	0.005
Bi　Bismuth	0.05
Ca　Calcium	0.002
Cd　Cadmium	0.002
Cu　Copper	0.08
Fe　Iron	0.002
S　Sulphur	0.002
Sb　Antimony	0.005
Sn　Tin	0.005
Zn　Zinc	0.002
Other(for example Te),each	0.001

序号 No.	税则号列 Tariff Item	货品名称	最惠国税率(%) M.F.N.	协定税率(%)Agreement 东盟 ASEAN	曼谷协定 BA	香港 H.K.	澳门 Macao	普通税率(%) Gen.	Article Description
	78.01	**未锻轧铅:**							**Unwrought lead:**
5143	7801.1000	-精炼铅	3					20	-Refined lead
		-其他:							-Other:
5144	7801.9100	--按重量计所含其他元素是以锑为主的	3					20	--Containing by weight antimony as the principal other elenent
5145	7801.9900	--其他	3					20	--Other
	78.02	**铅废碎料:**							**Lead waste and scrap:**
5146	7802.0000	铅废碎料	1.5					10	Lead waste and scrap
	78.03	**铅条、杆、型材及异型材或丝:**							**Lead bars,rods,profiles and wire:**
5147	7803.0000	铅条、杆、型材及异型材或丝	6					30	Lead bars,rods,profiles and wire
	78.04	**铅板、片、带、箔;铅粉及片状粉末:**							**Lead plates sheets,strip and foil;lead powders and flakes:**
		-板、片、带、箔:							-Plates,sheets,strip and foil:
5148	7804.1100	--片、带及厚度(衬背除外)不超过0.2毫米的箔	6					30	--Sheets,strip and foil of a thickness (excluding any backing)not exceeding 0.2mm
5149	7804.1900	--其他	6					30	--Other
5150	7804.2000	-粉末及片状粉末	6					35	-Powders and flakes
	78.05	**铅管及管子附件(例如,接头、肘管、管套):**							**Lead tubes,pipes and tube or pipe fittings (for example,couplings,elbows,sleeves):**
5151	7805.0010	---铅管	6					30	---Lead tubes or pipes
5152	7805.0020	---铅制管子附件(例如,接头、肘管、管套)	6					30	---Lead tube or pipe fittings (for example,couplings,elbows,sleeves)
	78.06	**其他铅制品:**							**Other articles of lead:**
5153	7806.0010	---工业用	6					40	---For technical use
5154	7806.0090	---其他	6					80	---Other

第七十九章
锌及其制品

注释：

本章所用名词解释如下：

一、条、杆

轧、挤、拔或锻制的实心产品，非成卷的，其全长截面均为圆形、椭圆形、矩形（包括正方形）、等边三角形或规则外凸多边形（包括相对两边为弧拱形，另外两边为等长平行直线的“扁圆形”及“变形矩形”）。对于矩形（包括正方形）、三角形或多边形截面的产品，其全长边角可经磨圆。矩形（包括“变形矩形”）截面的产品，其厚度应大于宽度的十分之一。所述条、杆也包括同样形状及尺寸的铸造或烧结产品。该产品在铸造或烧结后再经加工（简单剪修或去氧化皮的除外），但不具有其他税号所列制品或产品的特征。

二、型材及异型材

轧、挤、拔、锻制的产品或其他成型产品，不论是否成卷，其全长截面相同，但与条、杆、丝、板、片、带、箔、管的定义不相符合。同时也包括同样形状的铸造或烧结产品。该产品在铸造或烧结后再经加工（简单剪修或去氧化皮的除外），但不具有其他税号所列制品或产品的特征。

三、丝

盘卷的轧、挤或拔制实心产品，其全长截面均为圆形、椭圆形、矩形（包括正方形）、等边三角形或规则外凸多边形（包括相对两边为弧拱形，另外两边为等长平行直线的“扁圆形”及“变形矩形”）。对于矩形（包括正方形）、三角形或多边形截面的产品，其全长边角可经磨圆。矩形（包括“变形矩形”）截面的产品，其厚度应大于宽度的十分之一。

Chapter 79
Zinc and articles thereof

Notes:

In this Chapter the following expressions have the meanings hereby assigned to them:

1. Bars and rods

Rolled, extruded, drawn or forged products, not in coils, which have a uniform solid cross-section along their whole length in the shape of circles, ovals, rectangle (including squares), equilateral triangles or regular convex polygons (including "flattened circles" and "modified rectangles", of which two opposite sides are convex arcs, the other two sides being straight, of equal length and parallel). Products with a rectangular (including square), triangular or polygonal cross-section may have corners rounded alongtheir whole length. The thickness of such products which have a rectangular (including "modified rectangular") cross-section exceeds one-tenth of the width. The expression also covers cast or sintered products, of the same forms and dimensions, which have been subsequently worked after production (otherwise than by simple trimming or de-scaling), provided that they have not thereby assumed the character of articles or products of other headings.

2. Profiles

Rolled, extruded, drawn, forged or formed products, coiled or not, of a uniform cross-section along their whole length, which do not conform to any of the definitions of bars, rods, wire, plates, sheets, strip, foil, tubes or pipes. The expression also covers cast or sintered products, of the same forms, which have been subsequently worked after production (otherwise than by simple trimming or de-scaling), provided that they have not thereby assumed the character of articles or products of other headings.

3. Wire

Rolled, extruded or drawn products, in coils, which have a uniform solid cross-section along their whole length in the shape of circles, ovals, rectangles (including squares), equilateral triangles or regular convex polygons (including "flattened circles" and "modified rectangles", of which two opposite sides are convex arcs, the other two sides being straight, of equal length and parallel). Products with a rectangular (including square), triangular or polygonal cross-section may have corners rounded along their whole length. The thickness of such products which have a rectangular (including "modified rectangular") cross-section

exceeds one-tenth of the width.

四、板、片、带、箔

成卷或非成卷的平面产品（税号79.01的未锻轧产品除外），截面均为厚度相同的实心矩形（不包括正方形），不论边角是否磨圆（包括相对两边为弧拱形，另外两边为等长平行直线的"变形矩形"），并且符合以下规格：

1. 矩形（包括正方形）的，厚度不超过宽度的十分之一；
2. 矩形或正方形以外形状的，任何尺寸，但不具有其他税号所列制品或产品的特征。

税号79.05还适用于具有花样（例如，凹槽、肋条形，格槽、珠粒及菱形）的板、片、带、箔以及穿孔、抛光、涂层或制成瓦楞形的这类产品，但不具有其他税号所列制品或产品的特征。

4. Plates, sheets, strip and foil

Flat-surfaced products (other than the unwrought products of heading No. 79. 01), coiled or not, of solid rectangular (other than square) cross-section with or without rounded corners (including "modified rectangles" of which two opposite sides are convex arcs, the other two sides being straight, of equal length and parallel) of a uniform thickness, which are:

-of rectangular (including square) shape with a thickness not exceeding one-tenth of the width,

-of a shape other than rectangular or square, of any size, provided that they do not assume the character of articles or products of other headings.

Heading No. 79.05 applies, *inter alia*, to plates, sheets, strip and foil with patterns (for example, grooves, ribs chequers, tears, buttons, lozenges) and to such products which have been perforated, corrugated, polished or coated, provided that they do not thereby assume the character of articles or products of other headings.

五、管

全长截面及管壁厚度相同并只有一个闭合空间的空心产品，成卷或非成卷的，其截面为圆形、椭圆形、矩形（包括正方形）、等边三角形或规则外凸多边形。对于截面为矩形（包括正方形）、等边三角形或规则外凸多边形的产品，不论全长边角是否磨圆，只要其内外截面为同一圆心并为同样形状及同一轴向，也可视为管子。上述截面的管子可经抛光、涂层、弯曲、攻丝、钻孔、缩腰、胀口、成锥形或装法兰、颈圈或套环。

5. Tubes and pipes

Hollow products, coiled or not, which have a uniform cross-section with only one enclosed void along their whole length in the shape of circles, ovals, rectangles (including squares), equilateral triangles or regular convex polygons, and which have a uniform wall thickness. Products with a rectangular (including square), equilateral triangular or regular convex polygonal cross-section, which may have corners rounded along their whole length, are also to be considered as tubes and pipes provided the inner and outer cross-sections are concentric and have the same form and orientation. Tubes and pipes of the foregoing cross-sections may be polished, coated, bent, threaded, drilled, waisted, expanded, cone-shaped or fitted with flanges, collars or rings.

子目注释：

本章所用有关名词解释如下：

一、非合金锌

按重量计含锌量至少为97.5%的金属。

二、锌合金

按重量计含锌量大于其他元素单项含量的金属物质，但按重量计其他元素的总含量超过2.5%。

Subheading Notes:

In this Chapter the following expressions have the meanings hereby assigned to them:

1. Zinc, not alloyed

Metal containing by weight at least 97.5% of zinc.

2. Zinc alloys

Metallic substances in which zinc predominates by weight over each of the other elements, provided that the total content by weight of such other elements exceeds 2.5%.

三、锌末

冷凝锌雾所得的锌末。该产品由球形微粒组成，比锌粉更为精细，按重量计至少80%的微粒可以通过孔径为63微米的筛子，而且必须含有按重量计至少为85%的金属锌。

3. Zinc dust

Dust obtained by condensation of zinc vapour, consisting of spherical particles which are finer than zinc powders. At least 80% by weight of the particles pass through a sieve with 63 micrometres (microns) mesh. It must contain at least 85% by weight of metallic zinc.

序号 No.	税则号列 Tariff Item	货品名称	最惠国税率(%) M.F.N.	协定税率(%)Agreement 东盟 ASEAN	曼谷协定 BA	香港 H.K.	澳门 Macao	普通税率(%) Gen.	Article Description
	79.01	**未锻轧锌：**							**Unwrought zinc:**
		-非合金锌：							-Zinc, not alloyed:
5155	7901.1100	--按重量计含锌量在99.99%及以上	3					20	--Containing by weight 99.99% or more of zinc
5156	7901.1200	--按重量计含锌量低于99.99%	3					20	--Containing by weight less than 99.99% of zinc
5157	7901.2000	-锌合金	3					20	-Zinc alloys
	79.02	**锌废碎料：**							**Zinc waste and scrap:**
5158	7902.0000	锌废碎料	1.5					20	Zinc waste and scrap
	79.03	**锌末、锌粉及片状粉末：**							**Zinc dust, powders and flakes:**
5159	7903.1000	-锌末	6					20	-Zinc dust
5160	7903.9000	-其他	6					20	-Other
	79.04	**锌条、杆、型材及异型材或丝：**							**Zinc bars, rods, profiles and wire:**
5161	7904.0000	锌条、杆、型材及异型材或丝	6					30	Zinc bars, rods, profiles and wire
	79.05	**锌板、片、带、箔：**							**Zinc plates, sheets, strip and foil:**
5162	7905.0000	锌板、片、带、箔	6					30	Zinc plates, sheets, strip and foil
	79.06	**锌管及管子附件（例如，接头、肘管、管套）：**							**Zinc tubes, pipes and tube or pipe fittings (for example, couplings, elbows, sleeves):**
5163	7906.0010	---锌管	6					30	---Zinc tubes or pipes
5164	7906.0020	---锌制管子附件（例如，接头、肘管、管套）	6					30	---Zinc tube or pipe fittings (for example, couplings, elbows, sleeves)
	79.07	**其他锌制品：**							**Other articles of zinc:**
		---工业用：							---For industrial use:
5165	7907.0011	----电池壳体坯料（锌饼）	6					40	----Cellpacking blanks (zinc biscuits)
5166	7907.0019	----其他	6					40	----Other
5167	7907.0090	---非工业用	6					80	---Not for industrial uses

第八十章
锡及其制品

注释：

本章所用有关名词解释如下：

一、条、杆

轧、挤、拔或锻制的实心产品，非成卷的，其全长截面均为圆形、椭圆形、矩形（包括正方形)、等边三角形或规则外凸多边形（包括相对两边为弧拱形，另外两边为等长平行直线的“扁圆形”及“变形矩形”)。对于矩形（包括正方形)、三角形或多边形截面的产品，其全长边角可经磨圆。矩形（包括“变形矩形”）截面的产品，其厚度应大于宽度的十分之一。所述条、杆也包括同样形状及尺寸的铸造或烧结产品。该产品在铸造或烧结后再经加工（简单剪修或去氧化皮的除外)，但不具有其他税号所列制品或产品的特征。

二、型材及异型材

轧、挤、拔、锻制的产品或其他成型产品，不论是否成卷，其全长截面相同，但与条、杆、丝、板、片、带、箔、管的定义不相符合。同时也包括同样形状的铸造或烧结产品。该产品在铸造或烧结后再经加工（简单剪修或去氧化皮的除外)，但不具有其他税号所列制品或产品的特征。

三、丝

盘卷的轧、挤或拔制实心产品，其全长截面均为圆形、椭圆形、矩形（包括正方形)、等边三角形或规则外凸多边形（包括相对两边为弧拱形，另外两边为等长平行直线的“扁圆形”及“变形矩形”)。对于矩形（包括正方形)、三角形或多边形截面的产品，其全长边角可经磨圆。矩形（包括“变形矩形”）截面的产品，其厚度应大于宽度的十分之一。

Chapter 80
Tin and articles thereof

Notes:

In this Chapter the following expressions have the meanings hereby assigned to them:

1. Bars and rods

Rolled, extruded, drawn or forged products, not in coils, which have a uniform solid cross-section along their whole length in the shape of circles, ovals, rectangles (including squares), equilateral triangles or regular convex polygons (including "flattened circles" and "modified rectangles", of which two opposite sides are convex arcs, the other two sides being straight, of equal length and parallel). Products with a rectangular (including square), triangular or polygonal cross-section may have corners rounded alongtheir whole length. The thickness of such products which have a rectangular (including "modified rectangular") cross-section exceeds one-tenth of the width. The expression also covers cast or sintered products, of the same forms and dimensions, which have been subsequently worked after production (otherwise than by simple trimming or de-scaling), provided that they have not thereby assumed the character of articles or products of other headings.

2. Profiles

Rolled, extruded, drawn, forged or formed products, coiled or not, of a uniform cross-section along their whole length, which do not conform to any of the definitions of bars, rods, wire, plates, sheets, strip, foil, tubes or pipes. The expression also covers cast or sintered products, of the same forms, which have been subsequently worked after production (otherwise than by simple trimming or de-scaling), provided that they have not thereby assumed the character of articles or products of other headings.

3. Wire

Rolled, extruded or drawn products, in coils, which have a uniform solid cross-section along their whole length in the shape of circles, ovals, rectanges (including squares), equilateral triangles or regular convex polygons (including "flattened circles" and "modified rectangles", of which two opposite sides are convex arcs, the other two sides being straight, of equal length and parallel). Products with a rectangular (including square), triangular or polygonal cross-section may have corners rounded along their whole length. The thickness of such products which have a rectangular (including "modified rectangular") cross-section

exceeds one-tenth of the width.

四、板、片、带、箔

成卷或非成卷的平面产品（税号 80.01 的未锻轧产品除外），截面均为厚度相同的实心矩形（不包括正方形），不论边角是否磨圆（包括相对两边为弧拱形，另外两边为等长平行直线的"变形矩形"），并且符合以下规格：

1. 矩形（包括正方形）的，厚度不超过宽度的十分之一；
2. 矩形或正方形以外形状的，任何尺寸，但不具有其他税号所列制品或产品的特征。

税号 80.04 和 80.05 还适用于具有花样（例如，凹槽、肋条形、格槽、珠粒及菱形）的板、片、带、箔以及穿孔、抛光、涂层或制成瓦楞形的这类产品，但不具有其他税号所列制品或产品的特征。

4. Plates, sheets, strip and foil

Flat-surfaced products (other than the unwrought products of heading No. 80.01), coiled or not, of solid rectangular (other than square) cross-section with or without rounded corners (including "modified rectangles" of which two opposite sides are convex arcs, the other two sides being straight, of equal length and parallel) of a uniform thickness, which are:

-of rectangular (including square) shape with a thickness not exceeding one-tenth of the width,

-of a shape other than rectangular or square, of any size, provided that they do not assume the character of articles or products of other headings.

Headings Nos. 80.04 and 80.05 apply, *inter alia*, to plates, sheets, strip and foil with patterns (for example, grooves, ribs, chequers, tears, buttons, lozenges) and to such products which have been perforated, corrugated, polished or coated, provided that they do not thereby assume the character of articles or products of other headings.

五、管

全长截面及管壁厚度相同并只有一个闭合空间的空心产品，成卷或非成卷的，其截面为圆形、椭圆形、矩形（包括正方形）、等边三角形或规则外凸多边形。对于截面为矩形（包括正方形）、等边三角形或规则外凸多边形的产品，不论全长边角是否磨圆，只要其内外截面为同一圆心并为同样形状及同一轴向，也可视为管子。上述截面的管子可经抛光、涂层、弯曲、攻丝、钻孔、缩腰、胀口、成锥形或装法兰、颈圈或套环。

5. Tubes and pipes

Hollow products, coiled or not, which have a uniform cross-section with only one enclosed void along their whole length in the shape of circles, ovals, rectangles (including squares), equilateral triangles or regular convex polygons, and which have a uniform wall thickness. Products with a rectangular (including square), equilateral triangular or regular convex polygonal cross-section, which may have corners rounded along their whole length, are also to be considered as tubes and pipes provided the inner and outer cross-sections are concentric and have the same form and orientation. Tubes and pipes of the foregoing cross-sections may be polished, coated, bent, threaded, drilled, waisted, expanded, cone-shaped or fitted with flanges, collars or rings.

子目注释：

本章所用有关名词解释如下：

一、非合金锡

按重量计含锡量至少为 99% 的金属，但含铋量或含铜量不超过下表中规定的限量：

Subheading Notes:

In this Chapter the following expressions have the meanings hereby assigned to them:

1. Tin, not alloyed.

Metal containing by weight at least 99% of tin, provided that the content by weight of any bismuth or copper is less than the limit specified in the following table:

其他元素表

元　　素	所含重量百分比
Bi　铋	0.1
Cu　铜	0.4

TABLE—Other elements

Element	Limiting content % by weight
Bi　Bismuth	0.1
Cu　Copper	0.4

二、锡合金

按重量计含锡量大于其他元素单项含量的金属物质，但：

（一）按重量计其他元素的总含量超过1%；或

（二）按重量计含铋量或含铜量应等于或大于上表中规定的限量。

2. Tin alloys

Metallic substances in which tin predominates by weight over each of the other elements, provided that:

(a) the total content by weight of such other elements exceeds 1%; or

(b) the content by weight of either bismuth or copper is equal to or greater than the limit specified in the foregoing table.

序号 No.	税则号列 Tariff Item	货品名称	最惠国税率(%) M.F.N.	协定税率(%) Agreement 东盟 ASEAN	曼谷协定 BA	香港 H.K.	澳门 Macao	普通税率(%) Gen.	Article Description
	80.01	**未锻轧锡：**							**Unwrought tin:**
5168	8001.1000	-非合金锡	3					20	-Tin, not alloyed
		-锡合金：							-Tin alloys:
5169	8001.2010	---锡基巴毕脱合金	3					20	---Babbitt metal
5170	8001.2020	---焊锡	3					30	---solder
5171	8001.2090	---其他	3					30	---Other
	80.02	**锡废碎料：**							**Tin waste and scrap:**
5172	8002.0000	锡废碎料	1.5					30	Tin waste and scrap
	80.03	**锡条、杆、型材及异型材或丝：**							**Tin bars, rods, profiles and wire:**
5173	8003.0000	锡条、杆、型材及异型材或丝	8			0		40	Tin bars, rods, profiles and wire
	80.04	**锡板、片及带，厚度超过0.2毫米：**							**Tin plates, sheets and strip, of a thickness exceeding 0.2 mm:**
5174	8004.0000	锡板、片及带，厚度超过0.2毫米	8					40	Tin plates, sheets and strip, of a thickness exceeding 0.2 mm
	80.05	**锡箔（不论是否印花或用纸、纸板、塑料或类似材料衬背），厚度（衬背除外）不超过0.2毫米；锡粉及片状粉末：**							**Tin foil (whether or not printed or backed with paper, paperboard, plastics or similar backing materials), of a thickness (excluding any backing) not exceeding 0.2 mm; tin powders and flakes**
5175	8005.0000	锡箔（不论是否印花或用纸、纸板、塑料或类似材料衬背），厚度（衬背除外）不超过0.2毫米；锡粉及片状粉末	8					40	Tin foil (whether or not printed or backed with paper, paperboard, plastics or similar backing materials), of a thickness (excluding any backing) not exceeding 0.2 mm; tin powders and flakes
	80.06	**锡管及管子附件（例如，接头、肘管、管套）：**							**Tin tubes, pipes and tube or pipe fittings (for example, couplings, elbows, sleeves):**

序号 No.	税则号列 Tariff Item	货品名称	最惠国税率(%) M.F.N.	协定税率(%) Agreement 东盟 ASEAN	曼谷协定 BA	香港 H.K.	澳门 Macao	普通税率(%) Gen.	Article Description
5176	8006.0000	锡管及管子附件(例如,接头、肘管、管套)	8					45	Tin tubes, pipes and tube or pipe fittings (for example, couplings, elbows, sleeves)
	80.07	**其他锡制品:**							**Other articles of tin:**
5177	8007.0010	---工业用	8					40	---For technical use
5178	8007.0090	---其他	18					80	---Other

第八十一章
其他贱金属、金属陶瓷及其制品

Chapter 81
Other base metals; cermets; artiles thereof

子目注释：

第七十四章注释中有关“条、杆”、“型材及异型材”、“丝”及“板、片、带、箔”的规定也适用于本章。

Subheading Note:

Note 1 to Chapter 74, defining "bars and rods", "profiles", "wire" and "plates, sheets, strip and foil" applies, *mutatis mutandis*, to this Chapter.

序号 No.	税则号列 Tariff Item	货品名称	最惠国税率(%) M.F.N.	协定税率(%) Agreement				普通税率(%) Gen.	Article Description
				东盟 ASEAN	曼谷协定 BA	香港 H.K.	澳门 Macao		
	81.01	**钨及其制品，包括废碎料：**							**Tungsten (wolfram) and articles thereof, including waste and scrap:**
5179	8101.1000	-粉末	6					20	-Powders
		-其他：							-Other:
5180	8101.9400	--未锻轧钨，包括简单烧结而成的条、杆	3					20	--Unwrought tungsten, including bars and rods obtained simply by sintering
5181	8101.9500	--条、杆，但简单烧结而成的除外；型材及异型材、板、片、带、箔	5					30	--Bars and rods, other than those obtained simply by sintering, profiles, plates, sheets, strip and foil
5182	8101.9600	--丝	8					20	--Wire
5183	8101.9700	--废碎料	3					20	--Waste and scrap
5184	8101.9900	--其他	8					70	--Other
	81.02	**钼及其制品，包括废碎料：**							**Molybdenum and articles thereof, including waste and scrap:**
5185	8102.1000	-粉末	6					20	-Powders
		-其他：							-Other:
5186	8102.9400	--未锻轧钼，包括简单烧结而成的条、杆	3					20	--Unwrought molybdenum, including bars and rods obtained simply by sintering
5187	8102.9500	--条、杆，但简单烧结而成的除外；型材及异型材、板、片、带、箔	8					30	--Bars and rods, other than those obtained simply by sintering, profiles, plates, sheets, strip and foil
5188	8102.9600	--丝	8					20	--Wire
5189	8102.9700	--废碎料	3					20	--Waste and scrap
5190	8102.9900	--其他	8					70	--Other
	81.03	**钽及其制品，包括废碎料：**							**Tantalum and articles thereof, including waste and scrap:**
5191	8103.2000	-未锻轧钽，包括简单烧结而成的条、杆；粉末	6					14	-Unwrought tantalum, including bars and rods obtained simply by sintering; powders
5192	8103.3000	-废碎料	6					14	-Waste and scrap
5193	8103.9000	-其他	8					30	-Other
	81.04	**镁及其制品，包括废碎料：**							**Magnesium and articles thereof, including waste and scrap:**
		-未锻轧镁：							-Unwrought magnesium:
5194	8104.1100	--按重量计含镁量至少为99.8%	6					20	--Containing at least 99.8% by weight of magnesium

序号 No.	税则号列 Tariff Item	货品名称	最惠国税率(%) M.F.N.	协定税率(%) Agreement 东盟 ASEAN	曼谷协定 BA	香港 H.K.	澳门 Macao	普通税率(%) Gen.	Article Description
5195	8104.1900	--其他	6					20	--Other
5196	8104.2000	-废碎料	1.5					20	-Waste and scrap
5197	8104.3000	-锉屑、车屑及颗粒,已按规格分级的;粉末	8					30	-Raspings, turnings and granules, graded according to size; powders
		-其他:							-Other:
5198	8104.9010	---锻轧镁	8					30	---Wrought magnesium
5199	8104.9020	---镁制品	8.4					70	---Magnesium articles
	81.05	**钴锍及其他冶炼钴时所得的中间产品;钴及其制品,包括废碎料:**							**Cobalt mattes and other intermediate products of cobalt matallurgy; cobalt and articles thereof, including waste and scrap:**
5200	8105.2000	-钴锍及其他冶炼钴时所得的中间产品;未锻轧钴;粉末	4					14	-Cobalt mattes and other intermediate products of cobalt metallurgy; unwrought cobalt; powders
5201	8105.3000	-废碎料	4					14	-waste and scrap
5202	8105.9000	-其他	8					30	-Other
	81.06	**铋及其制品,包括废碎料:**							**Bismuth and articles thereof, including waste and scrap:**
5203	8106.0010	---未锻轧铋;废碎料;粉末	3					20	---Unwroght bismuth; waste and scrap; powders
5204	8106.0090	---其他	8					30	---Other
	81.07	**镉及其制品,包括废碎料:**							**Cadmium and articles thereof, including waste and scrap:**
5205	8107.2000	-未锻轧镉;粉末	3					14	-Unwrought cadmium; powders
5206	8107.3000	-废碎料	3					14	-waste and scrap
5207	8107.9000	-其他	8					30	-Other
	81.08	**钛及其制品,包括废碎料:**							**Titanium and articles thereof, including waste and scrap:**
		-未锻轧钛;粉末:							-Unwrought titanium; powders
5208	8108.2010	---海绵钛	3					14	---Titanium sponge
5209	8108.2090	---其他	3					14	---Other
5210	8108.3000	-废碎料	3					14	-Waste and scrap
		-其他:							-Other:
5211	8108.9010	---条、杆、型材及异型材	8					30	---Bars、rods、shapes and sections
5212	8108.9020	---丝	8					30	---Wire
		---板、片、带、箔:							---Plates, sheets, strap, foil:
5213	8108.9031	----厚度不超过0.8毫米	8					30	---Of a thickness not more than 0.8mm
5214	8108.9032	----厚度超过0.8毫米	8					30	----Of a thickness more than 0.8mm
5215	8108.9040	---管	8					30	---Tubes or pipes
5216	8108.9090	---其他	8					30	---Other
	81.09	**锆及其制品,包括废碎料:**							**Zirconium and articles thereof, including waste and scrap:**
5217	8109.2000	-未锻轧锆;粉末	3					20	-Unwrought zirconium; powders
5218	8109.3000	-废碎料	3					20	-waste and scrap

序号 No.	税则号列 Tariff Item	货品名称	最惠国税率(%) M.F.N.	协定税率(%)Agreement 东盟 ASEAN	曼谷协定 BA	香港 H.K.	澳门 Macao	普通税率(%) Gen.	Article Description
5219	8109.9000	-其他	8					30	-Other
	81.10	**锑及其制品,包括废碎料:**							**Antimony and articles thereof, including waste and scrap:**
		-未锻轧锑;粉末:							-Unwrought antimony; powders:
5220	8110.1010	---未锻轧锑	3					30	---Unwrought antimony
5221	8110.1020	---粉末	3					30	---powders
5222	8110.2000	-废碎料	3					30	-Antimony waste and scrap
5223	8110.9000	-其他	8					40	-Other
	81.11	**锰及其制品,包括废碎料:**							**Manganese and articles thereof, including waste and scrap:**
5224	8111.0010	---未锻轧锰;废碎料;粉末	3					20	---Unwrought manganese; waste and scrap; powders
5225	8111.0090	---其他	8					30	---Other
	81.12	**铍、铬、锗、钒、镓、铪、铟、铼、铌、铊及其制品,包括废碎料:**							**Beryllium, chromium, germanium, vanadium, gallium, hafnium, indium, niobium (columbium), rhenium and thallium, and articles of these metals, including waste and scrap:**
		-铍:							-Beryllium:
5226	8112.1200	--未锻轧铍;粉末	3					30	--Unwrought; powders
5227	8112.1300	--废碎料	3					30	--waste and scrap
5228	8112.1900	--其他	8					30	--Other
		-铬:							-Chromium:
5229	8112.2100	--未锻轧铬;粉末	3					20	--Unwrought; powders
5230	8112.2200	--废碎料	3					20	--Waste and Scrap
5231	8112.2900	--其他	3					20	--Other
5232	8112.3000	-锗	3					20	-Germanium
5233	8112.4000	-钒	3					20	-Vanadium
		-铊:							-Thallium:
5234	8112.5100	--未锻轧铊;粉末	3					20	--Unwrought; powders
5235	8112.5200	--废碎料	3					20	--Waste and Scrap
5236	8112.5900	--其他	8					30	--Other
		-其他:							-Other:
5237	8112.9200	--未锻轧;废碎料;粉末	3					20	--Unwrought; waste and scrap; powders
5238	8112.9900	--其他	8					30	--Other
	81.13	**金属陶瓷及其制品,包括废碎料:**							**Cermets and articles thereof, including waste and scrap:**
5239	8113.0000	金属陶瓷及其制品,包括废碎料	8.4					30	Cermets and articles thereof, including waste and scrap

第八十二章
贱金属工具、器具、利口器、餐匙、餐叉及其零件

Chapter 82
Tools, implements, cutlery, spoons and forks, of base metal

注释：

一、除喷灯、轻便锻炉、带支架的砂轮、修指甲和修脚用器具及税号82.09的货品外，本章仅包括带有用下列材料制成的刀片、工作刃、工作面或其他工作部件的物品：

(一) 贱金属；

(二) 硬质合金或金属陶瓷；

(三) 装于贱金属、硬质合金或金属陶瓷底座上的宝石或半宝石（天然、合成或再造）；

(四) 附于贱金属底座上的磨料，当附上磨料后，所具有的切齿、沟、槽或类似结构仍保持其特性及功能。

二、本章所列物品的贱金属零件，应与该制品归入同一税号，但具体列名的零件及手工工具的工具夹具（税号84.66）除外。第十五类注释二所述的通用零件，均不归入本章。

电动剃须刀及电动毛发推剪的刀头、刀片应归入税号85.10。

三、由税号82.11的一把或多把刀具与税号82.15至少数量相同的物品构成的成套货品应归入税号82.15。

Notes:

1. Apart from blow lamps, portable forges, grinding wheels with frameworks, manicure or pedicure sets, and goods of heading No. 82.09, this Chapter covers only articles with a blade, working edge, working surface or other working part of:

(a) Base metal;

(b) Metal carbides or cermets;

(c) Precious or semi-precious stones (natural, synthetic or reconstructed) on a support of base metal, metal carbide or cermet; or

(d) Abrasive materials on a support of base metal, provided that the articles have cutting teeth, flutes, grooves, or the like, of base metal, which retain their identity and function after the application of the abrasive.

2. Parts of base metal of the articles of this Chapter are to be classified with the articles of which they are parts, except parts separately specified as such and tool-holders for hand tools (heading No. 84.66). However, parts of general use as defined in Note 2 to Section XV are in all cases excluded from this Chapter. Heads, blades and cutting plates for electric shavers or electric hair clippers are to be classified in heading No. 85.10.

3. Sets consisting of one or more knives of heading No. 82.11 and at least an equal number of articles of heading No. 82.15 are to be classified in heading No. 82.15.

序号 No.	税则号列 Tariff Item	货品名称	最惠国税率(%) M.F.N.	协定税率(%)Agreement 东盟 ASEAN	曼谷协定 BA	香港 H.K.	澳门 Macao	普通税率(%) Gen.	Article Description
	82.01	**锹、铲、镐、锄、叉及耙；斧子、钩刀及类似砍伐工具；各种修枝用剪刀；镰刀、秣刀、树篱剪、伐木楔子及其他农业、园艺或林业用手工工具：**							**Hand tools, the following: spades, shovels, mattocks, picks, hoes, forks and rakes; axes, bill hooks and similar hewing tools; secateurs and prundrs of any kind; scythes, hay knives, hedge shears, timber wedges and other tools of a kind used in agriculture, horticulture or forestry:**
5240	8201.1000	-锹及铲	8					50	-Spades and shovels
5241	8201.2000	-叉	8					50	-Forks
5242	8201.3000	-镐、锄及耙	8					50	-Mattocks, picks, hoes and rakes

序号 No.	税则号列 Tariff Item	货品名称	最惠国税率(%) M.F.N.	协定税率(%) Agreement 东盟 ASEAN	曼谷协定 BA	香港 H.K.	澳门 Macao	普通税率(%) Gen.	Article Description
5243	8201.4000	-斧子、钩刀及类似砍伐工具	8					50	-Axes, bill hooks and similar hewing tools
5244	8201.5000	-修枝剪及类似的单手操作剪刀(包括家禽剪)	8					50	-Secateurs and similar one-handed pruners and shears (including poultry shears)
5245	8201.6000	-树篱剪、双手修枝剪及类似的双手操作剪刀	8					50	-Hedge shears, two-handed pruning shears and similar two-handed shears
5246	8201.9000	-用于农业、园艺或林业的其他手工工具	8					50	-Other hand tools of a kind used in agriculture, horticulture or forestry
	82.02	**手工锯;各种锯的锯片(包括切条、切槽或无齿锯片):**							**Hand saws; blades for saws of all kinds (including slitting, slotting or toothless saw blades):**
5247	8202.1000	-手工锯	8.4					50	-Hand saws
5248	8202.2000	-带锯片	8					20	-Band saw blades
		-圆锯片(包括切条或切槽锯片):							--Circular saw blades(including slitting or slotting saw blades):
5249	8202.3100	--带有钢制工作部件	8					20	--With working part of steel
5250	8202.3900	--其他,包括部件	8					20	--Other, including parts
5251	8202.4000	-链锯片	8					20	-Chain saw blades
		-其他锯片:							-Other saw blades:
		--直锯片,加工金属用:							--Straight saw blades, for working metal:
5252	8202.9110	---机械锯用	8		7			20	---For sawing machines
5253	8202.9190	---其他	8					50	---Other
		--其他:							--Other:
5254	8202.9910	---机械锯用	8.4					20	---For sawing machines
5255	8202.9990	---其他	10.5					50	---Other
	82.03	**钢锉、木锉、钳子(包括剪钳)、镊子、白铁剪、切管器、螺栓切头器、打孔冲子及类似手工工具:**							**Files, rasps, pliers (including cutting pliers), pincers, tweezers, metal cutting shears, pipe-cutters, bolt croppers, perforating punches and similar hand tools:**
5256	8203.1000	-钢锉、木锉及类似工具	10.5					50	-Files, rasps and similar tools
5257	8203.2000	-钳子(包括剪钳)、镊子及类似工具	10.5					50	-Pliers (including cutting pliers), pincers, tweezers and similar tools
5258	8203.3000	-白铁剪及类似工具	10.5					50	-Metal cutting shears and similar tools
5259	8203.4000	-切管器、螺栓切头器、打孔冲子及类似工具	10.5					50	-Pipe-cutters, bolt croppers, perforating punches and similar tools
	82.04	**手动扳手及扳钳(包括转矩扳手,但不包括丝锥扳手);可互换的扳手套筒,不论是否带手柄:**							**Hand-operated spanners and wrenches (including torque meter wrenches but not including tap wrenches); interchangeable spanner sockets, with or without handles:**
		-手动扳手及扳钳:							-Hand-operated spanners and wrenches:
5260	8204.1100	--固定的	10.5					50	--Non-adjustable
5261	8204.1200	--可调的	10					50	--Adjustable

序号 No.	税则号列 Tariff Item	货品名称	最惠国税率(%) M.F.N.	协定税率(%) Agreement 东盟 ASEAN	曼谷协定 BA	香港 H.K.	澳门 Macao	普通税率(%) Gen.	Article Description
5262	8204.2000	-可互换的扳手套筒,不论是否带手柄	10					50	-Interchangeable spanner sockets, with or without handles
	82.05	**其他税号未列名的手工工具(包括玻璃刀);喷灯;台钳、夹钳及类似品,但作为机床附件或零件的除外;砧;轻便锻炉;带支架的手摇或脚踏砂轮:**							**Hand tools (including glaziers' diamonds), not elsewhere specified or included; blow lamps; vices, clamps and the like, other than accessories for and parts of, machine tools; anvils; portable forges; hand or pedal operated grinding wheels with frame works:**
5263	8205.1000	-钻孔或攻丝工具	10					50	-Drilling, threading or tapping tools
5264	8205.2000	-锤子	10					50	-Hammers and sledge hammers
5265	8205.3000	-木工用刨子、凿子及类似切削工具	10.5					50	-Planes, chisels, gouges and similar cutting tools for working wood
5266	8205.4000	-螺丝刀	10.5					50	-Screwdrivers
		-其他手工工具(包括玻璃刀):							-Other hand tools (including glaziers' diamonds):
5267	8205.5100	--家用工具	10.5					50	--Household tools
5268	8205.5900	--其他	10					50	--Other
5269	8205.6000	-喷灯	10					50	-Blow lamps
5270	8205.7000	-台钳、夹钳及类似品	10.5					50	-Vices, clamps and the like
5271	8205.8000	-砧、轻便锻炉;带支架的手摇或脚踏砂轮	10.5					50	-Anvils; portable forges; hand or pedal operated grindding wheels with frameworks
5272	8205.9000	-由上述两个或多个子目所列物品组成的成套货品	10.5					50	-Sets of articles of two or more of the foregoing subheadings
	82.06	**由税号82.02至82.05中两个或多个税目所列工具组成的零售包装成套货品:**							**Tools of two or more of the headings Nos. 82.02 to 82.05, put up in sets for retail sale:**
5273	8206.0000	由税号82.02至82.05中两个或多个税目所列工具组成的零售包装成套货品	10.5					50	Tools of two or more of the headings Nos. 82.02 to 82.05, put up in sets for retail sale
	82.07	**手工工具(不论是否有动力装置)及机床(例如,锻压、冲压、攻丝、钻孔、镗孔、铰孔及铣削、车削或上螺丝用的机器)的可互换工具,包括金属拉拔或挤压用模以及凿岩或钻探工具:**							**Interchangeable tools for hand tools, whether or not power-operated, or for machine-tools (for example, for pressing, stamping, punching, tapping, threading, drilling, boring, broaching, milling, turning or screw driv-ing), including dies for drawing or extruding metal, and rock drilling or earth boring tools:**
		-凿岩或钻探工具:							-Rock drilling or earth boring tools:
5274	8207.1300	--带有金属陶瓷制的工作部件	8					20	--With working part of cermets

序号 No.	税则号列 Tariff Item	货品名称	最惠国税率(%) M.F.N.	协定税率(%) Agreement 东盟 ASEAN	曼谷协定 BA	香港 H.K.	澳门 Macao	普通税率(%) Gen.	Article Description
		--其他,包括部件:							--Other,including parts:
5275	8207.1910	---带有天然或合成金刚石、立方氮化硼制的工作部件	8					20	---With working part of natural or synthetic diamonds or cubic boron nitride
5276	8207.1990	---其他	8					20	---Other
		-金属拉拔或挤压用模:							-Dies for drawing or extruding metal:
5277	8207.2010	---带有天然或合成金刚石、立方氮化硼制的工作部件	8					20	---With working part of natural or synthetic diamonds or cubic boron nitride
5278	8207.2090	---其他	8					20	---Other
5279	8207.3000	-锻压或冲压工具	8			0		20	-Tools for pressing, stamping or punching
5280	8207.4000	-攻丝工具	8					20	-Tools for tapping or threading
		-钻孔工具,但凿岩及钻探用的除外:							-Tools for drilling,other than for rock drilling:
5281	8207.5010	---带有天然或合成金刚石、立方氮化硼制的工作部件	8					20	---With working part of natural or synthetic diamonds or cubic boron nitride
5282	8207.5090	---其他	8					20	---Other
		-镗孔或铰孔工具:							-Tools for boring or broaching:
5283	8207.6010	---带有天然或合成金刚石、立方氮化硼制的工作部件	8					20	---With working part of natural or synthetic diamonds or cubic boron nitride
5284	8207.6090	---其他	8					20	---Other
5285	8207.7000	-铣削工具	8					20	-Tools for milling
5286	8207.8000	-车削工具	8					20	-Tools for turning
		-其他可互换工具:							-Other interchangeable tools:
5287	8207.9010	---带有天然或合成金刚石、立方氮化硼制的工作部件	8					20	---with working part of natural or synthetic diamonds or cubic boron nitride
5288	8207.9090	---其他	8					20	---Other
	82.08	**机器或机械器具的刀及刀片:**							**Knives and cutting blades, for machines or for mechanical appliances:**
5289	8208.1000	-金属加工用	8					20	-For metal working
5290	8208.2000	-木器加工用	8					20	-For wood working
5291	8208.3000	-厨房器具或食品工业机器用	8					20	-For kitchen appliances or for machines used by the food industry
5292	8208.4000	-农业、园艺或林业机器用	8					20	-For agricultural, horticultural or forestry machines
5293	8208.9000	-其他	8					20	-Other
	82.09	**未装配的工具用金属陶瓷板、杆、刀头及类似品:**							**Plates, sticks, tips and the like for tools, unmounted, of cermets:**
5294	8209.0000	未装配的工具用金属陶瓷板、杆、刀头及类似品	8					20	Plates, sticks, tips and the like for tools, unmounted, of cermets

序号 No.	税则号列 Tariff Item	货品名称	最惠国税率(%) M.F.N.	协定税率(%)Agreement 东盟 ASEAN	曼谷协定 BA	香港 H.K.	澳门 Macao	普通税率(%) Gen.	Article Description
	82.10	**用于加工或调制食品或饮料的手动机械器具,重量不超过10公斤:**							**Hand-operated mechanical appliances, weighing 10kg or less, used in the preparation, conditioning or serving of food or drink:**
5295	8210.0000	用于加工或调制食品或饮料的手动机械器具,重量不超过10公斤	18					80	Hand-operated mechanical appliances, weighing 10kg or less, used in the preparation, conditioning or serving of food or drink
	82.11	**有刃口的刀及其刀片,不论是否有锯齿(包括整枝刀),但税号82.08的刀除外:**							**Knives with cutting blades, serrated or not (including pruning knives), other than knives of heading No.82.08, and blades therefor:**
5296	8211.1000	-成套货品	18					80	-Sets of assorted articles
		-其他:							-Other:
5297	8211.9100	--刃面固定的餐刀	18					80	--Table knives having fixed blades
5298	8211.9200	--刃面固定的其他刀	12					80	--Other knives having fixed blades
5299	8211.9300	--可换刃面的刀	18					80	--Knives having other than fixed blades
5300	8211.9400	--刀片	14					80	--Blades
5301	8211.9500	--贱金属制的刀柄	12					80	--Handles of base metal
	82.12	**剃刀及其刀片(包括未分开的刀片条):**							**Razors and razor blades(including razor blade blanks in strips):**
5302	8212.1000	-剃刀	12					80	-Razors
5303	8212.2000	-安全刀片,包括未分开的刀片条	14					80	-Safety razor blades, including razor blade blanks in strips
5304	8212.9000	-其他零件	12					80	-Other parts
	82.13	**剪刀、裁缝剪刀及类似品、剪刀片:**							**Scissors, tailors' shears and similar shears, and blades therefor:**
5305	8213.0000	剪刀、裁缝剪刀及类似品、剪刀片	12					80	Scissors, tailors' shears and similar shears, and blades therefor
	82.14	**其他利口器(例如,理发推剪、屠刀、砍骨刀、切肉刀、切菜刀、裁纸刀);修指甲及修脚用具(包括指甲锉):**							**Other articles of cutlery(for example, hair clippers, butchers' or kitchen cleavers, choppers and mincing knives, paper knives); manicure or pedicure sets and instruments(including nail files):**
5306	8214.1000	-裁纸刀、开信刀、改错刀、铅笔刀及其刀片	12					80	-Paper knives, letter openers, erasing knives, pencil sharpeners and blades therefor
5307	8214.2000	-修指甲及修脚用具(包括指甲锉)	18					90	-Manicure or pedicure sets and instruments(including nail files)
5308	8214.9000	-其他	18					80	-Other
	82.15	**餐匙、餐叉、长柄勺、漏勺、糕点夹、鱼刀、黄油刀、糖块夹及类似的厨房或餐桌用具:**							**Spoons, forks, ladles, skimmers, cake-serv-ers, fish-knives, butter-knives, sugar tongs and similar kitchen or tableware:**

序号 No.	税则号列 Tariff Item	货品名称	最惠国税率(%) M.F.N.	协定税率(%)Agreement 东盟 ASEAN	曼谷协定 BA	香港 H.K.	澳门 Macao	普通税率(%) Gen.	Article Description
5309	8215.1000	-成套货品,至少其中一件物品是镀贵金属的	18					80	-Sets of assorted articles containing at least one article plated with precious metal
5310	8215.2000	-其他成套货品	18					80	-Other sets of assorted articles
		-其他:							-Other:
5311	8215.9100	--镀贵金属的	18					80	--Plated with precious metal
5312	8215.9900	--其他	18			0		80	--Other

第八十三章
贱金属杂项制品

Chapter 83
Miscellaneous articles of base metal

注释：

一、在本章，贱金属零件应与制品一同归类。但税号 73.12、73.15、73.17、73.18 及 73.20 的钢铁制品或其他贱金属（第七十四章至第七十六章及第七十八章至第八十一章）制的类似物品不应视为本章制品的零件。

二、税号 83.02 所称"脚轮"，是指直径（对于有胎的，连胎计算在内，下同）不超过 75 毫米的或直径虽超过 75 毫米，但所装轮或胎的宽度必须小于 30 毫米的脚轮。

Notes:

1. For the purposes of this Chapter, parts of base metal are to be classified with their parent artcles. However, articles of iron or steel of heading No. 73.12, 73.15, 73.17, 73.18 or 73.20, or similar articles of other base metal (Chapters 74 to 76 and 78 to 81) are not to be taken as parts of articles of this Chapter.

2. For the purposes of heading No. 83.02, the word "castors" means those having a diameter (including, where appropriate, tyres) not exceeding 75mm, or those having a diameter (including, where appropriate, tyres) exceeding 75mm provided that the width of the wheel or tyre fitted thereto is less than 30mm.

序号 No.	税则号列 Tariff Item	货品名称	最惠国税率(%) M.F.N.	协定税率(%)Agreement 东盟 ASEAN	曼谷协定 BA	香港 H.K.	澳门 Macao	普通税率(%) Gen.	Article Description
	83.01	**贱金属制的锁(钥匙锁、数码锁及电动锁);贱金属制带锁的扣环及扣环框架;上述锁的贱金属制钥匙:**							**Padlocks and locks(key, combination or electrically operated), of base metal; clasps and frames with clasps, incorporating locks, of base metal; keys for any of the foregoing articles, of base metal:**
5313	8301.1000	-挂锁	14					80	-Padlocks
		-机动车用锁:							-Locks of a kind used for motor vehicles:
5314	8301.2010	---中央控制门锁	10					80	---Central control door lock
5315	8301.2090	---其他	10					80	---Other
5316	8301.3000	-家具用锁	14					80	-Locks of a kind used for furniture
5317	8301.4000	-其他锁	14					80	-Other locks
5318	8301.5000	-带锁的扣环及扣环框架	14			0		80	-Clasps and frames with clasps, incorporating locks
5319	8301.6000	-零件	12					80	-Parts
5320	8301.7000	-钥匙	10					80	-Keys presented separately
	83.02	**用于家具、门窗、楼梯、百叶窗、车厢、鞍具、衣箱、盒子及类似品的贱金属附件及架座;贱金属制帽架、帽钩、托架及类似品;用贱金属做支架的小脚轮;贱金属制的自动闭门器:**							**Base metal mountings, fittings and similar articles suitable for furniture, doors, staircases, windows, blinds, coachwork, saddlery, trunks, chests, caskets, or the like; base metal hat-racks, hat-pegs, brackets and similar fixtures; castors with mountings of base metal; automatic door closers of base metal:**
5321	8302.1000	-铰链(折叶)	10			0		80	-Hinges
5322	8302.2000	-小脚轮	12					80	-Castors

序号 No.	税则号列 Tariff Item	货品名称	最惠国税率(%) M.F.N.	协定税率(%) Agreement 东盟 ASEAN	曼谷协定 BA	香港 H.K.	澳门 Macao	普通税率(%) Gen.	Article Description
5323	8302.3000	-机动车辆用的其他附件及架座	10					80	-Other mountings, fittings and similar articles suitable for motor vehicles
		-其他附件及架座:							-Other mountings, fittings and similar articles:
5324	8302.4100	--建筑用	14					80	--Suitable for buildings
5325	8302.4200	--其他,家具用	12					80	--Other, suitable for furniture
5326	8302.4900	--其他	12					80	--Other
5327	8302.5000	-帽架、帽钩、托架及类似品	14					80	-Hat-racks, hat-pegs, brackets and similar fixtures
5328	8302.6000	-自动闭门器	12					80	-Automatic door closers
	83.03	**装甲或加强的贱金属制保险箱、保险柜及保险库的门和带锁保险储存橱、钱箱、契约箱及类似品:**							**Armoured or reinforced safes, strong-boxes and doors and safe deposit lockers for strong-rooms, cash or deed boxes and the like, of base metal:**
5329	8303.0000	装甲或加强的贱金属制保险箱、保险柜及保险库的门和带锁保险储存橱、钱箱、契约箱及类似品	14					50	Armoured or reinforced safes, strong-boxes and doors and safe deposit lockers for strong-rooms, cash or deed boxes and the like, of base metal
	83.04	**贱金属制的档案柜、卡片索引柜、文件盘、文件篮、笔盘、公章架及类似的办公用具,但税号94.03的办公室家具除外:**							**Filing cabinets, card-index cabinets, paper trays, paper rests, pen trays, office-stamp stands and similar office or desk equipment, of base metal, other than office furniture of heading No. 94.03:**
5330	8304.0000	贱金属制的档案柜、卡片索引柜、文件盘、文件篮、笔盘、公章架及类似的办公用具,但税号94.03的办公室家具除外	10.5			0		80	Filing cabinets, card-index cabinets, paper trays, paper rests, pen trays, office-stamp stands and similar office or desk equipment, of base metal, other than office furniture of heading No. 94.03
	83.05	**活页夹、卷宗夹的贱金属附件,贱金属制的信夹、信角、文件夹、索引标签及类似的办公用品;贱金属制的成条钉书钉(例如,供办公室、室内装饰或包装用):**							**Fittings for loose-leaf binders or files, letter clips, letter corners, paper clips, indexing tags and similar office articles, of base metal; staples in strips (for example, for offices, upholstery, packaging), of base metal:**
5331	8305.1000	-活页夹或卷宗夹的附件	10.5					80	-Fittings for loose-leaf binders of files
5332	8305.2000	-成条钉书钉	10.5					80	-Staples in strips
5333	8305.9000	-其他,包括零件	10.5			0		80	-Other, including parts

序号 No.	税则号列 Tariff Item	货品名称	最惠国税率(%) M.F.N.	协定税率(%)Agreement 东盟 ASEAN	曼谷协定 BA	香港 H.K.	澳门 Macao	普通税率(%) Gen.	Article Description
	83.06	**非电动的贱金属铃、钟、锣及类似品；贱金属雕塑像及其他装饰品；贱金属相框或画框及类似框架；贱金属镜子：**							**Bells, gongs and the like, non-electric, of base metal; statuettes and other ornaments, of base metal; photograph, picture or similar frames, of base metal; mirrors of base metal:**
5334	8306.1000	-铃、钟、锣及类似品	8					80	-Bells, gongs and the like
		-雕塑像及其他装饰品：							-Statuettes and other ornaments:
5335	8306.2100	--镀贵金属的	8					100	--Plated with precious metal
		--其他：							-Other:
5336	8306.2910	---景泰蓝的	8					100	---Cloisonne
5337	8306.2990	---其他	8					100	---Other
5338	8306.3000	-相框、画框及类似框架；镜子	8					100	-Photograph, picture or similar frames; mirrors
	83.07	**贱金属软管，不论是否有附件：**							**Flexible tubing of base metal, with or without fittings:**
5339	8307.1000	-钢铁制	8.4					35	-Of iron or steel
5340	8307.9000	-其他贱金属制	8.4					35	-Of other base metal
	83.08	**贱金属制的扣、钩、环、眼及类似品，用于衣着、鞋靴、天篷、提包、旅行用品或其他制成品；贱金属制的管形铆钉及开口铆钉；贱金属制的珠子及亮晶片：**							**Clasps, frames with clasps, buckles, buckle-clasps, hooks, eyes, eyelets and the like, of base metal, of a kind used for clothing, footwear, awnings, handbags, travel goods or other made up articles; tubular or bifurcated rivets, of base metal; beads and spangles, of base metal:**
5341	8308.1000	-钩、环及眼	10.5			0		80	-Hooks, eyes and eyelets
5342	8308.2000	-管形铆钉及开口铆钉	10.5			0		80	-Tubular or bifurcated rivets
5343	8308.9000	-其他，包括零件	10.5			0		80	-Other, including parts
	83.09	**贱金属制的塞子、盖子（包括冠形瓶塞、螺口盖及倒水塞）、瓶帽、螺口塞、塞子帽、封志及其他包装用附件：**							**Stoppers, caps and lids (including crown corks, screw caps and pouring stoppers), capsules for bottles, threaded bungs, bung covers, seals and other packing accessories, of base metal:**
5344	8309.1000	-冠形瓶塞	18			0		90	-Crown corks
5345	8309.9000	-其他	12			0		80	-Other
	83.10	**贱金属制的标志牌、铭牌、地名牌及类似品、号码、字母及类似标志，但税号 94.05 的货品除外：**							**Sign-plates, name-plates, address-plates and similar plates, numbers, letters and other symbols, of base metal, excluding those of heading No. 94.05:**
5346	8310.0000	贱金属制的标志牌、铭牌、地名牌及类似品、号码、字母及类似标志，但税号 94.05 的货品除外	18					80	Sign-plates, name-plates, address-plates and similar plates, numbers, letters and other sym-bols, of base metal, excluding those of heading No. 94.05

序号 No.	税则号列 Tariff Item	货品名称	最惠国税率(%) M.F.N.	协定税率(%) Agreement 东盟 ASEAN	曼谷协定 BA	香港 H.K.	澳门 Macao	普通税率(%) Gen.	Article Description
	83.11	**贱金属或硬质合金制的丝、条、管、板、电极及类似品,以焊剂涂面或以焊剂为芯,用于焊接或沉积金属、硬质合金;贱金属粉粘聚而成的丝或条,供金属喷镀用:**							**Wire, rods, tubes, plates, electrodes and similar products, of base metal or of metal carbides, coated or cored with flux material, of a kind used for soldering, brazing, welding or deposition of metal or of metal carbides; wire and rods, of agglomerated base metal powder, used for metal spraying:**
5347	8311.1000	-以焊剂涂面的贱金属制电极,电弧焊用	8					30	-Coated electrodes of base metal, for electric arc-welding
5348	8311.2000	-以焊剂为芯的贱金属制焊丝,电弧焊用	8					30	-Cored wire of base metal, for electric arcwelding
5349	8311.3000	-以焊剂涂面或以焊剂为芯的贱金属条或丝,钎焊或气焊用	8					30	-Coated rods and cored wire, of base metal, for soldering, brazing or welding by flame
5350	8311.9000	-其他,包括零件	8					30	-Other, including parts

第十六类

机器、机械器具、电气设备及其零件；录音机及放声机、电视图像、声音的录制和重放设备及其零件、附件

注释：

一、本类不包括：

（一）第三十九章的塑料或税号 40.10 的硫化橡胶制的传动带、输送带；除硬质橡胶以外的硫化橡胶制的机器、机械器具、电气器具或其他专门技术用途的物品（税号 40.16）；

（二）机器、机械器具或其他专门技术用途的皮革、再生皮革（税号 42.04）或毛皮（税号 43.03）的制品；

（三）各种材料（例如，第三十九章、第四十章、第四十四章、第四十八章及第十五类的材料）制的筒管、卷轴、纡子、锥形筒管、芯子、线轴及类似品；

（四）提花机及类似机器用的穿孔卡片（例如，归入第三十九章、第四十八章或第十五类的）；

（五）纺织材料制的传动带、输送带或带料（税号 59.10）或专门技术用途的其他纺织材料制品（税号 59.11）；

（六）税号 71.02 至 71.04 的宝石或半宝石（天然、合成或再造）或税号 71.16 的完全以宝石或半宝石制成的物品，但已加工未装配的唱针用蓝宝石和钻石除外（税号 85.22）；

（七）第十五类注释二所规定的贱金属制通用零件（第十五类）及塑料制的类似品（第三十九章）；

（八）钻管（税号 73.04）；

（九）金属丝、带制的环形带（第十五类）；

（十）第八十二章或第八十三章的物品；

（十一）第十七类的物品；

（十二）第九十章的物品；

（十三）第九十一章的钟、表及其他物品；

（十四）税号 82.07 的可互换工具及作为机器

SECTION XVI

MACHINERY AND MECHANICAL APPLIANCES; ELECTRICAL EQUIPMENT; PARTS THEREOF; SOUND RECORDERS AND REPRODUCERS, TELEVISION IMAGE AND SOUND RECORDERS AND REPRODUCERS, AND PARTS AND ACCESSORIES OF SUCH ARTICLES

Notes:

1. This Section does not cover:

(a) Transmission or conveyor belts or belting, of plastics of Chapter 39, or of vulcanized rubber (heading No. 40.10); or other articles of a kind used in machinery or mechanical or electrical appliances or for other technical uses, of vulcanized rubber other than hard rubber (heading No. 40.16);

(b) Articles of leather or of composition leather (heading No. 42.04) or of furskin (heading No. 43.03), of a kind used in machinery or mechanical appliances or for other technical uses;

(c) Bobbins, spools, cops, cones, cores, reels or similar supports of any material (for example, Chapter 39, 40, 44 or 48 or Section XV);

(d) Perforated cards for Jacquard or similar machines (for example, Chapter 39 or 48 Section XV);

(e) Transmission or conveyor belts or belting of textile material (heading No. 59.10) or other articles of textile material for technical uses (heading No. 59.11);

(f) Precious or semi-precious stones (natural, synthetic or reconstructed) of headings Nos. 71.02 to 71.04, or articles wholly of such stones of heading No. 71.16, except unmounted worked sapphires and diamonds for styli (heading No. 85.22);

(g) Parts of general use, as defined in Note 2 to Section XV, of base metal (Section XV), or similar goods of plastics (Chapter 39);

(h) Drill pipe (heading No. 73.04);

(ij) Endless belts of metal wire or strip (Section XV);

(k) Articles of Chapter 82 or 83;

(l) Articles of Section XVII;

(m) Articles of Chapter 90;

(n) Clocks, watches or other articles of Chapter 91;

(o) Interchangeable tools of heading No. 82.07 brushes of

零件的刷子（税号96.03）；类似的可互换工具应按其构成工作部件的材料归类（例如，归入第四十章、第四十二章、第四十三章、第四十五章、第五十九章或税号68.04、69.09）；

（十五）第九十五章的物品。

（十六）打字机色带或类似色带，不论是否装轴或装盒（按其材料属性归类；如已上油或经其他方法处理能着色的，应归入税号96.12）。

二、除本类注释一、第八十四章注释一及第八十五章注释一另有规定的以外，机器零件（不属于税号84.84、85.44、85.45、85.46或85.47所列物品的零件）应按下列规定归类：

（一）凡在第八十四章、第八十五章的税号（税号84.09、84.31、84.48、84.66、84.73、84.85、85.03、85.22、85.29、85.38及85.48除外）列名的货品，均应归入该两章的相应税号；

（二）专用于或主要用于某一种机器或同一税号的多种机器（包括税号84.79或85.43的机器）的零件，应与该种机器一并归类，或酌情归入税号84.09、84.31、84.48、84.66、84.73、85.03、85.22、85.29或85.38。但能同时主要用于税号85.17和85.25至85.28所列机器的零件，应归入税号85.17；

（三）所有其他零件应酌情归入税号84.09、84.31、84.48、84.66、84.73、85.03、85.22、85.29或85.38，如不能归入上述税号，则应归入税号84.85或85.48。

三、由两部及两部以上机器装配在一起形成的组合式机器，或具有两种及两种以上互补或交替功能的机器，除条文另有规定的以外，应按具有主要功能的机器归类。

四、由不同独立部件（不论是否分开或由管道、传动装置、电缆或其他装置连接）组成的机器（包括机组），如果组合后明显具有一种第八十四章或第八十五章某个税号所列功能，则全部机器应按其功能归入有关税号。

五、上述各注释所称“机器”，是指第八十四章或第八十五章各税号所列的各种机器、设备、装置及器具。

a kind used as parts of machines (heading No. 96.03), similar interchangable tools are to be classified according to the material of working parts (for example, Chapter 40, 42, 43, 45, 59 or heading No. 68.04 or 69.09);

(p) Articles of Chapter 95;

(q) Typewriter or similar ribbons, whether or not on spools or in cartridges (classified according to their constituent material, or in heading 96.12 if inked or otherwise prepared for giving impressions).

2. Subject to Note 1 to this Section, Note 1 to Chapter 84 and Note 1 to Chapter 85, parts of machines (not being parts of the articles of heading No. 84.84, 85.44, 85.45, 85.46 or 85.47) are to be classified according to the following rules:

(a) Parts which are goods included in any of the headings of Chapter 84 or 85 (other than headings Nos. 84.09, 84.31, 84.48, 84.66, 84.73, 84.85, 85.03, 85.22, 85.29, 85.38, and 85.48) are in all cases to be classified in their respective headings;

(b) Other parts, if suitable for use solely or principally with a particular kind of machine, or with a number of machines of the same heading (including machines of heading No. 84.79 or 85.43) are to be classified with the machines of that kind or in heading No. 84.09, 84.31, 84.48, 84.66, 84.73, 85.03, 85.22, 85.29 or 85.38 as appropriate. However, parts which are equally suitable for use principally with the goods of headings Nos. 85.17 and 85.25 to 85.28 are to be classified in heading No.85.17;

(c) All other parts are to be classified in heading No.84.09, 84.31, 84.48, 84.66, 84.73, 85.03, 85.22, 85.29 or 85.38 as appropriate or, failing that, in heading No. 84.85 or 85.48.

3. Unless the context otherwise requires, composite machines consisting of two or more machines fitted together to form a whole and other machines adapted for the purpose of performing two or more complementary or alternative functions are to be classified as if consisting only of that component or as being that machine which performs the principal function.

4. Where a machine (including a combination of machines) consists of individual components (whether separate or interconnected by piping, by transmission devices, by electric cables or by other devices) intended to contribute together to a clearly defined function covered by one of the headings in Chapter 84 or Chapter 85, then the whole falls to be classified in the heading appropriate to that function.

5. For the purposes of these Notes, the expression "machine" means any machine, machinery, plant, equipment, apparatus or appliance cited in the headings of Chapter 84 or 85.

第八十四章
核反应堆、锅炉、机器、机械器具及其零件

Chapter 84
Nuclear reactors，boilers，machinery and mechanical appliances；parts thereof

注释：

一、本章不包括：

（一）第六十八章的石磨、石碾及其他物品；

（二）陶瓷材料制的机器或器具（例如，泵）及供任何材料制的机器或器具用的陶瓷零件（第六十九章）；

（三）实验室用玻璃器（税号 70.17）；玻璃制的机器、器具或其他专门技术用途的物品及其零件（税号 70.19 或 70.20）；

（四）税号 73.21 或 73.22 的物品或其他贱金属制的类似物品（第七十四章至第七十六章或第七十八章至第八十一章）；

（五）税号 85.09 的家用电动器具；或税号 85.25 的数字照相机；

（六）非机动的手工操作地板清扫器（税号 96.03）。

二、除第十六类注释三另有规定的以外，如果某种机器或器具既符合税号 84.01 至 84.24 中一个或几个税号的规定，又符合税号 84.25 至 84.80 中一个或几个税号的规定，则应归入税号 84.01 至 84.24 中的相应税号，而不归入税号 84.25 至 84.80 中的有关税号。

但税号 84.19 不包括：

（一）催芽装置、孵卵器或育雏器（税号 84.36）；

（二）谷物调湿机（税号 84.37）；

（三）萃取糖汁的浸提装置（税号 84.38）；

（四）纱线、织物及纺织制品的热处理机器（税号 84.51）；

（五）温度变化（即使必不可少）仅作为辅助功能的机器设备。

税号 84.22 不包括：

（一）缝合袋子或类似品用的缝纫机（税号 84.52）；

（二）税号 84.72 的办公室用机器。

税号 84.24 不包括：

喷墨印刷（打印）机器（税号 84.43 或 84.71）。

三、如果用于加工各种材料的某种机床既符合税号 84.56 的规定，又符合税号 84.57、84.58、84.59、84.60、84.61、84.64 或 84.65 的规定，则应归入税号 84.56。

Notes：

1．This Chapter does not cover：

（a）Millstones，grindstones or other articles of Chapter 68；

（b）Machinery or appliances （for example，pumps） of ceramic material and ceramic parts of machinery or appliances of any material （Chapter 69）；

（c）Laboratory glassware（heading No. 70.17）；machinery，appliances or other articles for technical uses or parts thereof，of glass（heading No. 70.19 or 70.20）；

（d）Articles of heading No. 73.21 or 73.22 or similar articles of other base metals （Chapters 74 to 76 or 78 to 81）；

（e）Elctro-mechanical domestic appliances of heading No. 85.09；or digital cameras of heading 85.25； or

（f）Hand-operated mechanical floor sweepers，not motorized （heading No. 96.03）.

2．Subject to the operation of Note 3 to Section XVI，a machine or appliance which answers to a description in one or more of the headings Nos. 84.01 to 84.24 and at the same time to a description in one or other of the headings Nos. 84.25 to 84.80 is to be classified under the appropriate heading of the former group and not the latter.

Heading No. 84.19 does not，however，cover：

（a）Germination plant，incubators or brooders （heading No. 84.36）；

（b）Grain dampening machines （heading No. 84.37）；

（c）Diffusing apparatus for sugar juice extraction （heading No. 84.38）；

（d）Machinery for the heat-treatment of textile yarns，fabrics or made up textile articles（heading No. 84.51）；or

（e）Machinery or plant，designed for mechanical operation，in which a change of temperature，even if necessary，is subsidiary.

Heading No. 84.22 does not cover：

（a）Sewing machines for closing bags or similar containers （heading No. 84.52）； or

（b）Office machinery of heading No. 84.72.

Heading No. 84.24 does not cover：

Ink-jet printing machines（heading No. 84.43 or 84.71）.

3．A machine-tool for working any material which answers to a description in heading No. 84.56 and at the same time to a description in heading No. 8 4.57，84.58，84.59，84.60，84.61，84.64 or 84.65 is to be classified in heading No.

四、税号 84.57 仅适用于可以完成下列不同形式机械操作的金属加工机床，但车床（包括车削中心）除外：

（一）按照机械加工程序从刀具库中自动更换刀具（加工中心）；

（二）同时或顺序地自动使用不同的动力头对固定不动的工件进行加工（单工位组合机床）；

（三）自动将工件送向不同的动力头（多工位组合机床）。

五、（一）税号 84.71 所称"自动数据处理设备"，是指：

1. 数字式计算机，该机能够（1）存储处理程序和执行程序直接需要的起码的数据；（2）按照用户的要求随意编辑程序；（3）按照用户指令进行算术计算；以及（4）在运行过程中，可不需人为干预而通过逻辑判断，执行一个处理程序，这个处理程序可改变计算机指令的执行。

2. 模拟式计算机，即能够模拟数学模型并且至少包括模拟部件、控制部件和程序部件的机器。

3. 混合式计算机，即具有模拟部件的数字式计算机或具有数字部件的模拟式计算机。

（二）自动数据处理设备可以是一套由若干单独部件所组成的系统。除本条注释（五）另有规定的以外，一个部件如果符合下列所有规定，即可视为整套系统的一部分：

1. 专用于或主要用于自动数据处理系统；

2. 可以直接或通过一个或几个其他部件同中央处理器相联接；

3. 能够以本系统所使用的方式（代码或信号）接收或传送数据。

（三）自动数据处理设备的部件如果单独进口或出口，应归入税号 84.71。

（四）打印机、键盘、X—Y 座标输入装置及盘（片）式存储部件，只要符合上述注

84.56.

4. Heading No. 84.57 applies only to machine-tools for working metal other than lathes (including turning centres), which can carry out different types of machining operations either:

(a) by automatic tool change from a magazine or the like in conformity with a machining programme (machining centres),

(b) by the automatic use, simultaneously or sequentially, of different unit heads working on a fixed position workpiece (unit construction machines, single station), or

(c) by the automatic transfer of the workpiece to different unit heads (multi-station transfer machines).

5. (a) For the purposes of heading No. 84.71, the expression "automatic data processing machines" means:

(a) Digital machines, capable of (1) storing the processing program or programs and at least the data immediately necessary for the execution of the program; (2) being freely programmed in accordance with the requirements of the user; (3) performing arthmetical computations specified by the user; and (4) executing, without human intervention, a processing program which requires them to modify their execution, by logical decision during the processing run.

(b) Analogue machines capable of simulating mathematical models and comprising at least: analogue elements, control elements and programming elements.

(c) Hybrid machines consisting of either a digital machine with analogue elements or an analogue machine with digital elements.

(b) Automatic data processing machines may be in the form of systems consisting of a variable number of separate units. Subject to paragraph (E) below, a unit is to be regarded as being a part of a complete system if it meets all of the following conditions:

(a) It is of a kind solely or principally used in an automatic data processing system;

(b) It is connectable to the central processing unit either directly or through one or more other units; and

(c) It is able to accept or deliver data in a form (codes or signals) which can be used by the system.

(c) Separately presented units of an automatic data processing machine are to be classified in heading No.84.71.

(d) Printers, keyboards, X-Y co-ordinate input devices and disk storage units which satisfy the conditions of

释（二）2、3 所列的规定，应一律作为税号 84.71 的部件归类。

（五）装有自动数据处理装置或与自动数据处理设备连接使用，但却从事数据处理以外的某项专门功能的机器，应按其功能归入相应的税号，对于无法按功能归类的，应归入未列名税号。

六、税号 84.82 还包括最大直径及最小直径与标称直径相差均不超过 1% 或 0.05 毫米（以相差数值较小的为准）的抛光钢珠，其他钢珠归入税号 73.26。

七、具有一种以上用途的机器在归类时，其主要用途可作为唯一的用途对待。除本章注释二、第十六类注释三另有规定的以外，凡任何税号都未列明其主要用途的机器，以及没有哪一种用途是主要用途的机器，均应归入税号 84.79。税号 84.79 还包括将金属丝、纺织纱线或其他各种材料以及它们的混合材料制成绳、缆的机器（例如，捻股机、绞扭机、制缆机）。

八、税号 84.70 所称“袖珍式”，仅适用于外形尺寸不超过 170 毫米×100 毫米×45 毫米的机器。

子目注释：

一、子目号 8471.49 所称“系统”，是指各部件符合第八十四章注释五（二）所列条件，并且至少由一个中央处理部件、一个输入部件（例如，键盘或扫描器）及一个输出部件（例如，视频显示器或打印机）组成的自动数据处理设备。

二、子目号 8482.40 仅包括滚柱直径相同，最大不超过 5 毫米，且长度至少是直径三倍的圆滚柱轴承，滚柱的两端可以磨圆。

paragraphs (B) (b) and (B) (c) above, are in all cases to be classified as units of heading No. 84.71.

(e) Machines performing a specific function other than data processing and incorporating or working in conjunction with an automatic data processing machine are to be classified in the headings appropriate to their respective functions or, failing that, in residual headings.

6. Heading No. 84.82 applies, *inter alia*, to polished steel balls, the maximum and minimum diameters of which do not differ from the nominal diameter by more than 1% or by more than 0.05mm, whichever is less. Other steel balls are to be classified in heading No. 73.26.

7. A machine which is used for more than one purpose is, for the purposes of classification, to be treated as if its principal purpose were its sole purpose. Subject to Note 2 to this Chapter and Note 3 to Section XVI, a machine, the principal purpose of which is not described in any heading or for which no one purpose is the principal purpose, unless the context otherwise requires, is to be classified in heading No. 84.79. Heading No. 84.79 also covers machines for making rope or cable (for example, stranding, twisting or cabling machines) from metal wire, textile yarn or any other material or from a combination of such materials.

8. For the purposes of heading No. 84.70, the term "pocket-size" applies only to machines the dimensions of which do not exceed 170mm×100mm×45mm.

Subheading Notes:

1. For the purposes of subheading No.8471.49, the term "systems" means automatic data processing machines whose units satisfy the conditions laid down in Note 5 (B) to Chapter 84 and which comprise at least a central processing unit, one input unit (e. g., a keyboard or a scanner), and one output unit (e. g., a visual display unit or a printer).

2. Subheading No. 8482.40 applies only to bearings with cylindrical rollers of a uniform diameter not exceeding 5mm and having a length which is at least three times the diameter. The ends of the rollers may be rounded.

序号 No.	税则号列 Tariff Item	货品名称	最惠国税率(%) M.F.N.	协定税率(%)Agreement 东盟 ASEAN	曼谷协定 BA	香港 H.K.	澳门 Macao	普通税率(%) Gen.	Article Description
	84.01	**核反应堆;核反应堆的未辐照燃料元件(释热元件);同位素分离机器及装置:**							**Nuclear reactors; fuel elements (cartridges), non-irradiated, for nuclear reactors; machinery and apparatus for isotopic separation:**
5351	8401.1000	-核反应堆	2					8	-Nuclear reactors
5352	8401.2000	-同位素分离机器、装置及其零件	1					8	-Machinery and apparatus for isotopic separation, and parts thereof
		-未辐照燃料元件(释热元件):							-Fuel elements(cartridges), non-irradiated:
5353	8401.3010	---未辐照燃料元件	2					8	---Fuel elements, non-irrdadiated
5354	8401.3090	---未辐照燃料元件的零件	1					8	---Parts for fuel elements non-irradiated
		-核反应堆零件:							-Parts of nuclear reactors:
5355	8401.4010	---未辐照相关组件	1					8	---Non-irradiated Associated Assembly
5356	8401.4020	---堆内构件	1					8	---Reactor internals
5357	8401.4090	---其他	1					8	---Other
	84.02	**蒸汽锅炉(能产生低压水蒸汽的集中供暖用的热水锅炉除外);过热水锅炉:**							**Steam or other vapour generating boilers(other than central heating hot water boilers capable also of producting low pressure steam); super-heated water boilers:**
		-蒸汽锅炉:							-Steam or other vapour generating boilers:
		--蒸发量超过45吨/时的水管锅炉:							--Watertube boilers with a steam production exceeding 45t per hour:
5358	8402.1110	---蒸发量在900吨/时及以上的发电用锅炉	3					11	---Boilers for generating electricity with a steam production 900t or more per hour
5359	8402.1190	---其他	14					35	---Other
5360	8402.1200	--蒸发量不超过45吨/时的水管锅炉	5		3.9			35	--Watertube boilers with a steam production not exceeding 45t per hour
5361	8402.1900	--其他蒸汽锅炉,包括混合式锅炉	5					35	--Other vapour generating boilers, including hybrid boilers
5362	8402.2000	-过热水锅炉	16					35	-Super-heated water boilers
5363	8402.9000	-零件	2					11	-Parts
	84.03	**集中供暖用的热水锅炉,但税号84.02的货品除外:**							**Central heating boilers other than those of heading No.84.02:**
5364	8403.1010	-家用型	10					80	-Household type
5365	8403.1090	-其他	10					80	-Other
5366	8403.9000	-零件	6					80	-Parts
	84.04	**税号84.02或84.03所列锅炉的辅助设备(例如,节热器、过热器、除灰器、气体回收器);水蒸汽或其他蒸汽动力装置的冷凝器:**							**Auxiliary plant for use with boilers of heading No. 84.02 or 84.03 (for example, economizers, super-heaters, soot removers, gas recoverers); condensers for steam or other vapour power units:**

序号 No.	税则号列 Tariff Item	货品名称	最惠国税率(%) M.F.N.	协定税率(%)Agreement 东盟 ASEAN	曼谷协定 BA	香港 H.K.	澳门 Macao	普通税率(%) Gen.	Article Description
		-税号84.02或84.03所列锅炉的辅助设备:							-Auxiliary plant for use with boilers of heading No.84.02 or 84.03:
5367	8404.1010	---税号84.02所列锅炉的辅助设备	7					35	---For use with boilers of heading No. 84.02
5368	8404.1020	---税号84.03所列锅炉的辅助设备	10					80	---For use with boilers of heading No. 84.03
5369	8404.2000	-水蒸汽或其他蒸汽动力装置的冷凝器	14					35	-Condensers for steam or other vapour power units
		-零件:							-Parts:
5370	8404.9010	---子目号8404.1020所列设备的零件	10					80	---Of the auxiliary plant of subheading No.8404.1020
5371	8404.9090	---其他	7					35	---Other
	84.05	**煤气发生器,不论有无净化器;乙炔发生器及类似的水解气体发生器,不论有无净化器:**							**Producer gas or water gas generators, with or without their purifiers; acetylene gas generators and similar water process gas generators, with or without their purifiers:**
5372	8405.1000	-煤气发生器,不论有无净化器;乙炔发生器及类似的水解气体发生器,不论有无净化器	14					30	-Producer gas or water gas generators, with or without their purifiers; acetylene gas generators and similar water process gas generators, with or without their purifiers
5373	8405.9000	-零件	8					30	-Parts
	84.06	**汽轮机:**							**Steam turbines and other vapour turbines:**
5374	8406.1000	-船舶动力用汽轮机	5					35	-Turbines for marine propulsion
		-其他汽轮机:							-Other turbines:
		--输出功率超过40兆瓦的:							--Of an output exceeding 40 MW:
5375	8406.8110	---输出功率不超过100兆瓦的	5					35	---Of an output not exceeding 100 MW
5376	8406.8120	---输出功率超过100兆瓦,但不超过350兆瓦的	5					35	---Of an output exceeding 100 MW but not exceeding 350 MW
5377	8406.8130	---输出功率超过350兆瓦的	6					11	---Of an output exceeding 350 MW
5378	8406.8200	--输出功率不超过40兆瓦的	5					35	--Of an output not exceeding 40 MW
5379	8406.9000	-零件	2					11	-Parts
	84.07	**点燃往复式或旋转式活塞内燃发动机:**							**Spark-ignition reciprocating or rotary internal combustion piston engines:**
		-航空器发动机:							-Aircraft engines:
5380	8407.1010	---输出功率不超过298千瓦	2					11	---Of an output not exceeding 298kW
5381	8407.1020	---输出功率超过298千瓦	2					11	---Of an output exceeding 298kW

序号 No.	税则号列 Tariff Item	货品名称	最惠国税率(%) M.F.N.	协定税率(%) Agreement 东盟 ASEAN	曼谷协定 BA	香港 H.K.	澳门 Macao	普通税率(%) Gen.	Article Description
		-船舶发动机:							-Marine propulsion engines:
5382	8407.2100	--舷外发动机	8					35	--Outboard motors
5383	8407.2900	--其他	8					20	--Other
		-用于第八十七章所列车辆的往复式活塞发动机:							-Reciprocating piston engines of a kind used for the propulsion of vehicles of Chapter 87:
5384	8407.3100	---气缸容量(排气量)不超过50毫升	10					35	--Of a cylinder capacity not exceeding 50cc
5385	8407.3200	--气缸容量(排气量)超过50毫升,但不超过250毫升	10					35	--Of a cylinder capacity exceeding 50cc but not exceeding 250cc
5386	8407.3300	--气缸容量(排气量)超过250毫升,但不超过1000毫升	17.5					70	--Of a cylinder capacity exceeding 250cc but not exceeding 1000cc
		--气缸容量(排气量)超过1000毫升:							--Of a cylinder capacity exceeding 1000cc:
5387	8407.3410	---气缸容量(排气量)超过1000毫升,但不超过3000毫升	10					70	---Of a cylinder capacity exceeding 1000cc but not exceeding 3000cc
5388	8407.3420	---气缸容量(排气量)超过3000毫升	10					35	---Of a cylinder capacity exceeding 3000cc
		-其他发动机:							-Other engines:
5389	8407.9010	---沼气发动机	12					35	---Firedamp engines
5390	8407.9090	---其他	18					35	---Other
	84.08	**压燃式活塞内燃发动机(柴油或半柴油发动机):**							**Compression-ignition internal combustion piston engines(diesel or semi-diesel engines):**
5391	8408.1000	-船舶发动机	5					11	-Marine propulsion engines
		-用于第八十七章所列车辆的发动机:							-Engines of a kind used for the propulsion of vehicles of Chapter 87:
5392	8408.2010	---输出功率在132.39千瓦(180马力)及以上	9					14	---Of an output of 132.39kW(180H.P.)or more
5393	8408.2090	---其他	25					35	---Other
		-其他发动机:							-Other engines:
5394	8408.9010	---机车发动机	6					11	---Locomotive engines
		---其他:							---Other:
5395	8408.9091	----输出功率不超过14千瓦	5					35	----Of an output not exceeding 14kW
5396	8408.9092	----输出功率超过14千瓦,但小于132.39千瓦(180马力)	8.4					35	----Of an output exceeding 14kW but not exceeding 132.39kW(180H.P.)
5397	8408.9093	----输出功率在132.39千瓦(180马力)及以上	5					14	----Of an output of 132.39kW(180H.P.)or more

序号 No.	税则号列 Tariff Item	货品名称	最惠国税率(%) M.F.N.	协定税率(%)Agreement 东盟 ASEAN	曼谷协定 BA	香港 H.K.	澳门 Macao	普通税率(%) Gen.	Article Description
	84.09	**专用于或主要用于税号84.07或84.08所列发动机的零件:**							**Parts suitable for use solely or principally with the engines of heading No. 84.07 or 84.08:**
5398	8409.1000	-航空器发动机用	2					11	-For aircraft engines
		-其他:							-Other:
		--专用于或主要用于点燃式活塞内燃发动机的:							--Suitable for use solely or principally with spark-ignition internal combustion piston engines:
5399	8409.9110	---船舶发动机用	6					17	---For marine propulsion engines
		---其他:							---Other:
5400	8409.9191	----电控燃油喷射装置	5					35	----Electric fuel injection devices
5401	8409.9199	----其他	5					35	----Other
		--其他:							--Other:
5402	8409.9910	---船舶发动机用	5					11	---For marine propulsion engines
5403	8409.9920	---机车发动机用	2					11	---For locomotive engines
		---其他:							---Other:
5404	8409.9991	----输出功率在132.39千瓦(180马力)及以上的发动机用	2					11	----For engines with an output of 132.39kW(180H.P.)or more
5405	8409.9999	----其他	8.4					35	----Other
	84.10	**水轮机、水轮及其调节器:**							**Hydraulic turbines, water wheels, and regulators therefor:**
		-水轮机及水轮:							-Hydraulic turbines and water wheels:
5406	8410.1100	--功率不超过1000千瓦	10					35	--Of a power not exceeding 1000kW
5407	8410.1200	--功率超过1000千瓦,但不超过10000千瓦	10					35	--Of a power exceeding 1000kW but not exceeding 10000kW
		--功率超过10000千瓦:							--Of a power exceeding 10000kW:
5408	8410.1310	---功率超过30000千瓦的冲击式水轮机及水轮	10					35	---Impulse hydraulic turbines and water wheels of a power>30000kW
5409	8410.1320	---功率超过35000千瓦的贯流式水轮机及水轮	10					35	---Radial hydraulic turbines and water wheels of a power>35000kW
5410	8410.1330	---功率超过200000千瓦的水泵水轮机及水轮	10					35	---Pumping hydraulic turbines and water wheels of a power>200000kW
5411	8410.1390	---其他	10					35	---Other
		-零件,包括调节器:							-Parts, including regulators:
5412	8410.9010	---调节器	6					35	---Regulators
5413	8410.9090	---其他	6					35	---Other
	84.11	**涡轮喷气发动机,涡轮螺桨发动机及其他燃气轮机:**							**Turbo-jets, turbo-propellers and other gas turbines:**
		-涡轮喷气发动机:							-Turbo-jets:

序号 No.	税则号列 Tariff Item	货品名称	最惠国税率(%) M.F.N.	协定税率(%)Agreement 东盟 ASEAN	曼谷协定 BA	香港 H.K.	澳门 Macao	普通税率(%) Gen.	Article Description
		--推力不超过25千牛顿:							--Of a thrust not exceeding 25kN:
5414	8411.1110	---涡轮风扇发动机	1					11	---Turbofan engines
5415	8411.1190	---其他	1					11	---Other
		--推力超过25千牛顿:							--Of a thrust exceeding 25kN:
5416	8411.1210	---涡轮风扇发动机	1					11	---Turbofan engines
5417	8411.1290	---其他	1					11	---Other
		-涡轮螺桨发动机:							-Turbo-propellers:
5418	8411.2100	--功率不超过1100千瓦	2					11	--Of a power not exceeding 1100kW
		--功率超过1100千瓦:							--Of a power exceeding 1100kW:
5419	8411.2210	---功率超过1100千瓦,但不超过2238千瓦	2					11	---Of a power exceeding 1100kW but not exceeding 2238kW
5420	8411.2220	---功率超过2238千瓦,但不超过3730千瓦	2					11	---Of a power exceeding 2238kW but not exceeding 3730kW
5421	8411.2230	---功率超过3730千瓦	2					11	---Of a power exceeding 3730kW
		-其他燃气轮机:							-Other gas turbines:
5422	8411.8100	--功率不超过5000千瓦	15					35	--Of a power not exceeding 5000kW
5423	8411.8200	--功率超过5000千瓦	3					35	--Of a power exceeding 5000kW
		-零件:							-Parts:
5424	8411.9100	--涡轮喷气发动机或涡轮螺桨发动机用	1					11	--Of turbo-jets or turbo-propellers
		--其他:							--Other:
5425	8411.9910	---涡轮轴发动机用	5					35	---Of turboshaft engines
5426	8411.9990	---其他	5					35	---Other
	84.12	**其他发动机及动力装置:**							**Other engines and motors:**
		-喷气发动机,但涡轮喷气发动机除外:							-Jet engines other than turbo-jets:
5427	8412.1010	---航空器及航天器用	3					11	---For aircraft or spacecraft
5428	8412.1090	---其他	10					35	---Other
		-液压动力装置:							-Hydraulic power engines and motors:
5429	8412.2100	--直线作用(液压缸)的	12					35	--Linear acting(cylinders)
		--其他:							--Other:
5430	8412.2910	---液压马达	10					35	---Hydraulic motors
5431	8412.2990	---其他	14					35	---Other
		-气压动力装置:							-Pneumatic power engines and motors:
5432	8412.3100	--直线作用(气压缸)的	14					35	--Linear acting(cylinders)
5433	8412.3900	--其他	14					35	--Other
5434	8412.8000	-其他	10					35	-Other
		-零件:							-Parts:
5435	8412.9010	---子目号8412.1010所列机器的零件	2					11	---For machines of subheading No. 8412.1010
5436	8412.9090	---其他	8					35	---Other

序号 No.	税则号列 Tariff Item	货品名称	最惠国税率(%) M.F.N.	协定税率(%)Agreement 东盟 ASEAN	曼谷协定 BA	香港 H.K.	澳门 Macao	普通税率(%) Gen.	Article Description
	84.13	**液体泵,不论是否装有计量装置;液体提升机:**							**Pumps for liquids, whether or not fitted with a measuring device; liquid elevators:**
		-装有或可装计量装置的泵:							-Pumps fitted or designed to be fitted with a measuring device:
5437	8413.1100	--分装燃料或润滑油的泵,用于加油站或车库	10					30	--Pumps for dispensing fuel or lubricants, of the type used in filling-stations or in garages
5438	8413.1900	--其他	10					30	--Other
5439	8413.2000	-手泵,但子目号8413.11或8413.19的货品除外	10					30	-Hand pumps, other than those of subheading No. 8413.11 or 8413.19
		-活塞式内燃发动机用的燃油泵、润滑油泵或冷却剂泵:							-Fuel, lubricating or cooling medium pumps for internal combustion piston engines:
		---燃油泵:							---Fuel pumps:
5440	8413.3021	----输出功率在132.39千瓦(180马力)及以上的发动机用燃油泵	3					30	----Fuel pumps for engines of an output of 132.39kW(180hp) or more
5441	8413.3029	----其他	3					30	----Other
5442	8413.3030	---润滑油泵	3					30	---Lubricating oil pumps
5443	8413.3090	---其他	3					30	---Other
5444	8413.4000	-混凝土泵	8			0		30	-Concrete pumps
		-其他往复式排液泵:							-Other reciprocating positive displacement pumps:
5445	8413.5010	---气动式	10					40	---Pneumatic
5446	8413.5020	---电动式	10					40	---Electric
5447	8413.5030	---液压式	10					40	---Hydraulic
5448	8413.5090	---其他	10					40	---Other
		-其他回转式排液泵:							-Other rotary positive displacement pumps:
5449	8413.6010	---电动潜油泵及潜水电泵	10					40	---Electric submersible oil pumps and electric submersible pumps
5450	8413.6090	---其他	10					40	---Other
		-其他离心泵:							-Other centrifugal pumps:
5451	8413.7010	---转速在10000转/分及以上	8					40	---Rotational speed no less than 10000r/min
5452	8413.7090	---其他	8					40	---Other
		-其他泵;液体提升机:							-Other pumps; liquid elevators:
5453	8413.8100	--泵	8					40	--Pumps
5454	8413.8200	--液体提升机	8					30	--Liquid elevators
		-零件:							-Parts:
5455	8413.9100	--泵用	5			0		30	--Of pumps
5456	8413.9200	--液体提升机用	6					30	--Of liquid elevators

序号 No.	税则号列 Tariff Item	货品名称	最惠国税率(%) M.F.N.	协定税率(%)Agreement				普通税率(%) Gen.	Article Description
				东盟 ASEAN	曼谷协定 BA	香港 H.K.	澳门 Macao		
	84.14	**空气泵或真空泵、空气及其他气体压缩机、风机、风扇；装有风扇的通风罩或循环气罩，不论是否装有过滤器：**							**Air or vacuum pumps, air or other gas compressors and fans; ventilating or recycling hoods incorporating a fan, whether or not fitted with filters:**
5457	8414.1000	-真空泵	8					30	-Vacuum pumps
5458	8414.2000	-手动或脚踏式空气泵	8					30	-Hand-or foot-operated pumps
		-用于制冷设备的压缩机：							-Compressors of a kind used in refrigerating equipment:
		---电动机驱动的压缩机：							---Driven by a motor:
5459	8414.3011	----冷藏箱或冷冻箱用，电动机额定功率不超过0.4千瓦	8		5.9	0		80	----For refrigerators or freezers, of a motor power not exceeding 0.4kW
5460	8414.3012	----冷藏箱或冷冻箱用，电动机额定功率超过0.4千瓦，但不超过5千瓦	10		8.5	0		80	----For refrigerators or freezers, of a motor power exceeding 0.4kW but not exceeding 5kW
5461	8414.3013	----空气调节器用，电动机额定功率超过0.4千瓦，但不超过5千瓦	10		8.5	0		80	----For air conditioning machines, of a motor power exceeding 0.4kW but not exceeding 5kW
5462	8414.3014	----空气调节器用，电动机额定功率超过5千瓦	10		8.5	0		80	----For air conditioning machines, of a motor power exceeding 5kW
5463	8414.3015	----冷冻或冷藏设备用，电动机额定功率超过5千瓦	10		9.2	0		30	----For refrigerators or freezers, of a motor power exceeding 5kw
5464	8414.3019	----其他	10		9.2	0		30	----Other
5465	8414.3090	---非电动机驱动的压缩机	9		8.8	0		80	---Driven by a non-motor
5466	8414.4000	-装在拖车底盘上的空气压缩机	8					30	-Air compressors mounted on a wheeled chassis for towing
		-风机、风扇：							-Fans:
		--台扇、落地扇、壁扇、换气扇或吊扇，包括风机，本身装有一个输出功率不超过125瓦的电动机：							--Table, floor, wall, window, ceiling or roof fans, with a self-contained electric motor of an output not exceeding 125W:
5467	8414.5110	---吊扇	20					130	---Ceiling or roof fans
5468	8414.5120	换气扇	20					130	---Window fans
5469	8414.5130	---具有旋转导风轮的风扇	12					130	---Repeating front louver fan
		---其他：							---Other:
5470	8414.5191	----台扇	10					130	----Table fans
5471	8414.5192	----落地扇	10					130	----Floor fans
5472	8414.5193	----壁扇	10					130	----Wall fans

序号 No.	税则号列 Tariff Item	货品名称	最惠国税率(%) M.F.N.	协定税率(%)Agreement 东盟 ASEAN	曼谷协定 BA	香港 H.K.	澳门 Macao	普通税率(%) Gen.	Article Description
5473	8414.5199	----其他:	10					130	----Other
		--其他:							--Other:
5474	8414.5910	---吊扇	8					30	---Ceiling or roof fans
5475	8414.5920	---换气扇	8					30	---Window fans
5476	8414.5930	---离心通风机	10					30	---Centrifugal ventilation fans
5477	8414.5990	---其他	8					30	---Other
		-罩的平面最大边长不超过120厘米的通风罩或循环气罩:							-Hoods having a maximum horizontal side not exceeding 120cm:
5478	8414.6010	---抽油烟机	10					130	---Range hoods
5479	8414.6090	---其他	10					130	---Other
		-其他:							-Other:
5480	8414.8010	---燃气轮机用的自由活塞式发生器	8					50	---Free piston generators for gas turbines
5481	8414.8020	---二氧化碳压缩机	7					30	---CO_2 compressors
5482	8414.8090	---其他:	7					30	---Other
		-零件:							-Parts:
		---子目号8414.3011至8414.3014及8414.3090所列机器的零件:							---Of the machines of subheadings Nos. 8414.3011 to 8414.3014 and 8414.3090:
5483	8414.9011	----压缩机进、排气阀片	8					80	----In take valve leaf or discharge valve leaf
5484	8414.9019	----其他	8					80	----Other
5485	8414.9020	---子目号8414.5110至8414.5199及8414.6000所列机器的零件	12					130	---Of the machines of subheadings Nos. 8414.5110 to 8414.5199 or 8414.6000
5486	8414.9090	---其他	7			0		30	---Other
	84.15	**空气调节器,装有电扇及调温、调湿装置,包括不能单独调湿的空调器:**							**Air conditioning machines, comprising a motor-driven fan and elements for changing the temperature and humidity, including those machines in which the humidity cannot be separately regulated:**
		-窗式或壁式,独立的或分体的:							-Window or wall types, self – contained or "split – system":
5487	8415.1010	---独立式	15					130	---Self – contained
		---分体式							---Split – systerm:
5488	8415.1021	----制冷量不超过4000大卡/时	15					130	----Of a refirigerating effect not exceeding 4000 Cal per hour
5489	8415.1022	----制冷量超过4000大卡/时	15					90	----Of a refirigerating effect exceeding 4000 Cal per hour
5490	8415.2000	-机动车辆上供人使用的	20			0△		110	-Of a kind used for persons, in motor vehicles
		-其他:							-Other:

序号 No.	税则号列 Tariff Item	货品名称	最惠国税率(%) M.F.N.	协定税率(%) Agreement 东盟 ASEAN	曼谷协定 BA	香港 H.K.	澳门 Macao	普通税率(%) Gen.	Article Description
		--装有制冷装置及一个冷热循环换向阀的(可逆式热泵):							--Incorporating a refrigerating unit and a valve for reversal of the cooling/heat cycle (reversible heat pumps):
5491	8415.8110	---制冷量不超过4000大卡/时	15					130	---Of a refrigerating effect not exceeding 4000 Cal per hour
5492	8415.8120	---制冷量超过4000大卡/时	20					90	---Of a refrigerating effect exceeding 4000 Cal per hour
		--其他,装有制冷装置的:							--Other, incorporating a refrigerating unit:
5493	8415.8210	---制冷量不超过4000大卡/时	15					130	---Of a refrigerating effect not exceeding 4000 Cal per hour
5494	8415.8220	---制冷量超过4000大卡/时	20					90	---Of a refrigerating effect exceeding 4000 Cal per hour
5495	8415.8300	--未装有制冷装置的	10					90	--Not incorporating a refrigerating unit
		-零件:							-Parts:
5496	8415.9010	---子目号8415.1010、8415.1021、8415.8110及8415.8210所列设备的零件	10		8	0		130	---Of the machines of subheading No. 8415.1010, 8415.1021, 8415.8110 or 8415.8210
5497	8415.9090	---其他	10		8	0		90	---Other
	84.16	**使用液体燃料、粉状固体燃料或气体燃料的炉用燃烧器;机械加煤机,包括其机械炉篦、机械出灰器及类似装置:**							**Furnace burners for liquid fuel, for pulverzied solid fuel or for gas; mechanical stokers, including their mechanical grates, mechanical ash dischargers and similar appliances:**
5498	8416.1000	-使用液体燃料的炉用燃烧器	10					35	-Furnace burners for liquid fuel
		-其他炉用燃烧器,包括复式燃烧器:							-Other furnace burners, including combination burners:
		---气体的:							---For gas:
5499	8416.2011	----使用天然气的	10.5					35	----Of using natural gas
5500	8416.2019	----其他	10.5					35	----Other
5501	8416.2090	---其他	10.5					35	---Other
5502	8416.3000	-机械加煤机,包括其机械炉篦、机械出灰器及类似装置	8.4					35	-Mechanical stokers, including their mechanical grates, mechanical ash dischargers and similar appliances
5503	8416.9000	-零件	6					35	-Parts
	84.17	**非电热的工业或实验室用炉及烘箱,包括焚烧炉:**							**Industrial or laboratory furnaces and ovens, including incinerators, non-elec-tric:**
5504	8417.1000	-矿砂、黄铁矿或金属的焙烧、熔化或其他热处理用炉及烘箱	10					35	-Furnaces and ovens for the roasting, melting or other heat-treatment of ores, pyrites or of metals
5505	8417.2000	-面包房用烤炉及烘箱,包括做饼干用的	10					35	-Bakery ovens, including biscuit ovens

序号 No.	税则号列 Tariff Item	货品名称	最惠国税率(%) M.F.N.	协定税率(%) Agreement 东盟 ASEAN	曼谷协定 BA	香港 H.K.	澳门 Macao	普通税率(%) Gen.	Article Description
		-其他：							-Other:
5506	8417.8010	---炼焦炉	10					35	---Coke ovens
5507	8417.8020	---放射性废物焚烧炉	5					35	---Burn furnaces for radioactive waste
5508	8417.8030	---水泥回转窑	10					35	---Cement rotary kilns
5509	8417.8040	---石灰石分解炉	10					35	---Limestone decomposition furnace
5510	8417.8090	---其他	10					35	---Other
		-零件：							-Parts:
5511	8417.9010	---海绵铁回转窑用	7					35	---For sponge iron rotary kiln
5512	8417.9020	---炼焦炉用	7					35	---For coke ovens
5513	8417.9090	---其他	7					35	---Other
	84.18	**电气或非电气的冷藏箱、冷冻箱及其他制冷设备；热泵，但税号84.15的空气调节器除外：**							**Refrigerators, freezers and other refrigerating or freezing equipment, electric or other; heat pumps other than air conditioning machines of heading No.84.15:**
		-冷藏-冷冻组合机，各自装有单独外门的：							-Combined refrigerator-freezers, fitted with separate external doors:
5514	8418.1010	---容积超过500升	10					100	---Of a capacity exceeding 500 l
5515	8418.1020	---容积超过200升，但不超过500升	15					130	---Of a capacity exceeding 200 l, not exceeding 500 l
5516	8418.1030	---容积不超过200升	15					130	---Of a capacity not exceeding 200l
		-家用型冷藏箱：							-Refrigerators, household type:
		--压缩式：							--Compression-type:
5517	8418.2110	---容积超过150升	10					130	---Of a capacity exceeding 150 l
5518	8418.2120	---容积超过50升，但不超过150升	10					130	---Of a capacity exceeding 50L, not exceeding 150L
5519	8418.2130	---容积不超过50升	10					130	---Of a capacity not exceeding 50L
5520	8418.2200	--电气吸收式	15					130	--Absorption-type, electrical
5521	8418.2900	--其他	30					130	--Other
		-柜式冷冻箱，容积不超过800升：							-Freezers of the chest type, not exceeding 800L capacity:
5522	8418.3010	---制冷温度在－40℃及以下	9				$0^{\triangle}$	50	---Of a refrigerating temperature of －40℃ or lower
		---制冷温度在－40℃以上：							---Of a refrigerating temperature higher than －40℃:
5523	8418.3021	----容积超过500升	23				$0^{\triangle}$	100	----Of a capacity exceeding 500L
5524	8418.3029	----其他	30				$0^{\triangle}$	130	----Other
		-立式冷冻箱，容积不超过900升：							-Freezers of the upright type, not exceeding 900L capacity:
5525	8418.4010	---制冷温度在－40℃及以下	9				$0^{\triangle}$	50	---Of a refrigerating temperature of －40℃ or lower
		---制冷温度在－40℃以上：							---Of a refrigerating temperature higher than －40℃:
5526	8418.4021	----容积超过500升	15				$0^{\triangle}$	100	----Of a capacity exceeding 500L
5527	8418.4029	----其他	30				$0^{\triangle}$	130	----Other

序号 No.	税则号列 Tariff Item	货品名称	最惠国税率(%) M.F.N.	协定税率(%) Agreement 东盟 ASEAN	曼谷协定 BA	香港 H.K.	澳门 Macao	普通税率(%) Gen.	Article Description
5528	8418.5000	-其他冷藏或冷冻柜、箱、展示台、陈列箱及类似的冷藏或冷冻设备	10			0	0△	100	-Other refrigerating or freezing chests, cabinets, display counters, show-cases and similar refrigerating or freezing furniture
		-其他制冷设备;热泵:							-Other refrigerating or freezing equipment; heat pumps:
		--冷凝器为热交换器的压缩式设备:							--Compression type units whose condensers are heat exchangers:
5529	8418.6110	---制冷机组;热泵	10					90	---Refrigerating units; heat pumps
5530	8418.6190	---其他	10				0△	130	---Other
		--其他:							--Other:
5531	8418.6910	---制冷机组;热泵	15					90	---Refrigerating units; heat pumps
5532	8418.6990	---其他	15					130	---Other
		-零件:							-Parts:
5533	8418.9100	--冷藏或冷冻设备专用的特制家具	18					130	--Furniture designed to receive refrigerating or freezing equipment
		--其他:							--Other:
5534	8418.9910	---制冷机组及热泵用	10					90	--Of refrigerating units and heat pumps
		---其他:							---Other:
5535	8418.9991	----制冷温度在-40℃及以下的冷冻设备用	9.5				0△	50	----Of freezing equipment of a refrigerating temperature of -40℃ or lower
5536	8418.9992	----制冷温度在-40℃以上,但容积超过500升的冷藏或冷冻设备用	10				0△	100	----Of refrigerating or freezing equipment of a refrigerating temperature higher than -40℃ and a capacity exceeding 500L
5537	8418.9999	----其他	10				0△	130	----Other
	84.19	**利用温度变化处理材料的机器、装置及类似的实验室设备,例如,加热、烹煮、烘炒、蒸馏、精馏、消毒、灭菌、汽蒸、干燥、蒸发、气化、冷凝、冷却的机器设备,不论是否电热的(不包括税目8514的炉、烘箱及其他设备),但家用的除外;非电热的快速热水器或贮备式热水器:**							**Machinery, plant or laboratory equipment, whether or not electrically heated (excluding furnaces, ovens and other equipment of heading 85.14), for the treatment of materials by a process involving a change of temperature such as heating, cooking, roasting, distilling, rectifying, sterilizing, pasteurizing, steaming, drying, evaporating, vaporizing, condensing or cooling, other than machinery or plant of a kind used for domestic purposes; instantaneous or storage water heaters, non-electric:**
		-非电热的快速热水器或贮备式热水器:							-Instantaneous or storage water heaters, non-electric:
5538	8419.1100	--燃气快速热水器	35					100	--Instantaneous gas water heaters
5539	8419.1900	--其他	35					100	--Other

序号 No.	税则号列 Tariff Item	货品名称	最惠国税率(%) M.F.N.	协定税率(%) Agreement 东盟 ASEAN	曼谷协定 BA	香港 H.K.	澳门 Macao	普通税率(%) Gen.	Article Description
5540	8419.2000	-医用或实验室用消毒器具	4					30	-Medical, surgical or laboratory sterilizers
		-干燥器:							-Dryers:
5541	8419.3100	--农产品干燥用	8					30	--For agricultural products
5542	8419.3200	--木材、纸浆、纸或纸板干燥用	9					30	--For wood, paper pulp, paper or paperboard
		--其他:							--Other:
5543	8419.3910	---微空气流动陶瓷坯件干燥器	9					30	---Breeze pottery blanks dryers
5544	8419.3990	---其他	9					30	---Other
		-蒸馏或精馏设备:							-Distilling or rectifying plant:
5545	8419.4010	---提净塔	10					30	---Stripping towers
5546	8419.4020	---精馏塔	10					30	---Rectifying towers
5547	8419.4090	---其他	10					30	---Other
5548	8419.5000	-热交换装置	10			0△		30	-Heat exchange units
		-液化空气或其他气体的机器:							-Machinery for liquefying air or other gases:
		---制氧机:							---Oxygen producers:
5549	8419.6011	----制氧量在15000立方米/小时及以上	12			0△		30	----Oxygen preparation volume no less than 15000m^3/h
5550	8419.6019	----其他	13			0△		30	----Other
5551	8419.6090	---其他	10			0△		30	---Other
		-其他机器设备:							-Other machinery, plant and equipment:
5552	8419.8100	--加工热饮料或烹调、加热食品用	10					30	--For making hot drinks of for cooking or heating food
		--其他:							--Other:
5553	8419.8910	---加氢反应器	0		0			30	---Hydroformer vessels
5554	8419.8990	---其他	0		0			30	---Other
		-零件:							-Parts:
5555	8419.9010	---热水器用	0					100	---Of water heaters
5556	8419.9090	---其他	4			0		30	---Other
	84.20	**研光机或其他滚压机器及其滚筒,但加工金属或玻璃用的除外:**							**Calendering or other rolling machines, other than for metals or glass, and cylinders therefor:**
5557	8420.1000	-研光机或其他滚压机器	8.4					30	-Calendering or other rolling machines
		-零件:							-Parts:
5558	8420.9100	--滚筒	8					30	--Cylinders
5559	8420.9900	--其他	8					30	--Other
	84.21	**离心机,包括离心干燥机;液体或气体的过滤、净化机器及装置:**							**Centrifuges, including centrifugal dryers; filtering or purifying machinery and apparatus, for liquids or gases:**

序号 No.	税则号列 Tariff Item	货品名称	最惠国税率(%) M.F.N.	协定税率(%)Agreement 东盟 ASEAN	曼谷协定 BA	香港 H.K.	澳门 Macao	普通税率(%) Gen.	Article Description
		-离心机,包括离心干燥机:							-Centrifuges, including centrifugal dryers:
5560	8421.1100	--奶油分离器	8.4					30	--Cream separators
		--干衣机:							--Clothes-dryers:
5561	8421.1210	---干衣量不超过10公斤	17.5					70	---Of a dry linen capacity not exceeding 10kg
5562	8421.1290	---其他	8					30	---Other
		--其他:							--Other:
5563	8421.1910	---脱水机	10					30	---Dewaterers
5564	8421.1920	---固液分离机	10					30	---Solid-liquor separators
5565	8421.1990	---其他	10					30	---Other
		-液体的过滤、净化机器及装置:							-Filtering or purifying machinery and apparatus for liquids:
		--过滤或净化水用:							--For filtering or purifying water:
5566	8421.2110	---家用型	25					63	---Of the household type
5567	8421.2190	---其他	5					50	---Other
5568	8421.2200	--过滤或净化饮料(水除外)用	12					40	---For filtering or purifying beverages other than water
5569	8421.2300	--内燃发动机的滤油器	10			0		40	--Oil or petrol-filters for internal combustion engines
		--其他:							--Other:
5570	8421.2910	---压滤机	5			0		40	---Press filters
5571	8421.2990	---其他	5			0		40	---Other
		-气体的过滤、净化机器及装置:							-Filtering or purifying machinery and apparatus for gases:
5572	8421.3100	--内燃发动机的进气过滤器	10			0		40	--Intake air filters for internal combustion engines
		--其他:							--Other:
5573	8421.3910	---家用型	15			0△		100	---Of the household type
		---工业用除尘器:							---Dust collectors for industry uses:
5574	8421.3921	----静电除尘器	5			0△		40	----Electrostatic
5575	8421.3922	----袋式除尘器	5			0△		40	----Baghoused
5576	8421.3923	----旋风式除尘器	5			0△		40	----Cyclone
5577	8421.3929	----其他	5			0△		40	----Other
5578	8421.3990	---其他	5			0△		40	---Other
		-零件:							-Parts:
		--离心机用,包括离心干燥机用:							--Of centrifuges, including centrifugal dryers:
5579	8421.9110	---干衣量不超过10公斤的干衣机用	0					70	---Of clothes-dryers of a dry linen capacity not exceeding 10kg
5580	8421.9190	---其他	0					30	---Other
		--其他:							--Other:
5581	8421.9910	---家用型过滤、净化装置用	10			0		100	---Of household-type filtering or purifying machines
5582	8421.9990	---其他	5			0		40	---Other

序号 No.	税则号列 Tariff Item	货品名称	最惠国税率(%) M.F.N.	协定税率(%)Agreement 东盟 ASEAN	曼谷协定 BA	香港 H.K.	澳门 Macao	普通税率(%) Gen.	Article Description
	84.22	**洗碟机;瓶子及其他容器的洗涤或干燥机器;瓶、罐、箱、袋或其他容器装填、封口、密封、贴标签的机器;瓶、罐、管、筒或类似容器的包封机器;其他包装或打包机器(包括热缩包装机器);饮料充气机:**							**Dish washing machines;machinery for cleaning or drying bottles or other containers;machinery for filling,closing,sealing or labelling bottles,cans,boxes,bags or other containers;machinery for capsuling bootles,jars,tubes and similar containers other packing or wrapping machinery(including heat-shrink wrapping machinery);machinery for aerating beverages:**
		-洗碟机:							-Dish washing machines:
5583	8422.1100	--家用型	10					90	--Of the household type
5584	8422.1900	--其他	14					90	--Other
5585	8422.2000	-瓶子或其他容器的洗涤或干燥机器	10					35	-Machinery for cleaning or drying bottles or other containers
		-瓶、罐、箱、袋或其他容器的装填、封口、密封、贴标签的机器;瓶、罐、管、筒或类似容器的包封机器;饮料充气机:							-Machinery for filling,closing,sealing,or labelling bottles,cans,boxes,bags or other containers;machinery for capsuling bottles,jars,tubes and similar containers;machinery for aerating beverages:
		---饮料及液体食品灌装设备:							---Bottling or canning machinery for beverages or liquid food:
5586	8422.3011	----电动手提式	10			0		45	----Electro－mechanical hand－lifting
5587	8422.3019	----其他	12			0		45	----Other
		---水泥包装机:							---Machinery for packing cement:
5588	8422.3021	----全自动灌包机	12			0		45	----Automatic filling and sacking machines
5589	8422.3029	----其他	12			0		45	----Other
5590	8422.3030	---其他包装机	10			0		35	----Other packing machines
5591	8422.3090	---其他	10			0		35	---Other
5592	8422.4000	-其他包装或打包机器(包括热缩包装机器)	10			0△		35	-Other packing or wrapping machinery (including heat-shrink wrapping machinery)
		-零件:							-Parts:
5593	8422.9010	---洗碟机用	10.5					90	---Of dish washing machines
5594	8422.9020	---饮料及液体食品灌装设备用	8.5					45	---Of bottling or canning machinery for beverages or liquid food
5595	8422.9090	---其他	8.5			0		35	---Other
	84.23	**衡器(感量为50毫克或更精密的天平除外),包括计数或检验用的衡器;衡器用的各种砝码、秤砣:**							**Weighing machinery(excluding balances of a sensitivity of 50 mg or better),including weight operated counting or checking machines; weighing machine weights of all kinds:**
5596	8423.1000	-体重计,包括婴儿秤;家用秤	10.5					80	-Personal weighing machines, including baby scales;household scales

序号 No.	税则号列 Tariff Item	货品名称	最惠国税率(%) M.F.N.	协定税率(%) Agreement 东盟 ASEAN	曼谷协定 BA	香港 H.K.	澳门 Macao	普通税率(%) Gen.	Article Description
		-输送带上连续称货的秤:							-Scales for continuous weighing of goods on conveyors:
5597	8423.2010	---电子皮带秤	10					80	---Electronic belt weighing machines
5598	8423.2090	---其他	10					80	---Other
		-恒定秤、物料定量装袋或装容器用的秤,包括库秤:							-Constant weight scales and scales for discharging a predetermined weight of material into a bag or container, including hopper scales:
5599	8423.3010	---定量包装秤	10.5					80	---Rationed packing scales
5600	8423.3020	---定量分选秤	10.5					80	---Rationed sorting scales
5601	8423.3030	---配料秤	10.5					80	---Proporating scales
5602	8423.3090	---其他	10.5					80	---Other
		-其他衡器:							-Other weighing machinery:
		--最大称量不超过30公斤:							--Having a maximum weighing capacity not exceeding 30kg:
5603	8423.8110	---计价秤	10.5					80	---Account balances
5604	8423.8120	---弹簧秤	10.5					80	---Spring balances
5605	8423.8190	---其他	10.5					80	---Other
		--最大称量超过30公斤,但不超过5000公斤:							--Having a maximum weighing capacity exceeding 30kg but not exceeding 5000kg:
5606	8423.8210	---地中衡	10.5					80	---Weighbridges
5607	8423.8290	---其他	10.5					80	---Other
		--其他:							--Other:
5608	8423.8910	---地中衡	10					80	---Weighbridges
5609	8423.8920	---轨道衡	10					80	---Track scales
5610	8423.8930	---吊秤	10					80	---Hanging scales
5611	8423.8990	---其他	10					80	---Other
5612	8423.9000	-衡器用的各种砝码、秤砣;衡器的零件	10					80	-Weighing machine weights of all kinds; parts of weighing machinery
	84.24	**液体或粉末的喷射、散布或喷雾的机械器具(不论是否手工操作);灭火器,不论是否装药;喷枪及类似器具;喷汽机、喷砂机及类似的喷射机器:**							**Mechanical appliances (whether or not hand-operated) for projecting, dispersing or spraying liquids or powders; fire extinguishers, whether or not charged; spray guns and similar appliances; steam or sand blasting machines and similar jet projecting machines:**
5613	8424.1000	-灭火器,不论是否装药	8.4					70	-Fire extinguishers, whether or not charged
5614	8424.2000	-喷枪及类似器具	8.4					40	-Spray guns and similar appliances
5615	8424.3000	-喷汽机、喷砂机及类似的喷射机器	8.4					40	-Steam or sand blasting machines and similar jet projecting machines
		-其他器具:							-Other appliances:
5616	8424.8100	--农业或园艺用	8			0		30	--Agricultural or horticultural
		--其他:							--Other:
5517	8424.8910	---家用型	0					80	---Of the household type

序号 No.	税则号列 Tariff Item	货品名称	最惠国税率(%) M.F.N.	协定税率(%) Agreement				普通税率(%) Gen.	Article Description
				东盟 ASEAN	曼谷协定 BA	香港 H.K.	澳门 Macao		
		---其他:							---Other:
5618	8424.8991	----船用洗舱机	0					30	----Marine cabinet washer
5619	8424.8999	----其他	0					30	----Other
		-零件:							-Parts:
5620	8424.9010	---子目号 8424.1000 所列器具的零件	0					70	---Of the apparatus of subheading No. 8424.1000
5621	8424.9020	---子目号 8424.8910 所列器具的零件	0					80	---Of the apparatus of subheading No. 8424.8910
5622	8424.9090	---其他	0			0		30	---Other
	84.25	**滑车及提升机,但倒卸式提升机除外;卷扬机及绞盘;千斤顶:**							**Pulley tackle and hoists other than skip hoists; winches and capstans; jacks:**
		-滑车及提升机,但倒卸式提升机及提升车辆用的提升机除外:							-Pulley tackle and hoists other than skip hoists or hoists of a kind used for raising vehicles:
5623	8425.1100	--电动的	6					30	--Powered by electric motor
5624	8425.1900	--其他	5					30	--Other
		-矿井口卷扬装置;专为井下使用设计的卷扬机:							-Pit-head winding gear; winches specially designed for use underground:
		---矿井卷扬机:							---Mine hoists:
5625	8425.2011	----圆筒直径在 2 米及以上	10					30	----Reel diameter no less than 2m
5626	8425.2019	----其他	10					30	----Other
5627	8425.2090	----其他	10					30	---Other
		-其他卷扬机;绞盘:							-Other winches; capstans:
5628	8425.3100	--电动的	5					30	--Powered by electric motor
5629	8425.3900	--其他	5			0		30	--Other
		-千斤顶;提升车辆用的提升机:							-Jacks; hoists of a kind used for raising vehicles:
5630	8425.4100	--车库中使用的固定千斤顶系统	3					30	--Built-in jacking systems of a type used in garages
		--其他液压千斤顶及提升机:							--Other jacks and hoists, hydraulic:
5631	8425.4210	---液压千斤顶	3					30	---Hydraulic jacks
5632	8425.4290	---其他	5					30	---Other
		--其他:							--Other:
5633	8425.4910	---其他千斤顶	5					30	---Other jacks
5634	8425.4990	---其他	10					30	---Other
	84.26	**船用桅杆式起重机;起重机,包括缆式起重机;移动式吊运架、跨运车及装有起重机的工作车:**							**Ships derricks; cranes, including cable cranes; mobile lifting frames, straddle carriers and works trucks fitted with a crane:**
		-高架移动式起重机、桁架桥式起重机、龙门起重机、桥式起重机、移动式吊运架及跨运车:							-Overhead travelling cranes, transporter cranes, gantry cranes, bridge cranes, mobile lifting frames and straddle carriers:

序号 No.	税则号列 Tariff Item	货品名称	最惠国税率(%) M.F.N.	协定税率(%)Agreement 东盟 ASEAN	曼谷协定 BA	香港 H.K.	澳门 Macao	普通税率(%) Gen.	Article Description
		--固定支架的高架移动式起重机:							--Overhead travelling cranes on fixed support:
5635	8426.1120	---通用桥式起重机	8					30	---Bridge cranes, all-purpose
5636	8426.1190	---其他	8					30	---Other:
5637	8426.1200	--带胶轮的移动式吊运架及跨运车	6					30	--Mobile lifting frames on tyres and straddle carriers
		--其他:							--Other:
5638	8426.1910	---装船机	5					30	---Ship loading cranes
		---卸船机:							---Ship unloading cranes:
5639	8426.1921	----抓斗式	5					30	---Grab ship unloading cranes
5640	8426.1929	----其他	5					30	---Other
5641	8426.1930	---龙门式起重机	10					30	---Gantry cranes
		---装卸桥:							---Loading and unlcading bridges:
5642	8426.1941	----门式装卸桥	10					30	----Frame loading and unloading bridges
5643	8426.1942	----集装箱装卸桥	10					30	----Container loading and unloading bridges
5644	8426.1943	----其他动臂式装卸桥	10					30	----Derrick loading and unloading bridges
5645	8426.1949	----其他	10					30	----Other
5646	8426.1990	---其他	10					30	---Other
5647	8426.2000	-塔式起重机	10					30	-Tower cranes
5648	8426.3000	-门座式起重机及座式旋臂起重机	6					30	-Portal or pedestal jib cranes
		-其他自推进机械:							-Other machinery, self-propelled:
		--带胶轮的:							--On tyres:
5649	8426.4110	---轮胎式起重机	5					30	---Wheel-mounted cranes
5650	8426.4190	---其他	5					30	---Other
		--其他:							--Other:
5651	8426.4910	---履带式起重机	8					30	---Crawler cranes
5652	8426.4990	---其他	13					30	---Other
		-其他机械:							-Other machinery:
5653	8426.9100	--供装于公路车辆的	10					30	--Designed for mounting on road vehicles
5654	8426.9900	--其他	6					30	--Other
	84.27	**叉车;其他装有升降或搬运装置的工作车:**							**Fork-lift trucks; other works trucks fitted with lifting or handling equipment:**
		-电动机推进的机动车:							-Self-propelled trucks powered by an electric motor:
5655	8427.1010	---有轨巷道堆垛机	9					30	---Track alleyway stackers
5656	8427.1020	---无轨巷道堆垛机	9					30	---Trackless alleyway stackers
5657	8427.1090	---其他	9					30	---Other
		-其他机动车:							-Other self-propelled trucks:
5658	8427.2010	---集装箱叉车	9					30	---Fork-lift trucks cranes
5659	8427.2090	---其他	9					30	---Other
5660	8427.9000	-其他车	9					30	-Other trucks

序号 No.	税则号列 Tariff Item	货品名称	最惠国税率(%) M.F.N.	协定税率(%) Agreement 东盟 ASEAN	曼谷协定 BA	香港 H.K.	澳门 Macao	普通税率(%) Gen.	Article Description
	84.28	**其他升降、搬运、装卸机械(例如,升降机、自动梯、输送机、缆车):**							**Other lifting, handling, loading or unloading machinery (for example, lifts, escalators, conveyors, teleferics):**
		-升降机及倒卸式起重机:							-Lifts and skip hoists:
5661	8428.1010	---载客电梯	8			0		30	---Designed for the transport of persons
5662	8428.1090	---其他	6			0		30	---Other
5663	8428.2000	-气压升降机及输送机	5					30	-Pneumatic elevators and conveyors
		-其他用于连续运送货物或材料的升降机及输送机:							-Other continuous-action elevators and conveyors, for goods or materials:
5664	8428.3100	--地下专用的	5					30	--Specially designed for underground use
5665	8428.3200	--其他,斗式	5					30	--Other, bucket type
5666	8428.3300	--其他,带式	5		4.3			30	--Other, belt type
		--其他:							--Other:
5667	8428.3910	---链式	5					30	---Chain type
5668	8428.3920	---辊式	5					30	---Roller type
5669	8428.3990	---其他	5					30	---Other
5670	8428.4000	-自动梯及自动人行道	5					30	-Escalators and moving walkways
5671	8428.5000	-矿车推动机、铁道机车或货车的转车台、货车倾卸装置及类似的铁道货车搬运装置	10					30	-Mine wagon pushers, locmotive or wagon traversers, wagon tippers and similar railway wagon handling equipment
		-缆车、座式升降机、滑雪拉索、索道用牵引装置:							-Teleferics, chair-lifts, ski-draglines; traction mechanisms for funiculars:
5672	8428.6010	---货运架空索道	8					30	---Cargo aerial cableways
		---客运架空索道:							---Passanger aerial cableways:
5673	8428.6021	----单线循环式	8					30	----Monocable endless
5674	8428.6029	----其他	8					30	----Other
5675	8428.6090	---其他	8					30	---Other
5676	8428.9000	-其他机械	5					30	-Other machinery
	84.29	**机动推土机、侧铲推土机、筑路机、平地机、铲运机、机械铲、挖掘机、机铲装载机、捣固机械及压路机:**							**Self-propelled bulldozers, angledozers, graders, levellers, scrapers, mechanical shovels, excavators, shovel loaders, tamping machines and road rollers:**
		-推土机及侧铲推土机:							-Bulldozers and angledozers:
		--履带式:							--Track laying:
5677	8429.1110	---发动机输出功率超过235.36千瓦(320马力)的	7					17	---With an engine of an output exceeding 235.36kW(320HP)
5678	8429.1190	---其他	7					30	---Other
		--其他:							--Other:

序号 No.	税则号列 Tariff Item	货品名称	最惠国税率(%) M.F.N.	协定税率(%) Agreement 东盟 ASEAN	曼谷协定 BA	香港 H.K.	澳门 Macao	普通税率(%) Gen.	Article Description
5679	8429.1910	---发动机输出功率超过235.36千瓦(320马力)的	7					17	---With an engine of an output exceeding 235.36kW(320HP)
5680	8429.1990	---其他	7					30	---Other
		-筑路机及平地机:							-Graders and levellers:
5681	8429.2010	---发动机输出功率超过235.36千瓦(320马力)的	5					17	---With an engine of an output exceeding 235.36kW(320HP)
5682	8429.2090	---其他	5					30	---Other:
		-铲运机:							-Scrapers:
5683	8429.3010	---斗容量超过10立方米的	3					17	---Having a capacity of shovel exceeding $10m^3$
5684	8429.3090	---其他	5					30	---Other
		-捣固机械及压路机:							-Tamping machines and road rollers:
		---机动压路机:							---Self-propelled road rollers:
5685	8429.4011	----机重18吨及以上的振动压路机	7					20	----Vibration type, of a deadweight of 18t or more
5686	8429.4019	----其他	8					40	----Other
5687	8429.4090	---其他	6					30	---Other
		-机械铲、挖掘机及机铲装载机:							-Mechanical shovels, excavators and shovel loaders:
5688	8429.5100	--前铲装载机	5					30	--Front-end shovel loaders
		--上部结构可旋转360度的机械:							--Machinery with a 360° revolving superstructure
		---挖掘机:							---Excavators:
5689	8429.5211	----轮胎式	8		7.2			30	----Tyre-mounted
5690	8429.5212	----履带式	8		8			30	----Track-mounted
5691	8429.5219	----其他	8		7.2			30	----Other
5692	8429.5290	---其他	8		7.2			30	---Other
5693	8429.5900	--其他	8					30	--Other
	84.30	**泥土、矿物或矿石的运送、平整、铲运、挖掘、捣固、压实、开采或钻探机械;打桩机及拔桩机;扫雪机及吹雪机:**							**Other moving, grading, levelling, scraping, excavating, tamping, compacting, extracting or boring machinery, for earth, minerals or ores; pile-drivers and pile-extractors; snow-ploughs and snow-blowers:**
5694	8430.1000	-打桩机及拔桩机	10					30	-Pile-drivers and pile-extranctors
5695	8430.2000	-扫雪机及吹雪机	10					30	-Snow-ploughs and snow-blowers
		-截煤机、凿岩机及隧道掘进机:							-Coal or rock cutters and tunnelling machinery:
5696	8430.3100	--自推进的	10					30	--Self-propelle
5697	8430.3900	--其他	6					30	--Other
		-其他钻探或凿井机械:							-Other boring or sinking machinery:
		--自推进的:							--Self-propelled:
		---石油及天然气钻探机:							---Oil and natural gas drilling machinery:

序号 No.	税则号列 Tariff Item	货品名称	最惠国税率(%) M.F.N.	协定税率(%)Agreement 东盟 ASEAN	曼谷协定 BA	香港 H.K.	澳门 Macao	普通税率(%) Gen.	Article Description
5698	8430.4111	----钻探深度在6000米及以上的	5			0		11	----Of drilling depth of 6000m or more
5699	8430.4119	----其他	5			0		17	----Other
		---其他钻探机:							---Other drilling machinery:
5700	8430.4121	----钻探深度在6000米及以上的	5			0		11	----Of drilling depth of 6000m or more
5701	8430.4122	----钻探深度在6000米以下的履带式自推进钻机	5			0		17	----Crawler boring machinery of drilling depth＜6000m
5702	8430.4129	----钻探深度在6000米以下的其他钻探机	5			0		17	----Other boring machinery of drilling depth＜6000m
5703	8430.4190	---其他	5			0		30	---Other
5704	8430.4900	--其他	5					30	--Other
		-其他自推进机械:							-Other machinery, self-propelled:
5705	8430.5010	---其他采油机械	3					17	---For oil production
5706	8430.5020	---矿用电铲	7					30	---Mining power shovels
		---采矿钻机:							---Mining drills:
5707	8430.5031	----牙轮直径380毫米及以上	5					30	----Gear wheel diameter more than 380mm
5708	8430.5039	----其他	5					30	----Other
5709	8430.5090	---其他	5					30	---Other
		-其他非自推进机械:							-Other machinery, not self-propelled:
5710	8430.6100	--捣固或压实机械	6					30	--Tamping or compacting machinery
		--其他:							--Other:
		---工程钻机:							---Engineering drills:
5711	8430.6911	----钻筒直径3米及以上	6					30	----Boring casing diameter more than 3m
5712	8430.6919	----其他	6					30	----Other
5713	8430.6920	---铲运机	6					30	---Scrapers
5714	8430.6990	---其他	6					30	---Other
	84.31	**专用于或主要用于税号84.25至84.30所列机械的零件:**							**Parts suitable for use solely or principally with the machinery of headings Nos.84.25 to 84.30:**
5715	8431.1000	-税号84.25所列机械的零件	3					30	-Of machinery of heading No.84.25
5716	8431.2000	-税号84.27所列机械的零件	6					30	-Of machinery of heading No.84.27
		-税号84.28所列机械的零件:							-Of machinery of heading No.84.28:
5717	8431.3100	--升降机、倒卸式起重机或自动梯的零件	3					30	--Of lifts, skip hoists or escalators
5718	8431.3900	--其他	5					30	--Other
		-税号84.26、84.29、或84.30所列机械的零件:							-Of machinery of heading No.84.26, 84.29 or 84.30:
5719	8431.4100	--戽斗、铲斗、抓斗及夹斗	6					17	--Buckets, shovels, grabs and grips

序号 No.	税则号列 Tariff Item	货品名称	最惠国税率(%) M.F.N.	协定税率(%) Agreement 东盟 ASEAN	曼谷协定 BA	香港 H.K.	澳门 Macao	普通税率(%) Gen.	Article Description
5720	8431.4200	--推土机或侧铲推土机用铲	6					17	--Bulldozer or angledozer blades
		--子目号 8430.41 或 8430.49 所列钻探或凿井机械的零件:							--Parts of boring or sinking machinery of subheading No. 8430.41 or 8430.49:
5721	8431.4310	---石油或天然气钻探机用	4			0		11	---Of oil and natural gas drilling machinery
5722	8431.4320	---其他钻探机用	4			0		11	---Of other drilling machinery
5723	8431.4390	---其他	5			0		17	---Of sinking machinery
		--其他:							--Other:
5724	8431.4910	---矿用电铲用	5					17	---For mining power shovels
5725	8431.4990	---其他	5					17	--Other
	84.32	**农业、园艺及林业用整地或耕作机械;草坪及运动场地滚压机:**							**Agricultural, horticultural or forestry machinery for soil preparation or cultivation; lawn or sports-ground rollers:**
5726	8432.1000	-犁	5					30	-Ploughs
		-耙、松土机、中耕机、除草机及耕耘机:							-Harrows, scarifiers, cultivators, weeders and hoes:
5727	8432.2100	--圆盘耙	5					30	--Disc harrows
5728	8432.2900	--其他	4					30	--Other
5729	8432.3000	-播种机、种植机及移植机	4					30	-Seeders, planters and transplanters
5730	8432.4000	-施肥机	4					30	-Manure spreaders and fertilizer distributors
		-其他机械:							-Other machinery:
5731	8432.8010	---草坪及运动场地滚压机	7					40	---Lawn or sports-ground rollers
5732	8432.8090	---其他	4					30	---Other
5733	8432.9000	-零件	4					17	-Parts
	84.33	**收割机、脱粒机,包括草料打包机;割草机;蛋类、水果或其他农产品的清洁、分选、分级机器,但税号 84.37 的机器除外:**							**Harvesting or threshing machinery, including straw or fodder balers; grass or hay mowers; machines for cleaning, sorting or grading eggs, fruit or other agricultural produce, other than machinery of heading No. 84.37:**
		-草坪、公园或运动场地用的割草机:							-Mowers for lawns, parks or sports grounds:
5734	8433.1100	--机动的,切割装置在同一水平面上旋转的	6					30	--Powered, with the cutting device rotating in a horizontal plane
5735	8433.1900	--其他	6					30	--Other
5736	8433.2000	-其他割草机,包括牵引装置用的刀具杆	4					30	-Other mowers, including cutter bars for tractor mounting
5737	8433.3000	-其他干草切割、翻晒机器	5					30	-Other haymaking machinery

序号 No.	税则号列 Tariff Item	货品名称	最惠国税率(%) M.F.N.	协定税率(%)Agreement 东盟 ASEAN	曼谷协定 BA	香港 H.K.	澳门 Macao	普通税率(%) Gen.	Article Description
5738	8433.4000	-草料打包机,包括收集打包机	5					30	-Straw or fodder balers, including pick-up balers
		-其他收割机;脱粒机:							-Other harvesting machinery; threshing machinery:
5739	8433.5100	--联合收割机	8					17	--Combine harvester-threshers
5740	8433.5200	--其他脱粒机	8					30	--Other threshing machinery
5741	8433.5300	--根茎或块茎收获机	8					30	--Root or tuber harvesting machines
		--其他:							--Other:
5742	8433.5910	---甘蔗收获机	8					30	---Sugarcane harvesters
5743	8433.5990	---其他	8					30	---Other
5744	8433.6000	-蛋类、水果或其他农产品的清洁、分选、分级机器	5					30	-Machines for cleaning, sorting or grading eggs, fruit or other agricultural produce
		-零件:							-Parts:
5745	8433.9010	---联合收割机用	5					11	---Of combine harvester-thrashers
5746	8433.9090	---其他	3					17	---Other
	84.34	**挤奶机及乳品加工机器:**							**Milking machines and dairy machinery:**
5747	8434.1000	-挤奶机	10					20	-Milking machines
5748	8434.2000	-乳品加工机器	6					30	-Dairy machinery
5749	8434.9000	-零件	5					17	-Parts
	84.35	**制酒、制果汁或制类似饮料用的压榨机、轧碎机及类似机器:**							**Presses, crushers and similar machinery used in the manufacture of wine, cider, fruit juices or similar beverages:**
5750	8435.1000	-机器	10					30	-Machinery
5751	8435.9000	-零件	6					30	-Parts
	84.36	**农业、园艺、林业、家禽饲养业或养蜂业用的其他机器,包括装有机械或热力装置的催芽设备;家禽孵卵器及育雏器:**							**Other agricultural, horticultural, forestry, poultry-keeping or bee-keeping machinery, including germination plant fitted with mechanical or thermal equipment; poultry incubators and brooders:**
5752	8436.1000	-动物饲料配制机	7					30	-Machinery for preparing animal feeding stuffs
		-家禽饲养用的机器;家禽孵卵器及育雏器:							-Poultry-keeping machinery; poultry incubators and brooders:
5753	8436.2100	--家禽孵卵器及育雏器	5					30	--Poultry incubators and brooders
5754	8436.2900	--其他	10					30	--Other
5755	8436.8000	-其他机器	10					30	-Other machinery
		-零件:							-Parts:
5756	8436.9100	--家禽饲养用机器的零件或家禽孵卵器及育雏器的零件	6					17	--Of poultry-keeping machinery or poultry incubators and brooders
5757	8436.9900	--其他	6					17	--Other

序号 No.	税则号列 Tariff Item	货品名称	最惠国税率(%) M.F.N.	协定税率(%)Agreement 东盟 ASEAN	曼谷协定 BA	香港 H.K.	澳门 Macao	普通税率(%) Gen.	Article Description
	84.37	**种子、谷物或干豆的清洁、分选或分级机器；谷物磨粉业加工机器或谷物、干豆加工机器，但农业用机器除外：**							**Machines for cleaning, sorting or grading seed, grain or dried leguminous vegetables; machinery used in the milling industry or for the working of cereals or dried leguminous vegetables, other than farm-type machinery:**
5758	8437.1000	-种子、谷物或干豆的清洁、分选或分级机器	10					30	-Machines for cleaning, sorting or grading seed, grain or dried leguminous vegetables
5759	8437.8000	-其他机器	10					30	-Other machinery
5760	8437.9000	-零件	6					30	-Parts
	84.38	**本章其他税号未列名的食品、饮料工业用的生产或加工机器，但提取、加工动物油脂或植物固定油脂的机器除外：**							**Machinery, not specified or included elsewhere in this chapter, for the industrial preparation or manufacture of food or drink, other than machinery for the extraction or preparation of animal or fixed vegetable fats or oils:**
5761	8438.1000	-糕点加工机器及生产通心粉、面条或类似产品的机器	7					30	-Bakery machinery and machinery for the manufacture of macaroni, spaghetti or similar products
5762	8438.2000	-生产糖果、可可粉、巧克力的机器	8					30	-Machinery for the manufacture of confectionery, cocoa or chocolate
5763	8438.3000	-制糖机器	10					30	-Machinery for sugar manufacture
5764	8438.4000	-酿酒机器	7					30	-Brewery machinery
5765	8438.5000	-肉类或家禽加工机器	7					30	-Machinery for the preparation of meat or poultry
5766	8438.6000	-水果、坚果或蔬菜加工机器	10					30	-Machinery for the preparation of fruits, nuts or vegetables
5767	8438.8000	-其他机器	8.5					30	-Other machinery
5768	8438.9000	-零件	5					30	-Parts
	84.39	**纤维素纸浆、纸及纸板的制造或整理机器：**							**Machinery for making pulp of fibrous cellulosic material or for making or finishing paper or paperboard:**
5769	8439.1000	-制造纤维素纸浆的机器	8.4					30	-Machinery for making pulp of fibrous cellulosic material
5770	8439.2000	-纸或纸板的抄造机器	8.4					30	-Machinery for making paper or paper board
5771	8439.3000	-纸或纸板的整理机器	8.4					30	-Machinery for finishing paper or paper board
		-零件：							-Parts:
5772	8439.9100	--制造纤维素纸浆的机器用	6					30	--Of machinery for making pulp of fibrous cellulosic material
5773	8439.9900	--其他	6					30	--Other
	84.40	**书本装订机器，包括锁线订书机：**							**Book-binding machinery, including booksewing machines:**

序号 No.	税则号列 Tariff Item	货品名称	最惠国税率(%) M.F.N.	协定税率(%) Agreement 东盟 ASEAN	曼谷协定 BA	香港 H.K.	澳门 Macao	普通税率(%) Gen.	Article Description
		-机器:							-Machinery:
5774	8440.1010	---锁线装订机	10					35	---Sewing bookbinders
5775	8440.1020	---胶订机	12					35	---Glueing bookbinders
5776	8440.1090	---其他	12					35	---Other
5777	8440.9000	-零件	8					35	-Parts
	84.41	**其他制造纸浆制品、纸制品或纸板制品的机器,包括各种切纸机:**							**Other machinery for making up paper pulp, paper or paperboard, including cutting machines of all kinds:**
5778	8441.1000	-切纸机	12					50	-Cutting machines
5779	8441.2000	-制造包、袋或信封的机器	12					30	-Machines for making bags, sacks or envelopes
		-制造箱、盒、管、桶或类似容器的机器,但模制成型机器除外:							-Machines for making cartons, boxes, cases, tubes, drums or similar containers, other than by moulding:
5780	8441.3010	---制造纸塑铝复合罐的生产设备	13.5					30	---Machines for paper, plastic and aluminium composite can manufacture
5781	8441.3090	---其他	13.5					30	---Other
5782	8441.4000	-纸浆、纸或纸板制品模制成型机器	12					30	-Machines for moulding articles in paper pulp, paper or paperboard
		-其他机器:							-Other machinery:
5783	8441.8010	---制造纸塑铝软包装的生产设备	12					30	---Machines for paper plastic and aluminium flexible packaging manufacture
5784	8441.8090	---其他	12					30	---Other
		-零件:							-Parts:
5785	8441.9010	---切纸机用	8					50	---Of cutting machines
5786	8441.9090	---其他	8.4					30	---Other
	84.42	**铸字、排字或制版用的机器、器具及设备(税号84.56至84.65的机床除外);活字、印刷用版、片、滚筒及其他部件;制成供印刷用(例如,刨平、压纹或抛光)的板、片、滚筒及石板:**							**Machinery, apparatus and equipment (other than the machine-tools of headings Nos.84.56 to 84.65), for type-founding or type-setting, for preparing or making printing blocks, plates, cylinders or other printing components; printing type, blocks, plates, cylinders and other printing components; blocks, plates, cylinders and lithographic stones, prepared for printing purposes (for example, planed, grained or polished):**
5787	8442.1000	-照像排版及排字机器	9					35	-Phototype-setting and composing machines
5788	8442.2000	-其他方法排字的机器、器具及设备,不论是否有铸字装置	9					35	-Machinery, apparatus and equipment for type-setting or composing by other processes, with or without founding device
		-其他机器、器具及设备:							-Other machinery, apparatus and equipment:

序号 No.	税则号列 Tariff Item	货品名称	最惠国税率(%) M.F.N.	协定税率(%) Agreement 东盟 ASEAN	曼谷协定 BA	香港 H.K.	澳门 Macao	普通税率(%) Gen.	Article Description
5789	8442.3010	---铸字机	9					35	---Type casters
5790	8442.3020	---制版机器、器具及设备	9					35	---Other machinery, apparatus and equipment for typesetting
5791	8442.3090	---其他	9					35	---Other
5792	8442.4000	-上述机器、器具及设备的零件	7					20	-Parts of the foregoing machinery, apparatus or equipment
5793	8442.5000	-活字、印刷用版、片、滚筒及其他部件;制成供印刷用(例如,刨平、压纹或抛光)的板、片、滚筒及石板	7			0		35	-Printing type, blocks, plates, cylinders and other printing components; blocks, plates, cylinders and lithographic stones, prepared for printing purposes(for example, planed, grained or polished)
	84.43	**用于税目 84.42 的活字、印刷用板、片、滚筒及其他印刷部件进行印刷的机器;喷墨印刷机,但税号 84.71 的货品除外;印刷用辅助机器:**							**Printing machinery used for printing by means of the printing type, blocks, plates, cylinders and other printing components of heading 84.42; ink-jet printing machines, other than those of heading 84.71; machines for uses ancillary to printing:**
		-胶印机:							Offset printing machinery:
5794	8443.1100	--卷取进料式	10					35	--Reel fed
5795	8443.1200	--办公室用片取进料式(片尺寸不超过 22×36 厘米)	12					35	--Sheet fed, office type(sheet size not exceeding 22×36cm)
		--其他:							--Other:
5796	8443.1910	---平张纸进料式	10		8.9			35	---Sheet fed
5797	8443.1990	---其他	10		8.9			35	---Other
		-凸版印刷机,但不包括苯胺印刷机:							-Letterpress printing machinery, excluding flexographic printing:
5798	8443.2100	--卷取进料式	12					35	--Reel fed
5799	8443.2900	--其他	12					35	--Other
5800	8443.3000	-苯胺印刷机	10					35	-Flexographic printing machinery
5801	8443.4000	-凹版印刷机	18		17			35	-Gravure printing machinery
		-其他印刷机:							-Other printing machinery:
5802	8443.5100	--喷墨印刷机	8					30	--Ink-jet printing machines
		--其他:							--Other:
		---网式印刷机:							---Screen printing machinery:
5803	8443.5911	----圆网印刷机	10					35	----Cylinder screen press
5804	8443.5912	----平网印刷机	10					35	----Platen screen press
5805	8443.5919	----其他	10					35	----Other
5806	8443.5990	---其他	8			0		35	---Other
5807	8443.6000	-印刷用辅助机器	12					35	-Machines for uses ancillary to printing
5808	8443.9000	-零件	6					20	-Parts
	84.44	**化学纺织纤维挤压、拉伸、变形或切割机器:**							**Machines for extruding, drawing, texturing or cutting man-made textile materials:**

序号 No.	税则号列 Tariff Item	货品名称	最惠国税率(%) M.F.N.	协定税率(%) Agreement 东盟 ASEAN	曼谷协定 BA	香港 H.K.	澳门 Macao	普通税率(%) Gen.	Article Description
5809	8444.0010	---合成纤维长丝纺丝机	10		9			30	---Synthetic filaments spinning jets
5810	8444.0020	---合成纤维短纤纺丝机	10		9			30	---Synthetic staple fibres spinning jets
5811	8444.0030	---人造纤维纺丝机	10		9			30	---Artificial fibres spinning jets
5812	8444.0040	---化学纤维变形机	10		9			30	---Man-made filaments crimping machinery
5813	8444.0050	---化学纤维切断机	10		9			30	---Man-made filaments cutting machinery
5814	8444.0090	---其他	10		9			30	---Other
	84.45	**纺织纤维的预处理机器;纺纱机、并线机、加捻机及其他生产纺织纱线的机器;摇纱机、络纱机(包括卷纬机)及处理税号84.46或84.47所列机器用的纺织纱线的机器:**							**Machines for preparing textile fibres; spinning, doubling or twisting machines and other machinery for producing textile yarns; textile reeling or winding(including weft-winding) machines and machines for preparing textile yarns for use on the machines of heading No.84.46 or 84.47:**
		-纺织纤维的预处理机器:							-Machines for preparing textile fibres:
		--梳理机:							--Carding machines:
		---棉纤维型:							---For cotton type fibres:
5815	8445.1111	----清梳联合机	10		9			30	----Blowing-carding Machinery
5816	8445.1112	----自动抓棉机	10		9			30	----Bale Plucker
5817	8445.1113	----梳棉机	10		9			30	----Card or Carding Mackine
5818	8445.1119	----其他	10		9			30	----Other
5819	8445.1120	---毛纤维型	10		9			30	---For wool type fibres
5820	8445.1190	---其他	10		9			30	---Other
		--精梳机:							--Combing machines:
5821	8445.1210	---棉精梳机	10					30	----Cotton Comber
5822	8445.1220	---毛精梳机	10					30	----Worsted Comber
5823	8445.1290	---其他	10					30	----Other
		--拉伸机或粗纱机:							--Drawing or roving machines:
5824	8445.1310	---拉伸机	10					30	---Drawing machines
		---粗纱机:							---Roving machines:
5825	8445.1321	----棉纺粗纱机	10					30	----Cotton Roving Frames
5826	8445.1322	----毛纺粗纱机	10					30	----Worsted Roving Machines
5827	8445.1329	----其他	10					30	----Other
5828	8445.1900	--其他	10		9			30	--Other
		-纺纱机:							Textile spinning machines:
		---自由端纺纱机:							---Open-end spinner:
5829	8445.2031	----转杯纺纱机	10		9			30	----Rotor Spinning Machine
5830	8445.2032	----喷气纺纱机	10					30	----Jet spinner
5831	8445.2039	----其他	10					30	----Other
		---环锭细纱机:							---Ring spinning frames
5832	8445.2041	----棉细纱机	10.5		10			40	----Cotton Ring Spinning Frame

序号 No.	税则号列 Tariff Item	货品名称	最惠国税率(%) M.F.N.	协定税率(%) Agreement 东盟 ASEAN	曼谷协定 BA	香港 H.K.	澳门 Macao	普通税率(%) Gen.	Article Description
5833	8445.2042	----毛细纱机	10					40	----Worsted Ring Spinning Frame
5834	8445.2049	----其他	10					40	----Other
5835	8445.2090	---其他	10		9			30	---Other
5836	8445.3000	-并线机或加捻机	10		9			30	-Textile doubling or twisting machines
		-络纱机(包括卷纬机)或摇纱机:							-Textile winding(including weft-winding)or reeling machines:
5837	8445.4010	---自动络筒机	10		9			30	---Automatic bobbin winders
5838	8445.4090	---其他	10		9			30	---Other
		-其他:							-Other:
5839	8445.9010	---整经机	10		9			30	---Warping machines
5840	8445.9020	---浆纱机	10		9			30	---Sizing machines
5841	8445.9090	---其他	10		9			30	---Other
	84.46	**织机:**							**Weaving machines(looms):**
5842	8446.1000	-所织织物宽度不超过30厘米的织机	8					30	-For weaving fabrics of a width not exceeding 30cm
		-所织织物宽度超过30厘米的梭织机:							-For weaving fabrics of width exceeding 30cm,shuttle type:
		--动力织机:							--Power looms:
5843	8446.2110	---地毯织机	12		11			35	---For making carpets or rugs
5844	8446.2190	---其他	10		9			30	---Other
5845	8446.2900	--其他	10					30	--Other
		-所织织物宽度超过30厘米的无梭织机:							-For weaving fabrics of a width exceeding 30cm,shuttleless type:
5846	8446.3020	---剑杆织机	8		7			30	---Rapier looms
5847	8446.3030	---片梭织机	8					30	---Carrier looms
5848	8446.3040	---喷水织机	8		7			30	----Water jet looms
5849	8446.3050	---喷气织机	8		7			30	----Air jet looms
5850	8446.3090	---其他	8		7			30	---Other
	84.47	**针织机、缝编机及制粗松螺旋花线、网眼薄纱、花边、刺绣品、装饰带、编织带或网的机器及簇绒机:**							**Knitting machines, stitch-bonding machines and machines for making gimped yarn, tulle lace, embroidery, trimmings, braid or net and machines for tufting:**
		-圆型针织机:							-Circular knitting machines:
5851	8447.1100	--圆筒直径不超过165毫米	8		7			30	--With cylinder diameter not exceeding 165mm
5852	8447.1200	--圆筒直径超过165毫米	8					30	--With cylinder diameter exceeding 165mm
		-平型针织机;缝编机:							-Flat knitting machines;stitch-bonding machines:
5853	8447.2010	---经编机	8		7			30	---Warp knitting machines
5854	8447.2020	---其他平型针织机	8		7			30	---Other flat knitting machines
5855	8447.2030	---缝编机	8		7			30	---Stitch-bonding machines
		-其他:							-Other:
		---簇绒机:							---Tufting machines:
5856	8447.9011	----地毯织机	7		6			35	----For making carpets or rugs

序号 No.	税则号列 Tariff Item	货品名称	最惠国税率(%) M.F.N.	协定税率(%)Agreement 东盟 ASEAN	曼谷协定 BA	香港 H.K.	澳门 Macao	普通税率(%) Gen.	Article Description
5857	8447.9019	----其他	8		7			30	----Other
5858	8447.9020	---绣花机	8		7			30	---Embroidery machines
5859	8447.9090	---其他	10		9			30	---Other
	84.48	**税号84.44、84.45、84.46或84.47所列机器的辅助机器(例如,多臂机、提花机、自停装置及换梭装置);专用于或主要用于税号84.44、84.45、84.46或84.47所列机器的零件、附件(例如,锭子、锭壳、钢丝针布、梳、喷丝头、梭子、综丝、综框、针织机用针):**							**Auxiliary machinery for use with machines of heading No. 84.44, 84.45, 84.46 or 84.47 (for example, dobbies, Jacquards, automatic stop motions, shuttle changing mechanisms); parts and accessories suitable for use solely or principally with the machines of this heading or of heading No. 84.44, 84.45, 84.46 or 84.47 (for example, spindles and spindle flyers, card clothing, combs, extruding nipples, shuttles, healds and heald-frames, hosiery needles):**
		-税号84.44、84.45、84.46或84.47所列机器的辅助机器:							-Auxiliary machinery for machines of heading No. 84.44, 84.45, 84.46 or 84.47:
5860	8448.1100	--多臂机或提花机及其所用的卡片缩小、复制、穿孔或汇编机器	8					20	--Dobbies and Jacquards; card reducing, copying, punching or assembling machines for use therewith
5861	8448.1900	--其他	8					20	--Other
		-税号84.44所列机器及其辅助机器的零件、附件:							-Parts and accessories of machines of heading No. 84.44 or of their auxiliary machinery:
5862	8448.2020	---喷丝头或喷丝板	6					14	---Extruding nipples or spinnerets
5863	8448.2090	---其他	6					17	---Other
		-税号84.45所列机器及其辅助机器的零件、附件:							-Parts and accessories of machines of heading No. 84.45 or of their auxiliary machinery:
5864	8448.3100	--钢丝针布	6					17	--Card clothing
5865	8448.3200	--纺织纤维预处理机器的零件、附件,但钢丝针布除外	6					17	--Of machines for preparing textile fibres, other than card clothing
		--锭子、锭壳、纺丝环、钢丝圈:							--Spindles, spindle flyers, spinning rings and ring travellers:
5866	8448.3310	---络筒锭	6					17	---Winding spindle
5867	8448.3390	---其他	6					17	---Other
		--其他:							--Other:
5868	8448.3910	---气流杯	6					14	---Open-end rotors
5869	8448.3920	---电子清纱器	6					17	---Electronic yarn clearers
5870	8448.3930	---空气捻接器	6					17	---Air twisting devices
5871	8448.3990	---其他	6					17	---Other
		-织机及其辅助机器的零件、附件:							-Parts and accessories of weaving machines(looms)or of their auxiliary machinery:

序号 No.	税则号列 Tariff Item	货品名称	最惠国税率(%) M.F.N.	协定税率(%) Agreement 东盟 ASEAN	曼谷协定 BA	香港 H.K.	澳门 Macao	普通税率(%) Gen.	Article Description
5872	8448.4100	--梭子	6					50	--Shuttles
5873	8448.4200	--织机用筘、综丝及综框	6					50	--Reeds for looms, healds and heald-frames
		--其他:							--Other:
5874	8448.4910	---接、投梭箱	6					17	---Catching and throwing shuttle boxes
5875	8448.4920	---引纬、送经装置	6					17	---Weft insertion and let-off motions
5876	8448.4990	---其他	6					17	---Other
		-税号 84.47 所列机器及其辅助机器的零件、附件:							-Parts and accessories of machines of heading No. 84.47 or of their auxiliary machinery:
		--沉降片、织针及其他成圈机件:							--Sinkers, needles and other articles used in forming stitches:
5877	8448.5120	---针织机用 28 号以下的弹簧针、钩针及复合针	6					50	---Barbered needles, crotchet hooks and complex needles for knitting machines, smaler than gauge No. 28
5878	8448.5190	---其他	6					17	---Other
5879	8448.5900	--其他	6					17	--Other
	84.49	**成匹、成形的毡呢或无纺织物制造或整理机器,包括制毡呢帽机器;帽模:**							**Machinery for the manufacture or finishing of felt or nonwovens in the piece or in shapes, including machinery for making felt hats; blocks for making hats:**
5880	8449.0000	成匹、成形的毡呢或无纺织物制造或整理机器,包括制毡呢帽机器;帽模	8					30	Machinery for the manufacture or finishing of felt or nonwovens in the piece or in shapes, including machinery for making felt hats; blocks for making hats
	84.50	**家用型或洗衣房用洗衣机,包括洗涤干燥两用机:**							**Household or laundry-type washing machines, including machines which both wash and dry:**
		-干衣量不超过 10 公斤的洗衣机:							-Machines, each of a dry linen capacity not exceeding 10kg:
		--全自动的:							--Fully-automatic machines
5881	8450.1110	---波轮式	10		8.7			130	---Of the continuously rotating impeller
5882	8450.1120	---滚筒式	10		8.7			130	---Of the drum type
5883	8450.1190	---其他	10		8.7			130	---Other
5884	8450.1200	--其他机器,装有离心甩干机	30		25			130	--Other machines, with built-in centrifugal drier
5885	8450.1900	--其他	30					130	--Other
5886	8450.2000	-干衣量超过 10 公斤的洗衣机	10					80	-Machines, each of a dry linen capacity exceeding 10kg
		-零件:							-Parts:
5887	8450.9010	---干衣量不超过 10 公斤的洗衣机用	5		4.4			130	---Of the machines of subheadings Nos. 8450.1110 to 8450.1900

序号 No.	税则号列 Tariff Item	货品名称	最惠国税率(%) M.F.N.	协定税率(%)Agreement 东盟 ASEAN	曼谷协定 BA	香港 H.K.	澳门 Macao	普通税率(%) Gen.	Article Description
5888	8450.9090	---其他	16					80	---Other
	84.51	**纱线、织物及纺织制品的洗涤、清洁、绞拧、干燥、熨烫、挤压(包括熔压)、漂白、染色、上浆、整理、涂布或浸渍机器(税号84.50的机器除外);列诺伦(亚麻油地毡)及类似铺地制品的布基或其他底布的浆料涂布机器;纺织物的卷绕、退绕、折叠、剪切或剪齿边机器:**							**Machinery (other than machines of heading No. 84. 50) for washing, cleaning, wringing, drying, ironing, pressing (including fusing presses), bleaching, dyeing, dressing, finishing, coating or impregnating textile yarns, fabrics or made up textile articles and machines for applying the paste to the base fabric or other support used in the manufacture of floor coverings such as linoleum; machines for reeling, unreeling, folding, cutting or pinking textile fabrics:**
5889	8451.1000	-干洗机	21		16			80	-Dry-cleaning machines
		-干燥机:							-Drying machines:
5890	8451.2100	--干衣量不超过10公斤	15					80	--Each of a dry linen capacity not exceeding 10kg
5891	8451.2900	--其他	8					30	--Other
5892	8451.3000	-熨烫机及挤压机(包括熔压机)	8					30	-Ironing machines and presses(including fusing presses)
5893	8451.4000	-洗涤、漂白或染色机器	8.4		7	0	0	20	-Washing, bleaching or dyeing machines
5894	8451.5000	-纺织物的卷绕、退绕、折叠、剪切或剪齿边机器	8					20	-Machines for reeling, unreeling, folding, cutting or pinking textile fabrics
5895	8451.8000	-其他机器	12			0		30	-Other machinery
5896	8451.9000	-零件	8					20	-Parts
	84.52	**缝纫机,但税号84.40的锁线订书机除外;缝纫机专用的特制家具、底座及罩盖;缝纫机针:**							**Sewing machines, other than book-sewing machines of heading No. 84. 40; furniture, bases and covers specially designed for sewing machines; sewing machine needles:**
5897	8452.1000	-家用型缝纫机	21		17				-Sewing machines of household type
		-其他缝纫机:							-Other sewing machines:
		--自动的:							--Automatic units:
5898	8452.2110	---平缝机	12		10.7			40	---Flatseam
5899	8452.2190	---其他	12		10.7			40	---Other
5900	8452.2900	--其他	12		11	0		40	--Other
5901	8452.3000	-缝纫机针	14					100	-Sewing machine needles
5902	8452.4000	-缝纫机专用的特制家具、底座和罩盖及其零件	14					100	-Furniture, bases and covers for sewing machines and parts thereof
		-缝纫机的其他零件:							-Other parts of sewing machines:
		---家用型缝纫机用:							---Of sewing machines of the household type:

序号 No.	税则号列 Tariff Item	货品名称	最惠国税率(%) M.F.N.	协定税率(%)Agreement 东盟 ASEAN	曼谷协定 BA	香港 H.K.	澳门 Macao	普通税率(%) Gen.	Article Description
5903	8452.9011	----旋梭	14		13			80	----Rotating shuttles
5904	8452.9019	----其他	14		13			80	----Other
		---其他:							---Other:
5905	8452.9091	----旋梭	14					40	----Rotating shuttles
5906	8452.9099	----其他	14					40	----Other
	84.53	**生皮、皮革的处理、鞣制或加工机器,鞋靴、毛皮及其他皮革制品的制作或修理机器,但缝纫机除外:**							**Machinery for preparing, tanning or working hides, skins or leather or for making or repairing footwear or other articles of hides, skins or leather, other than sewing machines:**
5907	8453.1000	-生皮、皮革的处理、鞣制或加工机器	8.4		8			30	-Machinery for preparing, tanning or working hides, skins or leather
5908	8453.2000	-鞋靴制作或修理机器	8.4		8			30	-Machinery for making or repairing footwear
5909	8453.8000	-其他机器	8.4					30	-Other machinery
5910	8453.9000	-零件	8					30	-Parts
	84.54	**金属冶炼及铸造用的转炉、浇包、锭模及铸造机:**							**Converters, ladles, ingot moulds and casting machines, of a kind used in metallurgy or in metal foundries:**
5911	8454.1000	-转炉	8.4					35	-Converters
		-锭模及浇包:							-Ingot moulds and ladles:
5912	8454.2010	---炉外精炼设备	8.4					35	---Fining equipments, outside of converters
5913	8454.2090	---其他	8.4					35	---Other
		-铸造机:							-Casting machines:
5914	8454.3010	---冷室压铸机	12					35	---Cold chamber die-casting machines
		---钢坯连铸机:							---Ingot continuous casting machines:
5915	8454.3021	----方坯连铸机	10					35	----Ingot block
5916	8454.3022	----板坯连铸机	12					35	----Ingot slab
5917	8454.3029	----其他	12					35	----Other
5918	8454.3090	---其他	12					35	---Other
		-零件:							-Parts:
5919	8454.9010	---炉外精炼设备用	8					20	---For the fining equipments outside of converters
		---钢坯连铸机用:							---For ingot continuous casting machines:
5920	8454.9021	----结晶器	8					20	----Crystallizers
5921	8454.9022	----振动装置	8					20	----Vibrating devices
5922	8454.9029	----其他	8					20	----Other
5923	8454.9090	---其他	8					20	---Other
	84.55	**金属轧机及其轧辊:**							**Metal-rolling mills and rolls therefor:**
		-轧管机:							-Tube mills:
5924	8455.1010	---热轧管机	12					35	---Tube mills, for hot-rolled
5925	8455.1020	---冷轧管机	12					35	---Tube mills for cold-rolled
5926	8455.1030	---定减径轧管机	12					35	---Fixed and reduced tube mills
5927	8455.1090	---其他	12					35	---Other

序号 No.	税则号列 Tariff Item	货品名称	最惠国税率(%) M.F.N.	协定税率(%) Agreement 东盟 ASEAN	曼谷协定 BA	香港 H.K.	澳门 Macao	普通税率(%) Gen.	Article Description
		-其他轧机:							-Other rolling mills:
		--热轧机或冷热联合轧机:							--Hot or combination hot and cold:
5928	8455.2110	---板材热轧机	15					35	---Sheet mills, hot-rolled
5929	8455.2120	---型钢轧机	15					35	---Rolled-steel section mills
5930	8455.2130	---线材轧机	15					35	---Wire mills
5931	8455.2190	---其他	15					35	---Other
		--冷轧机:							--Cold:
5932	8455.2210	---板材冷轧机	10					35	---Sheet mills
5933	8455.2290	---其他	15					35	---Other
5934	8455.3000	-轧机用轧辊	8.4					20	-Rolls for rolling mills
5935	8455.9000	-其他零件	8					20	-Other parts
	84.56	**用激光、其他光、光子束、超声波、放电、电化学法、电子束、离子束或等离子弧处理各种材料的加工机床:**							**Machine-tools for working any material by removal of material, by laser or other light or photon beam, ultrasonic, electro-discharge, electro-chemical, electron beam, ionic-beam or plasma arc processes:**
5936	8456.1000	-用激光、其他光或光子束处理的	0					30	-Operated by laser or other light or photon beam processes
5937	8456.2000	-用超声波处理的	10					30	-Operated by ultrasonic processes
		-用放电处理的:							-Operated by electro-discharge processes:
5938	8456.3010	---数控的	9.7					30	---Numerically controlled
5939	8456.3090	---其他	10					30	---Other
		-其他:							-Other:
5940	8456.9100	--用干法蚀刻半导体材料的	0					30	--For dry-etching patterns on semiconductor materials
		--其他:							--Other:
5941	8456.9910	---等离子切割机	0					30	---Cutting machines of plasma arc
5942	8456.9990	---其他	0					30	---Other
	84.57	**加工金属的加工中心、单工位组合机床及多工位组合机床:**							**Machining centres, unit construction machines (single station) and multi-station transfer machines, for working metal:**
		-加工中心:							-Machining centres:
5943	8457.1010	---立式	9.7					20	---Vertical
5944	8457.1020	---卧式	9.7					20	---Horizontal
5945	8457.1030	---龙门式	9.7					20	---Plano
5946	8457.1090	---其他	9.7					20	---Other
5947	8457.2000	--单工位组合机床	8					20	--Unit construction machines (single station)
5948	8457.3000	-多工位组合机床	5					20	-Multi-station transfer machines
	84.58	**切削金属的车床(包括车削中心):**							**Lathes (including turning centres) for removing metal:**
		-卧式车床:							-Horizontal lathes:

序号 No.	税则号列 Tariff Item	货品名称	最惠国税率(%) M.F.N.	协定税率(%) Agreement 东盟 ASEAN	曼谷协定 BA	香港 H.K.	澳门 Macao	普通税率(%) Gen.	Article Description
5949	8458.1100	--数控的	9.7					20	--Numerically controlled
5950	8458.1900	--其他	12					50	--Other
		-其他车床:							-Other lathes:
5951	8458.9100	--数控的	5					20	--Numerically controlled
5952	8458.9900	--其他	12					50	--Other
	84.59	**切削金属的钻床、镗床、铣床、攻丝机床(包括直线移动式动力头钻床),但税号84.58的车床(包括车削中心)除外:**							**Machine-tools (including way-type unit head machines) for drilling, boring, milling, threading or tapping by removing metal, other than lathes (including turning centres) of heading No.84.58:**
5953	8459.1000	-直线移动式动力头钻床	15					50	-Way-type unit head machines
		-其他钻床:							-Other drilling machines:
5954	8459.2100	--数控的	9.7					20	--Numerically controlled
5955	8459.2900	--其他	15					50	--Other
		-其他镗铣机床:							-Other boring-milling machines:
5956	8459.3100	--数控的	9.7					20	--Numerically controlled
5957	8459.3900	--其他	10					50	--Other
		-其他镗床:							-Other boring machines:
5958	8459.4010	---数控的	9.7					20	---Numerically controlled
5959	8459.4090	---其他	15					50	---Other
		-升降台式铣床:							-Milling machines, knee-type:
5960	8459.5100	--数控的	9.7					20	--Numerically controlled
5961	8459.5900	--其他	15					50	--Other
		-其他铣床:							-Other milling machines:
		--数控的:							--Numerically controlled:
5962	8459.6110	---龙门铣床	5					20	---Planomilling machines
5963	8459.6190	---其他	5					20	---Other
		--其他:							--Other:
5964	8459.6910	---龙门铣床	12		10			50	---Planomilling machines
5965	8459.6990	---其他	12		11			50	---Other
5966	8459.7000	-其他攻丝机床	12					50	-Other threading or tapping machines
	84.60	**用磨石、磨料或抛光材料对金属或金属陶瓷进行去毛刺、刃磨、磨削、珩磨、研磨、抛光或其他精加工的机床,但税号84.61的切齿机、齿轮磨床或齿轮精加工机床除外:**							**Machine-tools for deburring, sharpening, grinding, honing, lapping, polishing or otherwise finishing metal or cermets by means of grinding stones, abrasives or polishing products, other than gear cutting, gear grinding or gear finishing machines of heading No.84.61:**
		-平面磨床,在任一坐标的定位精度至少是0.01毫米:							-Flat-surface grinding machines, in which the positioning in any one axis can be set up to an accuracy of at least 0.01mm:
5967	8460.1100	--数控的	9.7					20	--Numerically controlled

序号 No.	税则号列 Tariff Item	货品名称	最惠国税率(%) M.F.N.	协定税率(%) Agreement 东盟 ASEAN	曼谷协定 BA	香港 H.K.	澳门 Macao	普通税率(%) Gen.	Article Description
5968	8460.1900	--其他	15					50	--Other
		-其他磨床,在任一坐标的定位精度至少是0.01毫米:							-Other grinding machines, in which the positioning in any one axis can be set up to an accuracy of at least 0.01mm:
		--数控的:							--Numerically controlled:
5969	8460.2110	---外圆磨床	9.7					20	---Cylindrical grinding machines
5970	8460.2120	---内圆磨床	9.7					20	---Internal grinding machines
5971	8460.2190	---其他	9.7					20	---Other
		--其他:							--Other:
5972	8460.2910	---外圆磨床	15					50	---Cylindrical grinding machines
5973	8460.2920	---内圆磨床	15					50	---Internal grinding machines
5974	8460.2930	---轧辊磨床	13					50	---Grinding machines of roll
5975	8460.2990	---其他	13					50	---Other
		-刃磨(工具或刀具)机床:							-Sharpening (tool or cutter grinding) machines:
5976	8460.3100	--数控的	9.7					20	--Numerically controlled
5977	8460.3900	--其他	15					50	--Other
		-珩磨或研磨机床:							-Honing or lapping machines:
5978	8460.4010	---珩磨	13					50	---Honing
5979	8460.4020	---研磨	13					50	---Lapping
		-其他:							-Other:
5980	8460.9010	---砂轮机	15					50	---Grinding wheel machines
5981	8460.9020	---抛光机床	15					50	---Polishing machines
5982	8460.9090	---其他	15					50	---Other
	84.61	**切削金属或金属陶瓷的刨床、牛头刨床、插床、拉床、切齿机、齿轮磨床或齿轮精加工机床、锯床、切断机及其他税号未列名的切削机床:**							**Machine-tools for planing, shaping, slotting, broaching, gear cutting, gear grinding or gear finishing, sawing, cutting-off and other machine-tools working by removing metal or cermets, not elsewhere specified or included:**
		-牛头刨床或插床:							-Shaping or slotting machines:
5983	8461.2010	---牛头刨床	15					50	---Shaping machines
5984	8461.2020	---插床	15					50	---Slotting machines
5985	8461.3000	-拉床	12					50	-Broaching machines
		-切齿机、齿轮磨床或齿轮精加工机床:							-Gear cutting, gear grinding or gear finishing machines:
5986	8461.4010	---数控的	9.7					20	---Numerically controlled
5987	8461.4090	---其他	15					50	---Other
5988	8461.5000	-锯床或切断机	12					50	-Sawing or cutting-off machines
		-其他:							-Other:
		---刨床:							---Planing machines:
5989	8461.9011	----龙门刨床	15					50	----Double - column (open - side) planing machines
5990	8461.9019	----其他	15					50	----Other

序号 No.	税则号列 Tariff Item	货品名称	最惠国税率(%) M.F.N.	协定税率(%)Agreement				普通税率(%) Gen.	Article Description
				东盟 ASEAN	曼谷协定 BA	香港 H.K.	澳门 Macao		
5991	8461.9090	---其他	12					50	---Other
	84.62	**加工金属的锻造(包括模锻)或冲压机床;加工金属的弯曲、折叠、矫直、矫平、剪切、冲孔或开槽机床;其他加工金属或硬质合金的压力机:**							**Machine-tools (including presses) for working metal by forging, hammering or die-stamping; machine-tools (including presses) for working metal by bending, folding, straightening, flattening, shearing, punching or notching; presses for working metal or metal carbides, not specified above:**
		-锻造(包括模锻)或冲压机床及锻锤:							-Forging or die-stamping machines(including presses)and hammers:
5992	8462.1010	---数控的	9.7					20	---Numerically controlled
5993	8462.1090	---其他	12					50	---Other
		-弯曲、折叠、矫直或矫平机床:							-Bending, folding, straightening or flattening machines (including presses):
		--数控的:							--Numerically controlled:
5994	8462.2110	---矫直机	9.7					20	---Straightening machines
5995	8462.2190	---其他	9.7					20	---Other
		--其他:							--Other:
5996	8462.2910	---矫直机	10					50	---Straightening machines
5997	8462.2990	---其他	10					50	---Other
		-剪切机床,但冲剪两用机除外:							-Shearing machines (including presses), other than combined punching and shearing machines:
		--数控的:							--Numerically controlled:
5998	8462.3110	---板带纵剪机	7					20	---Shearing lengthwise
5999	8462.3120	---板带横剪机	7					20	---Shearing transverse
6000	8462.3190	---其他	7					20	---Other
		--其他:							--Other:
6001	8462.3910	---板带纵剪机	10					50	---Shearing lengthwise
6002	8462.3920	---板带横剪机	10					50	---Shearing transverse
6003	8462.3990	---其他	10					50	---Other
		-冲孔或开槽机床,包括冲剪两用机:							-Punching or notching machines (including presses), including combined punching and shearing machines:
		--数控的:							--Numerically controlled:
		---冲床:							---Punch press:
6004	8462.4111	----自动模式数控步冲压力机	9.7						----CNC automatic tool change punch press
6005	8462.4119	----其他	9.7						----Other
6006	8462.4190	---其他	9.7						---Other
6007	8462.4900	--其他	10					50	--Other
		-其他:							-Other:
		--液压压力机:							--Hydraulic presses:
6008	8462.9110	---金属型材挤压机	10		9.2			50	---Metal section squeezing machine

序号 No.	税则号列 Tariff Item	货品名称	最惠国税率(%) M.F.N.	协定税率(%) Agreement 东盟 ASEAN	曼谷协定 BA	香港 H.K.	澳门 Macao	普通税率(%) Gen.	Article Description
6009	8462.9190	---其他	10		9.2			50	---Other
		--其他:							--Other:
6010	8462.9910	---机械压力机	10					50	---Mechanical presses
6011	8462.9990	---其他	10					50	---Other
	84.63	**金属或金属陶瓷的其他非切削加工机床:**							**Other machine-tools for working metal or cermets, without removing material:**
		-杆、管、型材、异型材、丝及类似品的拉拔机:							-Draw-benches for bars, tubes, profiles, wire or the like:
		---冷拔管机:							---Cold-drawing tube benches:
6012	8463.1011	----300吨及以下	10					50	----With 300t or less
6013	8463.1019	----其他	10					50	----Other
6014	8463.1020	---拔丝机	10					50	---Wiredrawing machines
6015	8463.1090	---其他	10					50	---Other
6016	8463.2000	-螺纹滚轧机	15					50	-Thread rolling machines
6017	8463.3000	-金属丝加工机	10					50	-Machines for working wire
6018	8463.9000	-其他	10					50	-Other
	84.64	**石料、陶瓷、混凝土、石棉水泥或类似矿物材料的加工机床、玻璃冷加工机床:**							**Machine-tools for working stone, ceramics, concrete, asbestos-cement or like mineral materials or for cold working glass:**
		-锯床:							-Sawing machines:
6019	8464.1010	---圆盘锯	0					30	---Of disk saw
6020	8464.1020	---钢丝锯	0					30	---Of scroll saw
6021	8464.1090	---其他	0					30	---Other
		-研磨或抛光机床:							-Grinding or polishing machines:
6022	8464.2010	---玻璃研磨或抛光机床	0					30	---Machines for grinding or polishing glass or glassware
6023	8464.2090	---其他	0					30	---Other
		-其他:							-Other:
		---玻璃的其他冷加工机床:							---Other machines for cold-working glass or glassware:
6024	8464.9011	----切割机	0					30	----Cutting-off machines
6025	8464.9012	----刻花机	0					30	----Carving machines
6026	8464.9019	----其他	0					30	----Other
6027	8464.9090	---其他	0					30	---Other
	84.65	**木材、软木、骨、硬质橡胶、硬质塑料或类似硬质材料的加工机床(包括用打钉或打U形钉、胶粘或其他方法组合前述材料的机器):**							**Machine-tools (including machines for nailing, stapling, glueing or otherwise assembl-ing) for working wood, cork, bone, hard rubber, hard plastics or similar hard materials:**
6028	8465.1000	-不需更换工具即可进行不同机械加工的机器	10					30	-Machines which can carry out different types of machining operations without tool change between such operations

序号 No.	税则号列 Tariff Item	货 品 名 称	最惠国税率(%) M.F.N.	协定税率(%)Agreement 东盟 ASEAN	曼谷协定 BA	香港 H.K.	澳门 Macao	普通税率(%) Gen.	Article Description
		-其他:							-Other:
6029	8465.9100	--锯床	10			0		30	--Sawing machines
6030	8465.9200	--刨、铣或切削成形机器	10					30	--Planing, milling or moulding (by cutting) machines
6031	8465.9300	--研磨、砂磨或抛光机器	10					30	--Grinding, sanding or polishing machines
6032	8465.9400	--弯曲或装配机器	10					30	--Bending or assembling machines
6033	8465.9500	--钻孔或凿榫机器	10					30	--Drilling or mortising machines
6034	8465.9600	--剖开、切片或刮削机器	10					30	--Splitting, slicing or paring machines
6035	8465.9900	--其他	10					30	--Other
	84.66	**专用于或主要用于税号 84.56 至 84.65 所列机器的零件、附件,包括工件或工具的夹具、自启板牙切头、分度头及其他专用于机床的附件;各种手提工具的工具夹具:**							**Parts and accessories suitable for use solely or principally with the machines of headings Nos. 84.56 to 84.65, including work or tool holders, self-opening dieheads, dividing heads and other special attachments for machine-tools; tool holders for any type of tool for working in the hand:**
6036	8466.1000	-工具夹具及自启板牙切头	7					17	-Tool holders and self-opening dieheads
6037	8466.2000	-工件夹具	7					17	-Work holders
6038	8466.3000	-分度头及其他专用于机床的附件	7					17	-Dividing heads and other special attachments for machine-tools
		-其他:							-Other:
6039	8466.9100	--税号 84.64 所列机器用	0					17	--For machines of heading No. 84.64
6040	8466.9200	--税号 84.65 所列机器用	6					17	--For machines of heading No. 84.65
6041	8466.9300	--税号 84.56 至 84.61 所列机器用	0					17	--For machines of headings Nos. 84.56 to 84.61
6042	8466.9400	--税号 84.62 或 84.63 所列机器用	6					17	--For machines of heading No. 84.62 or 84.63
	84.67	**手提式风动或液压工具及本身装有电动或非电动动力装置的手提式工具:**							**Tools for working in the hand, pneumatic hydraulic or with self-contained electric or non-electric motor:**
		-风动的:							-Pneumatlic:
6043	8467.1100	--旋转式(包括旋转冲击式的)	8					30	--Rotary type (including combined rotarypercussion)
6044	8467.1900	--其他	8					30	--Other
		-本身装有电动动力装置的:							-With self − contained electric motor:
6045	8467.2100	--各种钻	10		9			30	--Drills of all kinds
		--锯:							--Saws:
6046	8467.2210	---链锯	10		9			30	---Chain saws

序号 No.	税则号列 Tariff Item	货品名称	最惠国税率(%) M.F.N.	协定税率(%)Agreement 东盟 ASEAN	曼谷协定 BA	香港 H.K.	澳门 Macao	普通税率(%) Gen.	Article Description
6047	8467.2290	---其他	10		9			30	---Other
		--其他:							--Other tools:
6048	8467.2910	---砂磨工具(包括磨光机,砂光机,砂轮机等)	10		8	0△		30	---Grinding tools(induding burnisher, belt sander, wheel - sander)
6049	8467.2920	---电刨	10		8	0△		30	---Planings
6050	8467.2990	---其他	10		8	0△		30	---Other
		-其他工具:							-Other tools:
6051	8467.8100	--链锯	8					30	--Chain saws
6052	8467.8900	--其他	8					30	--Other
		-零件:							-Parts:
		--链锯用:							--Of chain saws:
6053	8467.9110	---电动的	6		4.8	0		30	---With self - contained electric motor
6054	8467.9190	---其他	6		6			30	---Other
6055	8467.9200	--风动工具用	6					30	--Of pneumatic tools
		--其他:							--Other:
6056	8467.9910	---电动工具用	10		9	0		30	---With self - contained electric motor
6057	8467.9990	---其他	6					30	---Other
	84.68	**焊接机器及装置,不论是否兼有切割功能,但税号 85.15 的货品除外;气体加温表面回火机器及装置:**							**Machinery and apparatus for soldering, brazing or welding, whether or not capable of cutting, other than those of heading No. 85.15; gas-operated surface tempering machines and appliances:**
6058	8468.1000	-手提喷焊器	12					30	-Hand-held blow pipes
6059	8468.2000	-其他气体焊接或表面回火机器及装置	12					30	-Other gas-operated machinery and apparatus
6060	8468.8000	--其他机器及装置	12					30	--Other machinery and apparatus
6061	8468.9000	-零件	7					30	-Parts
	84.69	**打字机,但税号 84.71 的打印机除外;文字处理机:**							**Typewriters other than printers of heading No. 84. 71; word-processing machines:**
		-自动打字机及文字处理机:							-Automatic typewriters and word-processing machines:
6062	8469.1100	--文字处理机	0					40	--Word-processing machines
6063	8469.1200	--自动打字机	12					40	--Automatic typewriters
6064	8469.2000	-其他电动打字机	12					40	-Other typewriters, electric
6065	8469.3000	-其他非电动打字机	12					40	-Other typewriters, non-electric
	84.70	**计算机器及具有计算功能的袖珍式数据记录、重现及显示机器;装有计算装置的会计计算机、邮资盖戳机、售票机及类似机器;现金出纳机:**							**Calculating machines and pocket-size data recording, reproducing and displaying machines with calculating functions; accounting machines, postage-franking machines, ticket-issuing machines and similar machines, incorporating a calculating device; cash registers:**

序号 No.	税则号列 Tariff Item	货品名称	最惠国税率(%) M.F.N.	协定税率(%) Agreement 东盟 ASEAN	曼谷协定 BA	香港 H.K.	澳门 Macao	普通税率(%) Gen.	Article Description
6066	8470.1000	-不需外接电源的电子计算器及具有计算功能的袖珍式数据记录、重现及显示机器	0					80	-Electronic calculators capable of operation without an external source of electric power and pocket-size data recording, reproducing and displaying machines with caluclating functions
		-其他电子计算器：							-Other electronic calculating machines:
6067	8470.2100	--装有打印装置的	0					80	--Incorporating a printing device
6068	8470.2900	--其他	0					80	--Other
6069	8470.3000	-其他计算机器	0					40	-Other calculating machines
6070	8470.4000	-会计计算机	0					40	-Accounting machines
		-现金出纳机：							-Cash registers:
6071	8470.5010	---销售点终端出纳机	0					40	---Terminal registers for market
6072	8470.5090	---其他	0					40	---Other
6073	8470.9000	-其他	0					40	-Other
	84.71	**自动数据处理设备及其部件；其他税号未列名的磁性或光学阅读机、将数据以代码形式转录到数据记录媒体的机器及处理这些数据的机器：**							**Automatic data processing machines and units thereof; magnetic or optical readers, machines for transcribing data onto data media in coded form and machines for processing such data, not elsewhere specified or included:**
6074	8471.1000	-模拟式或混合式自动数据处理设备	0					14	-Analogue or hybrid automatic data processing machines
6075	8471.3000	-重量不超过10公斤的便携数字式自动数据处理设备，至少由一个中央处理部件、一个键盘及一个显示器组成	0					70	-Portable digital automatic data processing machines, weighing not more than 10kg, consisting of at least a central processing unit, a keyboard and a display
		-其他数字式自动数据处理设备：							-Other digital automatic data processing machines:
		--同一机壳内至少有一个中央处理部件及一个输入和输出部件，不论是否组合式：							--Comprising in the same housing at least a central processing unit and an input and output unit, whether or not combined:
6076	8471.4110	---巨型机、大型机及中型机	0					14	---Mainframes
6077	8471.4120	---小型机	0					14	---Mini-computers
6078	8471.4140	---微型机	0					70	---Microprocessings
6079	8471.4190	---其他	0					70	---Other
		--其他，以系统形式进口或出口的：							Other, presented in the form of systems:
6080	8471.4910	---巨型机、大型机及中型机	0					29	---Mainframes
6081	8471.4920	---小型机	0					29	---Mini-computers
6082	8471.4940	---微型机	0					70	---Microprocessings

序号 No.	税则号列 Tariff Item	货品名称	最惠国税率(%) M.F.N.	协定税率(%) Agreement				普通税率(%) Gen.	Article Description
				东盟 ASEAN	曼谷协定 BA	香港 H.K.	澳门 Macao		
		---其他:							---Other:
6083	8471.4991	----分散型工业过程控制设备	0					70	----Processing machines for the distributed control system
6084	8471.4999	----其他	0					70	----Other
		-子目号 8471.41 或 8471.49 所列以外的数字式处理部件,不论是否在同一机壳内有一个或两个下列部件:存储部件、输入部件、输出部件:							-Digital processing units other than those of subheading 8471.41 or 8471.49, whether or not containing in the same housing one or two of the following types of unit: storage units, input units, output units:
6085	8471.5010	---巨型机、大型机及中型机的	0					14	---Mainframes
6086	8471.5020	---小型机的	0					14	---Mini-computers
6087	8471.5040	---微型机的	0					70	---Microprocessings
6088	8471.5090	---其他	0					70	---Other
		-输入或输出部件,不论是否在同一机壳内有存储部件:							-Input or output units, whether or not containing storage units in the same housing:
		---显示器:							---Displays:
6089	8471.6011	----液晶的	0						----With liquid crystal display panel
6090	8471.6012	----阴极射线管的	0						----With cathode-ray tubes
6091	8471.6019	----其他	0						----Other
		---打印机:							---Printers:
6092	8471.6031	----针式打印机	0					40	---Stylus printers
6093	8471.6032	----激光打印机	0					14	----Laser printers
6094	8471.6033	----喷墨打印机	0					14	----Ink-jet printers
6095	8471.6039	----其他	0					14	----Other
6096	8471.6040	---巨型机、大型机、中型机及小型机用终端	0					14	---Terminating machines for the huge computers, mainframes and mini-computers
6097	8471.6050	---扫描仪	0					14	---Scanner
6098	8471.6060	---数字化仪	0					14	---Digitizer
6099	8471.6070	---键盘、鼠标器	0		0			40	---Keyboards, mouses
6100	8471.6090	---其他	0					14	---Other
		-存储部件:							-Storage units:
6101	8471.7010	---硬盘驱动器	0					14	---Rigid disk drivers
6102	8471.7020	---软盘驱动器	0					14	---Floppy disk drivers
6103	8471.7030	---光盘驱动器	0					14	---CD drivers
6104	8471.7090	---其他	0					14	---Other
		-自动数据处理设备的其他部件:							-Other units of automatic data processing machines:
6105	8471.8010	---集线器	0						---Network concentrators
6106	8471.8020	---路由器	0						---Network path-control devices

序号 No.	税则号列 Tariff Item	货品名称	最惠国税率(%) M.F.N.	协定税率(%) Agreement 东盟 ASEAN	曼谷协定 BA	香港 H.K.	澳门 Macao	普通税率(%) Gen.	Article Description
6107	8471.8090	---其他	0						---Other
6108	8471.9000	-其他	0					40	-Other
	84.72	**其他办公室用机器(例如,胶版复印机、油印机、地址印写机、自动付钞机、硬币分类、计数及包装机、削铅笔机、打洞机或订书机):**							**Other office machines (for example, hectograph or stencil duplicating machines, addressing machines, automatic banknote dispensers, coin-sorting machines, coincounting or wrapping machines, pencil-sharpening machines, perforating or stapling machines):**
6109	8472.1000	-胶版复印机、油印机	14					40	-Duplicating machines
6110	8472.2000	-地址印写机及地址铭牌压印机	14					40	-Addressing machines and address plate embossing machines
		-信件分类或折叠机或信件装封机、信件开封或闭封机、粘贴或盖销邮票机:							-Machines for sorting or folding mail or for inserting mail in envelopes or bands, machines for opening, closing or sealing mail and machines for affixing or cancelling postage stamps:
6111	8472.3010	---邮政信件分拣及封装设备	10					40	---Machines for sorting or banding mail
6112	8472.3090	---其他	14					40	---Other
		-其他:							-Other:
6113	8472.9010	---自动柜员机	0					40	---Automated teller
		---装订用机器:							---Stapling machines:
6114	8472.9021	----打洞机	0					40	----Perforator
6115	8472.9022	----订书机	0					40	----Stapler
6116	8472.9029	----其他	0					40	----Other
6117	8472.9030	---碎纸机	0					40	----Paper shrudders
6118	8472.9090	---其他	0					40	---Other
	84.73	**专用于或主要用于税号84.69至84.72所列机器的零件、附件(罩套、提箱及类似品除外):**							**Parts and accessories (other than covers, carrying cases and the like) suitable for use solely or principally with machines of headings Nos. 84.69 to 84.72:**
6119	8473.1000	-税号84.69所列机器的零件、附件	8					35	-Parts and accessories of the machines of heading No. 84.69
		-税号84.70所列机器的零件、附件:							-Parts and accessories of the machines of heading No. 84.70:
6120	8473.2100	--子目号8470.10、8470.21或8470.29所列电子计算器的零件、附件	0					50	--Of the electronic calculating machines of subheading No. 8470.10, 8470.21 or 8470.29
6121	8473.2900	--其他	0					35	--Other
		-税号84.71所列机器的零件、附件:							-Parts and accessories of the machines of heading No. 84.71:

序号 No.	税则号列 Tariff Item	货品名称	最惠国税率(%) M.F.N.	协定税率(%) Agreement 东盟 ASEAN	曼谷协定 BA	香港 H.K.	澳门 Macao	普通税率(%) Gen.	Article Description
6122	8473.3010	---子目号8471.1000、8471.4110、8471.4120、8471.4910、8471.4920、8471.5010、8471.5020、8471.6090、8471.7010、8471.7020、8471.7030及8471.7090所列机器及装置的零件、附件	0					14	---Of the machines of subheading No. 8471.1000, 8471.4110, 8471.4120, 8471.4910, 8471.4920, 8471.5010, 8471.5020, 8471.6090, 8471.7019, 8471.7020, 8471.7030 or 8471.7090
		---子目号8471.6031、8471.6032、8471.6033、8471.6039所列机器的零件、附件:							---Parts and accessories of the machines of subheading No.8471.6031、8471.6032、8471.6033、8471.6039:
6123	8473.3021	----针式打印机打印头	0		0			35	----Print element of stylus printers
6124	8473.3029	----其他	0		0			35	----Other
6125	8473.3090	---其他	0		0			40	---Other
		-税号84.72所列机器的零件、附件:							-Parts and accessories of the machines of heading No.84.72:
6126	8473.4010	---自动柜员机用出钞器	10.5					35	---Banknote dispenser of automated teller
6127	8473.4090	---其他	10.5					35	---Other
6128	8473.5000	-同样适用于税号84.69至84.72中两个或两个以上税号所列机器的零件、附件	0		0			35	-Parts and accessories equally suitable for use with machines of two or more of the headings No.84.69 to 84.72
	84.74	**泥土、石料、矿石或其他固体(包括粉状、浆状)矿物质的分类、筛选、分离、洗涤、破碎、磨粉、混合或搅拌机器;固体矿物燃料、陶瓷坯泥、未硬化水泥、石膏材料或其他粉状、浆状矿产品的粘聚或成形机器;铸造用砂模的成形机器:**							**Machinery for sorting, screening, separating, washing, crushing, grinding, mixing or kneading earth, stone, ores or other mineral substances, in solid (including powder or paste) form; machinery for agglomerating, shaping or moulding solid mineral fuels, ceramic paste, unhardened cements, plastering materials or other mineral products in powder or paste form; machines for forming foundry moulds of sand:**
6129	8474.1000	-分类、筛选、分离或洗涤机器	5					30	-Sorting, screening, separating or washing machines
		-破碎或磨粉机器:							-Crushing or grinding machines:
6130	8474.2010	---齿辊式	5					30	---Toothing roller type
6131	8474.2020	---球磨式	5					30	---Em-Peters type
6132	8474.2090	---其他	5					30	---Other
		-混合或搅拌机器:							-Mixing or kneading machines:
6133	8474.3100	--混凝土或砂浆混合机器	7					30	--Concrete or mortar mixers
6134	8474.3200	--矿物与沥青的混合机器	7					30	--Machines for mixing mineral substances with bitumen
6135	8474.3900	--其他	5					30	--Other
		-其他机器:							-Other machinery:
6136	8474.8010	---辊压成型机	5					30	---Rolling forming machines

序号 No.	税则号列 Tariff Item	货品名称	最惠国税率(%) M.F.N.	协定税率(%) Agreement 东盟 ASEAN	曼谷协定 BA	香港 H.K.	澳门 Macao	普通税率(%) Gen.	Article Description
6137	8474.8090	---其他	5					30	---Other
6138	8474.9000	-零件	5					30	-Parts
	84.75	**白炽灯泡、灯管、放电灯管、电子管、闪光灯泡及类似品的封装机器；玻璃或玻璃制品的制造或热加工机器：**							**Machines for assembling electric or electronic lamps, tubes or valves or flashbulbs, in glass envelopes; machines for manufacturing or hot working glass or glassware:**
6139	8475.1000	-白炽灯泡、灯管、放电灯管、电子管、闪光灯泡及类似品的封装机器	8					30	-Machines for assembling electric or electronic lamps, tubes or valves or flashbulbs, in glass envelopes
		-玻璃或玻璃制品的制造或热加工机器：							-Machines for manufacturing or hot working glass or glassware:
6140	8475.2100	--制造光导纤维及其预制棒的机器	10					30	--Machines for making optical fibres and preforms thereof
		--其他：							--Other:
		---玻璃的热加工设备：							---Equipments for hot working glass or glasswares:
6141	8475.2911	----连续式玻璃热弯炉	10					30	----Continuous hot bending furnaces
6142	8475.2912	----玻璃纤维拉丝机(光纤拉丝机除外)	10					30	----Fiber glass winder(excluding Opticaefiber winder)
6143	8475.2919	----其他	10					30	----Other
6144	8475.2990	---其他	10					30	---Other
6145	8475.9000	-零件	8					30	-Parts
	84.76	**自动售货机(例如，出售邮票、香烟、食品或饮料的机器)，包括钱币兑换机：**							**Automatic goods-vending machines (for example, postage stamp, cigarette, food or beverage machines), including money-changing machines:**
		-饮料自动销售机：							-Automatic beverage-vending machines:
6146	8476.2100	--装有加热或制冷装置的	14					50	--Incorporating heating or refrigerating devices
6147	8476.2900	--其他	15					50	--Other
		-其他机器：							-Other machines:
6148	8476.8100	--装有加热或制冷装置的	14					50	--Incorporating heating or refrigerating devices
6149	8476.8900	--其他	15					50	--Other
6150	8476.9000	-零件	10					50	-Parts
	84.77	**本章其他税号未列名的橡胶或塑料及其产品的加工机器：**							**Machinery for working rubber or plastics or for the manufacture of products from these materials, not specified or included elsewhere in this Chapter:**
		-注射机：							-Injection-moulding machines:
6151	8477.1010	---注塑机	0		0			45	---For working plastics
6152	8477.1090	---其他	0					30	---Other

序号 No.	税则号列 Tariff Item	货品名称	最惠国税率(%) M.F.N.	协定税率(%)Agreement 东盟 ASEAN	曼谷协定 BA	香港 H.K.	澳门 Macao	普通税率(%) Gen.	Article Description
		-挤出机:							-Extruders:
6153	8477.2010	---塑料造粒机	5			0		30	---Plastic pelletizers
6154	8477.2090	---其他	5			0		30	---Other
6155	8477.3000	-吹塑机	5					30	-Blow moulding machines
		-真空模塑机器及其他热成型机器:							-Vacuum moulding machines and other thermoforming machines:
6156	8477.4010	---塑料中空成型机	5					30	---Plastics brideg-die-forming machines
6157	8477.4020	---塑料压延成型机	5					30	---Plastics calender-forming machines
6158	8477.4090	---其他	5					30	---Other
		-其他模塑或成型机器:							-Other machinery for moulding or otherwise forming:
6159	8477.5100	--用于充气轮胎模塑或翻新的机器及内胎模塑或用其他方法成型的机器	5					30	--For moulding or retreading pneumatic tyres or for moulding or otherwise forming inner tubes
6160	8477.5900	--其他	5					30	--Other
6161	8477.8000	-其他机器	5			0		30	-Other machinery
6162	8477.9000	-零件	0					30	-Parts
	84.78	**本章其他税号未列名的烟草加工及制作机器:**							**Machinery for preparing or making up tobacco, not specified or included elsewhere in this Chapter:**
6163	8478.1000	-机器	5					30	-Machinery
6164	8478.9000	-零件	10					30	-Parts
	84.79	**本章其他税号未列名的具有独立功能的机器及机械器具:**							**Machines and mechanical appliances having individual functions,not specified or included elsewhere in this Chapter:**
		-公共工程用机器:							-Machinery for public works, building or the like:
		---摊铺机:							---Spreading machines:
6165	8479.1021	----沥青混凝土摊铺机	8					30	----Machines for spreading bituminous concrete
6166	8479.1022	----稳定土摊铺机	8					30	----Stabilizer spreading machines
6167	8479.1029	----其他	8					30	----Other
6168	8479.1090	---其他	8					30	---Other
6169	8479.2000	-提取、加工动物油脂或植物固定油脂的机器	10					30	-Machinery for the extraction or preparation of animal or fixed vegetable fats or oils
6170	8479.3000	-木碎料板或木纤维板的挤压机及其他木材或软木处理机	10					30	-Presses for the manufacture of particle board or fibre building board or wood or other ligneous materials and other machinery for treating wood or cork
6171	8479.4000	-绳或缆的制造机器	7					30	-Rope or cable-making machines
		-未列名工业机器人:							Industrial robots,not elsewhere specified or included:
6172	8479.5010	---多功能工业机器人	0		0			20	---Industrial robots for multiple uses

序号 No.	税则号列 Tariff Item	货品名称	最惠国税率(%) M.F.N.	协定税率(%)Agreement 东盟 ASEAN	曼谷协定 BA	香港 H.K.	澳门 Macao	普通税率(%) Gen.	Article Description
6173	8479.5090	---其他	0		0			30	---Other
6174	8479.6000	-蒸发式空气冷却器	10		9			30	-Evaporative air coolers
		-其他机器及机械器具：							-Other machines and mechanical appliances:
		--处理金属的机械，包括线圈绕线机：							--For treating metal, including electric wire coil-winders:
6175	8479.8110	---绕线机	9.5			0△		30	---Filament winding machines
6176	8479.8190	---其他	9.5			0△		30	---Other
6177	8479.8200	--混合、搅拌、轧碎、研磨、筛选、均化或乳化机器	7					30	--Mixing, kneading, crushing, grinding, screening, sifting, homogenizing, emulsifying or stirring machines
		--其他：							--Other:
6178	8479.8910	---船舶用舵机及陀螺稳定器	0		0			14	---Steering and rudder equipment or gyroscopic stabilizers for ships
6179	8479.8920	---空气增湿器及减湿器	0		0			70	---Air humidifiers or dehumidifiers
6180	8479.8940	---邮政用包裹、印刷品分拣设备	0		0			30	---Bundle and printed matter sorting machines used in post offices
6181	8479.8950	---放射性废物压实机	0		0			30	---Presses for radioactive waste material
		---在印刷电路电路板上装配元器件的机器：							---Machines for assemblying elements on printed circuit boards:
6182	8479.8961	----自动插件机	0					30	----Automatic plug-in machines
6183	8479.8962	----自动贴片机	0			0		30	----Automatic coreslice adhering machines
6184	8479.8969	----其他	0					30	----Other
6185	8479.8990	---其他	0		0			30	---Other
		-零件：							-Parts:
6186	8479.9010	---船舶用舵机及陀螺稳定器用	0					14	---Of the machines of subheading No. 8479.8910
6187	8479.9020	---空气增湿器及减湿器用	0					70	---Of the machines of subheading No. 8479.8920
6188	8479.9090	---其他	0					20	---Other
	84.80	**金属铸造用型箱；型模底板；阳模；金属用型模(锭模除外)、硬质合金、玻璃、矿物材料、橡胶或塑料用型模：**							**Moulding boxes for metal foundry; mould bases; moulding patterns; moulds for metal (other than ingot moulds), metal carbides, glass, mineral materials, rubber or plastics:**
6189	8480.1000	-金属铸造用型箱	10					20	-Moulding boxes for metal foundry
6190	8480.2000	-型模底板	8					20	-Mould bases
6191	8480.3000	-阳模	10					20	-Moulding patterns
		-金属、硬质合金用型模：							-Moulds for metal or metal carbides:
6192	8480.4100	--注模或压模	8			0		20	--Injection or compression types
6193	8480.4900	--其他	8					20	--Other

序号 No.	税则号列 Tariff Item	货品名称	最惠国税率(%) M.F.N.	协定税率(%)Agreement 东盟 ASEAN	曼谷协定 BA	香港 H.K.	澳门 Macao	普通税率(%) Gen.	Article Description
6194	8480.5000	-玻璃用型模	8.4					20	-Moulds for glass
6195	8480.6000	-矿物材料用型模	8.4					20	-Moulds for mineral materials
		-塑料或橡胶用型模：							-Moulds for rubber or plastics：
6196	8480.7100	--注模或压模	0					20	--Injection or compression types
6197	8480.7900	--其他	5			0		20	--Other
	84.81	**用于管道、锅炉、罐、桶或类似品的龙头、旋塞、阀门及类似装置，包括减压阀及恒温控制阀：**							**Taps, cocks, valves and similar appliances for pipes, boiler shells, tanks, vats or the like, including pressure-reducing valves and thermostatically controlled valves:**
6198	8481.1000	-减压阀	5					30	-Pressure-reducing valves
		-油压或气压传动阀：							-Valves for oleohydraulic or pneumatic transmissions：
6199	8481.2010	---油压的	5					30	---For oleohydraulic transmissions
6200	8481.2020	---气压的	5					30	---For pneumatic transmissions
6201	8481.3000	-止回阀	5					30	-Check (nonreturn) valves
6202	8481.4000	-安全阀或溢流阀	5					30	-Safety or relief valves
		-其他器具：							-Other appliances：
6203	8481.8010	---其他阀门	7			0			---Other valves
6204	8481.8090	---其他	5			0△		50	---Other
		-零件：							-Parts：
6205	8481.9010	---阀门用	8					30	---Of valves
6206	8481.9090	---其他	8					50	---Other
	84.82	**滚动轴承：**							**Ball or roller bearings:**
6207	8482.1000	-滚珠轴承	8			0		20	-Ball bearings
6208	8482.2000	-锥形滚子轴承，包括锥形滚子组件	8					20	-Tapered roller bearings, including cone and tapered roller assemblies
6209	8482.3000	-鼓形滚子轴承	8					20	-Spherical roller bearings
6210	8482.4000	-滚针轴承	8					20	-Needle roller bearings
6211	8482.5000	-其他圆柱形滚子轴承	8					20	-Other cylindrical roller bearings
6212	8482.8000	-其他，包括球、柱混合轴承	8					20	-Other, including combined ball/roller bearings
		-零件：							-Parts：
6213	8482.9100	--滚珠、滚针及滚柱	8					20	--Balls, needles and rollers
6214	8482.9900	--其他	6					20	--Other
	84.83	**传动轴(包括凸轮轴及曲柄轴)及曲柄；轴承座及滑动轴承；齿轮及齿轮传动装置；滚珠或滚子螺杆传动装置；齿轮箱及其他变速装置，包括扭矩变换器；飞轮及滑轮，包括滑轮组；离合器及联轴器(包括万向节)：**							**Transmission shafts (including cam shafts and crank shafts) and cranks; bearing housings and plain shaft bearings; gears and gearing ball or roller screws; gear boxes and other speed changers, including torque converters; flywheels and pulleys, including pulley blocks; clutches and shaft couplings (including universal joints):**
		-传动轴(包括凸轮轴及曲柄轴)及曲柄：							-Transmission shafts (including cam shafts and crank shafts) and cranks：

序号 No.	税则号列 Tariff Item	货品名称	最惠国税率(%) M.F.N.	协定税率(%) Agreement 东盟 ASEAN	曼谷协定 BA	香港 H.K.	澳门 Macao	普通税率(%) Gen.	Article Description
6215	8483.1010	---船舶用传动轴	6			0		14	---Transmission shafts for ships
6216	8483.1090	---其他	6			0		30	---Other
6217	8483.2000	-装有滚珠或滚子轴承的轴承座	6					30	-Bearing housings, incorporating ball or roller bearings
6218	8483.3000	-未装有滚珠或滚子轴承的轴承座;滑动轴承	6					30	-Bearing housings, not incorporating ball or roller bearings; plain shaft bearings
		-齿轮及齿轮传动装置,但单独进口或出口的带齿的轮、链轮及其他传动元件除外;滚珠或滚子螺杆传动装置;齿轮箱及其他变速装置,包括扭矩变换器:							-Gears and gearing, other than toothed wheels, chain sprockets and other transmission elements presented separately; ball or roller screws; gear boxes and other speed changers, including torque converters:
6219	8483.4010	---滚子螺杆传动装置	8			0		30	---Roller Screws
6220	8483.4020	---行星齿轮减速器	8			0		30	---Planet decelerators
6221	8483.4090	---其他	8			0		30	---Other
6222	8483.5000	-飞轮及滑轮,包括滑轮组	8					30	-Flywheels and pulleys, including pulley blocks
6223	8483.6000	-离合器及联轴器(包括万向节)	8					30	-Clutches and shaft couplings (including universal joints)
6224	8483.9000	-单独报验的带齿的轮、链轮及其他传动元件;零件	8			0		30	-Toothed wheels, chain sprockets and other transmission elements presented separately; parts
	84.84	**密封垫或类似接合衬垫,用金属片与其他材料制成或用双层或多层金属片制成;成套或各种不同材料的密封垫或类似接合衬垫,装于袋、套或类似包装内;机械密封件:**							**Gaskets and similar joints of metal sheeting combined with other material or of two or more layers of metal; sets or assortments of gaskets and similar joints, dissimilar in composition, put up in pouches, envelopes or similar packings mechanical seals:**
6225	8484.1000	-密封垫或类似接合衬垫,用金属片与其他材料制成或用双层或多层金属片制成	8					30	-Gaskets and similar joints of metal sheeting combined with other material or of two or more layers of metal
6226	8484.2000	-机械密封件	8					30	-Mechanical seals
6227	8484.9000	-其他	8					30	-Other
	84.85	**本章其他税号未列名的机器零件,不具有电气接插件、绝缘体、线圈、触点或其他电气器材特征的:**							**Machinery parts, not containing electrical connectors, insulators, coils, contacts or other electrical features, not specified or included elsewhere in this Chapter:**
6228	8485.1000	-船用推进器及桨叶	6					14	-Ships' or boats' propellers and blades therefor
6229	8485.9000	-其他	8			0		30	-Other

第八十五章
电机、电气设备及其零件；录音机及放声机、电视图像、声音的录制和重放设备及其零件、附件

注释：

一、本章不包括：

（一）电暖的毯子、褥子、足套及类似品，电暖的衣服、靴、鞋、耳套或其他供人穿戴的电暖物品；

（二）税号70.11的玻璃制品；

（三）第九十四章的电热家具。

二、税号85.01至85.04不适用于税号85.11、85.12、85.40、85.41或85.42的货品，但金属槽汞弧整流器仍归入税号85.04。

三、税号85.09仅包括通常供家用的下列电动器具：

（一）任何重量的真空吸尘器包括干式及湿式真空吸尘器、地板打蜡机、食品研磨机及食品搅拌器，水果或蔬菜的榨汁器；

（二）重量不超过20公斤的其他机器。

但该税号不适用于风机、风扇或装有风扇的通风罩及循环气罩（不论是否装有过滤器）（税号84.14）、离心干衣机（税号84.21）、洗碟机（税号84.22）、家用洗衣机（税号84.50）、滚筒式或其他形式的熨烫机器（税号84.20或84.51）、缝纫机（税号84.52）、电剪子（税号84. 67）或电热器具（税号85.16）。

四、税号85.34所称“印刷电路”，是指采用各种印制方法（例如，压印、覆镀、腐蚀）或采用“膜电路”工艺，将导线、接点或其他印制元件（例如，电感器、电阻器、电容器）按预定的图形单独或互相连接地印制在绝缘基片上的电路，但能够产生、整流、调制或放大电信号的元件（例如，半导体元件）除外。

Chapter 85
Electrical machinery and equipment and parts thereof; sound recorders and reproducers, television image and sound recorders and reproducers, and parts and accessories of such articles

Notes:

1. This Chapter does not cover:

(a) Electrically warmed blankets, bed pads, foot-muffs or the like; electrically warmed clothing, footwear or ear pads or other electrically warmed articles worn on or about the person;

(b) Articles of glass of heading No.70.11; or

(c) Electrically heated furniture of Chapter 94.

2. Headings No.85.01 to 85.04 do not apply to goods described in heading No.85.11, 85.12, 85.40, 85.41 or 85.42.However, metal tank mercury arc rectifiers remain classified in heading No.85.04.

3. Heading No.85.09 covers only the following electro mechanical machines of the kind commonly used for domestic purposes:

(a) Vacuum cleaners, including dry and wet vacuum cleaners, floor polishers, food grinders and mixers, and fruit or vegetable juice extractors, of any weight;

(b) Other machines provided the weight of such machines does not exceed 20 kg.

The heading does not, however, apply to fans or ventilating or recycling hoods incorporating a fan, whether or not fitted with filters (heading No.84.14), centrifugal clothes-dryers (heading No.84.21), dish washing machines (heading No.84.22), household washing machines (heading No.84.50), roller or other ironing machines (heading No.84.20 or 84.51), sewing machines (heading No.84.52), electric scissors (heading No.84.67) or to electrothermic appliances (heading No.85.16).

4. For the purposes of heading No.85.34 "printed circuits" are circuits obtained by forming on an insulating base, by any printing process (for example, embossing, plating up, etching) or by the "film circuit" technique, conductor elements, contacts or other printed components (for example, inductances, resistors, capacitors) alone or interconnected according to a preestablished pattern, other than elements which can produce, rectify, modulate or amplify an electrical signal (for example, semiconductor elements).

所称“印刷电路”，不包括装有非印制元件的电路，也不包括单个的分立式电阻器、电容器及电感器。但印刷电路可配有非经印刷的连接元件。

用同样工艺制得的无源元件及有源元件组成的薄膜电路或厚膜电路应归入税号85.42。

五、税号85.41及85.42所称：

(一)“二极管、晶体管及类似的半导体器件”，是指那些依靠外加电场引起电阻率的变化而进行工作的半导体器件。

(二)“集成电路及微电子组件”，是指：

1. 单片集成电路，即电路元件（二极管、晶体管、电阻器、电容器、连接线等）主要整体制作在一片半导体材料（例如掺杂硅）基片的表面，并不可分割地连接在一起的电路。

2. 混合集成电路，即通过薄膜或厚膜工艺制得的无源元件（电阻器、电容器、连接线等）和通过半导体工艺制得的有源元件（二极管、晶体管、单片集成电路等）不可分割地组合在同一绝缘基片（玻璃、陶瓷等）上的电路。这种电路也可包括分立元件。

3. 模制组件、微模组件或类似组件，即由分立元件、有源元件或有源元件及无源元件互相结合并连接在一起所构成的微型组合件。

本注释所述物品在归类时，即使本目录其他税号涉及到上述的物品，尤其是物品的功能，仍应优先考虑归入税号85.41及85.42。

六、税号85.23及85.24的唱片、磁带及其他记录媒体，当与所用设备一同报验时，仍应归入上述税号。

本注释不适用于与所用设备以外的物品一同报验的上述媒体。

七、税号85.48所称“废原电池、废原电池组及废蓄电池”，是指因破损、拆解、耗尽或其他原因而不能再使用，也不能再充电的电池。

The expression “printed circuits” does not cover circuits combined with elements other than those obtained during the printing process, nor does it cover individual, discrete resistors, capacitors or inductances.

Thinor thick-film circuits comprising passive and active elements obtained during the same technological process are to be classified in heading No.85.42.

5. For the purposes of headings Nos.85.41 and 85.42:

(a) “Diodes, transistors and similar semiconductor devices” are semiconductor devices the operation of which depends on variations in resistivity on the application of an electric field;

(b) “Electronic integrated circuits and microassemblies” are:

(1) Monolithic integrated circuits in which the circuit elements (diodes, transistors, resistors, capacitors, interconnections, etc.) are created in the mass (essentially) and on the surface of a semiconductor material (doped silicon, for example) and are inseparably associated;

(2) Hybrid integrated circuits in which passive elements (resistors, capacitors, interconnections, etc.), obtained by thin-or thick-film technology, and active elements (diodes, transistors, monolithic integrated circuits, etc.), obtained by semiconductor technology, are combined to all intents and purposes indivisibly, on a single insulating substrate (glass, ceramic, etc.). These circuits may also include discrete components;

(3) Microassemblies of the moulded module, micromodule or similar types, consisting of discrete, active or both active and passive, components which are combined and interconnected.

For the classification of the articles defined in this Note, headings No.85.41 and 85.42 shall take precedence over any other heading in the Nomenclature which might cover them by reference to, in particular, their function.

6. Records, tapes and other media of heading 85.23 or 85.24 remain classified in those headings when presented with the apparatus for which they are intended.

This Note does not apply to such media when they are presented with articles other than the apparatus for which they are intended.

7. For the purposes of heading No.85.48, spent primary cells, “spent primary batteries and spent electric accumulators” are those which are neither usable as such because of breakage, cutting-up, wear or other reasons, nor capable of being recharged.

子目注释：

一、子目号8519.92和8527.12仅包括有内置放大器但无内置扬声器的盒式磁带放声机，不需外接电源即能工作，且外形尺寸不超过170毫米×100毫米×45毫米。

二、子目8542.10所称“智能卡”是指装有任何类型集成电路（微处理器）芯片的卡，不论是否装有磁条。

Subheading Notes：

1. Subheadings Nos.8519.92 and 8527.12 cover only cassette-players with built-in amplifier, without built-in loudspeaker, capable of operating without an external source of electric power and the dimensions of which do not exceed 170mm × 100mm × 45mm.
2. For the purposes of subheading 8542.10, the term “smart cards” means cards which have embedded in them an electronic integrated circuit (microprocessor) of any type in the form of a chip and which may or may not have a magnetic stripe.

序号 No.	税则号列 Tariff Item	货品名称	最惠国税率(%) M.F.N.	协定税率(%)Agreement 东盟 ASEAN	曼谷协定 BA	香港 H.K.	澳门 Macao	普通税率(%) Gen.	Article Description
	85.01	**电动机及发电机(不包括发电机组)：**							**Electric motors and generators (excluding generating sets):**
		-输出功率不超过37.5瓦的电动机：							-Motors of an output not exceeding 37.5W：
6230	8501.1010	---玩具用	24.5			0		80	---For use in toys
		---其他：							---Other：
6231	8501.1091	----微电机，机座尺寸在20毫米及以上，但不超过39毫米	9			0		70	----Micromotors with a housing size of 20mm or more but not exceeding 39mm
6232	8501.1099	----其他	9			0		35	----Other
6233	8501.2000	-交直流两用电动机，输出功率超过37.5瓦	12					35	-Universal AC/DC motors of an output exceeding 37.5 W
		-其他直流电动机；直流发电机：							-Other DC motors;DC generators：
6234	8501.3100	--输出功率不超过750瓦	12		11	0	0	35	--Of an output not exceeding 750W
6235	8501.3200	--输出功率超过750瓦，但不超过75千瓦	10					35	--Of an output exceeding 750 W but not exceeding 75 kW
6236	8501.3300	--输出功率超过75千瓦，但不超过375千瓦	5					35	--Of an output exceeding 75 kW but not exceeding 375 kW
6237	8501.3400	--输出功率超过375千瓦	12					35	--Of an output exceeding 375 kW
6238	8501.4000	-其他单相交流电动机	12					35	-Other AC motors,single-phase
		-其他多相交流电动机：							-Other AC motors,multi-phase：
6239	8501.5100	--输出功率不超过750瓦	5					35	--Of an output not exceeding 750W
6240	8501.5200	--输出功率超过750瓦，但不超过75千瓦	10					35	--Of an output exceeding 750 W but not exceeding 75 kW
6241	8501.5300	--输出功率超过75千瓦	12					35	--Of an output exceeding 75 kW
		-交流发电机：							-AC generators(alternators)：
6242	8501.6100	--输出功率不超过75千伏安	5					30	--Of an output not exceeding 75 kVA

序号 No.	税则号列 Tariff Item	货品名称	最惠国税率(%) M.F.N.	协定税率(%)Agreement 东盟 ASEAN	曼谷协定 BA	香港 H.K.	澳门 Macao	普通税率(%) Gen.	Article Description
6243	8501.6200	--输出功率超过75千伏安,但不超过375千伏安	12					30	--Of an output exceeding 75 kVA but not exceeding 375 kVA
6244	8501.6300	--输出功率超过375千伏安,但不超过750千伏安	12					30	--Of an output exceeding 375 kVA but not exceeding 750 kVA
		--输出功率超过750千伏安:							--Of an output exceeding 750 kVA:
6245	8501.6410	---输出功率超过750千伏安,但不超过350兆伏安	10					30	---Of an output exceeding 750 kVA but not exceeding 350 MVA
6246	8501.6420	---输出功率超过350兆伏安,但不超过665兆伏安	5.8					14	---Of an output exceeding 350 MVA but not exceeding 665 MVA
6247	8501.6430	---输出功率超过665兆伏安	6					11	---Of an output exceeding 665 MVA
	85.02	**发电机组及旋转式变流机:**							**Electric generating sets and rotary converters:**
		-装有压燃式活塞内燃发动机(柴油或半柴油发动机)的发电机组:							-Generating sets with compression-ignition internal combustion piston engines(diesel or semi-diesel engines):
6248	8502.1100	--输出功率不超过75千伏安	10					45	--Of an output not exceeding 75 kVA
6249	8502.1200	--输出功率超过75千伏安,但不超过375千伏安	10					45	--Of an output exceeding 75 kVA but not exceeding 375 kVA
		--输出功率超过375千伏安:							--Of an output exceeding 375 kVA:
6250	8502.1310	---输出功率超过375千伏安,但不超过2兆伏安	10					45	---Of an output exceeding 375 kVA but not exceeding 2 MVA
6251	8502.1320	---输出功率超过2兆伏安	10					30	---Of an output exceeding 2 MVA
6252	8502.2000	-装有点燃式活塞内燃发动机的发电机组	10					45	-Generating sets with spark-ignition internal combustion piston engines
		-其他发电机组:							-Other generating sets:
6253	8502.3100	--风力驱动的	8					30	--Wind-powered
6254	8502.3900	--其他	10					30	--Other
6255	8502.4000	-旋转式变流机	10					30	-Electric rotary converters
	85.03	**专用于或主要用于税号85.01或85.02所列机器的零件:**							**Parts suitable for use solely or principally with the machines of heading No.85.01 or 85.02:**
6256	8503.0010	---子目号8501.1010及8501.1091所列电动机用	12			0		70	---Of the motors of subheading No. 8501.1010 or 8501.1091

序号 No.	税则号列 Tariff Item	货品名称	最惠国税率(%) M.F.N.	协定税率(%) Agreement 东盟 ASEAN	曼谷协定 BA	香港 H.K.	澳门 Macao	普通税率(%) Gen.	Article Description
6257	8503.0020	---子目号 8501.6420 及 8501.6430 所列发电机用	3			0		11	---Of the generators of subheading No.8501.6420 or 8501.6430
6258	8503.0030	---子目号 8502.3100 所列发电机组用	3			0		30	---Of the generating sets of subheading No.8502.3100
6259	8503.0090	---其他	8			0		30	---Other
	85.04	**变压器、静止式变流器(例如整流器)及电感器:**							**Electrical transformers, static converters (for example, rectifiers) and inductors:**
		-放电灯或放电管用镇流器:							-Ballasts for discharge lamps or tubes:
6260	8504.1010	---电子镇流器	10				0	35	---Electronic ballats
6261	8504.1090	---其他	10				0	35	---Other
		-液体介质变压器:							-Liquid dielectric transformers:
6262	8504.2100	--额定容量不超过 650 千伏安	10.5					50	--Having a power handling capacity not exceeding 650 kVA
6263	8504.2200	--额定容量超过 650 千伏安,但不超过 10 兆伏安	12.6					50	--Having a power handling capacity exceeding 650 kVA but not exceeding 10 MVA
		--额定容量超过 10 兆伏安:							--Having a power handling capacity exceeding 10 MVA:
		---额定容量超过 10 兆伏安,但小于 400 兆伏安:							---Having a power handling capacity exceeding 10 MVA but less than 400 MVA:
6264	8504.2311	----额定容量超过 10 兆伏安,但小于 220 兆伏安	10					50	----Having a power handing capacity exceeding 10MVA but less than 220MVA
6265	8504.2312	----额定容量在 220 兆伏安及以上,但小于 330 兆伏安	10					50	----Having a power handling capacity exceeding 220MVA but less than 330 MVA
6266	8504.2313	----额定容量在 330 兆伏安及以上,但小于 400 兆伏安	10					50	----Having a power handling capacity exceeding 330MVA but less than 400 MVA
		---额定容量在 400 兆伏安及以上:							---Having a power handling capacity of 400 MVA or more:
6267	8504.2321	----额定容量在 400 兆伏安及以上,但小于 500 兆伏安	6					11	----Having a power handling capacity exceeding 400MVA but less than 500 MVA
6268	8504.2329	----其他	6					11	----Other
		-其他变压器:							-Other transformers:
		--额定容量不超过 1 千伏安:							--Having a power handling capacity not exceeding 1 kVA:
6269	8504.3110	---互感器	5				0	50	---Mutual inductor
6270	8504.3190	---其他	5			0	0	50	---Other

序号 No.	税则号列 Tariff Item	货品名称	最惠国税率(%) M.F.N.	协定税率(%) Agreement 东盟 ASEAN	曼谷协定 BA	香港 H.K.	澳门 Macao	普通税率(%) Gen.	Article Description
		--额定容量超过1千伏安,但不超过16千伏安:							--Having a power handling capacity exceeding 1 kVA but not exceeding 16 kVA:
6271	8504.3210	---互感器	5				0	50	---Mutual inductor
6272	8504.3290	---其他	5				0	50	---Other
6273	8504.3300	--额定容量超过16千伏安,但不超过500千伏安	5				0	50	--Having a power handling capacity exceeding 16 kVA but not exceeding 500 kVA
6274	8504.3400	--额定容量超过500千伏安	14				0	50	--Having a power handling capacity exceeding 500 kVA
		-静止式变流器:							-Static converters:
		---稳压电源:							---Voltage-stabilized suppliers:
6275	8504.4013	----税号84.71所列机器用	0					40	----Of the machines of heading No. 84.71
6276	8504.4014	----其他直流稳压电源,功率小于1千瓦,精度低于万分之一	7					80	----Other DC Voltage-stabilized suppliers, of a power of less than 1 kW and an accuracy of not better than 0.0001
6277	8504.4015	----其他交流稳压电源,功率小于10千瓦,精度低于千分之一	0					80	----Other AC voltage-stabilized suppliers, of a power of less than 10 kW and an accuracy of not better than 0.001
6278	8504.4019	----其他	0					50	----Other
6279	8504.4020	---不间断供电电源	10			0△		50	---Uninterrupted power suppliers
		---其他							---Other
6280	8504.4091	----具有变流功能的半导体模块	10			0	0	30	----Semiconductor modules with converting function
6281	8504.4099	----其他	10			0	0	30	----Other
6282	8504.5000	-其他电感器	0					35	-Other inductors
		-零件:							-Parts:
		---变压器用:							---Of transformers:
6283	8504.9011	----子目号8504.2321,8504.2329所列变压器用	5		4.5	0		11	----Of the transformers of subheading No.8504.2321,8504,2329
6284	8504.9019	----其他	8		4.5	0		50	----Other
6285	8504.9020	---稳压电源及不间断供电电源用	8			0		50	---Of voltage-stabilized suppliers and uninterrupted power suppliers
6286	8504.9090	---其他	8			0		30	---Other
	85.05	**电磁铁;永磁铁及磁化后准备制永磁铁的物品;电磁铁或永磁铁卡盘、夹具及类似的工件夹具;电磁联轴节、离合器及制动器;电磁起重吸盘:**							**Electro-magnets; permanent magnets and articles intended to become permanent magnets after magnetization; electro-magnetic or permanent magnet chucks, clamps and similar holding devices; electro-magnetic couplings, clutches and brakes; electro-magnetic lifting heads:**

序号 No.	税则号列 Tariff Item	货品名称	最惠国税率(%) M.F.N.	协定税率(%)Agreement 东盟 ASEAN	曼谷协定 BA	香港 H.K.	澳门 Macao	普通税率(%) Gen.	Article Description
		-永磁铁及磁化后准备制永磁铁的物品：							-Permanent magnets and articles intended to become permanent magnets after magnetization:
		--金属的：							--Of metal:
6287	8505.1110	---稀土永磁体	7					20	---Of rare-earth metals
6288	8505.1190	---其他	7			0		20	---Other
6289	8505.1900	--其他	7			0		20	--Other
6290	8505.2000	-电磁联轴节、离合器及制动器	8					20	-Electro-magnetic couplings, clutches and brakes
6291	8505.3000	-电磁起重吸盘	8					20	-Electro-magnetic lifting heads
6292	8505.9000	-其他，包括零件	8					20	-Other, including parts
	85.06	**原电池及原电池组：**							**Primary cells and primary batteries:**
		-二氧化锰的：							-Manganese dioxide
6293	8506.1010	---碱性锌锰的	20			0		80	---Alkaline zinc – manganese dioxide
6294	8506.1090	---其他	20			0		80	---Other
6295	8506.3000	-氧化汞的	14					40	-Mercuric oxide
6296	8506.4000	-氧化银的	14					40	-Silver oxide
6297	8506.5000	-锂的	14					40	-Lithium
6298	8506.6000	-锌空气的	14					40	-Air-zinc
6299	8506.8000	-其他原电池及原电池组	14			0		40	-Other primary cells and primary batteries
		-零件：							-Parts:
6300	8506.9010	---子目号8506.1000所列电池用	14					80	---Of the cells of subheading No. 8506.1000
6301	8506.9090	---其他	10					40	---Other
	85.07	**蓄电池，包括隔板，不论是否矩形(包括正方形)：**							**Electric accumulators, including separators therefor, whether or not rectangular (including square):**
6302	8507.1000	-铅酸蓄电池，用于起动活塞式发动机	10		6.9			90	-Lead-acid, of a kind used for starting piston engines
6303	8507.2000	-其他铅酸蓄电池	10		6.9			90	-Other lead-acid accumulators
6304	8507.3000	-镍镉蓄电池	10		8	0		40	-Nickel-cadmium
6305	8507.4000	-镍铁蓄电池	12		10			40	-Nickel-iron
		-其他蓄电池：							-Other accumulators:
6306	8507.8010	---镍氢电池	12		10	0		40	---Nickel-hydrogen
6307	8507.8020	---锂离子电池	12		10	0		40	---Lithium ion
6308	8507.8090	---其他	12		9	0		40	---Other
		-零件：							-Parts:
6309	8507.9010	---铅酸蓄电池用	10			0		90	---Of lead-acid accumulators
6310	8507.9090	---其他	8			0		40	---Other
	85.09	**家用电动器具：**							**Electro-mechanical domestic appliances, with self-contained electric motor:**
6311	8509.1000	-真空吸尘器，包括干式及湿式真空吸尘器	10		8.2			130	-Vacuum cleaners, including dry and wet vacuum cleaners
6312	8509.2000	-地板打蜡机	30					100	-Floor polishers

序号 No.	税则号列 Tariff Item	货品名称	最惠国税率(%) M.F.N.	协定税率(%) Agreement 东盟 ASEAN	曼谷协定 BA	香港 H.K.	澳门 Macao	普通税率(%) Gen.	Article Description
6313	8509.3000	-厨房废物处理器	20					100	-Kitchen waste disposers
6314		-食品研磨机及搅拌器；水果或蔬菜的榨汁机：							-Food grinders and mixers; fruit or vegetable juice extractors:
6315	8509.4010	---水果或蔬菜的榨汁机	10					100	---Fruit or vegetable juice extractors
	8509.4090	---其他	10					100	---Other
6316	8509.8000	-其他器具	30					100	-Other appliances
6317	8509.9000	-零件	12			0		100	-Parts
	85.10	**电动剃须刀、电动毛发推剪及电动脱毛器：**							**Shavers, hair clippers and hair-removing appliances, with self-contained electric motor:**
6318	8510.1000	-剃须刀	30					100	-Shavers
6319	8510.2000	-毛发推剪	30			0△		100	-Hair clippers
6320	8510.3000	-脱毛器	20					100	-Hair-removing appliances
6321	8510.9000	-零件	24.5					100	-Parts
	85.11	**点燃式或压燃式内燃发动机用的电点火及电起动装置（例如，点火磁电机、永磁直流发电机、点火线圈、火花塞、电热塞及起动电机）；附属于上述内燃发动机的发电机（例如，直流发电机、交流发电机）及断流器：**							**Electrical ignition or starting equipment of a kind used for spark-ignition or compression-ignition internal combustion engines (for example, ignition magnetos, magnetodynamos, ignition coils, sparking plugs and glow plugs, starter motors); generators (for example, dynamos, alternators) and cut-outs of a kind used in conjunction with such engines:**
6322	8511.1000	-火花塞	10					30	-Sparking plugs
		-点火磁电机；永磁直流发电机；磁飞轮：							-Ignition magnetos; magneto-dynamos; magnetic flywheels:
6323	8511.2010	---机车、航空器及船舶用	5					11	---For locomotives, aircraft or ships
6324	8511.2090	---其他	10					30	---Other
		--分电器；点火线圈：							--Distributors; ignition coils:
6325	8511.3010	---机车、航空器及船舶用	5					11	---For locomotives, aircraft or ships
6326	8511.3090	---其他	8.4					30	---Other
		-起动电机及两用启动发电机：							-Starter motors and dual purpose startergenerators:
6327	8511.4010	---机车、航空器及船舶用	5					11	---For locomotives, aircraft or ships
		---其他：							---Other:
6328	8511.4091	----输出功率在132.39千瓦(180马力)及以上的发动机用启动电机	8.4					30	----Starter motors for engines of an output of 132.39 KW(180HP) or more
6329	8511.4099	----其他	8.4					30	----Other

序号 No.	税则号列 Tariff Item	货品名称	最惠国税率(%) M.F.N.	协定税率(%) Agreement 东盟 ASEAN	曼谷协定 BA	香港 H.K.	澳门 Macao	普通税率(%) Gen.	Article Description
		-其他发电机:							-Other generators:
6330	8511.5010	----机车、航空器及船舶用	5					11	---For locomotives, aircraft or ships
6331	8511.5090	---其他	8.4					30	---Other
6332	8511.8000	-其他装置	8.4					30	-Other equipment
		-零件:							-Parts:
6333	8511.9010	---本税号所列供机车、航空器及船舶用的各种装置的零件	4.5					11	---Of the equipment of heading No. 85.11 used for locomotives, aircraft or ships
6334	8511.9090	---其他	5					30	---Other
	85.12	**自行车或机动车辆用的电气照明或信号装置(税号85.39的物品除外)、风挡刮水器、除霜器及去雾器:**							**Electrical lighting or signalling equipment (excluding articles of heading No. 85.39), windscreen wipers, defrosters and demisters, of a kind used for cycles or motor vehicles:**
6335	8512.1000	-自行车用照明或视觉信号装置	10.5					45	-Lighting or visual signalling equipment of a kind used on bicycles
		-其他照明或视觉信号装置:							-Other lighting or visual signalling equipment:
6336	8512.2010	---机动车辆用照明装置	10					45	---Lighting equipment of a kind used for motor vehicles
6337	8512.2090	---其他	10					45	---Other
		-音响信号装置:							-Sound signalling equipment:
		---机动车辆用:							---For motor Vehicles:
6338	8512.3011	----喇叭、蜂鸣器	10					45	----Loudspeaker, buzzers
6339	8512.3019	----其他	10					45	----other
6340	8512.3090	---其他	10					45	---other
6341	8512.4000	-风挡刮水器、除霜器及去雾器	10					45	-Windscreen wipers, defrosters and demisters
6342	8512.9000	-零件	8					45	-Parts
	85.13	**自供能源(例如,使用干电池、蓄电池、永磁发电机)的手提式电灯,但税号85.12的照明装置除外:**							**Portable electric lamps designed to function by their own source of energy (for example, dry batteries, accumulators, magnetos), other than lighting equipment of heading No.85.12:**
		-灯:							-Lamps:
6343	8513.1010	---手电筒	15		13.2		0	100	---Portable electric torches designed to function by dry batteries
6344	8513.1090	---其他	17.5				0	70	---Other
		-零件:							-Parts:
6345	8513.9010	---手电筒用	14			0	0	100	---Of the torches of subheading No. 8513.1010
6346	8513.9090	---其他	14			0	0	70	---Other

序号 No.	税则号列 Tariff Item	货品名称	最惠国税率(%) M.F.N.	协定税率(%)Agreement 东盟 ASEAN	曼谷协定 BA	香港 H.K.	澳门 Macao	普通税率(%) Gen.	Article Description
	85.14	**工业或实验室用电炉及电烘箱(包括感应的或电介质的);工业或实验室用其他通过感应或介质损耗对材料进行热处理的设备:**							**Industrial or laboratory electric furnaces and ovens(including those functioning by induction or dielectric loss);other industrial or laboratory equipment for the heat treatment of materials by induction or dielectric loss:**
		-电阻加热的炉及烘箱:							-Resistance heated furnaces and ovens:
6347	8514.1010	---可控气氛热处理炉	0					30	---Furnaces for heat treatment, atmosphere controllable
6348	8514.1090	---其他	0					30	---Other
6349	8514.2000	-通过感应或介质损耗工作的炉及烘箱	0					30	-Furnaces and ovens functioning by induction or dielectric loss
6350	8514.3000	-其他炉及烘箱	0					30	-Other furnaces and ovens
6351	8514.4000	-其他通过感应或介质损耗对材料进行热处理的设备	10					30	-Other equipment for the heat treatment of materials by induction or dielectric loss
		-零件:							-Parts:
6352	8514.9010	---炼钢电炉用	8					30	---Of steel making electric furnaces
6353	8514.9090	---其他	0					30	---Other
	85.15	**电气(包括电热气体)、激光、其他光、光子束、超声波、电子束、磁脉冲或等离子弧焊接机器及装置,不论是否兼有切割功能;用于热喷金属或金属陶瓷的电气机器及装置:**							**Electric (including electrically heated gas), laser or other light or photon beam, ultrasonic, electron beam, magnetic pulse or plasma are soldering, brazing or welding machines and apparatus, whether or not capable of cutting;electric machines and apparatus for hot spraying of metals or cermets:**
		-钎焊机器及装置:							-Brazing or soldering machines and apparatus:
6354	8515.1100	--烙铁及焊枪	10					30	--Soldering irons and guns
6355	8515.1900	--其他	10					30	--Other
		-电阻焊接机器及装置:							-Machines and apparatus for resistance welding of metals:
		--全自动或半自动的:							--Fully or partly automatic:
6356	8515.2110	---直缝焊管机	10			0		30	---Aligning tube welding machines
6357	8515.2190	---其他	10			0		30	---Other
6358	8515.2900	--其他	10					30	--Other
		-电弧(包括等离子弧)焊接机器及装置:							-Machines and apparatus for are (including plasma arc)welding of metals:
		--全自动或半自动的:							--Fully or partly automatic:
6359	8515.3110	---螺旋焊管机	10					30	---Spiralling tube welding machines
6360	8515.3190	---其他	10					30	---Other
6361	8515.3900	--其他	10					30	--Other
6362	8515.8000	-其他机器及装置	8			0		30	-Other machines and apparatus
6363	8515.9000	-零件	6			0		30	-Parts

序号 No.	税则号列 Tariff Item	货品名称	最惠国税率(%) M.F.N.	协定税率(%) Agreement 东盟 ASEAN	曼谷协定 BA	香港 H.K.	澳门 Macao	普通税率(%) Gen.	Article Description
	85.16	**电热的快速热水器、储存式热水器、浸入式液体加热器；电气空间加热器及土壤加热器；电热的理发器具(例如，电吹风机、电卷发器、电热发钳)及干手器；电熨斗；其他家用电热器具；加热电阻器，但税号85.45的货品除外：**							**Electric instantaneous or storage water heaters and immersion heaters; electric space heating apparatus and soil heating apparatus; electro-thermic hair-dressing apparatus (for example, hair dryers, hair curlers, curling tong heaters) and hand dryers; electric smoothing irons; other electro-thermic appliances of a kind used for domestic purposes; electric heating resistors, other than those of heading No. 85.45:**
6364	8516.1000	-电热的快速热水器、储存式热水器、浸入式液体加热器	10					100	-Electric instantaneous or storage waterheaters and immersion heaters
		-电气空间加热器及土壤加热器：							-Electric space heating apparatus and electric soil heating apparatus:
6365	8516.2100	--储存式散热器	35				0△	100	--Storage heating radiators
		--其他：							--Other:
6366	8516.2910	---土壤加热器	10			0△	0△	40	---Electric soil heating apparatus
6367	8516.2990	---其他	10			0△	0△	100	---Other
		-电热的理发器具及干手器：							-Electro-thermic hair-dressing or hand-drying apparatus:
6368	8516.3100	--吹风机	10				0	100	--Hair dryers
6369	8516.3200	--其他理发器具	35				0	100	--Other hair-dressing apparatus
6370	8516.3300	--干手器	35				0	100	--Hand-drying apparatus
6371	8516.4000	-电熨斗	35				0	100	-Electric smoothing irons
6372	8516.5000	-微波炉	15		13.9			130	-Microwave ovens
		-其他炉；电锅、电热板、加热环、烧烤炉及烘烤器：							-Other ovens; cookers, cooking plates, boiling rings, grillers and roasters:
6373	8516.6010	---电磁炉	15				0	130	---Electromagnetic ovens
6374	8516.6030	---电饭锅	15				0	130	---Electric rice cookers
6375	8516.6040	---电炒锅	15				0	130	---Electric frying pans
6376	8516.6090	---其他	15				0	130	---Other
		-其他电热器具：							-Other electro-thermic appliances:
6377	8516.7100	--咖啡壶或茶壶	32				0	130	---Coffee or tea makers
6378	8516.7200	--烤面包器	32				0	130	--Toasters
		--其他：							--Other:
6379	8516.7910	---电热饮水机	18.3				0	100	---Electro－thermic water dispensers
6380	8516.7990	---其他	18.3				0	100	---Other
6381	8516.8000	-加热电阻器	10					40	-Electric heating resistors
		-零件：							-Parts:

序号 No.	税则号列 Tariff Item	货品名称	最惠国税率(%) M.F.N.	协定税率(%)Agreement 东盟 ASEAN	曼谷协定 BA	香港 H.K.	澳门 Macao	普通税率(%) Gen.	Article Description
6382	8516.9010	---土壤加热器及加热电阻器用	8			0		40	---Of apparatus of subheading No. 8516.2910 or 8516.8000
6383	8516.9090	---其他	12			0		100	---Other
	85.17	**有线电话、电报设备，包括无绳电话机、有线载波通信设备及有线数字通信设备；可视电话：**							**Electrical apparatus for line telephony or line telegraphy, including line telephone sets with cordless handsets and telecommunication apparatus for carrier-current line systems or for digital line systems; videophones:**
		-电话机；可视电视：							-Telephone sets; videophones:
6384	8517.1100	--无绳电话机	0					30	--Line telephone sets with cordless handsets
		--其他：							--Other:
6385	8517.1910	---可视电话机	0					30	---Videophones
6386	8517.1990	---其他	0					30	---Other
		-传真机及电传打字机：							-Facsimile machines and teleprinters:
6387	8517.2100	--传真机	0					17	--Facsimile machines
6388	8517.2200	--电传打字机	0					30	--Teleprinters
		-电话或电报交换机：							-Telephonic or telegraphic switching apparatus:
		---数字式程控电话或电报交换机：							---Digital program-controlled switching systems:
6389	8517.3011	----容量在5000门及以上的局用电话交换机；长途电话交换机；电报交换机	0		0			17	----Public telephonic switching systems with a capacity exceeding 5000 lines, toll telephonic or telegraphic switching systems
6390	8517.3013	----移动通信交换机	0		0			40	----Mobile communication switching system
6391	8517.3019	----其他电话交换机	0		0			40	----Other telephonic switching systems
		---其他：							---Other:
6392	8517.3091	----模拟式移动通信交换机	0		0			30	----Anological mobile communication switching system
6393	8517.3099	----其他	0		0			30	----Other
		-有线载波通信或有线数字通信用的其他设备：							-Other apparatus, for carrier-current line systems or for digital line systems:
		---光通讯设备：							---Optical communication equipments:
6394	8517.5021	----光端机及脉冲编码调制设备(PCM)	0					17	----Optical line terminal equipments and pulse code modulation mutilexers
6395	8517.5022	----波分复用光传输设备	0					30	----Optical transmission equipments for wave-division multiplexing
6396	8517.5029	----其他	0					30	----Other
		---其他有线数字通信设备：							---Other telecommunication apparatus for digital line system:

序号 No.	税则号列 Tariff Item	货品名称	最惠国税率(%) M.F.N.	协定税率(%) Agreement 东盟 ASEAN	曼谷协定 BA	香港 H.K.	澳门 Macao	普通税率(%) Gen.	Article Description
6397	8517.5031	----通信网络时钟同步设备	0					30	----Communiction network synchronizing equipments
6398	8517.5032	----以太网络交换机	0					30	----Ethernet exchangers
6399	8517.5033	----IP 电话信号转换设备	0					30	----IP telephone signal converters
6400	8517.5036	----调制解调器	0					30	----Modem
6401	8517.5039	----其他	0					30	----Other
6402	8517.5090	---其他	0					30	---Other
6403	8517.8000	-其他设备	0					30	-Other apparatus
		-零件：							-Parts:
6404	8517.9010	---数字式程控电话或电报交换机用	0					14	---Of the apparatus of subheading No. 8517.3011, 8517.3013 or 8517.3019
6405	8517.9020	---光端机及脉冲编码调制设备(PCM)用	0					14	---Of the equipment of subheading No.8517.5021
		---传真机用：							---Of facsimile apparatus:
6406	8517.9031	----热敏记录头	0					14	----Thermograph recording heads
6407	8517.9032	----接触式图像传感器	0					14	----Contact image sensors
6408	8517.9039	----其他	0					14	----Other
6409	8517.9090	---其他	0					30	---Other
	85.18	**传声器(麦克风)及其座架；扬声器，不论是否装成音箱；耳机、耳塞机，不论是否装有传声器，由传声器及一个或多个扬声器组成的组合机；音频扩大器；电气扩音机组：**							**Microphones and stands therefor; loudspeakers, whether or not mounted in their enclosures; headphones, earphones, whether or not combined with a microphone, and sets consisting of a microphone and one or more loudspeakers; audio - frequency electric amplifiers; electric sound amplifier sets:**
6410	8518.1000	-传声器(麦克风)及其座架	10			0		40	-Microphones and stands therefor
		-扬声器，不论是否装成音箱：							-Loudspeakers, whether or not mounted in their enclosures:
6411	8518.2100	--单喇叭音箱	10					40	--Single loudspeakers, mounted in their enclosures
6412	8518.2200	--多喇叭音箱	10					40	--Multiple loudspeakers, mounted in the same enclosure
6413	8518.2900	--其他	0					40	--Other
6414	8518.3000	-耳机、耳塞机，不论是否装有传声器，由传声器及一个或多个扬声器组成的组合机	0					40	-Headphones and earphones, whether or not combined with a microphone, and sets consisting of a microphone and one or more loudspeakers
6415	8518.4000	-音频扩大器	12					40	-Audio-frequency electric amplifiers
6416	8518.5000	-电气扩音机组	10					40	-Electric sound amplifier sets
6417	8518.9000	-零件	10.5			0		40	-Parts

序号 No.	税则号列 Tariff Item	货品名称	最惠国税率(%) M.F.N.	协定税率(%)Agreement 东盟 ASEAN	曼谷协定 BA	香港 H.K.	澳门 Macao	普通税率(%) Gen.	Article Description
	85.19	**转盘(唱机唱盘)、唱机、盒式磁带放声机及其他声音重放设备,未装有声音录制装置:**							**Turntables (record-decks), record-players, cassette-players and other sound reproducing apparatus, not incorporating a sound recording device:**
6418	8519.1000	-投币式唱机	30					130	-Coin-or disc-operated record-players
		-其他唱机:							-Other record-players:
6419	8519.2100	--不带扬声器的	30					130	--Without loudspeaker
6420	8519.2900	--其他	30					130	--Other
		-转盘(唱机唱盘):							-Turntables(record-decks):
6421	8519.3100	--装有自动换片装置	30					130	--With automatic record changing mechanism
6422	8519.3900	--其他	30					130	--Other
6423	8519.4000	-编辑节目用放声机	20					80	-Transcribing machines
		-其他声音重放设备:							-Other sound reproducing apparatus:
6424	8519.9200	--袖珍盒式磁带放声机	17					130	--Pocket-size cassette-players
6425	8519.9300	--其他,盒式磁带型	20					130	--Other, cassette-type
		--其他:							--Other:
6426	8519.9910	---激光唱机	30		26			80	---Compact disc players
6427	8519.9990	---其他	20		16.4			80	---Other
	85.20	**磁带录音机及其他声音录制设备,不论是否装有声音重放装置:**							**Magnetic tape recorders and other sound recording apparatus, whether or not incorporating a sound reproducing device:**
6428	8520.1000	-需外接电源的口授记录机	25					80	-Dictating machines not capable of operating without an external source of power
6429	8520.2000	-电话自动应答机	0					80	-Telephone answering machines
		-装有声音重放装置的其他磁带录音机:							-Other magnetic tape recorders incorporating sound reproducing apparatus:
		--数字音频式:							---Digital audio type:
6430	8520.3210	---盒式磁带型	30					130	---Cassette-type
6431	8520.3290	---其他	25					80	---Other
6432	8520.3300	--其他,盒式磁带型	30					130	--Other cassette-type
		--其他:							--Other:
6433	8520.3910	---开盘式录音机	25					80	---Open-reel recorders
6434	8520.3990	---其他	20					80	---Other
		-其他							-Other
6435	8520.9010	---闪速存储器型	20					80	---Flash memory type
6436	8520.9090	---其他	20					80	---Other
	85.21	**视频信号录制或重放设备,不论是否装有高频调谐器:**							**Video recording or reproducing apparatus, whether or not incorporating a video tuner:**
		-磁带型:							-Magnetic tape-type:
		---录像机							---Video tape recorders
6437	8521.1011	----广播级	见表4 Table 4		见尾注①			见表4 Table	----Broadcast quality

序号 No.	税则号列 Tariff Item	货品名称	最惠国税率(%) M.F.N.	协定税率(%) Agreement 东盟 ASEAN	曼谷协定 BA	香港 H.K.	澳门 Macao	普通税率(%) Gen.	Article Description
6438	8521.1019	----其他	见表4		见尾注②			见表4	----Other
6439	8521.1020	---放像机	见表4		见尾注③			见表4	----Video tape reproducers
		-其他:							-Other:
		---激光视盘机:							---Laser video compact disk player:
6440	8521.9011	----视频高密光盘(VCD)播放机	20					130	----Video Compact Disc player
6441	8521.9012	----数字化视频光盘(DVD)播放机	20					130	----Digital Video Disc player
6442	8521.9019	----其他	20					130	----Other
6443	8521.9090	---其他	20					130	---Other
	85.22	**专用于或主要用于税号85.19至85.21所列设备的零件、附件:**							**Parts and accessories suitable for use solely or principally with the apparatus of headings Nos.85.19 to 85.21:**
6444	8522.1000	-拾音头	35			0		130	-Pick-up cartridges
		-其他:							-Other:
6445	8522.9010	---转盘或唱机用	25			0		130	---Of turntables (record decks) or recordplayers
		---盒式磁带录音机或放声机用:							---Of cassette magnetic tape recorders or reproducers:
6446	8522.9021	----走带机构(机芯),不论是否装有磁头	25			0		100	----Transport mechanisms, whether or not incorporating a magnetic head
6447	8522.9022	----磁头	25			0		100	----Magnetic heads
6448	8522.9023	----磁头零件	20			0		100	----Parts of magnetic heads
6449	8522.9029	----其他	30			0		100	----Other
		---视频信号录制或重放设备用:							---Of video recording or reproducing apparatus:
6450	8522.9031	----激光视盘机的机芯	30			0		100	----Movements for Laser video compact disk player
6451	8522.9039	----其他	30			0		100	----other
6452	8522.9090	---其他	20			0		80	---Other
	85.23	**制成供灌(录)音或录制其他信息用的未录制媒体,但第三十七章的产品除外:**							**Prepared unrecorded media for sound recording or similar recording of other phenomena, other than products of Chapter 37:**
		-磁带:							-Magnetic tapes:
6453	8523.1100	--宽度不超过4毫米	0					130	--Of a width not exceeding 4 mm
6454	8523.1200	--宽度超过4毫米,但不超过6.5毫米	0					130	--Of a width exceeding 4 mm but not exceeding 6.5 mm
		--宽度超过6.5毫米:							--Of a width exceeding 6.5 mm:
6455	8523.1310	---税号84.71所列机器用	0					20	---For the machines of heading No. 84.71
6456	8523.1320	---录音带	0					130	---Sound-recording tapes
6457	8523.1330	---录像带	0					130	---Video tapes
6458	8523.1390	---其他	0					70	---Other
		-磁盘:							-Magnetic discs:

序号 No.	税则号列 Tariff Item	货品名称	最惠国税率(%) M.F.N.	协定税率(%) Agreement 东盟 ASEAN	曼谷协定 BA	香港 H.K.	澳门 Macao	普通税率(%) Gen.	Article Description
6459	8523.2010	---税号 84.71 所列机器用	0					20	---For the machines of heading No. 84.71
6460	8523.2090	---其他	0				0	70	---Other
6461	8523.3000	-磁条卡	17.5			0		70	-Cards incorporating a magnetic stripe
6462	8523.9000	-其他	0			0	0	70	-Other
	85.24	**已灌(录)音或录制其他信息的唱片、磁带及其他媒体,包括供复制用的母片及母带,但不包括第三十七章的产品:**							**Records, tapes and other recorded media for sound or other similarly recorded phenommena, including matrices and masters for the production of records, but excluding products of Chapter 37:**
		-唱片:							-Gramophone records:
6463	8524.1010	---教学用	0					0	---For deucational use
6464	8524.1090	---其他	15					130	---Other
		-用于激光阅读系统的光盘:							-Discs for laser reading systems:
6465	8524.3100	--用于重放声音或图像以外信息的	0					14	--For reproducing phenomena other than sound or image
		--仅用于重放声音的:							--For reproducing sound only:
6466	8524.3210	---教学用	0					0	---For educational use
6467	8524.3290	---其他	10			0		130	---Other
		--其他:							--Other:
6468	8524.3910	---教学用	0					0	---For educational use
6469	8524.3920	---税号 84.71 所列机器用	0					14	---For the machines of heading No. 84.71
6470	8524.3990	---其他	0					130	---Other
		-用于重放声音或图像以外信息的磁带:							-Magnetic tapes for reproducing phenomena other than sound or image:
6471	8524.4010	---教学用	0					0	---For educational use
		---其他:							---Other:
6472	8524.4091	----宽度不超过 6.5 毫米	0					14	----Of a width not exceeding 6.5 mm
6473	8524.4099	----其他	0					14	----Other
		-其他磁带:							-Other magnetic tapes:
		--宽度不超过 4 毫米:							--Of a width not exceeding 4 mm:
6474	8524.5110	---教学用	0					0	---For educational use
6475	8524.5190	---其他	15					130	---Other
		--宽度超过 4 毫米,但不超过6.5毫米:							--Of a width exceeding 4 mm but not exceeding 6.5 mm:
6476	8524.5210	---教学用	0					0	---For educational use
6477	8524.5290	---其他	10					130	---Other
		--宽度超过 6.5 毫米:							--Of a width exceeding 6.5 mm:
6478	8524.5310	---教学用	0					0	---For educational use
6479	8524.5390	---其他	10					130	---Other
6480	8524.6000	-磁条卡	15			0		130	-Cards incorporating a magnetic stripe

序号 No.	税则号列 Tariff Item	货品名称	最惠国税率(%) M.F.N.	协定税率(%) Agreement 东盟 ASEAN	曼谷协定 BA	香港 H.K.	澳门 Macao	普通税率(%) Gen.	Article Description
		-其他:							-Other:
		--用于重放声音或图像以外信息的:							--For reproducing phenomena other than sound or image:
6481	8524.9110	---教学用	0					0	---For educational use
6482	8524.9120	---税号 84.71 所列机器用	0					14	---For the machines of heading No. 84.71
6483	8524.9190	---其他	0					130	---Other
		--其他:							--Other:
6484	8524.9910	---教学用	0					0	---For educational use
6485	8524.9920	---税号 84.71 所列机器用	0					14	---For the machines of heading No. 84.71
6486	8524.9990	---其他	0					130	---Other
	85.25	**无线电话、电报、无线电广播、电视发送设备,不论是否装有接收装置或声音的录制、重放装置;电视摄像机;静像视频摄像机及其他视频摄录一体机;数字照相机:**							**Transmission apparatus for radio-telephony, radio-telegraphy, radio-broadcasting or television, whether or not incorporating reception apparatus or sound recording or reproducing apparatus; teleyision cameras; still image video cameras and other video camera recorders; digital cameras:**
		-发送设备:							-Transmission apparatus:
6487	8525.1010	---广播电视用	0					30	---For broadcast or television
6488	8525.1090	---其他	0					14	---Other
		-装有接收装置的发送设备:							-Transmission apparatus incorporating reception apparatus:
		---卫星地面站设备:							---Satellite earth station:
6489	8525.2011	----电视用	0					30	----For television
6490	8525.2019	----其他	0					14	----Other
		---可移动通信设备:							---Mobile communication equipment:
6491	8525.2022	----手持(包括车载)式无线电话机	0					20	----Radio telephone handsets(including vehicle installed)
6492	8525.2023	----对讲机	0					17	----Walkie-talkie
6493	8525.2029	----其他	0					14	----Other
		---其他:							---Other:
6494	8525.2091	----广播电视用	0					30	----For broadcast or television
6495	8525.2092	----移动通信基地站	0					14	----Mobile communication base station
6496	8525.2093	----无线用户接入网设备	0					14	----Wireless subscriber communicating equipments
6497	8525.2099	----其他	0					14	----Other
		-电视摄像机:							-Television cameras:
6498	8525.3010	---特种用途的	10					17	---For special purposes
		---其他							---Other
6499	8525.3091	----广播级	见表 4 Table 4					见表 4 Table 4	Table 4----Broadcast quality
6500	8525.3099	----其他	见表 4 Table 4					见表 4 Table 4	----Other
		-静像视频摄像机及其他视频摄录一体机;数字照相机:							-Still image video cameras and other video camera recorders; digital cameras:

序号 No.	税则号列 Tariff Item	货品名称	最惠国税率(%) M.F.N.	协定税率(%) Agreement 东盟 ASEAN	曼谷协定 BA	香港 H.K.	澳门 Macao	普通税率(%) Gen.	Article Description
6501	8525.4010	---特种用途的	0					17	---For special purposes
		---其他静像视频摄像机及其他视频摄录一体机:							---other still image video cameras and other video camera recorders:
6502	8525.4041	----广播级	见表4					见表4	----Broadcast quality
6503	8525.4042	----家用型摄录一体机	0					130	----Household video camera recorders
6504	8525.4049	----其他	见表4					见表4	----other
6505	8525.4050	---其他数字照相机	见表4					见表4	---other digital cameras
	85.26	**雷达设备、无线电导航设备及无线电遥控设备:**							**Radar apparatus, radio navigational aid apparatus and radio remote control apparatus:**
		-雷达设备:							-Radar apparatus:
6506	8526.1010	---导航用	2					8	---For navigational aid
6507	8526.1090	---其他	5					14	---Other
		-其他:							-Other:
		--无线电导航设备							--Radio navigational aid apparatus
6508	8526.9110	---机动车辆用	2					8	---For motor vehicles
6509	8526.9190	---其他	2					8	---Other
6510	8526.9200	--无线电遥控设备	5					14	--Radio remote control apparatus
	85.27	**无线电话、电报、无线电广播接收设备,不论是否与声音的录制、重放装置或时钟组合在同一机壳内:**							**Reception apparatus for radio-telephony, radio-telegraphy or radio-broadcasting, whether or not combined, in the same housing, with sound recording or reproducing apparatus or a clock:**
		-不需外接电源的无线电收音机,包括兼可接收无线电话、电报的设备:							-Radio-broadcast receivers capable of operating without an external source of power, including apparatus capable of receiving also radio-telephony or radio-telegraphy:
6511	8527.1200	--袖珍盒式磁带收放机	20					130	--Pocket-size radio cassette-players
6512	8527.1300	--其他收录(放)音组合机	15					130	--Other apparatus combined with sound recording or reproducing apparatus
6513	8527.1900	--其他	15					130	--Other
		-需外接电源的汽车用无线电收音机,包括兼可接收无线电话、电报的设备:							-Radio-broadcast receivers not capable of operating without an external source of power, of a kind used in motor vehicles, including apparatus capable of receiving also radio-telephony or radio-telegraphy:
6514	8527.2100	--收录(放)音组合机	15					130	--Combined with sound recording or reproducing apparatus
6515	8527.2900	--其他	15					130	--Other

序号 No.	税则号列 Tariff Item	货品名称	最惠国税率(%) M.F.N.	协定税率(%) Agreement 东盟 ASEAN	曼谷协定 BA	香港 H.K.	澳门 Macao	普通税率(%) Gen.	Article Description
		-其他无线电收音机,包括兼可接收无线电话、电报的设备:							-Other radio-broadcast receivers, including apparatus capable of receiving also radio-telephony or radio-telegraphy:
6516	8527.3100	--收录(放)音组合机	15					130	--Combined with sound recording or reproducing apparatus
6517	8527.3200	--带时钟的收音机	15					130	--Not combined with sound recording or reproducing apparatus but combined with a clock
6518	8527.3900	--其他	27					130	--Other
		-其他设备:							-Other apparatus:
6519	8527.9010	---无线寻呼机	0					35	---Radio paging receivers
6520	8527.9090	---其他	9					14	---Other
	85.28	**电视接收装置,不论是否装有无线电收音装置或声音、图像的录制或重放装置;视频监视器及视频投影机:**							**Reception apparatus for television, whether or not incorporating radio-broadcast receivers or sound or video recording or reproducing apparatus; video monitors and video projectors:**
		-电视接收装置,不论是否装有无线电收音装置或声音、图像的录制或重放装置:							-Reception apparatus for television, whether or not incorporating radio-broadcast receivers or sound or video recording or reproducing apparatus:
		--彩色的:							--Colour:
6521	8528.1210	---卫星电视接收机	30		26			130	---Satellite television receivers
		---阴极射线显像管的:							---With cathode-ray picture tube:
6522	8528.1221	----显示屏幕不超过42厘米的	30		26			130	----With a diagonal measurement of the screen not exceeding 42 cm
6523	8528.1222	----显示屏幕超过42厘米不超过52厘米的	30		26			130	----With a diagonal measurement of the screen exceeding 42 cm but not exceeding 52 cm
6524	8528.1223	----显示屏幕超过52厘米不超过74厘米的	30		29.1			130	----With a diagonal measurement of the screen exceeding 52 cm but not exceeding 74 cm
6525	8528.1224	----显示屏幕超过74厘米的	30		29.1			130	----With a diagonal measurement of the screen exceeding 74 cm
		---液晶显示器的:							---With liquid crystal display:
6526	8528.1238	----显示屏幕超过52厘米的	30		29.1		0△	130	----With a diagonal measurement of the screen exceeding 52 cm
6527	8528.1239	----其他	30		26		0△	130	----Other
		---等离子显示器的:							---With plasma Display Panels:
6528	8528.1248	----显示屏幕超过52厘米的	30		29.1		0△	130	----With a diagonal measurement of the screen exceeding 52 cm
6529	8528.1249	----其他	30		26		0△	130	----Other
6530	8528.1290	---其他	30					130	---Other
		--黑白的或其他单色的:							--Black and white or other monochrome:

序号 No.	税则号列 Tariff Item	货品名称	最惠国税率(%) M.F.N.	协定税率(%)Agreement 东盟 ASEAN	曼谷协定 BA	香港 H.K.	澳门 Macao	普通税率(%) Gen.	Article Description
6531	8528.1310	---显示屏幕尺寸不超过16厘米	15				0△	100	---With a diagonal measurement of the screen not exceeding 16 cm
6532	8528.1320	---显示屏幕尺寸超过16厘米,但不超过42厘米	15				0△	100	---With a diagonal measurement of the screen exceeding 16 cm but not exceeding 42 cm
6533	8528.1330	---显示屏幕尺寸超过42厘米,但不超过52厘米	15				0△	100	---With a diagonal measurement of the screen exceeding 42 cm but not exceeding 52 cm
6534	8528.1340	---显示屏幕尺寸超过52厘米	15				0△	100	---With a diagonal measurement of the screen exceeding 52 cm
		-视频监视器:							-Video monitors:
6535	8528.2100	--彩色的	30		26		0△	130	--Colour
6536	8528.2200	--黑色的或其他单色的	19				0△	100	--Black and white or other monochrome
		-视频投影机:							-Video projectors:
6537	8528.3010	---彩色的	30		26		0△	130	---Colour
6538	8528.3020	---黑白的或其他单色的	15				0△	100	---Black and white or other monochrome
	85.29	**专用于或主要用于税号85.25至85.28所列装置或设备的零件:**							**Parts suitable for use solely or principally with the apparatus of headings No3.85.25 to 85.28:**
		-各种天线或天线反射器及其零件:							-Aerials and aerial reflectors of all kinds; parts suitable for use therewith:
6539	8529.1010	---雷达设备及无线电导航设备用	1.5			0		8	---For radar apparatus and radio navigational aid apparatus
6540	8529.1020	---无线电收音机及其组合机、电视接收机用	0					90	---For radio-broadcast receivers and their combinations, television receivers
6541	8529.1090	---其他	2					20	---Other
		-其他:							-Other:
6542	8529.9010	---电视发送、差转设备及卫星电视地面接收转播设备用	0		0			30	---Of television transmission or translation apparatus, satellite television ground receiving and relaying apparatus
6543	8529.9020	---手持式无线电话机用	0		0			17	---Of radio telephone handsets
6544	8529.9030	---对讲机用	8		7	0		20	---Of walkie-talkie
		---电视摄像机、静像视频摄像机及其他视频摄录一体机、数字照相机用:							---Of television cameras, still image video cameras, other video camera recorders and digital cameras:
6545	8529.9041	----特种用途的	8		7	0		17	----Of special purpose
6546	8529.9042	----非特种用途的取像模块	12		11	0		100	----Camera modules without special purpose
6547	8529.9049	----其他	12		11	0		100	----Other

序号 No.	税则号列 Tariff Item	货品名称	最惠国税率(%) M.F.N.	协定税率(%) Agreement 东盟 ASEAN	曼谷协定 BA	香港 H.K.	澳门 Macao	普通税率(%) Gen.	Article Description
6548	8529.9050	---雷达设备及无线电导航设备用	1.5		1.5			8	---Of radar apparatus and radio navigational aid apparatus
6549	8529.9060	---无线电收音机及其组合机用	15		12	0		130	---Of radio-broadcast receivers and their combinations
6550	8529.9070	---无线寻呼机用	0		0			20	---Of radio paging receivers
		---电视接收机用(高频调谐器除外):							---Of television receivers (Other than H.F. turners):
6551	8529.9081	----彩色电视接收机用(等离子显像组件及其零件除外)	15		12			80	----Of colour television receivers (Other than plasma display modules or parts thereof)
6552	8529.9082	----等离子显像组件及其零件	15		12			80	----Plasma display modules and part thereof
6553	8529.9089	----其他	0		0			50	----Other
6554	8529.9090	---其他:	0		0	0		57	---Other
	85.30	**铁道、电车道、道路或内河航道、停车场、港口或机场用的电气信号、安全或交通管理设备(税号 86.08 的货品除外):**							**Electrical signalling, safety or traffic control equipment for railways, tramways, roads, inland waterways, parking facilities, port installations or airfields (other than those of heading No.86.08):**
6555	8530.1000	-铁道或电车道用的设备	10					20	-Equipment for railways or tramways
6556	8530.8000	-其他设备	8					20	-Other equipment
6557	8530.9000	-零件	8					20	-Parts
	85.31	**电气音响或视觉信号装置(例如,电铃、电笛、显示板、防盗或防火报警器),但税号 85.12 或 85.30 的货品除外**							**Electric sound or visual signalling apparatus (for example, bells, sirens, indicatorpanels, burglar or fire alarms), other than those of heading No.85.12 or 85.30:**
		-防盗或防火报警器及类似装置							-Burglar or fire alarms and similar apparatus
6558	8531.1010	---机动车辆防盗装置	10					40	---Burglar alarm for motor vehicles
6559	8531.1090	---其他	10			0△		40	---Other
6560	8531.2000	-装有液晶装置(LCD)或发光二极管(LED)的显示板	0					70	-Indicator panels incorporating liquid crystal devices (LCD) or light emitting diodes (LED)
		-其他装置:							-Other apparatus:
6561	8531.8010	---蜂鸣器	15			0		70	---Buzzers
6562	8531.8090	---其他	10			0△		70	---Other
		-零件:							-Parts:
6563	8531.9010	---防盗或防火报警器及类似装置用	0					40	---Of burglar or fire alarms and similar apparatus
6564	8531.9090	---其他	0					70	---Other
	85.32	**固定、可变或可调(微调)电容器:**							**Electrical capacitors, fixed, variable or adjustable (pre-set):**

序号 No.	税则号列 Tariff Item	货品名称	最惠国税率(%) M.F.N.	协定税率(%) Agreement 东盟 ASEAN	曼谷协定 BA	香港 H.K.	澳门 Macao	普通税率(%) Gen.	Article Description
6565	8532.1000	-固定电容器,用于50/60赫兹电路,其额定无功功率不低于0.5千乏(电力电容器)	0					20	-Fixed capacitors designed for use in 50/60 Hz circuits and having a reactive power handling capacity of not less than 0.5 kVar(power capacitors)
		-其他固定电容器:							-Other fixed capacitors:
		--钽电容器:							--Tantalum:
6566	8532.2110	---片式	0					35	---Laminate
6567	8532.2190	---其他	0					35	---Other
6568	8532.2200	--铝电解电容器	0					35	--Aluminium electrolytic
6569	8532.2300	--单层瓷介电容器	0					35	--Ceramic dielectric,single layer
		--多层瓷介电容器:							--Ceramic dielectric,multilayer:
6570	8532.2410	---片式	0					35	---Laminate
6571	8532.2490	---其他	0					35	---Other
		--纸介质或塑料介质电容器:							--Dielectric of paper or plastics:
6572	8532.2510	---片式	0					35	---Laminate
6573	8532.2590	---其他	0					35	---Other
6574	8532.2900	--其他	0					35	--Other
6575	8532.3000	-可变或可调(微调)电容器	0					35	-Variable or adjustable (pre-set)capacitors
		-零件:							-Parts:
6576	8532.9010	---子目号8532.1000所列电容器用	0					20	---Of the capacitors of subheading No. 8532.1000
6577	8532.9090	---其他	0					35	---Other
	85.33	**电阻器(包括变阻器及电位器),但加热电阻器除外:**							**Electrical resistors (including rheostats and potentiometers), other than heating resistors:**
6578	8533.1000	-固定碳质电阻器,合成或薄膜式	0					50	-Fixed carbon resistors,composition or film types
		-其他固定电阻器:							-Other fixed resistors:
		--额定功率不超过20瓦:							--For a power handling capacity not exceeding 20 W:
6579	8533.2110	---片式	0					50	---Laminate
6580	8533.2190	---其他	0					50	---Other
6581	8533.2900	--其他	0					50	--Other
		-线绕可变电阻器,包括变阻器及电位器:							-Wirewound variable resistors, including rheostats and potentiometers:
6582	8533.3100	--额定功率不超过20瓦	0					50	--For a power handling capacity not exceedint 20 W
6583	8533.3900	--其他	0					50	--Other
6584	8533.4000	-其他可变电阻器,包括变阻器及电位器	0					50	-Other variable resistors, including rheostats and potentiometers

序号 No.	税则号列 Tariff Item	货品名称	最惠国税率(%) M.F.N.	协定税率(%)Agreement 东盟 ASEAN	曼谷协定 BA	香港 H.K.	澳门 Macao	普通税率(%) Gen.	Article Description
6585	8533.9000	-零件	0					50	-Parts
	85.34	**印刷电路:**							**Printed circuits:**
6586	8534.0010	---4层以上的	0					35	---Of more than 4 layers
6587	8534.0090	---其他	0					50	---Other
	85.35	**电路的开关、保护或连接用的电气装置(例如,开关、熔断器、避雷器、电压限幅器、电涌抑制器、插头、接线盒),用于电压超过1000伏的线路:**							**Electrical apparatus for switching or protecting electrical circuits, or for making connections to or in electrical circuits (for example, switches, fuses, lightning arresters, voltage limiters, surge suppressors, plugs, junction boxes), for a voltage exceeding 1000 volts:**
6588	8535.1000	-熔断器	14					50	-Fuses
		-自动断路器:							-Automatic circuit breakers:
6589	8535.2100	--用于电压低于72.5千伏的线路	14					50	--For a voltage of less than 72.5 kV
6590	8535.2900	--其他	10					50	--Other
6591	8535.3000	-隔离开关及断续开关	10					50	-Isolating switches and make-and-break switches
6592	8535.4000	-避雷器、电压限幅器及电涌抑制器	18			0		50	-Lightning arresters, voltage limiters and surge suppressors
6593	8535.9000	-其他	10			0△		50	-Other
	85.36	**电路的开关、保护或连接用的电气装置(例如,开关、继电器、熔断器、电涌抑制器、插头、插座、灯座、接线盒),用于电压不超过1000伏的线路:**							**Electrical apparatus for switching or protecting electrical circuits, or for making connections to or in electrical-circuits(for example, switches, relays, fuses, surge suppressors, plugs, sockets, lamp-holders, junction boxes), for a voltage not exceeding 1000 volts:**
6594	8536.1000	-熔断器	10			0		50	-Fuses
6595	8536.2000	-自动断路器	9			0		50	-Automatic circuit breakers
6596	8536.3000	-其他电路保护装置	9					50	-Other apparatus for protecting electrical circuits
		-继电器:							-Relays:
6597	8536.4100	--用于电压不超过60伏的线路	10					50	--For a voltage not exceeding 60V
6598	8536.4900	--其他	10			0		50	--Other
6599	8536.5000	-其他开关	0					50	-Other switches
		-灯座、插头及插座:							-Lamp-holders, plugs and sockets:
6600	8536.6100	--灯座	10					50	--Lamp-holders
6601	8536.6900	--其他	0					50	--Other
6602	8536.9000	-其他装置	0					50	-Other apparatus

序号 No.	税则号列 Tariff Item	货品名称	最惠国税率(%) M.F.N.	协定税率(%) Agreement 东盟 ASEAN	曼谷协定 BA	香港 H.K.	澳门 Macao	普通税率(%) Gen.	Article Description
	85.37	**用于电气控制或电力分配的盘、板、台、柜及其他基座，装有两个或多个税号85.35或85.36所列的装置，包括装有第九十章所列的仪器或装置，以及数控装置，但税号85.17的交换机除外：**							**Boards, panels, consoles, desks, cabinets and other bases, equipped with two or more apparatus of heading No. 85.35 or 85.36, for electric control or the distribution of electricity, including those incorporating instruments or apparatus of Chapter 90, and numerical control apparatus, other than switching apparatus of heading No. 85.17:**
		-用于电压不超过1000伏的线路：							-For a voltage not exceeding 1 000 V:
6603	8537.1010	---数控装置	5			0		14	---Numerical control panels
6604	8537.1090	---其他	8.4			0	0	50	---Other
		-用于电压超过1000伏的线路：							-For a voltage exceeding 1 000 V:
6605	8537.2010	---全封闭组合式高压开关装置，用于电压在500千伏及以上的线路	8.4					30	---Gas insulated switchgear, for a voltage of 500 kV or more
6606	8537.2090	---其他	8.4			0		50	---Other
	85.38	**专用于或主要用于税号85.35、85.36或85.37所列装置的零件：**							**Parts suitable for use solely or principally with the apparatus of heading No. 85.35, 85.36 or 85.37:**
		-税号85.37所列货品用的盘、板、台、柜及其他基座，但未装有关装置：							-Boards, panels, consoles, desks, cabinets and other bases for the goods of heading No. 85.37, not equipped with their apparatus:
6607	8538.1010	---税号8537.2010所列货品用	8.4					50	---For the goods of heading No. 8537.2010
6608	8538.1090	---其他	7					50	---Other
6609	8538.9000	其他	7			0		50	-Other
	85.39	**白炽灯泡、放电灯管，包括封闭式聚光灯及紫外线灯管或红外线灯泡；弧光灯：**							**Electric filament or discharge lamps, including sealed beam lamp units and ultraviolet or infra-red lamps; arc-lamps:**
6610	8539.1000	-封闭式聚光灯	10					45	-Sealed beam lamp units
		-其他白炽灯泡，但不包括紫外线灯管或红外线灯泡：							-Other filament lamps, excluding ultraviolet or infra-red lamps:
		--卤钨灯：							--Tungsten halogen:
6611	8539.2110	---科研、医疗专用	8					20	---For scientific or medical uses only
6612	8539.2120	---火车、航空器及船舶用	8					20	---For locomotives and rolling-stock, aircraft or ships
6613	8539.2130	---机动车辆用	10					45	---For motor vehicles

序号 No.	税则号列 Tariff Item	货品名称	最惠国税率(%) M.F.N.	协定税率(%) Agreement 东盟 ASEAN	曼谷协定 BA	香港 H.K.	澳门 Macao	普通税率(%) Gen.	Article Description
6614	8539.2190	---其他	10.5					70	---Other
		--其他灯，功率不超过200瓦，但额定电压超过100伏：							--Other, of a power not exceeding 200 W and for a voltage exceeding 100 V:
6615	8539.2210	---科研、医疗专用	10.5					20	---For scientific or medical uses only
6616	8539.2290	---其他	5					70	---Other
		--其他：							--Other:
6617	8539.2910	---科研、医疗专用	5					20	---For scientific or medical uses only
6618	8539.2920	---火车、航空器及船舶用	10.5					20	---For locomotives and rolling-stock, aircraft or ships
6619	8539.2930	---机动车辆用	5					45	---For motor vehicles
		---其他：							---Other:
6620	8539.2991	----12伏及以下的	12			0		70	----Of a voltage 12 V or less
6621	8539.2999	----其他	12					70	----Other
		-放电灯管，但紫外线灯管除外：							-Discharge lamps, other than ultra-violet lamps:
		--热阴极荧光灯：							--Fluorescent, hot cathode:
6622	8539.3110	---科研、医疗专用	8					20	---For scientific or medical uses only
6623	8539.3120	---火车、航空器及船舶用	8					20	---For locomotives and rolling-stock, aircraft or ships
6624	8539.3190	---其他	8					70	---Other
		--汞或钠蒸汽灯；金属卤化物灯：							--Mercury or sodium vapour lamps; metal halide lamps:
6625	8539.3210	---科研、医疗专用	8					20	---For scientific or medical uses only
6626	8539.3220	---火车、航空器及船舶用	8					20	---For locomotives and rolling-stock, aircraft or ships
6627	8539.3290	---其他	8					70	---Other
		--其他：							--Other:
6628	8539.3910	---科研、医疗专用	8					20	---For scientific or medical uses only
6629	8539.3920	---火车、航空器及船舶用	8					20	---For locomotives and rolling-stock, aircraft or ships
6630	8539.3990	---其他	8					70	---Other
		-紫外线灯管或红外线灯泡；弧光灯：							-Ultraviolet or infra-red lamps; arc-lamps:
6631	8539.4100	--弧光灯	8					20	--Arc-lamps
6632	8539.4900	--其他	8					20	--Other
6633	8539.9000	-零件	8					20	-Parts
	85.40	**热电子管、冷阴极管或光阴极管（例如，真空管或充气管、汞弧整流管、阴极射线管、电视摄像管）：**							**Thermionic, cold cathode or photo-cathode valves and tubes (for example, vacuum or vapour or gas filled valves and tubes, mercury arc rectifying valves and tubes, cathode-ray tubes, television camera tubes):**
		-阴极射线电视显像管，包括视频监视器用阴极射线管：							-Cathode-ray television picture tubes, including video monitor cathode-ray tubes:

序号 No.	税则号列 Tariff Item	货品名称	最惠国税率(%) M.F.N.	协定税率(%)Agreement 东盟 ASEAN	曼谷协定 BA	香港 H.K.	澳门 Macao	普通税率(%) Gen.	Article Description
6634	8540.1100	--彩色的	12					40	--Colour
6635	8540.1200	--黑白的或其他单色的	15					40	--Black and white or other monochrome
		-电视摄像管;变像管及图像增强管;其他光阴极管:							-Television camera tubes; image converters and intensifiers; other photocathode tubes:
6636	8540.2010	---电视摄像管	12					35	---Television camera tubes
6637	8540.2090	---其他	8					17	---Other
6638	8540.4000	-彩色的数据/图形显示管,屏幕荧光点间距小于0.4毫米	8			0		17	-Data/graphic display tubes, colour, with a phosphor dot screen pitch smaller than 0.4 mm
6639	8540.5000	-黑白的或其他单色的数据/图形显示管	8					17	-Data/graphic display tubes, black and white or other monochrome
		-其他阴极射线管:							-Other cathode-ray tubes:
6640	8540.6010	---雷达显示管	6					14	---Radar display tubes
6641	8540.6090	---其他	8					17	---Other
		-微波管(例如,磁控管、速调管、行波管、返波管),但不包括栅控管:							-Microwave tubes (for example, magnetrons, klystrons, travelling wave tubes, carcinotrons), excluding grid-controlled tubes:
6642	8540.7100	--磁控管	8					17	--Magnetrons
6643	8540.7200	--速调管	8					17	--Klystrons
6644	8540.7900	--其他	8					17	--Other
		-其他管:							-Other valves and tubes:
6645	8540.8100	--接收管或放大管	8					17	--Receiver or amplifier valves and tubes
6646	8540.8900	--其他	8					17	--Other
		-零件:							-Parts:
		--阴极射线管用:							--Of cathode-ray tubes:
6647	8540.9110	---电视显像管用	6					40	---Of television picture tubes
6648	8540.9120	---雷达显示管用	5					14	---Of radar display tubes
6649	8540.9190	---其他	8					17	---Other
		--其他:							--Other:
6650	8540.9910	---电视摄像管用	8					35	---Of television camera tubes
6651	8540.9990	---其他	8					17	---Other
	85.41	**二极管、晶体管及类似的半导体器件;光敏半导体器件,包括不论是否装在组件内或组装成块的光电池;发光二极管;已装配的压电晶体:**							**Diodes, transistors and similar semiconductor devices; photosensitive semi-conductor devices, including photovoltaic cells whether or not assembled in modules or made up into panels; light emitting diodes; mounted piezo-electric crystals:**
6652	8541.1000	-二极管,但光敏二极管或发光二极管除外	0					30	-Diodes, other than photosensitive or light emitting diodes

序号 No.	税则号列 Tariff Item	货品名称	最惠国税率(%) M.F.N.	协定税率(%)Agreement 东盟 ASEAN	曼谷协定 BA	香港 H.K.	澳门 Macao	普通税率(%) Gen.	Article Description
		-晶体管,但光敏晶体管除外:							-Transistors, other than photosensitive transistors:
6653	8541.2100	--耗散功率小于1瓦的	0					30	--With a dissipation rate of less than 1 W
6654	8541.2900	--其他	0					30	--Other
6655	8541.3000	-半导体开关元件、两端交流开关元件及三端双向可控硅开关元件,但光敏器件除外	0					30	-Thyristors, diacs and triacs, other than photosensitive devices
6656	8541.4000	-光敏半导体器件,包括不论是否装在组件内或组装成块的光电池;发光二极管	0					30	-Photosensitive semiconductor devices, including photovoltaic cells whether or not assembled in modules or made up into panels; light emitting diodes
6657	8541.5000	-其他半导体器件	0					30	-Other semiconductor devices
6658	8541.6000	-已装配的压电晶体	0					30	-Mounted piezo-electric crystals
6659	8541.9000	-零件	0					30	-Parts
	85.42	**集成电路及微电子组件:**							**Electronic integrated circuits and microassemblies:**
6660	8542.1000	-装有集成电路的卡(智能卡)	0					21	-Cards incorporating an electronic integrated circuits("smart"cards)
		-单片集成电路:							-Monolithic integrated circuits
		--数字式:							--Digital:
		---线宽不超过0.18微米:							---Line width not exceeding 0.18 micron:
6661	8542.2111	----圆片(未经切割加工)	0					24	----Wafer(not cut-processed)
6662	8542.2119	----其他	0					24	----Other
		---线宽超过0.18微米,但不超过0.35微米:							---Line width exceeding 0.18 micron but not exceeding 0.35 micron:
6663	8542.2121	----圆片(未经切割加工)	0					24	----Wafer(not cut-processed)
6664	8542.2129	----其他	0					24	----Other
		---线宽超过0.35微米:							---Line width exceeding 0.35micron:
6665	8542.2191	----圆片(未经切割加工)	0					24	----Wafer(not cut-processed)
6666	8542.2199	----其他	0					24	----Other
6667	8542.2900	--其他	0					24	--Other
6668	8542.6000	-混合集成电路	0					30	-Hybrid integrated circuits
		-微电子组件:							-Electronic microassemblies:
6669	8542.7010	---光通信设备的激光收发模块	0					30	---Laser transmitting and receiving unit of laser communication equipment

序号 No.	税则号列 Tariff Item	货品名称	最惠国税率(%) M.F.N.	协定税率(%) Agreement 东盟 ASEAN	曼谷协定 BA	香港 H.K.	澳门 Macao	普通税率(%) Gen.	Article Description
6670	8542.7090	---其他	0					30	---Other
6671	8542.9000	-零件	0					30	-Parts
	85.43	**本章其他税号未列名的具有独立功能的电气设备及装置:**							**Electrical machines and apparatus, having individual functions, not specified or included elsewhere in this Chapter:**
		-粒子加速器:							-Particle accelerators:
6672	8543.1100	--用于对半导体材料掺杂的离子注入机	0					11	--Ion implanters for doping semiconductor materials
6673	8543.1900	--其他	5					11	--Other
		-信号发生器:							-Signal generators:
6674	8543.2010	---输出信号频率在1500兆赫兹以下的通用信号发生器	15					80	---Universal signal generators, with a frequency range of less than 1500 MHz
6675	8543.2090	---其他	8					20	---Other
6676	8543.3000	-电镀、电解或电泳设备及装置	0					35	-Machines and apparatus for electroplating, electrolysis or electrophoresis
6677	8543.4000	-电篱网激发器	10					35	-Electric fence energizers
		-其他设备及装置:							-Other machines and apparatus:
6678	8543.8100	--邻近卡及牌	0					35	--Proximity cards and tags
		--其他:							--Other:
6679	8543.8910	---金属、矿藏探测器	0					17	---Metal or mine detectors
6680	8543.8920	---高、中频放大器	0					17	---High or intermediate frequency amplifiers
6681	8543.8990	---其他	0					35	---Other
		-零件:							-Parts:
6682	8543.9010	---粒子加速器用	0					11	---Of particle accelerators
		---信号发生器用:							---Of signal generators:
6683	8543.9021	----输出信号频率在1500兆赫兹以下的通用信号发生器用	0					80	----Of the generators of subheading No.8543.2010
6684	8543.9029	----其他	0					20	----Other
6685	8543.9030	---金属、矿藏探测器用	0					17	---Of metal or mine detectors
6686	8543.9040	---高、中频放大器用	0					17	---Of high or intermediate frequency amplifiers
6687	8543.9090	---其他	0					35	---Other
	85.44	**绝缘(包括漆包或阳极化处理)电线、电缆(包括同轴电缆)及其他绝缘电导体,不论是否有接头;由每根被覆光纤组成的光缆,不论是否与电导体装配或装有接头:**							**Insulated (including enamelled or anodized) wire, cable (including co-axial cable) and other insulated electric conductors, whether or not ritted with connectors; optical fibre cables, made up of individually sheathed fibres, whether or not assembled with electric conductors or fitted with connectors:**

序号 No.	税则号列 Tariff Item	货品名称	最惠国税率(%) M.F.N.	协定税率(%)Agreement 东盟 ASEAN	曼谷协定 BA	香港 H.K.	澳门 Macao	普通税率(%) Gen.	Article Description
		-绕组电线:							-Winding wire:
6688	8544.1100	--铜制	10			0	0	70	--Of copper
6689	8544.1900	--其他	20					70	--Other
6690	8544.2000	-同轴电缆及其他同轴电导体	10		9			20	-Co-axial cable and other co-axial electric conductors
		-车辆、航空器、船舶用点火布线组及其他布线组:							-Ignition wiring sets and other wiring sets of a kind used in vehicles, aircraft or ships:
6691	8544.3020	---机动车辆用	10					20	---For motor vehicles
6692	8544.3090	---其他	5					70	---Other
		-其他电导体,额定电压不超过80伏							-Other electric conductors, for a voltage not exceeding 80 V:
		--有接头:							--Fitted with connectors:
6693	8544.4110	---电缆	0					20	---Electric cable
6694	8544.4190	---其他	0			0		70	---Other
		--其他:							--Other:
6695	8544.4910	---电缆	0					20	---Electric cable
6696	8544.4990	---其他	0					70	---Other
		-其他电导体,额定电压超过80伏,但不超过1000伏:							-Other electric conductors, for a voltage exceeding 80 V but not exceeding 1000 V:
		---有接头:							---Fitted with connectors:
6697	8544.5110	---电缆	0					20	---Electric cable
6698	8544.5190	---其他	0			0	0	70	---Other
		--其他:							--Other:
6699	8544.5910	---电缆	6			0		20	---Electric cable
6700	8544.5990	---其他	12			0	0	70	---Other
		-其他电导体,额定电压超过1000伏:							-Other electric conductors, for a voltage exceeding 1000 V:
		---电缆:							---Electric cable:
6701	8544.6011	----额定电压超过35千伏	8.4		8	0		20	----For a voltage excedding 35 kV
6702	8544.6019	----其他	10		8.9	0		50	----Other
6703	8544.6090	---其他	21		20	0		70	---Other
6704	8544.7000	-光缆	0		0			20	-Optical fibre cables
	85.45	**碳电极、碳刷、灯碳棒、电池碳棒及电气设备用的其他石墨或碳精制品,不论是否带金属:**							**Carbon electrodes, carbon brushes, lamp carbons, battery carbons and other articles of graphite or other carbon, with or without metal, of a kind used for electrical purposes:**
		-碳电极:							-Electrodes:
6705	8545.1100	--炉用	8					35	--Of a kind used for furnaces
6706	8545.1900	--其他	10.5					35	--Other
6707	8545.2000	-碳刷	10.5					35	-Brushes

序号 No.	税则号列 Tariff Item	货品名称	最惠国税率(%) M.F.N.	协定税率(%)Agreement 东盟 ASEAN	曼谷协定 BA	香港 H.K.	澳门 Macao	普通税率(%) Gen.	Article Description
6708	8545.9000	-其他	10.5					35	-Other
	85.46	**各种材料制的绝缘子:**							**Electrical insulators of any material:**
6709	8546.1000	-玻璃制	10.5					35	-Of glass
		-陶瓷制:							-Of ceramics:
6710	8546.2010	---输变电线路绝缘瓷套管	6					35	---Power transmission and converting ceramic bushings
6711	8546.2090	---其他	12					35	---Other
6712	8546.9000	-其他	10					35	-Other
	85.47	**电气机器、器具或设备用的绝缘零件,除了为装配需要而在模制时装入的小金属零件(例如螺纹孔)以外,全部用绝缘材料制成,但税号85.46的绝缘子除外;内衬绝缘材料的贱金属制线路导管及其接头:**							**Insulating fittings for electrical machines, appliances or equipment, being fittings wholly of insulating material apart from any minor components of metal (for example, threaded sockets) incorporated during moulding solely for purposes of assembly, other than insulators of heading No. 85.46; electrical conduit tubing and joints therefor, of base metal lined with insulating material:**
6713	8547.1000	-陶瓷制绝缘零件	8					35	-Insulating fittings of ceramics
6714	8547.2000	-塑料制绝缘零件	8					35	-Insulating fittings of plastics
		-其他:							-Other
6715	8547.9010	---内衬绝缘材料的贱金属制线路导管及其接头	10					50	---Electrical conduit tubing and joints therefor, of base metal lined with insulating material
6716	8547.9090	---其他	8					35	---Other
	85.48	**原电池、原电池组及蓄电池的废碎料;废原电池、废原电池组及废蓄电池;机器或设备的本章其他税号未列名的电气零件:**							**Waste and scrap of primary cells, primary batteries and electric accumulators; spent primary cells, spent primary batteries and spent electric accumulators; electrical parts of machinery or apparatus, not specified or included elsewhere in this Chapter:**
6717	8548.1000	-原电池、原电池组及蓄电池的废碎料;废原电池、废原电池组及废蓄电池	8					36	-Waste and scrap of primary cells, primary batteries and electric accumulators; spent primary cells, spent primary batteries and spent electric accumulators
6718	8548.9000	-其他	12					40	-Other

注:①85211011 广播级录像机的协定税率:每台完税价格低于或等于 2000 美元,执行单一从价税,税率为 16%;每台完税价格高于 2000 美元,每台征收从量税,税额 2158 元,加上 3%的从价税;

②85211019 其他磁带录像机的协定税率:每台完税价格低于或等于 2000 美元,执行单一从价税,税率 25%;每台完税价格高于 2000 美元,每台征收从量税,税额 3652 元,加上 3%的从价税;

③85211020 放像机的协定税率:每台完税价格低于或等于 2000 美元,执行单一从价税,税率为 25%;每台完税价格高于 2000 美元,每台征收从量税,税额 3652 元,加上 3%的从价税。

Notes:

①85211011 If the Duty Paying Value(DPV)of each video tape recorder with broadcasting quality is less than or equal to US $ 2000, it should be imposed an Ad Valorem duty of 16%; If the DPV is higher than US $ 2000. it should be imposed a compound duty, i.e. a Specific duty of 2158 RMB plus an Ad Valorem duty of 3% for each one.

②85211019: If the DPV of each other magnetic tape-type video recorder is less than or equal to US $ 2000, it should be imposed an Ad Valorem duty of 25%; If the DPV is higher than US $ 2000, it should be imposed a compound duty, i.e. a Specific duty of 3652 RMB plus an Ad Valorem duty of 3% for each one.

③85211020: If the DPV of each video tape reproducer is less than or equal to US $ 2000, it should be imposed an Ad Valorem duty of 25%; If the DPV is higher than US $ 2000, it should be imposed a compound duty, i.e. a Specific duty of 3652 RMB plus an Ad Valorem duty of 3% for each one.

第十七类

车辆、航空器、船舶及有关运输设备

注释：

一、本类不包括税号95.01、95.03或95.08的物品以及税号95.06的长雪橇、平底雪橇及类似品。

二、本类所称“零件”及“零件、附件”，不适用于下列货品，不论其是否确定为供本类货品使用：

(一) 各种材料制的接头、垫圈或类似品（按其构成材料归类或归入税号84.84）或硫化橡胶（硬质橡胶除外）的其他制品（税号40.16）；

(二) 第十五类注释二所规定的贱金属制通用零件（第十五类）或塑料制的类似品（第三十九章）；

(三) 第八十二章的物品（工具）；

(四) 税号83.06的物品；

(五) 税号84.01至84.79的机器或装置及其零件；税号84.81或84.82的物品及税号84.83的物品（这些物品是构成发动机或其他动力装置所必需的）；

(六) 电机或电气设备（第八十五章）；

(七) 第九十章的物品；

(八) 第九十一章的物品；

(九) 武器（第九十三章）；

(十) 税号94.05的灯具或照明装置；

(十一) 作为车辆零件的刷子（税号96.03）。

三、第八十六章至第八十八章所称“零件”或“附件”，不适用于那些非专用于或非主要用于这几章所列物品的零件、附件。同时符合这几章内两个或两个以上税号规定的零件、附件，应按其主要用途归入相应的税号。

四、在本类中：

(一) 既可在道路上又可在轨道上行驶的特殊构造的车辆，应归入第八十七章的相应税号；

(二) 水陆两用的机动车辆，应归入第八十七章的相应税号；

SECTION XVII

VEHICLES, AIRCRAFT, VESSELS AND ASSOCIATED TRANSPORT EQUIPMENT

Notes:

1. This Section does not cover articles of heading No.95.01, 95.03 or 95.08, or bobsleighs, toboggans or the like of heading No.95.06.

2. The expressions “parts” and “parts and accessories” do not apply to the following articles, whether or not they are identifiable as for the goods of this Section:

(a) Joints, washers or the like of any material (classified according to their constituent material or in heading No.84.84) or other articles of vulcanized rubber other than hard rubber (heading No.40.16);

(b) Parts of general use, as defined in Note 2 to Section XV, of base metal (Section XV), or similar goods of plastics (Chapter 39);

(c) Articles of Chapter 82 (tools);

(d) Articles of heading No.83.06;

(e) Machines or apparatus of headings Nos.84.01 to 84.79, or parts thereof; articles of heading No.84.81 or 84.82 or, provided they constitute integral parts of engines or motors, articles of heading No.84.83;

(f) Electrical machinery or equipment (Chapter 85);

(g) Articles of Chapter 90;

(h) Articles of Chapter 91;

(ij) Arms (Chapter 93);

(k) Lamps or lighting fittings of heading No.94.05; or

(l) Brushes of a kind used as parts of vehicles (heading No.96.03).

3. References in Chapters 86 to 88 to “parts” or “accessories” do not apply to parts or accessories which are not suitable for use solely or principally with the articles of those Chapters. A part or accessory which answers to a description in two or more of the headings of those Chapters is to be classified under that heading which corresponds to the principal use of that part or accessory.

4. For the purposes of this Section:

(a) Vehicles specially constructed to travel on both road and rail are classified under the appropriate heading of Chapter 87;

(b) Amphibious motor vehicles are classified under the appropriate heading of Chapter 87;

（三）可兼作地面车辆使用的特殊构造的航空器，应归入第八十八章的相应税号。

(c) Aircraft specially constructed so that they can also be used as road vehicles are classified under the appropriate heading of Chapter 88.

五、气垫运输工具应按本类最相似的运输工具归类，其规定如下：
（一）在导轨上运行的（气垫火车），归入第八十六章；
（二）在陆地行驶或水陆两用的，归入第八十七章；
（三）在水上航行的，不论能否在海滩或浮码头登陆及能否在冰上行驶，一律归入第八十九章。
气垫运输工具的零件、附件，应按照上述规定，与最相类似的运输工具的零件、附件一并归类。

气垫火车的导轨固定装置及附件应与铁道轨道固定装置及附件一并归类。气垫火车运行系统的信号、安全或交通管理设备应与铁路的信号、安全或交通管理设备一并归类。

5. Air-cushion vehicles are to be classified within this Section with the vehicles to which they are most akin as follows:
(a) In Chapter 86 if designed to travel on a guide-track (hovertrains);
(b) In Chapter 87 if designed to travel over land or over both land and water;
(c) In Chapter 89 if designed to travel over water, whether or not able to land on beaches or landingstages or also able to travel over ice.
Parts and accessories of air-cushion vehicles are to be classified in the same way as those of vehicles of the heading in which the air-cushion vehicles are classified under the above provisions.

Hovertrain track fixtures and fittings are to be classified as railway track fixtures and fittings, and signalling, safety or traffic control equipment for hovertrain transport systems as signalling, safety or traffic control equipment for railways.

第八十六章
铁道及电车道机车、车辆及其零件；铁道及电车道轨道固定装置及其零件、附件；各种机械（包括电动机械）交通信号设备

Chapter 86
Railway or tramway locomotives, rolling-stock and parts thereof; railway or tramway track fixtures and fittings and parts thereof; mechanical (including electro-mechanical) traffic signalling equipment of all kinds

注释：

一、本章不包括：
（一）木制或混凝土制的铁道或电车道轨枕及气垫火车用的混凝土导轨（税号 44.06 或 68.10）；
（二）税号 73.02 的铁道及电车道铺轨用钢铁材料；
（三）税号 85.30 的电气信号、安全或交通管理设备。

二、税号 86.07 主要适用于：
（一）轴、轮、行走机构、金属轮箍、轮圈、毂及轮子的其他零件；
（二）车架、底架、转向架；
（三）轴箱；制动装置；
（四）车辆缓冲器；钩或其他联结器及车厢走廊联结装置；
（五）车身。

Notes:

1. This Chapter does not cover:
(a) Railway or tramway sleepers of wood or of concrete, or concrete guide-track sections for hovertrain (heading No.44.06 or 68.10);
(b) Railway or tramway track construction material of iron or steel of heading No.73.02; or
(c) Electrical signalling, safety of traffic control equipment of heading No.85.30.

2. Heading No.86.07 applies, *inter alia*, to:
(a) Axles, wheels, wheel sets (running gear), metal tyres, hoops and hubs and other parts of wheels;
(b) Frames, underframes, bogies and bissel-bogies;
(c) Axle boxes; brake gear;
(d) Buffers for rolling-stock; hooks and other coupling gear and corridor connections;
(e) Coachwork.

三、除上述注释一另有规定的以外，税号86.08包括：

(一) 已装配的轨道、转车台、站台缓冲器、量载规；

(二) 铁道及电车道、道路、内河航道、停车场、港口或机场用的臂板信号机、机械信号盘、平交道口控制器、信号及道岔控制器及其他机械（包括电动机械）信号、安全或交通管理设备，不论是否装有电力照明装置。

3. Subject to the provisions of Note 1 above, heading No.86.08 applies, *inter alia*, to:

(a) Assembled track, turntables, platform buffers, loading gauges;

(b) Semaphores, mechanical signal discs, level crossing control gear, signal and point controls, and other mechanical (including electro-mechanical) signalling, safety or traffic control equipment, whether or not fitted with electric lighting, for railways, tramways, roads, inland waterways, parking facilities, port installations or airfields.

序号 No.	税则号列 Tariff Item	货品名称	最惠国税率(%) M.F.N.	协定税率(%) Agreement 东盟 ASEAN	曼谷协定 BA	香港 H.K.	澳门 Macao	普通税率(%) Gen.	Article Description
	86.01	**铁道电力机车，由外部电力或蓄电池驱动：**							**Rail locomotives powered from an external source of electricity or by electric accumulators:**
		-由外部电力驱动：							-Powered from an external source of electricity:
		---直流电机驱动的：							---Drived by DC motors:
6719	8601.1011	----微型机控制的	3					11	----Controlled by microprocessings
6720	8601.1019	----其他	3					11	----Other
6721	8601.1020	---交流电机驱动的	3					11	---Drived by AC motors
6722	8601.1090	---其他	3					11	---Other
6723	8601.2000	-由蓄电池驱动	3					11	-Powered by electric accumulators
	86.02	**其他铁道机车；机车煤水车：**							**Other rail locomotives; locomotive tenders:**
		-柴油电力机车：							-Diesel-electric locomotives:
6724	8602.1010	---微型机控制的	3					11	---Controled by microprocessings
6725	8602.1090	---其他	3					11	---Other
6726	8602.9000	-其他	3					11	-Other
	86.03	**铁道及电车道用的机动客车、货车、敞车，但税号86.04的货品除外：**							**Self-propelled railway or tramway coaches, vans and trucks, other than those of heading No.86.04:**
6727	8603.1000	-由外部电力驱动	3					11	-Powered from an external source of electricity:
6728	8603.9000	-其他	3					11	-Other
	86.04	**铁道及电车道用的维修或服务车，不论是否机动（例如，工场车、起重机车、道碴捣固车、轨道校正车、检验车及查道车）：**							**Railway or tramway maintenance or service vehicles, whether or not self-propelled (for example, workshops, cranes, ballast tampers, trackliners, testing coaches and track inspection vehicles):**
		---检验车及查道车：							---Testing coaches and track inspection vehicles

序号 No.	税则号列 Tariff Item	货品名称	最惠国税率(%) M.F.N.	协定税率(%)Agreement 东盟 ASEAN	曼谷协定 BA	香港 H.K.	澳门 Macao	普通税率(%) Gen.	Article Description
6729	8604.0011	----隧道限界检查车	3					14	----Inspection vehicles for tunnel clearance
6730	8604.0012	----钢轨在线打磨列车	3					14	----Sanding vehicles for on-line rails
6731	8604.0019	----其他	5					14	----Other
		---其他:							---Other:
6732	8604.0091	----电气化接触网架线机(轨行式)	5					20	----Installing vehicles for suspension of contact wire (running on rails)
6733	8604.0099	----其他	7					20	----Other
	86.05	**铁道及电车道用的非机动客车;行李车、邮政车和其他铁道及电车道用的非机动特殊用途车辆(税号86.04的货品除外):**							**Railway or tramway passenger coaches, not self-propelled; luggage vans, post office coaches and other special purpose railway or tramway coaches, not self-propelled (excluding those of heading No.86.04):**
6734	8605.0010	---铁道客车	5					14	---Railway passenger coaches
6735	8605.0090	---其他	5					14	---Other
	86.06	**铁道及电车道用的非机动有篷及无篷货车:**							**Railway or tramway goods vans and wagons, not self-propelled:**
6736	8606.1000	-油罐货车及类似车	5					14	-Tank wagons and the like
6737	8606.2000	-保温或冷藏货车,但子目号8606.1000的货品除外	5					14	-Insulated or refrigerated vans and wagons, other than those of subheading No.8606.1000
6738	8606.3000	-自卸货车,但子目号8606.1000或8606.2000的货品除外	5					14	-Self-discharging vans and wagons, other than those of subheading No.8606.1000 or 8606.2000
		-其他:							-Other:
6739	8606.9100	--带篷及封闭的	5					14	--Covered and closed
6740	8606.9200	--敞篷的,厢壁固定且高度超过60厘米	5					14	--Open, with non-removable sides of a height exceeding 60 cm
6741	8606.9900	--其他	5					14	--Other
	86.07	**铁道及电车道机车或其他车辆的零件:**							**Parts of railway or tramway locomotives or rolling-stock:**
		-转向架、轴、轮及其零件:							-Bogies, bissel-bogies, axles and wheels, and parts thereof:
6742	8607.1100	--驾驶转向架	3					11	--Driving bogies and bissel-bogies
6743	8607.1200	--其他转向架	3					11	--Other bogies and bissel-bogies
		--其他,包括零件:							--Other, including parts:
6744	8607.1910	---轴	3					11	---Axles
6745	8607.1990	---其他	3					11	---Other
		-制动装置及其零件:							-Brakes and parts thereof:
6746	8607.2100	--空气制动器及其零件	3					11	--Air brakes and parts thereof
6747	8607.2900	--其他	3					11	--Other
6748	8607.3000	-钩、其他联结器、缓冲器及其零件	3					11	-Hooks and other coupling devices, buffers, and parts thereof
		-其他:							-Other:

序号 No.	税则号列 Tariff Item	货品名称	最惠国税率(%) M.F.N.	协定税率(%)Agreement 东盟 ASEAN	曼谷协定 BA	香港 H.K.	澳门 Macao	普通税率(%) Gen.	Article Description
6749	8607.9100	--机车用	3					11	--Of locomotives
6750	8607.9900	--其他	3					11	--Other
	86.08	**铁道及电车道轨道固定装置及附件;供铁道、电车道、道路、内河航道、停车场、港口或机场用的机械(包括电动机械)信号、安全或交通管理设备;上述货品的零件:**							**Railway or tramway track fixtures and fittings; mechanical (including electro-mechanical) signalling, safety or traffic control equipment for railways, tramways, roads, inland waterways, parking facilities, port installations or airfields; parts of the foregoing:**
6751	8608.0010	---轨道自动计轴设备	3					20	---Rail automatic axle counting equipments
6752	8608.0090	---其他	4					20	---Other
	86.09	**集装箱(包括运输液体的集装箱),经特殊设计、装备适用于各种运输方式:**							**Containers (including containers for the transport of fluids) specially designed and equipped for carriage by one or more modes of transport:**
6753	8609.0010	---20 英尺的	10.5					35	---Of 20 feet
6754	8609.0020	---40 英尺的	10.5					35	---Of 40 feet
6755	8409.0030	---45、48、53 英尺的	10.5					35	---Of 45、48、53 feet
6756	8609.0090	---其他	10.5					35	---Other

第八十七章
车辆及其零件、附件，但铁道及电车道车辆除外

Chapter 87
Vehicles other than railway or tramway rolling-stock, and parts and accessories thereof

注释：

一、本章不包括仅可在钢轨上运行的铁道及电车道车辆。

二、本章所称"牵引车、拖拉机"，是指主要为牵引或推动其他车辆、器具或重物的车辆。除了上述主要用途以外，不论其是否还具有装运工具、种子、肥料或其他货品的辅助装置。

用于安装在税号87.01的牵引车或拖拉机上，作为可替换设备的机器或作业工具，即使与牵引车或拖拉机一同进口或出口，不论其是否已安装在车（机）上，仍应归入其各自相应的税号。

三、装有驾驶室的机动车辆底盘，应归入税号87.02至87.04，而不归入税号87.06。

四、税号87.12包括所有儿童两轮车，其他儿童脚踏车归入税号95.01。

Notes:

1. This Chapter does not cover railway or tramway rollingstock designed solely for running on rails.

2. For the purposes of this Chapter, "tractors" means vehicles constructed essentially for hauling or pushing another vehicle, appliance or load, whether or not they contain subsidiary provision for the transport, in connection with the main use of the tractor, of tools, seeds, fertilizers or other goods.

Machines and working tools designed for fitting to tractors of heading No.87.01 as interchangeable equipment remain classified in their respective headings even if presented with the tractor, and whether or not mounted on it.

3. Motor chassis fitted with cabs fall in headings Nos.87.02 to 87.04, and not in heading No.87.06.

4. Heading No.87.12 includes all children's bicycles. Other children's cycles fall in heading No.95.01.

序号 No.	税则号列 Tariff Item	货品名称	最惠国税率(%) M.F.N.	协定税率(%) Agreement 东盟 ASEAN	曼谷协定 BA	香港 H.K.	澳门 Macao	普通税率(%) Gen.	Article Description
	87.01	**牵引车、拖拉机(税号87.09的牵引车除外)：**							**Tractors(other than tractors of heading No.87.09):**
6757	8701.1000	-手扶拖拉机	9					20	-Pedestrian controlled tractors
6758	8701.2000	-半挂车用的公路牵引车	6					20	-Road tractors for semi-trailers
6759	8701.3000	-履带式牵引车、拖拉机	6					20	-Track-laying tractors
		-其他：							-Other:
		---拖拉机：							---Tractors:
6760	8701.9011	----轮式	8					20	----Wheeled
6761	8701.9019	----其他	8					20	----Other
6762	8701.9090	---其他	8					20	---Other
	87.02	**客运机动车辆，10座及以上(包括驾驶座)：**							**Motor vehicles for the transport of ten or more persons, including the driver:**
		-装有压燃式活塞内燃发动机(柴油或半柴油发动机)的车辆：							-With compression-ignition internal combustion piston engine (diesel or semidiesel):
6763	8702.1020	---机坪客车	4					90	---Buses for transport passengers at airport

序号 No.	税则号列 Tariff Item	货品名称	最惠国税率(%) M.F.N.	协定税率(%) Agreement 东盟 ASEAN	曼谷协定 BA	香港 H.K.	澳门 Macao	普通税率(%) Gen.	Article Description
		---其他:							---Other:
6764	8702.1091	----30座及以上(大型客车)	25					90	----With 30 seats or more
6765	8702.1092	----20座及以上,但不超过29座	25					230	----With 20 seats or more, but not exceeding 29 seats
6766	8702.1093	----10座及以上,但不超过19座	25					230	----With 10 seats or more, but not exceeding 19 seats
		-其他:							-Other:
6767	8702.9010	---30座及以上(大型客车)	25					90	---With 30 seats or more
6768	8702.9020	---20座及以上,但不超过29座	25					230	---With 20 seats or more, but not exceeding 29 seats
6769	8702.9030	---10座及以上,但不超过19座	25					230	---With 10 seats or more, but not exceeding 19 seats
	87.03	**主要用于载人的机动车辆(税号87.02的货品除外),包括旅行小客车及赛车:**							**Motor cars and other motor vehicles principally designed for the transport of persons (other than those of heading No. 87.02), including station wagons and racing cars:**
6770	8703.1000	-雪地行走专用车;高尔夫球车及类似车辆	25					150	-Vehicles specially designed for travelling on snow; golf cars and similar vehicles
		-装有点燃往复式活塞内燃发动机的其他车辆:							-Other vehicles, with spark-ignition internal combustion reciprocating piston engine:
		--气缸容量(排气量)不超过1000毫升:							--Of a cylinder capacity not exceeding 1000 cc:
6771	8703.2130	---小轿车	30					230	---Saloon cars
6772	8703.2190	---其他	30					230	---Other
		--气缸容量(排气量)超过1000毫升,但不超过1500毫升:							--Of a cylinder capacity exceeding 1000 cc but not exceeding 1500 cc:
6773	8703.2230	---小轿车	30					230	---Saloon cars
6774	8703.2240	---越野车(4轮驱动)	30					230	---Cross-country cars(4WD)
6775	8703.2250	---9座及以下的小客车	30					230	---Station wagons(with 9 seats or less)
6776	8703.2290	---其他	30					230	---Other
		--气缸容量(排气量)超过1500毫升,但不超过3000毫升:							--Of a cylinder capacity exceeding 1500 cc but not exceeding 3000 cc:

序号 No.	税则号列 Tariff Item	货品名称	最惠国税率(%) M.F.N.	协定税率(%) Agreement 东盟 ASEAN	曼谷协定 BA	香港 H.K.	澳门 Macao	普通税率(%) Gen.	Article Description
		---气缸容量(排气量)超过1500毫升,但不超过2500毫升:							---Of a cylinder capacity exceeding 1500 cc but not exceeding 2500 cc:
6777	8703.2314	----小轿车	30					230	----Saloon cars
6778	8703.2315	----越野车(4轮驱动)	30					230	----Cross-country cars (4WD)
6779	8703.2316	----9座及以下的小客车	30					230	----Station wagons (with 9 seats or less)
6780	8703.2319	----其他	30					230	----Other
		---气缸容量(排气量)超过2500毫升,但不超过3000毫升:							---Of a cylinder capacity exceeding 2500 cc but not exceeding 3000 cc:
6781	8703.2334	----小轿车	30					270	----Saloon cars
6782	8703.2335	----越野车(4轮驱动)	30					270	----Cross-country cars (4WD)
6783	8703.2336	----9座及以下的小客车	30					270	----Station wagons (with 9 seats or less)
6784	8703.2339	----其他	30					270	----Other
		--气缸容量(排气量)超过3000毫升:							--Of a cylinder capacity exceeding 3000 cc:
6785	8703.2430	---小轿车	30					270	---Saloon cars
6786	8703.2440	---越野车(4轮驱动)	30					270	---Cross-country cars(4WD)
6787	8703.2450	---9座及以下的小客车	30					270	---Station wagons (with 9 seats or less)
6788	8703.2490	---其他	30					270	---Other
		-装有压燃式活塞内燃发动机(柴油或半柴油发动机)的其他车辆:							-Other vehicles, with compression-ignition internal combustion piston engine (diesel or semi-diesel):
		--气缸容量(排气量)不超过1500毫升:							--Of a cylinder capacity not exceeding 1500 cc:
6789	8703.3130	---小轿车	30					230	---Saloon cars
6790	8703.3140	---越野车(4轮驱动)	30					230	---Cross-country cars (4WD)
6791	8703.3150	---9座及以下的小客车	30					230	---Station wagons(with 9 seats or less)
6792	8703.3190	---其他	30					230	---Other
		--气缸容量(排气量)超过1500毫升,但不超过2500毫升:							--Of a cylinder capacity exceeding 1500 cc but not exceeding 2500 cc:
6793	8703.3230	---小轿车	30					230	---Saloon cars
6794	8703.3240	---越野车(4轮驱动)	30					230	---Cross-country cars(4WD)
6795	8703.3250	---9座及以下的小客车	30					230	---Station wagons(with 9 seats or less)
6796	8703.3290	---其他	30					230	---Other
		--气缸容量(排气量)超过2500毫升:							--Of a cylinder capacity exceeding 2500 cc:
6797	8703.3330	---小轿车	30					270	---Saloon cars
6798	8703.3340	---越野车(4轮驱动)	30					270	---Cross-country cars(4WD)
6799	8703.3350	---9座及以下的小客车	30					270	---Station wagons(with 9 seats or less)

序号 No.	税则号列 Tariff Item	货品名称	最惠国税率(%) M.F.N.	协定税率(%) Agreement 东盟 ASEAN	曼谷协定 BA	香港 H.K.	澳门 Macao	普通税率(%) Gen.	Article Description
6800	8703.3390	---其他	30					270	---Other
6801	8703.9000	-其他	30					230	-Other
	87.04	**货运机动车辆:**							**Motor vehicles for the transport of goods:**
		-非公路用自卸车:							-Dumpers designed for off-highway use:
6802	8704.1030	---电动轮货运自卸车	6					20	---Electromobile dumpers for the transport of goods
6803	8704.1090	---其他	6					20	---Other
		-装有压燃式活塞内燃发动机(柴油或半柴油发动机)的其他货车:							-Other, with compression-ignition internal combustion piston engine(diesel or semidiesesl):
6804	8704.2100	--车辆总重量不超过5吨	25					70	--g.v.w. not exceeding 5 tons
		--车辆总重量超过5吨,但不超过20吨:							--g.v.w. exceeding 5 tons but not exceeding 20 tons:
6805	8704.2230	---车辆总重量超过5吨,但小于14吨	20					70	---g.v.w. exceeding 5 tons but not exceeding 14 tons
6806	8704.2240	---车辆总重量在14吨及以上,但不超过20吨	20					40	---g.v.w. of 14 tons or more but not exceeding 20 tons
6807	8704.2300	--车辆总重量超过20吨	15					40	--g.v.w. exceeding 20 tons
		-装有点燃式活塞内燃发动机的其他货车:							-Other, with spark-ignition internal combustion piston engine:
6808	8704.3100	--车辆总重量不超过5吨	25					70	--g.v.w. not exceeding 5 tons
		--车辆总重量超过5吨:							--g.v.w. exceeding 5 tons:
6809	8704.3230	---车辆总重量超过5吨,但不超过8吨	20					70	---g.v.w. exceeding 5 tons, but not exceeding 8 tons.
6810	8704.3240	---车辆总重量超过8吨	22					70	---g.v.w. exceeding 8 tons
6811	8704.9000	-其他	25					70	-Other
	87.05	**特殊用途的机动车辆(例如,抢修车、起重车、救火车、混凝土搅拌车、道路清洁车、喷洒车、流动工场车及流动放射线检查车),但主要用于载人或运货的车辆除外:**							**Special purpose motor vehicles, other than those principally designed for the transport of persons or goods(for example, breakdown lorries, crane lorries, fire fighting vehicles, concrete-mixer lorries, road sweeper lorries, spraying lorries, mobile workshops, mobile radiological units):**
		-起重车:							-Crane lorries:
		---全路面起重车:							---All-road crane lorries:
6812	8705.1021	----最大起重重量不超过50吨	15					30	----Of maxium lifting capacity not more than 50 tons

序号 No.	税则号列 Tariff Item	货品名称	最惠国税率(%) M.F.N.	协定税率(%) Agreement 东盟 ASEAN	曼谷协定 BA	香港 H.K.	澳门 Macao	普通税率(%) Gen.	Article Description
6813	8705.1022	----最大起重重量超过50吨,但不超过100吨	10					30	----Of a maxium lifting capacity exceeding 50tons but not exceeding 100tons
6814	8705.1023	----最大起重重量超过100吨	10					30	----Of a maxium lifting capacity exceeding 100tons
		---其他:							---Other:
6815	8705.1091	----最大起重重量不超过50吨	15					30	----Of maxium lifting capacity not more than 50tons
6816	8705.1092	----最大起重重量超过50吨,但不超过100吨	10					30	----Of a maxium lifting capacity exceeding 50tons but not exceeding 100tons
6817	8705.1093	----最大起重重量超过100吨	10					30	----Of a maxium lifting capacity exceeding 100tonsp
6818	8705.2000	-钻探车	12					17	-Mobile drilling derricks
		-救火车:							-Fire fighting vehicles:
6819	8705.3010	---装有云梯的救火车	3					8	---Mounted with scaling ladder
6820	8705.3090	---其他	3					8	---Other
6821	8705.4000	-混凝土搅拌车	15					35	-Concrete-mixer lorries
		-其他:							-Other:
6822	8705.9010	---无线电通信车	9					35	---radio communication vans
6823	8705.9020	---放射线检查车	9					14	---Mobile radiological units
6824	8705.9030	---环境监测车	12					20	---Mobile environmental monitoring units
6825	8705.9040	---医疗车	12					30	---Mobile clinics
		---电源车:							---Mobile electric generator sets:
6826	8705.9051	----航空电源车(频率为400赫兹)	12					30	----Airplane charging vehicles (frequency 400Hz)
6827	8705.9059	----其他	12					30	----Other
6828	8705.9060	---飞机加油车、调温车、除冰车	12					35	---Mobile vehicles for aircraft refuelling, air-conditioning or deicing
6829	8705.9070	---道路(包括跑道)扫雪车	12					35	---Snow sweep vehicles for cleansing streets or airfield runways
6830	8705.9080	---石油测井车、压裂车、混沙车	12					35	---Petrolcum well logging trucks, fracturing unit trucks and mixing sand trucks
6831	8705.9090	---其他	12					35	---Other
	87.06	**装有发动机的机动车辆底盘,税号87.01至87.05所列车辆用:**							**Chassis fitted with engines, for the motor vehicles of headings Nos.87.01 to 87.05:**
6832	8706.0010	---非公路用自卸车底盘	8					14	---For the vehicles of subheading No.8704.1030 or 8704.1090
		---货车底盘:							---For the vehicles of subheading Nos.8704.2100 to 8704.9000:
6833	8706.0021	----车辆总重量在14吨及以上的	10					30	----For vehicles g.v.w of 14 tons or more

序号 No.	税则号列 Tariff Item	货 品 名 称	最惠国税率(%) M.F.N.	协定税率(%)Agreement 东盟 ASEAN	曼谷协定 BA	香港 H.K.	澳门 Macao	普通税率(%) Gen.	Article Description
6834	8706.0022	----车辆总重量在14吨以下的	12					45	----For vehicles g. v. w less than 14 tons
6835	8706.0030	---大型客车底盘	20					70	---For passenger motor vehicles with 30 seats or more
6836	8706.0040	---汽车起重机底盘	20					100	---For crane lorries
6837	8706.0090	---其他	20.7					100	---Other
	87.07	**机动车辆的车身(包括驾驶室),税号87.01至87.05所列车辆用:**							**Bodies (including cabs), for the motor vehicles of headings Nos. 87.01 to 87.05:**
6838	8707.1000	-税号87.03所列车辆用	22.9					100	-For the vehicles of heading No.87.03
		-其他:							-Other:
6839	8707.9010	---税号8702.1092、8702.1093、8702.9020及8702.9030所列车辆用	18.6					70	---For the vehicles of subheading No.8702.1092、8702.1093、8702.9020 or 8702.9030
6840	8707.9090	---其他	18.6					70	---Other
	87.08	**机动车辆的零件、附件,税号87.01至87.05所列车辆用:**							**Parts and accessories of the motor vehicles of headings Nos. 87.01 to 87.05:**
6841	8708.1000	-缓冲器(保险杠)及其零件	12.9		12.3			100	-Bumpers and parts thereof
		-车身(包括驾驶室)的其他零件、附件:							-Other parts and accessories of bodies (including cabs):
6842	8708.2100	--座椅安全带	12.9					100	--Safety seat belts
		--其他:							--Other:
6843	8708.2920	---安全气囊装置	18.6					100	---Safety air-sac equipment
6844	8708.2930	---车窗玻璃升降器	10					100	---Windowpane raiser
6845	8708.2990	---其他	12.9					100	---Other
		-制动器、助力制动器及其零件:							-Brakes and servo-brakes and parts thereof:
6846	8708.3100	--装在蹄片上的制动摩擦片	10					100	--Mounted brake linings
		--其他:							--Other:
6847	8708.3910	---税号87.01所列车辆用	6					14	---Of the vehicles of heading No.87.01
6848	8708.3920	---子目号8702.1091及8702.9010所列车辆用	12.9					70	---Of the vehicles of subheading No.8702.1091 or 8702.9010
6849	8708.3930	---子目号8704.1030及8704.1090所列车辆用	6					11	---Of the vehicles of subheading No.8704.1030 or 8704.1090
6850	8708.3940	---子目号8704.2100、8704.2230、8704.3100及8704.3230所列车辆用	12.9					45	---Of the vehicles of subheading No.8704.2100, 8704.2230, 8704.3100 or 8704.3230

序号 No.	税则号列 Tariff Item	货品名称	最惠国税率(%) M.F.N.	协定税率(%)Agreement 东盟 ASEAN	曼谷协定 BA	香港 H.K.	澳门 Macao	普通税率(%) Gen.	Article Description
6851	8708.3950	---子目号 8704.2240、8704.2300 及 8704.3240 所列车辆用	10					30	---Of the vehicles of subheading No.8704.2240, 8704.2300 or 8704.3240
6852	8708.3960	---税号 87.05 所列车辆用	10					100	---Of the vehicles of heading No.87.05
		---其他:							---Other:
6853	8708.3991	----防抱死制动系统	18.6					100	----Anti-skid brake systerm
6854	8708.3999	----其他	18.6					100	----Other
		-变速箱:							-Gear boxes:
6855	8708.4010	---税号 87.01 所列车辆用	6					14	---Of the vehicles of heading No.87.01
6856	8708.4020	---子目号 8702.1091 及 8702.9010 所列车辆用	12.9					70	---Of the vehicles of subheading No.8702.1091 or 8702.9010
6857	8708.4030	---子目号 8704.1030 及 8704.1090 所列车辆用	6					11	---Of the vehicles of subheading No.8704.1030 or 8704.1090
6858	8708.4040	---子目号 8704.2100、8704.2230、8704.3100 及 8704.3230 所列车辆用	12.9					45	---Of the vehicles of subheading No.8704.2100,8704.2230,8704.3100 or 8704.3230
6859	8708.4050	---子目号 8704.2240、8704.2300 及 8704.3240 所列车辆用	10					30	---Of the vehicles of subheading No.8704.2240, 8704.2300 or 8704.3240
6860	8708.4060	---税号 87.05 所列车辆用	10					100	---Of the vehicles of heading No.87.05
		---其他:							---Other:
6861	8708.4091	----小轿车用自动换档变速箱	18.6					100	----Automatic gearshift for saloon cars
6862	8708.4099	----其他	18.6					100	----Other
		-装有差速器的驱动桥,不论是否装有其他传动部件:							-Drive-axles with differential, whether or not provided with other transmission components:
6863	8708.5010	---税号 87.01 所列车辆用	6					14	---Of the vehicles of heading No.87.01
6864	8708.5020	---子目号 8702.1091 及 8702.9010 所列车辆用	12.9					70	---Of the vehicles of subheading No.8702.1091 or 8702.9010
6865	8708.5030	---子目号 8704.1030 及 8704.1090 所列车辆用	6					11	---Of the vehicles of subheading No.8704.1030 or 8704.1090
6866	8708.5040	---子目号 8704.2100、8704.2230、8704.3100 及 8704.3230 所列车辆用	12.9					45	---Of the vehicles of subheading No.8704.2100,8704.2230,8704.3100 or 8704.3230

序号 No.	税则号列 Tariff Item	货品名称	最惠国税率(%) M.F.N.	协定税率(%)Agreement 东盟 ASEAN	曼谷协定 BA	香港 H.K.	澳门 Macao	普通税率(%) Gen.	Article Description
6867	8708.5050	---子目号8704.2240、8704.2300及8704.3240所列车辆用	10					30	---Of the vehicles of subheading No.8704.2240, 8704.2300 or 8704.3240
6868	8708.5060	---税号87.05所列车辆用	10					100	---Of the vehicles of heading No.87.05
6869	8708.5090	---其他	13.5					100	---Other
		-非驱动的桥及其零件:							-Non-driving axles and parts thereof:
6870	8708.6010	---税号87.01所列车辆用	6					14	---Of the vehicles of heading No.87.01
6871	8708.6020	---子目号8702.1091及8702.9010所列车辆用	17.3					70	---Of the vehicles of subheading No.8702.1091 or 8702.9010
6872	8708.6030	---子目号8704.1030及8704.1090所列车辆用	6					11	---Of the vehicles of subheading No.8704.1030 or 8704.1090
6873	8708.6040	---子目号8704.2100、8704.2230、8704.3100及8704.3230所列车辆用	12.9					45	---Of the vehicles of subheading No.8704.2100, 8704.2230, 8704.3100 or 8704.3230
6874	8708.6050	---子目号8704.2240、8704.2300及8704.3240所列车辆用	10					30	---Of the vehicles of subheading No.8704.2240, 8704.2300 or 8704.3240
6875	8708.6060	---税号87.05所列车辆用	10					100	---Of the vehicles of heading No.87.05
6876	8708.6090	---其他	13.5					100	---Other
		-车轮及其零件、附件:							-Road wheels and parts and accessories thereof:
6877	8708.7010	---子目号87.01所列车辆用	6					14	---Of the vehicles of heading No.87.01
6878	8708.7020	---子目号8702.1091及8702.9010所列车辆用	12.9					70	---Of the vehicles of subheading No.8702.1091 or 8702.9010
6879	8708.7030	---子目号8704.1030及8704.1090所列车辆用	6					11	---Of the vehicles of subheading No.8704.1030 or 8704.1090
6880	8708.7040	---子目号8704.2100、8704.2230、8704.3100及8704.3230所列车辆用	12.9					45	---Of the vehicles of subheading No.8704.2100, 8704.2230, 8704.3100 or 8704.3230
6881	8708.7050	---子目号8704.2240、8704.2300及8704.3240所列车辆用	10					30	---Of the vehicles of subheading No.8704.2240, 8704.2300 or 8704.3240

序号 No.	税则号列 Tariff Item	货品名称	最惠国税率(%) M.F.N.	协定税率(%) Agreement 东盟 ASEAN	曼谷协定 BA	香港 H.K.	澳门 Macao	普通税率(%) Gen.	Article Description
6882	8708.7060	---税号87.05所列车辆用	10					100	---Of the vehicles of heading No.87.05
6883	8708.7090	---其他	13.5					100	---Other
		-悬挂减震器:							-Suspension shock-absorbers:
6884	8708.8010	---税号87.03所列车辆用	13.5					100	---Of the vehicles of heading No.87.03
6885	8708.8090	---其他	13.5					100	---Other
		-其他零件、附件:							-Other parts and accessories:
6886	8708.9100	--散热器(水箱)	12.9					100	--Radiators
6887	8708.9200	--消声器及排气管	12.9					100	--Silencers and exhaust pipes
		--离合器及其零件:							--Clutches and parts thereof:
6888	8708.9310	---税号87.01所列车辆用	6					14	---Of the vehicles of heading No.87.01
6889	8708.9320	---子目号8702.1091及8702.9010所列车辆用	12.9					70	---Of the vehicles of subheading No.8702.1091 or 8702.9010
6890	8708.9330	---子目号8704.1030及8704.1090所列车辆用	6					11	---Of the vehicles of subheading No.8704.1030 or 8704.1090
6891	8708.9340	---子目号8704.2100、8704.2230、8704.3100及8704.3230所列车辆用	12.9					45	---Of the vehicles of subheading No.8704.2100,8704.2230,8704.3100 or 8704.3230
6892	8708.9350	---子目号8704.2240、8704.2300及8704.3240所列车辆用	10					30	---Of the vehicles of subheading No.8704.2240, 8704. 2300 or 8704.3240
6893	8708.9360	---税号87.05所列车辆用	10					100	---Of the vehicles of heading No.87.05
6894	8708.9390	---其他	13.5					100	---Other
		--转向盘、转向柱及转向器:							--Steering wheels, steering columns andsteering boxes:
6895	8708.9410	---税号87.01所列车辆用	6					14	---Of the vehicles of heading No.87.01
6896	8708.9420	---子目号8702.1091及8702.9010所列车辆用	12.9					70	---Of the vehicles of subheading No.8702.1091 or 8702.9010
6897	8708.9430	---子目号8704.1030及8704.1090所列车辆用	6					11	---Of the vehicles of subheading No.8704.1030 or 8704.1090
6898	8708.9440	---子目号8704.2100、8704.2230、8704.3100及8704.3230所列车辆用	12.9					45	---Of the vehicles of subheading No.8704.2100,8704.2230,8704.3100 or 8704.3230

序号 No.	税则号列 Tariff Item	货品名称	最惠国税率(%) M.F.N.	协定税率(%)Agreement 东盟 ASEAN	曼谷协定 BA	香港 H.K.	澳门 Macao	普通税率(%) Gen.	Article Description
6899	8708.9450	---子目号 8704.2240、8704.2300 及 8704.3240 所列车辆用	10					30	---Of the vehicles of subheading No.8704.2240, 8704.2300 or 8704.3240
6900	8708.9460	---税号 87.05 所列车辆用转向器	10					100	---Of the vehicles of heading No.87.05
6901	8708.9490	---其他	10					100	---Other
		--其他：							--Other:
6902	8708.9910	---税号 87.01 所列车辆用	6					14	---Of the vehicles of heading No.87.01
		---子目号 8702.1091 及 8702.9010 所列车辆用：							---Of the vehicles of subheading No.8702.1091 or 8702.9010:
6903	8708.9921	----车架	25					70	----Frames
6904	8708.9929	----其它	25					70	----Other
		---子目号 8704.1030 及 8704.1090 所列车辆用：							---Of the vehicles of subheading No.8704.1030 or 8704.1090:2
6905	8708.9931	----车架	6					11	----Frames
6906	8708.9939	----其它	6					11	----Other
		---子目号 8704.2100、8704.2230、8704.3100 及 8704.3230 所列车辆用：							---Of the vehicles of subheading No.8704.2100, 8704.2230, 8704.3100 or 8704.3230:
6907	8708.9941	----车架	25					45	----Frames
6908	8708.9949	----其它	25					45	----Other
		---子目号 8704.2240、8704.2300 及 8704.3240 所列车辆用：							---Of the vehicles of subheading No.8704.2240, 8704.2300 or 8704.3240:
6909	8708.9951	----车架	10					30	----Frames
6910	8708.9959	----其它	10					30	----Other
6911	8708.9960	---税号 87.05 所列车辆用	15					100	---Of the vehicles of heading No.87.05
		---其他：							---Other:
6912	8708.9991	----车架	13.5					100	----Frames
6913	8708.9999	----其它	13.5					100	----Other
	87.09	**短距离运输货物的机动车辆，未装有提升或搬运设备，用于工厂、仓库、码头或机场；火车站台上用的牵引车；上述车辆的零件：**							**Works trucks, self-propelled, not fitted with lifting or handling equipment, of the type used in factories, warehouses, dock areas or airports for short distance transport of goods; tractors of the type used on railway station platforms; parts of the foregoing vehicles:**

序号 No.	税则号列 Tariff Item	货品名称	最惠国税率(%) M.F.N.	协定税率(%) Agreement 东盟 ASEAN	曼谷协定 BA	香港 H.K.	澳门 Macao	普通税率(%) Gen.	Article Description
		-车辆:							-Vehicles:
		--电动的:							--Electrical:
6914	8709.1110	---牵引车	10					30	---Tractors
6915	8709.1190	---其他	10					30	---Other
		--其他:							--Other:
6916	8709.1910	---牵引车	10.5					30	---Tractors
6917	8709.1990	---其他	10.5					30	---Other
6918	8709.9000	-零件	8.4					17	-Parts
	87.10	**坦克及其他机动装甲战斗车辆,不论是否装有武器;上述车辆的零件:**							**Tanks and other armoured fighting vehicles, motorized, whether or not fitted with weapons, and parts of such vehicles:**
6919	8710.0010	---整车	15					100	---Assembled
6920	8710.0090	---零件	15					100	---Parts and accessories
	87.11	**摩托车(包括机器脚踏两用车)及装有辅助发动机的脚踏车,不论有无边车;边车:**							**Motorcycles (including mopeds) and cycles fitted with an auxiliary motor, with or without side-cars; side-cars:**
6921	8711.1000	-装有往复式活塞内燃发动机,气缸容量(排气量)不超过50毫升	45					150	-With reciprocating internal combustion piston engine of a cylinder capacity not exceeding 50cc
		-装有往复式活塞内燃发动机,气缸容量(排气量)超过50毫升,但不超过250毫升:							-With reciprocating internal combustion piston engine of a cylinder capacity exceeding 50cc but not exceeding 250cc:
6922	8711.2010	---气缸容量超过50毫升,但不超过100毫升	45					150	---Of a cylinder capacity exceeding 50cc but not exceeding 100cc
6923	8711.2020	---气缸容量超过100毫升,但不超过125毫升	45					150	---Of a cylinder capacity exceeding 100cc but not exceeding 125cc
6924	8711.2030	---气缸容量超过125毫升,但不超过150毫升	45					150	---Of a cylinder capacity exceeding 125cc but not exceeding 150cc
6925	8711.2040	---气缸容量超过150毫升,但不超过200毫升	45					150	---Of a cylinder capacity exceeding 150cc but not exceeding 200cc
6926	8711.2050	---气缸容量超过200毫升,但不超过250毫升	45					150	---Of a cylinder capacity exceeding 200cc but not exceeding 250cc
		-装有往复式活塞内燃发动机,气缸容量(排气量)超过250毫升,但不超过500毫升:							-With reciprocating internal combustion piston engine of a cylinder capacity exceeding 250cc but not exceeding 500cc:

序号 No.	税则号列 Tariff Item	货品名称	最惠国税率(%) M.F.N.	协定税率(%) Agreement 东盟 ASEAN	曼谷协定 BA	香港 H.K.	澳门 Macao	普通税率(%) Gen.	Article Description
6927	8711.3010	---气缸容量超过250毫升,但不超过400毫升	45		32.8			150	---Of a cylinder capacity exceeding 250cc but not exceeding 400cc
6928	8711.3020	---气缸容量超过400毫升,但不超过500毫升	45		32.8			150	---Of a cylinder capacity exceeding 400cc but not exceeding 500cc
6929	8711.4000	-装有往复式活塞内燃发动机,气缸容量(排气量)超过500毫升,但不超过800毫升	40					150	-With reciprocating internal combustion piston engine of a cylinder capacity exceeding 500cc but not exceeding 800cc
6930	8711.5000	-装有往复式活塞内燃发动机,气缸容量(排气量)超过800毫升	30					150	-With reciprocating internal combustion piston engine of a cylinder capacity exceeding 800cc
6931	8711.9000	-其他	45					150	-Other
	87.12	**自行车及其他非机动脚踏车(包括运货三轮脚踏车):**							**Bicycles and other cycles (including delivery tricycles), not motorized:**
6932	8712.0020	---竞赛型自行车	13					130	---Racing bicycle
6933	8712.0030	---山地自行车	13					130	---Mountain bicycle
		---越野自行车:							---Cross-country bicycles:
6934	8712.0041	----16、18及20英寸	13					130	----16″,18″or 20″
6935	8712.0049	----其他	13					130	----Other
		---其他自行车:							---Other cycles:
6936	8712.0081	----16英寸及以下	13					130	----Not larger than 16″
6937	8712.0089	----其他	13					130	----Other
6938	8712.0090	---其他	23					130	---Other
	87.13	**残疾人用车,不论是否机动或其他机械驱动:**							**Carriages for disabled persons, whether or not motorized or otherwise mechanically propelled:**
6939	8713.1000	-非机械驱动	6					20	-Not mechanically propelled
6940	8713.9000	-其他	4					20	-Other
	87.14	**零件、附件、税号87.11至87.13所列车辆用:**							**Parts and accessoris of vehicles of headings Nos.87.11 to 87.13:**
		-摩托车(包括机器脚踏两用车)用:							-Of motorcycles (including mopeds):
6941	8714.1100	--鞍座	30					100	--Saddles
6942	8714.1900	--其他	30					100	--Other
6943	8714.2000	-残疾人车辆用	5					17	-Of carriages for disabled persons
		-其他:							Other:
6944	8714.9100	--车架、轮叉及其零件	12					80	--Frames and forks, and parts thereof
6945	8714.9200	--轮圈及辐条	12					80	--Wheel rims and spokes
		--轮毂(倒轮制动毂及毂闸除外);飞轮、链轮:							--Hubs, other than coaster braking hubs and hub brakes; and free-wheel, sprocket wheels

序号 No.	税则号列 Tariff Item	货品名称	最惠国税率(%) M.F.N.	协定税率(%) Agreement 东盟 ASEAN	曼谷协定 BA	香港 H.K.	澳门 Macao	普通税率(%) Gen.	Article Description
6946	8714.9310	---轮毂	12					80	---Hubs
6947	8714.9320	---飞轮	12					80	---Free wheel
6948	8714.9390	---其他	12					80	---Other
6949	8714.9400	--制动器(包括倒轮制动毂及毂闸)及其零件	12					80	--Brakes, including coaster braking hubs and hub brakes, and parts thereof
6950	8714.9500	--鞍座	12					80	--Saddles
		--脚蹬、曲柄链轮及其零件:							--Pedals and crank-gear, and parts thereof:
6951	8714.9610	---脚蹬及其零件	12					80	---Pedals and parts thereof
6952	8714.9620	---曲柄链轮及其零件	12					80	---Crank-gear and parts thereof
6953	8714.9900	--其他	12					80	--Other
	87.15	**婴孩车及其零件:**							**Baby carriages and parts thereof:**
6954	8715.0000	婴孩车及其零件	20					80	Baby carriages and parts thereof
	87.16	**挂车及半挂车或其他非机械驱动车辆及其零件:**							**Trailers and semi-trailers; other vehicles, not mechanically propelled; parts thereof:**
6955	8716.1000	-供居住或野营用厢式挂车及半挂车	10					35	-Trailers and semi-trailers of the caravan type, for housing or camping
6956	8716.2000	-农用自装或自卸式挂车及半挂车	10					35	-Self-loading or self-unloading trailers and semi-trailers for agricultural purposes
		-其他货运挂车及半挂车:							-Other trailers and semi-trailers for the transport of goods:
		--罐式挂车及半挂车:							--Tanker trailers and tanker semi-trailers:
6957	8716.3110	---油罐挂车及半挂车	10					20	---Oil tanker trailers and semi-trailers
6958	8716.3190	---其他	10					35	---Other
		--其他:							--Other:
6959	8716.3910	---货柜挂车及半挂车	10					20	---Van trailers and semi-trailers
6960	8716.3990	---其他	10					35	---Other
6961	8716.4000	-其他挂车及半挂车	10					35	-Other trailers and semi-trailers
6962	8716.8000	-其他车辆	10					80	-Other vehicles
6963	8716.9000	-零件	10					35	-Parts

第八十八章
航空器、航天器及其零件

Chapter 88
Aircraft, spacecraft and parts thereof

子目注释：

子目号8802.11至8802.40所称“空载重量”，是指航空器在正常飞行状态下，除去机组人员、燃料及非永久性安装设备后的重量。

Note:

For the purposes of subheadings Nos. 8802.11 to 8802.40, the expression "unladen weight" means the weight of the machine in normal flying order, excluding the weight of the crew and of fuel and equipment other than permanently fitted items of equipment.

序号 No.	税则号列 Tariff Item	货品名称	最惠国税率(%) M.F.N.	协定税率(%) Agreement 东盟 ASEAN	曼谷协定 BA	香港 H.K.	澳门 Macao	普通税率(%) Gen.	Article Description
	88.01	**气球及飞艇；滑翔机、悬挂滑翔机及其他无动力航空器：**							**Balloons and dirigibles; gliders, hang gliders and other non-powered aircraft:**
6964	8801.1000	-滑翔机及悬挂滑翔机	3					11	-Gliders and hang gliders
6965	8801.9000	-其他	3					11	-Other
	88.02	**其他航空器(例如，直升机、飞机)；航天器(包括卫星)及其运载工具、亚轨道运载工具：**							**Other aircraft (for example, helicopters, aeroplanes); spacecraft (including satellites) and suborbital and spacecraft launch vehicles:**
		-直升机：							-Helicopters:
6966	8802.1100	--空载重量不超过2000公斤	2					11	--Of an unladen weight not exceeding 2000kg
		--空载重量超过2000公斤：							--Of an unladen weight exceeding 2000kg:
6967	8802.1210	---空载重量超过2000公斤，但不超过7000公斤	2					11	---Of an unladen weight exceeding 2000kg but not exceeding 7000kg
6968	8802.1220	---空载重量超过7000公斤	2					11	---Of an unladen weight exceeding 7000kg
6969	8802.2000	-飞机及其他航空器，空载重量不超过2000公斤	5					11	-Aeroplanes and other aircraft, of an unladen weight not exceeding 2000kg
6970	8802.3000	-飞机及其他航空器，空载重量超过2000公斤，但不超过15000公斤	4					11	-Aeroplanes and other aircraft, of an unladen weight exceeding 2000kg but not exceeding 15000kg
		-飞机及其他航空器，空载重量超过15000公斤：							-Aeroplanes and other aircraft, of an unladen weight exceeding 15000kg:
6971	8802.4010	---空载重量超过15000公斤，但不超过45000公斤	5					11	---Of an unladen weight exceeding 15000kg but not exceeding 45000kg

序号 No.	税则号列 Tariff Item	货品名称	最惠国税率(%) M.F.N.	协定税率(%)Agreement 东盟 ASEAN	曼谷协定 BA	香港 H.K.	澳门 Macao	普通税率(%) Gen.	Article Description
6972	8802.4020	---空载重量超过45000公斤	1					11	---Of an unladen weight exceeding 45000kg
6973	8802.6000	-航天器(包括卫星)及其运载工具、亚轨道运载工具	2					11	-Spacecraft (including satellites) and suborbital and spacecraft launch vehicles
	88.03	**税号88.01或88.02所列货品的零件:**							**Parts of goods of heading No.88.01 or 88.02:**
6974	8803.1000	-推进器、水平旋翼及其零件	1					11	-Propellers and rotors and parts thereof
6975	8803.2000	-起落架及其零件	1					11	-Under-carriages and parts thereof
6976	8803.3000	-飞机及直升机的其他零件	1					11	-Other parts of aeroplanes or helicopters
6977	8803.9000	-其他	0					11	-Other
	88.04	**降落伞(包括可操纵降落伞及滑翔伞)、旋翼降落伞及其零件、附件:**							**Parachutes (including dirigible parachutes and paragliders) and rotochutes; parts thereof and accessories thereto:**
6978	8804.0000	降落伞(包括可操纵降落伞及滑翔伞)、旋翼降落伞及其零件、附件	2					11	Parachutes (including dirigible parachutes and paragliders) and rotochutes; parts thereof and accessories thereto
	88.05	**航空器的发射装置、甲板停机装置或类似装置和地面飞行训练器及其零件:**							**Aircraft launching gear; deck-arrestor or similar gear; ground flying trainers; parts of the foregoing articles:**
6979	8805.1000	-航空器的发射装置及其零件;甲板停机装置或类似装置及其零件	1.5					11	-Aircraft launching gear and parts thereof; deck-arrestor or similar gear and parts thereof
		-地面飞行训练器及其零件							-Ground flying trainers and parts thereof
6980	8805.2100	--空战模拟器及其零件	1.5					11	--Air combat simulators and parts thereof
6981	8805.2900	--其他	1.5					11	--Other

第八十九章
船舶及浮动结构体

Chapter 89
Ships, boats and floating structures

注释：

已装配、未装配或已拆卸的船体、未完工或不完整的船舶以及未装配或已拆卸的完整船舶，如果不具有某种船舶的基本特征，应归入税号89.06。

Notes:

1. A hull, an unfinished or incomplete vessel, assembled, unassembled or disassembled, or a complete vessel unassembled or disassembled, is to be classified in heading No.89.06 if it does not have the essential character of a vessel of a particular kind.

序号 No.	税则号列 Tariff Item	货品名称	最惠国税率(%) M.F.N.	协定税率(%) Agreement 东盟 ASEAN	曼谷协定 BA	香港 H.K.	澳门 Macao	普通税率(%) Gen.	Article Description
	89.01	**巡航船、游览船、渡船、货船、驳船及类似的客运或货运船舶：**							**Cruise ships, excursion boats, ferry-boats, cargo ships, barges and similar vessels for the transport of persons or goods:**
		-巡航船、游览船及主要用于客运的类似船舶；各式渡船：							-Cruise ships, excursion boats and similar vessels principally designed for the transport of persons; ferry-boats of all kinds:
6982	8901.1010	---机动船舶	5					14	---Motor vessels
6983	8901.1090	---非机动船舶	8					30	---Other
		-液货船：							-Tankers:
		---成品油船：							---Finished oil tankers:
6984	8901.2011	----载重量不超过10万吨	9					14	----Loading not exceeding 100000t
6985	8901.2012	----载重量超过10万吨，但不超过30万吨	9					14	----Loading exceeding 100000t, but not exceeding 300000t
6986	8901.2013	----载重量超过30万吨	6					14	----Loading exceeding 300000t
		---原油船：							---Crude oil tankers
6987	8901.2021	----载重量不超过15万吨	9					14	----Loading not exceeding 150000t
6988	8901.2022	----载重量超过15万吨，但不超过30万吨	9					14	----Loading exceeding 150000t, but not exceeding 300000t
6989	8901.2023	----载重量超过30万吨	6					14	----Loading exceeding 300000t
		---液化石油气船：							---Liquified petroleum gas carriers:
6990	8901.2031	----容积在20000立方米及以下	9					14	----Volume with 20000 m^3 or less
6991	8901.2032	----容积在20000立方米以上	6					14	----Volume more than 20000 m^3
		---液化天然气船：							---Liquified natural gas carriers:
6992	8901.2041	----容积在20000立方米及以下	9					14	----Volume with 20000 m^3 or less

序号 No.	税则号列 Tariff Item	货品名称	最惠国税率(%) M.F.N.	协定税率(%) Agreement 东盟 ASEAN	曼谷协定 BA	香港 H.K.	澳门 Macao	普通税率(%) Gen.	Article Description
6993	8901.2042	----容积在20000立方米以上	6					14	----Volume more than 20000 m^3
6994	8901.2090	---其他	9					14	---Other
6995	8901.3000	-冷藏船，但子目号8901.20的船舶除外	9					14	-Refrigerated vessels, other than those of subheading No.8901.20
		-其他货运船舶及其他客货兼运船舶：							-Other vessels for the transport of goods and other vessels for the transport of both persons and goods:
		---机动集装箱船：							---Motor container vessels:
6996	8901.9021	----可载标准集装箱在6000箱及以下	9					14	----Capable loading standard containers with 6000 or less
6997	8901.9022	----可载标准集装箱在6000箱以上	6					14	----Capable loading standard containers more than 6000
		---机动滚装船：							---Motor Ro-Ro carriers:
6998	8901.9031	----载重量在2万吨及以下	9					14	----Loading with 2000t or less
6999	8901.9032	----载重量在2万吨以上	6					14	----Loading more than 2000t
		---机动散货船：							---Motor bulk carriers:
7000	8901.9041	----载重量不超过15万吨	9					14	----Loading not exceeding 150000t
7001	8901.9042	----载重量超过15万吨，但不超过30万吨	9					14	----Loading exceeding 150000t, not exceeding 300000t
7002	8901.9043	----载重量超过30万吨	9					14	----Loading exceeding 300000t
7003	8901.9050	---机动多用途船	9					14	---Multi-purposes motor vessels
7004	8901.9080	---其他，机动的	9					14	---Other motor vessels
7005	8901.9090	---非机动的	8					30	---Other non-motor vessels
	89.02	**捕鱼船；加工船及其他加工保藏鱼类产品的船舶：**							**Fishing vessels; factory ships and other vessels for processing or preserving fishery products:**
7006	8902.0010	---机动船舶	7					14	---Motor vessels
7007	8902.0090	---非机动船舶	8					30	---Other
	89.03	**娱乐或运动用快艇及其他船舶；划艇及轻舟：**							**Yachts and other vessels for pleasure or sports; rowing boats and canoes:**
7008	8903.1000	-充气的	10					30	-Inflatable
		-其他：							-Other:
7009	8903.9100	--帆船，不论是否装有辅助发动机	8					30	--Sailboats, with or without auxiliary motor
7010	8903.9200	--汽艇，但装有舷外发动机的除外	10.5					30	--Motorboats, other than outboard motorboats
7011	8903.9900	--其他	10					30	--Other
	89.04	**拖轮及顶推船：**							**Tugs and pusher craft:**
7012	8904.0000	拖轮及顶推船	9					14	Tugs and pusher craft

序号 No.	税则号列 Tariff Item	货品名称	最惠国税率(%) M.F.N.	协定税率(%) Agreement 东盟 ASEAN	曼谷协定 BA	香港 H.K.	澳门 Macao	普通税率(%) Gen.	Article Description
	89.05	**灯船、消防船、挖泥船、起重船及其他不以航行为主要功能的船舶;浮船坞;浮动或潜水式钻探或生产平台:**							**Light-vessels, fire-floats, dredgers, floating cranes, and other vessels the navigability of which is subsidiary to their main function; floating docks; floating or submersible drilling or production platforms:**
7013	8905.1000	-挖泥船	3					11	-Dredgers
7014	8905.2000	-浮动或潜水式钻探或生产平台	6					11	-Floating or submersible drilling or production platforms
		-其他:							-Other:
7015	8905.9010	---浮船坞	8					30	---Floating docks
7016	8905.9090	---其他	3					11	---Other
	89.06	**其他船舶,包括军舰及救生船,但划艇除外:**							**Other vessels, including warships and lifeboats other than rowing boats:**
7017	8906.1000	-军舰	5					14	-Warships
		-其他:							-Other:
7018	8906.9010	---机动船舶	5					14	---Motor vessels
7019	8906.9090	---非机动船舶	8					30	---Other
	89.07	**其他浮动结构体(例如,筏、柜、潜水箱、浮码头、浮筒及航标):**							**Other floating structures (for example, rafts, tanks, coffer-dams, landing-stages, buoys and beacons):**
7020	8907.1000	-充气筏	8					30	-Inflatable rafts
7021	8907.9000	-其他	8					30	-Other
	89.08	**供拆卸的船舶及其他浮动结构体:**							**Vessels and other floating structures for breaking up:**
7022	8908.0000	供拆卸的船舶及其他浮动结构体	3					11	Vessels and other floating structures for breaking up

第十八类

光学、照相、电影、计量、检验、医疗或外科用仪器及设备、精密仪器及设备；钟表；乐器；上述物品的零件、附件

SECTION XVIII

OPTICAL, PHOTOGRAPHIC, CINEMATOGRAPHIC, MEASURING, CHECKING, PRECISION, MEDICAL OR SURGICAL INSTRUMENTS AND APPARATUS; CLOCKS AND WATCHES; MUSICAL INSTRUMENTS; PARTS AND ACCESSORIES THEREOF

第九十章
光学、照相、电影、计量、检验、医疗或外科用仪器及设备、精密仪器及设备；上述物品的零件、附件

Chapter 90
Optical, photographic, cinematographic, measuring, checking, precision, medical or surgical instruments and apparatus; parts and accessories thereof

注释：

一、本章不包括：

（一）机器、设备或其他专门技术用途的硫化橡胶（硬质橡胶除外）制品（税号40.16）、皮革或再生皮革制品（税号42.04）或纺织材料制品（税号59.11）；

（二）纺织材料制的承托带及其他承托物品，其承托器官的作用仅依靠自身的弹性（例如，孕妇用的承托带，用于胸部、腹部、关节或肌肉的承托绷带）（第十一类）；

（三）税号69.03的耐火材料制品；税号69.09的实验室、化学或其他专门技术用途的陶瓷器；

（四）税号70.09的未经光学加工的玻璃镜及税号83.06或第七十一章的非光学元件的贱金属或贵金属制的镜子；

（五）税号70.07、70.08、70.11、70.14、70.15或70.17的货品；

（六）第十五类注释二所规定的贱金属制通用零件（第十五类）或塑料制的类似品（第三十九章）；

Notes:

1. This Chapter does not cover:

(a) Articles of a kind used in machines, appliances or for other technical uses, of vulcanized rubber other than hard rubber (heading No.40.16), of leather or of composition leather (heading No.42.04) or of textile material (heading No.59.11);

(b) Supporting belts or other support articles of textile material, whose intended effect on the organ to besupported or held derives solely from their elasticity (for example, maternity belts, thoracic support bandages, abdominal support bandages, supports for joints or muscles) (Section XI);

(c) Refractory goods of heading No.69.03; ceramic wares for laboratory, chemical or other technical uses, of heading No.69.09;

(d) Glass mirrors, not optically worked, of heading No.70.09, or mirrors of base metal or of precious metal, not being optical elements (heading No.83.06 or Chapter 71);

(e) Goods of heading No.70.07, 70.08, 70.11, 70.14, 70.15 or 70.17;

(f) Parts of general use, as defined in Note 2 to Section XV, of base metal (Section XV) or similar goods of plastics (Chapter 39);

（七）税号 84.13 的装有计量装置的泵；计数和检验用的衡器或单独进口或出口的天平砝码（税号 84.23）；升降、起重及搬运机械（税号 84.25 至 84.28）；纸张或纸板的各种切割机器（税号 84.41）；税号 84.66 的用于机床上调整工件或工具的附件，包括具有读度用的光学装置的附件（例如，“光学”分度头），但其本身主要是光学仪器的除外（例如校直望远镜）；计算机器（税号 84.70）；税号 84.81 的阀门及其他装置；

（八）自行车或机动车辆用探照灯或聚光灯（税号 85.12）；税号 85.13 的手提式电灯；电影录音机、还音机及转录机（税号 85.19 或 85.20）；拾音头或录音头（税号 85.22）；静像视频摄像机、其他视频摄录一体机和数字照相机（税目 85.25）；雷达设备、无线电导航设备或无线电遥控设备（税目 85.26）；税目 85.37 的数控装置税号 85.39 的封闭式聚光灯；税号 85.44 的光缆；

（九）税号 94.05 的探照灯及聚光灯；

（十）第九十五章的物品；

（十一）容量的计量器具（按其构成的材料归类）；

（十二）卷轴、线轴及类似芯子（按其构成材料归类，例如，归入税号 39.23 或第十五类）。

二、除上述注释一另有规定的以外，本章各税号所列机器、设备、仪器或器具的零件、附件，应按下列规定归类：

（一）凡零件、附件本身已构成本章或第八十四章、第八十五章或第九十一章各税号（税号 84.85、85.48 或 90.33 除外）所包括的货品，应一律归入其相应的税号；

（二）其他零件、附件，如果专用于或主要用于某种或同一税号项下的多种机器、仪器或器具（包括税号 90.10、90.13 或 90.31 的机器、仪器或器具），应归入相应机器、仪器或器具的税号；

(g) Pumps incorporating measuring devices, of heading No.84.13; weight-operated counting or checking machinery, or separately presented weights for balances (heading No.84.23); lifting or handling machinery (headings Nos.84.25 to 84.28); paper or paperboard cutting machines of all kinds (heading No.84.41); fittings for adjusting work or tools on machine-tools, of heading No.84.66, including fittings with optical devices for reading the scale (for example, "optical" dividing heads) but not those which are in themselves essentially optical instruments (for example, alignment telescopes); calculating machines (heading No.84.70); valves or other appliances of heading No.84.81;

(h) Searchlights or spotlights of a kind used for cycles or motor vehicles (heading No.85.12); portable electric lamps of heading No.85.13; cinematographic sound recording, reproducing or re-recording apparatus (heading No.85.19 or 85.20); sound-heads (heading No.85.22); image video cameras, other video camera recorders and digital cameras (heading 85.25); radar apparatus, radio navigational aid apparatus or radio remote control apparatus (heading 85.26); numerical control apparatus of heading 85.37; sealed beam lamp units of heading No.85.39; optical fibre cables of heading No.85.44;

(ij) Searchlights or spotlights of heading No.94.05;

(k) Articles of Chapter 95;

(l) Capacity measures, which are to be classified according to their constituent material; or

(m) Spools, reels or similar supports (which are to be classified according to their constituent material, for example, in heading No.39.23 or Section XV).

2. Subject to Note 1 above, parts and accessories for machines, apparatus, instruments or articles of this Chapter are to be classified according to the following rules:

(a) Parts and accessories which are goods included in any of the headings of this Chapter or of Chapter 84, 85 or 91 (other than heading No.84.85, 85.48 or 90.33) are in all cases to be classified in their respective headings;

(b) Other parts and accessories, if suitable for use solely or principally with a particular kind of machine, instrument or apparatus, or with a number of machines, instruments or apparatus of the same heading (including a machine, instrument or apparatus of heading No.90.10, 90.13 or 90.31) are to be classified with the machines, instruments or apparatus of that kind;

（三）所有其他零件、附件均应归入税号 90.33。

三、第十六类的注释四也适用于本章。

四、税号 90.05 不包括武器用望远镜瞄准具、潜艇或坦克上的潜望镜式望远镜及本章或第十六类的机器、设备、仪器或器具用的望远镜；这类望远镜瞄准具及望远镜应归入税号 90.13。

五、计量或检验用的光学仪器、器具或机器，如果既可归入税号 90.13，又可归入税号 90.31，则应归入税号 90.31。

六、税目 90.21 所称"矫形器具"，是指下列用途的器具：
预防或矫正人体畸变；
生病、手术或受伤后人体部位支撑或固定；
矫形器具包括用于矫正畸形的鞋及特种鞋垫，但需符合下列任一条件：
（一）定制的
（二）成批生产的、单独报验、且不成双的、设计为左右两脚同样适用。

七、税号 90.32 仅适用于：
（一）液体或气体的流量、液位、压力或其他变化量的自动控制仪器及装置或温度自动控制装置，不论其是否依靠要被自控的因素所发生的电现象来进行工作，这些仪器或装置将被自控因素调到并保持在一设定值上，通过持续或定期测量实际值来保持稳定，修正偏差。

（二）电量自动调节器及自动控制非电量的仪器或装置，依靠要被控制的因素所发生的电现象来进行工作，这些仪器或装置将被控制的因素调到并保持在一设定值上，通过持续或定期测量实际来保持稳定，修正偏差。

(c) All other parts and accessories are to be classified in heading No.90.33.

3. The provisions of Note 4 to Section XVI apply also to this Chapter.

4. Heading No.90.05 does not apply to telescopic sights for fitting to arms, periscopic telescopes for fitting to submarines or tanks, or to telescopes for machines, appliances, instruments or apparatus of this Chapter of Section XVI ; such telescopic sights and telescopes are to be classified in heading No.90.13.

5. Measuring or checking optical instruments, appliances or machines which, but for this Note, could be classified both in heading No.90.13 and in heading No.90.31 are to be classified in heading No.90.31.

6. For the purposes of heading 90.21, the expression "orthopaedic appliances" means appliances for:
-Preventing or correcting bodily deformities; or
-Supporting or holding parts of the body following an illness, operation or injury.

Orthopaedic appliances include footwear and special insoles designed to correct orthopaedic conditions, provided that they are either (1) made to measure or (2) mass-produced, presented singly

7. Heading 90.32 applies only to:
(a) Instruments and apparatus for automatically controlling the flow, level, pressure or other variables of liquids or gases, or for automatically controlling temperature, whether or not their operation depends on an electrical phenomenon which varies according to the factor to be automatically controlled, which are designed to bring this factor to, and maintain it at, a desired value, stabilised against disturbances, by constantly or periodically measuring its actual value; and
(b) Automatic regulators of electrical quantities, and instruments or apparatus for automatically controlling non-electrical quantities the operation of which depends on an electrical phenomenon varying according to the factor to be controlled, which are designed to bring this factor to, and maintain it at, a desired value, stabilised against disturbances, by constantly or periodically measuring its actual value.

序号 No.	税则号列 Tariff Item	货品名称	最惠国税率(%) M.F.N.	协定税率(%) Agreement 东盟 ASEAN	曼谷协定 BA	香港 H.K.	澳门 Macao	普通税率(%) Gen.	Article Description
	90.01	**光导纤维及光导纤维束;光缆,但税号85.44的货品除外;偏振材料制的片及板;未装配的各种材料制透镜(包括隐形眼镜片)、棱镜、反射镜及其他光学元件,但未经光学加工的玻璃制上述元件除外:**							**Optical fibres and optical fibre bundles; optical fibre cables, other than those of heading No.85.44; sheets and plates of polarizing material; lenses (including contact lenses), prisms, mirrors and other optical elements, of any material, unmounted, other than such elements of glass not optically worked:**
7023	9001.1000	-光导纤维、光导纤维束及光缆	5			0		20	-Optical fibres, optical fibre bundles and cables
7024	9001.2000	-偏振材料制的片及板	8			0		20	-Sheets and plates of polarizing material
7025	9001.3000	-隐形眼镜片	10			0		70	-Contact lenses
		-玻璃制眼镜片:							-Spectacle lenses of glass:
7026	9001.4010	---变色镜片	20					90	---Photochromic
		---其他:							---Other:
7027	9001.4091	----太阳镜片	20					90	----For sunglasses
7028	9001.4099	----其他	20					70	----Other
		-其他材料制眼镜片:							-Spectacle lenses of other materials:
7029	9001.5010	---变色镜片	20			0		90	---Photochromic
		---其他:							---Other:
7030	9001.5091	----太阳镜片	20			0		90	----For sunglasses
7031	9001.5099	----其他	20			0		70	----Other
7032	9001.9000	-其他	8			0		20	-Other
	90.02	**已装配的各种材料制透镜、棱镜、反射镜及其他光学元件,作为仪器或装置的零件、配件,但未经光学加工的玻璃制上述元件除外:**							**Lenses, prisms, mirrors and other optical elements, of any material, mounted, being parts of or fittings for instruments or apparatus, other than such elements of glass not optically worked:**
		-物镜:							-Objective lenses:
		--照相机、投影仪、照片放大机及缩片机用:							--For cameras, projectors or photographic enlargers or reducers:
7033	9002.1110	---子目号9006.1010至9006.3000所列照相机用	8					14	---For the photographic cameras of subheadings Nos.9006.1010 to 9006.3000
7034	9002.1120	---缩微阅读机用	8					14	---For microfilm, microfiche or other microform readers
7035	9002.1190	---其他	15					80	---Other
		--其他:							--Other:
7036	9002.1910	---摄影机或放映机用	15					40	---For cinematographic cameras or projectors
7037	9002.1990	---其他	15			0		50	---Other
		-滤色镜:							-Filters:

序号 No.	税则号列 Tariff Item	货品名称	最惠国税率(%) M.F.N.	协定税率(%) Agreement 东盟 ASEAN	曼谷协定 BA	香港 H.K.	澳门 Macao	普通税率(%) Gen.	Article Description
7038	9002.2010	---照相机用	15					80	---For cameras
7039	9002.2090	---其他	15					40	---Other
		-其他：							-Other:
7040	9002.9010	---照相机用	15			0		80	---For cameras
7041	9002.9090	---其他	15			0		40	---Other
	90.03	**眼镜架及其零件：**							**Frames and mountings for spectacles, goggles or the like, and parts thereof:**
		-眼镜架：							-Frames and mountings:
7042	9003.1100	--塑料制	18					70	--Of plastics
7043	9003.1900	--其他材料制	10					70	--Of other materials
7044	9003.9000	-零件	10					70	-Parts
	90.04	**矫正视力、保护眼睛或其他用途的眼镜、挡风镜及类似品：**							**Spectacles, goggles and the like, corrective, protective or other:**
7045	9004.1000	-太阳镜	20					100	-Sunglasses
		-其他：							-Other:
7046	9004.9010	---变色镜	16					100	---Photochromic spectacles
7047	9004.9090	---其他	20					90	---Other
	90.05	**双筒望远镜、单筒望远镜、其他光学望远镜及其座架；其他天文仪器及其座架，但不包括射电天文仪器：**							**Binoculars, monoculars, other optical telescopes, and mountings therefor; other astronomical instruments and mountings therefor, but not including instruments for radio-astronomy:**
7048	9005.1000	-双筒望远镜	15					50	-Binoculars
		-其他仪器：							-Other instruments:
7049	9005.8010	---天文望远镜及其他天文仪器	3					8	---Astronomical telescopes and other astronomical instruments
7050	9005.8090	---其他	12					50	---Other
		-零件、附件(包括座架)：							-Parts and accessories(including mountings):
7051	9005.9010	---天文望远镜及其他天文仪器用	2					8	---Of instruments of subheading No.9005.8010
7052	9005.9090	---其他	8					30	---Other
	90.06	**照相机(电影摄影机除外)；照相闪光灯装置及闪光灯泡，但税号85.39的放电灯泡除外：**							**Photographic (other than cinematographic) cameras; photographic flashlight apparatus and flashbulbs, other than discharge lamps of heading No.85.39:**
		-制版照相机：							-Cameras of a kind used for preparing printing plates or cylinders:
7053	9006.1010	---电子分色机	12					20	---Electronic colour scanners
7054	9006.1090	---其他	10					20	---Other
7055	9006.2000	-缩微照相机，使用缩微胶卷、胶片或其他缩微品的	9					17	-Cameras of a kind used for recording documents on microfilm, microfiche or other microforms

序号 No.	税则号列 Tariff Item	货品名称	最惠国税率(%) M.F.N.	协定税率(%) Agreement 东盟 ASEAN	曼谷协定 BA	香港 H.K.	澳门 Macao	普通税率(%) Gen.	Article Description
7056	9006.3000	-水下、航空测量或体内器官检查用的特种照相机;法庭或犯罪学用的比较照相机	9					17	-Cameras specially designed for underwater use, for aerial survey or for medical or surgical examination of internal organs; comparison cameras for forensic or criminological purposes
7057	9006.4000	-一次成像照相机	5					70	-Instant print cameras
		-其他照相机:							-Other cameras:
7058	9006.5100	--通过镜头取景(单镜头反光式(SLR)),使用胶片宽度不超过35毫米	25					100	--With a through-the-lens viewfinder (single lens reflex (SLR)), for roll film of a width not exceeding 35 mm
7059	9006.5200	--其他,使用胶片宽度小于35毫米	25					100	--Other, for roll film of a width less than35 mm
7060	9006.5300	--其他,使用胶片宽度为35毫米	20					100	--Other, for roll film of a width of 35mm
		--其他:							--Other
7061	9006.5910	---激光照相排版设备	9					35	---Laser photo typesetting equipments
7062	9006.5990	---其他	25					100	---Other
		-照相闪光灯装置及闪光灯泡:							-Photographic flashlight apparatus and flashbulbs:
7063	9006.6100	--放电式(电子式)闪光灯装置	18					80	--Discharge lamp ("electronic") flashlight apparatus
7064	9006.6200	--闪光灯泡、方形闪光灯及类似品	18					80	--Flashbulbs, flashcubes and the like
7065	9006.6900	--其他	18					80	--Other
		-零件、附件:							-Parts and accessories:
		--照相机用:							--For cameras:
7066	9006.9110	---子目号9006.1010至9006.3000所列相机用	8			0		17	---For cameras of subheadings Nos.9006.1010 to 9006.3000
7067	9006.9120	---一次成像照相机用	5			0		100	---For instant print cameras
		---其他:							---Other:
7068	9006.9191	----自动调焦组件	10			0		100	----Automatic focal setting units
7069	9006.9192	----快门组件	10			0		100	----Shutter units
7070	9006.9199	----其他	10			0		100	----Other
7071	9006.9900	--其他	12					80	--Other
	90.07	**电影摄影机、放映机,不论是否带有声音的录制或重放装置:**							**Cinematographic cameras and projectors, whether or not incorporating sound recording or reproducing apparatus:**
		-摄影机:							-Cameras:
7072	9007.1100	--胶片宽度小于16毫米或双8毫米	14					40	--For film of less than 16 mm width or for double-8 mm film
		--其他:							--Other:
7073	9007.1910	---高速摄影机	14					40	---High speed cameras
7074	9007.1990	---其他	14					40	---Other
7075	9007.2000	-放映机	14					40	-Projectors
		-零件、附件:							-Parts and accessories:

序号 No.	税则号列 Tariff Item	货品名称	最惠国税率(%) M.F.N.	协定税率(%)Agreement 东盟 ASEAN	曼谷协定 BA	香港 H.K.	澳门 Macao	普通税率(%) Gen.	Article Description
7076	9007.9100	--摄影机用	8.4					40	--For cameras
7077	9007.9200	--放映机用	8.4					40	--For projectors
	90.08	**影像投影仪,但电影用除外;照片(电影片除外)放大机及缩片机:**							**Image projectors, other than cinematographic; photographic (other than cinematographic) enlargers and reducers:**
7078	9008.1000	-幻灯机	14					40	-Slide projectors
7079	9008.2000	-缩微胶卷、缩微胶片或其他缩微品的阅读机,不论是否可以进行复制	10					17	-Microfilm, microfiche or other microform readers, whether or not capable of producing copies
		-其他影像投影仪:							-Other image projectors:
7080	9008.3010	---正射投影仪	18					40	---Orthographical projectors
7081	9008.3090	---其他	18					40	---Other
7082	9008.4000	-照片(电影片除外)放大机及缩片机	20					80	-Photographic (other than cinematographic) enlargers and reducers
		-零件、附件:							-Parts and accessories:
7083	9008.9010	---缩微阅读机用	8					17	---Of microfilm, microfiche or other microform readers
7084	9008.9020	---照片放大机及缩片机用	14					80	---Of photographic enlargers and reducers
7085	9008.9090	---其他	14					40	---Other
	90.09	**装有光学系统的或接触式的感光复印设备及热敏复印设备:**							**Photo-copying apparatus incorporating an optical system or of the contact type and thermo-copying apparatus:**
		-静电感光复印设备:							-Electrostatic photo-copying apparatus:
		--将原件直接复印的(直接法):							--Operating by reproducing the original image directly onto the copy (direct process):
7086	9009.1110	---多色复印的	0					70	---Of colourful photocopying
7087	9009.1190	---其他	0					70	---Other
		--将原件通过中间体转印的(间接法):							--Operating by reproducing the original image via an intermediate onto the copy (indirect process):
7088	9009.1210	---多色复印的	10					70	---Of colourful photocopying
		---其他:							---Other:
7089	9009.1291	----多功能复印一体机	10					70	----Digital multifunctional machines with copying function
7090	9009.1299	----其他	10					70	----Other
		-其他感光复印设备:							-Other photocopying apparatus:
		--带有光学系统的:							--Incorporating an optical system:
7091	9009.2110	---多色复印的	0					70	---Of colourful photocopying

序号 No.	税则号列 Tariff Item	货品名称	最惠国税率(%) M.F.N.	协定税率(%) Agreement 东盟 ASEAN	曼谷协定 BA	香港 H.K.	澳门 Macao	普通税率(%) Gen.	Article Description
7092	9009.2190	---其他	0					70	---Other
		--接触式的：							--Of the contact type:
7093	9009.2210	---多色复印的	20					70	---Of colourful photocopying
7094	9009.2290	---其他	20					70	---Other
		-热敏复印设备：							-Thermo-copying apparatus:
7095	9009.3010	---多色复印的	20					70	---Of colourful photocopying
7096	9009.3090	---其他	20					70	---Other
		-零件、附件：							-Parts and accessories:
7097	9009.9100	--文件自动送入器	0					35	--Automatic document feeders
7098	9009.9200	--送纸器	0					35	--Paper feeders
7099	9009.9300	--分页器	0					35	--Sorters
		--其他							--other:
7100	9009.9910	---有机光导体感光鼓	0					35	---Organic photo－conductor
7101	9009.9990	---其他	0					70	---other
	90.10	**本章其他税号未列名的照相(包括电影)洗印用装置及设备(包括将电路图投影或绘制到感光半导体材料上的装置);负片显示器;银幕及其他投影屏幕：**							**Apparatus and equipment for photographic (including cinematographic) laboratories (including apparatus for the projection or drawing of circuit patterns on sensitized semiconductor materials), not specified or included elsewhere in this Chapter; negatoscopes; projection screens:**
		-照相(包括电影)胶卷或成卷感光纸的自动显影装置及设备或将已冲洗胶卷自动曝光到成卷感光纸上的装置及设备：							-Apparatus and equipment for automatically developing photographic (including cinematographic) film or paper in rolls or for automatically exposing developed film to rolls of photographic paper:
7102	9010.1010	---电影用	14			0		40	---Of a kind used in cinematographic film
7103	9010.1020	---特种照相用	8.4			0		20	---Of a kind used in special photographic film or paper
		---其他：							---Other:
7104	9010.1091	----彩色胶卷用	25			0		100	----For the colour photographic film in rolls
7105	9010.1099	----其他	15			0		100	----Other
		-将电路图投影或绘制到感光半导体材料上的装置：							-Apparatus for the projection or drawing of circuit patterns on sensitized semiconductor materials:
7106	9010.4100	--直接记录到晶片上的装置	0					100	--Direct write-on-wafer apparatus
7107	9010.4200	--分步重复光刻机	0					100	--Step and repeat aligners
7108	9010.4900	--其他	0					100	--Other

序号 No.	税则号列 Tariff Item	货品名称	最惠国税率(%) M.F.N.	协定税率(%) Agreement 东盟 ASEAN	曼谷协定 BA	香港 H.K.	澳门 Macao	普通税率(%) Gen.	Article Description
		-照相(包括电影)洗印用其他装置及设备;负片显示器:							-Other apparatus and equipment for photographic (including cinematographic) laboratories; negatoscope:
7109	9010.5010	---负片显示器	14					50	---Negatoscopes
		---其他:							---Other:
7110	9010.5021	----电影用	14					40	----Of a kind used in cinematographic film
7111	9010.5022	----特种照相用	8.4					20	----Of a kind used in special photographic film or paper
7112	9010.5029	----其他	17					100	----Other
7113	9010.6000	-银幕及其他投影屏幕	14					50	-Projection screens
		-零件、附件:							-Parts and accessories:
7114	9010.9010	---电影用	0					40	---Of a kind used in cinematographic film
7115	9010.9020	---特种照相用	0					20	---Of a kind used in special photographic film or paper
7116	9010.9090	---其他	0					100	---Other
	90.11	**复式光学显微镜,包括用于缩微照相、显微电影摄影及显微投影的:**							**Compound optical microscopes, including those for photomicrography, cinephotomi-crography or microprojection:**
7117	9011.1000	-立体显微镜	0					14	-Stereoscopic microscopes
7118	9011.2000	-缩微照相、显微电影摄影及显微投影用的其他显微镜	0					14	-Other microscopes, for photomicrography, cinephotomicrography or microprojection
7119	9011.8000	-其他显微镜	7					14	-Other microscopes
7120	9011.9000	-零件、附件	0					14	-Parts and accessories
	90.12	**显微镜,但光学显微镜除外;衍射设备:**							**Microscopes other than optical microscopes; diffraction apparatus:**
7121	9012.1000	-显微镜,但光学显微镜除外;衍射设备	0					14	-Microscopes other than optical microscopes; and diffraction apparatus
7122	9012.9000	-零件、附件	0					14	-Parts and accessories
	90.13	**其他税号未列名的液晶装置;激光器,但激光二极管除外;本章其他税号未列名的光学仪器及器具:**							**Liquid crystal devices not constituting articles provided for more specifically in other headings; lasers, other than laser diodes; other optical appliances and instruments, not specified or included elsewhere in this Chapter:**
7123	9013.1000	-武器用望远镜瞄准具;潜望镜式望远镜;作为本章或第十六类的机器、设备、仪器或器具部件的望远镜	8					14	-Telescopic sights for fitting to arms; periscopes; telescopes designed to formparts of machines, appliances, instruments or apparatus of this Chapter or Section XVI

序号 No.	税则号列 Tariff Item	货品名称	最惠国税率(%) M.F.N.	协定税率(%) Agreement 东盟 ASEAN	曼谷协定 BA	香港 H.K.	澳门 Macao	普通税率(%) Gen.	Article Description
7124	9013.2000	-激光器,但激光二极管除外	6				0	11	-Lasers, other than laser diodes
		-其他装置、仪器及器具:							-Other devices, appliances and instruments:
7125	9013.8010	---放大镜	12			0	0	50	---Hand magnifying glasses
7126	9013.8020	---光学门眼	12					50	---"Door eyes"
7127	9013.8030	---液晶显示板	5			0	0	50	---Liquid crystal display panel
7128	9013.8090	---其他	5			0	0	17	---Other
		-零件、附件:							-Parts and accessories:
7129	9013.9010	---子目录9013.1000及9013.2000所列货品用	6			0	0	11	---For goods of subheading No.9013.1000 or 9013.2000
7130	9013.9090	---其他	8			0	0	17	---Other
	90.14	**定向罗盘;其他导航仪器及装置:**							**Direction finding compasses; other navigational instruments and appliances:**
7131	9014.1000	-定向罗盘	2					8	-Direction finding compasses
7132	9014.2000	-航空或航天导航仪器及装置(罗盘除外)	2					8	-Instruments and appliances for aeronautical or space navigation (other than compasses)
7133	9014.8000	-其他仪器及装置	2					8	-Other instruments and appliances
7134	9014.9000	-零件、附件	1.5					8	-Parts and accessories
	90.15	**大地测量(包括摄影测量)、水道测量、海洋、水文、气象或地球物理用仪器及装置,不包括罗盘;测距仪:**							**Surveying (including photogrammetrical surveying), hydrographic, oceanographic, hydrological, meteorological or geophysical instruments and appliances, excluding compasses; rangefinders:**
7135	9015.1000	-测距仪	9					14	-Rangefinders
7136	9015.2000	-经纬仪及视距仪	9					14	-Theodolites and tachymeters (tacheometers)
7137	9015.3000	-水平仪	9					14	-Levels
7138	9015.4000	-摄影测量用仪器及装置	9					14	-Photogrammetrical surveying instruments and appliances
7139	9015.8000	-其他仪器及装置	5					14	-Other instruments and appliances
7140	9015.9000	-零件、附件	5					14	-Parts and accessories
	90.16	**感量为50毫克或更精密的天平,不论是否带有砝码:**							**Balances of a sensitivity of 50 mg or better, with or without weights:**
7141	9016.0010	---感量为0.1毫克或更精密的天平	9					14	---Of a sensitivity of 0.1 mg or better
7142	9016.0090	---其他	10.5					30	---Other

序号 No.	税则号列 Tariff Item	货品名称	最惠国税率(%) M.F.N.	协定税率(%)Agreement 东盟 ASEAN	曼谷协定 BA	香港 H.K.	澳门 Macao	普通税率(%) Gen.	Article Description
	90.17	**绘图、划线或数学计算仪器及器具(例如,绘图机、比例缩放仪、分度规、绘图工具、计算尺及盘式计算器);本章其他税号未列名的手用测量长度的器具(例如,量尺、量带、千分尺及卡尺):**							**Drawing, marking-out or mathematicalcalculating instruments (for example, drafting machines, pantographs, protractors, drawing sets, slide rules, disc calculators); instruments for measuring length, for use in the hand (for example, measuring rods and tapes, micrometers, callipers), not specified or included else-where in this Chapter:**
7143	9017.1000	-绘图台及绘图机,不论是否自动	8					20	-Drafting tables and machines, whether or not automatic
7144	9017.2000	-其他绘图、划线或数学计算器具	0					70	-Other drawing, marking-out or mathematical calculating instruments
7145	9017.3000	-千分尺、卡尺及量规	8					20	-Micrometers, callipers and gauges
7146	9017.8000	-其他仪器及器具	8					20	-Other instruments
7147	9017.9000	-零件、附件	0					20	-Parts and accessories
	90.18	**医疗、外科、牙科或兽医用仪器及器具,包括闪烁扫描装置、其他电气医疗装置及视力检查仪器:**							**Instruments and appliances used in medical, surgical, dental or veterinary sciences, including scintigraphic apparatus, other electro-medical apparatus and sight-testing instruments:**
		-电气诊断装置(包括功能检查或生理参数检查用装置):							-Electro-diagnostic apparatus (including apparatus for functional exploratory examination or for checking physiological parameters):
7148	9018.1100	--心电图记录仪	5					17	--Electro-cardiographs
		--超声波扫描装置:							--Ultrasonic scanning apparatus:
7149	9018.1210	---B型超声波诊断仪	7		6			35	---B-ultrasonic diagnostic equipment
		---其他:							---Other:
7150	9018.1291	----彩色超声波诊断仪	5					17	----Chromoscope ultrasonic diagnostic equipment
7151	9018.1299	----其他	5					17	----Other
7152	9018.1300	--核磁共振成像装置	4					17	--Magnetic resonance imaging apparatus
7153	9018.1400	--闪烁摄影装置	5					17	--Scintigraphic apparatus
		--其他:							--Other:
7154	9018.1930	---病员监护仪	4					17	---Patient monitors
7155	9018.1990	---其他	4					17	---Other
7156	9018.2000	-紫外线及红外线装置	4					17	-Ultra-violet or infra-red ray apparatus
		-注射器、针、导管、插管及类似品:							-Syringes, needles, catheters, cannulae and the like:
7157	9018.3100	--注射器,不论是否装有针头	8					50	--Syringes, with or without needles

序号 No.	税则号列 Tariff Item	货品名称	最惠国税率(%) M.F.N.	协定税率(%) Agreement 东盟 ASEAN	曼谷协定 BA	香港 H.K.	澳门 Macao	普通税率(%) Gen.	Article Description
		--管状金属针头及缝合用针:							--Tubular metal needles and needles for sutures:
7158	9018.3210	---管状金属针头	8		7			50	---Tubular metal needles
7159	9018.3220	---缝合用针	4					17	---Needles for sutures
7160	9018.3900	--其他	4					17	--Other
		-牙科用其他仪器及器具:							-Other instruments and appliances, used in dental sciences:
7161	9018.4100	--牙钻机,不论是否与其他牙科设备组装在同一底座上	4					17	--Dental drill engines, whether or not combined on a single base with other dental equipment
		--其他:							--Other:
7162	9018.4910	---装有牙科设备的牙科用椅	4					17	---Dentists'chairs incorporating dental equipment
7163	9018.4990	---其他	4					17	---Other
7164	9018.5000	-眼科用其他仪器及器具	4					17	-Other ophthalmic instruments and appliances
		-其他仪器及器具:							-Other instruments and appliances:
7165	9018.9010	---听诊器	4			0△		17	---Stethoscopes
7166	9018.9020	---血压测量仪器及器具	4			0△		17	---Sphygmomanometers
7167	9018.9030	---内窥镜	4			0△		17	---Endoscopes
7168	9018.9040	---肾脏透析设备(人工肾)	4			0△		17	---Artificial kidney(dialysis)apparatus
7169	9018.9050	---透热疗法设备	4			0△		17	---Diathermy apparatus
7170	9018.9060	---输血设备	4			0△		17	---Blood transfusion apparatus
7171	9018.9070	---麻醉设备	4			0△		17	---Anaesthetic apparatus and instruments
7172	9018.9080	---宫内节育器	4			0△		17	---Intrauterine contraceptive device
7173	9018.9090	---其他	4			0△		17	---Other
	90.19	**机械疗法器具;按摩器具;心理功能测验装置;臭氧治疗器;氧气治疗器、喷雾治疗器、人工呼吸器及其他治疗用呼吸器具:**							**Mechano-therapy appliances; massage ap-paratus; psychological aptitude-testing ap-paratus; ozone therapy, oxygen therapy, aerosol therapy, artificial respiration or other therapeutic respiration apparatus:**
		-机械疗法器具;按摩器具;心理功能测验装置:							-Mechano-therapy appliances; massage ap-paratus; psychological aptitude-testing ap-paratus:
7174	9019.1010	按摩器具	15			0△		40	---Massage apparatus
7175	9019.1090	---其他	4			0△		30	---Other
7176	9019.2000	-臭氧治疗器、氧气治疗器、喷雾治疗器、人工呼吸器及其他治疗用呼吸器具	4					17	-Ozone therapy, oxygen therapy, aerosol therapy, artificial respiration or other therapeutic respiration apparatus

序号 No.	税则号列 Tariff Item	货品名称	最惠国税率(%) M.F.N.	协定税率(%)Agreement 东盟 ASEAN	曼谷协定 BA	香港 H.K.	澳门 Macao	普通税率(%) Gen.	Article Description
	90.20	**其他呼吸器具及防毒面具,但不包括既无机械零件又无可互换过滤器的防护面具:**							**Other breathing appliances and gas masks, excluding protective masks having neither mechanical parts nor replaceable filters:**
7177	9020.0000	其他呼吸器具及防毒面具,但不包括既无机械零件又无可互换过滤器的防护面具	8				0	30	Other breathing appliances and gas masks, excluding protective masks having neither mechanical parts nor replaceable filters
	90.21	**矫形器具,包括支具、外科手术带、疝气带;夹板及其他骨折用具;人造的人体部分;助听器及为弥补生理缺陷或残疾而穿戴、携带或植入人体内的其他器具:**							**Orthopaedic appliances, including crutches, surgical belts and trusses; splints and other fracture appliances; artificial parts of the body; hearing aids and other appliances which are worn or carried, or implanted in the body, to compensate for a defect or disability:**
7178	9021.1000	-矫形或骨折用器具:	4					17	-Orthopaedic or fracture appliances
		-假牙及牙齿固定件:							-Artificial teeth and dental fittings:
7179	9021.2100	--假牙	4					17	--Artificial teeth
7180	9021.2900	--其他	4					17	--Other
		-其他人造的人体部分							-Other artificial parts of the body:
7181	9021.3100	--人造关节	4					17	--Artificial joints
7182	9021.3900	--其他	4					17	--Other
7183	9021.4000	-助听器,不包括零件、附件	4					17	-Hearing aids, excluding parts and accessories
7184	9021.5000	-心脏起搏器,不包括零件、附件	4					17	-Pacemakers for stimulating heart muscles, excluding parts and accessories
7185	9021.9000	-其他	4					17	-Other
	90.22	**X射线或α射线、β射线、γ射线的应用设备,不论是否用于医疗、外科、牙科或兽医,包括射线照相及射线治疗设备,X射线管及其他X射线发生器、高压发生器、控制板及控制台、荧光屏、检查或治疗用的桌、椅及类似品:**							**Apparatus based on the use of X-rays or of alpha, beta or gamma radiations, whether or not for medical, surgical, dental or veterinary uses, including radiography or radiotherapy apparatus, X-ray tubes and other X-ray generators, high tension generators, control panels and desks, screens, examination or treatment tables, chairs and the like:**
		-X射线的应用设备,不论是否用于医疗、外科、牙科或兽医,包括射线照相或射线治疗设备:							-Apparatus based on the use of X-rays, whether or not for medical, surgical, dental or veterinary uses, including radiography or radiotherapy apparatus:

序号 No.	税则号列 Tariff Item	货品名称	最惠国税率(%) M.F.N.	协定税率(%) Agreement 东盟 ASEAN	曼谷协定 BA	香港 H.K.	澳门 Macao	普通税率(%) Gen.	Article Description
7186	9022.1200	--X射线断层检查仪	4					11	--Computed tomography apparatus
7187	9022.1300	--其他,牙科用	4					11	--Other,for dental uses
7188	9022.1400	--其他,医疗、外科或兽医用	4					11	--Other,for medical,surgical or veterinary uses
		--其他:							--For other uses:
7189	9022.1910	---低剂量X射线安全检查设备	4					11	---Low dosage X-ray security inspecting equipment
7190	9022.1990	---其他	4					11	---Other
		-α射线、β射线、γ射线的应用设备,不论是否用于医疗、外科、牙科或兽医,包括射线照相或射线治疗设备:							-Apparatus based on the use of alpha, beta or gamma radiations, whether or not for medical, surgical, dental or veterinary uses, including radiography or radiothe rapy apparatus:
7191	9022.2100	--医疗、外科、牙科或兽医用	4					11	--For medical, surgical, dental or veterinary uses
7192	9022.2900	--其他	6					11	--For other uses
7193	9022.3000	-X射线管	2					11	-X-ray tubes
		-其他,包括零件、附件:							-Other, including parts and accessories:
7194	9022.9010	---X射线影像增强器	6					11	---X-ray intensifiers
7195	9022.9090	---其他	6					11	---Other
	90.23	**专供示范(例如,教学或展览)而无其他用途的仪器、装置及模型:**							**Instruments, apparatus and models, designed for demonstrational purposes (for example, in education or exhibitions), unsuitable for other uses:**
7196	9023.0000	专供示范(例如,教学或展览)而无其他用途的仪器、装置及模型	7					20	Instruments, apparatus and models, designed for demonstrational purposes (for example, in education or exhibitions), unsuitable for other uses
	90.24	**各种材料(例如,金属、木材、纺织材料、纸张、塑料)的硬度、强度、压缩性、弹性或其他机械性能的试验机器及器具:**							**Machines and appliances for testing the hardness, strength, compressibility, elasticity or other mechanical properties of materials (for example, metals, wood, textiles, paper, plastics):**
7197	9024.1000	-金属材料的试验用机器及器具	7					20	-Machines and appliances for testing metals
7198	9024.8000	-其他机器及器具	5					20	-Other machines and appliances
7199	9024.9000	-零件、附件	6					20	-Parts and accessories
	90.25	**记录式或非记录式的液体比重计及类似的浮子式仪器、温度计、高温计、气压计、温度计、干湿球温度计及其组合装置:**							**Hydrometers and similar floating instruments, thermometers and pyrometers, barometers, hygrometers and psychrometers, recording or not, and any combination of these instruments:**

序号 No.	税则号列 Tariff Item	货品名称	最惠国税率(%) M.F.N.	协定税率(%)Agreement 东盟 ASEAN	曼谷协定 BA	香港 H.K.	澳门 Macao	普通税率(%) Gen.	Article Description
		-温度计及高温计,未与其他仪器组合:							-Thermometers and pyrometers, not combined with other instruments:
7200	9025.1100	--液体温度计,可直接读数	4					40	--Liquid-filled,for direct reading
		--其他:							--Other:
7201	9025.1910	---工业用	8.4					20	---For technical use
7202	9025.1990	---其他	8.4					80	---Other
7203	9025.8000	-其他仪器	11					30	-Other instruments
7204	9025.9000	-零件、附件	8					20	-Parts and accessories
	90.26	**液体或气体的流量、液位、压力或其他变化量的测量或检验仪器及装置(例如,流量计、液位计、压力表、热量计),但不包括税号90.14、90.15、90.28或90.32的仪器及装置:**							**Instruments and apparatus for measuring or checking the flow, level, pressure or other variables of liquids or gases (for example, flow meters, level gauges, manometers, heat meters), excluding instruments and apparatus of heading No.90.14,90.15, 90.28 or 90.32:**
7205	9026.1000	-测量、检验液体流量或液位的仪器及装置	0					17	-For measuring or checking the flow or level of liquids
		-测量、检验压力的仪器及装置							-For measuring or checking pressure
7206	9026.2010	---压力/差压变送器	0					17	---Pressure/differential pressure transmitters
7207	9026.2090	---其他	0					17	---Other
7208	9026.8000	-其他仪器及装置	0					17	-Other instruments or apparatus
7209	9026.9000	-零件、附件	0					17	-Parts and accessories
	90.27	**理化分析仪器及装置(例如,偏振仪、折光仪、分光仪、气体或烟雾分析仪);测量或检验粘性、多孔性、膨胀性、表面张力及类似性能的仪器及装置;测量或检验热量、声量或光量的仪器及装置(包括曝光表);检镜切片机:**							**Instruments and apparatus for physical or chemical analysis(for example, polarimeters, refractometers, spectrometers,gas or smoke analysis apparatus);instruments and apparatus for measuring or checking viscosity, porosity,expansion,surface tension or the like; instruments and apparatus for measuring or checking quantities of heat, sound or light(including exposure meters);microtomes:**
7210	9027.1000	-气体或烟雾分析仪	7			0△		17	-Gas or smoke analysis apparatus
7211	9027.2000	-色谱仪及电泳仪	0					17	-Chromatographs and electrophoresis instruments
7212	9027.3000	-使用光学射线(紫外线、可见光、红外线)的分光仪、分光光度计及摄谱仪	0					17	-Spectrometers, spectrophotometers and spectrographs using optical radiations(UV,visible,IR)
7213	9027.4000	-曝光表	14					70	-Exposure meters

序号 No.	税则号列 Tariff Item	货品名称	最惠国税率(%) M.F.N.	协定税率(%) Agreement 东盟 ASEAN	曼谷协定 BA	香港 H.K.	澳门 Macao	普通税率(%) Gen.	Article Description
7214	9027.5000	-使用光学射线(紫外线、可见光、红外线)的其他仪器及装置	0					17	-Other instruments and apparatus using optical radiations(UV,visible,IR)
		-其他仪器及装置:							-Other instruments and apparatus:
		---质谱仪:							---Mass spectrograph:
7215	9027.8011	----集成电路生产用氦质谱检漏台	0					17	----Integrated circuit belium spectra leak detectors
7216	9027.8019	----其他	0					17	----Other
7217	9027.8090	---其他	0					17	---Other
7218	9027.9000	-检镜切片机;零件、附件	0					17	-Microtomes;parts and accessories
	90.28	**生产或供应气体、液体及电力用的计量仪表,包括它们的校准仪表:**							**Gas, liquid or electricity supply or production meters, including calibrating meters therefor:**
		-气量计:							-Gas meters:
7219	9028.1010	---煤气表	10					30	---Coal gas meters
7220	9028.1090	---其他	10					30	---Other
		-液量计:							-Liquid meters:
7221	9028.2010	---水表	10					30	---Water meters
7222	9028.2090	---其他	10					30	---Other
		-电量计:							-Electricity meters:
7223	9028.3010	---电度表	10			0△		30	---Watt-hour meter
7224	9028.3090	---其他	10			0△		30	---Other
		-零件、附件:							-Parts and accessories:
7225	9028.9010	---工业用	8.4					30	---For technical use
7226	9028.9090	---其他	8.4					50	---Other
	90.29	**转数计、产量计数器、车费计、里程计、步数计及类似仪表;速度计及转速表,税号90.14及90.15的仪表除外;频闪观测仪:**							**Revolution counters, production counters, taximeters, mileometers, pedometers and the like;speed indicators and tachometers, other than those of heading No.90.14 or 90.15;stroboscopes:**
		-转数计、产量计数器、车费计、里程计、步数计及类似仪表:							-Revolution counters, production counters, taximeters, mileometers, pedometers and the like:
7227	9029.1010	---转数计	15					50	---Revolution counters
7228	9029.1020	---车费计、里程计	15					35	---Taximeters and mileometers
7229	9029.1090	---其他	15					35	---Other
		-速度计及转速表,频闪观测仪:							-Speed indicators and tachometers; stroboscopes:
7230	9029.2010	---车辆用速度计	10					35	---Speed indicators for motor vehicles

序号 No.	税则号列 Tariff Item	货品名称	最惠国税率(%) M.F.N.	协定税率(%) Agreement 东盟 ASEAN	曼谷协定 BA	香港 H.K.	澳门 Macao	普通税率(%) Gen.	Article Description
7231	9029.2090	---其他	10					35	---Other
7232	9029.9000	-零件、附件	6					35	-Parts and accessories
	90.30	**示波器、频谱分析仪及其他用于电量测量或检验的仪器和装置,但不包括税号90.28的各种仪表;α射线、β射线、γ射线、X射线、宇宙射线或其他离子射线的测量或检验仪器及装置:**							**Oscilloscopes, spectrum analysers and other instruments and apparatus for measuring or checking eclectrical quantities, excluding meters of heading No.90.28; instruments and apparatus for measuring or detecting alpha, beta, gamma, X-ray, cosmic or other ionizing radiations:**
7233	9030.1000	-离子射线的测量或检验仪器及装置	5					20	-Instruments and apparatus for measuring or detecting ionizing radiations
		-阴极射线示波器:							-Cathode-ray oscilloscopes and cathoderay oscillographs:
7234	9030.2010	---测试频率在300兆赫兹以下的通用示波器	10					80	---For general use, of test frequency less than 300 MHz
7235	9030.2090	---其他	5					20	---Other
		-检测电压、电流、电阻或功率的其他仪器及装置,不带记录装置:							-Other instruments and apparatus, for measuring or checking voltage, current, resistance or power, without a recording device:
		--万用表:							--Multimeters:
7236	9030.3110	---量程在五位半及以下的数字万用表	15					130	---Digital, of measuring range of 5 1/2 or less
7237	9030.3190	---其他	5					20	---Other
		--其他:							--Other:
7238	9030.3910	---量程在五位半及以下的数字电流表、电压表	15			0△		130	---Digital ammeters or voltmeters, of measuring rang of 5 1/2 or less
7239	9030.3920	---电阻测试仪	14			0△		80	---Resistance measuring instruments
7240	9030.3990	---其他	9			0△		20	---Other
		-无线电通讯专用的其他仪器及装置(例如,串音测试器、增益测量仪、失真度表、噪声计):							-Other instruments and apparatus, specially designed for telecommunications (for example, cross-talk meters, gain measuring instruments, distortion factor meters, psophometers):
7241	9030.4010	---测试频率在12.4千兆赫兹以下的数字式频率计	0					80	---Digital frequency meters, of test frequency less than 12.4 GHz
7242	9030.4090	---其他	0					20	---Other
		-其他仪器及装置:							-Other instruments and apparatus:
7243	9030.8200	--测试或检验半导体晶片或器件用	0					20	--For measuring or checking semiconductor wafers or devices
		--其他,装有记录装置的:							--Other, with a recording device:

序号 No.	税则号列 Tariff Item	货品名称	最惠国税率(%) M.F.N.	协定税率(%) Agreement 东盟 ASEAN	曼谷协定 BA	香港 H.K.	澳门 Macao	普通税率(%) Gen.	Article Description
7244	9030.8310	---电感及电容测试仪	10					80	---For measuring inductances or capacitances
7245	9030.8390	---其他	8					20	---Other
		--其他:							--Other:
7246	9030.8910	---电感及电容测试仪	14					80	---For measuring inductances or capacitances
7247	9030.8990	---其他	8					20	---Other
7248	9030.9000	-零件、附件	7			0		17	-Parts and accessories
	90.31	**本章其他税号未列名的测量或检验仪器、器具及机器;轮廓投影仪:**							**Measuring or checking instruments, appliances and machines, not specified or included elsewhere in this Chapter; profile projectors:**
7249	9031.1000	-机械零件平衡试验机	7					17	-Machines for balancing mechanical parts
7250	9031.2000	-试验台	7					17	-Test benches
7251	9031.3000	-轮廓投影仪	10					20	-Profile projectors
		-其他光学仪器及器具:							-Other optical instruments and appliances:
7252	9031.4100	--制造半导体器件时检验半导体晶片、器件或检测光掩模或光栅用	0					17	--For inspecting semiconductor wafers or devices or for inspecting photomasks or reticles used in manufacturing semiconductor devices
7253	9031.4900	--其他	0					17	--Other
		-其他仪器、器具及机器:							-Other instruments, appliances and machines:
7254	9031.8010	---光纤通信及光纤性能测试仪	5			0		17	---Optical telecommunication and optical fibre performance testing instruments
7255	9031.8020	---坐标测量仪	5			0		17	---Coordinate measuring machine
7256	9031.8090	---其他	5			0		17	---Other
7257	9031.9000	-零件、附件	0					17	-Parts and accessories
	90.32	**自动调节或控制仪器及装置:**							**Automatic regulating or controlling instruments and apparatus:**
7258	9032.1000	-恒温器	7			0		17	-Thermostats
7259	9032.2000	-恒压器	7					17	-Manostats
		-其他仪器及装置:							-Other instruments and apparatus:
7260	9032.8100	--液压或气压的	7					17	--Hydraulic or pneumatic
7261	9032.8900	--其他	7					17	--Other
7262	9032.9000	-零件、附件	5					17	-Parts and accessories
	90.33	**第九十章所列机器、器具、仪器或装置用的本章其他税号未列名的零件、附件:**							**Parts and accessories (not specified or included elsewhere in this Chapter) for machines, appliances, instruments or apparatus of Chapter 90:**
7263	9033.0000	第九十章所列机器、器具、仪器或装置用的本章其他税号未列名的零件、附件	6					17	Parts and accessories (not specified or in cluded elsewhere in this Chapter) for machines, appliances, instruments or apparatus of Chapter 90

第九十一章
钟表及其零件

注释：

一、本章不包括：

（一）钟表玻璃及钟锤（按其构成材料归类）；

（二）表链（根据不同情况，归入税号 71.13 或 71.17）；

（三）第十五类注释二所规定的贱金属制通用零件（第十五类）、塑料制的类似品（第三十九章）及贵金属或包贵金属制的类似品（一般归入税号 71.15）；但钟、表发条则应作为钟、表的零件归类（税号 91.14）；

（四）轴承滚珠（根据不同情况，归入税号 73.26 或 84.82）；

（五）税号 84.12 的物品，不需擒纵器可以工作的；

（六）滚珠轴承（税号 84.82）；

（七）第八十五章的物品，本身未组装在或未与其他零件组装在钟、表机芯内，也未组装成专用于或主要用于钟、表机芯零件的（第八十五章）。

二、税号 91.01 仅包括表壳完全以贵金属或包贵金属制的表，以及用贵金属或包贵金属与税号 71.01 至 71.04 的天然、养殖珍珠或宝石、半宝石（天然、合成或再造）合制的表。用贱金属上镶嵌贵金属制成表壳的表应归入税号 91.02。

三、本章所称“表芯”，是指由摆轮及游丝、石英晶体或其他能确定时间间隔的装置来进行调节的机构，并带有显示器或可装机械指示器的系统。表芯的厚度不超过 12 毫米，长、宽或直径不超过 50 毫米。

四、除注释一另有规定的以外，钟、表的机芯及其他零件，既适用于钟或表，又适用于其他物品（例如精密仪器）的，均应归入本章。

Chapter 91
Clocks and watches and parts thereof

Notes:

1. This Chapter does not cover:

(a) Clock or watch glasses or weights (classified according to their constituent material);

(b) Watch chains (heading No. 71.13 or 71.17, as the case may be);

(c) Parts of general use defined in Note 2 to Section XV, of base metal (Section XV), or similar goods of plastics (Chapter 39) or of precious metal or metal clad with precious metal (generally heading No. 71.15); clock or watch springs are, however, to be classified as clock or watch parts (heading No. 91.14);

(d) Bearing balls (heading No. 73.26 or 84.82, as the case may be);

(e) Articles of heading No. 84.12 constructed to work without an escapement;

(f) Ball bearings (heading No. 84.82); or

(g) Articles of Chapter 85, not yet assembled together or with other components into watch or clock movements or into articles suitable for use solely or principally as parts of such movements (Chapter 85).

2. Heading No. 91.01 covers only watches with case wholly of precious metal or of metal clad with precious metal, or of the same materials combined with natural or cultured pearls, or precious or semiprecious stones (natural, synthetic or reconstructed) of headings No. 71.01 to 71.04. Watches with case of base metal inlaid with precious metal fall in heading No. 91.02.

3. For the purposes of this Chapter, the expression "watch movements" means devices regulated by a balance-wheel and hairspring, quartz crystal or any other system capable of determining intevals of time, with a display or a system to which a mechanical display can be incorporated. Such watch movements shall not exceed 12 mm in thickness and 50 mm in width, length or diameter.

4. Except as provided in Note 1, movements and other parts suitable for use both in clocks or watches and in other articles (for example, precision instruments) are to be classified in this Chapter.

序号 No.	税则号列 Tariff Item	货品名称	最惠国税率(%) M.F.N.	协定税率(%) Agreement 东盟 ASEAN	曼谷协定 BA	香港 H.K.	澳门 Macao	普通税率(%) Gen.	Article Description
	91.01	**手表、怀表及其他表,包括秒表,表壳用贵金属或包贵金属制成的:**							**Wrist-watches, pocket-watches and other watches, including stop-watches, with case of precious metal or of metal clad with pre-cious metal:**
		-电力驱动的手表,不论是否附有秒表装置:							-Wrist-watches, electrically operated whether or not incorporating a stop-watch facility:
7264	9101.1100	--仅有机械指示器的	11					100	--With mechanical display only
7265	9101.1200	--仅有光电显示器的	16					100	--With opto-electronic display only
7266	9101.1900	--其他	15					100	--Other
		-其他手表,不论是否附有秒表装置:							-Other wrist-watches, whether or not incorporating a stop -watch facility:
7267	9101.2100	--自动上弦的	11					80	--Automatic winding
7268	9101.2900	--其他	15					80	--Other
		-其他:							-Other:
7269	9101.9100	--电力驱动的	15					100	--Electrically operated
7270	9101.9900	--其他	20					80	--Other
	91.02	**手表、怀表及其他表,包括秒表,但税号91.01的货品除外:**							**Wrist-watches, pocket-watches and other watches, including stop-watches, other than those of heading No.91.01:**
		-电力驱动的手表,不论是否附有秒表装置:							-Wrist-watches, electrically operated, whether or not incorporating a stop-watch facility:
7271	9102.1100	--仅有机械指示器的	12.5			0	0	100	--With mechanical display only
7272	9102.1200	--仅有光电显示器的	23			0	0	100	--With opto-electronic display only
7273	9102.1900	--其他	15					100	--Other
		-其他手表,不论是否装有秒表装置:							-Other wrist-watches, whether or not incorporating a stop-watch facility:
7274	9102.2100	--自动上弦的	11					80	--Automatic winding
7275	9102.2900	--其他	15					80	--Other
		-其他:							-Other:
7276	9102.9100	--电力驱动的	15					100	--Electrically operated
7277	9102.9900	--其他	20					80	--Other
	91.03	**以表芯装成的钟,但不包括税号91.04的钟:**							**Clocks with watch movements, excluding clocks of heading No.91.04:**
7278	9103.1000	-电力驱动的	23			0	0	100	-Electrically operated
7279	9103.9000	-其他	20					100	-Other
	91.04	**仪表板钟及车辆、航空器、航大器或船舶用的类似钟:**							**Instrument panel clocks and clocks of a similar type for vehicles ,aircraft, spacecraft or vessels:**
7280	9104.0000	仪表板钟及车辆、航空器、航天器或船舶用的类似钟	10					100	Instrument panel clocks and clocks of a similar type for vehicles, aircraft, space-craft or vessels
	91.05	**其他钟:**							**Other clocks:**

序号 No.	税则号列 Tariff Item	货品名称	最惠国税率(%) M.F.N.	协定税率(%) Agreement 东盟 ASEAN	曼谷协定 BA	香港 H.K.	澳门 Macao	普通税率(%) Gen.	Article Description
		-闹钟:							-Alarm clocks:
7281	9105.1100	--电力驱动的	23			0	0	100	--Electrically operated
7282	9105.1900	--其他	20					100	--Other
		-挂钟:							-Wall clocks:
7283	9105.2100	--电力驱动的	23					100	--Electrically operated
7284	9105.2900	--其他	20					100	--Other
		-其他:							-Other:
		--电力驱动的:							--Electrically operated:
7285	9105.9110	---天文钟	3					8	---Astronomical chronometer
7286	9105.9190	---其他	23					100	---Other
7287	9105.9900	--其他	16					100	--Other
	91.06	**时间记录器以及测量、记录或指示时间间隔的装置,装有钟、表机芯或同步电动机的(例如,考勤钟、时刻记录器):**							**Time of day recording apparatus and apparatus for measuring, recording or otherwise indicating intervals of time, with clock or watch movement or with synchronous motor (for example, time-registers, time-recorders):**
7288	9106.1000	-考勤钟、时刻记录器	16					50	-Time-registers; time-recorders
7289	9106.2000	-停车计时表	16					50	-Parking meters
7290	9106.9000	-其他	16					50	-Other
	91.07	**装有钟、表机芯或同步电动机的定时开关:**							**Time switches with clock or watch movement or with synchronous motor:**
7291	9107.0000	装有钟、表机芯或同步电动机的定时开关	12					50	Time switches with clock or watch movement or with synchronous motor
	91.08	**已组装的完整表芯:**							**Watch movements, complete and assembled:**
		-电力驱动的:							-Electrically operated:
7292	9108.1100	--仅有机械指示器或有可装机械指示器的装置的	16			0	0	80	--With mechanical display only or with a device to which a mechanical display can be incorporated
7293	9108.1200	--仅有光电显示器的	16			0	0	0	--With opto-electronic display only
7294	9108.1900	--其他	16		12	0	0	80	--Other
7295	9108.2000	-自动上弦的	16					80	-Automatic winding
		-其他:							-Other:
7296	9108.9010	---表面尺寸在 33.8 毫米及以下	16					80	---Measuring 33.8 mm or less
7297	9108.9090	---其他	16			0	0	80	---Other
	91.09	**已组装的完整钟芯:**							**Clock movements, complete and assembled:**
		-电力驱动的:							-Eletrically operated:

序号 No.	税则号列 Tariff Item	货品名称	最惠国税率(%) M.F.N.	协定税率(%) Agreement 东盟 ASEAN	曼谷协定 BA	香港 H.K.	澳门 Macao	普通税率(%) Gen.	Article Description
7298	9109.1100	--闹钟芯	16					100	--Of alarm clocks
7299	9109.1900	--其他	16					100	--Other
7300	9109.9000	-其他	16					100	-Other
	91.10	**未组装或部分组装的完整钟、表机芯(机芯套装件);已组装的不完整钟、表机芯;未组装的不完整钟、表机芯:**							**Complete watch or clock movements, unassembled or partly assembled (movement sets); incomplete watch or clock movements, assembled; rough watch or clock movements:**
		-表的:							-Of watches:
7301	9110.1100	--未组装或部分组装的完整机芯(机芯套装件)	16					80	--Complete movements, unassembled or partly assembled (movement sets)
7302	9110.1200	--已组装的不完整机芯	16					70	--Incomplete movements, assembled
7303	9110.1900	--未组装的不完整机芯	16					70	--Rough movements
		-其他:							-Other:
7304	9110.9010	---未组装或部分组装的完整机芯	16					100	---Complete movements, unassembled or partly assembled
7305	9110.9090	---其他	16					80	---Other
	91.11	**表壳及其零件:**							**Watch cases and parts thereof:**
7306	9111.1000	-贵金属表壳或包贵金属表壳	14					80	-Cases of precious metal or of metal clad with precious metal
7307	9111.2000	-贱金属表壳,不论是否镀金或镀银	14		10	0	0	80	-Cases of base metal, whether or not goldor silver-plated
7308	9111.8000	-其他表壳	14					80	-Other cases
7309	9111.9000	-零件	14					80	-Parts
	91.12	**钟壳和本章所列其他货品的类似外壳及其零件:**							**Clock cases and cases of a similar type for other goods of this Chapter, and parts thereof:**
7310	9112.2000	-壳	14					80	-Cases
7311	9112.9000	-零件	12					80	-Parts
	91.13	**表带及其零件:**							**Watch straps, watch bands and watch bracelets, and parts thereof:**
7312	9113.1000	-贵金属或包贵金属制	20					130	-Of precious metal or of metal clad with pecious metal
7313	9113.2000	-贱金属制,不论是否镀金或镀银	14			0	0	100	-Of base metal, whether or not gold-plated or silver-plated
7314	9113.9000	-其他	14					100	-Other
	91.14	**钟、表的其他零件:**							**Other clock or watch parts:**
7315	9114.1000	-发条,包括游丝	14					50	-Springs, including hair-springs
7316	9114.2000	-宝石轴承	14					50	-Jewel bearings
7317	9114.3000	-钟面或表面	14					50	-Dials
7318	9114.4000	-夹板及横担(过桥)	14					50	-Plates and bridges
7319	9114.9000	-其他	14			0	0	70	-Other

第九十二章

乐器及其零件、附件

Chapter 92

Musical instruments; parts and accessories of such articles

注释：

一、本章不包括：

（一）第十五类注释二所规定的贱金属制通用零件（第十五类）或塑料制的类似品（第三十九章）；

（二）第八十五章或第九十章的传声器、扩大器、扬声器、耳机、开关、频闪观测仪及其他附属仪器、器具或设备，虽用于本章物品但未与该物品组成一体或安装在同一机壳内；

（三）玩具乐器或器具（税号 95.03）；

（四）清洁乐器用的刷子（税号 96.03）；

（五）收藏品或古物（税号 97.05 或 97.06）。

二、用于演奏税号 92.02、92.06 所列乐器的弓、槌及类似品，如果与该乐器一同进口或出口，数量合理，用途明确，应归入有关乐器的相应税号。

税号 92.09 的卡片、盘或卷，即使与乐器一同进口或出口，也不视为该乐器的组成部分，而应作为单独进口或出口的物品对待。

Notes:

1. This Chapter does not cover:

(a) Parts of general use, as defined in Note 2 to Section XV, of base metal (Section XV), or similar goods of plastics (Chapter 39);

(b) Microphones, amplifiers, loudspeakers, headphones, switches, stroboscopes or other accessory instruments, apparatus or equipment of Chapter 85 or 90, for use with but not incorporated in or housed in the same cabinet as instruments of this Chapter;

(c) Toy instruments or apparatus (heading No.95.03);

(d) Brushes for cleaning musical instruments (heading No.96.03); or

(e) Collectors' pieces or antiques (heading No.97.05 or 97.06).

2. Bows and sticks and similar devices used in playing the musical instruments of heading No.92.02 or 92.06 presented with such instruments in numbers normal thereto and clearly intended for use therewith, are to be classified in the same heading as the relative instruments.

Cards, discs and rolls of heading No.92.09 presented with an instrument are to be treated as separate articles and not as forming a part of such instrument.

序号 No.	税则号列 Tariff Item	货品名称	最惠国税率(%) M.F.N.	协定税率(%)Agreement 东盟 ASEAN	曼谷协定 BA	香港 H.K.	澳门 Macao	普通税率(%) Gen.	Article Description
	92.01	**钢琴，包括自动钢琴、拨弦古钢琴及其他键盘弦乐器：**							**Pianos, including automatic pianos; harpsichords and other keyboard stringed instruments:**
7320	9201.1000	-竖式钢琴	17.5				0	70	-Upright pianos
7321	9201.2000	-大钢琴	17.5					70	-Grand pianos
7322	9201.9000	-其他	17.5					70	-Other
	92.02	**其他弦乐器（例如，吉他、小提琴、竖琴）：**							**Other string musical instruments (for example, guitars, violins, harps):**
7323	9202.1000	-弓弦乐器	17.5					70	-Played with a bow
7324	9202.9000	-其他	17.5					70	-Other
	92.03	**键盘管风琴；簧风琴及类似的游离金属簧片键盘乐器：**							**Keyboard pipe organs; harmoniums and similar keyboard instruments with free metal reeds:**
7325	9203.0000	键盘管风琴；簧风琴及类似的游离金属簧片键盘乐器	20					80	Keyboard pipe organs; harmoniums and similar keyboard instruments with free metal reeds

序号 No.	税则号列 Tariff Item	货品名称	最惠国税率(%) M.F.N.	协定税率(%) Agreement 东盟 ASEAN	曼谷协定 BA	香港 H.K.	澳门 Macao	普通税率(%) Gen.	Article Description
	92.04	**手风琴及类似乐器;口琴:**							**Accordions and similar instruments; mouth organs:**
7326	9204.1000	-手风琴及类似乐器	21					80	-Accordions and similar instruments
7327	9204.2000	-口琴	21					80	-Mouth organs
	92.05	**其他管乐器(例如,单簧管、小号、风笛):**							**Other wind musical instruments (for example, clarinets, trumpets, bagpipes):**
7328	9205.1000	-铜管乐器	17.5					70	-Brass-wind instruments
7329	9205.9000	-其他	17.5					70	-Other
	92.06	**打击乐器(例如,鼓、木琴、铙、钹、响板、响葫芦):**							**Percussion musical instruments (for example, drums, xylophones, cymbals, castanets, maracas):**
7330	9206.0000	打击乐器(例如,鼓、木琴、铙、钹、响板、响葫芦)	17.5					70	Percussion musical instruments (for example, drums, xylophones, cymbals, castanets, maracas)
	92.07	**通过电产生或扩大声音的乐器(例如,电风琴、电吉他、电手风琴):**							**Musical instruments, the sound of which is produced or must be amplified electrically (for example, organs, guitars, accordions):**
7331	9207.1000	-键盘乐器,但手风琴除外	30					100	-Keyboard instruments, other than accordions
7332	9207.9000	-其他	30					100	-Other
	92.08	**百音盒、游艺场风琴、手摇风琴、机械鸣禽、乐锯及本章其他税号未列名的其他乐器;各种媒诱音响器、哨子、号角、口吹音响信号器:**							**Musical boxes, fairground organs, mechanical street organs, mechanical singing birds, musical saws and other musical instruments not falling within any other heading of this Chapter; decoy calls of all kinds; whistles, call horns and other mouth-blown sound signalling instruments:**
7333	9208.1000	-百音盒	22					80	-Musical boxes
7334	9208.9000	-其他	22					80	-Other
	92.09	**乐器的零件(例如百音盒的机械装置)、附件(例如,机械乐器用的卡片、盘及带卷),节拍器、音叉及各种定音管:**							**Parts (for example, mechanisms for musical boxes) and accessories (for example, cards, discs and rolls for mechanical instruments) of musical instruments; metronomes, tuning forks and pitch pipes of all kinds:**
7335	9209.1000	-节拍器、音叉及定音管	17.5					70	-Metronomes, tuning forks and pitch pipes
7336	9209.2000	-百音盒的机械装置	17.5					70	-Mechanisms for musical boxes
7337	9209.3000	-乐器用的弦	17.5					70	-Musical instrument strings
		-其他:							-Other:
7338	9209.9100	--钢琴的零件、附件	17.5					70	--Parts and accessories for pianos

序号 No.	税则号列 Tariff Item	货品名称	最惠国税率(%) M.F.N.	协定税率(%) Agreement 东盟 ASEAN	曼谷协定 BA	香港 H.K.	澳门 Macao	普通税率(%) Gen.	Article Description
7339	9209.9200	--税号 92.02 所列乐器的零件、附件	17.5					70	--Parts and accessories for the musical instruments of heanding No.92.02
7340	9209.9300	--税号 92.03 所列乐器的零件、附件	17.5					70	--Parts and accessories for the musical instruments of heading No.92.03
7341	9209.9400	--税号 92.07 所列乐器的零件、附件	17.5					70	--Parts and accessories for the musical instruments of heading No.92.07
7342	9209.9900	--其他	17.5					70	--Other

第十九类

武器、弹药及其零件、附件

SECTION XIX

ARMS AND AMMUNITION; PARTS AND ACCESSORIES THEREOF

第九十三章
武器、弹药及其零件、附件

Chapter 93
Arms and ammunition; parts and accessories thereof

注释:

一、本章不包括:

(一) 第三十六章的货品(例如,火帽、雷管、信号弹);

(二) 第十五类注释二所规定的贱金属制通用零件(第十五类)或塑料制的类似品(第三十九章);

(三) 装甲战斗车辆(税号 87.10);

(四) 武器用的望远镜瞄准具及其他光学装置(第九十章),但安装在武器上或与武器一同进口或出口以备安装在该武器上的除外;

(五) 弓、箭、钝头击剑或玩具(第九十五章);

(六) 收藏品及古物(税号 97.05 或 97.06)。

二、税号 93.06 所称“零件”,不包括税号 85.26 的无线电设备及雷达设备。

Notes:

1. This Chapter does not cover:

(a) Goods of Chapter 36 (for example, percussion caps, detonators, signalling flares);

(b) Parts of general use, as defined in Note 2 to Section XV, of base metal (Section XV), or similar goods of plastics (Chapter 39);

(c) Armoured fighting vehicles (heading No.87.10);

(d) Telescopic sights or other optical devices suitable for use with arms, unless mounted on a firearm or presented with the firearm on which they are designed to be mounted (Chapter 90);

(e) Bows, arrows, fencing foils or toys (Chapter 95); or

(f) Collectors' pieces or antiques (heading No.97.05 or 97.06).

2. In heading No.93.06, the reference to "parts thereof" does not include radio or radar apparatus of heading No.85.26.

序号 No.	税则号列 Tariff Item	货品名称	最惠国税率(%) M.F.N.	协定税率(%) Agreement 东盟 ASEAN	曼谷协定 BA	香港 H.K.	澳门 Macao	普通税率(%) Gen.	Article Description
	93.01	**军用武器,但左轮手枪、其他手枪及税目 93.07 的兵器除外:**							**Military weapons, other than revolvers, pistols and the arms of heading 93.07.**
		-火炮武器(例如,榴弹炮及迫击炮):							-Artillery weapons(for example, guns, howitzers and mortars):
7343	9301.1100	--自推进的	13					80	--Self-propelled
7344	9301.1900	--其他	13					80	--Other
7345	9301.2000	-火箭发射装置;火焰喷射器;手榴弹发射器;鱼雷发射管及类似发射装置自推进的	13					80	-Rocket launchers; flame-throwers; grenade launchers; torpedo tubes and similar projectors
7346	9301.9000	-其他	13					80	-Other.

序号 No.	税则号列 Tariff Item	货品名称	最惠国税率(%) M.F.N.	协定税率(%)Agreement 东盟 ASEAN	曼谷协定 BA	香港 H.K.	澳门 Macao	普通税率(%) Gen.	Article Description
7347	9302.0000	左轮手枪及其他手枪,但税号93.03或93.04的货品除外	13					80	Revolvers and pistols, other than those of heading No. 93.03 or 93.04
	93.03	**靠爆炸药发射的其他火器及类似装置(例如,运动用猎枪及步枪、前装枪、维利式信号枪及其他专为发射信号弹的装置、发射空包弹的左轮手枪和其他手枪、弩枪式无痛捕杀器、抛缆枪):**							**Other firearms and similar devices which operate by the firing of an explosive charge (for example, sporting shotguns and rifles, muzzle-loading firearms, Very pistols and other devices designed to project only signal flares, pistols and revolvers for firing blank ammunition, captive-bolt humane killers, linethrowing guns):**
7348	9303.1000	-前装枪	13					80	-Muzzle-loading firearms
7349	9303.2000	-其他运动、狩猎或打靶用猎枪,包括组合式滑膛来复枪	13					80	-Other sporting, hunting or target shooting shotguns, including combination shotgunrifles
7350	9303.3000	-其他运动、狩猎或打靶用步枪	13					80	-Other sporting, hunting or target shooting rifles
7351	9303.9000	-其他	13					80	-Other
	93.04	**其他武器(例如,弹簧枪、气枪、气手枪、警棍),但不包括税号93.07的货品:**							**Other arms (for example, spring, air or gas guns and pistols, truncheons), excluding those of heading No.93.07:**
7352	9304.0000	其他武器(例如,弹簧枪、气枪、气手枪、警棍),但不包括税号93.07的货品	13					80	Other arms(for example, spring, air or gas guns and pistols, truncheons,) excluding those of heading No.93.07
	93.05	**税号93.01至93.04所列物品的零件、附件:**							**Parts and accessories of articles of headings No.93.01 to 93.04:**
7353	9305.1000	-左轮手枪或其他手枪用	13					80	-Of revolvers or pistols
		-税号93.03的猎枪或步枪用:							-Of shotguns or rifles of heading No.93.03:
7354	9305.2100	--猎枪筒	13					80	--Shotgun barrels
7355	9305.2900	--其他	13					80	--Other
		-其他:							-Other:
7356	9305.9100	--税目93.01的军用武器用	13					80	--Of military weapons of heading 93.01
7357	9305.9900	--其他	13					80	--Other

序号 No.	税则号列 Tariff Item	货品名称	最惠国税率(%) M.F.N.	协定税率(%)Agreement 东盟 ASEAN	曼谷协定 BA	香港 H.K.	澳门 Macao	普通税率(%) Gen.	Article Description
	93.06	**炸弹、手榴弹、鱼雷、地雷、水雷、导弹及类似武器及其零件；子弹、其他弹药和射弹及其零件，包括弹丸及弹垫：**							**Bombs, grenades, torpedoes, mines, missiles, and similar munitions of war and parts thereof; cartridges and other ammunition and projectiles and parts thereof, including shot and cartridge wads:**
7358	9306.1000	-铆接机或类似工具用及弩枪式无痛捕杀器用子弹及其零件	13					80	-Cartridges for riveting or similar tools or for captive-bolt humane killers and parts thereof
		-猎枪子弹及其零件；气枪弹丸：							-Shotgun cartridges and parts thereof; air gun pellets:
7359	9306.2100	--猎枪子弹	13					80	--Cartridges
7360	9306.2900	--其他	13					80	--Other
7361	9306.3000	-其他子弹及其零件	13					80	-Other cartridges and parts thereof
7362	9306.9000	-其他	13					80	-Other
	93.07	**剑、短弯刀、刺刀、长矛和类似的武器及其零件；刀鞘、剑鞘：**							**Swords, cutlasses, bayonets, lances and similar arms and parts thereof and scabbards and sheaths therefor:**
7363	9307.0000	剑、短弯刀、刺刀、长矛和类似的武器及其零件；刀鞘、剑鞘	13					80	Swords, cutlasses, bayonets, lances and similar arms and parts thereof and scabbards and sheaths therefor

第二十类

杂 项 制 品

第九十四章
家具；寝具、褥垫、弹簧床垫、软座垫及类似的填充制品；未列名灯具及照明装置；发光标志、发光铭牌及类似品；活动房屋

注释：

一、本章不包括：

（一）第三十九章、第四十章或第六十三章的充气或充水的褥垫、枕头及座垫；

（二）落地镜〔例如税号 70.09 的试衣镜（旋转镜）〕；

（三）第七十一章的物品；

（四）第十五类注释二所规定的贱金属制通用零件（第十五类）、塑料制的类似品（第三十九章）或税号 83.03 的保险箱；

（五）冷藏或冷冻设备专用的特制家具（税号 84.18）；缝纫机专用的特制家具（税号 84.52）；

（六）第八十五章的灯具及照明装置；

（七）税号 85.18、85.19 至 85.21 或税号 85.25 至 85.28 所列装置专用的特制家具（应分别归入税号 85.18、85.22 或 85.29）；

（八）税号 87.14 的物品；

（九）装有税号 90.18 所列牙科用器具或漱口盂的牙科用椅（税号 90.18）；

（十）第九十一章的物品（例如，钟及钟壳）；

（十一）玩具家具、玩具灯或玩具照明装置（税号 95.03）、台球桌或其他供游戏用的特制家具（税号 95.04）、魔术用的特制家具或中国灯笼及类似的装饰品（电气彩灯串除外）（税号 95.05）。

SECTION XX

MISCELLANEOUS MANUFACTURED ARTICLES

Chapter 94
Furniture; bedding, mattresses, mattress supports, cushions and similar stuffed furnishings; lamps and lighting fittings, not elsewhere specified or included; illuminated signs, illuminated name-plates and the like; prefabricated buildings

Notes:

1. This Chapter does not cover:

(a) Pneumatic or water mattresses, pillows or cushions, of Chapter 39, 40 or 63;

(b) Mirrors designed for placing on the floor or ground (for example, cheval-glasses (swing-mirrors) of heading No.70.09);

(c) Articles of Chapter 71;

(d) Parts of general use as defined in Note 2 to Section XV, of base metal (Section XV), or similar goods of plastics (Chapter 39), or safes of heading No.83.03;

(e) Furniture specially designed as parts of refrigerating or freezing equipment of heading No.84.18; furniture specially designed for sewing machines (heading No. 84.52);

(f) Lamps or lighting fittings of Chapter 85;

(g) Furniture specially designed as parts of apparatus of heading No.85.18 (heading No.85.18), of headings No.85.19 to 85.21 (heading No.85.22) or of headings No.85.25 to 85.28 (heading No.85.29);

(h) Articles of heading No.87.14;

(ij) Dentists' chairs incorporating dental appliances of heading No.90.18 or dentists' spittoons (heading No.90.18);

(k) Articles of Chapter 91 (for example, clocks and clock-cases); or

(l) Toy furniture or toy lamps or lighting fittings (heading No.95.03), billiard tables or other furniture specially constructed for games (heading No.95.04), furnitute for conjuring tricks or decorations (other than electric garlands) such as Chinese lanterns (heading No.95.05).

二、税号 94.01 至 94.03 的物品（零件除外），只适用于落地式的物品。

2. The articles (other than parts) referred to in headings No.94.01 to 94.03 are to be classified in those headings only if they are designed for placing on the floor or ground.

对下列物品，即使是悬挂的、固定在墙壁上的或叠摞的，仍归入上述各税号：

（一）碗橱、书柜、架式家具及组合家具；

（二）坐具及床。

The following are, however, to be classified in the abovementioned headings even if they are designed to be hung, to be fixed to the wall or to stand one on the other:

(a) Cupboards, bookcases, other shelved furniture and unit furniture;

(b) Seats and beds.

三、（一）税号 94.01 至 94.03 所列货品的零件，不包括玻璃（包括镜子）、大理石或其他石料以及第六十八章及第六十九章所列任何其他材料的片、块（不论是否切割成形，但未与其他零件组装）。

（二）税号 94.04 的货品，如果单独进口或出口，不能作为税号 94.01、94.02 或 94.03 所列货品的零件归类。

3. (a) In headings No.94.01 to 94.03 references to parts of goods do not include references to sheets or slabs (whether or not cut to shape but not combined with other parts) of glass (including mirrors), marble or other stone or of any other material referred to in Chapter 68 or 69.

(b) Goods described in heading No.94.04, presented separately, are not to be classified in heading No.94.01, 94.02 or 94.03 as parts of goods.

四、税号 94.06 所称“活动房屋”，是指在工厂制成成品或制成部件并一同进口或出口，供以后在有关地点上组装的房屋，例如，工地用房、办公室、学校、店铺、工作棚、车房或类似的建筑物。

4. For the purposes of heading No. 94.06, the expression “prefabricated buildings” means buildings which are finished in the factory or put up as elements, presented together, to be assembled on site, such as housing or worksite accommodation, offices, schools, shops, sheds, garages or similar buildings.

序号 No.	税则号列 Tariff Item	货品名称	最惠国税率(%) M.F.N.	协定税率(%)Agreement 东盟 ASEAN	曼谷协定 BA	香港 H.K.	澳门 Macao	普通税率(%) Gen.	Article Description
	94.01	**坐具（包括能作床用的两用椅，但税号 94.02 的货品除外）及其零件：**							**Seats (other than those of heading No. 94.02), whether or not convertible into beds, and parts thereof:**
7364	9401.1000	-飞机用坐具	0					100	-Seats of a kind used for aircraft
		-机动车辆用坐具：							-Seats of a kind used formotor vehicles:
7365	9401.2010	---皮革或再生皮革面的	10					100	---With outer suvface of leather or composition leather
7366	9401.2090	---其他	10					100	---Other
7367	9401.3000	-可调高度的转动坐具	0			0△		100	-Swivel seats with variable height adjustment
		-能作床用的两用椅，但庭园坐具或野营设备除外：							-Seats other than garden seats or camping equipment, convertible into beds:
7368	9401.4010	---皮革或再生皮革面的	0			0△		100	---With outer suvface of leather or composition leather
7369	9401.4090	---其他	0			0△		100	---Other

序号 No.	税则号列 Tariff Item	货品名称	最惠国税率(%) M.F.N.	协定税率(%) Agreement 东盟 ASEAN	曼谷协定 BA	香港 H.K.	澳门 Macao	普通税率(%) Gen.	Article Description
7370	9401.5000	-藤、柳条、竹及类似材料制的坐具	0	见表1 Table 1				100	-Seats of cane, osier, bamboo or similar materials
		-木框架的其他坐具:							-Other seats, with wooden frames:
		--装软垫的:							--upholstered:
7371	9401.6110	---皮革或再生皮革面的	0			0△		100	---With outer suvface of leather or composition leather
7372	9401.6190	---其他	0			0△		100	---Other
7373	9401.6900	--其他	0			0△		100	--Other
		-金属框架的其他坐具:							-Other seats, with metal frames:
		--装软垫的:							--Upholstered:
7374	9401.7110	---皮革或再生皮革面的	0			0△		100	---With outer suvface of leather or composition leather
7375	9401.7190	---其他	0			0△		100	---Other
7376	9401.7900	--其他	0			0		100	--Other
		-其他坐具:							-Other seats:
7377	9401.8010	---石制的	0			0	0	100	---Of stone
7378	9401.8090	---其他	0			0		100	---Other
		-零件							-Parts
		---机动车辆用:							---Of the motor Vehicles:
7379	9401.9011	----座椅调角器	18.6			0△		100	----Seat angle regulating devices
7380	9401.9019	----其他	0			0△		100	----Other
7381	9401.9090	---其他	0			0△		100	---Other
	94.02	**医疗、外科、牙科或兽医用家具(例如,手术台、检查台、带机械装置的病床、牙科用椅);有旋转、倾斜、升降装置的理发用椅及类似椅;上述物品的零件:**							**Medical, surgical, dental or veterinary furniture (for example, operating tables, examination tables, hospital beds with mechanical fittings, dentists' chairs); barbers' chairs and similar chairs, having rotating as well as both reclining and elevating movements; parts of the foregoing articles:**
		-牙科、理发及类似用途的椅及其零件:							-Dentists', barbers' or similar chairs and parts thereof:
7382	9402.1010	---理发用椅及其零件	0					100	---Barbers' chairs and parts thereof
7383	9402.1090	---其他	0					30	---Other
7384	9402.9000	-其他	0					30	-Other
	94.03	**其他家具及其零件:**							**Other furniture and parts thereof:**
7385	9403.1000	-办公室用金属家具	0			0△		100	-Metal furniture of a kind used in offices
7386	9403.2000	-其他金属家具	0			0△		100	-Other metal furniture
7387	9403.3000	-办公室用木家具	0			0△		100	-Wooden furniture of a kind used in offices
7388	9403.4000	-厨房用木家具	0			0△		100	-Wooden furniture of a kind used in the kitchen
		-卧室用木家具:							-Wooden furniture of a kind used in the bedroom:
7389	9403.5010	---红木制	0			0△		100	---Of rose wood

序号 No.	税则号列 Tariff Item	货品名称	最惠国税率(%) M.F.N.	协定税率(%) Agreement 东盟 ASEAN	曼谷协定 BA	香港 H.K.	澳门 Macao	普通税率(%) Gen.	Article Description
		---其他:							---Other:
7390	9403.5091	----漆木家具	0			0△		100	----Of lacquered wood
7391	9403.5099	----其他	0			0△		100	----Other
		-其他木家具:							-Other wooden furniture:
7392	9403.6010	---红木制	0			0△		100	---Of rose wood
		---其他:							---Other:
7393	9403.6091	----漆木家具	0			0△		100	----Of lacquered wood
7394	9403.6099	----其他	0			0△		100	----Other
7395	9403.7000	-塑料家具	0			0		100	- Furniture of plastics
		-其他材料制的家具,包括藤、柳条、竹或类似材料制的:							-Furniture of other materials, including cane, osier, bamboo or similar materials:
7396	9403.8010	---藤、柳条、竹及类似材料制	0	见表1 Table 1				100	---Of cane, osier, bamboo or similar materials
		---其他:							---Other:
7397	9403.8091	----石制的	0				0	100	----Of stone
7398	9403.8099	----其他	0					100	----Other
7399	9403.9000	-零件	0			0△		100	-Parts
	94.04	**弹簧床垫;寝具及类似用品,装有弹簧、内部用任何材料填充、衬垫或用海绵橡胶、泡沫塑料制成,不论是否包面(例如,褥垫、棉被、羽绒被、靠垫、座垫及枕头):**							**Mattress supports; articles of bedding and similar furnishing (for example, mattresses, quilts, eiderdowns, cushions, pouffes and pillows) fitted with springs or stuffed or internally fitted with any material or of cellular rubber or plastics, whether or not covered:**
7400	9404.1000	-弹簧床垫	20					100	-Mattress supports
		-褥垫:							-Mattresses:
7401	9404.2100	--海绵橡胶或泡沫塑料制,不论是否包面	20			0△		100	--Of cellular rubber or plastics, whether or not covered
7402	9404.2900	--其他材料制	20			0△		100	--Of other materials
		-睡袋:							-Sleeping bags:
7403	9404.3010	---羽毛或羽绒填充的	20					130	---Stuffed with feathers or domn
7404	9404.3090	---其他	20					100	---Other
		-其他:							-Other:
7405	9404.9010	---羽毛或羽绒填充的	20			0△		130	---Stuffed with feathers or down
7406	9404.9020	---兽毛填充的	20			0△		130	---Stuffed with animal hair
7407	9404.9030	---丝棉填充的	20			0△		130	---Stuffed with silk wadding
7408	9404.9040	---化纤棉填充的	20			0△		130	---Stuffed with man – made fibres
7409	9404.9090	---其他	20			0△		100	---Other

序号 No.	税则号列 Tariff Item	货品名称	最惠国税率(%) M.F.N.	协定税率(%) Agreement 东盟 ASEAN	曼谷协定 BA	香港 H.K.	澳门 Macao	普通税率(%) Gen.	Article Description
	94.05	**其他税号未列名的灯具及照明装置,包括探照灯、聚光灯及其零件;装有固定光源的发光标志、发光铭牌及类似品,以及其他税号未列名的这些货品的零件:**							**Lamps and lighting fittings including searchlights and spotlights and parts thereof, not elsewhere specified or included; illuminated signs, illuminated name - plates and the like, having a permanently fixed light source, and parts thereof not elsewhere specified or included:**
7410	9405.1000	-枝形吊灯及天花板或墙壁上的其他电气照明装置,但不包括公共露天场所或街道上的电气照明装置	10					80	-Chandeliers and other electric ceiling or wall lighting fittings, excluding those of a kind used for lighting public open spaces or thorough-fares
7411	9405.2000	-电气的台灯、床头灯或落地灯	20					80	-Electric table, desk , bedeside or floorstanding lamps
7412	9405.3000	-圣诞树用的成套灯具	16					100	-Lighting sets of a kind used for Christmas trees
		-其他电灯及照明装置:							-Other electric lamps and lighting fittings:
7413	9405.4010	---探照灯	17.5				0	70	---Searchlights
7414	9405.4020	---聚光灯	17.5				0	70	---Spotlights
7415	9405.4090	---其他	10				0	80	---Other
7416	9405.5000	-非电气的灯具及照明装置	20					80	-Non - electrical lamps and lighting fittings
7417	9405.6000	-发光标志、发光铭牌及类似品	20					80	-Illuminated signs, illuminated name plates and the like
		-零件:							-Parts:
7418	9405.9100	--玻璃制	20					70	--Of glass
7419	9405.9200	--塑料制	20					70	--Of plastics
7420	9405.9900	--其他	20					70	--Other
	94.06	**活动房屋:**							**Prefabricated buildings:**
7421	9406.0000	活动房屋	10					70	Prefabricated buildings

第九十五章
玩具、游戏品、运动用品及其零件、附件

注释：

一、本章不包括：

(一) 圣诞树蜡烛 (税号 34.06)；

(二) 税号 36.04 的烟花、爆竹或其他烟火制品；

(三) 已切成一定长度但未制成钓鱼线的纱线、单丝、绳、肠线及类似品 (第三十九章、税号 42.06 或第十一类)；

(四) 税号 42.02、43.03 或 43.04 的运动用袋或其他容器；

(五) 第六十一章或第六十二章的纺织品制的运动服或化妆舞会服装；

(六) 第六十三章的纺织品制的旗帜及帆板或滑行车用帆；

(七) 第六十四章的运动鞋靴 (装有冰刀或滑轮的溜冰鞋除外) 或第六十五章的运动用帽；

(八) 手杖、鞭子、马鞭或类似品 (税号 66.02) 及其零件 (税号 66.03)；

(九) 税号 70.18 的未装配的玩偶或其他玩具用的玻璃假眼；

(十) 第十五类注释二所规定的贱金属制通用零件 (第十五类) 或塑料制的类似货品 (第三十九章)；

(十一) 税号 83.06 的铃、钟、锣及类似品；

(十二) 液体泵 (税号 84.13)、液体或气体的过滤、净化机器及装置 (税号 84.21)、电动机 (税号 85.01)、变压器 (税号 85.04) 或无线电遥控设备 (税号 85.26)；

(十三) 第十七类的运动用车辆 (长雪橇、平底雪橇及类似品除外)；

(十四) 儿童两轮车 (税号 87.12)；

(十五) 运动用船艇，例如，轻舟、赛艇 (第八十九章) 及其桨、橹和类似品 (木制的归入第四十四章)；

(十六) 运动及户外游戏用的眼镜、护目镜及类似品 (税号 90.04)；

(十七) 媒诱音响器及哨子 (税号 92.08)；

(十八) 第九十三章的武器及其他物品；

(十九) 各种电气彩灯串 (税号 94.05)；

(二十) 球拍线、帐篷或类似的野营用品、手套、棒球手套和露指手套 (按其构成材料归类)。

Chapter 95
Toys, games and sports requisites; parts and accessories thereof

Notes:

1. This Chapter does not cover:

(a) Christmas tree candles (heading No.34.06);

(b) Fireworks or other pyrotechnic articles of heading No.36.04;

(c) Yarns, monofilament, cords or gut or the like for fishing, cut to length but not made up into fishing lines, of Chapter 39, heading No.42.46 or Section XI;

(d) Sports bags or other containers of heading No.42.02, 43.03 or 43.04;

(e) Sports clothing or fancy dress, of textiles, of Chapter 61 or 62;

(f) Textile flags or bunting , or sails for boats , sailboards or land craft, of Chapter 63 ;

(g) Sports footwear (other than skating boots with ice or roller skates attached) of Chapter 64, or sports headgear of Chapter 65;

(h) Walking-sticks , whips, riding-crops or the like (heading No.66.02), or parts thereof (heading No.66.03);

(ij) Unmounted glass eyes for dolls or other toys , of heading No.70.18;

(k) Parts of general use, as defined in Note 2 to Section XV, of base metal (Section XV), or similar goods of plastics (Chapter 39);

(l) Bells, gongs or the like of heading No.83.06;

(m) Pumps for liquids (heading No.84.13), filtering or purifying machinery and apparatus for liquids or gases (heading No.84.21), electric motors (heading No.85.01), electric transformers (heading No.85.04) or radio remote control apparatus (heading No.85.26);

(n) Sports vehicles (other than bobsleighs, toboggans and the like) of Section XVII;

(o) Children's bicycles (heading No.87.12);

(p) Sports craft such as canoes and skiffs (Chapter 89), or their means of propulsion (Chapter 44 for such articles made of wood);

(q) Spectacles, goggles or the like, for sports or outdoor games (heading No.90.04);

(r) Decoy calls or whistles (heading No.92.08);

(s) Arms or other articles of Chapter 93;

(t) Electric garlands of all kinds (heading No.94.05); or

(u) Racket strings, tents or other camping goods, or gloves, mittens and mitts (classified according to their constituent material).

二、本章包括天然或养殖珍珠、宝石或半宝石（天然、合成或再造）、贵金属或包贵金属只作为小零件的物品。

2. This Chapter includes articles in which natural or cultured pearls, precious or semi-precious stones (natural, synthetic or reconstructed), precious metal or metal clad with precious metal constitute only minor consistuents.

三、除上述注释一另有规定的以外，凡专用于或主要用于本章各税号所列物品的零件、附件，应与有关物品一并归类。

3. Subject to Note 1 above, parts and accessories which are suitable for use solely or principally with articles of this Chapter are to be classified with those articles.

四、税目 95.03 不包括因其设计、形状或构成材料可确认为专供动物使用的物品，例如“宠物玩具”归入其相应的税目。

4. Heading No.95.03 does not cover articles which, on account of their design, shape or constituent material, are identifiable as intended exclusively for animals, e. g., “pet toys” (classification in their own appropriate heading).

序号 No.	税则号列 Tariff Item	货品名称	最惠国税率(%) M.F.N.	协定税率(%) Agreement 东盟 ASEAN	曼谷协定 BA	香港 H.K.	澳门 Macao	普通税率(%) Gen.	Article Description
	95.01	**供儿童乘骑的带轮玩具(例如,三轮车、踏板车、踏板汽车);玩偶车:**							**Wheeled toys designed to be ridden by children (for example, tricycles, scooters, pedal cars); dolls' carriages:**
7422	9501.0000	供儿童乘骑的带轮玩具(例如,三轮车、踏板车、踏板汽车);玩偶车	0					80	Wheeled toys designed to be ridden by children (for example, tricycles, scooters, pedal cars); dolls' carriages
	95.02	**玩偶:**							**Dolls representing only human beings:**
7423	9502.1000	-玩偶,不论是否着装	0					80	-Dolls, whether or not dressed
		-零件、附件:							-Parts and accessories:
7424	9502.9100	--服装及其附件、鞋、靴、帽	0					80	--Garments and accessories therefor, footwear and headgear
7425	9502.9900	--其他	0					80	--Other
	95.03	**其他玩具;缩小(按比例缩小)的模型及类似的娱乐用模型,不论是否活动;各种智力玩具:**							**Other toys; reduced-size ("scale") models and similar recreational models, working or not; puzzles of all kinds:**
7426	9503.1000	-电动火车,包括轨道、信号及其他附件	0					80	-Electric trains, including tracks, signals and other accessories therefor
7427	9503.2000	-缩小(按比例缩小)的全套模型组件,不论是否活动,但子目号 9503.10 的货品除外	0					80	-Reduced-size ("scale") model assembly kits, whether or not working models, excluding those of subheading No.9503.10
7428	9503.3000	-其他建筑套件及建筑玩具	0					80	-Other construction sets and construcional toys
		-玩具动物:							-Toys representing animals or non-human creatures:
7429	9503.4100	--填充的	0					80	--Stuffed

序号 No.	税则号列 Tariff Item	货品名称	最惠国税率(%) M.F.N.	协定税率(%)Agreement 东盟 ASEAN	曼谷协定 BA	香港 H.K.	澳门 Macao	普通税率(%) Gen.	Article Description
7430	9503.4900	--其他	0					80	--Other
7431	9503.5000	-玩具乐器	0					80	-Toy musical instruments and apparatus
7432	9503.6000	-智力玩具	0					80	-Puzzles
7433	9503.7000	-其他玩具,组装成套或全套的	0					80	-Other toys, put up in sets or outfits
7434	9503.8000	-其他带动力装置的玩具及模型	0					80	-Other toys and models, incorporating a motor
7435	9503.9000	-其他	0			0		80	-Other
	95.04	**游艺场所、桌上或室内游戏用品,包括弹球机、台球、娱乐专用桌及保龄球自动球道设备:**							**Articles for funfair, table or parlour games, including pintables, billiards, special tables for casino games and automatic bowling alley equipment:**
7436	9504.1000	-与电视接收机配套使用的电子游戏机	0					130	-Video games of a kind used with a television receiver
7437	9504.2000	-台球用品及附件	0					80	-Articles and accessories for billiards
		-使用硬币、钞票(纸币)、圆形代币及类似品的其他游戏用品,保龄球道设备除外:							-Other games, operated by coins, banknotes(paper currency), discs or other similar articles, other than bowling alley equipment
7438	9504.3010	---电子游戏机	0					130	---Video games
7439	9504.3090	---其他	0					80	---Other
7440	9504.4000	-扑克牌	0					80	-Playing cards
		-其他:							-Other:
7441	9504.9010	---其他电子游戏机	0					130	---Other video games
		---保龄球自动球道设备及器具:							---Automatic bowling alley equipments and appliances
7442	9504.9021	----保龄球自动分瓶机	0			0		80	----Automatic bowling pin distributing machines
7443	9504.9022	----保龄球	0			0		80	----Bowling balls
7444	9504.9023	----保龄球瓶	0			0		80	----Bowling pins
7445	9504.9029	----其他	0			0		80	----Other
7446	9504.9030	---中国象棋、国际象棋、跳棋等棋类用品	0			0		80	---Chess and other board games, including Chinese chess, international chess, Chinese cherkers and draughts
7447	9504.9040	---麻将及类似桌上游戏用品	0					80	---Mahjong and similiar table games
7448	9504.9090	---其他	0			0		80	---Other
	95.05	**节日(包括狂欢节)用品或其他娱乐用品,包括魔术道具及嬉戏品:**							**Festive, carnival or other entertainment articles, including conjuring tricks and novelty jokes:**
7449	9505.1000	-圣诞节用品	0					100	-Articles for Christmas festivities
7450	9505.9000	-其他	0					100	-Other

序号 No.	税则号列 Tariff Item	货品名称	最惠国税率(%) M.F.N.	协定税率(%) Agreement 东盟 ASEAN	曼谷协定 BA	香港 H.K.	澳门 Macao	普通税率(%) Gen.	Article Description
	95.06	**一般的体育活动、体操、竞技及其他运动(包括乒乓球运动)或户外游戏用的本章其他税号未列名用品及设备;游泳池或戏水池:**							**Articles and equipment for general physical exercise, gymnastics, athletics, other sports (including table-tennis) or out door games, not specified or included elsewhere in this Chapter; swimming pools and paddling pools:**
		-滑雪屐及其他滑雪用具:							-Snow-skis and other snow-ski equipment:
7451	9506.1100	--滑雪屐	14					50	--Skis
7452	9506.1200	--滑雪屐扣件(滑雪屐带)	14					50	--Ski-fastenings(ski-bindings)
7453	9506.1900	--其他	14					50	--Other
		-滑水板、冲浪板、帆板及其他水上运动用具:							-Water-skis, surf-boards, sailboards and other water-sport equipment:
7454	9506.2100	--帆板	12					50	--Sailboards
7455	9506.2900	--其他	14					50	--Other
		-高尔夫球棍及其他高尔夫球用具:							-Golf clubs and other golf equipment:
7456	9506.3100	--棍,全套	14					50	--Clubs, complete
7457	9506.3200	--球	12					50	--Balls
7458	9506.3900	--其他	14					50	--Other
		-乒乓球运动用品及器械:							-Articles and equipment for table-tennis:
7459	9506.4010	---乒乓球	12					50	---Table-tennis balls
7460	9506.4090	---其他	14					50	---Other
		-网球拍、羽毛球拍或类似的球拍,不论是否装弦:							-Tennis, badminton or similar rackets, whether or not strung:
7461	9506.5100	--草地网球拍,不论是否装弦	14					50	--Lawn-tennis rackets, whether or not strung
7462	9506.5900	--其他	14					50	--Other
		-球,但高尔夫球及乒乓球除外:							-Balls, other than golf balls and tabletennis balls:
7463	9506.6100	--草地网球	12					50	--Lawn-tennis balls
		--可充气的球:							--Inflatable:
7464	9506.6210	---篮球、足球、排球	12					50	---Basketballs, footballs or volley balls
7465	9506.6290	---其他	12					50	---Other
7466	9506.6900	--其他	12					50	--Other
		-溜冰鞋及旱冰鞋,包括装有冰刀的溜冰靴:							-Ice skates and roller skates, including skating boots with skates attached:
7467	9506.7010	---溜冰鞋	14		12			50	---Ice skates
7468	9506.7020	---旱冰鞋	14		12			50	---Roller skates
		-其他:							-Other:
		--一般的体育活动、体操或竞技用品及设备:							--Articles and equipment for general physical exercise, gymnastics or athletics:

序号 No.	税则号列 Tariff Item	货品名称	最惠国税率(%) M.F.N.	协定税率(%) Agreement 东盟 ASEAN	曼谷协定 BA	香港 H.K.	澳门 Macao	普通税率(%) Gen.	Article Description
7469	9506.9110	---健身及康复器械	12			0		50	---Equipment for exercise and recovery
7470	9506.9120	---滑板	12			0		50	---Skateboards
7471	9506.9190	---其他	12			0		50	---Other
7472	9506.9900	--其他	12			0	0	50	--Other
	95.07	**钓鱼竿、钓鱼钩及其他钓鱼用品；捞鱼网、捕蝶网及类似网；囮子"鸟"（税号92.08或97.05的货品除外）以及类似的狩猎用品：**							**Fishing rods, fish-hooks and other line fishing tackle; fish landing nets, butterfly nets and similar nets; decoy "birds" (other than those of heading No. 92.08 or 97.05) and similar hunting or shooting requisites:**
7473	9507.1000	-钩鱼竿	21					80	-Fishing rods
7474	9507.2000	-钓鱼钩，不论有无系钩丝	21					80	-Fish-hooks, whether or not snelled
7475	9507.3000	-钓线轮	21					80	-Fisning reels
7476	9507.9000	-其他	21					80	-Other
	95.08	**旋转木马、秋千、射击用靶及其他游乐场的娱乐设备；流动马戏团及流动动物园；流动剧团：**							**Roundabouts, swings, shooting galleries and other fairground amusements; travelling circuses and travelling menageries ; travelling theatres:**
7477	9508.1000	-流动马戏团及流动动物园	15					100	-Travelling circuses and travelling menageries
7478	9508.9000	-其他	15					100	-Other

第九十六章
杂 项 制 品

注释：

一、本章不包括：

（一）化妆盥洗用笔（第三十三章）；

（二）第六十六章的制品（例如，伞或手杖的零件）；

（三）仿首饰（税号 71.17）；

（四）第十五类注释二所规定的贱金属制通用零件（第十五类）或塑料制的类似品（第三十九章）；

（五）第八十二章的利口器及其他物品，其柄或其他零件是雕刻或模塑材料制的；但税号 96.01 或 96.02 适用于单独进口或出口的上述物品的柄或其他零件；

（六）第九十章的物品，例如，眼镜架（税号 90.03）、数学绘图笔（税号 90.17）、各种牙科、医疗、外科或兽医专用刷子（税号 90.18）；

（七）第九十一章的物品（例如，钟壳或表壳）；

（八）乐器及其零件、附件（第九十二章）；

（九）第九十三章的物品（武器及其零件）；

（十）第九十四章的物品（例如，家具、灯具及照明装置）；

（十一）第九十五章的物品（玩具、游戏品、运动用品）；

（十二）艺术品、收藏品及古物（第九十七章）。

二、税号 96.02 所称“植物质或矿物质雕刻材料”，是指：

（一）用于雕刻的硬种子、硬果核、硬果壳、坚果及类似植物材料（例如，象牙果及棕榈子）；

（二）琥珀、海泡石、粘聚琥珀、粘聚海泡石、黑玉及其矿物代用品。

三、税号 96.03 所称“制帚、制刷用成束、成簇的材料”，仅指未装配的成束、成簇的兽毛、植物纤维或其他材料。这些成束、成簇的材料无需分开即可安装在帚、刷之上，或只需经过简单加工（例如将顶端修剪成形）即可安装的。

Chapter 96
Miscellaneous manufactured articles

Notes:

1. This Chapter does not cover:

(a) Pencils for cosmetic or toilet uses (Chapter 33);

(b) Articles of Chapter 66 (for example, parts of umbrellas or walking-sticks);

(c) Imitation jewellery (heading No. 71. 17);

(d) Parts of general use, as defined in Note 2 to Section XV, of base metal (Section XV), or similar goods of plastics (Chapter 39);

(e) Cutlery or other articles of Chapter 82 with handles or other parts of carving or moulding materials; heading No. 96. 01 or 96. 02 applies, however, to separately pressented handles or other parts of such articles;

(f) Articles of Chapter 90, for example, spectacle frames (heading No. 90. 03), mathematical drawing pens (heading No. 90. 17), brushes of a kind specialized for use in dentistry or for medical, surgical or veterinary purposes (heading No. 90.18);

(g) Articles of Chapter 91 (for example, clock or watch cases);

(h) Musical instruments or parts or accessories thereof (Chapter 92);

(i) Articles of Chapter 93 (arms and parts thereof);

(j) Articles of Chapter 94 (for example, furniture, lamps and lighting fittings);

(k) Articles of Chapter 95 (toys, games, sports requisites); or

(l) Works of art, collectors' pieces or antiques (Chapter 97).

2. In heading No. 96.02 the expression "vegetable or mineral carving material" means:

(a) Hard seeds, pips, hulls and nuts and similar vegetable materials of a kind used for carving (for example, corozo and dom);

(b) Amber, meerschaum, agglomerated amber and agglomerated meerschaum, jet and mineral substitutes for jet.

3. In heading No. 96. 03 the expression "prepared knots and tufts for broom or brush making" applies only to unmounted knots and tufts of animal hair, vegetable fibre or other material, which are ready for incorporation without division in brooms or brushes, or which require only such further minor processes as trimming to shape at the top, to render them ready for such incorporation.

四、除税号96.01至96.06或96.15的货品以外，本章的物品还包括全部或部分用贵金属、包贵金属、天然或养殖珍珠、宝石或半宝石（天然、合成或再造）制成的物品。而且，税号96.01至96.06及96.15包括天然或养殖珍珠、宝石或半宝石（天然、合成或再造）、贵金属或包贵金属只作为小零件的物品。

4. Articles of this Chapter, other than those of headings No. 96.01 to 96.06 or 96.15, remain classified in the Chapter whether or not composed wholly or partly of precious metal or metal clad with precious metal, of natural or cultured pearls, or precious or semi-precious stones (natural, synthetic or reconstructed). However, headings No. 96.01 to 96.06 and 96.15 include articles in which natural or cultured pearls, precious or semi-pre-cious stones (natural, synthetic or reconstructed), precious metal or metal clad with precious metal constitute only minor constituents.

序号 No.	税则号列 Tariff Item	货品名称	最惠国税率(%) M.F.N.	协定税率(%) Agreement				普通税率(%) Gen.	Article Description
				东盟 ASEAN	曼谷协定 BA	香港 H.K.	澳门 Macao		
	96.01	**已加工的兽牙、骨、玳瑁壳、角、鹿角、珊瑚、珍珠母及其他动物质雕刻材料及其制品(包括模塑制品):**							**Worked ivory, bone, tortoise-shell, horn, antlers, coral, mother-of-pearl and other animal carving material and articles of these materials (including articles obtained by moulding):**
7479	9601.1000	-已加工的兽牙及其制品	20					100	-Worked ivory and articles of ivory
7480	9601.9000	-其他	20					100	-Other
	96.02	**已加工的植物质或矿物质雕刻材料及其制品;蜡、硬脂、天然树胶、天然树脂或塑型膏制成的模塑或雕刻制品以及其他税号未列名的模塑或雕刻制品;已加工的未硬化明胶(税号35.03的明胶除外)及未硬化明胶制品:**							**Worked vegetable or mineral carving mate-rial and articles of these materi-als; mould-ed or carved articles of wax, of stearin, of natural gums or natural resins or of mod-elling pastes, and other moulded or carved articles, not elsewhere specified or included; worked, unhardened gelatin (except gelatin of heading No.35.03) and ar-ticles of unhardened gelatin:**
7481	9602.0010	---装药用胶囊	10.5					40	---Pharmaceutical capsules
7482	9602.0090	---其他	25					100	---Other
	96.03	**帚、刷(包括作为机器、器具、车辆零件的刷)、非机动的手工操作地板清扫器、拖把及毛掸;供制帚、刷用的成束或成簇的材料;油漆块垫及滚筒;橡皮扫帚(橡皮辊除外):**							**Brooms, brushes (including brushes consti-tuting parts of machines, ap-pliances or ve-hicles), hand-operated mechanical floor sweepers, not mo-torized, mops and feather dusters; prepared knots and tufts for broom or brush making; paint pads and rollers; squeegees (other than roller aqueegees):**
7483	9603.1000	-用枝条或其他植物材料捆扎而成的帚及刷，不论是否有把	25				0△	100	-Brooms and brushes, consisting of twigs or other vegetable materials bound together, with or without han-dles

序号 No.	税则号列 Tariff Item	货品名称	最惠国税率(%) M.F.N.	协定税率(%)Agreement 东盟 ASEAN	曼谷协定 BA	香港 H.K.	澳门 Macao	普通税率(%) Gen.	Article Description
		-牙刷、剃须刷、发刷、指甲刷、睫毛刷及其他人体化妆用刷，包括作为器具零件的上述刷：							-Tooth brushes, shaving brushes, hair brushes, nail brushes, eyelash brushes and other toilet brushes for use on the person, including such brushes constituting parts of appliances:
7484	9603.2100	--牙刷，包括齿板刷	25					100	--Tooth brushes, including ventalplate brushes
7485	9603.2900	--其他	15					100	--Other
		-画笔、毛笔及化妆用的类似笔：							-Artists' brushes, writing brushes and similar brushes for the application of cosmetics:
7486	9603.3010	---画笔	25		15			100	---Artists' brushes
7487	9603.3020	---毛笔	20					100	---Writing brushes
7488	9603.3090	---其他	25					100	---Other
		-油漆刷、涂料刷、清漆刷及类似的刷（子目号9603.30 的货品除外）；油漆块垫及滚筒：							-Paint, distemper, varnish or similar brushes (other than brushes of subheading No. 9603.30); paint pads and rollers:
		---漆刷及类似刷：							---Paint, distemper, varnish or similar brushes:
7489	9603.4011	----猪鬃制	20					100	----Of pigs', hogs'or boars'bristle
7490	9603.4019	----其他	23					100	----Other
7491	9603.4020	---油漆块垫及滚筒	23					100	---Paint pads and rollers
		-作为机器、器具、车辆零件的刷：							-Other brushes constituting parts of ma-chines, appliances or vehicles:
		---金属丝刷：							---Brushes of metal wire:
7492	9603.5011	----作为机器、器具零件的刷	14					50	----Constituting parts of machines or appli-ances
7493	9603.5019	----其他	14					100	----Other
		---其他：							---Other:
7494	9603.5091	----作为机器、器具零件的刷	14					50	----Constituting parts of machines or appliances
7495	9603.5099	----其他	14					100	----Other
		-其他：							-Other:
7496	9603.9010	---羽毛掸	21					130	---Feather dusters
7497	9603.9090	---其他	15					100	---Other
	96.04	**手用粗筛、细筛：**							**Hand sieves and hand riddles:**
7498	9604.0000	手用粗筛、细筛	21					100	Hand sieves and hand riddles
	96.05	**个人梳妆、缝纫或清洁鞋靴、衣服用的成套旅行用具：**							**Travel sets for personal toilet, sewing or shoe or clothes cleaning**
7499	9605.0000	个人梳妆、缝纫或清洁鞋靴、衣服用的成套旅行用具	15					100	Travel sets for personal toilet, sewing or shoe or clothes cleaning

序号 No.	税则号列 Tariff Item	货 品 名 称	最惠国税率(%) M.F.N.	协定税率(%)Agreement 东盟 ASEAN	曼谷协定 BA	香港 H.K.	澳门 Macao	普通税率(%) Gen.	Article Description
	96.06	**钮扣、揿扣、钮扣芯及钮扣和揿扣的其他零件;钮扣坯:**							**Buttons, press-fasteners, snap-fasteners and press-studs, button moulds and other parts of these articles; button blanks:**
7500	9606.1000	-揿扣及其零件	21					100	-Press-fasteners, snap-fasteners and pressstuds and parts therefor
		-钮扣:							-Buttons:
7501	9606.2100	--塑料制,未用纺织材料包裹	21					100	--Of plastics, not covered with textile ma-terial
7502	9606.2200	--金属制,未用纺织材料包裹	15			0	0	100	--Of base metal, not covered with textile material
7503	9606.2900	--其他	15			0		100	--Other
7504	9606.3000	-钮扣芯及钮扣的其他零件;钮扣坯	15			0△		100	-Button moulds and other parts of buttons; button blanks
	96.07	**拉链及其零件:**							**Slide fasteners and parts thereof:**
		-拉链:							-Slide fasteners:
7505	9607.1100	--装有贱金属制咪牙齿的	21			0	0	130	--Fitted with chain scoops of base metal
7506	9607.1900	--其他	21			0	0	130	--Other
7507	9607.2000	-零件	21					130	-Parts
	96.08	**圆珠笔;毡尖和其他渗水式笔尖笔及唛头笔;自来水笔、铁笔型自来水笔及其他钢笔;蜡纸铁笔;活动铅笔;钢笔杆、铅笔套及类似的笔套;上述物品的零件(包括帽、夹),但税号96.09的货品除外:**							**Ball point pens; felt tipped and other porous-tipped pens and markers; fountain pens, stylograph pens and other pens; du-plicating stylos; propelling or sliding pen-cils; pen-holders, pencil-holders and similar holders; parts (including caps and clips) of the foregoing articles, other than those of heading No. 96.09:**
7508	9608.1000	-圆珠笔	15					80	- Ball point pens
7509	9608.2000	-毡尖和其他渗水式笔尖笔及唛头笔	21					80	-Felt tipped and other porous-tipped pens and markers
		-自来水笔、铁笔型自来水笔及其他钢笔:							-Fountain pens, stylograph pens and other pens:
7510	9608.3100	--墨汁画笔	21					80	--Indian ink drawing pens
		--其他:							--Other:
7511	9608.3910	---自来水笔	21					80	---Fountain pens
7512	9608.3990	---其他	21					80	---Other
7513	9608.4000	-活动铅笔	21					80	-Propelling or sliding pencils
7514	9608.5000	-由上述两个或多个子目所列物品组成的成套货品	21					80	-Sets of articles from two or more of the foregoing subheadings
7515	9608.6000	-圆珠笔芯,由圆珠笔头和墨芯构成	21					80	-Refills for ball point pens, comprising the ball point and ink-reservoir

序号 No.	税则号列 Tariff Item	货品名称	最惠国税率(%) M.F.N.	协定税率(%) Agreement 东盟 ASEAN	曼谷协定 BA	香港 H.K.	澳门 Macao	普通税率(%) Gen.	Article Description
		-其他:							-Other:
7516	9608.9100	--钢笔头及笔尖粒	12					70	--Pen nibs and nib points
		--其他:							--Other:
7517	9608.9910	---机器、仪器用笔	17.5					40	---Of a kind used on machines or instru-ments
7518	9608.9920	---蜡纸铁笔;钢笔杆、铅笔杆及类似的笔杆	21					80	---Duplicating stylos; pen-holders, pencil-holders and similar holders
7519	9608.9990	---其他	21					80	---Other
	96.09	**铅笔(税号96.08的铅笔除外)、颜色铅笔、铅笔芯、蜡笔、图画碳笔、书写或绘画用粉笔及裁缝划粉:**							**Pencils (other than pencils of heading No. 96.08), crayons, pencil leads, pastels, drawing charcoals, writing or drawing chalks and tailors' chalks:**
		-铅笔及颜色铅笔:							-Pencils and crayons, with leads encased in a rigid sheath:
7520	9609.1010	---铅笔	21					80	---Pencils
7521	9609.1020	---颜色铅笔	21					80	---Crayons
7522	9609.2000	-铅笔芯,黑的或其他颜色的	21					80	-Pencil leads, black or coloured
7523	9609.9000	-其他	15					80	-Other
	96.10	**具有书写或绘画面的石板、黑板及类似板,不论是否镶框:**							**Slates and boards, with writing or drawing surfaces, whether or not framed:**
7524	9610.0000	具有书写或绘画面的石板、黑板及类似板,不论是否镶框	15					80	Slates and boards, with writing or drawing surfaces, whether or not tramed
	96.11	**手用日期戳、封缄戳、编号戳及类似印戳(包括标签压印器);手工操作的排字盘及带有排字盘的手印器:**							**Daet, sealing or numbering stamps, and the like (including devices for printing or embossing labels), designed for operating in the hand; hand-operated composing sticks and hand printing sets incorporating such composing sticks:**
7525	9611.0000	手用日期戳、封缄戳、编号戳及类似印戳(包括标签压印器);手工操作的排字盘及带有排字盘的手印器	21					80	Date, sealing or numbering stamps, and the like (including devices for printing or embossing labels), designed for operating in the hand; hand-operated composing sticks and hand printing sets incorporating such composing sticks
	96.12	**打字机色带或类似色带,已上油或经其他方法处理能着色的,不论是否装轴或装盒;印台,不论是否已加印油或带盒子:**							**Typewriter or similar ribbons, inked or otherwise prepared for giving impressions, whether or not on spools or in cartridges; ink-pads, whether or not inked, with or without boxes:**

序号 No.	税则号列 Tariff Item	货品名称	最惠国税率(%) M.F.N.	协定税率(%)Agreement 东盟 ASEAN	曼谷协定 BA	香港 H.K.	澳门 Macao	普通税率(%) Gen.	Article Description
7526	9612.1000	-色带	10.5				0	35	-Inked ribbons
7527	9612.2000	-印台	25					100	-Ink-pads
	96.13	**香烟打火机和其他打火器(不论是机械的,还是电气的)及其零件,但打火石及打火机芯除外:**							**Cigarette lighters and other lighters, whether or not mechanical or electrical, and parts thereof other than flints and wicks:**
7528	9613.1000	-袖珍气体打火机,一次性的	25					130	-Pocket lighters, gas fuelled, non refillable
7529	9613.2000	-袖珍气体打火机,可充气的	25					130	-Pocket lighters, gas fuelled, refillable
7530	9613.8000	-其他打火器	25					130	-Other lighters
7531	9613.9000	-零件	25					130	-Parts
	96.14	**烟斗(包括烟斗头)和烟嘴及其零件:**							**Smoking pipes (including pipe bowls) and cigar or cigarette holders, and parts there-of:**
7532	9614.2000	-烟斗及烟斗头	25					130	-Pipes and pipe bowls
7533	9614.9000	-其他	25					130	-Other
	96.15	**梳子、发夹及类似品;发卡、卷发夹、卷发器或类似品及其零件,但税号85.16的货品除外:**							**Combs, hair-slides and the like; hair-pins, curling pins, curling grips, hair-curlers and the like, other than those of heading No.85.16, and parts thereof:**
		-梳子、发夹及类似品:							-Combs, hair-slides and the like:
7534	9615.1100	--硬质橡胶或塑料制	18					130	--Of hard rubber or plastics
7535	9615.1900	--其他	18					130	--Other
7536	9615.9000	-其他	18					130	-Other
	96.16	**香水喷雾器或类似的化妆用喷雾器及其座架、喷头;粉扑及粉拍,施敷脂粉或化妆品用:**							**Scent sprays and similar toilet sprays, and mounts and heads therefor; powderpuffs and pads for the application of cosmetics or toilet preparations:**
7537	9616.1000	-香水喷雾器或类似的化妆用喷雾器及其座架、喷头	18					130	-Scent sprays and similar toilet sprays, and mounts and heads therefor
7538	9616.2000	-粉扑及粉拍,施敷脂粉或化妆品用	18					130	-Powder-puffs and pads for the application of cosmetics or toilet preparations
	96.17	**带壳的保温瓶和其他真空容器及其零件,但玻璃瓶胆除外:**							**Vacuum flasks and other vacuum vessels, complete with cases; parts therof other than glass inners:**
7539	9617.0010	---保温瓶	24			0		130	---Vacuum flasks

序号 No.	税则号列 Tariff Item	货品名称	最惠国税率(%) M.F.N.	协定税率(%) Agreement				普通税率(%) Gen.	Article Description
				东盟 ASEAN	曼谷协定 BA	香港 H.K.	澳门 Macao		
7540	9617.0090	---其他	18			0		130	---Other
	96.18	**裁缝用人体模型及其他人体活动模型;橱窗装饰用的自动模型及其他活动陈列品:**							**Tailors' dummies and other lay figures; automata and other animated displays used for shop window dressing:**
7541	9618.0000	裁缝用人体模型及其他人体活动模型;橱窗装饰用的自动模型及其他活动陈列品	21					80	Tailors' dummies and other lay figures; automata and other animated displays used for shop window dressing

第二十一类

艺术品、收藏品及古物

SECTION XXI

WORKS OF ART, COLLECTORS' PIECES AND ANTIQUES

第九十七章
艺术品、收藏品及古物

Chapter 97
Works of art, collectors' pieces and antiques

注释：

一、本章不包括：

（一）在指运国流通或新发行的未经使用的邮票、印花税票、邮政信笺（印有邮票的纸品）及类似的票证（第四十九章）；

（二）作舞台、摄影的布景及类似用途的已绘制画布（税号59.07），但可归入税号97.06的除外；

（三）天然或养殖珍珠、宝石或半宝石（税号71.01至71.03）。

二、税号97.02所称"雕版画、印制画、石印画的原本"，是指以艺术家完全手工制作的单块或数块印版直接印制出来的黑白或彩色原本，不论艺术家使用何种方法或材料，但不包括使用机器或照相制版方法制作的。

三、税号97.03不适用于成批生产的复制品及具有商业性质的传统手工艺品，即使这些物品是艺术家设计或创造的。

四、（一）除上述注释一至三另有规定的以外，可归入本章各税号的物品，均应归入本章的相应税号而不归入本目录的其他税号；

（二）税号97.06不适用于可以归入本章其他各税号的物品。

五、已装框的油画、粉画及其他绘画、版画、拼贴画及类似装饰板，如果框架的种类及价值与作品相称，应与作品一并归类。如果框架的种类及价值与作品不相称，应分别归类。

Notes:

1. This Chapter does not cover:

(a) Unused postage or revenue stamps, postal stationery (stamped paper) or the like, of heading 49.07;

(b) Theatrical scenery, studio back-cloths or the like, of painted canvas (heading No.59.07) except if they may be classified in heading No.97.06; or

(c) Pearls, natural or cultured, or percious or semi-precious stones (headings No.71.01 to 71.03).

2. For the purposes of heading No.97.02, the expression "original engravings, prints and lithographs" means impressions produced directly, in black and white or in colour, of one or of several plates wholly executed by hand by the artist, irrespective of the process or of the material employed by him, but not including any mechanical or photomechanical process.

3. Heading No.97.03 does not apply to mass-produced reproductions or works of conventional craftsmanship of a commercial character, even if these articles are designed or created by artists.

4. (a) Subject to Notes 1 to 3 above, articles of this Chapter are to be classified in this Chapter and not in any other Chapter of the Nomenclature.

(b) Heading No.97.06 does not apply to articles of the preceding headings of this Chapter.

5. Frames around paintings, drawings, pastels, collages or similar decorative plaques, engravings, prints or lithographs are to be classified with those articles, provided they are of a kind and of a value normal to those articles.

Frames which are not of a kind or of a value normal to the articles referred to in this Note are to be classified separately.

序号 No.	税则号列 Tariff Item	货品名称	最惠国税率(%) M.F.N.	协定税率(%) Agreement 东盟 ASEAN	曼谷协定 BA	香港 H.K.	澳门 Macao	普通税率(%) Gen.	Article Description
	97.01	**油画、粉画及其他手绘画,但带有手工绘制及手工描饰的制品或税号49.06的图纸除外;拼贴画及类似装饰板:**							**Paintings, drawings and pastels, executed entirely by hand, other than drawings of heading No.49.06 and other than hand-painted or handdecorated manufactured articles; collages and similar decorative plaques:**
		-油画、粉画及其他手绘画:							-Paintings, drawings and pastels:
7542	9701.1010	---原件	12					50	---The originals
7543	9701.1020	---复制品	14					50	---Reproductions
7544	9701.9000	-其他	14					50	-Other
	97.02	**雕版画、印制画、石印画的原本:**							**Original engravings, prints and lithographs:**
7545	9702.0000	雕版画、印制画、石印画的原本:	12					50	Original engravings, prints and lithographs
	97.03	**各种材料制的雕塑品原件**							**Original sculptures and statuary, in any material:**
7546	9703.0000	各种材料制的雕塑品原件	12					50	Original sculptures and statuary, in any material
	97.04	**使用过或未使用过的邮票、印花税票、邮戳印记、首日封、邮政信笺(印有邮票的纸品)及类似品,但税目49.07的货品除外:**							**Used or unused postage or revenue stamps, stamp-post-marks, first-day covers, postal stationery (stamped paper), and the like, other than those of heading No.49.07:**
7547	9704.0010	---邮票	8					50	---Postage
7548	9704.0090	---其他	14					50	---Other
	97.05	**具有动物学、植物学、矿物学、解剖学、历史学、考古学、古生物学、人种学或钱币学意义的收集品及珍藏品:**							**Collections and collectors'pieces of zoological, botanical, mineralogical, anatomical, historical, archaeological, palaeontological, ethnographic or numismatic interest:**
7549	9705.0000	具有动物学、植物学、矿物学、解剖学、历史学、考古学、古生物学、人种学或钱币学意义的收集品及珍藏品	0					0	Collections and collectors'pieces of zoological, botanical, mineralogical, anatomical, historical, archaeolohical, palaeontological, ethnographic or numismatic interest
	97.06	**超过一百年的古物:**							**Antiques of an age exceeding one hundred years:**
7550	9706.0000	超过一百年的古物	0					0	Antiques of an age exceeding one hundred years

中华人民共和国出口税则

(2005 年 1 月 1 日起实施)

Customs Tariff of Export of the People's Republic of China (Enforced from January 1, 2005)

注释：

出口税则也采用《商品名称及编码协调制度》目录，进口税则的归类总规则同样适用于出口税则。

Note:

The Harmonized Commodity Description and Conding System is also adopted for this Export Tariff, and the General Rules For the Interpretation of the lmport Tariff are samely applicable to the Export Tariff.

序号 No.	税则号列 Tariff Item.	货品名称	出口税率 (%) Export Duty Rate	Article Description
	03.01			
1	03019210	鳗鱼苗	20	Eels (Anguilla spp.) fry
	05.06			
2	05060000	骨、角柱及他们的粉末、废料	40	Bones and horn-cores, powder and waste thereof
	26.07			
3	26070000	铅矿砂及其精矿	30	Lead ores and concentrates
	26.08			
4	26080000	锌矿砂及其精矿	30	Zinc ores and concentrates
	26.09			
5	26090000	锡矿砂及其精矿	50	Tin ores and concentrates
	26.11			
6	26110000	钨矿砂及其精矿	20	Tungsten ores and concen trates
	26.15			
7	26159000	钽铌矿砂及其精矿	30	Tantalum and Niobium concentrates
	26.17			
8	26171010	生锑（锑精矿，选矿产品）	20	Crude antimony (Antimony concentrates which are mineral products)
	28.04			
9	28047000	磷	20	Phosphorus
	28.26			
10	28269000	氟钽酸钾	30	Potassium fluotantalate
	29.02			
11	29022000	苯	40	Benzenne
	41.03			
12	41031010	山羊板皮	20	Raw hides and skins of goats
	72.01			
13	72011000	非合金生铁，含磷量小于或等于0.5%	20	Non-alloy pig iron containing by weigut 0.5% or less of phosphorus
14	72012000	非合金生铁，含磷量大于0.5%	20	Non-alloy pig iron, by weight more than 0.5% of phosphorus
15	72015000	合金生铁	20	Alloy pig iron
	72.02			
16	72021000	锰铁	20	Ferro-manganese
17	72022000	硅铁	25	Ferro-silicon
18	72023000	硅锰铁	20	Ferro-silico-manganese
19	72024000	铬铁	40	Ferro-chromium
	72.04			
20	72040000	钢铁废碎料；供再熔的碎料钢铁锭	40	Ferrous waste and scrap; remelting scrap ingots of iron or steel
	74.02			

序号 No.	税则号列 Tariff Item.	货品名称	出口税率 (%) Export Duty Rate	Article Description
21	74020000	未精炼铜，电解精炼用的铜阳极	30	Unrefined copper; copper anodes for electrolytic refining
	74.03			
22	74030000	未锻轧的精炼铜及铜合金	30	Refined copper and copper alloys, unwrought
	74.04			
23	7404000	铜废碎料	30	Copper waste and scrap
	74.07			
24	74070000	铜条、杆、型材及异型材	30	Copper bars, rods and profiles
	74.08			
25	74080000	铜丝	30	Copper wire
	74.09			
26	74090000	铜板、片及带，厚度>0.15mm	30	Copper plates, sheets and strip, of a thickness exceeding 0.15mm
	75.02			
27	75020000	未锻轧镍	40	Unwrought nickel
	75.08			
28	75089010	电镀用镍阳极	40	Electroplating nickel anodes
	76.01			
29	76010000	未锻轧铝	30	Unwrought aluminium
	76.02			
30	76020000	铝废粹料	30	Aluminum Waste and scrap
	76.04			
31	76040000	铝条、杆、型材及异型材	20	Aluminum bars, rods and profiles
	76.05			
32	76050000	铝丝	20	Aluminium wire
	76.06			
33	76060000	铝板、片及带，厚度>0.2mm	20	Aluminium plates, sheets and strip, of a thickness exceeding 0.2mm
	79.01			
34	79010000	未锻轧锌	20	Unwrought zinc
	81.10			
35	81101010	未锻轧锑	20	unwrought antimony
36	81101020	锑粉末	20	Antimony powders
37	81102000	锑废碎料	20	Antimony waste and scrap

附　　表

Appendix

表 1:《中国－东盟全面经济合作框架协议》早期收获安排附加表

Table 1: Annex to Early Havest Arrangement under the Framework on Comprehensive Economic Co－operation Between China and ASEAN

序号 No.	税则号列 Tariff Item	*货品名称 (简　称)	文莱 (%) Brunei	柬埔寨 (%) Cambodia	印尼 (%) Indonesia	老挝 (%) Laos	马来西亚 (%) Malaysia	缅甸 (%) Myanmar	新加坡 (%) Singapore	泰国 (%) Thailand	越南 (%) Vietnam	**Description (in brief)
1	01039200	重量≥50kg 的猪	0		0		0	0	0	0	0	Live swine weighing≥50kg, other than for pure－bred breeding
2	01051190	重量≤185g 的其他鸡	0	0	0	0	0	0	0	0		Live chicken ≤ 185g, other than pure－bred breeding
3	01059210	重量≤2000g 的改良种用鸡	0	0	0		0	0	0	0	0	Live chicken > 185g but ≤ 2000g, pure－bred breed
4	01059290	重量≤2000g 的其他鸡	0	0	0		0	0	0	0		Live chicken > 185g but≤2000g, other than pure－bred breed
5	01059310	重量>2000g 的改良种用鸡	0	0	0		0	0	0	0		Live chicken > 2000g, for pure－bred breeding
6	01059390	重量>2000g 的其他鸡	0	0	0		0	0	0	0		Live chicken > 2000g, other than pure－bred breeding
7	01059910	重量>185g 的其他改良种用家禽	0	0	0		0	0	0	0	0	Live ducks, geese, turkeys & guineafowls > 185g for pure bred breeding
8	01059991	重量>185g 的非改良种用鸭	0	0	0		0	0	0	0		Live ducks>185g, other than for pure－bred breeding
9	01059992	重量>185g 的非改良种用鹅	0	0	0		0	0	0	0		Live geese>185g, other than for pure－bred breeding

＊:此表中“货品名称”为简称,准确的货品名称以进口税则的货品名称描述为准。

＊＊:The *Article Description* in this table is a brief description of the article concerned. For the exact description of an article, the correspondent *Article Description in the Customs Import Tariff of the People's Republic of China* should be consulted.

序号 No.	税则号列 Tariff Item	*货品名称（简称）	文莱（%）Brunei	柬埔寨（%）Cambodia	印尼（%）Indonesia	老挝（%）Laos	马来西亚（%）Malaysia	缅甸（%）Myanmar	新加坡（%）Singapore	泰国（%）Thailand	越南（%）Vietnam	**Description (in brief)
10	01059993	重量＞185g 的非改良种用珍珠鸡	0	0	0		0	0	0	0		Live guines fowls > 185g, other than for pure－bred breeding
11	01059994	重量＞185g 的非改良种用火鸡	0	0	0		0	0	0	0		Live turkeys > 185g, other than for pure－bred breeding
12	01061110	改良种用灵长目动物	0	0	0		0	0	0	0	0	Primates, for pure－bred breeding
13	01061190	其他灵长目动物	0	0	0		0	0	0	0	0	Other primates, other than for pure－bred breeding
14	01061200	鲸、海豚及鼠海豚；海牛及儒艮	0	0	0		0	0	0	0	0	Whales, dolphins and propoises; manatees and dugongs
15	01061910	改良种用哺乳动物	0	0	0		0	0	0	0	0	Other mammals, for pure－bred breeding
16	01061920	食用哺乳动物	0	0	0		0	0	0	0	0	Other edible mammals, other than for pre－bred breeding
17	01061990	其他哺乳动物	0	0	0		0	0	0	0	0	Mammals, nes
18	01062011	改良种用鳄鱼苗	0	0	0		0	0	0	0	0	Crocodiles for cultivation, for pure－bred breeding
19	01062019	其他改良种用爬行动物	0	0	0		0	0	0	0	0	Reptiles, for pure－bred breeding
20	01062020	食用爬行动物	0	0	0		0	0	0	0	0	Edible reptiles
21	01062090	其他爬行动物	0	0	0		0	0	0	0	0	Reptiles, nes
22	01063110	改良种用猛禽	0	0	0		0	0	0	0	0	Birds, of prey, for pure－bred breeding
23	01063190	其他猛禽	0	0	0		0	0	0	0	0	Other birds of prey, other than for pure－bred breeding

序号 No.	税则号列 Tariff Item	*货品名称 （简 称）	文莱 （%） Brunei	柬埔寨 （%） Cambodia	印尼 （%） Indonesia	老挝 （%） Laos	马来西亚 （%） Malaysia	缅甸 （%） Myanmar	新加坡 （%） Singapore	泰国 （%） Thailand	越南 （%） Vietnam	**Description （in brief）
24	01063210	改良种用鹦形目鸟	0	0	0		0	0	0	0	0	Psittaciformes, for pure – bred breeding
25	01063290	其他鹦形目鸟	0	0	0		0	0	0	0	0	Other Psittaciformes, other than for pure – bred breeding
26	01063910	其他改良种用鸟	0	0	0		0	0	0	0	0	Other birds, for pure – bred breeding
27	01063921	乳鸽	0	0	0		0	0	0	0	0	Squabs, edible
28	01063922	鸵鸟	0	0	0		0	0	0	0	0	Ostrich, edible
29	01063923	野鸭	0	0	0		0	0	0	0	0	Teals, edible
30	01063929	其他食用鸟	0	0	0		0	0	0	0	0	other birds, edible
31	01063990	其他鸟	0	0	0		0	0	0	0	0	Birds, nes
32	01069011	改良种用蛙苗	0	0	0		0	0	0	0	0	Tadpole and young frogs for pure – bred breeding
33	01069019	其他改良种用动物	0	0	0		0	0	0	0	0	Other live animals, for pure – bred breeding
34	01069020	其他食用动物	0	0	0		0	0	0	0	0	Other live animals, edible
35	01069090	其他动物	0	0	0		0	0	0	0	0	Other live animals, not edible
36	02021000	冻的整头及半头牛肉	5	5	5		5	5	5	5	5	Frozen bovine carcasses & half carcasses
37	02022000	冻的带骨牛肉	5	5	5		5	5	5	5	5	Frozen unboned bovine meat (excl. carcasses)
38	02023000	冻的去骨牛肉	5	5	5		5	5	5	5	5	Frozen boneless bovine meat
39	02031110	鲜、冷的整头及半头乳猪肉	5	5	5		5	5	5	5	5	Fresh or chilled carcasses & half carcasses of sucking pig
40	02031190	其他鲜、冷的整头及半头猪肉	5	5	5		5	5	5	5	5	Fresh or chilled swine carcasses & half carcasses nes

序号 No.	税则号列 Tariff Item	*货品名称（简称）	文莱（%）Brunei	柬埔寨（%）Cambodia	印尼（%）Indonesia	老挝（%）Laos	马来西亚（%）Malaysia	缅甸（%）Myanmar	新加坡（%）Singapore	泰国（%）Thailand	越南（%）Vietnam	**Description (in brief)
41	02031200	鲜、冷的带骨猪前腿、后腿及其肉块	5	5	5		5	5	5	5	5	Fresh or chilled hams, shoulders & cuts with bone in thereof of swine meat
42	02031900	其他鲜、冷猪肉	5	5	5		5	5	5	5	5	Fresh or chilled swine meat, nes (unboned)
43	02032110	冻整头及半头乳猪肉	0	0	0		0	0	0	0	0	Frozen carcasses & half carcasses of sucking pig
44	02032190	其他冻整头及半头猪肉	0	0	0		0	0	0	0	0	Frozen swine carcasses & half carcasses , nes
45	02032200	冻的带骨猪前腿、后腿及其肉块	0	0	0		0	0	0	0	0	Frozen unboned hams, shoulders & cuts there-of of swine
46	02071100	整只，鲜或冷的鸡	5		5		5	5	5	5		Fresh or chilled whole chickens, not cut in pieces
47	02071200	整只，冻的鸡	见本表尾注		见本表尾注		见本表尾注	见本表尾注	见本表尾注	见本表尾注		Frozen whole chickens, not cut in pieces
48	02071311	鲜或冷的带骨鸡块	5		5		5	5	5	5		Fresh or chilled cuts & offal of chicken, with bone in
49	02071319	鲜或冷的其他鸡块	5		5		5	5	5	5		Fresh or chilled cuts of chicken, nes
50	02071321	鲜或冷的鸡翼(不包括翼尖)	5		5		5	5	5	5		Fresh or chilled wing of chicken (other than wiungtips)
51	02071329	鲜或冷的其他鸡杂碎	5		5		5	5	5	5		Fresh or chilled offal of chicken, nes
52	02071411	冻的的带骨鸡块	见本表尾注		见本表尾注	见本表尾注	见本表尾注	见本表尾注	见本表尾注	见本表尾注		Frozen cuts chicken, with bone im
53	02071419	冻的其他鸡块	见本表尾注		见本表尾注	见本表尾注	见本表尾注	见本表尾注	见本表尾注	见本表尾注		Frozen cuts chicken, nes
54	02071421	冻的鸡翼（不包括翼尖）	见本表尾注		见本表尾注	见本表尾注	见本表尾注	见本表尾注	见本表尾注	见本表尾注		Frozen wing of chicken (other than wingtips)

序号 No.	税则号列 Tariff Item	*货品名称 (简　称)	文莱 (%) Brunei	柬埔寨 (%) Cambodia	印尼 (%) Indonesia	老挝 (%) Laos	马来西亚 (%) Malaysia	缅甸 (%) Myanmar	新加坡 (%) Singapore	泰国 (%) Thailand	越南 (%) Vietnam	**Description (in brief)
55	02071422	冻的鸡爪	见本表尾注		见本表尾注	见本表尾注	见本表尾注	见本表尾注	见本表尾注	见本表尾注		Frozen chicken claw
56	02071429	冻的其他鸡杂碎	见本表尾注		见本表尾注	见本表尾注	见本表尾注	见本表尾注	见本表尾注	见本表尾注		Frozen offal of chicken, nes
57	02072400	整只,鲜或冷的火鸡	5	5	5		5	5	5	5	5	Fresh or chilled whole turkeys
58	02072500	整只,冻的火鸡	5	5	5		5	5	5	5	5	Frozen whole turkeys
59	02072600	鲜或冷的火鸡块及杂碎	5	5	5		5	5	5	5		Fresh or chilled cuts & offal of turkey
60	02072700	冻的火鸡块及杂碎	0	0	0	0	0	0	0	0		Frozen cuts & offal of turkey
61	02073210	整只,鲜或冷的鸭	5	5	5		5	5	5	5	5	Frozen or chilled whole ducks
62	02073220	整只,鲜或冷的鹅	5	5	5		5	5	5	5	5	Frozen or chilled whole geese
63	02073230	整只,鲜或冷的珍珠鸡	5	5	5		5	5	5	5	5	Frozen or chilled whole guinea fowls
64	02073310	整只,冻的鸭	5	5	5		5	5	5	5	5	Frozen whole ducks
65	02073320	整只,冻的鹅	5	5	5		5	5	5	5	5	Frozen whole geese
66	02073330	整只,冻的珍珠鸡	5	5	5		5	5	5	5	5	Frozen whole guinea fowls
67	02073400	鲜或冷的税目的0105所列家禽的肥肝	5	5	5		5	5	5	5	5	Fresh or chilled fatty livers of poultry
68	02073510	鲜或冷的鸭块及杂碎	5	5	5		5	5	5	5	5	Fresh or chilled cuts/offal of duck
69	02073520	鲜或冷的鹅块及杂碎	5	5	5		5	5	5	5	5	Fresh or chilled cuts/offal of geese
70	02073530	鲜或冷的珍珠鸡块及杂碎	5	5	5		5	5	5	5	5	Fresh or chilled cuts/offal ofguinea fowl
71	02101110	干、熏、盐制的带骨猪腿	5	5	5		5	5	5	5	5	Unboned swine hams&shoulders, salted, in brine driedd or smoked

序号 No.	税则号列 Tariff Item	*货品名称（简称）	文莱（%）Brunei	柬埔寨（%）Cambodia	印尼（%）Indonesia	老挝（%）Laos	马来西亚（%）Malaysia	缅甸（%）Myanmar	新加坡（%）Singapore	泰国（%）Thailand	越南（%）Vietnam	**Description (in brief)
72	02101190	干、熏、盐制的带骨猪腿肉块	5	5	5		5	5	5	5	5	Cuts of unboned swine hams&shoulders, salted, smoked or
73	02101200	干、熏、盐制的猪腹肉	5	5	5		5	5	5	5	5	Bellies & cuts thereof of swine, salted , in brine, dried, in brine drd or smoked
74	02101900	干、熏、盐制的其他猪肉	5	5	5		5	5	5	5	5	Meat of swine ,salted, in brine, dried or smoked,nes
75	02102000	干、熏、盐制的牛肉	5	5	5		5	5	5	5	5	Meat of bovine animals, salted, in brine, dried or smoked
76	02109100	干、熏、盐制的灵长目动物肉及食用杂碎	5	5	5		5	5	5	5	5	Primates, nes, saltd, dried or smokd; flours & meals of meat or offal
77	02109200	干、熏、盐制的鲸、海豚及鼠海豚;海牛及儒艮肉及食用杂碎	5	5	5		5	5	5	5	5	Whales, dophins, nes, saltd, dried or smokd; flours & meals of meat or offal
78	02109300	干、熏、盐制的爬行动物肉及食用杂碎	5	5	5		5	5	5	5	5	Reptiles, nes, saltd, dried or smokd; flours & meals of meat or offal
79	02109900	干、熏、盐制的其他肉及食用杂碎	5	5	5		5	5	5	5	5	Other meat , nes, saltd, dried or smokd; flours & meals of meat or offal
80	03019190	其他活鳟鱼	0	0	0		0	0	0	0	0	Live trout,excl.fry
81	03019310	鲤鱼苗	0		0	0	0	0	0	0	0	Live carp fry

序号 No.	税则号列 Tariff Item	*货品名称 （简　称）	文莱 （%） Brunei	柬埔寨 （%） Cambodia	印尼 （%） Indonesia	老挝 （%） Laos	马来西亚 （%） Malaysia	缅甸 （%） Myanmar	新加坡 （%） Singapore	泰国 （%） Thailand	越南 （%） Vietnam	**Description （in brief）
82	03019390	其他活鲤鱼	0		0	0	0	0	0	0	0	Live carp, excl. fry
83	04070010	种用禽蛋	0	0	0		0	0	0	0		Birds'eggs for hatching
84	04070021	带壳鲜鸡蛋	5	5	5		5	5	5	5		Hen's eggs, in shell, fresh, o/t for hatching
85	04070022	带壳鲜鸭蛋	5	5	5		5	5	5	5		Duck's eggs, in shell, fresh, o/t for hatching
86	04070023	带壳鲜鹅蛋	5	5	5		5	5	5	5		Goose's eggs, in shell, fresh, o/t for hatching
87	04070029	其他带壳鲜禽蛋	5	5	5		5	5	5	5		Birds' eggs nes, in shell, fresh, o/t for hatching
88	04070091	咸蛋	5	5	5		5	5	5	5		Birds' eggs, in shell, salted
89	04070092	皮蛋	5	5	5		5	5	5	5		Birds' eggs, in shell, lime－preserved
90	04070099	其他腌制或煮过的带壳禽蛋	5	5	5		5	5	5	5		Birds' eggs, in shell, preserved or cooked, nes
91	05040011	整个或切块的盐渍猪肠衣（猪大肠头除外）	5	5	5		5	5	5	5	5	Hog casings, salted
92	05040012	整个或切块的盐渍绵羊肠衣	5	5	5		5	5	5	5	5	Sheep casings, salted
93	05040013	整个或切块的盐渍山羊肠衣	5	5	5		5	5	5	5	5	Goat casings, salted
94	05040014	整个或切块的盐渍猪大肠头	5	5	5		5	5	5	5	5	Hog fat－ends, salted
95	05040019	整个或切块的其他动物肠衣	5	5	5		5	5	5	5	5	Sausage casings of animal guts, nes
96	05040021	冷和冻的鸡肫	见本表尾注	见本表尾注	见本表尾注		见本表尾注	见本表尾注	见本表尾注	见本表尾注	见本表尾注	Cold, frozen broiler gizzard
97	05040029	鲜、冷、冻、干、盐制的其他动物胃	5	5	5		5	5	5	5	5	Stomach of other animals, nes

序号 No.	税则号列 Tariff Item	*货品名称（简　称）	文莱（%）Brunei	柬埔寨（%）Cambodia	印尼（%）Indonesia	老挝（%）Laos	马来西亚（%）Malaysia	缅甸（%）Myanmar	新加坡（%）Singapore	泰国（%）Thailand	越南（%）Vietnam	**Description (in brief)
98	05040090	鲜、冷、冻、干、盐制的其他动物肠、膀胱、胃	5	5	5		5	5	5	5	5	Other guts, bladders and stomachs of animals, whole and pieces thereof, nes
99	07020000	鲜或冷藏的番茄	0		0		0	0	0	0	0	Tomatoes, fresh or chilled
100	07031010	鲜或冷的洋葱	0		0	0	0	0	0	0	0	Onions, fresh or chilled
101	07032010	鲜或冷藏的大蒜头	0		0	0	0	0	0	0	0	Bulbs of garlic, fresh or chilled
102	07032020	鲜或冷藏的大蒜蒜台及蒜苗（青蒜）	0		0	0	0	0	0	0	0	Stems or seedlins of garlic, fresh or chilled
103	07032090	其他鲜或冷藏的大蒜	0		0	0	0	0	0	0	0	Other parts of garlic, fresh or chilled
104	07041000	鲜或冷的菜花及硬花甘蓝	0		0	0	0	0	0	0	0	Cauliflowers & headed broccoli, fresh or chilled
105	07049000	鲜或冷的其他食用芥菜类蔬菜	0		0	0	0	0	0	0	0	White & red cabbages, kohlrabi, kale... etc, fresh or chilled
106	07051100	鲜或冷的结球莴苣（包心生菜）	0		0	0	0	0	0	0	0	Ccabbages lettuce, fresh or chilled
107	07051900	鲜或冷的其他莴苣	0		0	0	0	0	0	0	0	Lettuce, fresh or chilled, (excl. cabbage lettuce)
108	07061000	鲜或冷的胡萝卜及萝卜	0		0		0	0	0	0	0	Carrots & turnips, fresh or chilled
109	07069000	鲜或冷的小萝卜及类似食用根茎	0		0	0	0	0	0	0	0	Beetroot... radishes & other similar edible roots, fresh or chilled
110	07082000	鲜或冷的豇豆及菜豆	0		0		0	0	0	0	0	Beans, fresh or chilled
111	07093000	鲜或冷的茄子	0	0	0		0	0	0	0	0	Aubergines, fresh or chilled

序号 No.	税则号列 Tariff Item	*货品名称（简 称）	文莱（%） Brunei	柬埔寨（%） Cambodia	印尼（%） Indonesia	老挝（%） Laos	马来西亚（%） Malaysia	缅甸（%） Myanmar	新加坡（%） Singapore	泰国（%） Thailand	越南（%） Vietnam	**Description (in brief)
112	07095100	鲜或冷的其他伞菌属蘑菇	0	0	0		0	0	0	0	0	Mushrooms of the genus Agaricus, fresh or chilled
113	07096000	鲜或冷的辣椒，包括甜椒	0	0	0		0	0	0	0	0	Fruits of genus Capiscum or Pimenta, fresh or chilled
114	07099010	鲜或冷的竹笋	0		0		0	0	0	0	0	Bamboo shoots, fresh or chilled
115	07099090	鲜或冷的其他蔬菜	0		0		0	0	0	0	0	Other vegetables, fresh or chilled, nes
116	07102210	冷冻红小豆	0	0	0		0	0	0	0	0	Shelled or unshelled adzuki beans, frozen
117	07102290	其他冷冻豇豆及菜豆	0	0	0		0	0	0	0	0	Other shelled or unshelled beans, frozen
118	07102900	冷冻其他豆类蔬菜	0	0	0		0	0	0	0	0	Leguminous vegetables, shelled or unshelled nes, frozen
119	07104000	冷冻甜玉米	0	0	0		0	0	0	0	0	Sweet corn, frozen
120	07108010	冷冻松茸	0	0	0		0	0	0	0	0	Sungmo, frozen
121	07108020	其他蒜台及蒜苗（青蒜）	0	0	0		0	0	0	0	0	Garlic stems or seedlings, frozen
122	07108030	其他蒜头	0	0	0		0	0	0	0	0	Garlic bulbs
123	07108090	冷冻未列名蔬菜	0	0	0		0	0	0	0	0	Other vegetables, frozen
124	07109000	冷冻什锦蔬菜	0	0	0		0	0	0	0	0	Mixtures of vegetables, frozen
125	07114000	暂时保藏的黄瓜及小黄瓜	0	0	0		0	0	0	0	0	Cucumbers & gherkins provisionally preserved
126	07119031	盐水竹笋	0	0	0		0	0	0	0	0	Bamboo shoots in brine
127	07119034	盐水大蒜	0	0	0		0	0	0	0	0	Carlic, in brine
128	07119039	盐水其他蔬菜；什锦蔬菜	0	0	0		0	0	0	0	0	Other vegetables, in brine

序号 No.	税则号列 Tariff Item	*货品名称（简称）	文莱（%）Brunei	柬埔寨（%）Cambodia	印尼（%）Indonesia	老挝（%）Laos	马来西亚（%）Malaysia	缅甸（%）Myanmar	新加坡（%）Singapore	泰国（%）Thailand	越南（%）Vietnam	**Description (in brief)
129	07119090	暂时保藏的其他蔬菜；什锦蔬菜	0	0	0		0	0	0	0	0	Other vegetables provisionally preserved, nes
130	07122000	干制洋葱	0	0	0		0	0	0	0	0	Dried onions
131	07123100	干伞菌属蘑菇	0	0	0		0	0	0	0	0	Dried mushrooms of the genus Agaricus
132	07123910	干香菇	0	0	0		0	0	0	0	0	Dried shiitake
133	07123920	干金针菇	0	0	0		0	0	0	0	0	Dried winter mushroom
134	07123930	干草菇	0	0	0		0	0	0	0	0	Dried paddy straw mushroom
135	07123940	干口蘑	0	0	0		0	0	0	0	0	Dried tricholoma mongolicum lmai
136	07123950	干牛肝菌	0	0	0		0	0	0	0	0	Dried boletus; porcini
137	07123990	其他干制蘑菇及块菌	0	0	0		0	0	0	0	0	Other dried mushrooms and truffles
138	07129010	笋干丝	0	0	0		0	0	0	0	0	Dried bamboo shoots
139	07129020	紫萁（薇菜干）	0	0	0		0	0	0	0	0	Dried osmund
140	07129030	干金针菜（黄花菜）	0	0	0		0	0	0	0	0	Dried day lily flowers
141	07129040	蕨菜干	0	0	0		0	0	0	0	0	Dried chuch tsai
142	07129050	干制的大蒜	0	0	0		0	0	0	0	0	Dried garlic
143	07129060	干制的甜椒	0	0	0		0	0	0		0	Dried capsicum annuum var. grossum
144	07129090	干制的其他蔬菜及什锦蔬菜	0	0	0		0	0	0	0	0	Dried vegetables, nes
145	07133110	种用绿豆	0	0	0		0	0	0	0	0	Seed beans, dried, shelled
146	07133190	其他干绿豆	0	0	0		0	0	0	0	0	Other dried beans, shelled
147	07133211	种用红小豆	0	0	0		0	0	0	0	0	Small red (Adzuri) seed beans, dried shelled
148	07133219	其他种用赤豆	0	0	0		0	0	0	0	0	Other seed adzuki beans, dried shelled
149	07133290	其他干赤豆	0	0	0		0	0	0	0	0	Other dried adzuki beans, shelced

序号 No.	税则号列 Tariff Item	*货品名称（简称）	文莱（%）Brunei	柬埔寨（%）Cambodia	印尼（%）Indonesia	老挝（%）Laos	马来西亚（%）Malaysia	缅甸（%）Myanmar	新加坡（%）Singapore	泰国（%）Thailand	越南（%）Vietnam	**Description (in brief)
150	07133310	种用芸豆	0	0	0		0	0	0	0	0	Seed kidney beans, incl. white peabeans, dried, shelled
151	07133390	其他干芸豆	0	0	0		0	0	0	0	0	Other dried kidney beans, incl. white pea beans, shelled
152	07133900	干豇豆及菜豆	0	0	0		0	0	0	0	0	Dried beans, shelled, nes
153	07139010	种用干豆	0	0	0		0	0	0	0	0	Dried leguminous vegetables seed nes, shelled
154	07139090	其他干豆	0	0	0		0	0	0	0	0	Dried leguminous vegetables, shelled , nes
155	07141010	鲜木薯	0	0	0		0	0	0	0	0	Fresh manioc
156	07141020	干木薯	0	0	0		0	0	0	0	0	Dried manioc
157	07141030	冷或冻的木薯	0	0	0		0	0	0	0	0	Chilled or frozen manioc
158	07142011	种用鲜甘薯	0	0	0		0	0	0	0	0	Fresh sweet potatoes, for cultivatron
159	07142019	其他鲜甘薯	0	0	0		0	0	0	0	0	Other fresh sweet potatoes
160	07142020	干甘薯	0	0	0		0	0	0	0	0	Dried sweet potatoes
161	07142030	冷或冻的甘薯	0	0	0		0	0	0	0	0	Chilled or frozen sweet potatoes
162	07149010	鲜、干或冷、冻的荸荠	0	0	0		0	0	0	0	0	Waterchestnut, fresh or dried, chilled, frozen
163	07149021	种用藕	0	0	0		0	0	0	0	0	Lotus roots for cultivation, fresh or dried
164	07149029	其他藕	0	0	0		0	0	0	0	0	Lotus roots o/t for cultivation, fresh or dried
165	07149090	含有高淀粉或菊粉的其他类似根茎	0	0	0		0	0	0	0	0	Roots or tubers with high starch content, fresh or dried, nes
166	08011910	种用椰子	0		0	0	0	0	0	0	0	Coconuts, seedlings
167	08011990	其他鲜椰子	0		0	0	0	0	0	0	0	Other coconuts, fresh

序号 No.	税则号列 Tariff Item	*货品名称（简称）	文莱 (%) Brunei	柬埔寨 (%) Cambodia	印尼 (%) Indonesia	老挝 (%) Laos	马来西亚 (%) Malaysia	缅甸 (%) Myanmar	新加坡 (%) Singapore	泰国 (%) Thailand	越南 (%) Vietnam	**Description (in brief)
168	08013100	鲜或干的未去壳腰果	5	5	5	5	5	5	0	0	5	Cashew nuts, in shell, fresh or dried
169	08013200	鲜或干的去壳腰果	5	5	5	5	5	5	0	0	5	Cashew nuts, shelled, fresh or dried
170	08021100	鲜或干的未去壳巴旦杏	5	5	5		5	5	0	0	5	Almonds, in shell, fresh or dried
171	08021200	鲜或干的去壳巴旦杏	0	0	0		0	0	0	0	0	Almonds, without shells, fresh or dried
172	08022100	鲜或干的未去壳榛子	5	5	5		5	5	0	0	5	Hazlenuts in shell, fresh or dried
173	08022200	鲜或干的去壳榛子	0	0	0		0	0	0	0	0	Hazlenuts without shells, fresh or dried
174	08023100	鲜或干的未去壳核桃	5	5	5		5	5	0	0	5	Walnuts in shell, fresh or dried
175	08023200	鲜或干的去壳核桃	5	5	5		5	5	0	0	5	Walnuts without shells, fresh or dried
176	08024010	鲜或干板栗	5	5	5		5	5	0	0	5	Banli chestnut, fresh or dried
177	08024090	其他鲜或干栗子	5	5	5		5	5	0	0	5	Other chestnuts, fresh or dried
178	08025000	鲜或干的阿月浑子果（开心果）	0	0	0		0	0	0	0	0	Pistachio, fresh or dried
179	08029010	鲜或干的槟榔	0	0	0		0	0	0	0	0	Betel nuts, fresh or dried
180	08029020	鲜或干的白果	5	5	5		5	5	0	0	5	Gingko nuts, fresh or dried
181	08029030	鲜或干的松子仁	5	5	5		5	5	0	0	5	Pine – nuts, shelled, fresh or dried
182	08029041	鲜或干的种用夏威夷果	0	0	0		0	0	0	0	0	Hawaiian nuts seed, fresh or dried
183	08029049	鲜或干的其他夏威夷果	5	5	5		5	5	0	0	5	Hawaiian nuts excl. seed, fresh or dried

序号 No.	税则号列 Tariff Item	*货品名称（简称）	文莱 (%) Brunei	柬埔寨 (%) Cambodia	印尼 (%) Indonesia	老挝 (%) Laos	马来西亚 (%) Malaysia	缅甸 (%) Myanmar	新加坡 (%) Singapore	泰国 (%) Thailand	越南 (%) Vietnam	**Description (in brief)
184	08029090	其他鲜或干坚果	5	5	5		5	5	0	0	5	Other nuts, fresh or dried, nes
185	08042000	鲜或干的无花果	5	5	5	5	5	5	0	0	5	Figs, fresh or dried
186	08043000	鲜或干菠萝	0		0		0	0	0	0	0	Pineapples, fresh or dried
187	08044000	鲜或干鳄梨	5	5	5	5	5	5	0	0	5	Avocados, fresh or dried
188	08045010	鲜或干番石榴	5		5		5	5	0	0	5	Guavas, fresh or dried
189	08045020	鲜或干芒果	5		5		5	5	0	0	5	Mangoes, fresh or dried
190	08045030	鲜或干山竹果	5		5		5	5	0	0	5	Mangosteens, fresh or dried
191	08051000	鲜或干橙	5		5		5	5	0	0	5	Oranges, fresh or dried
192	08052010	鲜或干蕉柑	5	5	5	5	5	5	0	0	5	Chiao – Kan, fresh or dried
193	08052020	阔叶柑橘	5	5	5	5	5	5	0	0	5	Latifolia citrus, fresh or dried
194	08052090	其他鲜或干的柑桔及杂交柑桔	5	5	5	5	5	5	0	0	5	Mandarins, clementines, wilkings... etc, fresh or dried, nes
195	08054000	鲜或干柚	5	5	5	5	5	5	0	0		Grapefruit, fresh or dried
196	08055000	鲜或干的柠檬及酸橙	5	5	5		5	5	0	0		Lemons and limesgs, fresh or dried
197	08059000	其他鲜或干的柑桔属水果	5	5	5		5	5	0	0		Citrus fruit, fresh or dried, nes
198	08061000	鲜葡萄	5	5	5	5	5	5	0	0	5	Fresh grapes
199	08062000	葡萄干	5	5	5	5	5	5	0	0	5	Dried grapes
200	08071100	鲜西瓜	5		5		5	5	0	0	5	Watermelons, fresh
201	08071910	鲜哈密瓜	5		5		5	5	0	0	5	Hami melons, fresh

序号 No.	税则号列 Tariff Item	*货品名称（简称）	文莱（%）Brunei	柬埔寨（%）Cambodia	印尼（%）Indonesia	老挝（%）Laos	马来西亚（%）Malaysia	缅甸（%）Myanmar	新加坡（%）Singapore	泰国（%）Thailand	越南（%）Vietnam	**Description (in brief)
202	08071920	鲜罗马甜瓜及加勒比甜瓜	5		5		5	5	0	0	5	Cantaloupe and Galia melons, fresh
203	08071990	其他鲜甜瓜	5		5	5	5	5	0	0	5	Other melons, fresh
204	08072000	鲜木瓜	5	5	5		5	5	0	0	5	Papaws(papayas), fresh
205	08082012	鲜鸭梨、雪梨	5	5	5	5	5	5	0	0	5	Ya prars or Hsueh pears, fresh
206	08082013	鲜香梨	5	5	5	5	5	5	0	0	5	Xiang pears, fresh
207	08082020	鲜榅桲	5	5	5	5	5	5	0	0	5	Quinces, fresh
208	08091000	鲜杏	5	5	5	5	5	5	0	0	5	Apricots, fresh
209	08094000	鲜梅及李	5	5	5	5	5	5	0	0	5	Plums & sloes, fresh
210	08101000	鲜草莓	5	5	5	5	5	5	0	0	5	Strawberries, fresh
211	08102000	鲜的木莓、黑莓、桑椹及罗甘莓	5	5	5	5	5	5	0	0	5	Raspberries, blackberries, mulberries & loganberries, fresh
212	08103000	鲜的黑、白或红的醋栗及鹅莓	5	5	5	5	5	5	0	0	5	Black, white or red currants & gooseberries, fresh
213	08104000	鲜蔓越桔及越桔	5	5	5	5	5	5	0	0	5	Cranberries, bilberries...etc, fresh
214	08105000	鲜猕猴桃	5	5	5	5	5	5	0	0	5	Kiwifruit, fresh
215	08106000	鲜榴莲	5	5	5	5	5	5	0	0	5	Durian, fresh
216	08109010	鲜荔枝	5	5	5	5	5	5	0	0	5	Lychee, fresh
217	08109030	鲜龙眼	5		5	5	5	5	0	0	5	Longan, fresh
218	08109040	鲜红毛丹	5	5	5	5	5	5	0	0	5	Rambutan, fresh
219	08109050	鲜蕃荔枝	5	5	5	5	5	5	0	0	5	Sugar apple, fresh
220	08109060	鲜杨桃	5	5	5	5	5	5	0	0	5	Carambola, fresh
221	08109070	鲜莲雾	5	5	5	5	5	5	0	0	5	War apple, fresh
222	08109080	鲜火龙果	5	5	5	5	5	5	0	0	5	Dragon fruit, fresh
223	08109090	其他鲜果	5	5	5	5	5	5	0	0	5	Other fruit, fresh, nes
224	08111000	冷冻草莓	5	5	5	5	5	5	0	0	5	Strawberries, frozen
225	08112000	冷冻其他浆果	5	5	5	5	5	5	0	0	5	Raspberries, blackberveies..etc, frozen

序号 No.	税则号列 Tariff Item	*货品名称（简称）	文莱 (%) Brunei	柬埔寨 (%) Cambodia	印尼 (%) Indonesia	老挝 (%) Laos	马来西亚 (%) Malaysia	缅甸 (%) Myanmar	新加坡 (%) Singapore	泰国 (%) Thailand	越南 (%) Vietnam	**Description (in brief)
226	08119010	未去壳的冷冻栗子	5	5	5	5	5	5	0	0	5	Chestnuts in shell, frozen
227	08119090	其他冷冻水果及坚果	5	5	5	5	5	5	0	0	5	Other fruit & nuts, frozen, nes
228	08121000	暂时保藏的樱桃	5	5	5	5	5	5	0	0	5	Cherries, provisional provisional preserved, not for immediate consumptn
229	08129000	暂时保藏的其他水果及坚果	5	5	5	5	5	5	0	0	5	Fruit & nuts, provisional preserved, not for immediate consumptn
230	08131000	杏干	5	5	5	5	5	5	0	0	5	Dried apricots
231	08132000	梅干及李干	5	5	5	5	5	5	0	0	5	Dried prunes
232	08133000	苹果干	5	5	5	5	5	5	0	0	5	Dried apples
233	08134010	龙眼干、肉	5	5	5	5	5	5	0	0	5	Longans and longan pulps, dried
234	08134020	柿饼	5	5	5	5	5	5	0	0	5	Persimmons, dried
235	08134030	干红枣	5	5	5	5	5	5	0	0	5	Chinese dates (red jujubes), dried
236	08134040	荔枝干	5	5	5	5	5	5	0	0	5	Dried litchi
237	08134090	其他干果	5	5	5	5	5	5	0	0	5	Other fruit, dried, nes
238	08135000	本章的什锦坚果或干果	5	5	5	5	5	5	0	0	5	Mixtures of dried fruit & muts, nes
239	08140000	柑桔属水果或甜瓜的果皮	5	5	5	5	5	5	0	0	5	Peel of citrus fruit or melons, fresh, frozen, dried...etc.
240	09012200	已浸除咖啡碱的已焙炒咖啡			5		5		5			Roasted, decaffeinated coffee
241	15131100	初榨椰子油分离品	0		0		0		0			Crude coconut (copra) oil & fractions thereof

序号 No.	税则号列 Tariff Item	*货品名称（简称）	文莱（%）Brunei	柬埔寨（%）Cambodia	印尼（%）Indonesia	老挝（%）Laos	马来西亚（%）Malaysia	缅甸（%）Myanmar	新加坡（%）Singapore	泰国（%）Thailand	越南（%）Vietnam	**Description (in brief)
242	15131900	椰子油及其分离品	0		0				0			Cocout copra oii(excl. crude) & fractions thereof
243	15132100	初榨棕榈仁油或巴巴苏棕榈果油及其分离品	0		0		0		0			Crude palm kernel or babassu oil & fractions thereof
244	15132900	精制的棕榈仁油或巴巴苏棕榈果油及其分离品	0		0		0		0			Palm kernel or babassu oil (excl. crude) & fractions thereof
245	15162000	氢化、酯化或反油酸化植物油、脂及其分离品，但未进一步加工的	5		5		5		5			Vegetable fats & oils & their fractions, hydrogenated, etc, not further prepared
246	15179000	混合制成的食用油脂或制品	5		5		5		5			Edible prep.s of fats & oils, nes
247	18010000	生或焙炒的整颗或破碎的可可豆	0				0		0			Cocoa beans, whole or broken, raw or roasted
248	18031000	未脱脂可可膏	0				0		0			Cocoa paste, not defatted
249	18032000	全脱脂或部分脱脂的可可膏	0				0		0			Cocoa paste, wholly or parly defatted
250	18040000	可可油，可可脂	5				5		5			Cocoa butter, fat & oil
251	18050000	未加糖或其他甜物质的可可粉	5				5		5			Cocoa powder, with added sugar or other sweetening matter
252	18061000	含糖或其他甜物质的可可粉	0		0		0		0			Cocoa powder, without added sugar or other sweetening matter

序号 No.	税则号列 Tariff Item	*货品名称（简称）	文莱（%）Brunei	柬埔寨（%）Cambodia	印尼（%）Indonesia	老挝（%）Laos	马来西亚（%）Malaysia	缅甸（%）Myanmar	新加坡（%）Singapore	泰国（%）Thailand	越南（%）Vietnam	**Description (in brief)
253	27011100	未制成型的无烟煤，不论是否粉化	0				0		0	0		Anthracite, not agglonerated wheteher or not pulverized
254	27040010	煤制焦炭及半焦炭不论是否成型	0				0		0	0		Coke & semi – coke, whether or not agglomerated
255	34011990	其他用肥皂及有机表面活性产品，条状、块状或模制形状的，以及用肥皂或洗涤剂浸渍、涂面或包覆的纸、絮胎、毡呢及无纺织物			5		5		5			Soap and organic surface – active products & preparations for use as soap , in bars, cakes, moulded pieces or shapes; paper, wadding, felt & nonwovens, impregnated , coated or covered with soap or detergent, nes
256	34012000	其他形状的肥皂			5		5		5			Soap in other forms
257	38231100	硬脂酸	5		5		5		5			Industriac stearic acid
258	40169200	硫化橡胶制橡皮擦	5		5		5		5			Erasers of vulcanized rubber
259	70112000	阴极射线管用玻壳及零件	0		0		0		0			Glass envelopes(including bulbs/tubes) for cathode – ray tubes, nes
260	94015000	藤、柳条、竹及类似材料制的坐具			0				0			Seats of cane, osier, bamboo or similar materials
261	94038010	藤、柳条、竹及类似材料制家具			0				0			Furniture of cane, osier, bamboo or similar materials

注：

税号 02071200，02071411，02071419，02071421，02071422，02071429，05040021 按如下从量税执行：

02071200　0.33 元/公斤

02071411　0 元/公斤

02071419　**0** 元/公斤
02071421　**0** 元/公斤
02071422　**0** 元/公斤
02071429　**0** 元/公斤
05040021　**0**.**35** 元/公斤

abte:Specific Tariff Rates for items **02071200**, **02071411**, **02071419**, **02071421**, **02071422**,**02071429**, **05040021** are **0**.**33**RMB/KG, **0**RMB/KG, **0**RMB/KG, **0**RMB/KG, **0**RMB/KG, **0**RMB/KG, **0**.**35**RMB/KG respecitively.

表 2:进口关税特惠税率表

Table 2:Special Preferential Duty Rate on Imported Goods

序号 No.	税则号列 Tariff Item	商品名称(简称)	最惠国税率(%) M.F.N.	柬埔寨(%) Cambodia	缅甸(%) Myanmar	老挝(%) Laos	孟加拉(%) Bangladesh	非洲部分国家(%) Some Africa countries	Article Description (in brief)
1	01029000	其他牛	10	0	0	0			Live bovine animals, o/t pure - bred breeding
2	01039110	重量<10kg 的猪	10	0	0				Live swine weighing<10kg, o/t for pure - bred breeding
3	01039120	重量在 10 - 50kg 的猪	10	0	0				Live swine 10kg≤weighing <50kg, o/t for pure - bred breeding
4	01039200	重量≥50kg 的猪	10	0	0	0			Live swine weighing≥50kg, o/t for pure - bred breeding
5	01051190	重量≤185g 的其他鸡	10		0	0			Live chicken ≤ 185g, o/t pure - bred breeding
6	01051990	重量≤185g 的其他家禽	10			0			Live ducks, geese & guinea fowls ≤ 185g, o/t pure - bred breeding
7	01061920	食用哺乳动物	10		0				Other edible mammals, o/t for pre - bred breeding
8	01062020	食用爬行动物	10		0				Edible reptiles
9	01063921	乳鸽	10		0				Squabs, edible
10	01063922	鸵鸟	10		0				Ostrich, edible
11	01063923	野鸭	10		0				Teals, edible
12	01063929	其他食用鸟	10		0				Other birds, edible
13	01063990	未列名鸟	10					0	Birds, nes
14	01069020	其他食用动物	10		0				Other live animals, edible
15	01069090	其他非食用活动物	10					0	Other live animals, not edible
16	02011000	整头及半头鲜、冷牛肉	20	0		0			Fresh or chilled bovine carcasses&half carcasses
17	02012000	鲜、冷的带骨牛肉	12	0		0			Fresh or chilled unboned bovine meat (excl. carcasses)
18	02013000	鲜、冷的去骨牛肉	12	0					Fresh or chilled boneless bovine meat
19	02021000	冻的整头及半头牛肉	25	0		0			Frozen bovine carcasses & half carcasses
20	02022000	冻的带骨牛肉	12	0		0			Frozen unboned bovine meat (excl. carcasses)
21	02023000	冻的去骨牛肉	12	0					Frozen boneless bovine meat
22	02031110	鲜、冷的整头及半头乳猪肉	20	0		0			Fresh or chilled carcasses & half carcasses of sucking pig
23	02031190	其他鲜、冷的整头及半头猪肉	20	0		0			Fresh or chilled swine carcasses & half carcasses, nes

序号 No.	税则号列 Tariff Item	商品名称（简称）	最惠国税率(%) M.F.N.	柬埔寨(%) Cambodia	缅甸(%) Myanmar	老挝(%) Laos	孟加拉(%) Bangladesh	非洲部分国家(%) Some Africa countries	Article Description (in brief)
24	02031200	鲜、冷的带骨猪前腿、后腿及其肉块	20	0		0			Fresh or chilled unboned hams, shoulders & cuts thereof of swine
25	02031900	其他鲜、冷猪肉	20	0		0			Fresh or chilled swine meat, nes (unboned)
26	02032110	冻整头及半头乳猪肉	12	0		0			Frozen carcasses & half carcasses of sucking pig
27	02032190	其他冻整头及半头猪肉	12	0		0			Frozen swine carcasses & half carcasses, nes
28	02032200	冻的带骨猪前腿、后腿及其肉块	12	0		0			Frozen unboned hams, shoulders & cuts thereof of swine
29	02032900	其他冻猪肉	12	0					Frozen swine meat, nes
30	02061000	鲜、冷的牛杂碎	12	0		0			Fresh or chilled edible bovine offal
31	02062100	冻牛舌	12	0		0			Frozen bovine tongues
32	02062200	冻牛肝	12	0		0			Frozen bovine livers
33	02062900	其他冻牛杂碎	12	0		0			Frozen edible bovine offal (excl. tongues & livers)
34	02063000	鲜、冷的猪杂碎	20	0	0	0			Fresh or chilled edible swine offal
35	02064100	冻猪肝	20	0		0			Frozen swine livers
36	02064900	其他冻猪杂碎	12	0		0			Frozen edible swine offal (excl. livers)
37	02068000	鲜、冷的羊、马、驴、骡杂碎	20	0					Fresh or chilled edible offal of sheep, goats, horses…
38	02069000	冻的羊、马、驴、骡杂碎	18	0					Frozen edible offal of sheep, goats, horses…
39	02071100	整只，鲜或冷的鸡	20	0	0	0			Fresh or chilled whole chickens, not cut in pieces
40	02071200	整只，冻的鸡	20	见本表尾注	见本表尾注	见本表尾注			Frozen whole chickens, not cut in pieces
41	02071311	鲜或冷的带骨鸡块	20	0	0	0			Fresh or chilled cuts & offal of chicken, with bone in
42	02071319	鲜或冷的其他鸡块	20	0	0	0			Fresh or chilled cuts of chicken, nes
43	02071321	鲜或冷的鸡翼(不包括翼尖)	20	0	0	0			Fresh or chilled wing of chicken (other than wingtips)
44	02071329	鲜或冷的其他鸡杂碎	20	0	0	0			Fresh or chilled offal of chicken, nes
45	02071411	冻的的带骨鸡块	10	见本表尾注	见本表尾注	见本表尾注			Frozen cuts chicken, with bone in
46	02071419	冻的其他鸡块	10	见本表尾注	见本表尾注	见本表尾注			Frozen cuts chicken, nes

序号 No.	税则号列 Tariff Item	商品名称（简称）	最惠国税率（%）M.F.N.	柬埔寨（%）Cambodia	缅甸（%）Myanmar	老挝（%）Laos	孟加拉（%）Bangladesh	非洲部分国家（%）Some Africa countries	Article Description (in brief)
47	02071421	冻的鸡翼(不包括翼尖)	10	见本表尾注	见本表尾注	见本表尾注			Frozen wing of chicken(other than wingtips)
48	02071422	鲜、冷、冻鸡爪	10	见本表尾注	见本表尾注	见本表尾注			Fresh or chilled chicken claws
49	02071429	冻的其他鸡杂碎	10	见本表尾注	见本表尾注	见本表尾注			Frozen offal of chicken, nes
50	02072400	整只,鲜或冷的火鸡	20			0			Fresh or chilled whole turkeys
51	02072500	整只,冻的火鸡	20			0			Frozen whole turkeys
52	02072600	鲜或冷的火鸡块及杂碎	20			0			Fresh or chilled cuts & offal of turkey
53	02072700	冻的火鸡块及杂碎	10			0			Frozen cuts & offal of turkey
54	02073210	整只,鲜或冷的鸭	20	0	0	0			Fresh or chilled whole ducks
55	02073220	整只,鲜或冷的鹅	20	0	0	0			Fresh or chilled whole geese
56	02073230	整只,鲜或冷的珍珠鸡	20	0		0			Fresh or chilled whole guinea fowls
57	02073310	整只,冻的鸭	20	0	0	0			Frozen whole ducks
58	02073320	整只,冻的鹅	20	0	0	0			Frozen whole geese
59	02073330	整只,冻的珍珠鸡	20	0	0	0			Frozen whole guinea fowls
60	02073400	鲜或冷的肥肝	20	0		0			Fresh or chilled fatty livers of poultry
61	02073510	鲜或冷的鸭块及杂碎	20	0	0	0			Fresh or chilled cuts/offal of duck
62	02073520	鲜或冷的鹅块及杂碎	20	0	0	0			Fresh or chilled cuts/offal of geese
63	02073530	鲜或冷的珍珠鸡块及杂碎	20	0	0	0			Fresh or chilled cuts/offal of guinea fowl
64	02073610	冻的鸭块及杂碎	20	0	0	0			Frozen cuts/offal of duck
65	02073620	冻的鹅块及杂碎	20	0	0	0			Frozen cuts/offal of geese
66	02073630	冻的珍珠鸡块及杂碎	20	0	0	0			Frozen cuts/offal of guinea fowl
67	02101110	干、熏、盐制的带骨猪腿	25	0		0			Unboned swine hams&shoulders, salted, in brine, dried or smoked
68	02101190	干、熏、盐制的带骨猪腿肉块	25	0		0			Cuts of unboned swine hams & shoulders, salted, in brine drd of smoked
69	02101200	干、熏、盐制的猪腹肉	25	0		0			Bellies & cuts thereof of swine, salted, in brine, dried or smoked
70	02101900	干、熏、盐制的其他猪肉	25	0	0				Meat of swine, salted, in brine, dried or smoked, nes

序号 No.	税则号列 Tariff Item	商品名称（简称）	最惠国税率(%) M.F.N.	柬埔寨(%) Cambodia	缅甸(%) Myanmar	老挝(%) Laos	孟加拉(%) Bangladesh	非洲部分国家(%) Some Africa countries	Article Description (in brief)
71	02102000	干、熏、盐制的牛肉	25	0		0			Meat of bovine animals, salted, in brine, dried or smoked
72	02109100	干、熏、盐制的灵长目动物肉及食用杂碎	25	0		0			Primates, nes, saltd, dried or smokd; flours & meals of meat or offal
73	02109200	干、熏、盐制的鲸、海豚及鼠海豚;海牛及儒艮肉及食用杂碎	25	0		0			Whales, dophins, nes, saltd, dried or smokd; flours & meals of meat or offal
74	02109300	干、熏、盐制的爬行动物肉及食用杂碎	25	0		0			Reptiles, nes, saltd, dried or smokd; flours & meals of meat or offal
75	02109900	干、熏、盐制的其他肉及食用杂碎	25	0		0			Other meat, nes, saltd, dried or smokd; flours & meals of meat or offal
76	03019290	其他活鳗鱼	10	0	0				Live eels, excl. fry
77	03019390	其他活鲤鱼	10.5	0	0				Live carp, excl. fry
78	03019991	活罗非鱼	10.5	0	0				Live tilapia
79	03019992	活河鲀鱼	10.5	0	0				Live puffer fish
80	03019999	其他活鱼	10.5	0	0				Live fish nes, excl. Fry
81	03022900	其他鲜、冷比目鱼	12	0					Fresh or chilled flat fish (excl. halibut, plaice & sole)
82	03023100	鲜、冷长鳍金枪鱼	12	0					Fresh or chilled albacore or longfinned tunas
83	03023400	鲜、冷大眼金枪鱼	12	0	0				Fresh or chilled bigeye tunas
84	03023500	鲜、冷蓝鳍金枪鱼	12	0	0				Fresh or chilled bluefin tunas
85	03023600	鲜、冷南金枪鱼	12	0	0				Fresh or chilled southern bluefin tunas
86	03023900	其他鲜、冷金枪鱼	12	0	0				Fresh or chilled tunas, nes
87	03026600	鲜冷鳗鱼	12		0				Fresh or chilled eels
88	03026940	鲜冷罗非鱼	12	0	0				Fresh or chilled tilapia
89	03026950	鲜、冷河鲀鱼	12	0	0				Fresh or chilled puffer fish
90	03026990	其他鲜、冷鱼	12	0	0				Fresh or chilled fish, nes
91	03027000	鲜、冷鱼肝及鱼卵	12	0					Fresh or chilled fish livers & roes
92	03031100	冻红大马哈鱼,但鱼肝及鱼卵除外	10	0					Frozen sockeye salmon
93	03031900	其他冻大马哈鱼,但鱼肝及鱼卵除外	10	0					Other frozen pacific salmon, nes
94	03033300	冻鳎鱼	12					0	Frozen sole
95	03033900	其他冻比目鱼	10					0	Frozen flat fish (excl. halibut, plaice & sole)
96	03034400	冻大眼金枪鱼,但鱼肝及鱼卵除外	12	0	0				Frozen bigeye tunas
97	03034500	冻蓝鳍金枪鱼,但鱼肝及鱼卵除外	12	0	0				Frozen bluefin tunas

序号 No.	税则号列 Tariff Item	商品名称（简称）	最惠国税率（%）M.F.N.	柬埔寨（%）Cambodia	缅甸（%）Myanmar	老挝（%）Laos	孟加拉（%）Bangladesh	非洲部分国家（%）Some Africa countries	Article Description (in brief)
98	03034600	冻南金枪鱼，但鱼肝及鱼卵除外	12	0	0				Frozen southern bluefin
99	03034900	其他冻金枪鱼，但鱼肝及鱼卵除外	12	0	0				Frozen tunas, nes
100	03035000	冻鲱鱼（大西洋鲱鱼、太平洋鲱鱼）	10					0	Frozen herrings (excl. livers & roes)
101	03037100	冻沙丁鱼、黍鲱鱼	12					0	Frozen sardines, brisling or sprats
102	03037910	冻带鱼	10	0	0			0	Frozen scabber fish (trichurius)
103	03037920	冻黄鱼	10	0	0				Frozen yellow croaker(pseudosicaena)
104	03037930	冻鲳鱼	10	0	0				Frozen butterfish (pampus)
105	03037940	冻罗非鱼	10	0	0				Frozen tilapia
106	03037990	其他未列名冻鱼	10	0	0			0	Frozen fish, nes
107	03038000	冻鱼肝及鱼卵	10					0	Frozen fish livers & roes
108	03041000	鲜、冷的鱼片及其他鱼肉	12		0				Fresh or chilled fish fillets, and other fish meat
109	03042010	冻罗非鱼片	10	0	0				Frozen fillets of tilapia
110	03042090	其他冻鱼片	10	0	0				Frozen fillets, nes
111	03049000	其他冻鱼肉	10		0			0	Frozen fish meat (excl. fillets)
112	03053000	干或盐制的鱼片	10		0				Fish fillets, dried, salted or in brine, but not smoked
113	03054900	其他熏鱼及鱼片	14		0				Smoked fish (excl. salmon & herrings)
114	03055910	干海马、干海龙	2	0	0	2	2		Dried pipefish and hippocampi, not smoked
115	03055920	干鱼翅	15	0	0	5	5	0	Dried sharks' fins, not smoked
116	03055990	其他干鱼	16		0	5	5	0	Other dried fish, not smoked
117	03056910	盐腌及盐渍的带鱼	16		0	5	5		Scabber fish, salted or in brine but not dried or smoked
118	03056920	盐腌及盐渍的黄鱼	16		0	5	5		Yellow croaker, salted or in brine but not dried or smoked
119	03056930	盐腌及盐渍的鲳鱼	16		0	5	5		butterfish, salted or in brine but not dried or smoked
120	03056940	盐腌及盐渍的罗非鱼	16		0				Tilapia, salted but not dried nor smoked and fish in brine
121	03056990	盐腌及盐渍的其他鱼	16		0	5	5		Other fish, salted or in brine but not dried or smoked, nes
122	03061100	冻龙虾	10	0	0				Frozen rock lobster & other sea crawfish

序号 No.	税则号列 Tariff Item	商品名称（简称）	最惠国税率（%） M.F.N.	柬埔寨（%） Cambodia	缅甸（%） Myanmar	老挝（%） Laos	孟加拉（%） Bangladesh	非洲部分国家（%） Some Africa countries	Article Description (in brief)
123	03061200	冻大螯虾	10	0					Frozen lobsters
124	03061311	冻小虾仁	8	0	0				Frozen shelled shrimps
125	03061312	冻北方长额虾	5	0	0				Frozen northern pandalus
126	03061319	其他冻小虾	5	0	0			0	Frozen shrimps in shell
127	03061321	冻对虾仁	8	0	0				Frozen shelled prawns
128	03061329	其他冻对虾	5	0	0			0	Frozen prawns in shell
129	03061410	冻梭子蟹	10	0	0	5	5		Frozen fresh - water swimming crabs
130	03061490	其他冻蟹	10	0	0	5	5	0	Other frozen crabs, nes
131	03061911	冻淡水小龙虾仁	16	0					Frozen freshwater crawfish shelled
132	03061919	冻带壳淡水小龙虾	16	0					Frozen freshwater crawfish in shell
133	03061990	其他冻甲壳动物	16	0					Frozen crustaceans, nes, incl flours, meals, pellets for human consumption
134	03062190	未冻的龙虾	15	0	0				Not frozen lobster, nes
135	03062290	其他未冻的大螯虾	15	0	0				Unfrozen lobsters other than for cultivation
136	03062391	鲜、冷对虾	15	0	0				Prawns, fresh or chilled
137	03062399	其他小虾及对虾	12	0	0				Unfrozen shrimps and prawns, nes
138	03062491	未冻的中华绒毛蟹（大闸蟹）	14	0					Live fresh-water crabs
139	03062492	未冻的梭子蟹	14	0					Unfrozen swimming crabs
140	03062499	未冻的其他蟹	14	0					Unfrozen crabs, nes
141	03062990	其他带壳或去壳的未冻的甲壳动物	14	0	0				Unfrozen crustaceans, nes; incl flours, meals and pellets, fit for human consumption
142	03071090	其他牡蛎（蚝）	14		0				Oysters other than for cultivation
143	03072190	其他活、鲜、冷扇贝	14		0				Scallops, live, fresh or chilled, o/t for cultivation
144	03072900	其他冻、干、盐腌或盐渍的扇贝	14		0				Scallops, frozen, dried, saltd or in brine
145	03073190	其他活、鲜、冷贻贝	14	0	0				Mussels, live, fresh or chilled, o/t for cultivation
146	03073900	其他冻、干、盐制的贻贝	14	0	0				Mussels, frozen, dried, saltd or in brine
147	03074190	其他冷墨鱼及鱿鱼	12	0	0				Cuttle fish&squid, live, fresh or chilled, o/t cultivation
148	03074900	其他冻、干、盐制的墨鱼及鱿鱼	12	0	0			0	Cuttle fish & squid, frozen, dried, saltd or in brine

序号 No.	税则号列 Tariff Item	商品名称（简称）	最惠国税率(%) M.F.N.	柬埔寨(%) Cambodia	缅甸(%) Myanmar	老挝(%) Laos	孟加拉(%) Bangladesh	非洲部分国家(%) Some Africa countries	Article Description (in brief)
149	03075900	冻、干、盐腌或盐渍章鱼	17					0	Octopus, frozen, dried, salted or in brine
150	03076090	其他蜗牛及螺	14	0					Other snails o/t sea snails, excl for cultivation
151	03079191	活、鲜、冷鲍鱼	14	0	0				Abalones, live, fresh or chilled
152	03079192	活、鲜或冷的沙蚕	14	0	0				Clamworm
153	03079199	其他活、鲜、冷水生无脊椎动物	14	0	0				Other molluscs and aquatic invertibrates, nes, live, fresh or chilled
154	03079910	冻、干、盐腌或盐渍的鲍鱼	10	0	0				Abalone, frozen, dried, salted or in brine
155	03079920	冻、干、盐腌或盐渍的海参	10	0	0			0	Sea cucumbers, frozen, dried, salted or in brine
156	03079990	其他冻、干、盐制水生无脊椎动物	10	0	0				Other molluscs and aquatic invertibrats, nes, frz/dr saltd/br; inc 1 flours, meals and pellets, fit for human consumption
157	04070021	带壳鲜鸡蛋	20	0	0				Hen's eggs, in shell, fresh, o/t for hatching
158	04070022	带壳鲜鸭蛋	20	0	0				Duck's eggs, in shell, fresh, o/t for hatching
159	04070023	带壳鲜鹅蛋	20	0	0				Goose's eggs, in shell, fresh, o/t for hatching
160	04070029	其他带壳鲜禽蛋	20	0	0				Birds' eggs nes, in shell, fresh, o/t for hatching
161	04070091	咸蛋	20		0				Birds' eggs, in shell, salted
162	04090000	天然蜂蜜	15	0		0			Natural honey
163	04100010	燕窝	25	0	0				Edible bird's nests (salanganes' nests)
164	04100020	鲜蜂王浆	15		0				Pure royal jelly, fresh
165	04100030	鲜蜂王浆粉	15	0					Pure royal jelly, in powder
166	04100090	其他编号未列名的食用动物产品	20			0			Other edible products of animal origin, nes
167	05061000	经酸处理的骨胶原及骨	12			0			Ossein and bones treated with acid
168	05069011	含牛羊成分的骨粉及骨废料	12			0			Power and waste bones, containing bovine composition or sheep and goat's thereof
169	05069019	骨粉及骨废料	12			0			Other powder and waste of bones
170	05069090	其他骨及角柱	12			0			Bones and horn – cores, unworked, defatted, simply prepared, not cut to shape, nes

序号 No.	税则号列 Tariff Item	商品名称（简称）	最惠国税率(%) M.F.N.	柬埔寨(%) Cambodia	缅甸(%) Myanmar	老挝(%) Laos	孟加拉(%) Bangladesh	非洲部分国家(%) Some Africa countries	Article Description (in brief)
171	05071000	兽牙；兽牙粉末及废料	10					0	Ivory, its powder & waste, unworked
172	05080090	珊瑚等；软体、甲壳或棘皮动物壳，墨鱼骨	12					0	Coral; shells of molluscs, crustaceans, etc, not cut to shape
173	05119111	受精鱼卵	12					0	Fertilized fish eggs
174	05119119	鱼的其他产品	12					0	Other products of fish
175	05119190	甲壳动物、软体动物、其他水生无脊椎动物产品；第三章的死动物	12					0	Products of crustaceans, molluscs or other aquatic invertebrates; dead animals of chapter 3
176	06011010	休眠的番红花球茎	4			0			Dormant stigma croci corms
177	06011099	其他休眠的鳞茎、块茎、块根等	5			0			Dormant bulbs, tubers, corms, rhizomes, nes
178	06031000	鲜的插花及花蕾	10			0			Fresh cut flowers & buds
179	06039000	干的及经常染色等加工的插花及花蕾	23			0			Dried, dyed, bleached or otherwise prepared cut flowers & buds
180	06049900	其他制花束或装饰用不带花的植物枝、叶及草	10					0	Parts of plants, without flowers or buds, for ornamental purposes
181	07019000	其他鲜或冷的马铃薯	13	0					Other potatoes, fresh or chilled
182	07031010	鲜或冷的洋葱	13	0	0				Onions, fresh or chilled
183	07031020	鲜或冷的青葱	13	0	0				Shallots, fresh or chilled
184	07039010	鲜或冷的韭葱	13	0					Leeks, fresh or chilled
185	07039020	鲜或冷的大葱	13	0					Scallion, fresh or chilled
186	07039090	鲜或冷的其他葱属蔬菜	13	0					Otheralliaceous vegetables, fresh or chilled
187	07049000	鲜或冷的其他食用芥菜类蔬菜	13			0			White & red cabbages, kohlrabi, kale…etc, fresh or chilled
188	07070000	鲜或冷的黄瓜及小黄瓜	13	0					Cucumbers & gherkins, fresh or chilled
189	07081000	鲜或冷的豌豆	13	0					Peas, fresh or chilled
190	07082000	鲜或冷的豇豆及菜豆	13	0					Beans, fresh or chilled
191	07089000	鲜或冷的其他豆类蔬菜	13	0					Leguminous vegetables, fresh or chilled, nes
192	07095100	鲜或冷的伞菌属蘑菇	13	0					Mushrooms of the genus Agaricus, fresh or chilled
193	07095910	鲜或冷的松茸	13	0					Sungmo, fresh or chilled
194	07095920	鲜或冷的香菇	13	0					Shiitak, fresh or chilled
195	07095930	鲜或冷的金针菇	13	0					Winter mushroom, fresh or chilled

序号 No.	税则号列 Tariff Item	商品名称（简称）	最惠国税率(%) M.F.N.	柬埔寨(%) Cambodia	缅甸(%) Myanmar	老挝(%) Laos	孟加拉(%) Bangladesh	非洲部分国家(%) Some Africa countries	Article Description (in brief)
196	07095940	鲜或冷的草菇	13	0					Paddy straw mushroom, fresh or chilled
197	07095950	鲜或冷的口蘑	13	0					Trichiloma mongolicum I-mai, fresh or chilled
198	07095990	鲜或冷的其他蘑菇	13	0					Other mushrooms and truffles, nes, fresh or chilled
199	07096000	鲜或冷的辣椒,包括甜椒	13	0					Fruits of genus Capiscum or Pimenta, fresh or chilled
200	07099010	鲜或冷的竹笋	13	0					Bamboo shoots, fresh or chilled
201	07099090	鲜或冷的其他蔬菜	13	0					Other vegetables, fresh or chilled, nes
202	07102210	红小豆	13	0					Shelled or unshelled adzuki beans, frozen
203	07102290	其他冷冻豇豆及菜豆	13	0					Other shelled or unshelled beans, frozen
204	07102900	冷冻其他豆类蔬菜	13	0					Leguminous vegetables, shelled or unshelled, frozen, nes
205	07108010	冷冻松茸	13	0					Sungmo, frozen
206	07108020	其他蒜苔及蒜苗(青蒜)	13	0					Garlic stems or seedlings, frozen
207	07108030	蒜头	13	0					Garlic bulbs
208	07108090	冷冻未列名蔬菜	13	0					Other vegetables, frozen
209	07109000	冷冻什锦蔬菜	10	0					Mixtures of vegetables, frozen
210	07133190	其他干绿豆	3	0	0				Other dried beans, shelled
211	07133290	其他干赤豆	3	0					Other dried adzuki beans, shelled
212	07133390	其他干芸豆	7.5	0					Other dried kidney beans, incl. white pea beans, shelled
213	07133900	干豇豆及菜豆	7	0					Dried beans, shelled, nes
214	07141010	鲜木薯	10	0		0			Fresh manioc
215	07141020	干木薯	5	0		0			Dried manioc
216	07141030	冷或冻的木薯	10	0		0			Chilled or frozen manioc
217	07142011	种用鲜甘薯	0	0		0			Fresh sweet potatoes, for cultivatron
218	07142019	其他鲜甘薯	13	0		0			Other fresh sweet potatoes
219	07142020	干甘薯	13	0		0			Dried sweet potatoes
220	07142030	冷或冻的甘薯	13	0		0			Chilled or frozen sweet potatoes
221	07149010	干或冷、冻的荸荠	13	0		0			Waterchestnut, fresh or dried, chilled, frozen
222	07149029	其他藕	13	0		0			Lotus roots o/t for cultivation, fresh or dried
223	07149090	含有高淀粉或菊粉的其他类似根茎	13	0		0			Roots or tubers with high starch content, fresh or dried, nes

序号 No.	税则号列 Tariff Item	商品名称（简称）	最惠国税率(%) M.F.N.	柬埔寨(%) Cambodia	缅甸(%) Myanmar	老挝(%) Laos	孟加拉(%) Bangladesh	非洲部分国家(%) Some Africa countries	Article Description (in brief)
224	08011100	干的椰子	12	0					Desiccated coconuts
225	08011990	其他椰子	12	0					Other coconuts, fresh
226	08013100	鲜或干的未去壳腰果	20	0				0	Cashew nuts, in shell, fresh or dried
227	08013200	鲜或干的去壳腰果	10	0	0				Cashew nuts, shelled, fresh or dried
228	08030000	鲜或干的香蕉，包括芭蕉	10	0		0			Bananas, including plantains, fresh or dried
229	08043000	鲜或干菠萝	12	0					Pineapples, fresh or dried
230	08044000	鲜或干鳄梨	25	0					Avocados, fresh or dried
231	08072000	鲜木瓜	25	0					Papaws (papayas), fresh
232	08106000	鲜榴莲	20	0					Durian, fresh
233	08109030	鲜龙眼	12	0					Longan, fresh
234	08134010	龙眼干、肉	20	0					Longans and longan pulps, dried
235	08134020	柿饼	25	0					Persimmons, dried
236	08134030	干红枣	25	0					Chinese dates (red jujubes), dried
237	08134040	荔枝干	25	0					Preserved litchi
238	08134090	其他干果	25	0					Other fruit, dried, nes
239	09011100	未浸除咖啡碱的未焙炒咖啡	8	0		0		0	Coffee, not roasted or decaffeinated
240	09011200	已浸除咖啡碱的未焙炒咖啡	8	0		0		0	Decaffeinated coffee, not roasted
241	09012100	未浸除咖啡碱的已焙炒咖啡	15	0		0			Roasted coffee, not decaffeinated
242	09012200	已浸除咖啡碱的已焙炒咖啡	15	0		0			Roasted, decaffeinated coffee
243	09019010	咖啡豆荚及咖啡豆皮	10	0		0			Coffee husks & skins
244	09019020	含咖啡的咖啡代用品	30	0		0			Coffee substitutes containing coffee
245	09024010	每件净重＞3kg的乌龙茶	15			0			Oolong tea in packings ＞ 3kg
246	09024020	每件净重＞3kg的普洱茶	15			0			Pu－er tea in packings ＞ 3kg
247	09024090	每件净重＞3kg的其他红茶（已发酵）及关发酵茶	15			0			Other black tea & partly fermented tea nes, in packings＞ 3kg
248	09050000	香子兰豆	15					0	Vanilla
249	09070000	丁香（母丁香、公丁香及丁香梗）	3	0				0	Cloves (whole fruit, cloves & stems)
250	09081000	肉豆蔻	8	0		0			Nutmeg
251	09082000	肉豆蔻衣	8	0		0			Mace
252	09083000	豆蔻	3	0		0			Cardamoms
253	09101000	姜	15	0		0			Ginger
254	09103000	姜黄	15	0					Turmeric (curcuma)
255	10082000	谷子	2			0			Millet
256	10083000	加那利草子	2			0			Canary seed
257	10089090	其他谷物	3			0			Other cereal excl. seed, nes

序号 No.	税则号列 Tariff Item	商品名称（简称）	最惠国税率(%) M.F.N.	柬埔寨(%) Cambodia	缅甸(%) Myanmar	老挝(%) Laos	孟加拉(%) Bangladesh	非洲部分国家(%) Some Africa countries	Article Description (in brief)
258	11029000	其他谷物细粉	5	0					Other cereal flour, nes
259	11031990	其他谷物粗粒及粗粉	5	0					Cereal, meal, nes
260	11061000	干豆细粉、粗粉及粉末	10	0					Flour & meal of the dried leguminous vegetables of 07.13
261	11062000	西谷茎髓、植物根茎、块茎制成的细粉、粗粉、粉末	20	0					Flour & meal of sago, roots or tubers of 07.14
262	12010091	种用黄大豆	3	0		0			Yellow soya beans excl. seed
263	12010092	种用黑大豆	3	0		0			Black soya beans excl. seed
264	12010093	种用青大豆	3	0		0			Green soya beans excl. seed
265	12021090	种用其他未焙炒或未烹煮的带壳花生	15			0			Ground-nuts in shell, excl. seed, not cooked
266	12022000	未焙炒或烹煮的去壳花生	15			0			Ground – nuts, not roasted or otherwise cooked, shelled, whether or not broken
267	12030000	干椰子肉	15	0					Copra
268	12071090	其他油棕果及油棕仁	10	0					Palm nuts & kernels excl for sowing
269	12073090	其他蓖麻子	15	0		0			Castor oil seeds excl for sowing
270	12074090	其他芝麻	10	0	0	0		0	Sesamum seeds excl for sowing
271	12079991	牛油树果	20		0				Shea nuts (karite nuts)
272	12079999	其他含油子仁及果实	10		0				Other oil seeds & oleaginous fruits nes
273	12081000	大豆粉	9	0					Soya bean flour & meal
274	12089000	其他含油子仁或果实的细粉及粗粉	15		0				Other flours & meal of oil seeds or oleaginous fruit, nes
275	12113000	古柯叶	9	0					Coca leaf
276	12114000	罂粟秆	9	0					Poppy straw
277	12119011	鲜或干的当归	6	0					Radix angelicae sinensis, fresh or dried
278	12119012	鲜或干的田七	6	0					Radix pseudoginseng, used in pharmacy, fresh or dried
279	12119015	鲜或干的菊花	6	0					Flos chrysanthemi, used in pharmacy, fresh or dried
280	12119016	鲜或干的冬虫夏草	6	0					Cordyceps sinensis, used in pharmacy, fresh or dried
281	12119017	鲜或干的贝母	6	0					Bulbs of fritilariae thunbergii, used in pharmacy, fresh or dried
282	12119018	鲜或干的川芎	6	0					Rhizoma ligustici, used in pharmacy, fresh or dried

序号 No.	税则号列 Tariff Item	商品名称（简称）	最惠国税率(%) M.F.N.	柬埔寨(%) Cambodia	缅甸(%) Myanmar	老挝(%) Laos	孟加拉(%) Bangladesh	非洲部分国家(%) Some Africa countries	Article Description (in brief)
283	12119019	鲜或干的半夏	6	0					Rhizoma pinellia, used in pharmacy, fresh or dried
284	12119021	鲜或干的白芍	6	0					Radix paeoniae lactiflorae, used in pharmacy, fresh or dried
285	12119022	鲜或干的天麻	6	0					Rhizoma gastrodiae, used in pharmacy, fresh or dried
286	12119023	鲜或干的黄芪	6	0					Radix astragali, used in pharmacy, fresh or dried
287	12119024	鲜或干的大黄、籽黄	6	0					Rhubarb, used in pharmacy, fresh or dried
288	12119025	鲜或干的白术	6	0					Rhizoma atractylodis macrocephalae, used in pharmacy, fresh or dried
289	12119026	鲜或干的地黄	6	0					Radix rehmanniae, used in pharmacy, fresh or dried
290	12119027	鲜或干的槐米	6	0					Flos sophorae, used in pharmacy, fresh or dried
291	12119028	鲜或干的杜仲	6	0					Cortex eucommiae, used in pharmacy, fresh or dried
292	12119029	鲜或干的茯苓	6	0					Poria, used in pharmacy, fresh or dried
293	12119031	鲜或干的枸杞	6	0					Fructus lycii, used in pharmacy, fresh or dried
294	12119032	鲜或干的大海子	6	0					Bantaroi seeds, used in pharmacy, fresh or dried
295	12119033	鲜或干的沉香	3	0					Aloes wood, used in pharmacy, fresh or dried
296	12119034	沙参	6	0					Adenophora axilliflora, fresh or dried
297	12119035	青蒿	6	0					Southernwood
298	12119039	其他主要用作药料的植物	6	0					Plants and parts of plants, used in pharmacy, nes
299	12119050	主要用作香料的植物及其某部分	8	0				0	Plants and parts of plants, used in perfumery
300	12119091	鲜或干的鱼藤根、除虫菊	3	0					Dessis roots and pyrethrum, used for insecticides
301	12119099	其他鲜或干的杀虫、杀菌用植物	9	0					Plants or their parts, use for insecticides, nes
302	12122090	未列名海草及其他藻类	15					0	Edible seaweeds and other algae, fresh chilled frozen or dried nes
303	12130010	未经处理的稻草的茎、秆	12	0					Cereal straw, unprepared
304	12130090	其他未经处理的谷类植物茎、秆及谷壳	12	0					Other Cereal straw and husks
305	12141000	紫苜蓿粗粉及团粒	5	0					Lucerne (alfalfa) meal & pellets

序号 No.	税则号列 Tariff Item	商品名称（简称）	最惠国税率(%) M.F.N.	柬埔寨(%) Cambodia	缅甸(%) Myanmar	老挝(%) Laos	孟加拉(%) Bangladesh	非洲部分国家(%) Some Africa countries	Article Description (in brief)
306	12149000	芜菁甘蓝、饲料甜菜等其他植物饲料	9	0					Other forage products, nes
307	13012000	阿拉伯胶	15					0	Gum Arabic
308	13019020	乳香、没药及血竭	3					0	Olibanum, myrrh and dragon's blood
309	14011000	竹	10	0		0			Bamboos
310	14012000	藤	10	0		0			Rattans
311	14019010	谷类植物的茎秆(麦秸除外)	10	0					Cereal straw other than wheat straw
312	14019020	芦苇	10	0					Reeds
313	14019030	灯芯草属	10	0					Rushes
314	14019090	未列名主要用作编结用的植物材料	10					0	Other vegetable plaiting materials, nes
315	14020000	作填充或衬垫用植物材料	15	0					Vegetabl stuffing or padding materials
316	14030000	制帚或制刷用植物材料	15	0	0				Vegetable materials primarily for brooms or brushes, nes
317	14049000	未列名植物产品	15					0	Vegetable products, nes
318	15131900	其他椰子油及其分离品	9.5					0	Coconut copra oil (excl. crude) & fractions thereof
319	15152100	初榨玉米油的分离品	10	0		0			Crude maize (corn) oil
320	15152900	精制的玉米油及其分离品	10	0		0			Maize (corn) oil (excl. crude) & fractions
321	15153000	蓖麻油及其分离品	10	0		0			Castor oil & its fractions
322	15154000	桐油及其分离品	20	0					Tung oil & its fractions
323	15155000	芝麻油及其分离品	12	0	0				Sesame oil & fractions
324	15159010	希蒙得木油及其分离品	20	0					Jojoba oil & fractions
325	15159090	其他固定植物油、脂及其分离品	20	0					Other fixed vegetable fats & fractions, nes
326	16041300	制作或保藏的沙丁鱼、黍鲱鱼	5	0					Prepared or preserved sardines, sardinella, brisling or sprats (excl. minced)
327	16041400	制作或保藏的金枪鱼、鲣鱼	5	0					Prepared or preserved tuna, skip jack & bonito (sarda spp.)(excl. minced)
328	16041990	其他制作或保藏的鱼,整条或切块	12					0	Prepared or preserved fish (excl. minced), nes
329	16052000	制作或保藏的小虾及对虾	5	0					Shrimps & prawns, prepared or preserved
330	16053000	制作或保藏的龙虾	5	0					Lobster, prepared or preserved
331	16054011	制作或保藏的淡水小龙虾仁	5	0					Freshwater crawfish shelled, prepared or preserved
332	16054019	制作或保藏的带壳淡水小龙虾	5	0					Freshwater crawfish in shell, prepared or preserved

序号 No.	税则号列 Tariff Item	商品名称（简称）	最惠国税率(%) M.F.N.	柬埔寨(%) Cambodia	缅甸(%) Myanmar	老挝(%) Laos	孟加拉(%) Bangladesh	非洲部分国家(%) Some Africa countries	Article Description (in brief)
333	16054090	制作或保藏的其他甲壳动物	5	0					Crustaceans, nes, prepared or preserved
334	18010000	整颗或破碎的可可豆，生的或焙炒的	8					0	Cocoa beans, whole or broken, raw or roasted
335	18040000	可可脂、可可油	22					0	Cocoa butter, fat & oil
336	20041000	非醋方法制作或保藏的冷冻马铃薯	13					0	Potatoes, preserved o/t by vinegar or acetic acid, frozen
337	20060010	蜜枣	30			0			Ju jubes preserved by sugar
338	20060020	糖渍制橄榄	30		0				Olives preserved by sugar
339	20060090	其他糖渍蔬菜、水果、坚果、果皮	30		0				Other fruit, nuts or parts of plants preserved by sugar, nes
340	20079100	烹煮制成的柑桔属水果	30		0				Jams, fruit jellies, marmalades, etc, of citrus fruit
341	20081120	烘焙花生	30	0					Roasted ground－nuts
342	20081910	核桃仁罐头	20	0		0			Walnut meats, prepd, in airtight containers
343	20081920	其他果仁罐头	13	0		0			Other prepd nuts or seeds, in airtight containers
344	20081991	栗仁	10	0		0			Chestnut seed
345	20081999	其他坚果及子仁	10	0		0			Other nuts and seed
346	20082010	菠萝罐头	15			0			Pineapples prepd nes, in airtight containers
347	20082090	非用醋制作的其他菠萝	15			0			Pineapples prepard, nes
348	20083010	柑桔属水果罐头	20			0			Citrus fruit prepd nes, in airtight containers
349	20083090	非用醋制作的其他柑桔属水果	20			0			Citrus fruit prepared, nes
350	20084010	梨罐头	20			0			Pears prepd nes, in airtight containers
351	20084090	非用醋制作的其他梨	20			0			Pears prepared, nes
352	20087010	桃罐头	10			0			Peaches prepd nes, in airtight containers
353	20087090	非用醋制作的其他桃	20			0			Peaches, prepared or preserved, nes
354	20089100	非用醋制作的棕榈芯	5	0					Palm hearts, prepared or preserved, nes
355	20089200	非用醋制作的什锦果实	10	0					Mixtures of fruit, prepared or preserved, nes
356	20089920	龙眼罐头	15	0					Longan can
357	20089990	未列名制作或保藏的水果、坚果	15	0					Other fruit, etc, prepared or preserved, nes
358	20091100	冷冻的橙汁	7.5	0		0			Frozen orange juice, unfermented, not containing added spirit

序号 No.	税则号列 Tariff Item	商 品 名 称 （简 称）	最惠国税率(%) M.F.N.	柬埔寨 (%) Cambodia	缅甸 (%) Myanmar	老挝 (%) Laos	孟加拉 (%) Bangladesh	非洲部分国家(%) Some Africa countries	Article Description (in brief)
359	20091200	非冷冻的,白利糖度值不超过20的橙汁	30	0		0			Not frozen orange juice, of a Bris value not exceeding 20
360	20091900	其他橙汁	30	0		0			Unfrozen orange juice, unfermented, not containing added spirit
361	20092100	白利糖度值不超过20的柚汁	15			0			Grapefruit juice, of a Bris value not exceeding 20
362	20092900	其他柚汁	15			0			Grapefruit juice, unfermented, not containing added spirit, nes
363	20093110	白利糖度不超过20的柠檬汁	18	0					Lemon juice, of a Bris value not exceeding 20
364	20093190	其他未混合的白利糖度值不超过20的桔汁属水果汁	18	0					Single citrus fruit juice, of a Bris value not exceeding 20
365	20093910	白利糖度超过20的柠檬汁	18	0					Lemon juice, of a Bris value exceeding 20
366	20093990	其他未混合的桔汁属水果汁	18	0					Single citrus fruit juice, of a Bris value exceeding 20
367	20094100	白利糖度值不超过20的菠萝汁	10	0		0			Pineapple juice, of a Bris value not exceeding 20
368	20094900	其他菠萝汁	10	0		0			Pineapple juice, unfermented, not containing added spirit, nes
369	20095000	番茄汁	30			0			Tomato juice, unfermented, not containing added spirit
370	20098011	椰子汁	10		0	0			Coconut juice
371	20098012	芒果汁	20		0	0			Mango juice
372	20098013	西番莲果汁	20		0	0			Passion－fruit juice
373	20098014	番石榴果汁	20		0	0			Guava juice
374	20098019	其他未混合的水果汁	20		0	0			Other single fruit juice, unfermented, not containing added spirit
375	20098020	其他未混合的蔬菜汁	20		0				Other single vegetable juice, unfermented, not containing added spirit
376	20099010	混合水果汁	20	0		0			Mixtures of fruit juices, unfermented, not containing added spirit
377	20099090	混合蔬菜汁、水果与蔬菜的混合汁	20	0		0			Mixtures of veg juices, unfermented, not containing added spirit
378	21031000	酱油	28	0					Soya sauce
379	21032000	番茄沙司及其他番茄调味汁	15	0					Tomato ketchup & other tomato sauces

序号 No.	税则号列 Tariff Item	商品名称（简称）	最惠国税率(%) M.F.N.	柬埔寨(%) Cambodia	缅甸(%) Myanmar	老挝(%) Laos	孟加拉(%) Bangladesh	非洲部分国家(%) Some Africa countries	Article Description (in brief)
380	21033000	芥子粉及其调味品	15	0					Mustard flour & meal, prepared mustard
381	21069030	蜂王浆制剂	3			0			Royal jelly preparations
382	23022000	稻米糠、麸及其他残渣	5			0			Brans, sharps & other residues of rice
383	23062000	亚麻子油渣饼及固体残渣	5			0			Oil - cake & other solid residues of linseed
384	23065000	椰子或干椰肉油渣饼及固体残渣	5	0					Oil - cake & other solid residues of coconut or copra
385	23066000	油棕果或油棕仁油渣饼及固体残渣	5	0					Oil - cake & other solid residues of palm nuts & kernels
386	23067000	玉米胚芽油渣饼及固体残渣	5	0					Oil - cake & other solid residues of maize germ
387	23069000	其他油渣饼及固体残渣	5			0			Oil - cake & residues, of other vegetable fats & oils
388	25010020	纯氯化钠	3			0			Pure sodium chloride
389	25061000	石英	3					0	Quartz
390	25062100	原状或粗加修整石英岩	3					0	Crude or roughly trimmed quartzite
391	25062900	其他石英岩	3					0	Quartzite cut into blocks or slabs of a rectangular shape
392	25084000	其他粘土	3					0	Other clays, nes, whether or not calcined
393	25085000	红柱石、蓝晶石及硅线石	3					0	Andalusite, kyanite & sillimanite, whether or not calcined
394	25140000	板岩	3					0	Slate, whether or not roughly trimmed or cut into blocks or slabs of a rectangular (incl. square) shape
395	25151100	原状或粗加修整的大理石及石灰华	4					0	Marble & travertine crude or roughly trimmed
396	25161100	原状或粗加修整的花岗岩	4					0	Granite, crude or roughly trimmed
397	25162100	原状或粗加修整的砂岩	3					0	Sandstone, crude or roughly trimmed
398	25171000	卵石、砾石及碎石,圆石子及燧石	4					0	Pebbles, gravel, broken or crushed stone, commonly used for concrete aggregates, for road metalling or for railway or other ballast, shingle & flint, whether or not heat-treated

序号 No.	税则号列 Tariff Item	商品名称（简称）	最惠国税率(%) M.F.N.	柬埔寨(%) Cambodia	缅甸(%) Myanmar	老挝(%) Laos	孟加拉(%) Bangladesh	非洲部分国家(%) Some Africa countries	Article Description (in brief)
399	25174900	2515 及 2516 所列各种石料的碎粒、碎屑及粉末	3					0	Granules, chippings & powder, of stones (excl. marble) of heading 25.15 & 25.16, whether or not heat-treated
400	25201000	生石膏;硬石膏	5			0			Gypsum; anhydrite
401	25202010	牙科用熟石膏	5			0			Plasters for dentistry, whether or not coloured, with or without small quantities of accelerators or retarders
402	25202090	其他熟石膏	5			0			Plasters other than those for dentistry, whether or not coloured, with or without small quantities of accelerators or retarders
403	25309091	硅灰石	3					0	Wollastonite
404	25309099	其他税号未列名的矿产品	3					0	Mineral substances, nes
405	26203000	主要含铜矿灰及残渣	4					0	Ash & residues containing mainly copper & compound thereof
406	27011100	无烟煤	3			0			Anthracite, not agglomerated, whether or not pulverized
407	27011210	炼焦烟煤	3			0			Bituminous coking coal, not agglomerated, whether or not pulverized
408	27011290	其他烟煤	6			0			Other bituminous coal, other than coking coal, not agglomerated, whether or not pulverized
409	27011900	其他煤	5			0			Coal nes, not agglomerated, whether or not pulverized
410	27012000	煤砖、煤球及类似用煤制固体燃料	5			0			Briquettes, ovoids & similar solid fuels manufactured from coal
411	27101991	润滑油	6					0	Lubricating oils
412	27141000	沥青页岩、油页岩及焦油砂	6					0	Bituminous or oil shale & tar sands
413	29071990	未列名一元酚	5.5					0	Monophenols and their salts, nes
414	29095000	醚酚、醚醇酚及其卤、磺、硝或亚硝化衍生物	5.5					0	Ether-phenols, ether-alcohol-phenols and their halogenad, sulphonated, nitrated or nitrosated derivatives

序号 No.	税则号列 Tariff Item	商品名称（简称）	最惠国税率(%) M.F.N.	柬埔寨(%) Cambodia	缅甸(%) Myanmar	老挝(%) Laos	孟加拉(%) Bangladesh	非洲部分国家(%) Some Africa countries	Article Description (in brief)
415	29157090	棕榈酸及其盐和酯、硬脂酸盐和酯	5.5					0	Palmitic acid, its salts and esters; salts and esters of stearic acid
416	29392100	奎宁及其盐	4					0	Quinine and its salts
417	33011300	柠檬油	20					0	Essential oils of lemon (incl. concretes & absolutes)
418	33012990	未列名非柑桔属果实精油	15					0	Essential oils other than of citrus fruit, (incl. concretes & absolutes), nes
419	33019090	含浓缩精油的制品;其他萜烯副产品及精油液	20					0	Concentrates of essential oils in fats, in fixed oils, in waxes or the like; terpenic by-products of the deterpenation of essential oils (excl. citrus fruits); aqueous distillates & aqueous solutions of essential oils
420	33029000	其他工业用混合香料及香料为基本成份混合物	10					0	Odoriferous substances used as raw materials in other industry
421	34031900	矿物油<70%的润滑剂	10					0	Lubricating preparations, containing petroleum oils or oils obtained from bituminous minerals and their weight <70%
422	39152000	苯乙烯聚合物的废碎料及下脚料	9.7					0	Waste, parings & scrap, of polymers of styrene
423	39159010	聚对苯二甲酸乙二酯的塑料废碎料及下脚料	9.7					0	Waste, parings & scrap, of diglycol terephthalate of styrene
424	39159090	其他塑料废碎料及下脚料	9.7					0	Waste, parings & scrap, of other plastics, nes
425	39191099	其他宽≤20cm成卷塑料胶粘板、片、膜、箔等	6.5					0	Self-adhesive tape, plates, strip, sheet, film, foil & other flat shapes, of plastics, in rolls, width≤20cm, nes
426	39219090	未列名塑料板、片、膜、箔、扁条	6.5					0	Other plates ,strips ,sheet, film of plastics, nes
427	39221000	塑料浴缸、淋浴盘、洗涤槽及盥洗盆	10			0			Baths, shower-baths & wash-basins, of plastics
428	39222000	塑料马桶坐圈及盖	10			0			Lavatory seats & covers of plastics
429	39229000	塑料坐浴盆、便盆、抽水箱及类似卫生洁具	10					0	Bidets ,lavatory pans & other sanitary ware of plastics, nes

序号 No.	税则号列 Tariff Item	商品名称（简称）	最惠国税率(%) M.F.N.	柬埔寨(%) Cambodia	缅甸(%) Myanmar	老挝(%) Laos	孟加拉(%) Bangladesh	非洲部分国家(%) Some Africa countries	Article Description (in brief)
430	39269010	塑料制机器及仪器用零件	10					0	Machine or instruments parts of plastics
431	39269090	未列名塑料制品	10					0	Articles of plastics, nes
432	40111000	机动小客车用新的充气橡胶轮胎	10					0	New pneumatic tyres, of rubber of a kind used on motor cars
433	40169310	硫化橡胶制机器及仪器垫片、垫圈等密封垫	8					0	Gaskets , washers/seals of vulcanized rubber for machines
434	41012019	其他重量≤16公斤的牛皮	5	0		0			Whole hides and skins of bovine animals, of a weight per skin not exceeding 8kg when simply dried, 10kg when dry - salted, or 16kg when fresh, wet - salted or otherwise preserved, not tanned, parchment - dressed or further prepared, whether or not dehaired or s
435	41012020	重量≤16公斤马皮	5	0		0			Whole hides and skins of equine animals, of a weight per skin not exceeding 8kg when simply dried, 10kg when dry - salted, or 16kg when fresh, wet - salted or otherwise preserved, not tanned, parchment - dressed or further prepared, whether or not dehaired or s
436	41015019	其他重量>16公斤的牛皮	5	0		0			Whole hides and skins of bovine animals, of a weight exceeding 16 kg, not tanned, parchment - dressed or further prepared, whether or not dehaired or split, excl. have undergone a tanning which is reversible
437	41015020	重量>16公斤的马皮	5	0		0			Whole hides and skins of equine animals, of a weight exceeding 16 kg, not tanned, parchment - dressed or further prepared, whether or not dehaired or split

序号 No.	税则号列 Tariff Item	商品名称（简称）	最惠国税率(%) M.F.N.	柬埔寨(%) Cambodia	缅甸(%) Myanmar	老挝(%) Laos	孟加拉(%) Bangladesh	非洲部分国家(%) Some Africa countries	Article Description (in brief)
438	41019019	其他牛皮	5	0		0			Butts, bends and bellies and the other hides and skins of bovine animals, not tanned, parchment - dressed or further prepared, whether or not dehaired or split, excl. have undergone a tanning which is reversible
439	41019020	其他马皮	5	0		0			Butts, bends and bellies and the other hides and skins of equine animals, not tanned, parchment - dressed or further prepared, whether or not dehaired or split
440	41033000	生猪皮	9		0				Hides and skins of swine, fresh, or salted, dried, limed, pickled or otherwise preserved, but not tanned, parchment - dressed or further prepared, whether or not dehaired or split
441	41039000	其他生皮	9		0				Raw hides and skins, nes, fresh, or salted, dried, limed, pickled or otherwise preserved, but not tanned, parchment - dressed or further prepared, whether or not dehaired or split
442	41041111	全粒面未剖层及粒面剖层蓝湿牛皮	7					0	Chrome-tanned bovine leather (wetblue skin leather), full grains, unsplit, or grin splits, not further prepared
443	41041911	其他蓝湿牛皮	7					0	Wet blue bovine leather, not further prepared, nes
444	41051010	蓝湿绵羊或羔羊皮	14					0	Wet - blue sheep or lamb skin leather, without wool on, but not further prepared, whether or not split

序号 No.	税则号列 Tariff Item	商品名称（简称）	最惠国税率(%) M.F.N.	柬埔寨(%) Cambodia	缅甸(%) Myanmar	老挝(%) Laos	孟加拉(%) Bangladesh	非洲部分国家(%) Some Africa countries	Article Description (in brief)
445	41051090	其他绵羊或羔羊湿革	10					0	Sheep or lamb skin leather, in the wet state other than wet – blue, without wool on, but not further prepared, whether or not split
446	41053000	绵羊或羔羊干革(坯革)	8					0	Sheep or lamb skin leather, in the dry state (crust), without wool on, but not further prepared, whether or not split
447	41062100	山羊或小山羊湿革(包括蓝湿皮)	14					0	Goat or kid skin leather, in the wet state (incl. wet-blue), without hair on ,but not further prepared, whether or not split
448	41064000	经鞣制的不带毛爬行动物皮及其坯革	14					0	Leather of reptiles, tanned or crust, without hair on , but not further prepared, whether or not split
449	41079200	粒面剖层革(整张革除外)	5					0	Leather further prepared after tanning or crusting, including parchment-dressed leather, of bovine (including buffalo) or equine animals, without hair on ,whether or not split, other than leather of heading? 1. 14, other than whole hides and skins, grains s
450	41120000	鞣制或半硝后加工的不带毛的绵羊或羔羊皮革，但税目 41.14 的皮革除外	8					0	Leather further prepared after tanning or crusting, including parchment – dressed leather, of sheep or lamb, without wool on, whether or not split, other than leather of heading 41.14

序号 No.	税则号列 Tariff Item	商品名称（简称）	最惠国税率(%) M.F.N.	柬埔寨(%) Cambodia	缅甸(%) Myanmar	老挝(%) Laos	孟加拉(%) Bangladesh	非洲部分国家(%) Some Africa countries	Article Description (in brief)
451	41131000	经鞣制或半硝处理后加工的山羊或小山羊皮革，但税目 41.14 的皮革除外	14					0	Leather further prepared after tanning or crusting, including parchment - dressed leather, of goats or kids, without wool or hair on, whether or not split, other than leather of heading 41.14
452	41133000	经鞣制或半硝处理后加工的爬行动物皮革，但税目 41.14 的皮革除外	14					0	Leather further prepared after tanning or crusting, including parchment-dressed leather, of reptiles, without wool or hair on, whether or not split, other than leather of heading 41.14
453	42022200	塑料片或纺织材料作面的手提包	10					0	Handbags, whether or not with shoulder strap, incl. those without handle, with outer surface of plastic sheeting or of textile materials
454	42023100	皮革面通常置于口袋或手提包内的物品	10					0	Articles of a kind normally carried in the pocket or handbag, with outer surface of leather, of composition leather or of patent leather
455	42023200	塑料或纺织材料面的置于口袋或手提包内物品	20					0	Articles of a king normally carried in the pocket or handbag, with outer surface of plastic sheeting or of textile materials
456	42029200	塑料片或纺织材料作面的其他类似容器	10					0	Tool bags ,cutlery cases and containers nes, with outer surface of plastic sheeting or of textile materials
457	44201010	木刻及竹刻	0					0	Wood or bamboo carvings
458	44201090	其他木制小雕像及装饰品	0					0	Statuettes and other ornaments, of wood, nes
459	44209090	珠宝或刀具木盒及类似品；第 94 章以外木家具	0					0	Caskets and cases for jewellery or cutlery, and similar articles, of wood; wooden articles or furniture not falling in Chapter 94

序号 No.	税则号列 Tariff Item	商品名称（简称）	最惠国税率(%) M.F.N.	柬埔寨(%) Cambodia	缅甸(%) Myanmar	老挝(%) Laos	孟加拉(%) Bangladesh	非洲部分国家(%) Some Africa countries	Article Description (in brief)
460	46012010	藤制的席子、席料及帘子	9			0			Mats, matting and screens of rattans
461	46012021	灯心草属材料制的席子、席料及帘子	9			0			Mats, matting and screens of rushes
462	46012029	其他草制的席子、席料及帘子	9			0		0	Mats, matting and screens of grass or straw (other than rushes)
463	46012031	苇帘	9			0			Screens of reeds
464	46012039	芦苇制的席子、席料及帘子	9			0			Mats and matting of reeds
465	46012040	竹制的席子、席料、帘子	9			0			Mats,matting and screens of bamboo
466	46012090	其他植物材料制席子、席料及帘子	9			0			Mats, matting and screens of vegetable materials, nes
467	46019111	藤制的鞭条及其类似品	9			0			Plaits & similar products of plaiting materials, whether or not assembled into strips, of rattan
468	46019191	其他植物鞭条及其类似品	9			0			Plaits & similar products of plaiting materials, whether or not assembled into strips, of other vegetable plaiting material
469	46019199	未列名植物编结材料制其他平行连结或编结品	9					0	Plaits materials, plaits and similar products of plaiting materials, bond together in parallel strands or woven, in sheet form, whether or not being finished articles, of other vegetable plaiting material, nes
470	46019910	其他非植物鞭条及其类似品	9			0			Plaits & similar products of plaiting materials, whether or not assembled into strips, of nonvegetable plaiting materials
471	46021010	藤编制的篮筐及其他制品	9	0					Basketwork, wickerwork and other articles, of rattan
472	46021020	草编制的篮筐及其他制品	9	0				0	Basketwork, wickerwork and other articles, of grass or straw
473	46021030	竹编制的篮筐及其他制品	9	0					Basketwork, wickerwork and other articles, of bamboo

序号 No.	税则号列 Tariff Item	商品名称（简称）	最惠国税率(%) M.F.N.	柬埔寨(%) Cambodia	缅甸(%) Myanmar	老挝(%) Laos	孟加拉(%) Bangladesh	非洲部分国家(%) Some Africa countries	Article Description (in brief)
474	46021040	玉米皮编制的篮筐及其他制品	9	0					Basketwork, wickerwork and other articles, of maiz-shuck
475	46021050	柳条编制的篮筐及其他制品	9	0					Basketwork, wickerwork and other articles, of osier
476	46021090	其他植物材料编制篮筐及其他制品	9	0					Basketwork, wickerwork and other articles, of vegetable plaiting materials nes
477	46029000	其他编结材料制品及其他制品	9	0					Basketwork, wickerwork and other articles, of other non-vegetable plaiting materials; articles of loofah
478	49100000	印刷的各种日历,包括日历芯	7.5					0	Calendars of any kind, printed, including calendar blocks
479	50071010	未漂白或漂白的绸丝机织物	10			0			Unbleached or bleached woven fabrics of noil silk
480	50071090	其他绸丝机织物	10			0			Other woven fabrics of noil silk
481	50072011	未漂白或漂白的纯桑蚕丝机织物	10			0			Unbleachd or bleachd woven fabrics of mulberry silk, ≥85% silk
482	50072019	其他纯桑蚕丝机织物	10			0			Other woven fabrics of mulberry silk, containing≥85% silk
483	50072021	未漂白或漂白的纯柞蚕丝机织物	10			0			Unbleached or leached woven fabrics of tussah silk, ≥85% silk
484	50072029	其他纯柞蚕丝机织物	10			0			Other woven fabrics of tussah silk, containing≥85% silk
485	50072031	未漂白或漂白的纯绢丝机织物	10			0			Unbleached or bleached woven fabrics of silk waste, ≥85%, silk
486	50072039	其他纯绢丝机织物	10			0			Other woven fabrics of silk waste, containing ≥ 85% silk
487	50072090	其他纯丝机织物	10			0			Other woven fabrics of silk, containing≥85% silk, nes
488	50079010	未漂白或漂白其他丝机织物	10			0			Unbleached or bleached woven fabrics, containing < 85% silk

序号 No.	税则号列 Tariff Item	商品名称（简称）	最惠国税率(%) M.F.N.	柬埔寨(%) Cambodia	缅甸(%) Myanmar	老挝(%) Laos	孟加拉(%) Bangladesh	非洲部分国家(%) Some Africa countries	Article Description (in brief)
489	50079090	其他丝机织物	10			0			Other woven fabrics, nes, containing<85% silk
490	51081000	粗梳动物细毛纱线，非供零售用	5					0	Carded yarn of fine animal hair, not put up for retail sale
491	52041100	非供零售用全棉缝纫线	5			0			Cotton sewing thread, with≥85% cotton, not put up for retail sale
492	52042000	零售用棉制缝纫线	5			0			Cotton sewing thread, put up for retail sale
493	52052100	非零售精梳粗支纯棉单纱	5		0			0	Combed single cotton yarn, with≥85% cotton, nprs,≤14mn
494	52053200	棉≥85%，232≤单纱<714分特未精梳多股线	5					0	Uncombed cabled cotton yarn, with ≥85% cotton, nprs, >14mn but≤43mn
495	52054200	棉≥85%，232≤单纱<714分特精梳多股纱线	5					0	Combed cabled cotton yarn, with≥85% cotton, nprs, >14mn but≤43mn
496	52083200	棉≥85%染色平纹布，100g<平米重≤200g	10					0	Dyed plain cotton weave, with ≥ 85% cotton, > 100g/m2, ≤200g/m2
497	52085200	棉≥85%印花平纹布，100g<平米重≤200g	10					0	Printed plain cotton weave, with ≥ 85% cotton, > 100g/m2, ≤200g/m2
498	52091100	棉≥85%未漂平纹布，平米重>200g	10					0	Unbleached plain cotton weave, with≥85% cotton, >200g/m2
499	52105900	棉<85%与化纤纺印花其他布，平米重≤200g	10					0	Printed woven cotton fabrics, nes, with<85% cotton, ≤200g/m2
500	53031000	生或沤制黄麻，其他纺织用韧皮纤维	5	0		0	5		Jute, etc (excl. flax, true hemp & ramie), raw or retted
501	53039000	经加工、未纺的黄麻及其他纺织用韧皮纤维	5	0	0	0			Jute, etc (excl. flax, true hemp & ramie), nes; tow & waste…
502	53041000	生的纺织用龙舌兰类纤维	5					0	Sisal & other textile fibres of the genus Agave, raw
503	53049000	其他加工但未纺龙舌兰类纤维及其短纤和废麻	5					0	Sisal, etc (excl. raw), not spun; tow & waste of these fibres

序号 No.	税则号列 Tariff Item	商品名称（简称）	最惠国税率(%) M.F.N.	柬埔寨(%) Cambodia	缅甸(%) Myanmar	老挝(%) Laos	孟加拉(%) Bangladesh	非洲部分国家(%) Some Africa countries	Article Description (in brief)
504	53071000	黄麻及其他纺织用韧皮纤维单纱	6	0		5	5		Single yarn of jute or of other textile bast fibres of 53.03
505	53072000	黄麻及其他纺织用韧皮纤维多股纱或缆线	6	0		5	5		Multiple or cabled yarn of textile bast fibres of 53.03
506	53082000	大麻纱线	6					0	True hemp yarn
507	53101000	未漂白黄麻或其他韧皮纤维织物	10	0		5	5		Unbleached woven fabrics of jute or of other textile bast fibre
508	53109000	其他黄麻机织物	10	0		5	5		Woven fabrics of jute or oth textile bast fibres (excl. unbleached)
509	54076100	含聚酯非变形长丝≥85%的机织物	10					0	Other woven fabrics of synthetic yarn, ≥85% non-textured polyester
510	54079200	其他合成纤维长丝染色布	10					0	Dyed woven fabrics of synthetic filament yarn, nes
511	55081000	合成纤维短纤缝纫线	5					0	Sewing thread of synthetic staple fibres
512	56071000	黄麻或韧皮纤维纺制线、绳、索、缆	5			5	5		Twine, cordage, ropes & cables,of jute or other textile bast fibres
513	57023900	未制成其他纺织材料起绒铺地制品	14			5	5		Pile floor coveringsof other textiles, woven, not made up, nes
514	58012200	棉制割绒的灯芯绒	10					0	Cut corduroy of cotton
515	58071000	机织的纺织材料制非绣制标签、徽章及类似品	10					0	Labels, badges… of textiles, woven, in piece…, not embriodered
516	61031900	其他纺织材料制针织或钩编男式西服套装	17.5			0			Men's or boys' suits of other textiles, nes, knitted or crocheted
517	61032200	棉制针织或钩编男式便服套装	20			0			Men's or boys' ensembles of cotton, knitted or crocheted
518	61032900	其他纺织材料制针织或钩编男式便服套装	25			0			Men's or boys' ensembles of other textiles, nes, knitted/crocheted
519	61033200	棉制针织或钩编的男式上衣	16					0	Men's or boys' jackets & blazers of cotton, knitted or crocheted
520	61033300	合纤制针织或钩编男式上衣	19	0					Men's or boys' jackets… of synthetic fibres, knitted or crocheted

序号 No.	税则号列 Tariff Item	商品名称（简称）	最惠国税率(%) M.F.N.	柬埔寨(%) Cambodia	缅甸(%) Myanmar	老挝(%) Laos	孟加拉(%) Bangladesh	非洲部分国家(%) Some Africa countries	Article Description (in brief)
521	61033900	其他纺织材料制针织或钩编男式上衣	16	0		0			Men's or boys' jackdts… of other textiles, nes, knitted/crocheted
522	61034300	合纤制针织或钩编男长裤等	17.5	0					Men's or boys' trousers, etc, of synthetic fibres, knitted/crocheted
523	61034900	其他纺织材料制针织或钩编男长裤等	16	0		0			Men's or boys' trousers, etc, of other textiles, knitted/crocheted
524	61041200	棉制针织或钩编女式西服套装	17.5			0			Women's or girls' suits of cotton, knitted or crocheted
525	61041900	其他纺织材料制针织或钩编女式西服套装	17.5			0			Women's or girls' suits of other textiles, nes, knitted/crocheted
526	61042200	棉制针织或钩编女式便服套装	17.5			0			Women's or girls' ensembles, of cotton, knitted or crocheted
527	61043200	棉制针织或钩编的女式上衣	16					0	Women's or girls' jackets, of cotton, knitted or crocheted
528	61043300	合纤制针织女上衣	19	0					Women's or girls' jackets, of synthetic fibres, knitted/crocheted
529	61043900	其他纺织材料制针织女上衣	16	0					Woman's or girls' jackets, of other textiles, knitted or crocheted
530	61044300	合纤制针织或钩编连衣裙	17.5	0					Dresses of synthetic fibres, knitted or crocheted
531	61044400	人纤制针织或钩编连衣裙	16	0					Dresses of artificial fibres, knitted or crocheted
532	61044900	其他纺织材料制针织或钩编连衣裙	16	0					Dresses of other textile material, nes, knitted or crocheted
533	61045300	合纤制针织或钩编裙子及裙裤	16	0					Skirts & divided skirts of synthetic fibres, knitted or crocheted
534	61061000	棉制针织或钩编的女衬衫	16					0	Women's or girls' blouses, etc, of cotton, knitted or crocheted
535	61091000	棉制针织或钩编的T恤衫、汗衫、背心	14					0	T-shirts, singlets & other vests, of cotton, knitted or crocheted

序号 No.	税则号列 Tariff Item	商品名称（简称）	最惠国税率(%) M.F.N.	柬埔寨(%) Cambodia	缅甸(%) Myanmar	老挝(%) Laos	孟加拉(%) Bangladesh	非洲部分国家(%) Some Africa countries	Article Description (in brief)
536	61099090	未列名纺材制针织或钩编T恤衫、汗衫、背心	14					0	T-shirts, singlets, etc, of other textiles, nes, knitted/crocheted
537	61101100	羊毛制针织钩编套头衫开襟衫外穿背心及似品	14					0	Jerseys, pullovers, etc, of wool, knitted or crocheted
538	61102000	棉制针织钩编的套头衫、开襟衫、外穿背心等	14					0	Jerseys, pullovers, etc, of cotton, knitted or crocheted
539	61178000	针织或钩编的其他衣着附件	14					0	Other clothing accessories, knitted or crocheted, nes
540	62019900	其他纺织材料制男式防寒短上衣、防风衣	16		0				Men's or boy's anoraks, windcheaters, etc, of other textiles, nes
541	62031100	毛制男式西服套装	17.5					0	Men's or boys' suits of wool or fine animal hair
542	62033200	棉制男式上衣	16					0	Men's or boys' jackets & blazers of cotton
543	62033300	合成纤维制男式上衣	17.5					0	Men's or boys' jackets & blazers of synthetic fibres
544	62033990	未列名纺织材料制男式上衣	16					0	Men's or boys' jackets & blazers of other textl materi-als, nes
545	62034990	其他纺织材料制其他男裤	16					0	Men's or boys' trousers, breeches, nes, of oth textile fibres
546	62043100	毛制女式上衣	16					0	Women's or girls' jackets & blazers of wool or fine animal hair
547	62045990	未列名纺织材料制女式裙子及裙裤	14					0	Women's or girls' skirts of other textile materials, nes
548	62046200	棉制女裤	16					0	Women's or girls' trousers, breeches, etc, of cotton
549	62052000	棉制男衬衫	16					0	Men's or boys' shirts of cot-ton
550	62053000	化纤制男衬衫	16					0	Men's or boys' shirts of man-made fibres
551	62069000	其他纺织材料制女衬衫	16					0	Women's or girls' blouses, shirts, etc, of other textiles, nes
552	62079100	棉制其他男内衣、浴衣、晨衣及类似品	14					0	Men's or boys' singlets, dressing gowns, etc, of cot-ton
553	62131010	丝制刺绣手帕	14			0			Handkerchiefs of silk or silk waste, embroidered

序号 No.	税则号列 Tariff Item	商品名称（简称）	最惠国税率(%) M.F.N.	柬埔寨(%) Cambodia	缅甸(%) Myanmar	老挝(%) Laos	孟加拉(%) Bangladesh	非洲部分国家(%) Some Africa countries	Article Description (in brief)
554	62131090	其他丝及绢丝制手帕	14			0			Handkerchiefs of silk or silk waste, nes
555	62132010	棉制刺绣手帕	14			0			Handkerchiefs of cotton, embroidered
556	62132090	其他棉制手帕	14			0			Handkerchiefs of cotton, nes
557	62139010	其他纺织材料制刺绣手帕	14			0			Handkerchiefs of other textiles, embroidered
558	62139090	其他纺织材料制手帕	14			0			Handkerchiefs of other textiles, nes
559	62159000	其他纺织材料制领带及领结	14					0	Ties, bow ties & cravats of other textiles, nes
560	62179000	服装或衣着附件的零件	14					0	Parts of garments or of clothing accessories, nes
561	63051000	黄麻或其他韧皮纤维制货物包装袋	10			0	0		Sacks & bags, used for packing goods, of jute, etc
562	64031200	皮革制鞋面的滑雪靴	24	0					Ski – boots, etc, with rubber/plastics/leather.. soles, leather uppers
563	64031900	皮革制鞋面的其他运动鞋靴	15	0					Sports footwear, with rubber/plastics/leather.. soles, leather uppers
564	64032000	皮革条带为鞋面的皮底鞋	24	0					Sandles, with leather soles & straps (over instep, around big toe)
565	64033000	不带内底或金属护鞋头的皮面木屐	24	0					Footwear with a wood base, no inner soles or caps, leather uppers
566	64034000	装有金属护鞋头的其他皮革面鞋靴	24	0					Footwear, with a metal toecap, leather uppers
567	64035100	皮革制外底的皮革面短统靴(过踝)	10	0					Footwear with leather soles & uppers, covering the ankle
568	64035900	皮革制外底的皮革面其他鞋靴	10	0					Footwear with leather soles & uppers, not covering the ankle
569	64039100	其他皮革制面的短统靴(过踝)	10	0					Footwear with rubber … soles & leather uppers, covering the ankle
570	64039900	皮革制面的其他鞋靴	10	0					Footwear with rubber … soles, leather uppers, not coverng the ankle
571	68021010	大理石制表面积＜7cm×7cm砖瓦;染色石粉粒	24					0	Tiles etc of marble, side＜7cm; artificial colored granuls, powder

序号 No.	税则号列 Tariff Item	商品名称（简称）	最惠国税率(%) M.F.N.	柬埔寨(%) Cambodia	缅甸(%) Myanmar	老挝(%) Laos	孟加拉(%) Bangladesh	非洲部分国家(%) Some Africa countries	Article Description (in brief)
572	68029990	未列名已加工石制品	24					0	Worked building stone and articles thereof, nes
573	68159900	未列名石制品及其他矿物制品	17.5					0	Articles of stone or of other mineral substances nes
574	69031000	含＞50%石墨其他耐火陶瓷制品	8			0			Refractory ceramic goods nes, ＞50% graphite/oth forms of carbon etc
575	69032000	含＞50%氧化铝其他耐火陶瓷制品	8			0			Refractory ceramic goods nes, ＞50% A1203 or mix/compds of Si02
576	69039000	其他耐火陶瓷制品	8			0			Refractory ceramic goods nes
577	71023100	未加工或简单锯开、劈开或粗磨的非工业钻石	3					0	Diamonds non-industrial unworked or simply sawn, cleaved or bruted
578	71031000	未加工宝石或半宝石	3			0		0	Prec or semi－prec stones (o/t diamonds) unwkd or smp. sawn or rough shpd
579	71039100	经其他加工的红、蓝、绿宝石	8			0			Rubies, sapphires & emeralds further worked than sawn or rough shaped
580	71049099	经其他加工非工业用合成再造其他宝石半宝石	8					0	Syn/reconstr prec/semi-prec stones, furthr wkd, not for techncl use
581	71070000	以贱金属为底的包银材料	10.5			0			Base metals clad with silver, semi－manufactured
582	71081300	非货币用半制成金	0			0			Gold in oth semi－manufactured forms, non－monetary
583	71162000	宝石或半宝石(天然、合成或再造)制品	35					0	Articles of precious or semi-precious stones
584	71171900	贱金属制其他仿首饰	17					0	Imitation jewellery nes of base metal
585	72022900	硅铁，按重量计含硅量在55%及以下	2					0	Ferro-silicon, nes
586	73089000	其他钢铁结构体；钢结构体用部件及加工钢材	4					0	Structure/parts nes, prepd plate, rods etc for struct, i/s
587	73181500	其他螺钉及螺栓，不论是否带有螺母或垫圈	8					0	Bolts' screws nes, with' without their nuts' washers, i's
588	74011000	铜锍	2			0		0	Cu mattes
589	74012000	沉积铜(泥铜)	2			0			Cement Cu
590	74020000	未精炼铜；电解精炼用的铜阳极	2					0	Cu unrefined, Cu anodes for electrolytic refining
591	74031100	未锻轧的精炼铜阴极及阴极型材	2					0	Cu cathodes & sections of cathodes unwrought

序号 No.	税则号列 Tariff Item	商品名称（简称）	最惠国税率(%) M.F.N.	柬埔寨(%) Cambodia	缅甸(%) Myanmar	老挝(%) Laos	孟加拉(%) Bangladesh	非洲部分国家(%) Some Africa countries	Article Description (in brief)
592	74040000	铜废碎料	1.5			0		0	Waste & scrap, Cu′Cu alloy
593	74199990	未列名铜制品	20					0	Articles of Cu, nes, not for technical use
594	75022000	未锻轧的镍合金	3					0	Ni unwrought, alloyed
595	76011000	未锻轧的非合金铝	5					0	Al unwrought, not alloyed
596	76020000	铝废碎料	1.5					0	Waste & scrap, Al
597	81052000	钴锍等冶炼钴时所得中间产品;未锻轧钴;钴粉	4					0	Cobalt, unwrought, intermediate products, powders
598	81059000	锻轧钴及钴制品	8					0	Cobalt wrought & articles thereof
599	83024900	鞍具、衣箱、盒子等其他贱金属制附件及架座	12					0	Mountings, fittings & similar articles of base metal, nes
600	83081000	贱金属制钩、环、眼	10.5					0	Hooks, eyes & eyelets of base metal
601	83089000	贱金属制扣等;珠子及亮晶片,包括零件	10.5					0	Claps, buckles & like, beads & spangles of base metal
602	84139100	液体泵零件	5					0	Parts of pumps for liquids
603	84145110	功率≤125 瓦的吊扇	20			0			Ceiling/roof fans, with a built－in electric motor of output≤125 W
604	84145120	功率≤125 瓦的换气扇	20			0			Window fans, with a built－in electric motor of output≤125W
605	84145130	功率≤125 瓦,有旋转导风轮的风扇	12			0			Repeating front louver fans, with built-in electric motor, ≤125W
606	84145191	功率≤125 瓦的台扇	10			0			Table fans, with a built－in electric motor of output ≤125W
607	84145192	功率≤125 瓦的落地扇	10			0			Floor fans, with a built－in electric motor of output ≤125W
608	84145193	功率≤125 瓦的壁扇	10			0			Wall fans, with a built－in electric motor of output ≤125W
609	84145199	功率≤125 瓦其他风扇、风机	10			0			Fans nes, with a built-in electric motor of output ≤125W
610	84146010	抽油烟机	10			0			Range hoods
611	84146090	其他罩的平面最大边长不超过 120 厘米的通风罩或循环气罩	10			0			Other hoods having a maximum horizontal side ≤120cm
612	84149011	用于制冷设备的压缩机进、排气阀片	8			0			Air inlet on discharge valve plates for compressoors
613	84193100	农产品干燥器	8			0			Dryers for agricultural products

序号 No.	税则号列 Tariff Item	商品名称（简称）	最惠国税率(%) M.F.N.	柬埔寨(%) Cambodia	缅甸(%) Myanmar	老挝(%) Laos	孟加拉(%) Bangladesh	非洲部分国家(%) Some Africa countries	Article Description (in brief)
614	84193200	木材、纸浆、纸或纸板用干燥器	9			0			Dryers for wood, paper pulp, paper' paperboard
615	84195000	热交换装置	10			0			Heat exchange units, non - domestic, non - electric
616	84198100	加工热饮料、烹调、加热食品的机器	10			0			Machinery for makng hot drinks/cooking or heating food, non domestc
617	84198910	加氢反应器	0			0			Hydroformer vessels
618	84198990	其他利用温度变化处理材料的机器	0			0			Other machinery, plant & equip for treat of mat. by a chg of temp, nes
619	84314310	石油或天然气钻机的零件	4					0	Parts of oil/gas boring machinery
620	84339090	其他 8433 所列机械的零件	3					0	Parts of other machines of heading 84.33
621	84701000	不需外接电源的电子计算器	0					0	Electronic calculators, operating without external source of power
622	84821000	滚珠轴承	8					0	Bearings, ball
623	84829900	滚动轴承的其他零件	6					0	Bearing parts, nes
624	84831090	其他传动轴及曲柄	6					0	Transmission shafts not for ships; cranks
625	84834090	齿轮及其他变速、传动装置;滚珠螺杆传动轴	8					0	Gears/gearing, ball screws, gear boxes, speed changers, etc
626	84839000	单独报验的带齿的轮等;84.83 货品的其他零件	8					0	Parts of applianced of heading No.84.83
627	84849000	已包装成套或不同材料密封垫或类似接合衬垫	8					0	Gasket sets consisting of gaskets of different materials
628	85011099	其他电动机,P≤37.5W	9					0	Electric motors of an output ≤37.5 W, nes
629	85021100	压燃式内燃机发电机组,P≤75KVA	10					0	Generating sets, diesel or semidiesel engines, output≤ 75 KVA
630	85079010	铅酸蓄电池零件	10			5	5		Parts of lead - acid electric accumulators
631	85079090	其他蓄电池零件	8			5	5		Parts of other electric accumulators, nes
632	85139010	手电筒零件	14					0	Parts of torches of subheading No. 8513.1010
633	85189000	8518 所列货品的零件	10.5					0	Parts of mics, loudspeakers, headphones, earphones & elec sound ampli
634	85245390	其他已录制非教学用磁带,宽>6.5mm	10					0	Other recorded magnetic tapes, width>6.5mm, nes

序号 No.	税则号列 Tariff Item	商品名称（简称）	最惠国税率(%) M.F.N.	柬埔寨(%) Cambodia	缅甸(%) Myanmar	老挝(%) Laos	孟加拉(%) Bangladesh	非洲部分国家(%) Some Africa countries	Article Description (in brief)
635	85389000	8535、8536所列装置的其他零件	7					0	Parts for switches, fuses, panels and etc, nes
636	85481000	电池废碎料及废电池	8			5	5		Waste/scrap of primary cells/batteries/elec accumulators
637	87120041	16、18、20英寸越野自行车	13					0	Cross-country bicycles, 16″, 18″, 20″
638	90183100	注射器,不论是否装有针头	8					0	Syringes, with or without needles
639	90229090	X光发生器等、检查用家具等;9022设备零件	6					0	Parts & accessories of apparatus of 90.22, nes
640	90230000	专供示范而无其他用途的仪器、装置及模型	7					0	Instruments/apparatus/models, for demonstrational purposes
641	90328900	其他自动调节或控制仪器及装置	7					0	Automatic regulating or controlling instruments & apparatus, nes
642	91051100	电动的闹钟	23					0	Electric alarm clocks
643	92099200	9202所列乐器的零件、附件	17.5					0	Parts & accessories for musical instruments of heading No 92.02
644	94015000	藤、柳条、竹及类似材料制的坐具	0			0			Seats of cane, osier, bamboo or similar materials
645	94016900	其他木框架坐具	0					0	Seats with wooden frames, nes
646	94033000	办公室用木家具	0	0		0			Office furniture, wooden, nes
647	94034000	厨房用木家具	0	0					Kitchen furniture, wooden, nes
648	94035010	卧室用红木制家具	0	0					Bedroom furniture of padauk wood
649	94035091	卧室用漆木家具	0	0					Bedroom furniture of lacquered wood
650	94035099	卧室用其他木家具	0	0				0	Bedroom furniture, wooden, nes
651	94036010	其他红木制家具	0	0					Other furniture of padauk wood
652	94036091	其他漆木家具	0	0					Other furniture of lacquered wood
653	94036099	其他木家具	0	0				0	Furniture, wooden, nes
654	94038010	藤、柳条、竹及类似材料制家具	0	0					Furniture of cane, osier, bamboo or similar materials

序号 No.	税则号列 Tariff Item	商品名称（简称）	最惠国税率(%) M.F.N.	柬埔寨(%) Cambodia	缅甸(%) Myanmar	老挝(%) Laos	孟加拉(%) Bangladesh	非洲部分国家(%) Some Africa countries	Article Description (in brief)
655	96020090	已加工植物质或矿物质雕刻材料及其制品等	25					0	Workd veg/mineral carving material/artcls, mould/carved articles
656	96071100	装有贱金属齿的拉链	21					0	Slide fasteners fitted with chain scoops of base metal
657	96081000	圆珠笔	15					0	Ball point pens
658	97030000	各种材料制的雕塑品原件	12					0	Original sculptures & statuary, in any material

注：

税目 02071200,02071411,02071419,02071421,02071422,02071429,对应的税率为 0 元/公斤

Note:

Specific Tariff rates for items of 02071200,02071411,02071419,02071421,02071422,02071429 are all o RMB/kg.

表 3:中国—巴基斯坦优惠关税安排税目、税率表

Table 3:Duty Rate under the China-Pakistan Preferential Tariff Arrangement

序号 No.	税则号列 Tariff Item	*货 品 名 称 (简 称)	对巴基斯坦税率 (%) Pakistan	*Article Description (in brief)
1	01069020	其他食用动物	9	Other live animals, edible
2	01069090	其他动物	9	Other live animals, not edible
3	03019190	其他活鳟鱼	8	Live trout, excl. fry
4	03019290	其他活鳗鱼	6.7	Live eels, excl. fry
5	03019390	其他活鲤鱼	8	Live carp, excl. fry
6	03019992	其他活鲀	8	Live puffer fish, excl. Fry
7	03019999	其他活鱼	8	Live fish nes, excl. Fry
8	03021900	其他鲜、冷鲑鱼	8	Fresh or chilled salmonidae (excl. 0302. 11 & 0302.12)
9	03022100	鲜、冷庸鲽鱼	9	Fresh or chilled halibut
10	03022200	鲜、冷鲽鱼	9	Fresh or chilled plaice
11	03022300	鲜、冷鳎鱼	9	Fresh or chilled sole
12	03022900	其他鲜、冷比目鱼	9	Fresh or chilled flat fish (excl. halibut, plaice & sole)
13	03023100	鲜、冷长鳍金枪鱼	9	Fresh or chilled albacore or longfinned tunas
14	03023200	鲜、冷黄鳍金枪鱼	9	Fresh or chilled yellowfin tunas
15	03023300	鲜冷鲣鱼	8	Fresh or chilled skipjack or stripe – bellied bonito
16	03023900	其他鲜、冷金枪鱼	8	Fresh or chilled tunas, nes
17	03024000	鲜、冷鲱鱼	8	Fresh or chilled herrings (excl. livers & roes)
18	03025000	鲜、冷鳕鱼	8	Fresh or chilled cod (excl. livers & roes)
19	03026100	鲜、冷沙丁鱼、黍鲱鱼	8	Fresh or chilled sardines, brisling or sparts
20	03026200	鲜、冷黑线鳕鱼	8	Fresh or chilled haddock
21	03026300	鲜、冷绿青鳕鱼	8	Fresh or chilled coalfish
22	03026400	鲜、冷鲭鱼	8	Fresh or chilled mackerel
23	03026500	鲜、冷角鲨及其他鲨鱼	9	Fresh or chilled dogfish & other sharks
24	03026600	鲜冷鳗鱼	8	Fresh or chilled eels
25	03026910	鲜冷带鱼	8	Fresh or chilled scabber fish (trichurius)
26	03026920	鲜冷黄鱼	8	Fresh or chilled yellow croaker (pseudosicaena)
27	03026930	鲜冷鲳鱼	8	Fresh or chilled butterfish (pamus)
28	03026950	鲜冷鲀	8	Fresh or chilled puffer fish
29	03026990	其他鲜、冷鱼	8	Fresh or chilled fish, nes
30	03031100	冻红大马哈鱼,但鱼肝及鱼卵除外	6.7	Frozen sockeye salmon
31	03031900	其他冻大马哈鱼,但鱼肝及鱼卵除外	6.7	Other frozen pacific salmon, nes
32	03032900	其他冻鲑鱼	6.7	Frozen salmonidae,nes
33	03033110	冻格陵兰庸鲽鱼	6.7	Frozen Greenland halibut
34	03033190	其他冻庸鲽鱼	6.7	Other frozen halibut
35	03033200	冻鲽鱼	8	Frozen plaice
36	03033300	冻鳎鱼	8	Frozen sole

* :此表中"货品名称"为简称,准确的货品名称以进口税则的货品名称描述为准。

* :The *Article Description* in this table is a brief description of the article concerned. For the exact description of an article, the correspondent *Article Description* in *the Customs Import Tariff of the People's Republic of China* should be consulted.

序号 No.	税则号列 Tariff Item	*货 品 名 称 (简 称)	对巴基斯坦税率 (%) Pakistan	*Article Description (in brief)
37	03033900	其他冻比目鱼	8	Frozen flat fish (excl. halibut, plaice & sole)
38	03034100	冻长鳍金枪鱼	9	Frozen albacore or longfinned tunas
39	03034200	冻黄鳍金枪鱼	9	Frozen yellowfin tunas
40	03034300	冻鲣鱼	9	Frozen skipjack or stripe - bellied bonito
41	03034900	其他冻金枪鱼,但鱼肝及鱼卵除外	9	Frozen tunas, nes
42	03035000	冻鲱鱼(大西洋、太平洋鲱鱼),但鱼肝及鱼卵除外	6.7	Frozen herrings (excl. livers & roes)
43	03036000	冻鳕鱼	6.7	Frozen cod (excl. livers & roes)
44	03037100	冻沙丁鱼、黍鲱鱼	8	Frozen sardines, brisling or sprats
45	03037200	冻黑线鳕鱼	8	Frozen haddock
46	03037300	冻绿青鳕鱼	8	Frozen coalfish
47	03037400	冻鲭鱼	6.7	Frozen mackerel
48	03037500	冻角鲨及其他鲨鱼	9	Frozen dogfish & sharks
49	03037600	冻鳗鱼	8	Frozen eels
50	03037700	冻尖吻鲈鱼	8	Frozen sea - bass
51	03037910	冻带鱼	6.7	Frozen scabber fish (trichurius)
52	03037920	冻黄鱼	6.7	Frozen yellow croaker(pseudosicaena)
53	03037930	冻鲳鱼	6.7	Frozen butterfish (pampus)
54	03037990	其他未列名冻鱼	6.7	Frozen fish, nes
55	03038000	冻鱼肝及鱼卵	9	Frozen fish livers & roes
56	03041000	鲜、冷的鱼片及其他鱼肉	9	Fresh or chilled fish fillets, and other fish meat
57	03053000	干或盐制的鱼片	7.8	Fish fillets, dried, salted or in brine, but not smoked
58	03056100	盐腌及盐渍的鲱鱼(大西洋、太平洋鲱鱼)	11.1	Herrings, salted or in brine but not dried or smoked
59	03056200	盐腌及盐渍的鳕鱼	12	Cod, salted or in brine but not dried or smoked
60	03056300	盐腌及盐渍的 Anchovies(醍)鱼	12	Anchovies, salted or in brine but not dried or smoked
61	03061200	冻大螯虾	7.2	Frozen lobsters
62	03061311	冻小虾仁	5	Frozen shelled shrimps
63	03061319	其他冻小虾	4	Frozen shrimps in shell
64	03061321	冻对虾仁	5	Frozen shelled prawns
65	03061329	其他冻对虾	4	Frozen prawns in shell
66	03074900	其他冻、干、盐制的墨鱼及鱿鱼	10	Cuttle fish & squid, frozen, dried, salted or in brine
67	04052000	乳酱	8.1	Dairy spreads
68	05051000	填充用羽毛;羽绒	7.5	Raw feathers for stuffing; down
69	06029010	蘑菇菌丝	0	Mushroom spawn
70	07019000	其他鲜或冷的马铃薯	9	Other potatoes, fresh or chilled
71	07123100	干伞菌属蘑菇	9	Dried mushrooms of the genus Agaricus
72	07123910	干香菇	9	Dried shiitake
73	07123920	干金针菇	9	Dried winter mushroom
74	07123930	干草菇	9	Dried paddy straw mushroom
75	07123940	干口蘑	9	Dried tricholoma mongolicum lmai
76	07123950	干牛肝菌	9	Dried porcini mushroom

序号 No.	税则号列 Tariff Item	*货 品 名 称（简 称）	对巴基斯坦税率（%）Pakistan	*Article Description (in brief)
77	07123990	其他干制蘑菇及块菌	9	Other dried mushroom and truffles
78	08011100	干的椰子	10.5	Desiccated coconuts
79	08011910	种用椰子	0	Coconuts, seedlings
80	08011990	其他鲜椰子	10.5	Other coconuts, fresh
81	08030000	鲜或干的香蕉,包括芭蕉	6.9	Bananas, including plantains, fresh or dried
82	08043000	鲜或干菠萝	7.9	Pineapples, fresh or dried
83	08045010	鲜或干番石榴	10.6	Guavas, fresh or dried
84	08045020	鲜或干芒果	10.6	Mangoes, fresh or dried
85	08045030	鲜或干山竹果	10.6	Mangosteens, fresh or dried
86	08082012	鲜鸭梨、雪梨	10	Ya pears or Hsueh pears, fresh
87	08082013	鲜香梨	10	Xiang pears, fresh
88	08105000	鲜猕猴桃	16.5	Kiwifruit, fresh
89	08109010	鲜荔枝	20	Lychee, fresh
90	08109070	鲜莲雾	16.4	Wax apple, fresh
91	08109080	鲜火龙果	16.4	Dragon fruit, fresh
92	08109090	其他鲜果	16.4	Other fruit, fresh, nes
93	09021010	每件净重≤3kg 的花茶	12.5	Flavoured green tea in packings≤3kg
94	09021090	每件净重≤3kg 的其他绿茶	12.5	Unflavoured green tea in packings≤3kg
95	09022010	每件净重>3kg 的花茶	12.5	Flavoured green tea in packings > 3kg
96	09022090	每件净重>3kg 的其他绿茶	12.5	Unflavoured green tea in packings > 3kg
97	09023010	每件净重≤3kg 的乌龙茶	12.5	Oolong tea in packings≤3kg
98	09023090	每件净重≤3kg 的其他发酵、半发酵红茶	12.5	Other black tea & partly fermented tea nes, in packings≤3kg
99	09024010	每件净重>3kg 的乌龙茶	12.5	Oolong tea in packings > 3kg
100	09024090	每件净重>3kg 的其他红茶(已发酵)及半发酵茶	12.5	Other black tea & partly fermented tea nes, in packings > 3kg
101	12074010	种用芝麻	0	Sesamum seeds for sowing
102	12074090	其他芝麻	9	Sesamum seeds excl for sowing
103	12112020	鲜或干的野山参(西洋参除外)	16.4	Wild ginseng other than american ginseng, fresh or dried
104	12122020	鲜、冷、冻或干的发菜	16.2	Black moss, fresh chilled frozen or dried
105	12122090	鲜、冷、冻或干的其他海草及其他藻类	13.1	Edible seaweeds and other algae, fresh chilled frozen or dried nes
106	13021990	其他植物液汁及浸膏	15	Vegetable saps and extracts, nes, used in pesticide
107	13023200	刺槐豆、刺槐豆子或瓜尔豆制得的胶液及增稠剂	10	Mucilages&thickeners from locust beans/guar seeds
108	14041000	主要供染料或鞣料用的植物原料	4.3	Raw vegetable materials primarily for dyeing or tanning
109	16041910	制作或保藏的(河)鳗鱼,整条或切块,但未绞碎	9.9	Prepared or preserved river eel (excl. minced)
110	16041990	制作或保藏的其他鱼,整条或切块,但未绞碎	9.9	Prepared or preserved fish (excl. minced), nes
111	16042011	鱼翅罐头	9.9	Shark fin in airtight container
112	16042019	其他制作或保藏的鱼罐头	9.9	Prepared or preserved fish in airtight container
113	16042091	鱼翅	9.9	Shark fin
114	16042099	其他制作或保藏的鱼	9.9	Other prepared or preserved fish
115	16059090	其他制作或保藏的软件动物及其他水生无脊椎动物	3.9	Molluscs & other aquatic invertebrates, nes, prepared or preserved

序号 No.	税则号列 Tariff Item	*货　品　名　称（简　称）	对巴基斯坦税率（%）Pakistan	*Article Description (in brief)
116	17041000	口香糖，不论是否裹糖	9.5	Chewing gum
117	17049000	其他不含可可的糖食	8.2	Sugar confectionery not containing cocoa, nes
118	18062000	每件净重＞2kg的含可可食品	7.7	Chocolate, etc, in blocks, slabs or bars＞2kg
119	18063100	其他夹心块状或条状的含可可食品	6.4	Chocolate, etc, in blocks, slabs or bars, filled
120	18063200	其他不夹心块状或条状含可可食品	7.7	Chocolate, etc, in blocks, slabs or bars, not filled
121	18069000	其他巧克力及含可可的食品	6.4	Chocolate, etc, not in blocks, slabs or bars, nes
122	19023030	即食或快熟面条	13.1	Instant noodle
123	19023090	其他面食	13.1	Other pasta, nes
124	19053100	甜饼干	12.4	Sweet biscuits
125	19053200	华夫饼干及圣餐饼	12.4	Waffles & wafers
126	19059000	其他面包、糕点、饼干及其焙烘糕饼	17.1	Other bread, etc, nes; communion wafers, rice paper, etc
127	20093110	白利糖度不超过20的柠檬汁	16.8	Lemon juice, of a Brix value not exceeding 20
128	20093190	其他未混合的白利糖度不超过20的柑桔属水果汁	16.8	Juice of any other single citrus fruit , of a Brix value not exceeding 20
129	20093910	白利糖度超过20的柠檬汁	16.8	Lemon juice, of a Brix value exceeding 20
130	20093990	其他未混合的白利糖度超过20的柑桔属水果汁	16.8	Juice of any other single citrus fruit, of a Brix value exceeding 20
131	20098011	椰子汁	9	Coconut juice
132	20098012	芒果汁	17.4	Mango juice
133	20098013	西番莲果汁	17.4	Passion-fruit juice
134	20098014	番石榴果汁	17.4	Guava juice
135	20098019	其他未混合的水果汁	17.4	Other single fruit juice, unfermented, not containing added spirit
136	20099010	混合水果汁	17.4	Mixtures of fruit juices, unfermented, not containing added spirit
137	21039010	味精	18.2	Gourmet powder
138	21039090	其他调味品	18.4	Other sauces, mixed condiments or sea-sonings, nes
139	21069030	蜂王浆制剂	3	Royal jelly preparations
140	21069091	调味紫菜	18.4	Seasoned laver
141	21069099	其他税号未列名的食品	18.4	Other food prep.s, nes
142	22029000	其他无酒精饮料	29.5	Other non－alcoholic beverages, nes
143	22030000	麦芽酿造的啤酒	见本表尾注	Beer made from malt
144	22086000	伏特加酒	8.9	Voldka
145	22087000	利口酒及柯迪尔酒	8.9	Liqueurs and cordials
146	22089010	龙舌兰酒	8.9	Tequila, Mezcal
147	22089090	其他蒸馏酒及酒精饮料	8.9	Spirituous beverages, nes
148	24011010	未去梗的烤烟	9.4	Flue-cured tobacco, not stemmed/stripped
149	24031000	供吸用的烟丝	50	Smoking tobacco with or without tobacco substitutes
150	27074000	萘	6	Naphthalene

序号 No.	税则号列 Tariff Item	*货　品　名　称 (简　称)	对巴基斯坦税率 (%) Pakistan	*Article Description (in brief)
151	28151100	固体氢氧化钠	8	Sodium hydroxide (caustic soda), solid
152	28371120	氧氰化钠	5	Sodium cyanide oxide
153	29025000	苯乙烯	1.4	Styrene
154	29031300	氯仿(三氯甲烷)	9	Chloroform (trichloromethane)
155	29051300	正丁醇	5.5	Butan－1－ol (n－butyl alcohol)
156	29071110	苯酚	5.5	Phenol
157	29071190	苯酚的盐	5.5	Salts of phenol
158	29072300	4,4'－异亚丙基联苯酚(双酚A)及其盐	5.5	4,4'－Isopropylidenediphenol (bisphenol A, diphenylolpropane) and its salts
159	29091900	其他无环醚及其卤化、磺化、硝化、或亚硝化的等衍生物	5.5	Acyclic ethers and their halogenated, sulphonated, nitrated or nitrosated derivatives, nes
160	29152110	冰乙酸(冰醋酸)	5.5	Acetic acid, glacial
161	29152190	其他乙酸	5.5	Acetic acid, (other than glacial)
162	29161200	丙烯酸酯	6.5	Esters of acrylic acid
163	29173200	邻苯二甲酸二辛酯	6.5	Dioctyl orthophthalates
164	29173500	邻苯二甲酸酐(苯酐)	6.5	Phthalic anhydride
165	29173610	对苯二甲酸	9.1	Terephthalic acid
166	29173690	对苯二甲酸盐	6.5	Terephthalic acid salts
167	29224210	谷氨酸	8.6	Glutamic acid
168	29241100	甲丙氨酯	6.5	Meprobamate(INN)
169	29241910	二甲基甲酰胺(N,N－dimethylformamide)	6.5	N,N-dimethylformamide
170	29241990	其他无环酰胺及其衍生物,及它们的盐	6.5	Other acyclic amides and their derivatives; salts there of
171	29291010	2,4和2,6甲苯二异氰酸酯混合物(甲苯二异氰酸酯TDI)	6.5	Toluene diisdocyanate
172	29291020	二甲苯二异氰酸酯(TODI)	6.5	o－Xylene diisdocyanate
173	29291030	二苯基甲烷二异氰酸酯(纯MDI)	6.5	Diphenylmethane diisdocyanate
174	29291040	六亚甲基二异氰酸酯	6.5	Hexamethelene diisdocyanate
175	29291090	其他异氰酸酯	6.5	Isocyanates, nes
176	29331100	二甲基苯基吡唑酮(安替比林)及其衍生物	6	Phenazone (antipyrin) and its derivatives
177	29334910	环丙氟哌酸	6	Ciprofloxacin
178	29341000	结构上含有非稠合噻唑环的化合物(不论是否氢化)	6.5	Compounds containing an unfused thiazole ring (whether or not hydrogenated) in the structure
179	29349910	磺内酯及磺内酰胺	6.5	Sultones and sultams
180	29411011	氨苄青霉素	5	Ampicillin
181	29411012	氨苄青霉素三水酸	5	Ampicillin trihydrate
182	29411019	其他氨苄青霉素盐	5	Ampicillin salts, nes
183	29411091	羟氨苄青霉素	4	Amoxycillin
184	29411092	羟氨苄青霉素三水酸	4	Amoxycillin trihydrate
185	29411093	6氨基青霉烷酸(6APA)	4	6－Aminopenicillanic acid
186	29411094	青霉素V	4	Penicillins V
187	29411095	磺苄青霉素	4	Sulfobenzylpenicillins
188	29411096	邻氯青霉素	4	Cloxacillin
189	29411099	其他青霉素及其衍生物,及它们的盐	4	Other penicillins and derivatives with a penicillin acid structure; salts thereof

序号 No.	税则号列 Tariff Item	*货　品　名　称 （简　称）	对巴基斯坦税率 （%） Pakistan	*Article Description (in brief)
190	29412000	链霉素及其衍生物，及它们的盐	4	Streptomycins and their derivatives; salts thereof
191	29413011	四环素	4	Tetracyclines
192	29413012	四环素盐	4	Salts of tetracyclines
193	29413020	四环素衍生物及其盐	4	Tetracyclines derivatives and their salts
194	29414000	氯霉素及其衍生物，及它们的盐	4	Chloramphenicol and its derivatives; salts thereof
195	29415000	红霉素及其衍生物，及它们的盐	4	Erythromycin and its derivatives; salts thereof
196	29419010	庆大霉素及其衍生物，及它们的盐	4	Gentamicin and its derivatives; salts thereof
197	29419020	卡那霉素及其衍生物，及它们的盐	4	Kanamycin and its derivatives; salts thereof
198	29419030	利福平及其衍生物，及它们的盐	4	Rifampicin and its derivatives; salts thereof
199	29419040	林可霉素及其衍生物，及它们的盐	4	Lincomycin and its derivatives; salts thereof
200	29419051	7氨基头孢烷酸，7氨基脱乙酰氧基头孢烷酸	5	7 − Aminocephalosporianic acid, 7 − Aminodeacetoxycefanoic acid
201	29419052	头孢氨苄及其盐	5	Cefalexin and its salts
202	29419053	头孢唑啉及其盐	5	Cefazolin and its salts
203	29419054	头孢拉啶及其盐	5	Cefradine and its salts
204	29419055	头孢三嗪（头孢曲松）及其盐	5	Ceftriaxone and its salts
205	29419056	头孢哌酮及其盐	5	Cefoperazone and its salts
206	29419057	头孢噻肟及其盐	5	Cefotaxime and its salts
207	29419058	头孢克罗及其盐	5	Cefaclor and its salts
208	29419059	其他先锋霉素及其衍生物，及它们的盐	5	Cephamycin and its derivatives; salts thereof, nes
209	29419060	麦迪霉素及其衍生物，及它们的盐	5	Midecamycin and its derivatives; salts thereof
210	29419070	乙酰螺旋霉素及其衍生物，及它们的盐	4	Acetyl spiramycin and its derivatives; salts thereof
211	29419090	其他抗菌素	5	Antibiotics, nes
212	30041011	氨苄青霉素制剂（混合，治病或防病用，已配定剂量或制成零售包装）	5	Medicaments containing ampicillin, consisting of mixed or unmixed products for therapeutic or prophylactic uses, put up in measured doses or in forms of packing for retail sale
213	30041012	羟氨苄青霉素制剂（两种或两种以上成分混合而成的，治病或防病用，已配定剂量或制成零售包装）	5	Medicaments containing amoxycillin, consisting of mixed or unmixed products for therapeutic or prophylactic uses, put up in measured doses or in forms of packing for retail sale
214	30041013	青霉素V制剂（两种或两种以上成分混合而成的，治病或防病用，已配定剂量或制成零售包装）	5	Medicaments containing penicillins V, consisting of mixed or unmixed products for therapeutic or prophylactic uses, put up in measured doses or in forms of packing for retail sale

序号 No.	税则号列 Tariff Item	*货 品 名 称 (简 称)	对巴基斯坦税率 (%) Pakistan	*Article Description (in brief)
215	30041019	其他青霉素制剂(混合或非混合,治病或防病用,已配定剂量或制成零售包装)	5	Medicaments containing penicillins, nes, consisting of mixed or unmixed products for therapeutic or prophylactic uses, put up in measured doses or in forms of packing for retail sale
216	30041090	含有其他青霉素及具有青霉烷酸结构的青霉素衍生物或链霉素及其衍生物的药品(混合或非混合,治病或防病用,已配定剂量或制成零售包装)	5	Medicaments containing penicillins or derivatives thereof nes, or streptomycins or their derivatives, consisting of mixed or unmixed products for therapeutic or prophylactic uses, put up in measured doses or in forms of packing for retail sale
217	30042011	头孢噻肟制剂(混合或非混合,治病或防病用已配定剂量或制成零售包装)	5	Medicaments containing cefotaxime, consisting of mixed or unmixed products for therapeutic or prophylactic uses, put up in measured doses or in forms of packing for retail sale
218	30042012	头孢他啶制剂(混合或非混合,治病或防病用,已配定剂量或制成零售包装)	5	Medicaments containing ceftazidime, consisting of mixed or unmixed products for therapeutic or prophylactic uses, put up in measured doses or in forms of packing for retail sale
219	30042013	头孢西丁制剂(混合或非混合,治病或防病用,已配定剂量或制成零售包装)	5	Medicaments containing cefoxitin, consisting of mixed or unmixed products for therapeutic or prophylactic uses, put up in measured doses or in forms of packing for retail sale
220	30042014	头孢替唑制剂(混合或非混合,治病或防病用,已配定剂量或制成零售包装)	5	Medicaments containing ceftezole, consisting of mixed or unmixed products for therapeutic or prophylactic uses, put up in measured doses or in forms of packing for retail sale
221	30042015	头孢克罗制剂(混合或非混合,治病或防病用,已配定剂量或制成零售包装)	5	Medicaments containing cefaclor, consisting of mixed or unmixed products for therapeutic or prophylactic uses, put up in measured doses or in forms of packing for retail sale
222	30042016	头孢呋辛制剂(混合或非混合,治病或防病用,已配定剂量或制成零售包装)	5	Medicaments containing cefuroxime, consisting of mixed or unmixed products for therapeutic or prophylactic uses, put up in measured doses or in forms of packing for retail sale
223	30042017	头孢三嗪(头孢曲松)制剂(混合或非混合,治病或防病用,已配定剂量或制成零售包装)	5	Medicaments containing ceftriaxone, consisting of mixed or unmixed products for therapeutic or prophylactic uses, put up in measured doses or in forms of packing for retail sale

序号 No.	税则号列 Tariff Item	*货 品 名 称 （简 称）	对巴基斯坦税率 (%) Pakistan	*Article Description (in brief)
224	30042018	头孢哌酮制剂(混合或非混合,治病或防病用,已配定剂量或制成零售包装)	5	Medicaments containing cefopenazone, consisting of mixed or unmixed products for therapeutic or prophylactic uses, put up in measured doses or in forms of packing for retail sale
225	30042019	含有其他头孢菌素制剂(混合或非混合,治病或防病用,已配定剂量或制成零售包装)	5	Medicaments containing other cephamycines, consisting of mixed or unmixed products for therapeutic or prophylactic uses, put up in measured doses or in forms of packing for retail sale, nes
226	30042090	含有其他抗菌素的药品(混合或非混合,治病或防病用,已配定剂量或制成零售包装)	5	Medicaments containing other antibiotics, consisting of mixed or unmixed products for therapeutic or prophylactic uses, put up in measured doses or in forms of packing for retail sale, nes
227	30043100	含有胰岛素但不含抗菌素的药品(混合或非混合,治病或防病用,已配定剂量或零售包装)	4	Medicaments containing insulin, consisting of mixed or unmixed products for therapeutic or prophylactic uses, put up in measured doses or in forms of packing for retail sale
228	30043200	含肾上腺皮混合或非混合质激素但不含抗菌素的药品(混合或半混合,治病或防病用,已配定剂量或零售包装)	4	Medicaments containing of adrenal cortical hormones, consisting of mixed or unmixed products for therapeutic or prophylactic uses, put up in measured does or in forms of packing for retail sale
229	30043900	含有税目 29.37 其他产品但不含抗菌素的药品(混合或非混合,治病或防病用,已配定剂量或零售包装)	4	Medicaments containing other hormones, consisting of mixed or unmixed products for therapeutic or prophylactic uses, put up in measured doses or in forms of packing for retail sale, nes
230	30044010	含有奎宁或其盐,但不含抗菌素及税目 29.37 的产品的药品(混合或非混合,治病或防病用,已配定剂量或零售包装)	4	Medicaments containing quinine or its salts, but not containing antibiotics or products of heading 29.37, consisting of mixed or unmixed products for therapeutic or prophylactic uses, put up in measured doses or in forms of packing for retail sale
231	30044090	含有其他生物碱及其衍生物,但不含抗菌素及税目 29.37 的产品的药品(混合或非混合,治病或防病用,已配定剂量或零售包装)	5	Medicaments containing alkaloids or their derivatives, but not containing hormones or other products of heading 29.37 or antibiotics, consisting of mixed or unmixed products put up in measured doses or in forms of packing for retail sale, nes
232	30045000	含有维生素或税目 29.36 其他产品的其他药品(混合或非混合,治病或防病用,已配定剂量或零售包装)	5	Other medicaments containing vitamins or other products of 29.36, consisting of mixed or unmixed products for therapeutic or prophylactic uses, put up in measured doses or in forms of packing for retail sale

序号 No.	税则号列 Tariff Item	*货 品 名 称 （简 称）	对巴基斯坦税率 （%） Pakistan	*Article Description （in brief ）
233	30049010	含有磺胺的药品（两种或两种以上成分混合而成的，治病或防病用，已配定剂量或零售包装）	5	Medicaments containing sulfa drugs, consisting of mixed or unmixed products for therapeutic or prophylactic uses, put up in measured doses or in forms of packing for retail sale, nes
234	30049051	中药酒（混合或非混合，治病或防病用，已配定剂量或零售包装）	2	Chinese medicated liquors or wines, consisting of mixed or unmixed products for therapeutic or prophylactic uses, put up in measured doses or in forms of packing for retail sale
235	30049052	片仔癀（混合或非混合，治病或防病用，已配定剂量或零售包装）	2	Pien Tzu Huang, consisting of mixed or unmixed products for therapeutic or prophylactic uses, put up in measured doses or in forms of packing for retail sale
236	30049053	白药（混合或非混合，治病或防病用，已配定剂量或零售包装）	2	Bai Yao, consisting of mixed or unmixed products for therapeutic or prophylactic uses, put up in measured doses or in forms of packing for retail sale
237	30049054	清凉油（混合或非混合，治病或防病用，已配定剂量或零售包装）	2	Essential balm, consisting of mixed or unmixed products for therapeutic or prophylactic uses, put up in measured doses or in forms of packing for retail sale
238	30049059	其他中式成药（混合或非混合，治病或防病用已配定剂量或零售包装）	2	Medicaments of Chinese type, consisting of mixed or unmixed products for therapeutic or prophylactic uses, put up in measured doses or in forms of packing for retail sale, nes
239	30049060	已配剂量含有青蒿素及其衍生物的药品	4	Medicaments containing ar-temisinins and their derivatives, put up in measured doses
240	30049090	其他药品（混合或非混合，治病或防病用，已配定剂量或零售包装）	4	Medicaments of products for therapeutic, prophylactic or diagnostic uses, consisting of mixed or unmixed products for therapeutic or prophylactic uses, put up in measured doses or in forms of packing for retail sale, nes
241	30059010	药棉、纱布、绷带（经药物浸涂或制定零售包装供医疗、外科、牙科或兽医用）	3	Absorbent cotton, bandage, gauzes, impregnated or coated with pharmaceutical substances or put up in forms of packing for retail sale for medical ,surgical ,dental or veterinary purposes
242	31021000	尿素（不论是否水溶液）	40	Urea, whether or not in aqueous solution
243	31043000	硫酸钾	3	Potassium sulphate
244	32041100	分散染料及以其为基本成分的制品（不论是否已有化学定义）	7.7	Disperse dyes & preparations based thereon, whether or not chemically defined
245	32041200	酸性染料（不论是否预金属络合）及以其为基本成分的制品（不论是否已有化学定义）；媒染染料及以其为基本成分的制品（不论是否已有化学定义）	7.7	Acid dyes, whether or not premetallized, & preparations based thereon whether or not chemically defined; mordant dyes & preparations based thereon whether or not chemically defined

序号 No.	税则号列 Tariff Item	*货 品 名 称 （简 称）	对巴基斯坦税率 (%) Pakistan	*Article Description (in brief)
246	32041300	碱性染料及以其为基本成分的制品(不论是否已有化学定义)	6	Basic dyes & preparations based thereon, whether or not chemically defined
247	32041400	直接染料及以其为基本成分的制品(不论是否已有化学定义)	6	Direct dyes & preparations based thereon, whether or not chemically defined
248	32041510	合成靛蓝(还原靛蓝)(不论是否已有化学定义)	6	Synthetic indigo (reduction indigo), whether or not chemically defined
249	32041590	其他瓮染料(包括颜料用的)及以其为基本成分的制品(不论是否已有化学定义)	6	Vat dyes (incl. those usable in that state as pigments) & preparations based thereon, whether or not chemically defined, excl. synthetic indigo
250	32041600	活性染料及以其为基本成分的制品(不论是否已有化学定义)	7.7	Reactive dyes & preparations based thereon, whether or not chemically defined
251	32041700	颜料及以其为基本成分的制品(不论是否已有化学定义)	6	Pigments & preparations based thereon, whether or not chemically defined
252	32041911	硫化黑(硫化青)及以其为基本成分的制品(不论是否已有化学定义)	6	Sulphur black & preparations based thereon, whether or not chemically defined
253	32041919	其他硫化染料及以其为基本成分的制品(不论是否已有化学定义)	6	Sulphur dyes & preparations based thereon, nes, whether or not chemically defined
254	32041990	由子目号 3204.11 至 3204.19 中两个或多个子目所列着色料组成的混合物(不论是否已有化学定义)	6	Synthetic organic colouring matters, incl. mixtures of colouring matter of two or more of the subheadings No.3204. 11 to 3204.19, nes
255	32042000	用作萤光增白剂的有机合成产品(不论是否已有化学定义)	6	Synthetic organic products used as fluorescent brightening agents, whether or not chemically defined
256	32049010	生物染色剂及染料指示剂(不论是否已有化学定义)	6	Biological stains and dye indicators, whether or not chemically defined
257	32049090	其他用作发光体的有机合成产品(不论是否已有化学定义)	6	Synthetic organic products used as luminophores, nes, whether or not chemically defined
258	32081000	分散或溶于非水介质的聚酯油漆及清漆等	9	Paints & varnishes based on polyesters, dispersed or dissolved in a non – aqueous medium
259	32082010	分散或溶于非水介质的丙烯酸聚合物油漆及清漆	9	Paints & varnishes based on acrylic polymers, dispersed or dissolved in a non – aqueous medium
260	32082020	分散或溶于非水介质的乙烯聚合物油漆及清漆	9	Paints & varnishes based on vinyl polymers, dispersed or dissolved in a non – aqueous medium
261	32089010	分散或溶于非水介质的聚胺酯类油漆及清漆	9	Paints & varnishes based on polyurethane polymers, dispersed or dissolved in a non – aqueous medium
262	32089090	分散或溶于非水介质其他油漆、清漆溶液	9	Paints & varnishes, dispersed or dissolved in a non – aqueous medium, nes

序号 No.	税则号列 Tariff Item	*货 品 名 称 (简 称)	对巴基斯坦税率 (%) Pakistan	*Article Description (in brief)
263	32100000	其他油漆及清漆(包括瓷漆\大漆及水浆涂料);皮革用水性颜料	9	Paints and varnishes (incl. enamels, lacquers & distemper), nes; prepared water pigments for leather finishing
264	32139000	非成套颜料、调色料及类似品(艺术家、学生和广告美工用的,片状、管装、罐装、瓶装、扁盒装等类似形状或包装的)	9	Colours, modifying tints & the like used by artists, students, signboard painters, in tables, tubes, jars, bottles, pans or in similar forms of packings, nes
265	33012500	其他薄荷油(包括浸膏及净油)	14	Essential oils of mints (incl. concretes & absolutes), nes
266	33019010	提取的油树脂	19	Extracted oleoresins
267	33019090	用花香吸取法或浸渍法制定的含浓缩精油的脂肪、固定油、蜡及类似品;精油脱萜所得的萜烯副产品(柑桔属果实的除外);精油水溶液及水馏液	18.1	Concentrates of essential oils in fats, in fixed oils , in waxes or the like; terpenic by - products of the deterpenation of essential oils (excl. citrus fruits); aqueous distillates & aqueous solutions of essential oils
268	33021010	生产饮料用的混合香料以及以香料为基本成分的制品,按容量计酒精浓度≤0.5%	12.8	Preparations based on odoriferous substances for the manufacture of beverages, alcohol strength by volume≤0.5%
269	33030000	香水及花露水	8.2	Perfumes & toilet waters
270	33051000	洗发剂(香波)	10.5	Shampoos
271	33062000	牙线	9.2	Dental floss
272	34011100	盥洗用肥皂及有机表面活性产品,条状、块状或模制形状的,以及用肥皂或洗涤剂浸渍、涂面或包覆的纸、絮胎、毡呢及无纺织物	8.3	Soap and organic surface - active products and preparations for use as soap, in bars, cakes, moulded pieces or shapes; paper , wadding, felt and nonwovens, impregnated, coated or covered with soap or detergent
273	34012000	其他形状的肥皂	12.4	Soap in other forms
274	34021100	阴离子型有机表面活性剂	6	Anionic surface - active agents, (excl. soap), whether or not put up for retail sale
275	34021200	阳离子型有机表面活性剂	6	Cationic surface - active agents, (excl. soap), whether or not put up for retail sale
276	34021300	非离子型有机表面活性剂	6	Non - ionic surface - active agents, (excl. soap), whether or not put up for retail sale
277	34021900	其他有机表面活性剂	6	Organic surface - active agents, (excl. soap), nes, whether or not put up for retail sale
278	34022010	零售包装的合成洗涤粉	9.4	Synthetic detergent powder, put up for retail sale
279	34022090	零售包装有机表面活性剂制品(合成洗涤粉除外)	8.9	Surface - active preparations, washing & cleaning preparations, put up for retail sale, excl. synthetic detergent

序号 No.	税则号列 Tariff Item	*货 品 名 称 (简 称)	对巴基斯坦税率 (%) Pakistan	*Article Description (in brief)
280	34029000	非零售包装有机表面活性剂制品、洗涤剂及清洁剂	7.8	Surface – active preparations, washing preparations & cleaning preparations, not put up for retail sale
281	35061000	适于作胶或粘合剂的产品,零售包装每件净重≤1kg	9.2	Products suitable for use as glues or adhesives, put up for retail sale, net weight ≤ 1kg
282	35069900	其他调制胶、粘合剂	8.6	Prepared glues & other prepared adhesives, nes
283	38021000	活性碳	6.4	Activated carbon
284	38140000	其他税号未列名的有机复合溶剂及稀释剂;除漆剂	9	Organic composite solvents & thinners, nes; prepared paint or varnish removers
285	38170000	混合烷基苯及混合烷基萘,但税目27.07及29.02的产品除外	6.5	Mixed alkylbenzenes and mixed alkylnaphthalenes, other than those of heading 27.07 or 29.02
286	38220010	附于衬背上的诊断或实验用试剂,但税目32.02,32.06的货品除外	4	Diagnostic or laboratory reagents on a backing, other than those of heading 30.02 or 30.06
287	38247100	仅含氟、氯的无环烃全卤化衍生物的混合物	6.5	Mixtures containing acyclic hydrocarbons perhalogenated only with fluorine and chlorine
288	38247900	其他含有两种或两种以上不同卤素的无环烃全卤化衍生物的混合物	6.5	Other mixtures containing perhalogenated derivatives of acyclic hydrocarbons containing two or more different halognes
289	38249010	杂醇油	5.5	Fusel oils
290	38249020	除墨剂、蜡纸改正液及类似品	8.3	Ink – removers, stencil correctors and the like
291	38249040	多次甲基多苯基异氰酸酯(聚合 MDI)	6	Polymethane polyphenyl Isocyanate (Polymetric MDI or Crude MDI)
292	38249090	其他税目未列名的化学工业及其相关工业的化学产品及配制品	6	Chemical products and preparations of the chemical or allied industries, nes
293	39011000	初级形状比重<0.94 的聚乙烯	9.9	Polyethylene having a specific gravity < 0.94, in primary forms
294	39012000	初级形状比重≥0.94 的聚乙烯	9.9	Polyethylene having a specific gravity ≥ 0.94, in primary forms
295	39013000	初级形状乙烯–乙酸乙烯酯共聚物	6.2	Ethylene – vinyl acetate copolymers, in primary forms
296	39019090	其他初级形状的乙烯聚合物	6.3	Polymers of ethylene, in primary forms, nes
297	39021000	初级形状的聚丙烯	9.7	Polypropylene, in primary forms
298	39023010	初级形状的乙烯丙烯共聚物(乙丙橡胶丙烯单体单元的含量大于乙烯单体单元)	9.1	Ethylene – propylene copolymers, in primary forms, monomer of propylene by weight more than of ethylene
299	39023090	初级形状的其他丙烯共聚物	9.1	Propylene copolymers, in primary forms, nes
300	39031100	初级形状的可发性聚苯乙烯	9.1	Expansible polystyrene, in primary forms
301	39031900	初级形状的其他聚苯乙烯	9.1	Polystyrene (excl. expansible), in primary forms

序号 No.	税则号列 Tariff Item	*货　品　名　称 （简　称）	对巴基斯坦税率 （%） Pakistan	*Article Description (in brief)
302	39033000	丙烯腈－丁二烯－苯乙烯共聚物	9.1	Acrylonitrile－butadiene－styrene (ABS) copolymers, in primary forms
303	39039000	初级形状的其他苯乙烯聚合物	9.4	Polymers of styrene, in primary forms, nes
304	39041000	初级形状的纯聚氯乙烯	9.1	Polyvinyl chloride, not mixed with other substances, in primary forms
305	39079900	初级形状的其他聚酯	6.2	Polyesters, in primary forms, nes
306	39206900	其他聚酯板、片、膜、箔及扁条	9	Plate/foil/strip/sheet/film of other polyesters, not reinforced, laminated, supported or similarly combined with other materials, non－cellular, nes
307	39211100	泡沫聚苯乙烯板、片、带、箔及扁条	9	Cellular plates, strips, sheet, film of polymers of styrene
308	39264000	塑料制小雕塑品及其他装饰品	8.3	Statuettes & other ornamental articles of plastics
309	39269010	塑料制机器及仪器用零件	9	Machine or instruments parts of plastics
310	39269090	其他塑料制品	9.2	Articles of plastics, nes
311	40012100	天然橡胶烟胶片	17	Smoked sheets of natural rubber
312	40012900	其他初级形状的天然橡胶	17	Natural rubber, in primary forms or in plates, sheets or strip, nes
313	40022090	丁二烯橡胶板、片、带	7	Butadiene rubber, in plates, sheets or strips
314	40111000	机动小客车用新的充气橡胶轮胎	9.5	New pneumatic tyres, of rubber of a kind used on motor cars
315	40112000	客或货运车用新的充气橡胶轮胎	9.5	New pneumatic tyres, of rubber of a kind used on buses or lorries
316	40131000	汽车用橡胶内胎	13	Inner tubes, of rubber of a kind used on motor cars, buses or lorries
317	41012011	规定重量范围内的整张生牛皮，经逆鞣处理的	6	Whole hides and skins of bovine animals, of a weight per skin not exceeding 8kg when simply dried, 10kg when dry－salted, or 16kg when fresh, wet－salted or otherwise preserved, not tanned, parchment-dressed or further prepared, whether or not dehaired or split
318	41015011	经逆鞣处理的重量＞16公斤的整张生牛皮	7	Whole hides and skins of bovine animals, of a weight exceeding 16 kg, not tanned, parchment-dressed or further prepared, whether or not dehaired or split, have undergone a tanning which is reversible
319	41019011	其他（包括整张或半张的背皮及腹皮）经逆鞣处理的生牛皮	7	Butts, bends and bellies and the other hides and skins of bovine animals, not tanned, parchment-dressed or further prepared, whether or not dehaired or split, have undergone a tanning which is reversible

序号 No.	税则号列 Tariff Item	*货 品 名 称 (简 称)	对巴基斯坦税率 (%) Pakistan	*Article Description (in brief)
320	41022190	浸酸的不带毛绵羊或羔羊生皮,经逆鞣处理的除外	8	Raw skins of sheep or lambs, pickled, but not tanned, parchment - dressed or further prepared, without wool on, excl. have undergone a tanning which is reversible
321	41022990	其他不带毛的绵羊或羔羊生皮,经逆鞣处理的除外	6	Raw skins of sheep or lambs, fresh, or salted, dried, limed or otherwise preserved, without wool on, nes
322	41041111	全粒面未剖层或粒面剖层蓝湿牛皮	6.8	Chrome - tanned bovine leather (wet blue skin leather), full grains, unsplit, or grain splits, not further prepared
323	41041119	其他全粒面未剖层或粒面剖层湿牛皮革	6	Bovine leather, wet state other than wet - blue, full grains, unsplit, or grain splits, not further prepared
324	41041120	全粒面未剖层或粒面剖层马皮革	4	Equine leather, wet state, full grains, unsplit, or grain splits, not further prepared
325	41041911	其他蓝湿牛皮	6.8	Wet blue bovine leather, not further prepared, nes
326	41041919	其他湿牛皮革	6	Bovine leather, wet state other than wet - blue, nes, not further prepared
327	41041920	其他湿马皮革	5	Equine leather, wet state, not further prepared, nes
328	41044100	全粒面未剖层或粒面剖层干革(坯革)	4	Bovine or equine leather, without hair on, dry state(crust), full grains, unsplit, or grain splits, not further prepared
329	41044910	其他机器带用干革(坯革)	4	Bovine or equine leather, without hair on, dry state(crust) , not further prepared, for machinery belting, nes
330	41044990	其他干革(坯革)	6	Bovine or equine leather, without hair on, dry state(crust) , not further prepared, nes
331	41051010	蓝湿绵羊或羔羊皮	10	Wet - blue sheep or lamb skin leather, without wool on, but not further prepared, whether or not split
332	41051090	其他绵羊或羔羊湿皮革	7	Sheep or lamb skin leather, in the wet state other than wet - blue, without wool on, but not further prepared, whether or not split
333	41053000	绵羊或羔羊干革(坯革)	7	Sheep or lamb skin leather, in the dry state(crust), without wool on, but not further prepared, whether or not split
334	41062100	山羊或小山羊皮湿革	12	Goat or kid skin leather, in the wet state (incl. wet - blue), without hair on, but not further prepared, whether or not split
335	41062200	山羊或小山羊皮干革(坯革)	12	Goat or kid skin leather, in the dry state (crust), without hair on, but not further prepared, whether or not split

序号 No.	税则号列 Tariff Item	*货 品 名 称 (简 称)	对巴基斯坦税率 (%) Pakistan	*Article Description (in brief)
336	41120000	已鞣进一步加工的不带毛绵羊或羔羊皮革	7	Leather further prepared after tanning or crusting, including parchment - dressed leather, of sheep or lamb, without wool on, whether or not split, other than leather of heading 41.14
337	41131000	已鞣进一步加工的不带毛山羊或小山羊皮革	13	Leather further prepared after tanning or crusting, including parchment - dressed leather, of goats or kids, without wool or hair on, whether or not split, other than leather of heading 41.14
338	41142000	漆皮及层压漆皮;镀金属皮革	9	Patent leather and patent laminated leather; metallized leather
339	42010000	各种材料制成的鞍具及挽具,适合各种动物用	12	Saddlery and harness for any animal (incl. traces, leads, knees pads, muzzles, saddle cloths, saddle bags, dog coats and the like), of any material
340	42022100	以皮革、再生皮革、漆皮作面的手提包	6.9	Handbags, whether or not with shoulder strap, incl. those without handle, with outer surface of leather, of composition leather or of patent leather
341	42022200	以塑料片或纺织材料作面的手提包	8.2	Handbags, whether or not with shoulder strap, incl. those without handle, with outer surface of plastic sheeting or of textile materials
342	42022900	以钢纸或纸板作面的手提包	14	Handbags, whether or not with shoulder strap, incl. those without handle, nes (for example, with outer surface of vulcanized fibre or of paperboard)
343	42023100	以皮革、再生皮革作面的钱包等物品	6.9	Articles of a kind normally carried in the pocket or handbag, with outer surface of leather, of composition leather or of patent leather
344	42023200	以塑料或纺织品作面的钱包等物品	14	Articles of a kind normally carried in the pocket or handbag, with outer surface of plastic sheeting or of textile materials
345	42023900	以钢纸或纸板作面的钱包等物品	14	Articles of a kind normally carried in the pocket or handbag, nes (for example, with outer surface of vulcanized fibre or of paperboard)
346	44112100	未经机械加工或盖面的中密度木纤维板	4	Fibreboard of wood or other ligneous materials, of a density exceeding 0.5g/cm^3 but not exceeding 0.8g/cm^3, whether or not bonded with resins or other organic substances, not mechanically worked or surface covered

序号 No.	税则号列 Tariff Item	*货　品　名　称（简　称）	对巴基斯坦税率（%） Pakistan	*Article Description (in brief)
347	44112900	其他中密度纤维板	4	Fibreboard of wood or other ligneous materials, of a density exceeding 0.5g/cm^3 but not exceeding 0.8g/cm^3, whether or not bonded with resins or other organic substances, mechanically worked or surface covered
348	44113100	未机械加工或盖面的低密木纤维板	7.5	Fibreboard of wood or other ligneous materials, of a density exceeding 0.35g/cm^3 but not exceeding 0.5g/cm^3, whether or not bonded with resins or other organic substances, not mechanically worked or surface covered
349	44113900	其他低密度木纤维板	7.5	Fibreboard of wood or other ligneous materials, of a density exceeding 0.35g/cm^3 but not exceeding 0.5g/cm^3, whether or not bonded with resins or other organic substances, mechanically worked or surface covered
350	44119100	其他未机械加工或盖面的木纤维板	7.5	Fibreboard of wood or other ligneous materials, of a density not exceeding 0.35g/cm^3, whether or not bonded with resins or other organic substances, not mechanically worked or surface covered
351	44119900	其他木纤维板	4	Fibreboard of wood or other ligneous materials, of a density not exceeding 0.35g/cm^3, whether or not bonded with resins or other organic substances, mechanically worked or surface covered
352	44121300	至少有一表层为热带木薄板制的胶合板	12	Plywood consisting solely of sheets of wood, each ply not exceeding 6mm thickness, with at least one outer ply of tropical wood specified
353	44121410	至少有一表层为温带非针叶木单板制的胶合板	4	Plywood consisting solely of sheets of wood, each ply not exceeding 6mm thickness, with at least one outer ply of temperate non-coniferous wood
354	44121420	由两层或两层以上竹板层叠胶合而成的其他多层板	4	Multilayer board with two or more plies of bamboo
355	44121490	其他至少有一表层为非针叶木单板制的胶合板	4	Plywood consisting solely of sheets of wood, each ply not exceeding 6mm thickness, other, with at least one outer ply of non-coniferous wood
356	44121900	其他仅由薄木板制胶合板	4	Plywood consisting solely of sheets of wood, each ply not exceeding 6mm thickness, nes
357	44122200	至少一层为热带木的非针叶木面多层板	10	Plywood, veneered panels and similar laminated wood, with at least one outer ply of non – coniferous and at least one ply of tropical wood specified

序号 No.	税则号列 Tariff Item	*货　品　名　称（简　称）	对巴基斯坦税率（%）Pakistan	*Article Description (in brief)
358	44122300	至少一层为木碎板的非针叶木面多层板	10	Plywood, veneered panels and similar laminated wood, with at least one outer ply of non－coniferous and, containing at least one layer of particle board
359	44122910	至少有一层是温带非针叶木面多层板	10	Plywood, veneered panels and similar laminated wood, with at least one ply of temperate non-coniferous wood
360	44122990	其他非针叶木面多层板	10	plywood, veneered panels and similar laminated wood, other, with at least one outer-ply of non-coniferous wood
361	44129200	至少一层为热带木的针叶木面多层板	8	Plywood, veneered panels and similar laminated wood, with at least one ply of tropical wood specified, nes
362	44129300	至少一层为木碎板的针叶木面多层板	10	Plywood, veneered panels and similar laminated wood, containing at least one layer of particle board, nes
363	44129910	其他至少有一表层是温带非针叶木面多层板	4	Plywood, veneered panels and similar laminated wood, other, with at least one ply of temperate non-coniferous wood
364	44129990	其他的胶合板、单板饰面及类似的多层板	4	Other plywood, veneered panels and similar laminated wood
365	46029000	其他编结材料制品及其他制品	8	Basketwork, wickerwork and other articles, of other non－vegetable plaiting materials; articles of loofah
366	48041100	成卷或成张的未经涂布未漂白的牛皮挂面纸	5	Unbleached kraftliner, uncoated, in rolls or sheets
367	48041900	成卷或成张的未经涂布的漂白的牛皮挂面纸	5	Kraftliner (excl. unbleached), uncoated, in rolls or sheets
368	48042100	未漂白的袋用牛皮纸	5	Unbleached sack kraft paper, uncoated, in rolls or sheets
369	48042900	漂白的袋用牛皮纸	5	Sack kraft paper (excl. unbleached), uncoated, in rolls or sheets
370	48043100	未漂白的其他薄牛皮纸及纸板	2	Unbleached kraft paper and paperboard, nes, weighing $\leqslant 150g/m^2$, uncoated, in rolls or sheets
371	48043900	漂白的薄牛皮纸及纸板	2	Kraft paper and paperboard (excl. unbleached), nes, weighing $\leqslant 150g/m^2$, uncoated, in rolls or sheets
372	48044100	未漂白的其他中厚牛皮纸及纸板	2	Unbleached kraft paper and paperboard, nes, $150g/m^2 <$ weighing $< 225g/m^2$, uncoated, in rolls or sheets
373	48044200	本体均匀漂白的中厚牛皮纸及纸板	5	Kraft paper and paperboard, bleached uniformly throughout the mass and of which more than 95% by weight of the total fibre content consists of wood fibres obtained by a chemical process, nes, $150g/m^2 <$ weighing $< 225g/m^2$, uncoated, in rolls or sheets

序号 No.	税则号列 Tariff Item	*货品名称（简称）	对巴基斯坦税率（%）Pakistan	*Article Description (in brief)
374	48044900	其他漂白的中厚牛皮纸及纸板	2	Kraft paper and paperboard (excl. unbleached), nes, $150g/m^2 <$ weighing $< 225g/m^2$, uncoated, in rolls or sheets
375	48045100	未漂白的其他厚牛皮纸及纸板	2	Unbleached kraft paper and paperboard, nes, weighing $\geqslant 225g/m^2$, uncoated, in rolls or sheets
376	48045200	本体均匀漂白的厚牛皮纸及纸板	5	Kraft paper and paperboard, bleached uniformly throughout the mass and of which more than 95% by weight of the total fibre content consists of wood fibres obtained by a chemical process, nes, weighing $\geqslant 225g/m^2$, uncoated, in rolls or sheets
377	48045900	其他漂白的厚牛皮纸及纸板	2	Kraft paper and paperboard (excl. unbleached), nes, weighing $\geqslant 225g/m^2$, uncoated, in rolls or sheets
378	48101300	涂无机物书写(印刷)纸(板)，不含用机械方法制得的纤维或所含前述纤维不超过全部纤维重量的10%，成卷的	5	Paper/paperboard of a kind writing/printing/other graphic purposes, fibres obtained by a mechanical process$\leqslant$10% by weight of total fibre, weighing $\leqslant 150g/m^2$, coated with kaolin or other inorganic substances, and with no other coating, in rolls
379	48101400	涂无机物书写(印刷)纸(板)，不含用机械方法制得的纤维或所含前述纤维不超过全部纤维重量的10%，成张的，一边≤435mm，另一边≤297mm(以未折叠计)	5	Paper/paperboard of a kind writing/printing/other graphic purposes, fibres obtained by a mechanical process$\leqslant$10% by weight of total fibre, weighing $\leqslant 150g/m^2$, coated with kaolin or other inorganic substances, and with no other coating, in sheets with one side not exceeding 435mm and the other side not exceeding 297mm in the unfolded state
380	48101900	其他涂无机物书写(印刷)纸(板)，不含用机械方法制得的纤维或所含前述纤维不超过全部纤维重量的10%，成张的	5	Paper/paperboard of a kind writing/printing/other graphic purposes, fibres obtained by a mechanical process$\leqslant$10% by weight of total fibre, weighing $\leqslant 150g/m^2$, coated with kaolin or other inorganic substances, and with no other coating, in sheets, nes
381	48239020	神纸及类似用品	7.5	Joss paper and the like
382	48239030	纸扇	7.5	Paper fans
383	48239090	其他纸及纸制品	7.5	Paper, paperboard, cellulose wadding and webs of cellulose fibres, nes, cut to size; articles of paper pulp, paper, paperboard, cellulose wadding or webs of cellulose fibres, nes

序号 No.	税则号列 Tariff Item	*货　品　名　称（简　称）	对巴基斯坦税率（%）Pakistan	*Article Description (in brief)
384	50072011	未漂白或漂白的纯桑蚕丝机织物	9	Unbleached or bleached woven fabrics of mulberry silk, ≥85% silk
385	50072019	其他纯桑蚕丝机织物	9	Other woven fabrics of mulberry silk, containing≥85% silk
386	50072021	未漂白或漂白的纯柞蚕丝机织物	9	Unbleached or bleached woven fabrics of tussah silk, ≥85% silk
387	50072029	其他纯柞蚕丝机织物	9	Other woven fabrics of tussah silk, containing≥85% silk
388	50072031	未漂白或漂白的纯绢丝机织物	9	Unbleached or bleached woven fabrics of silk waste,≥85% silk
389	50072039	其他纯绢丝机织物	9	Other woven fabrics of silk waste, containing≥85% silk
390	50072090	其他纯丝机织物	9	Other woven fabrics of silk, containing≥85% silk, nes
391	51121100	重量≤200g/平米精梳全毛布	8.9	Woven fabrics with≥85% combed wool or fine animal hair,≤200g/m^2
392	52053100	非零售粗梳粗支纯棉多股纱	4.5	Uncombed cabled cotton yarn, with≥85% cotton, nprs,≤14mn
393	52054100	非零售精梳粗支纯棉多股纱	4.5	Combed cabled cotton yarn, with≥85% cotton, nprs,≤14mn
394	52054200	非零售精梳中支纯棉多股纱	4.5	Combed cabled cotton yarn, with≥85% cotton, nprs,>14mn but≤43mn
395	52054600	非零售精梳较特细支纯棉多股纱	4.5	Combed cabled cotton yarn, with≥85% cotton, nprs,>80mn but≤94mn
396	52054700	非零售精梳特细支纯棉多股纱	4.5	Combed cabled cotton yarn, with≥85% cotton, nprs,>94mn but≤120mn
397	52054800	非零售精梳超细支纯棉多股纱	4.5	Combed cabled cotton yarn, with≥85% cotton, nprs,>120mn
398	52062100	非零售精梳粗支混纺棉单纱	4.5	Combed single cotton yarn, with< 85% cotton, nprs,≤14mn
399	52071000	供零售用纯棉纱线	5	Cotton yarn (excl. sewing), put up for retail sale, with≥85% cotton
400	52083900	染色的轻质其他全棉机织物	9	Dyed woven cotton fabrics, with≥85% cotton, nes
401	52084900	色织的轻质其他全棉机织物	9	Coloured woven cotton fabrics, with≥85% cotton, nes
402	52093100	染色的重质全棉平纹布	9.3	Dyed plain cotton weave, with≥85% cotton,>200g/m^2
403	52094300	色织的重质全棉三、四线斜纹布	9.3	Coloured 3 or 4 - thread twill, with≥85% cotton,>200g/m^2
404	52095100	印花的重质全棉平纹布	9.3	Printed plain cotton weave, with≥85% cotton,>200g/m^2
405	52101100	与化纤混纺未漂白轻质平纹棉布	11	Unbleached plain cotton weave, with< 85% cotton,≤200g/m^2
406	53051100	生的椰壳纤维	4	Raw coconut fibres
407	53089011	未漂白或漂白的全苎麻纱线	6	Unbleached or bleached ramie yarn, with ≥85% ramie

序号 No.	税则号列 Tariff Item	*货 品 名 称 （简 称）	对巴基斯坦税率 （%） Pakistan	*Article Description （in brief ）
408	53089012	全苎麻色纱线	6	Coloured ramie yarn, with≥85% ramie
409	53089013	未漂白或漂白的混纺苎麻纱线	6	Unbleached or bleached ramie yarn, with <85% ramie
410	53089014	混纺苎麻色纱线	6	Coloured ramie yarn, with <85% ramie
411	53089091	纸纱线	6	Paper yarn
412	53089099	其他植物纺织纤维纱线	6	Yarn of other vegetable textile fibres, nes
413	53091900	其他全亚麻机织物	9.3	Other woven fabrics of flax, with≥85% flax, nes
414	53110090	其他纺织用植物纤维机织物	9.3	Woven fabrics of other vegetable textile fibres, nes
415	54023290	非零售其他粗尼龙变形纱线	4.5	Other textured yarn of polyamides, psy, > 50tex, not for retail sale
416	54024110	非零售用聚酰胺－6 纺制的未捻单纱	4.7	Single yarn of nylon－6, with≤50turns/m, not for retail sale
417	54024120	非零售用聚酰胺－66 纺制的未捻单纱	4.7	Single yarn of nylon－6, 6, with≤50turns/m, not for retail sale
418	54024130	非零售用芳香族聚酰胺纺制的未捻单纱	4.7	Single yarn of aramides, with≤50turns/m, not for retail sale
419	54024190	非零售其他尼龙未捻单纱	4.7	Single yarn of other polyamides nes, with ≤50turns/m, not for retail sale
420	54024300	非零售未捻的其他聚酯单纱	4.9	Single yarn of polyesters, nes, with≤50turns/m, not for retail sale
421	54025200	非零售加捻的其他聚酯加捻单纱	4.3	Single yarn of polyesters, with>50turns/m, not for retail sale
422	54074200	染色的纯尼龙布	9.7	Dyed woven fabrics of synthetic filament yarn, ≥85% nylon...
423	54075200	染色的纯聚酯变形长丝布	9.7	Dyed woven fabrics of synthetic filament yarn, ≥85% textured polyester
424	54076100	其他纯聚酯非变形长丝布	9.7	Other woven fabrics of synthetic yarn, ≥85% non－textured polyester
425	54076900	其他纯聚酯长丝布	9.7	Other woven fabrics of synthetic yarn, ≥85% polyester, nes
426	54077100	未漂白或漂白其他纯合成纤维长丝布	8.8	Unbleached or bleached woven fabrics, ≥85% synthetic filaments, nes
427	54077200	染色的其他纯合成纤维长丝布	9.7	Dyed woven fabrics, ≥85% synthetic filaments, nes
428	54077400	印花的其他纯合成纤维长丝布	9.7	Printed woven fabrics, ≥85% synthetic filaments, nes
429	54083200	染色的人纤长丝混纺布	9.9	Dyed woven fabrics of artificial filament yarn, nes
430	55013000	聚丙烯腈长丝丝束	4.5	Synthetic filament tow, acrylic or modacrylic
431	55032000	未梳的聚酯合成纤维短纤	4.5	Synthetic staple fibres, of polyesters, not carded, etc
432	55033000	未梳的聚丙烯腈合成纤维短纤	4.5	Acrylic or modacrylic synthetic staple fibres, not carded, etc

序号 No.	税则号列 Tariff Item	*货　品　名　称（简　称）	对巴基斯坦税率（%）Pakistan	*Article Description (in brief)
433	55062000	已梳的聚酯纤维短纤	4.5	Synthetic staple fibres, of polyesters, carded, etc
434	55063000	已梳的聚丙烯腈及其变性纤维短纤	4.5	Acrylic or modacrylic synthetic staple fibres, carded, etc
435	55081000	合成纤维短纤纺制的缝纫线	4.6	Sewing thread of synthetic staple fibres
436	55095300	非零售与棉混纺聚酯短纤纱线	4.3	Yarn, < 85% polyester staple fibres, mixed with cotton, nprs
437	55101100	非零售纯人造纤维短纤单纱	4.5	Single yarn, with≥85% artificial staple fibres, nprs
438	55109000	非零售与其他混纺人造纤维短纤纱线	4.5	Yarn, with<85% artificial staple fibres, nes, nprs
439	55122900	其他纯腈纶布	9.3	Printed, dyed or coloured woven fabrics, ≥85% acrylic.. staple fibres
440	55132100	与棉混纺染色的轻质聚酯平纹布	9.1	Dyed plain weave fabrics, < 85% polyester fibres + cotton, ≤170g/m^2
441	55133200	与棉混纺色织的轻质聚酯斜纹布	8.8	Coloured 3 or 4 − thread twill, < 85% polyester fibres + cotton, ≤170g/m^2
442	55141110	与棉混纺未漂白的重质聚酯平纹布	14.4	Unbleached plain weave fabric, < 85% polyster staple fibres + cotton, >170g/m^2
443	55151100	与粘胶纤维短纤混纺的聚酯布	8.8	Woven fabrics, < 85% polyester staple fibres with viscose rayon fibres
444	55162200	与化纤长丝混纺的染色人造纤维布	9.9	Dyed woven fabrics, < 85% artificial fibres, with man − made filaments
445	55169400	与其他纤维混纺的印花人造纤维布	9.4	Printed woven fabrics, artificial staple fibres, nes
446	56031110	每平米≤25g经浸渍化纤长丝无纺织物	9.7	Nonwovens of man − made filament, ≤25g/m^2, coated, etc.
447	56031210	25g<每平米≤70g浸渍化纤长丝无纺织物	9.7	Nonwovens of man − made filament, >25g/m^2 but≤70g/m^2, coated, etc.
448	56031310	70g<每平米≤150g浸渍化纤长丝无纺织物	9.7	Nonwovens of man − made filament, >70g/m^2 but≤150g/m^2, coated, etc.
449	56031410	每平米>150g经浸渍化纤长丝无纺织物	9.7	Nonwovens of man − made filament, >150g/m^2, coated, etc.
450	56039110	每平米≤25g经浸渍其他无纺织物	9.7	Nonwovens of other materials, ≤25g/m^2, coated, etc.
451	56039190	每平米≤25g的其他无纺织物	9.3	Nonwovens of other materials, ≤25g/m^2, not coated, etc.
452	56039210	25g<每平米≤70g浸渍其他无纺织物	9.7	Nonwovens of other materials, >25g/m^2 but≤70g/m^2, coated, etc.
453	56039290	25g<每平米≤70g其他无纺织物	9.3	Nonwovens of other materials, >25g/m^2 but≤70g/m^2, not coated, etc.
454	56039310	70g<每平米≤150g浸渍其他无纺织物	9.7	Nonwovens of other materials, >70g/m^2 but≤150g/m^2, coated, etc.
455	56039390	70g<每平米≤150g的其他无纺织物	9.3	Nonwovens of other materials, >70g/m^2 but≤150g/m^2, not coated, etc.

序号 No.	税则号列 Tariff Item	*货 品 名 称 (简 称)	对巴基斯坦税率 (%) Pakistan	*Article Description (in brief)
456	56039410	每平米>150g经浸渍其他无纺织物	9.7	Nonwovens of other materials, > 150g/m^2, coated, etc.
457	56039490	每平米>150g的其他无纺织物	9.3	Nonwovens of other materials, > 150g/m^2, not coated, etc.
458	57024900	制成的其他纺织材料起绒铺地制品	9.3	Pile floor coverings of other textiles, woven, made up, nes
459	58061090	其他材料狭幅起绒织物及绳绒织物	9.1	Narrow woven pile/chenille fabrics of other textile materials, nes
460	58071000	机织非绣制纺织材料标签、徽章等	9.5	Labels, badges... of textiles, woven, in piece..., not embriodered
461	59031010	用聚氯乙烯浸、涂的绝缘布或带	9	Textile insulating cloth or tape treated with polyvinyl chloride
462	59031020	用聚氯乙烯浸、涂的人造革	9.2	Textile imitation leather treated with polyvinyl chloride
463	59031090	用聚氯乙烯浸、涂的其他纺织物	9.1	Other textile fabrics treated with polyvinyl chloride
464	59032010	用聚氨基甲酸酯浸、涂的绝缘布或带	9	Textile insulating cloth or tape treated with polyurethane
465	59032020	用聚氨基甲酸酯浸、涂的人造革	9.2	Textile imitation leather treated with polyurethane
466	59032090	用聚氨基甲酸酯浸、涂的其他纺织物	9.1	Other textile fabrics treated with polyurethane
467	59039010	用其他塑料浸、涂的绝缘布或带	9	Textile insulating cloth or tape treated with other plastics, nes
468	59039020	用其他塑料浸、涂的人造革	9.2	Textile imitation leather treated with other plastics, nes
469	59039090	用其他塑料浸、涂的其他纺织物	9.1	Other textile fabrics treated with other plastics, nes
470	59111010	包覆纺锤用浸胶的起绒狭幅织物	7.2	Narrow fabrics of velvet impregnated with rubber for covering weaving beams
471	60011000	针织或钩编的长毛绒织物	9.4	Long pile fabrics, knitted or crocheted
472	60019200	化纤制针织或钩编起绒织物	9.4	Pile fabrics of man - made fibres, nes, knitted or crocheted
473	60033000	宽≤30cm合成纤维制的针织、钩编织物	9.4	Knitted/crocheted fabrics of synthetic fibres, wid≤30cm, <5%
474	60034000	宽≤30cm人造纤维制的针织、钩编织物	9.4	Knitted/crocheted fabrics of artificial fibres, wid≤30cm, <5%
475	60053100	未漂白或漂白合成纤维制的其他经编织物	9.4	Other warp knit fabrics of unbleached or bleached synthetic fibres
476	60053200	染色合成纤维制的其他经编织物	9.4	Other warp knit fabrics of dyed synthetic fibres
477	60053300	色织合成纤维制的其他经编织物	9.4	Other warp knit fabrics of synthetic fibres of yarns of different colours
478	60053400	印花合成纤维制的其他经编织物	9.4	Other warp knit fabrics of printed synthetic fibres
479	60054100	未漂白或漂白人造纤维制的其他经编织物	9.4	Other warp knit fabrics of unbleached or bleached artificial fibres

序号 No.	税则号列 Tariff Item	*货　品　名　称 （简　称）	对巴基斯坦税率 （%） Pakistan	*Article Description （in brief ）
480	60054200	染色人造纤维制的其他经编织物	9.4	Other warp knit fabrics of dyed artificial fibres
481	60054300	色织人造纤维制的其他经编织物	9.4	Other warp knit fabrics of artificial fibres of yarns of different colours
482	60054400	印花人造纤维制的其他经编织物	9.4	Other warp knit fabrics of printed artificial fibres
483	60063100	未漂白或漂白合成纤维制的其他针织、钩编织物	9.4	Other knitted/crocheted fabrics of unbleached/bleached synthetic fibres
484	60063200	染色合成纤维制的其他针织、钩编织物	9.4	Other knitted/crocheted fabrics of dyed synthetic fibres, nes
485	60063300	色织合成纤维制的其他针织、钩编织物	9.4	Other knitted/crocheted fabrics of synthetic fibres of different colors
486	60063400	印花合成纤维制的其他针织、钩编织物	9.4	Other knitted/crocheted fabrics of printed synthetic fibres, nes
487	60064100	未漂白或漂白人造纤维制的其他针织、钩编织物	9.4	Other knitted/crocheted fabrics of unbleached/bleached artificial fibres
488	60064200	染色人造纤维制的其他针织、钩编织物	9.4	Other knitted/crocheted fabrics of dyed artificial fibres, nes
489	60064300	色织人造纤维制的其他针织、钩编织物	9.4	Other knitted/crocheted fabrics of artificial fibres of different colors
490	60064400	印花人造纤维制的其他针织、钩编织物	9.4	Other knitted or crocheted fabrics of printed artificial fibres
491	60069000	未列名针织、钩编织物	10.9	Other knitted or crocheted fabrics, nes
492	61011000	毛制针织或钩编男式大衣、防风衣	18	Men's or boys' overcoats, etc, of wool..., knitted or crocheted
493	61012000	棉制针织或钩编男式大衣、防风衣	14	Men's or boys' overcoats, etc, of cotton, knitted or crocheted
494	61013000	化纤制针织或钩编男式大衣等	12.4	Men's or boys' overcoats, etc, of man-made fibres, knitted or crocheted
495	61019000	其他纺织材料制针织或钩编男式大衣、防风衣	12.4	Men's or boys' overcoats, etc, of other textiles, knitted or crocheted
496	61021000	毛制针织或钩编女式大衣、防风衣	18	Women's or girls' overcoats, etc, of wool..., knitted or crocheted
497	61022000	棉制针织或钩编女式大衣、防风衣	14	Women's or girls' overcoats, etc, of cotton, knitted or crocheted
498	61023000	化纤制针织或钩编女式大衣等	12.4	Women's or girls' overcoats, etc, of man-made fibres, knitted/crocheted
499	61029000	其他纺织材料制针织或钩编女式大衣、防风衣	13.8	Women's or girls' overcoats, etc, of other textiles, knitted/crocheted
500	61031100	毛制针织或钩编男式西服套装	18	Men's or boys' suits of wool or fine animal hair, knitted/crocheted
501	61031200	合纤制针织或钩编男西服套装	18	Men's or boys' suits of synthetic fibres, knitted or crocheted
502	61031900	其他纺织材料制针织或钩编男式西服套装	12.4	Men's or boys' suits of other textiles, nes, knitted or crocheted
503	61032100	毛制针织或钩编男式便服套装	18	Men's or boys' ensembles of wool or fine hair, knitted or crocheted

序号 No.	税则号列 Tariff Item	*货品名称（简称）	对巴基斯坦税率（%）Pakistan	*Article Description（in brief）
504	61032200	棉制针织或钩编男式便服套装	14.8	Men's or boys' ensembles of cotton, knitted or crocheted
505	61032300	合纤制针织或钩编男便服套装	18	Men's or boys' ensembles of synthetic fibres, knitted or crocheted
506	61032900	其他纺织材料制针织或钩编男式便服套装	18	Men's or boys' ensembles of other textiles, nes, knitted/crocheted
507	61033100	毛制针织或钩编男式上衣	11.7	Men's or boys' jackets & blazers of wool..., knitted or crocheted
508	61033200	棉制针织或钩编男式上衣	12	Men's or boys' jackets & blazers of cotton, knitted or crocheted
509	61033300	合纤制针织或钩编男式上衣	13.6	Men's or boys' jackets... of synthetic fibres, knitted or crocheted
510	61033900	其他纺织材料制针织或钩编男式上衣	11.7	Men's or boys' jackets... of other textiles, nes, knitted/crocheted
511	61034100	毛制针织或钩编男长裤、工装裤等	11.7	Men's or boys' trousers, etc, of wool..., knitted or crocheted
512	61034200	棉制针织或钩编男长裤、工装裤等	13.6	Men's or boys' trousers, etc, of cotton, knitted or crocheted
513	61034300	合纤制针织或钩编男长裤等	12.4	Men's or boys' trousers, etc, of synthetic fibres, knitted/crocheted
514	61034900	其他纺织材料制针织或钩编男长裤等	11.7	Men's or boys' trousers, etc, of other textiles, knitted/crocheted
515	61041100	毛制针织或钩编女式西服套装	12.4	Women's or girls' suits of wool or fine hair, knitted or crocheted
516	61041200	棉制针织或钩编女式西服套装	14	Women's or girls' suits of cotton, knitted or crocheted
517	61041300	合纤制针织或钩编女西服套装	18	Women's or girls' suits of synthetic fibres, knitted or crocheted
518	61041900	其他纺织材料制针织或钩编女式西服套装	12.4	Women's or girls' suits of other textiles, nes, knitted/crocheted
519	61042100	毛制针织或钩编女式便服套装	12.4	Women's or girls' ensembles, of wool..., knitted or crocheted
520	61042200	棉制针织或钩编女式便服套装	14	Women's or girls' ensembles, of cotton, knitted or crocheted
521	61042300	合纤制针织或钩编女便服套装	18	Women's or girls' ensembles, of synthetic fibres, knitted/crocheted
522	61042900	其他纺织材料制针织或钩编女式便服套装	10.4	Women's or girls' ensembles, of other textiles, knitted/crocheted
523	61043100	毛制针织女式上衣	11.7	Women's or girls' jackets, of wool..., knitted or crocheted
524	61043200	棉制针织女式上衣	13.6	Women's or girls' jackets, of cotton, knitted or crocheted
525	61043300	合纤制针织女上衣	13.6	Women's or girls' jackets, of synthetic fibres, knitted/crocheted
526	61043900	其他纺织材料制针织女上衣	11.7	Women's or girls' jackets, of other textiles, knitted or crocheted

序号 No.	税则号列 Tariff Item	*货 品 名 称 (简 称)	对巴基斯坦税率 (%) Pakistan	*Article Description (in brief)
527	61044100	毛制针织或钩编连衣裙	11.7	Dresses of wool or fine animal hair, knitted or crocheted
528	61044200	棉制针织或钩编连衣裙	13.6	Dresses of cotton, knitted or crocheted
529	61044300	合纤制针织或钩编连衣裙	12.4	Dresses of synthetic fibres, knitted or crocheted
530	61044400	人纤制针织或钩编连衣裙	11.7	Dresses of artificial fibres, knitted or crocheted
531	61044900	其他纺织材料制针织或钩编连衣裙	11.7	Dresses of other textile material, nes, knitted or crocheted
532	61045100	毛制针织或钩编裙子及裙裤	10.8	Skirts & divided skirts of wool or fine hair, knitted or crocheted
533	61045200	棉制针织裙子及裙裤	10.3	Skirts & divided skirts of cotton, knitted or crocheted
534	61045300	合纤制针织或钩编裙子及裙裤	11.7	Skirts & divided skirts of synthetic fibres, knitted or crocheted
535	61045900	其他纺织材料制针织或钩编裙子及裙裤	10.8	Skirts & divided skirts of other textiles, nes, knitted or crocheted
536	61046100	毛制针织或钩编女长裤、工装裤等	11.7	Women's or girls' trousers, etc, of wool..., knitted or crocheted
537	61046200	棉制针织或钩编女长裤、工装裤等	13.6	Women's or girls' trousers, etc, of cotton, knitted or crocheted
538	61046300	合纤制针织或钩编女长裤等	12.4	Women's or girls' trousers, etc, of synthetic fibres, knitted or crocheted
539	61046900	其他纺织材料制针织或钩编女长裤等	12	Women's or girls' trousers, etc, of other textile, knitted/crocheted
540	61051000	棉制针织或钩编男衬衫	13.6	Men's or boys' shirts of cotton, knitted or crocheted
541	61052000	化纤制针织或钩编男衬衫	12.4	Men's or boys' shirts of man – made fibres, knitted or crocheted
542	61059000	其他纺织材料制针织或钩编男衬衫	11.7	Men's or boys' shirts of other textiles, nes, knitted or crocheted
543	61061000	棉制针织或钩编女衬衫	13.6	Women's or girls' blouses, etc, of cotton, knitted or crocheted
544	61062000	化纤制针织或钩编女衬衫	12.4	Women's or girls' blouses, etc, of man – made fibres, knitted/crocheted
545	61069000	其他纺织材料制针织或钩编女衬衫	11.7	Women's or girls' blouses, etc, of other textiles, knitted/crocheted
546	61071200	化纤制针织或钩编男内裤及三角裤	10.9	Men's or boys' underpants, etc, of man – made fibres, knitted/crocheted
547	61071910	丝及绢丝制针织或钩编男内裤及三角裤	9.3	Men's or boys' underpants, etc, of silk or silk waste, knitted/crocheted
548	61071990	其他纺织材料制针织或钩编男内裤及三角裤	9.3	Men's or boys' underpants, etc, of other textiles, knitted/crocheted
549	61072200	化纤制针织或钩编男睡衣裤	10.9	Men's or boys' pyjamas of man – made fibres, knitted or crocheted

序号 No.	税则号列 Tariff Item	*货　品　名　称（简　称）	对巴基斯坦税率 (%) Pakistan	*Article Description (in brief)
550	61072910	丝及绢丝制针织或钩编男长睡衣及睡衣裤	9.3	Men's or boys' pyjamas of silk or silk waste, knitted or crocheted
551	61072990	其他纺织材料制针织或钩编男长睡衣及睡衣裤	9.3	Men's or boys' pyjamas of other textiles, nes, knitted or crocheted
552	61079200	化纤制针织或钩编男浴衣、晨衣	10.9	Men's or boys' dressing gowns, of man-made fibres, knitted/crocheted
553	61079900	其他纺织材料制针织或钩编男浴衣、晨衣	9.3	Men's or boys' dressing gowns, of other textiles, knitted/crocheted
554	61081100	化纤制针织或钩编长衬裙及衬裙	10.9	Women's/girls' slips, etc, of man-made fibres, knitted or crocheted
555	61081990	其他纺织材料制针织或钩编女式长衬裙及衬裙	9.3	Women's/girls' slips, etc, of other textile materials, knitted/crocheted
556	61082200	化纤制针织或钩编女三角裤及短衬裤	10.9	Women's or girls' briefs, etc, of man-made fibres, knitted/crocheted
557	61082910	丝及绢丝制针织或钩编女三角裤及短衬裤	9.3	Women's or girls' briefs, etc, of silk or silk waste, knitted/crocheted
558	61082990	其他纺织材料制针织或钩编女三角裤及短衬裤	9.3	Women's or girls' briefs, etc, of other textiles, knitted/crocheted
559	61083200	化纤制针织或钩编女睡衣及睡衣裤	10.9	Women's or girls' pyjamas, etc, of man-made fibres, knitted/crocheted
560	61083910	丝及绢丝制针织或钩编女睡衣及睡衣裤	9.3	Women's or girls' pyjamas, etc, of silk or silk waste, knitted/crocheted
561	61083990	其他纺织材料制针织或钩编女睡衣及睡衣裤	9.3	Women's or girls' nighties, etc, of other textiles, knitted/crocheted
562	61089200	化纤制针织或钩编女浴衣、晨衣	10.9	Women's or girls' dressing gowns of man-made fibre, knitted/crocheted
563	61089900	其他纺织材料制针织或钩编女浴衣、晨衣	9.3	Women's or girls' dressing gowns of other textiles, knitted/crocheted
564	61091000	棉制针织或钩编T恤衫、汗衫等	9.4	T-shirts, singlets & other vests, of cotton, knitted or crocheted
565	61099010	丝及绢丝制针织或钩编T恤衫、汗衫等	9.3	T-shirts, singlets, etc, of silk or silk waste, knitted/crocheted
566	61099090	其他纺织材料制针织或钩编T恤衫、汗衫等	9.3	T-shirts, singlets, etc, of other textiles, nes, knitted/crocheted
567	61101100	羊毛制针织或钩编套头衫等	9.3	Jerseys, pullovers, etc, of wool, knitted or crocheted
568	61101200	喀什米尔山羊细毛制针织或钩编套头衫等	9.3	Jerseys, pullovers, etc, of cashmere, knitted or crocheted
569	61101910	其他山羊细毛制针织或钩编套头衫等	9.3	Jerseys, pullovers, etc, of other goats hair, knitted/crocheted
570	61101920	兔毛制针织或钩编套头衫等	9.3	Jerseys, pullovers, etc, of rabbit or hare hair, knitted/crocheted
571	61101990	其他毛制针织或钩编套头衫等	9.3	Jerseys, pullovers, etc, of other animal fine hair, nes, knitted/crocheted
572	61102000	棉制针织或钩编套头衫等	11	Jerseys, pullovers, etc, of cotton, knitted or crocheted

序号 No.	税则号列 Tariff Item	*货品名称 (简称)	对巴基斯坦税率 (%) Pakistan	*Article Description (in brief)
573	61103000	化纤制针织或钩编套头衫等	10.9	Jerseys, pullovers, etc, of man－made fibres, knitted or crocheted
574	61109010	丝及绢丝制针织或钩编套头衫等	9.3	Jerseys, pullovers, etc, of silk or silk waste, knitted or crocheted
575	61109090	其他纺织材料制针织或钩编套头衫等	9.3	Jerseys, pullovers, etc, of other textiles, knitted or crocheted
576	61111000	毛制针织或钩编婴儿服装及附件	9.3	Babies' garments, etc, of wool or fine animal hair, knitted/crocheted
577	61113000	合纤制针织婴儿服装及附件	10.9	Babies' garments, etc, of synthetic fibres, knitted or crocheted
578	61119000	其他纺织材料制针织或钩编婴儿服装及附件	9.3	Babies' garments, etc, of other textiles, knitted or crocheted
579	61121100	棉制针织或钩编运动服	13.6	Track－suits of cotton, knitted or crocheted
580	61121200	合纤制针织或钩编运动服	12.4	Track－suits of synthetic fibres, knitted or crocheted
581	61121900	其他纺织材料制针织或钩编运动服	11.7	Track－suits of other textiles, knitted or crocheted
582	61122010	棉制针织或钩编滑雪服	13.6	Ski－suits, of cotton, knitted or crocheted
583	61122090	其他纺织材料制针织或钩编滑雪服	13.6	Ski-suits, of other textile materials, knitted or crocheted
584	61123100	合纤制针织或钩编男式游泳服	12.4	Men's or boys' swimwear of synthetic fibres, knitted or crocheted
585	61123900	其他纺织材料制针织或钩编男式游泳服	11.7	Men's or boys' swimwear of other textiles, knitted or crocheted
586	61124100	合纤制针织或钩编女式游泳服	12.4	Women's or girls' swimwear of synthetic fibres, knitted or crocheted
587	61124900	其他纺织材料制针织或钩编女式游泳服	11.7	Women's or girls' swimwear of other textiles, knitted or crocheted
588	61130000	涂层经处理针织或钩编织物制服装	11.7	Garments of knitted or crocheted fabrics of 59.03, 59.06, 59.07
589	61151200	单丝≥67分特合纤制连裤袜等	10.9	Panty hose, etc, of synthetic fibres, ≥67decitex, knitted/crocheted
590	61151990	其他纺织材料制针织连裤袜及紧身裤袜	9.3	Panty hose & tights of other textile materials nes, knitted/crocheted
591	61152000	单丝＜67分特制针织或钩编女统袜	9.3	Women's hosiery of〈67decitex, knitted/crocheted
592	61159300	合纤制针织或钩编短袜及其他袜类	10.9	Hosiery & footwear, of synthetic fibres, knitted or crocheted, nes
593	61159900	其他纺织材料制针织或钩编短袜及其他袜类	9.3	Hosiery & footwear, of other textiles, knitted or crocheted, nes
594	61169300	合纤制其他针织或钩编手套	10.9	Gloves, mittens & mitts, of synthetic fibres, knitted or crocheted
595	61169900	其他纺织材料制针织或钩编手套	9.3	Gloves, mittens & mitts, of other textiles, knitted or crocheted
596	61171000	针织或钩编披巾、头巾等	9.3	Shawls, scarves, mufflers, mantillas, veils, etc, knitted/crocheted

序号 No.	税则号列 Tariff Item	*货品名称 （简称）	对巴基斯坦税率 （%） Pakistan	*Article Description （in brief ）
597	61172000	针织或钩编领带及领结	9.3	Ties, bow ties & cravats, knitted or crocheted
598	61178000	针织或钩编其他衣着附件	9.3	Other clothing accessories, knitted or crocheted, nes
599	61179000	其他针织或钩编衣着零件	9.3	Parts of garments or clothing accessories, knitted or crocheted
600	62011100	毛制男式大衣、斗篷及类似品	11.7	Men's or boys' overcoats, etc, of wool or fine animal hair
601	62011310	化纤制男式羽绒服	12.4	Men's or boys' overcoats, etc, of man - made fibres, down stuffed
602	62011390	化纤制男式大衣、斗篷及类似品	12.4	Men's or boys' overcoats, etc, of man - made fibres, nes
603	62011900	其他纺织材料制男式大衣、斗篷及类似品	11.7	Men's or boys' overcoats, etc, of other textiles, nes
604	62019100	毛制男式带风帽防寒短上衣、防风衣	11.7	Men's or boys' anoraks, wind - cheaters, etc, of wool/fine animal hair
605	62019310	化纤制男式其他羽绒服	12.4	Men's/boys' anoraks, wind - cheaters.., of man - made fibres, down stuffed
606	62019390	化纤制男式防寒短上衣、防风衣	12.4	Men's or boys' anoraks, wind - cheaters, etc, of man - made fibres, nes
607	62019900	其他纺织材料制男式防寒短上衣、防风衣	11.7	Men's or boys' anoraks, wind - cheaters, etc, of other textiles, nes
608	62021100	毛制女式大衣、斗篷及类似品等	11.7	Women's or girls' overcoats, etc, of wool or fine animal hair
609	62021310	化纤制女式羽绒服	13.6	Women's or girls' overcoats, etc, of man - made fibres, down stuffed
610	62021390	化纤制女式大衣、斗篷及类似品	13.6	Women's or girls' overcoats, etc, of man - made fibres, nes
611	62021900	其他纺织材料制女式大衣、斗篷及类似品	11.7	Women's or girls' overcoats, etc, of other textiles, nes
612	62029100	毛制女式带风帽防寒短上衣、防风衣	11.7	Women's or girls' anoraks, wind - cheaters, etc, of wool...
613	62029310	化纤制女式其他羽绒服	12.4	Women's/girls' anoraks, wind - cheaters.., of man - made fibres, down stuffed
614	62029390	化纤制女式防风衣等	12.4	Women's or girls' anoraks, wind - cheaters.., of man - made fibres, nes
615	62029900	其他纺织材料制防风衣、防风短上衣等	11.7	Women's or girls' anoraks, wind - cheaters, etc, of other textiles, nes
616	62031100	毛制男式西服套装	12.4	Men's or boys' suits of wool or fine animal hair
617	62031200	合纤制男式西服套装	12.4	Men's or boys' suits of synthetic fibres
618	62031910	丝及绢丝制男式西服套装	12.4	Men's or boys' suits of silk or silk waste
619	62031990	其他纺织材料制男式西服套装	12.4	Men's or boys' suits of other textiles, nes
620	62032100	毛制男式便服套装	12.4	Men's or boys' ensembles of wool or fine animal hair
621	62032300	合纤制男式便服套装	12.4	Men's or boys' ensembles of synthetic fibres

序号 No.	税则号列 Tariff Item	*货 品 名 称 (简 称)	对巴基斯坦税率 (%) Pakistan	*Article Description (in brief)
622	62032910	丝及绢丝制男式便服套装	12.4	Men's or boys' ensembles of silk or silk waste
623	62032990	其他纺织材料制男式便服套装	12.4	Men's or boys' ensembles of other textiles, nes
624	62033100	毛制男式上衣	11.7	Men's or boys' jackets & blazers of wool or fine animal hair
625	62033300	合纤制男式上衣	12.4	Men's or boys' jackets & blazers of synthetic fibres
626	62033910	丝及绢丝制男式上衣	11.7	Men's or boys' jackets & blazers of silk or silk waste
627	62033990	其他纺织材料制男式上衣	11.7	Men's or boys' jackets & blazers of other textile materials, nes
628	62034100	毛制男式长裤、工装裤等	11.7	Men's or boys' trousers, breeches, etc, of wool or fine animal hair
629	62034210	棉制男式阿拉伯裤	12.9	Men's or boys' arabian trousers, breeches, of cotton
630	62034290	棉制男式长裤、工装裤等	12.9	Men's or boys' trousers, breeches, nes, of cotton
631	62034310	合成纤维制男式阿拉伯裤	13.1	Men's or boys' arabian trousers, breeches of synthetic fibres
632	62034390	合纤制男式长裤、工装裤等	13.1	Men's or boys' trousers, breeches, nes, of synthetic fibres
633	62034910	其他纺织材料制男式阿拉伯裤	11.7	Men's or boys' arabian trousers, breeches of other textile fibres
634	62034990	其他纺织材料制男童裤、工装裤	11.7	Men's or boys' trousers, breeches, nes, of other textile fibres
635	62041100	毛制女式西服套装	12.4	Women's or girls' suits of wool or fine animal hair
636	62041300	合纤制女式西服套装	12.4	Women's or girls' suits of synthetic fibres
637	62041910	丝及绢丝制女式西服套装	12.4	Women's or girls' suits of silk or silk waste
638	62041990	其他纺织材料制女式西服套装	12.4	Women's or girls' suits of other textiles, nes
639	62042100	毛制女式便服套装	12.4	Women's or girls' ensembles of wool or fine animal hair
640	62042300	合纤制女式便服套装	13.8	Women's or girls' ensembles of synthetic fibres
641	62042910	丝及绢丝制女式便服套装	13.8	Women's or girls' ensembles of silk or silk waste
642	62042990	其他纺织材料制女式便服套装	9.9	Women's or girls' ensembles of other textile materials, nes
643	62043100	毛制女式上衣	11.7	Women's or girls' jackets & blazers of wool or fine animal hair
644	62043300	合纤制女式上衣	12.4	Women's or girls' jackets & blazers of synthetic fibres
645	62043910	丝及绢丝制女式上衣	11.7	Women's or girls' jackets of silk or silk waste

序号 No.	税则号列 Tariff Item	*货 品 名 称 （简 称）	对巴基斯坦税率 （%） Pakistan	*Article Description （in brief）
646	62043990	其他纺织材料制女式上衣	11.7	Women's or girls' jackets of other textiles, nes
647	62044100	毛制连衣裙	11.7	Dresses of wool or fine animal hair
648	62044300	合纤制女式连衣裙	12.4	Dresses of synthetic fibres
649	62044400	人纤制女式连衣裙	11.7	Dresses of artificial fibres
650	62044910	丝及绢丝制连衣裙	11.7	Women's or girls' dresses of silk or silk waste
651	62044990	其他纺织材料制连衣裙	11.7	Women's or girls' dresses of other textile materials, nes
652	62045100	毛制裙子及裙裤	9.3	Skirts & divided skirts of wool or fine animal hair
653	62045300	合纤制裙子及裙裤	11.7	Skirts & divided skirts of synthetic fibres
654	62045910	丝及绢丝制裙子及裙裤	9.3	Women's or girls' skirts of silk or silk waste
655	62045990	其他纺织材料制裙子及裙裤	9.3	Women's or girls' skirts of other textile materials, nes
656	62046100	毛制女式长裤、工装裤等	11.7	Women's or girls' trousers/breeches, etc, of wool/fine animal hair
657	62046300	合纤制女式长裤、工装裤等	12.4	Women's or girls' trousers, breeches, etc, of synthetic fibres
658	62046900	其他纺织材料制女式长裤、工装裤等	12	Women's or girls' trousers, breeches, etc, of other textiles, nes
659	62051000	毛制男衬衫	11.7	Men's or boys' shirts of wool or fine animal hair
660	62052000	棉制男衬衫	13.6	Men's or boys' shirts of cotton
661	62053000	化纤制男衬衫	11.7	Men's or boys' shirts of man-made fibres
662	62059010	丝及绢丝制男衬衫	11.7	Men's or boys' shirts of silk or silk waste
663	62059090	其他纺织材料制男衬衫	11.7	Men's or boys' shirts of other textile materials, nes
664	62061000	丝及绢丝制女式衬衫	11.7	Women's or girls' blouses, shirts, etc, of silk or silk waste
665	62062000	毛制女衬衫	11.7	Women's or girls' blouses, shirts, etc, of wool or fine animal hair
666	62063000	棉制女衬衫	13.6	Women's or girls' blouses, shirts, etc, of cotton
667	62064000	化纤制女衬衫	12.4	Women's or girls' blouses, shirts, etc, of man-made fibres
668	62069000	其他纺织材料制女衬衫	11.7	Women's or girls' blouses, shirts, etc, of other textiles, nes
669	62071100	棉制男式内裤及三角裤	10.3	Men's or boys' underpants & briefs of cotton
670	62082100	棉制女式睡衣及睡衣裤	10.3	Women's or girls' nightdresses & pyjamas of cotton
671	62101010	毛制毡呢或无纺织物服装	11.7	Garments of felt or nonwoven, of wool or fine animal hair

序号 No.	税则号列 Tariff Item	*货品名称（简称）	对巴基斯坦税率（%）Pakistan	*Article Description (in brief)
672	62101030	化纤制毡呢或无纺织物服装	12.4	Garments of felt or nonwoven, of man-made fibres
673	62102000	用塑料、橡胶等处理的织物制男大衣等	11.7	Garments of 6201.11 to 19, from fabrics of 59.03, 59.06 or 59.07
674	62103000	用塑料、橡胶等处理的织物制女大衣等	11.7	Garments of 6202.11 to 19, from fabrics of 59.03, 59.06 or 59.07
675	62104000	用塑料、橡胶等处理的织物制的其他男式服装	11.7	Men's or boys' garments of fabrics of 59.03, 59.06 or 59.07
676	62105000	用塑料、橡胶等处理的织物制的其他女式服装	11.7	Women's or girls' garments of fabrics of 59.03, 59.06 or 59.07
677	62111100	男式游泳服	11.7	Men's or boys' swimwear
678	62111200	女式游泳服	11.7	Women's or girls' swimwear
679	62112090	其他纺织材料制滑雪服	13.6	Ski suits of other textile materials
680	62113100	毛制男式运动服及其他服装	11.7	Men's or boys' garments of wool or fine animal hair, nes
681	62113310	化纤制男式阿拉伯袍	12.4	Men's or boys' arabian robes of man-made fibres
682	62113390	化纤制男式运动服及其他服装	12.4	Men's or boys' garments of man-made fibres, nes
683	62113910	丝及绢丝制男式运动服及其他服装	11.7	Men's or boys' garments of silk or silk waste, nes
684	62113990	其他纺织材料制男式运动服及其他服装	11.7	Men's or boys' garments of other textiles, nes
685	62114100	毛制女式运动服及其他服装	11.7	Women's or girls' garments of wool or fine animal hair
686	62114300	化纤制女式运动服及其他服装	12.4	Women's or girls' garments of man-made fibres, nes
687	62114910	丝及绢丝制女式运动服及其他服装	11.7	Women's or girls' garments of silk or silk waste, nes
688	62114990	其他纺织材料制女式运动服及其他服装	11.7	Women's or girls' garments of other textiles, nes
689	62171010	非针织非钩编袜子及袜套	12.2	Stocking, socks and sockettes
690	62171020	非针织非钩编和服腰带	12.2	Kimono belts
691	62171090	非针织非钩编服装或衣着附件	12.2	Other made up clothing accessories, nes
692	64061000	鞋面及其零件,硬衬除外	13.2	Uppers & parts thereof (excl. stiffeners)
693	64069900	其他材料制鞋靴、护腿等零件	13.2	Non-wood parts of footwear (excl. uppers, outer soles & heels)
694	67030000	经梳理、稀疏等方法加工的人发及假发材料	18	Human hair, dressed, etc; animal hair & synthetic materials for wigs
695	68022300	具有一个平面的花岗岩及制品	9.2	Monumental/building stone, cut/sawn flat or even, granite
696	68029390	其他加工形式花岗岩制品	9.2	Worked building stone nes, of granite
697	69081000	上釉的小陶瓷砖、瓦、块及类似品	10.4	Tiles, cubes & sim＜7 cm rect or not etc, glazed ceramics
698	69111010	瓷餐具	10	Tableware of porcelain or china
699	69111020	瓷厨房器具	12.5	Kitchenware of porcelain or china

序号 No.	税则号列 Tariff Item	*货品名称（简称）	对巴基斯坦税率（%）Pakistan	*Article Description (in brief)
700	69119000	其他家用或盥洗用瓷器	20	Household articles nes & toilet articles of porcelain or china
701	71023900	其他非工业用钻石	5	Diamonds non – industrial nes excluding mounted or set diamonds
702	71031000	未加工宝石或半宝石	2.8	Prec or semi – prec stones (o/t diamonds) unworked or smp. sawn or rough shaped
703	71039100	经其他加工的红、蓝、绿宝石	7	Rubies, sapphires & emeralds further worked than sawn or rough shaped
704	71039910	经其他加工的翡翠	7	Jadeites, further worked than sawn or rough shaped
705	71039990	经其他加工的其他宝石或半宝石	7	Other precious or semi – precious stones, nes, further worked
706	71131110	镶嵌钻石的银首饰及其零件	16.5	Jewellery & parts thereof of silver, plated with diamond mounted or set
707	71131190	其他银首饰及其零件	16.5	Jewellery & parts thereof, of silver, nes
708	71131911	镶嵌钻石的黄金制首饰及其零件	16.5	Jewellery & parts thereof of gold, plated with diamond mounted
709	71131919	其他黄金制首饰及其零件	16.5	Jewellery & parts thereof of gold, nes
710	71131991	镶嵌钻石的其他贵金属制首饰及其零件	30	Jewellery& parts thereof of precious metal, plated with diamond mounted
711	71131999	其他贵金属制首饰及其零件	30	Jewellery& parts thereof of precious metal, nes
712	71132010	镶嵌钻石的以贱金属为底的包贵金属制首饰	30	Jewellery& parts of base metal clad with precious metal, plated with diamond
713	71132090	其他以贱金属为底的包贵金属制首饰	30	Jewellery & parts thereof, of base metal clad with precious metal, nes
714	71179000	未列名材料制仿首饰	30	Imitation jewellery nes
715	72091810	厚度＜0.3mm 的冷轧卷材	6	Flat-rolled products of iron or non-alloy steel, of a width of 600mm or more, in coils, not further worked than cold-rolled (cold-reduced), not clad, plated or coated, thickness＜0.3mm
716	72091890	其他厚度＜0.5mm 的冷轧卷材	6	Flat-rolled products of iron or non-alloy steel, of a width of 600mm or more, in coils, not further worked than cold-rolled (cold-reduced), not clad, plated or coated, 0.3mm≤thickness≤0.5mm
717	72092800	厚度＜0.5mm 的冷轧非卷材	6	Flat rlld prod, i/nas, nt in coil, cr, w≥600mm, thickness＜0.5mm
718	72101100	镀锡的铁或非合金钢厚宽平板轧材	10	Flat rlld prod, i/nas, plated/coated with tin, w≥600mm, thickness≥0.5mm
719	72101200	镀锡的铁或非合金钢薄宽平板轧材	5	Flat rlld prod, i/nas, plated or coated with tin, w≥600mm, thickness＜0.5mm
720	72102000	镀铅的铁或非合金钢宽平板轧材	4	Flat rlld prod, i/nas, platd/coatd with lead, w≥600mm, incl terne – plate
721	72103000	电镀锌的铁或非合金钢宽板材	8	Flat rlld prod, i/nas, electrocly platd/coatd with zinc, w≥600mm

序号 No.	税则号列 Tariff Item	*货　品　名　称（简　称）	对巴基斯坦税率（%）Pakistan	*Article Description (in brief)
722	72104100	镀锌的瓦楞形铁或非合金钢宽板材	8	Flat rlld prod, i/nas, pltd or ctd w zinc, corrugated, w≥600mm, nes
723	72104900	镀锌的其他形铁或非合金钢宽板材	4	Flat rlld prod, i/nas, plated or coated with zinc, w≥600mm, nes
724	72105000	镀氧化铬的铁或非合金钢宽板材	8	Flat rld prod, i/nas, pltd/ctd w chrm oxids/chrom w chrm oxide, w≥600mm
725	72106100	镀或涂铝锌合金的铁宽平板轧材	8	Flat rlld prod, i/nas, platd/coatd wth aluminium - zinc alloy, w≥600mm
726	72106900	其他镀或涂铝的铁宽平板轧材	8	Flat rlld prod, i/nas, plated or coated with aluminium, w≥600mm, nes
727	72107000	涂漆或涂塑的铁或非合金钢宽板材	4	Flat rlld prod, i/nas, painted, varnish./coated with plastics, w≥600mm
728	72109000	涂镀其他材料铁或非合金钢宽板材	8	Flat rlld prod, i/nas, clad, plated or coated, w≥600mm, nes
729	72142000	热加工带有轧制花纹的条、杆	3	Bars&rods, i/nas, hr, hd or he, cntg indent, ribs, etc, prod dur rp/tar, nes
730	72191100	厚度＞10mm 热轧不锈钢卷板	4	Flat rlld prod, stainless steel, hr, in coil, w≥600mm, thickness＞10mm
731	72191200	4.75mm≤厚度≤10mm 热轧不锈钢卷板	4	Flat rlld prod, stainls steel, hr, in coil, w≥600mm, 4.75mm≤thickness≤10mm
732	72191310	未经酸洗的 3mm≤厚度＜4.75mm 热轧不锈钢卷板	4	Not acid pickled flat rlld prod, stainls steel, hr in coil, w≥600mm, 3mm≤thickness＜4.75mm
733	72191320	经酸洗的 3mm≤厚度＜4.75mm 热轧不锈钢卷板	4	Acid pickled flat rlld prod, stainls steel, hr in coil, w≥600mm, 3mm≤thickness＜4.75mm
734	72191410	未经酸洗的厚度＜3mm 热轧不锈钢卷板	4	Not acid pickled flat rlld prod, stainls steel, hr in coil, w≥600mm, thickness＜3mm
735	72191420	经酸洗的厚度＜3mm 热轧不锈钢卷板	4	Acid pickled flat rlld prod, stainls steel, hr in coil, w≥600mm, thickness＜3mm
736	72192100	厚度＞10mm 热轧不锈钢平板	9.3	Flat rlld prod, stainless steel, hr, nic, w≥600mm, thickness＞10mm
737	72192200	4.75mm≤厚度≤10mm 热轧不锈钢平板	9.3	Flat rlld prod, stainless steel, hr, nic, w≥600mm, 4.75mm≤thickness≤10mm
738	72192300	3mm≤厚度＜4.75mm 热轧不锈钢平板	9.3	Flat rlld prod, stainless steel, hr, nic, w≥600mm, 3mm≤thickness＜4.75mm
739	72192410	1mm＜厚度＜3mm 热轧不锈钢平板	9.3	Flat-rolled products of stainless steel, of a width of 600mm or more, not further worked than hot-rolled, not in coils, of a thickness exceeding 1mm but less than 3mm
740	72192420	0.5mm≤厚度≤1mm 热轧不锈钢平板	9.3	Flat-rolled products of stainless steel, of a width of 600mm or more, not further worked than hot-rolled, not in coils, of a thickness of 0.5mm or more but not exceeding 1mm

序号 No.	税则号列 Tariff Item	*货 品 名 称 (简 称)	对巴基斯坦税率 (%) Pakistan	*Article Description (in brief)
741	72192430	厚度<0.5mm 热轧不锈钢平板	9.3	Flat-rolled products of stainless steel, of a width of 600mm or more, not further worked than hot-rolled, not in coils, of a thickness of less than 0.5mm
742	72193100	厚度≥4.75mm 冷轧不锈钢板	10	Flat rlld prod, stainless steel, cr, w≥600mm, thickness≥4.75mm
743	72193200	3mm≤厚度<4.75mm 冷轧不锈钢板材	10	Flat rlld prod, stainless steel, cr, w≥600mm, 3mm≤thickness<4.75mm
744	72193300	1mm<厚度<3mm 冷轧不锈钢板材	10	Flat rlld prod, stainless steel, cr, w≥600mm, 1mm<thickness<3mm
745	72193400	0.5mm≤厚度≤1mm 冷轧不锈钢板材	10	Flat rlld prod, stainless steel, cr, w≥600mm, 0.5mm≤thickness≤1mm
746	72193500	厚度<0.5mm 冷轧不锈钢板材	10	Flat rlld prod, stainless steel, cr, w≥600mm, thickness<0.5mm
747	72199000	其他不锈钢冷轧板材	10	Flat rlld prod, stainless steel, 600mm or more wide, nes
748	72210000	不锈钢热轧条、杆	8	Bars/rods, stainless steel, hot rlld in irregularly wound coils
749	72221100	热加工的圆形截面不锈钢条、杆	9.0	Bars/rods, stainls stl, nfw th ht rl/drwn/extrud, of circ cros sectn
750	72221900	热加工其他截面形状不锈钢条杆	9.0	Bars/rods, stainless steel, nfw than hot rld/drawn/extrud, nes
751	72223000	其他不锈钢条、杆	8.9	Bars/rods, stainless steel, nes
752	76061220	厚度<0.28mm 的铝合金制矩形铝板片带	5	Plate, sheet′ strip, Al alloy, rect′ sq, 0.28mm<thickness<0.28mm
753	76061230	0.28mm≤厚度≤0.35mm 的铝合金制矩形铝板片带	5	Plate, sheet′ strip, Al alloy, rect′ sq, 0.28mm≤thickness≤0.35mm
754	76061240	厚度>0.35mm 的铝合金制矩形铝板片带	5.5	Plate, sheet′strip, Al alloy, rect′sq, thick >0.35mm
755	76071110	厚度≤0.007mm,轧制后未经进一步加工的无衬背铝箔	5.7	Foil, Al, not backed, rolled but not further worked, thickness≤0.007mm
756	76071190	0.007mm<厚度≤0.2mm,轧制后未经进一步加工的无衬背铝箔	5.7	Foil, Al, not backed, rolled but not further worked, 0.007mm<thickness≤0.2mm
757	82029110	加工金属用的机械锯的直锯片	7	Straight saw blades for working metal, for use wiwth mechnical saws
758	84021200	蒸发量≤45t/h 的水管锅炉	3.9	Watertube boilers with a steam production not>45T per hour
759	84143011	功率≤0.4kw 的冷藏、冷冻箱用压缩机	5.9	Compressors for refrige/freezer, motor power≤0.4kw
760	84143012	功率 0.4-5kw 的冷藏、冷冻箱用压缩机	8.5	Compressors for refrige′freezer, 0.4kw< motor power≤5 kw
761	84143013	功率 0.4-5kw 的空气调节器用压缩机	8.5	Compressors for airconditioner, 0.4kw < motor power ≤5kw
762	84143014	功率>5kw 的空气调节器用压缩机	8.5	Compressors for airconditioner, motor power>5 kw
763	84143015	冷冻或冷藏设备用,电动机额定功率超过 5 千瓦的电动机驱动压缩机	9.2	Compressors driven by a motor, for refrigerators or freezers, of a motor power exceeding 5KW

序号 No.	税则号列 Tariff Item	*货 品 名 称 (简 称)	对巴基斯坦税率 (%) Pakistan	*Article Description (in brief)
764	84143019	其他制冷设备用压缩机	9.2	Compressors driven by a motor, nes
765	84143090	非电机驱动的压缩机	8.8	Compressors driven by a non－motor
766	84159010	制冷量≤4千大卡/时等空调的零件	8	Parts of air conditners of subheadngs 8415.1000,8415.8110,8415.8210
767	84159090	制冷量>4千大卡/时等空调的零件	8	Parts of air conditioners of other subheadings of 84.15
768	84198910	加氢反应器	0	Hydroformer vessels
769	84198990	其他利用温度变化处理材料的机器	0	Other machinery, plant & equip for treat of mat. by a chg of temp, nes
770	84283300	其他带式连续运货升降、输送机	4.3	Continous－action elevators/conveyors for goods nes, belt type
771	84295211	轮胎式挖掘机	7.2	Excavators, tyre－mounted, with a 3600 revolving superstructure
772	84295212	履带式挖掘机	8	Excavators, track － mounted, with a 3600 revolving superstructure
773	84295219	其他挖掘机	7.2	Other excavators, with a 360° revolving superstructure, nes
774	84295290	其他上部结构可转360度的挖掘机类似机械	7.2	Other shovels and machineries with a 360° revolving superstructure, nes
775	84431910	平张纸进料式胶印机	8.9	Sheet fed offset printing machinery nes
776	84431990	其他胶印机	8.9	Offset printing machinery nes
777	84434000	照像凹版印刷机	17	Gravure printing machinery
778	84440010	合成纤维长丝纺丝机	9	Synthetic filament spinning jets
779	84440020	合成纤维短丝纺丝机	9	Synthetic staple fibres spinning jets
780	84440030	人造纤维纺丝机	9	Artificial fibres spinning jets
781	84440040	化学纤维变形机	9	Man－made filament crimping machinery
782	84440050	化学纤维切断机	9	Man－made filament cutting machinery
783	84440090	其他化学纤维挤压、拉伸、变形或切割机器	9	Machines for extrud/draw/texture/cut man－made textile mat. nes
784	84451111	棉纤维清梳联合机	9	Blowing-carding machinery for cotton type fibres
785	84451112	棉纤维自动抓棉机	9	Bale plucker for cotton type fibres
786	84451113	棉纤维梳棉机	9	Card or carding machine for cotton type fibres
787	84451119	其他棉纤维梳理机	9	Other carding machines for cotton type fibres
788	84451120	毛纤维梳理机	9	Carding machines, for wool type fibres
789	84451190	其他纺织纤维梳理机	9	Other carding machines, nes
790	84451900	纺织纤维的其他预处理机器	9	Textile preparing machines, nes
791	84452031	转杯纺纱机	9	Rotor spinning machine
792	84452041	棉环锭细纱机	10	Cotton ring spinning frame
793	84452090	其他纺纱机	9	Other textile spinning machines, nes
794	84453000	并线机或加捻机	9	Textile doubling'twisting machines
795	84454010	自动络筒机	9	Automatic bobbin winders
796	84454090	其他络纱机(包括卷纬机)或摇纱机	9	Other textile winding or reeling machines
797	84459010	整经机	9	Warping machines
798	84459020	浆纱机	9	Sizing machines

序号 No.	税则号列 Tariff Item	*货品名称（简称）	对巴基斯坦税率（%）Pakistan	*Article Description (in brief)
799	84459090	其他生产及处理纺织纱线的机器	9	Other machines for producing' preparing textile yarn, nes
800	84462110	所织织物宽度＞30cm 的梭织动力地毯织机	11	Power looms for weaving carpets/rugs of width30cm, shuttle type
801	84462190	所织织物宽度＞30cm 的其他梭织动力织机	9	Power looms for weaving fabrics of width ＞30 cm, nes, shuttle type
802	84463020	所织织物宽度＞30cm 的剑杆织机	7	Rapier looms for weaving fabrics of a width＞30cm, shuttless
803	84463040	所织织物宽度＞30cm 的喷水织机	7	Water jet looms for weaving fabrics of a width＞30cm, shuttless
804	84463050	所织织物宽度＞30cm 的喷气织机	7	Air jet looms for weaving fabrics of a width＞30cm, shuttless
805	84463090	所织织物宽度＞30cm 的其他无梭织机	7	Machines for weaving fabrics of a width＞30cm, shuttleless nes
806	84471100	圆筒直径≤165mm 的圆型针织机	7	Circular knitting machines with cylinder diameter≤165 mm
807	84472010	经编机	7	Warp knitting machines
808	84472020	其他平型针织机	7	Other flat knitting machines
809	84472030	缝编机	7	Stitch – bonding machines
810	84479011	地毯织机	6	Tufting machines for making carpets or rugs
811	84479019	其他簇绒机	7	Other tufting machines
812	84479020	绣花机	7	Embroidery machines
813	84479090	税号 84.47 其他未列名机器	9	Machines for making gimped yarn/tulle/lace/trimmings/braid/net
814	84501110	干衣量≤10kg 的波轮式全自动洗衣机	8.7	Full – auto machines of the continuously rotating impeller, of a dry linen capacity≤10 kg
815	84501120	干衣量≤10kg 的滚筒式全自动洗衣机	8.7	Full – auto machines of the drum type, of a dry linen capacity≤10 kg
816	84501190	干衣量≤10kg 的其他全自动洗衣机	8.7	Full – automatic washing machines, of a dry linen capacity ≤10 kg
817	84501200	装有离心甩干机的非全自动洗衣机	25	Other washing mach, dry linen capacity≤10kg, with built – in dryer
818	84509010	干衣量≤10kg 的洗衣机零件	4.4	Parts of washing machines, dry linen capacity≤10kg
819	84511000	干洗机	16	Dry – cleaning machines o/t hdg No 84.50
820	84514000	洗涤、漂白或染色机器	7	Washing, bleaching'dyeing machines
820	84521000	家用型缝纫机	17	Sewing machines of household type
821	84522110	平缝机	10.7	Non – household type automatic sewing machines, flatseam
823	84522190	其他自动缝纫机	10.7	Other non – household type automatic sewing machines nes
824	84522900	其他非自动缝纫机	11	Sewing machines, other than book – sewing machines, nes

序号 No.	税则号列 Tariff Item	*货 品 名 称 （简 称）	对巴基斯坦税率 （%） Pakistan	*Article Description (in brief)
825	84529011	家用型缝纫机用旋梭	13	Rotating shuttles of household sewing machines
826	84529019	家用型缝纫机用其他零件	13	Parts of household sewing machines, nes
827	84531000	生皮、皮革的处理、鞣制或加工机器	8	Machinery for preparing, tanning′working hides, skins′leather
828	84532000	鞋靴制作或修理机器	8	Machinery for making′repairing footwear
829	84596910	切削金属的非数控龙门铣床	10	Planoilling machines, not numerically controlled
830	84596990	切削金属的其他铣床	11	Milling machines nes, not numerically controlled
831	84629110	金属型材挤压机	9.2	Metal section squeezing machine, hydraulic
832	84629190	其他液压压力机	9.2	Hydraulic presses nes
833	84672100	电动钻	9	Drills of all kinds
834	84672210	电动链锯	9	Chain saws
835	84672290	其他电动锯	9	Electric saws, working in the hand
836	84672910	电动砂磨工具	8	Grinding tools
837	84672920	电刨	8	Planings
838	84672990	其他电动工具	8	Other tolls for working in the hand, nes
839	84679110	电动链锯用零件	4.8	Parts of chain saw, with self − contained electri motor
840	84679190	其他链锯用零件	6	Parts of chain saw, nes
841	84679910	其他手提式电动工具用零件	9	Parts of self − contained electric motor of heading 8467, nes
842	84716070	键盘及鼠标	0	Keyboards and mouses
843	84733021	针式打印机打印头	0	Printing element of stylus printers of 8471.6031
844	84733029	针式打印机其他零件	0	Parts/accessories of printers of 8471.6032 to 8471.6039
845	84733090	税号 84.71 所列计算机的其他零附件	0	Parts/accessories of other machines of heading 84.71
846	84735000	税号 84.69－84.72 中所列机器零附件	0	Parts/accessr equaly suitbl for use with ≥ 2 machs of 84.69 to 84.72
847	84771010	注塑机	0	Injection − moulding mach for working plastics
848	84795010	多功能工业机器人	0	Industrial robots for multiples uses
849	84795090	其他工业机器人	0	Industrial robots, nes
850	84796000	蒸发式空气冷却器	9	Evaporative air coolers
851	84798910	船舶用舵机及陀螺稳定器	0	Steering/rudder equipt or gyrostabilizers for ships
852	84798920	空气增湿器及减湿器	0	Air humidifiers or dehumidifiers
853	84798940	邮政用包裹、印刷品分拣设备	0	Machines for sorting mailed packages or printed matters
854	84798950	放射性废物压实机	0	Machines for squeezing radioactive waste
855	84798990	本章其他税号未列名机器及机械器具	0	Mach & mechanical appliances having individual functions, nes

序号 No.	税则号列 Tariff Item	*货　品　名　称 （简　称）	对巴基斯坦税率 （%） Pakistan	*Article Description （in brief）
856	85013100	输出功率≤750瓦直流电动机、发电机	11	DC motors, DC generators, of an output not exceeding 750 W
857	85049011	额定容量>400KVA液体介质变压器零件	4.5	Parts of transformers of subheading No. 8504.2320
858	85049019	其他变压器零件	4.5	Parts of other transformers, nes
859	85071000	起动活塞式发动机用铅酸蓄电池	6.9	Pb – acid electric accumulators for starting piston engines
860	85072000	其他铅酸蓄电池	6.9	Lead – acid electric accumulators nes
861	85073000	镍镉蓄电池	8	Nickel – cadmium electric accumulators
862	85074000	镍铁蓄电池	10	Nickel – iron electric accumultors
863	85078010	镍氢电池	10	Nickel – hydrogen
864	85078020	锂离子电池	10	Lithium ion
865	85078090	其他蓄电池	9	Other electric accumulators, nes
866	85091000	真空吸尘器;包括干式及湿式	8.2	Domestic vacuum cleaners
867	85131010	手电筒	13.2	Portable electric torches functioned by dry batteries
868	85165000	微波炉	13.9	Microwave ovens
869	85173011	容量≥5千门局用电话交换机	0	Digtl pub tele switchg systm≥5000 lines, toll tele switg systm
870	85173013	数字移动通信交换机	0	Mobile tele switching system
871	85173019	其他数字式程控电话交换机	0	Other digital tele switching systems, nes
872	85173091	模拟式移动通信交换机	0	Anological mobile tele switching system
873	85173099	其他电话或电报交换机	0	Telephonic or telegraphic switching apparatus, nes
874	85199910	激光唱机	26	Compact disc players, sound
875	85199990	其他声音重放设备	16.4	Sound reproducing apparatus without a sound recorder, nes
876	85211011	广播级录像机	见本表尾注	Broadcast quality, magnetic video tape recorders
877	85211019	其他磁带录像机	见本表尾注	Other magnetic video tape recorders
878	85211020	磁带放像机	见本表尾注	Magnetic video tape reproducers
879	85281210	彩色卫星电视接收机	26	Colour satellite TV receivers
880	85281221	显示屏幕不超过42厘米的阴极射线显像管的彩色电视接收装置	26	Reception apparatus for television, whether or not incorporating radio-broadcast receivers or sound or video recording or reproducing apparaus; video monitors and video projectors, with cathode-ray picture tube, with a diagonal measurement of the screen not exceeding 42cm
881	85281222	显示屏幕超过42厘米不超过52厘米的阴极射线显像管的彩色电视接收装置	26	Reception apparatus for television, whether or not incorporating radio-broadcast receivers or sound or video recording or reproducing apparaus; video monitors and video projectors, with cathode-ray picture tube, with a diagonal measurement of the screen exceeding 42cm but not exceeding 52cm

序号 No.	税则号列 Tariff Item	*货 品 名 称（简 称）	对巴基斯坦税率 (%) Pakistan	*Article Description (in brief)
882	85281223	显示屏幕超过52厘米不超过74厘米的阴极射线显像管的彩色电视接收装置	29.1	Reception apparatus for television, whether or not incorporating radio-broadcast receivers or sound or video recording or reproducing apparaus; video monitors and video projectors, with cathode-ray picture tube, with a diagonal measurement of the screen exceeding 52cm but not exceeding 74cm
883	85281224	显示屏幕超过74厘米的阴极射线显像管的彩色电视接收装置	29.1	Reception apparatus for television, whether or not incorporating radio-broadcast receivers or sound or video recording or reproducing apparaus; video monitors and video projectors, with cathode-ray picture tube, with a diagonal measurement of the screen exceeding 74cm
884	85281238	显示屏幕超过52厘米的液晶显示器的彩色电视接收装置	29.1	Reception apparatus for television, whether or not incorporating radio-broadcast receivers or sound or video recording or reproducing apparaus; video monitors and video projectors, with liquid crystal display, with a diagonal measurement of the screen exceeding 52cm
885	85281239	其他液晶显示器的彩色电视接收装置	26	Reception apparatus for television, whether or not incorporating radio-broadcast receivers or sound or video recording or reproducing apparaus; video monitors and video projectors, other, with liquid crystal display
886	85281248	显示屏幕超过52厘米等离子显示器的彩色电视接收装置	29.1	Reception apparatus for television, whether or not incorporating radio-broadcast receivers or sound or video recording or reproducing apparaus; video monitors and video projectors, with plasma display panels, with a diagonal measurement of the screen exceeding 52cm
887	85281249	其他等离子显示器的彩色电视接收装置	26	Reception apparatus for television, whether or not incorporating radio-broadcast receivers or sound or video recording or reproducing apparaus; video monitors and video projectors, other, with plasma display panels
888	85282100	彩色视频监视器	26	Colour video monitors
889	85283010	彩色视频投影机	26	Colour video projectors
890	85299010	电视发送、差转等设备零件	0	Parts for TV transmisn/translat/satelite earth rece/relay apparatus
891	85299020	手持式无线电话机零件	0	Parts for portable radio telephone sets
892	85299030	对讲机零件	7	Parts for walkie－talkies

序号 No.	税则号列 Tariff Item	*货 品 名 称 （简 称）	对巴基斯坦税率 （%） Pakistan	*Article Description （in brief ）
893	85299041	特种用途电视摄像机、静像视频摄像机及其他视频摄录一体机，数字相机零件	7	Parts for TV comeras sevring special purposes
894	85299042	非特种用途的取像模块	11	Camera modules without special purpose
895	85299049	其他电视摄像机、静像视频摄像机及其他视频摄录一体机、数字相机零件	11	Parts for other TV cameras
896	85299050	雷达及无线电导航设备零件	1.5	Parts for radar or radio－navigational aid apparatus
897	85299060	收音机及其组合机的其他零件	12	Parts for radio broacast receivers and their combinations
898	85299070	无线寻呼机零件	0	Parts for radio paging receivers
899	85299081	彩色电视机零件	12	Parts for colour TV receivers
900	85299082	等离子显像组件及其零件	12	Plasma display modules and parts thereof
901	85299089	其他电视机零件	0	Parts for monochroms TV receivers
902	85299090	税号85.25至85.28所列设备的零件	0	Parts for use with apparatus of 85.25 to 85.28, nes
903	85442000	同轴电缆及其他同轴电导体	9	Co－axial cable & other co－axial electric conductors
904	85446011	耐压＞35KV的电缆	8	Electric cable, for a voltage ＞ 35 KV
905	85446019	耐压在1－35KV之间的电缆	8.9	Electric cable, 1,000V＜voltage≤35 KV
906	85446090	耐压＞1KV的其他电导体	20	Electric conductor nes, for a voltage ＞1,000 V
907	85447000	光缆	0	Optical fibre cables, made up of individually sheathed fibres
908	87081000	缓冲器（保险杠）及其零件	12.3	Bumpers & parts of motor vehicles
909	87113010	汽油型中小马力摩托车及脚踏两用车	32.8	Motorcycles with, 250 cc＜ piston engine ≤400cc
910	87113020	汽油型中大马力摩托车及脚踏两用车	32.8	Motorcycles with, 400 cc＜ piston engine ≤500cc
911	90181210	B型超声波诊断仪	6	B－ ultrasonic diagnostic apparatus
912	90183210	管状金属针头	7	Tubular metal needles
913	91081900	其他已组装的完整电子表芯	12	Electric watch movements, assembled, nes
914	91112000	贱金属制的表壳	10	Watch cases of base metal
915	95067010	溜冰鞋	12	Ice skates, including skating boots with skates attached
916	95067020	旱冰鞋	12	Roller skates, including skating boots with skates attached
917	96033010	画笔	15	Artists' brushes

注：

22030000　麦芽酿造的啤酒的中国/巴基斯坦协定税率：0元/升。

85211011 广播级录像机：每台完税价格低于或等于2000美元：执行单一从价税，税率为16%；每台完税价格高于2000美元：每台征收从量税，税额2158元，加上3%从价税。

85211019 其他磁带录像机：每台完税价格低于或等于2000美元：执行单一从价税，税率为25%；每台完税价格高于2000美元：每台征收从量税，税额3652元，加上3%从价税。

85211020 磁带放像机：每台完税价格低于或等于2000美元：执行单一从价税，税率为25%；每台完税价格高于2000美元：每台征收从量税，税额3652元，加上3%从价税。

Note:

22030000: Specific Tariff Rate of Beer made from malt under China/Pakistan agreement is 0 RMB/Liter.

85211011: If the Duty Paying Value(DPV) of each video tape recorder with broadcasting quality is less than or equal to US $ 2000, it should be imposed an Ad Valorem duty of 16%; If the DPV is higher than US $ 2000, it should be imposed a compound duty, i.e. a Specific duty of 2158 RMB plus an Ad Valorem duty of 3% for each one.

85211019: If the DPV of each other magnetic tape-type video recorder is less than or equal to US $ 2000, it should be imposed an Ad Valorem duty of 25%; If the DPV is higher than US $ 2000, it should be imposed a compound duty, i.e. a Specific duty of 3652 RMB plus an Ad Valorem duty of 3% for each one.

85211020: If the DPV of each video tape reproducer is less than or equal to US $ 2000, it should be imposed an Ad Valorem duty of 25%; If the DPV is higher than US $ 2000, it should be imposed a compound duty, i.e. a Specific duty of 3652 RMB plus an Ad Valorem duty of 3% for each one.

表 4:进口商品从量税、复合税税目税率表

Table 4:Specific and Compound Duty Rate on Imported Goods

序号 No.	税则号列 Tariff Item	货　品　名　称	普通税率 Gen.(RMB/unit)	最惠国税率 M.F.N.(RMB/unit)	Article Description
1	02071200	冻的整只鸡	5.6 元/千克	1.3 元/千克	Frozen whole chickens, not cut in pieces
2	02071411	冻的带骨鸡块(包括鸡胸脯、鸡大腿等)	4.2 元/千克	0.6 元/千克	Frozen chicken cuts、breasts、legs, with bone in
3	02071419	冻的不带骨鸡块(包括鸡胸脯、鸡大腿等)	9.5 元/千克	1.0 元/千克	Frozen boneless chicken cuts、breasts、legs
4	02071421	冻的鸡翼(不包括翼尖)	8.1 元/千克	0.9 元/千克	Frozen wing of chicken(other than wingtips)
5	02071422	冻的鸡爪	3.2 元/千克	0.5 元/千克	Frozen claw of chicken
6	02071429	冻的其他鸡杂碎(包括鸡翼尖、鸡爪、鸡肝等)	3.2 元/千克	0.5 元/千克	Frozen offal of chicken, nes
7	05040021	冷、冻的鸡肫(即鸡胃)	7.7 元/千克	1.4 元/千克	Cold, frozen gizzard
8	22030000	麦芽酿造的啤酒	7.5 元/升	0	Beer made from malt
9	27090000	石油原油(包括从沥青矿物提取的原油)	85 元/吨	0	Petroleum oils & oils obtained from bituminous minerals, crude
10	37013021	未曝光照相制版用激光照排片(任何一边>255 毫米)	70 元/平方米	4.3 元/平方米	Laser phototypesetting film, unexposed, of any material other than paper, paperboard or textiles, any side exceeding 255mm
11	37013022	未曝光照相制版用 PS 版(任何一边>255 毫米)	70 元/平方米	2.9 元/平方米	Unexposed P. S phototypesetting plates, any side exceeding 255mm
12	37013023	未曝光照相制版用 CTP 版(任何一边>255 毫米)	70 元/平方米	6.8 元/平方米	Unexposed CTP phototypesetting plates, any side exceeding 255mm
13	37013029	其他未曝光照相制版用感光硬软片(任何一边>255 毫米)	70 元/平方米	6.8 元/平方米	Other photographic plates and film, for preparing printing plates or cylinders, in the flat, sensitized, unexposed, of any material other than paper, paperboard or textiles, any side exceeding 255mm
14	37023100	未曝光无齿孔彩色窄胶卷(窄胶卷指宽度≤105 毫米,彩色摄影用)	433 元/平方米	67.7 元/平方米	Photographic film rolls, unexposed, without perforations, for colour photography, of any material other than paper, paperboard or textiles, width≤105mm

序号 No.	税则号列 Tariff Item	货　品　名　称	普通税率 Gen.(RMB/unit)	最惠国税率 M.F.N.(RMB/unit)	Article Description
15	37023220	照相制版涂卤化银液无齿孔窄胶卷(成卷未曝光感光胶片,窄胶卷指宽度≤105毫米)	104元/平方米	5.3元/平方米	Photographic film rolls, unexposed, without perforations, with silver halide emulsion, for preparing printing plates or cylinders, of any material other than paper, paperboard or textiles, width≤105mm
16	37023290	其他涂卤化银乳液无齿孔窄胶卷(成卷未曝光感光胶片,窄胶卷指宽度≤105毫米)	202元/平方米	36.8元/平方米	Photographic film rolls, unexposed, without perforations, with silver halide emulsion , of any material other than paper, paperboard or textiles, width≤105mm, nes
17	37023920	照相制版用其他无齿孔窄感光胶卷(成卷未曝光感光胶片,窄胶卷指宽度≤105毫米)	104元/平方米	3.7元/平方米	Photographic film rolls, unexposed, without perforations and silver halide emulsion, of any material other than paper, paperboard or textiles, width≤105mm, for preparing printing plates or cylinders
18	37023990	其他用无齿孔窄感光胶卷(成卷未曝光感光胶片,窄胶卷指宽度≤105毫米)	202元/平方米	36.9元/平方米	Photographic film rolls, unexposed, without perforations and silver halide emulsion, of any material other than paper, paperboard or textiles, width≤105mm, nes
19	37024100	未曝光无齿孔宽长彩色胶卷(宽长胶卷指宽度>610毫米,长度>200米)	202元/平方米	30.7/平方米	Film rolls, for colour photography, unexposed, without perforations, of any material other than paper, paperboard or textiles, width > 610mm, length>200m
20	37024221	印刷电路板制造用光致抗蚀干膜(指宽度>610毫米,长度>200米)	110元/平方米	2.1元/平方米	Wide anticorrosive photographic plate for printed circuit processing, unexposed, without perforations, of any material other than paper, paperboard or textiles, width > 610mm, length>200m

序号 No.	税则号列 Tariff Item	货 品 名 称	普通税率 Gen. (RMB/unit)	最惠国税率 M.F.N. (RMB/unit)	Article Description
21	37024229	照相制版其他未曝光无齿宽长胶卷(宽长指宽度>610毫米,长度>200米)	110元/平方米	3.7元/平方米	Film rolls for preparing printing plates or cylinders, unexposed, without perforations, of any material other than paper, paperboard or textiles, width > 610mm, length > 200m, excl. wide anticorrosive photographic place for printed circuit processing
22	37024291	红外或氦氖激光胶片	213元/平方米	34元/平方米	Infa-red or neon-nelium laser film
23	37024299	黑白其他未曝光无齿孔宽长胶卷(宽长胶卷指宽度>610毫米,长度>200米	213元/平方米	34元/平方米	Unexposed monochrome photographic film in rolls, without perforations, W > 610mm, L >200m
24	37024321	照相制版用激光照排片(宽度>610毫米,长度≤200米)	104元/平方米	2.7元/平方米	Laser phototypesetting film, unexposed, without perforations, width > 610mm, length ≤200m, of any material other than paper, paperboard or textiles
25	37024329	其他照相制版用未曝光无齿孔胶卷(指宽度>610毫米,长度≤200米)	104元/平方米	3.7元/平方米	Film rolls for preparing printing plates or cylinders, unexposed, without perforations, width > 610mm, length ≤ 200m, of any material other than paper, paperboard or textile, excl. laser phototypesetting film
26	37024390	彩色或黑白其他用未曝光无齿孔中长胶卷(中长胶卷指宽度>610毫米,长度≤200米)	202元/平方米	33.2元/平方米	Film rolls, unexposed, without perforations, width > 610mm, length≤200m, of any material other tham paper, paperboard or textiles, nes
27	37024421	照相制版用未曝光激光照排片(宽度大于105毫米,小于等于610毫米)	115元/平方米	2.7元/平方米	Laser phototypesetting film, unexposed, without perforations, 105mm < width ≤ 610mm, of any material other than paper, paperboard or textiles

序号 No.	税则号列 Tariff Item	货品名称	普通税率 Gen.(RMB/unit)	最惠国税率 M.F.N.(RMB/unit)	Article Description
28	37024422	印刷电路板制造用光致抗蚀干膜(宽度大于 105 毫米,小于等于 610 毫米)	115 元/平方米	2.1 元/平方米	Narrow anticorrosive photographic plate for printed circuit processing, unexposed, without perforations, 105mm < width ≤ 610mm, of any material other than paper, paperboard or textiles
29	37024429	其他照相制版用无齿孔未曝光胶卷(宽度大于 105 毫米,小于等于 610 毫米)	115 元/平方米	3.7 元/平方米	Film rolls for preparing printing plates or cylinders, unexposed, without perforations, 105mm< width≤610mm, of any material other than paper, paperboard or textiles
30	37024490	彩色或黑白其他用无齿孔未曝光中宽胶卷(中宽胶卷指宽度>105 毫米,但≤610 毫米)	202 元/平方米	33.5 元/平方米	Film rolls, unexposed, without perforations, 105mm < width ≤ 610mm, of any material other than paper, paperboard or textiles, nes
31	37025100	未曝光窄短彩色胶卷(窄短胶卷指宽度≤16 毫米,长度≤14 米)	433 元/平方米	128.6 元/平方米	Film rolls for colour photography, unexposed, width ≤ 16mm, & length ≤ 14m, of any material othre than paper, paperboard or textiles nes
32	37025200	未曝光中窄彩色胶卷(中窄胶卷指宽度≤16 毫米,长度>14 米)	433 元/平方米	95.6 元/平方米	Film rolls for colour photography, unexposed, width ≤ 16mm, length > 14m, of any material other than paper, paperboard or textiles, nes
33	37025300	幻灯片用未曝光彩色摄影胶卷(宽度>16 毫米,但≤35 毫米,长度≤30 米)	433 元/平方米	128.6 元/平方米	Slide film rolls for colour photography, unexposed, 16mm< width≤35mm, length< 30m, of any material other than paper, paperboard or textiles
34	37025410	非幻灯片用彩色摄影胶卷(宽度=35 毫米,长度≤2 米)	433 元/平方米	96 元/平方米	Film rolls for colour photography other than for slides, unexposed, width = 35mm and length≤ 2m, of any material other than paper, paperboard or textiles

序号 No.	税则号列 Tariff Item	货品名称	普通税率 Gen.(RMB/unit)	最惠国税率 M.F.N.(RMB/unit)	Article Description
35	37025490	其他非幻灯片用彩色摄影胶卷(宽度>16毫米,但≤35毫米,长度≤30米)	433元/平方米	65.6元/平方米	Film rolls for colour photography other than for slides, unexposed, 16mm < width < 35mm, 2m < length ≤ 30m, of any material other than paper, paperboard or textiles
36	37025520	未曝光的窄长彩色电影胶卷(窄长胶卷指宽度>16毫米,但≤35毫米,长度>30米)	232元/平方米	13.8元/平方米	Colour cinematographic film rolls, unexposed, 16mm < width ≤ 35mm, length > 30m, of any material other than paper, paperboard or textiles
37	37025590	其他未曝光窄长彩色胶卷窄长胶卷指宽度>16毫米,但≤35毫米,长度>30米)	433元/平方米	81.4元/平方米	Colour film rolls for colour photography, unexposed, 16mm < width ≤ 35mm, length > 30m, excl. cinematographic film
38	37025620	未曝光的中宽彩色电影胶卷(中宽胶卷指宽度>35毫米)	232元/平方米	13.8元/平方米	Colour cinematographic film rolls, unexposed, width > 35mm, of any material other than paper, paperboard or textiles
39	37025690	其他未曝光的中宽彩色胶卷(中宽胶卷指宽度>35毫米)	433元/平方米	78.2元/平方米	Film in rolls for colour photography, unexposed, width > 35mm, of any material other than paper, paperboard or textiles, excl. cinematographic film
40	37029100	未曝光窄短黑白胶卷(窄短胶卷指宽度≤16毫米)	210元/平方米	54.7元/平方米	Film rolls of neutral colour, unexposed, width ≤ 16mm, length ≤ 14m, of any material other than paper, paperboard or textiles
41	37029310	未曝光中长黑白胶卷(宽度=35毫米,长度≤2米)	210元/平方米	65元/平方米	Film rolls of neutral colour, unexposed, width = 35mm, lenght ≤ 2m, of any material other than paper, paperboard or textiles
42	37029390	其他未曝光中长黑白胶卷(中长胶卷指宽度>16毫米,但≤35毫米,长度≤30米)	210元/平方米	54.7元/平方米	Film rolls of neutral colour, unexposed, 16mm < width < 35mm, 2m < lenght ≤ 30m, of any material other than paper, paperboard or textiles

序号 No.	税则号列 Tariff Item	货　品　名　称	普通税率 Gen.(RMB/unit)	最惠国税率 M.F.N.(RMB/unit)	Article Description
43	37029420	未曝光的窄长黑白电影胶卷(窄长胶卷指宽度＞16毫米,但≤35毫米,长度＞30米)	210元/平方米	9元/平方米	Cinematographic film rolls of neutral colour, unexposed, 16mm ＜ width ≤ 35mm, length＞30mm, of any material other than paper, paperboard or textiles
44	37029490	其他用未曝光窄长黑白胶卷(窄长胶卷指宽度＞16毫米,但≤35毫米,长度＞30米)	210元/平方米	49.2元/平方米	Film rolls of neutral colour, unexposed, 16mm＜width≤35mm, length＞30m, of any material other than paper, paperboard or textiles, excl. cinematographic film
45	37029520	未曝光的中宽黑白电影胶卷(中宽胶卷指宽度＞35毫米)	210元/平方米	10.4元/平方米	Cine matographic film rolls of neutral colour, unexposed, width＞35mm, of any material other than paper, paperboard or textiles
46	37029590	其他用未曝光的中宽黑白胶卷(中宽胶卷指宽度＞35毫米)	210元/平方米	35.1元/平方米	Film rolls of neutral colour, unexposed, width＞35mm, of any material other than paper, paperboard or textiles, excl. cinematographic film
47	85211011	广播级录像机	每台完税价格低于或等于2000美元:执行单一从价税,税率为130%;每台完税价格高于2000美元:每台征收从量税,税额20600元,加上6%从价税	每台完税价格低于或等于2000美元:执行单一从价税,税率为30%;每台完税价格高于2000美元:每台征收从量税,税额4482元,加上3%从价税	Broadcast quality, magnetic video tape recorders
48	85211019	其他录像机	每台完税价格低于或等于2000美元:执行单一从价税,税率为130%;每台完税价格高于2000美元:每台征收从量税,税额20600元,加上6%从价税	每台完税价格低于或等于2000美元:执行单一从价税,税率为30%;每台完税价格高于2000美元:每台征收从量税,税额4482元,加上3%从价税	Other magnetic video tape recorders

序号 No.	税则号列 Tariff Item	货　品　名　称	普通税率 Gen.(RMB/unit)	最惠国税率 M.F.N.(RMB/unit)	Article Description
49	85211020	放像机	每台完税价格低于或等于2000美元:执行单一从价税,税率为130%;每台完税价格高于2000美元:每台征收从量税,税额20600元,加上6%从价税	每台完税价格低于或等于2000美元:执行单一从价税,税率为30%;每台完税价格高于2000美元:每台征收从量税,税额4482元,加上3%从价税	Magnetic video tape reproducers
50	85253091	非特种用途的广播级电视摄像机	每台完吮价格低于或等于5000美元:执行单一从价税,税率为130%;每台完税价格高于5000美元:每台征收从量税,税额51500元,加上6%从价税	每台完税价格低于或等于5000美元,执行单一从价税,税率:35%;每台完税价格高于5000美元:每台征收从量税,税额13280元,加上3%从价税	Broadcast quality , television cameras, nes
51	85253099	非特种用途的其他电视摄像机	每台完税价格低于或等于5000美元:执行单一从价税,税率为130%;每台完税价格高于5000美元:每台征收从量税,税额51500元,加上6%从价税	每台完税价格低于或等于5000美元,执行单一从价税,税率:35%;每台完税价格高于5000美元:每台征收从量税,税额13280元,加上3%从价税	Other television cameras, nes
52	85254041 85254049	非特种用途的静像视频摄像机及其他视频摄像一体机,家用型摄录一体机除外	每台完税价格低于或等于5000美元:执行单一从价税,税率为130%;每台完税价格高于5000美元:每台征收从量税,税额51500元,加上6%从价税	0	Video camera recorders, not for specific uses, other than household video camera recorder
53	85254050	数字方式存储图像的照相机	每台完税价格低于或等于5000美元:执行单一从价税,税率为130%;每台完税价格高于5000美元:每台征收从量税,税额51500元,加上6%从价税	0	Other digital cameras

表 5:关税配额商品税目、税率表

Table 5:Tariff Quota Rate on Imported Goods

序号 No.	商品类别 Article Description	税则号列 Tariff Item	普通税率(%) General	最惠国税率(%) M.F.N.	关税配额税率(%) In Quota Duty Rate
1	小麦 Wheat	10011000	180	65	1
		10019010	180	65	1
		10019090	180	65	1
		11010000	130	65	6
		11031100	130	65	9
		11032010	180	65	10
2	玉米 Maize(Corn)	10051000	180	20	1
		10059000	180	65	1
		11022000	130	40	9
		11031300	130	65	9
		11042300	180	65	10
3	稻谷和大米 Rice, whether or not busked	10061011	180	65	1
		10061019	180	65	1
		10061091	180	65	1
		10061099	180	65	1
		10062010	180	65	1
		10062090	180	65	1
		10063010	180	65	1
		10063090	180	65	1
		10064010	180	65	1
		10064090	180	65	1
		11023010	130	40	9
		11023090	130	40	9
		11031921	70	10	9
		11031929	70	10	9
4	豆油 Soybean oil	15071000	190	19.9	9
		15079000	190	19.9	9
5	棕榈油 Palm oil	15111000	60	19.9	9
		15119010	60	19.9	9
		15119090	60	19.9	9
6	菜子油 Rape oil	15141100	170	19.9	9
		15141900	170	19.9	9
		15149110	170	19.9	9
		15149190	170	19.9	9
		15149900	170	19.9	9
7	糖 Sugar	17011100	125	50	15
		17011200	125	50	15
		17019100	125	50	15
		17019910	125	50	15
		17019920	125	50	15
		17019990	125	50	15

序号 No.	商品类别 Article Description	税则号列 Tariff Item	普通税率(%) General	最惠国税率(%) M.F.N.	关税配额税率(%) In Quota Duty Rate
8	羊毛 Wool	51011100	50	38	1
		51011900	50	38	1
		51012100	50	38	1
		51012900	50	38	1
		51013000	50	38	1
		51031010	50	38	1
9	毛条 Wool Tops	51051000	50	38	3
		51052100	50	38	3
		51052900	50	38	3
10	棉花 Cotton	52010000	125	40	1
		52030000	125	40	1
11	化肥 Chemical Fertilizer	31021000	150	50	4
		31052000	150	50	4
		31053000	150	50	4

表 6:进口商品暂定税率表

Table 6: Interim Duty Rate on Imported Goods

序号 No.	税则号列 Tariff Item	商 品 名 称	最惠国税率(%) M.F.N.	暂定最惠国税率(%) Interim Duty Rate	Article Description
1	03033110	格陵兰庸鲽鱼	10	5	Greenland halibut
2	05119111	受精鱼卵	12	0	Fertilized fish eggs
3	18040000	可可脂	22	15	Cocoa butter
4	23099090	已配制的饲料	6.7	4	Feeding preparations
5	24039100	再造烟草	57	40	Reconwtituted tobacco
6	25030000	硫磺	3	1.5	Sulphur of all kinds, other than sublimed sulphur, precipitated sulphur and colloidal sulphur
7	25151100	原状或粗加修正的大理石	4	1	Marble, crude or roughly trimmed
8	25161100	原状或粗加修正的花岗石	4	1	Granite, crude or roughly trimmed
9	25162100	原状或粗加修正的砂岩	3	1	Sandstone, crude or roughly trimmed
10	25281000	天然硼砂及其精矿	3	1	Natural sodium borates and concentrates thereof (H_3BO_3>85%)
11	25289000	天然粗硼酸,含硼酸干重不超过 85%	5	1	Natural sodium borates, crude (H_3BO_3<85%)
12	27011210	炼焦煤	3	0	Bituminous coking coal, not agglomerated, whether or not pulverized
13	27073000	粗二甲苯	6	3	Xylole
14	27075000	芳烃混合物	7	3	Aromatic hydrocarbon mixtures
15	27076000	混合甲酚	7	3	(o;m;p.)cresol
16	27101191	壬烯(碳九混合异构体含量高于 95%)	9	4	Nonene
17	27101199	异戊烯同分异构体混合物	9	5	isopentene
18	27101929	350℃以下馏出物体积百分比小于 20%,550℃以下馏出物体积百分比大于 80%的蜡油	6	0	Paraffin oils: 350° Cistillage < 20%,550° C distillage>80%
19	27101919	正构烷烃(C9~C13)	6	4	n-Alkanes(C9 – C13)
20	28020000	升华硫磺、沉淀硫磺、胶态硫磺	5.5	1.5	Sulphur, sublimed or precipitated; colloidal sulphur
21	28046190	多晶硅	4	2	Polysilicon
22	28092010	磷酸及偏磷酸、焦磷酸	1	0	Phosphoric acid, metaphosphoric acid and pyrophosphiric acid
23	28121010	氯化亚砜	5.5	2	Sulphoxide chloride
24	28419000	钴酸锂	5.5	2	Lithium cobaltate
25	28369930	碳酸钴	5.5	2	Cobaltous carbonate
26	28269000	六氟磷酸锂	5.5	2	Lithium hexafluorophosphate
27	29012100	乙烯	2	0	Ethylene
28	29022000	纯苯	2	1	Benzene
29	29024100	邻二甲苯	2	1	o-Xylene

序号 No.	税则号列 Tariff Item	商 品 名 称	最惠国税率(%) M.F.N.	暂定最惠国税率(%) Interim Duty Rate	Article Description
30	29031500	1,2－二氯乙烷	5.5	1	1,2 － Dichloroethane (ethylene dichloride)
31	29032100	氯乙烯	5.5	1	chloroethylene
32	29036110	邻二氯苯	5.5	3	o－Dichlorobenzene
33	29042020	邻、对硝基甲苯	5.5	3	(O－,p－)Nitrotoluene
34	29051210	正丙醇	5.5	3	Propan－l－ol (propyl alcohol)
35	29051400	叔丁醇	5.5	3	Tert-Butanol
36	29053200	丙二醇	5.5	3	Propylene glycol
37	29071212	邻甲酚	5.5	3	o-Cresols
38	29071910	邻仲丁基酚、邻异丙基酚	4	2	o-Sec-butyl phenol, o-Isopropyl phenol
39	29094100	二甘醇	5.5	3	2,2'－Oxydiethanol (diethylene glycol,digol)
40	29094990	2－正丙氧基乙醇	5.5	3	2－propoxyphene
41	29121900	正丙醛	5.5	3	n-propionaldehyde
42	29121900	乙二醛	5.5	3	Glyoxal(Ethanedial)
43	29141200	丁酮	5.5	3	Butanone
44	29141300	甲基异丁基甲酮	5.5	3	4 － Methyl － 2 － pentanone (isobutyl methyl ketone)
45	29155010	丙酸	5.5	3	Propionic acid
46	29162010	DV 菊酸甲酯、二溴菊酸	4	2	DV － chrysanthemic monomethylate, Dibromochrysanthemic acid
47	29162090	二氯菊酰氯(DV 菊酰氯)	6.5	2	Dichlorochrysanoyl chloride (DV －chrysanoyl chloride)
48	29172010	四氢苯酐	4	2	Tetrahydro benzonic anhydride
49	29173610	对苯二甲酸	9.7	6.5	Terephthalic acid
50	29211910	二正丙胺	4	2	Di－n－propylamines
51	29211920	异丙胺	6.5	2	Isopropyl amine
52	29211990	三乙胺、月桂胺、一乙胺、正丁胺	6.5	3	Triethylamine, Laurylamine, ethylamine,n-butylamine
53	29212110	乙二胺	6.5	3	Ethylenediamine
54	29212900	二乙烯三胺	6.5	3	Diethylenetriamine
55	29214300	邻甲苯胺	6.5	3	o-Tuluidine
56	29214920	2,4、2,6－二甲基苯胺	6.5	2	2,4、2,6－Dimethylanilines
57	29214940	2,6－二乙基苯胺	6.5	2	2,6－Diethylaniline

序号 No.	税则号列 Tariff Item	商 品 名 称	最惠国税率(%) M.F.N.	暂定最惠国税率(%) Interim Duty Rate	Article Description
58	29221320	芳基聚氧乙烯磷酸酯	6.5	3	Aromatic polyoxyethylene phosphate
59	29224210	谷氨酸	10	5	Glutamic acid
60	29261000	丙烯腈	6.5	4	Acrylonitrile
61	29269090	已二腈	6.5	1	Hexanedinitrile
62	29304000	DL－羟基蛋氨酸	6.5	5	DL-hydroxy-methionine
63	29309090	乙硫醇	6.5	3	Ethyl mercaptan
64	29333210	六氢吡啶	4	2	Hexahydropyridine
65	29336910	三聚氰氯	6	2	Cyanuric chloride
66	29419051	7－氨基脱乙酰氧基头孢烷酸、7－氨基头孢烷酸	6	4	7－Aminocephalosporianic acid, 7－Aminodeacetoxycefanoic acid
67	32061110	金红石型钛白粉	6.5	5	Rutile Titanium white
68	32082010	光导纤维用涂料(主要成分为聚胺酯丙烯酸酯类化合物)	10	6	Paint for optial fibers (based on polyaminate acrylate resin polymers)
69	32089010	光导纤维用涂料(主要成分为聚胺酯丙烯酸酯类化合物)	10	6	Paint for optial fibers (based on polyaminate acrylate resin polymers)
70	33012990	黄樟油	15	7	Sassafras oil
71	33013010	鸢尾凝脂(香膏类)	20	10	Balsam of irises
72	34029000	十二烷基苯磺酸钙甲醇溶液(十二烷基苯磺酸钙含量应高于70%)	9	7	Calcium dodecyl benzosulfonate methanol solution (Calcium dodecyl benzosulfonate content ＞ 70%)
73	35030010	明胶	12	5	Gelatin
74	37019920	超微粒干版	10	8	Ultramicron dry plate
75	37024221	印刷电路板制造用光制抗蚀干膜	2.1元/平方米	1.0元/平方米	Photo resist dry film
76	37024291	红外或氦氖激光胶片，宽度＞80cm，长度大于1000m	34元/平方米	1.05元/平方米	Infra-red or helium neon laser film width＞80cm, length＞1000m
77	37024422	印刷电路板制造用光制抗蚀干膜	2.1元/平方米	1.0元/平方米	Photo resist dry film
78	37025520	未曝光的窄长彩色电影胶卷(正片)	13.8元/平方米	6元/平方米	Unexposed polychrome cinematographic film in rolls
79	38011000	人造石墨	6.5	3	Artificial graphite
80	38170000	混合烷基苯	6.5	2	Mixed alkylbenzenes
81	38237000	工业用脂肪醇	13	9	Aliphatic alcohols for Industrial uses

序号 No.	税则号列 Tariff Item	商品名称	最惠国税率(%) M.F.N.	暂定最惠国税率(%) Interim Duty Rate	Article Description
82	38249090	电极浆料(主要成分为银、镍成铜和有机溶剂,用于生产片式陶瓷电容器)	6.5	3	Pastes for electrodes (mainly containing silver、nickle or copper and organic solvent, for flat figuline capacitor manufacture)
83	39011000	聚乙烯(进口 CIF 价高于 1500 美元/吨)	10.3	3	Polyethylene (import CIF ≥ 1500usd/t)
84	39012000	聚乙烯(进口 CIF 价高于 1500 美元/吨)	10.3	3	Polyethylene (import CIF ≥ 1500usd/t)
85	39021000	电工级初级形状聚丙烯树脂(灰分含量不大于 30ppm)	9.7	5	Electrotechonical polypropylene resin in primary forms(Ash content not more than 30ppm)
86	39073000	溴的重量百分含量在 18%以上的环氧树脂(如溶于溶剂,以纯环氧树脂折算溴的百分含量)	9.7	4	Epoxyresins containing by weight more than 18% Bromine (base on pure Epoxyresins if dissolved in solvent)
87	39081019	尼龙 11、尼龙 12	9.7	7	Nylon 11, Nylon 12
88	39119000	偏苯三酸酐和异氰酸预缩聚物	6.5	3	Precondensed polymer of trimeutic anhydride isocyanate
89	39119000	芳基酸与芳基胺预缩聚物	6.5	3	Precondensed polymer of aromatic acid and aromatic amide
90	39119000	改性三羟乙基脲酸酯类预缩聚物	6.5	3	Precondensed polymer of modified trihydroxy acetate
91	39119000	聚苯硫醚	6.5	6	Polyphenyl sulfide
92	39121100	未塑化二、三醋酸纤维素	6.5	1	Non-plasticized cellulose diacetate, nonplasticized cellulose triacetate
93	39201090	乙烯一四氟乙烯膜(四氟乙烯单体含量约 50%,比重 $1.7-1.75g/cm^3$)	6.5	1	Ethylene tetrafluoroethylene copolymer (containg about 50% tetrafluoroethylene monomer, density is $1.7-1.75g/cm^3$)
94	40011000	天然胶乳	20	10	Natural rubber latex
95	40112000	断面宽度 24 英寸及以上的巨型轮胎	10	6	Tyres, having a fracture surface of a width of 24 inches or more
96	40116100	断面宽度 24 英寸及以上的巨型轮胎	17.5	6	Tyres, having a fracture surface of a width of 25 inches or more
97	40116300	断面宽度 24 英寸及以上的巨型轮胎	17.5	6	Tyres, having a fracture surface of a width of 26 inches or more
98	40116900	断面宽度 24 英寸及以上的巨型轮胎	17.5	6	Tyres, having a fracture surface of a width of 27 inches or more
99	40119200	断面宽度 24 英寸及以上的巨型轮胎	25	6	Tyres, having a fracture surface of a width of 28 inches or more

序号 No.	税则号列 Tariff Item	商品名称	最惠国税率(%) M.F.N.	暂定最惠国税率(%) Interim Duty Rate	Article Description
100	40119400	断面宽度24英寸及以上的巨型轮胎	25	6	Tyres, having a fracture surface of a width of 29 inches or more
101	40119900	断面宽度24英寸及以上的巨型轮胎	25	6	Tyres, having a fracure surface of a width of 30 inches or more
102	40129010	航空用实心或半实心橡胶轮胎	3	1	Aircraft solid or cushion rubber tyres
103	40139010	航空用橡胶内胎	3	1	Aircraft rubber inner tubes
104	40169310	橡胶密封圈	8	5	Rubber's seal
105	48022010	照相原纸	7.5	5	Photo paper base
106	56013000	由两种或两种以上有机聚合物纺制的纤维(横载面为皮芯结构或并列结构或海岛结构),长度不超过5毫米	10	5	Of two or more kinds of polymers (with cross section of skin-core or juxtapose or island structure), length not more than 5mm
107	68159900	碳纤维纱线(碳元素含量大于90%)	17.5	8	Carbon fibre yarn (containg more than 90% Carbon)
108	69071000	瓷砖、陶瓷等产品,未打磨和上釉的陶瓷(表面最宽小于7厘米)	24.5	22	Tiles, cubes and sim(7 cm rect or not etc, unglazed ceramics)
109	70023900	光通信用微光组件的玻璃毛细管、定位管(外径小于3mm)	12	3	Micro capillavy and glass tube for optical communication (out diameter<3mm)
110	70071110	空载重量25吨及以上飞机的挡风玻璃	2	1	Windshiled for airplane unloaden weight≥25t
111	70071900	低铁钢化太阳能电池组件封装专用玻璃(最大含铁量0.02% Fe_2O_3,玻璃厚度2.5mm-3.5mm)	14	10	ferrless toughen glass for solar energy battery encapsulation (max Fe 0.02Fe_2O_3, thichness of glass: 2.5-3.5mm)
112	70112090	显示管玻壳及其零件	10	6	glass envelopes and parts there of for display tubes
113	70195900	覆铜箔板玻璃纤维布	12	4	Glass fiber cloth, covered with copper foil
114	70200011	导电玻璃	10.5	7	Conductive glass
115	70200012	绝缘子用玻璃伞盘	10.5	4	Glass umbrella for insulator
116	71011011	天然黑珍珠	21	0	Tahitian pearls
117	71011091	天然黑珍珠	21	0	Tahitian pearls

序号 No.	税则号列 Tariff Item	商　品　名　称	最惠国税率(%) M.F.N.	暂定最惠国税率(%) Interim Duty Rate	Article Description
118	71159010	工业或试验室用贵或包金属制品	3	0	'Other articles of precious metal, for technical or lab use
119	72026000	镍铁	2	1	Ferro-nickel
120	72029300	铁钽铌合金(钽含量<10%)	2	1	FeNbTa alloy(Ta content<10%)
121	72029300	铌铁	2	1	Ferro-niobium
122	72031000	热压铁块	2	0	Hot Briguetted Iron(HBI)
123	72091810	厚度<0.3毫米冷扎卷板	6	3	Coils of a thickness of less than 0.3 mm, cold rolled(cold-reduced)
124	72269900	铁镍合金带材(生产集成电路框架用)	7	4	Fe-Ni alloy strip (production of the frame for electronic integrated circuits)
125	73044110	锅炉管(包括内螺纹)	10	5	Boiler seamless tubes and pipes (inside thread), of stainless steel, cold draw or cold rolled
126	73044910	锅炉管(包括内螺纹)	10	5	Boiler seamless tubes and pipes (inside thread), of stainless steel, cold draw or cold rolled
127	74101100	覆铜板及印刷线路板用铜箔	4	1	Foil of refined Cu, not backed, for printed circuits
128	75011000	镍锍	3	1	Nickel matte
129	75012000	氧化镍烧结物及镍冶炼的中间产品	3	1	Ni oxide sinters & other intermediate products of Ni metallurgy
130	75021000	金属镍(未煅轧)	3	1	Ni unwrought, not alloyed
131	76061120	厚度在0.3毫米及以上,但不超过0.36毫米的铝板	6	2	Aluminium plates, of a thickness of 0.3mm or more but not exceeding 0.36mm
132	81124000	钒氮合金	3	0	Vanadic-nitrogen
133	84079010	沼气发动机	12	6	Firedamp engines
134	84079090	立式输出汽油发动机	18	5	Petrol engine with a vertical crankshaft
135	84079090	转速<3600r/min的发动机用汽油发动机、税号8426-8430所列机械用转速<4650r/min的汽油发动机	18	8	Petrol engines for generators, rotational speed less than 3600r/min. Petrol engines for engineering mechanism of headings No. 8426 to 8430, rotational speed<4650r/min
136	84099999	电控柴油喷射装置	8.4	3	Electric diesel in jection devices

序号 No.	税则号列 Tariff Item	商品名称	最惠国税率(%) M.F.N.	暂定最惠国税率(%) Interim Duty Rate	Article Description
137	84119910	涡轮轴航空发动机用零件	5	1	Parts of turboshaft engines
138	84122990	载重4万吨及以上船舶用舱口盖液压装置	14	8	Hatch hydraulic device of vessels moer than 40000t
139	84148090	轿车用增压强,增压压力为1.5－1.6Bar,流量为320立方米/小时	7	3.4	Suppercharger for use with car, supercharged preesure: 1.5Bar－1.6Bar, flowrate: 320m^3/h
140	84335100	功率≥160马力的联合收割机	8	5	Combine harvester-threshers, power≥160H.P
141	84335910	甘蔗收获机	8	5	Sugarcane harvesters
142	84335990	棉花收获机、自走式青储饲料收获机	8	5	Cotton harvesting machinery mobile silage harvester
143	84435911	纺织用圆网印花机	10	8	Gylinder screen woven fabricprinting range
144	84435912	纺织用平网印花机	10	8	Flate screen woven fabric printing range
145	84481100	多臂机或提花机　转速指标:500转/分以上	8	4	Dobbies or Jacquards, rotational speed > 500 r/min
146	84483200	精梳机钳板、顶梳装置,精梳机锡林部件、精梳机车头凸轮传动、行星齿轮部件,精梳联合梳理装置,清梳联合给棉装置,清梳联合机盖板清洁	6	3	Nippers, topcombs, cylinders, head cams, epicyclic gears of combers; combing devices, drawing devices, flat cleaning devices of lapping and combing combines
147	84483310	络筒锭	6	3	Winding spindle
148	84483920	电子清纱器	6	3	Electronic yarn cleaner
149	84483930	空气捻接器	6	3	Air twisting devices
150	84484910	接、投梭箱	6	3	Catching and throwing shuttle boxes
151	84484920	引纬、送经装置	6	3	Weft insertion and warp left-off motion
152	84484990	自动寻纬、补纬装置、打纬装置	6	3	Automatic searching weft for the gripper looms, automatic mending weft for the air jet looms
153	84518000	柔软整理机	12	10	Supple finishing machine
154	84518000	磨毛机、丝光机	12	10	Napping grinder, mercerizing range
155	84518000	定型机、涂层机	12	10	Boarding machine, coating machine

序号 No.	税则号列 Tariff Item	商品名称	最惠国税率(%) M.F.N.	暂定最惠国税率(%) Interim Duty Rate	Article Description
156	84518000	罐蒸机、精炼机	12	10	Potting steamer, refining machine
157	84518000	剪绒、洗缩联合机	12	10	Pile shearing and shrinking combine
158	84518000	预缩机	12	10	Sanforizer
159	84518000	服装液氨整理机	12	10	Dress aqua ammoniae finishing range
160	84518000	服装定型焙烘炉	12	10	Dress shaping cabinet dryer
161	84518000	剪毛联合机	12	10	Sheepshearing combination machine
162	84518000	织物轧光机	12	6	Woven fabrics calender
163	84734010	自动柜员机出钞器	10.5	1	Banknote dispenser of automated teller
164	84734090	钞票清分机零附件	10.5	5	Parts of banknote processing system
165	84789000	烟草加工及制作机器的零件	10	5	Parts of machinery for preparing or making up tobacco nes
166	84798110	吐丝机	9.5	7	Metal filament blowing machine
167	84834020	磨煤机用行星齿轮减速器(由螺旋伞齿轮加行星齿轮二级立式减速机构组成,转盘外圆直径为1300mm-2040mm)	8	4	Planetary decelerators for use with coal mill,(composed of helical bevel gear and planetary gear two-stage vertical reducing mechanism, rotary table outline diameter: 1300mm-2040mm)
168	85011091	激光视盘机机芯精密微型电机	9	6	Minitype precise motor for mechanism of laser disc player
169	85011099	功率不大于0.5瓦(圆柱型:直径不大于6毫米,高不大于25毫米;扁园型:直径不大于15毫米,厚不大于5毫米)非用于激光视盘机机芯的微型电机	9	1	Micromotors with a cylindrical housing diameter not exceeding 6mm and height not exceesing 25mm or a oblate-cylindrical housing diameter not exceeding 15mm and thickness not exceeding 5mm,(Not including minitype motor for mechanism of laser disc player)
170	85015300	高速(200km/h)电力机车的交流异步牵引电动机	12	3	AC asyncronous traction motor for high speed electric locomotive (200KM/h)
171	85023100	风力发电设备	8	5	Windmill generating equipments

序号 No.	税则号列 Tariff Item	商品名称	最惠国税率(%) M.F.N.	暂定最惠国税率(%) Interim Duty Rate	Article Description
172	85030030	风力发电设备的零件	3	1	Parts of wind-powerd generating sets
173	85044091	具有变流功能的半导体模块	10	6	Semiconductor modules with converting function
174	85044099	高速(200km/h)电力机车的牵引变流器	10	3	Traction converter for highspeed electric locomotive(200KM/h)
175	85078020	手机专用锂离子电池芯,包括手机时钟控制用钮扣锂离子电池	12	9	Lithium ion battery core for use with mobile phone, including buttony Lithium ion battery use for time-contrl in mobile phone
176	85144000	焊缝中频退火装置	10	5	Intermediate frequency annealing device for weld
177	85229031	具有刻录功能的激光视盘机机芯	30	15	Mechanism of driver of laser disc player, for repeating photo and etching disc
178	85229039	视频信号录制或重放设备的零件	30	10	Parts for video recording or reproducing apparatus
179	85229039	激光视盘机的激光收发装置(激光头)	30	6	Laser transmitting and receiving device of laser disc player
180	85229039	激光视盘机激光收发装置用的零件	30	3	Parts of Laser transmitting and receiving device of laser disc player
181	85253099	非广播级由镜头+CCD/CMOS+数字信号处理电路三部分构成的摄像组件	35	25	Camera subassembly made up of lens, CCD/CMOS sensor and digtial signal processing circuit, not broadcast quality
182	85261090	飞机机载雷达(包括气象雷达、地形雷达和空中交通管制应答系统)	5	1	Airborne radars (including aero-radars, terrain − following radars and air − traffic − control − interrogator − responder systems)
183	85299042	静像视频摄像机、摄录一体机、数码相机用取像模块	12	6	Camera modules for still image video cameras, video camera recorders and digital cameras
184	85299049	摄像机、摄录一体机、数码相机的其他零件	12	4	Parts for cameras and video camera recorders
185	85299081	投影电视机用光机	15	7	Optical engine of projection
186	85299082	等离子显像组件及其零件	15	3	Plasma display modules and parts thereof
187	85318010	音量不超过110dB的小型蜂鸣器	15	7.5	Minitype electric buzzers, maximum volume not exceeding 110dB
188	85371010	机床用数控装置	5	3	Numerical control panels for machine tools

序号 No.	税则号列 Tariff Item	商 品 名 称	最惠国税率(%) M.F.N.	暂定最惠国税率(%) Interim Duty Rate	Article Description
189	85393290	彩色液晶投影机的照明光源	8	5	Parts of lens of colour liquid-crystal projector
190	85404000	彩色的数据/图形显示管，屏幕荧光点间距小于0.4毫米	8	4	Color data/graphic display tubes of phosphor dot screen pitch smaller than 0.4mm
191	85451100	直径为610mm±3mm的炉用碳电极	8	4	Carbon or graphite electrodes for furnaces, with a diameter of 610±3mm
192	85462090	输变电架空线路用长棒形瓷绝缘子瓷件(单支长度为1-2米,实芯)	12	6	Long solid rod ceramic insulator body for power overhead transmission and converting lines(each rod is 1-2 meters in length)
193	87019011	功率大于150马力的轮式拖拉机	8	5	Wheeled tractors with power more than 150H.P.
194	87019019	功率大于150马力的其他拖拉机	8	5	Other tractors with power more than 150H.P.
195	87042300	装有驾驶室的混凝土泵车用底盘	15	10	Chassis (incoraporating cabs) of lorries mounted with concrete pumps
196	87042300	装有驾驶室的固井水泥车、压裂车、混砂车用底盘(车辆总重量超过35吨)	15	10	Chassis (incoraporating cabs) of cementing unit trucks, fractruring unit trucks and mixing sand trucks
197	87042300	装有驾驶室和发动机的重型公路自卸车底盘散件(底盘车架的前宽超过1050mm)	15	8	Knocked-down kits for chassis fitted with cabs and engines, for heavy-duty dumpers designed for highway use (front end width of the truck chassis' frame exceeding 1050mm)
198	87042300	起重55吨及以上的汽车起重机用底盘	15	8	Chassis (incoraporating cabs) of crane lorries, eifting capacity ≥ 55t
199	87059090	跑道除冰车	12	10	Runway deicing trucks
200	87083999	液压调节器(ABS用),由ECU控制模块、电磁阀、电机、传感器等组成	18.6	7.4	Hydraulic Regulator (used in ABS) composed of ECU, Electromagnetic valve, electromotor and sensor etc.
201	87083999	驱动防滑(ASR)装置(ABS用),结构与ABS液压调节器相似,具有驱动防滑的功能	18.6	7.4	Anti-slip Driving System (Acceleration Slip Regulation) (uesd in ABS) is structrally similar with hydraulic regulator used in ABS, use for with preventing slip during accleration

序号 No.	税则号列 Tariff Item	商品名称	最惠国税率(%) M.F.N.	暂定最惠国税率(%) Interim Duty Rate	Article Description
202	87083999	制动力分配(EBD)装置(ABS用),结构与ABS液压调节器相似,具有调整车轮制动力分配功能	18.6	7.4	Electronic Brake Force Distribution Device(used in ABS)is structrally similar with hydraulic regulator used in ABS, use for with regulating the dirstibution of brake force
203	87084050	扭距≥90kgm变速箱、分动箱	10	6	Gear boxes and transfer case, torque ≥ 90kgm
204	87085020	轴荷≥10t的中后驱动桥	12.9	8	Middle and rear drive-axles, axle capacity ≥ 10t
205	87085050	轴荷≥6吨的前驱动桥	10	8	Front drive-axles, axle capacity ≥ 6t
206	87086020	30座及以上的客车用非驱动桥及其零件	17.3	10	Non-driving axles & parts thereof for buses with 30 seats or more
207	87088090	30座及以上的客车用悬挂减震器	13.5	8	Shock-absorbers for buses with 30 seats or more
208	87089320	座位≥30客车用离合器及零件	12.9	8	Clutches and parts thereof for buses with 30 seats or more
209	87089350	总重≥14吨柴油型货车用离合器及零件	10	8	Clutches and parts thereof for diesel trucks of g.v.w≥14t
210	87089420	30座及以上客车用转向器	12.9	8	Steering boxes of buses with 30 seats or more
211	87089920	轴荷≥10t的中后驱动桥的零件	25	8	Parts of middle and rear drive-axles, axle capacity ≥10
212	87089920	30座及以上客车用转向器的零件	25	8	Parts of steering boxes of buses with 30 seats of more
213	87089930	非公路用自卸车未列名零部件	6	3	Not specified parts of dumpers designed for off-highway use
214	87089950	总重≥14吨柴油型货车转向器的零件	10	8	Parts of steering boxes for diesel trucks of g.v.w≥14t
215	87089950	扭距≥90kgm变速箱、分动箱的零件	10	6	Parts of gear boxes and transfer case, torque ≥ 90kgm
216	88024010	空载重量在25吨及以上,但重量不超过45吨的客运飞机	5	1	Passenger aeroplanes of an unladen weight exceeding 25t but not exceeding 45t
217	90019000	背投电视机显屏(包括菲涅耳透镜屏幕、双透镜屏幕和保护屏)	8	6	Project TV screen(induding Fresnel lens screen, lenticular screen, screen shields)

序号 No.	税则号列 Tariff Item	商品名称	最惠国税率(%) M.F.N.	暂定最惠国税率(%) Interim Duty Rate	Article Description
218	90019000	光通信用微光组件的光学元件(包括工作波长为800nm～1700nm的薄膜滤光片、自聚焦透镜、法拉第旋转片)	8	4	Micro-optic component for optical communication (including thin film filter, gradient index lens and faraday Rotator, with working wavelength between 800nm and 1700nm)
219	90021190	彩色液晶投影机的镜头零件及数码相机的镜头	15	6	Part of lens of color liquid－crystal projector
220	90091291	多功能复印一体机	10	3	Digital multifunctional machines with copying function
221	90132000	2.5G B/S及以上SDH、波分复用光传输设备的980纳米、1480纳米的泵浦激光器	6	3	980nm and 1480nm pump laster of SDH≥2.5GB/s, optical transmission equipment for wave-divison multiplexing
222	90138030	液晶显示板	5	2.5	Liquid crystal display panel
223	90019000	激光视盘机激光收发装置用的微型镜片	6	3	Mini lens use for with laster transmitting and receving device of laster disc player
224	90142000	航空惯性导航仪	2	1	Inertial navigation system
225	90189080	宫内节育器	4	0	Intrauterine device(IUD)
226	90229090	射线发生器的零部件	6	3	Parts of ray generators
227	90318090	跑道摩擦系数测试仪	5	3	Runway friction coefficient tester
228	90328900	飞机自动驾驶系统(包括自动驾驶、电子控制飞行、自动故障分析、警告系统配平系统及推力监控设备及其相关仪表)	7	1	Automatic pilotting systems for aircraft(including automatic pilotting, electronic flight contral, automatic failure analysis, warning systems, trimming systems, thrust monitoring equipments and the meters therof)
229	90330000	飞机自动驾驶系统(包括自动驾驶、电子控制飞行、自动故障分析、警告系统配平系统及推力监控设备及其相关仪表)的零件	6	1	Parts of automatic pilotting systems for aircraft (including automatic pilotting, electronic flight contral, automatic failure analysis, warning systems, trimming systems, thrust monitoring equipments and the meters therof)
230	92012000	完税价格50000美元及以上的大钢琴	17.5	1	Grand pianos, the duty-paying value≥50000USD
231	92021000	完税价格15000美元及以上的弓弦乐器	17.5	1	Played with a bow, the duty-paying value≥15000USD
232	92051000	完税价格2000美元及以上的铜管乐器	17.5	1	Brass-wind instruments, the duty-paying ualue≥2000USD
233	92059000	完税价格10000美元及以上的其他管乐器	17.5	1	Other wind musical instruments, the duty-paying value≥10000USD

表 7:出口商品暂定税率表

Table 7: Interim Duty Rate on Exported Goods

序号 No.	税则号列 Tariff Item	货品名称	最惠国税率 (%) M.F.N.	暂定最惠国税率 (%) Interim Duty Rate	Article Description
1	03019210	鳗鱼苗	20	10	Eels(Anguilla Spp.)fry
2	05069090	已脱胶骨、角柱	40	0	Degelatinized bones
3	26080000	灰色饲料氧化锌(氧化锌 ZnO 含量大于 80%)	30	0	Gray feed grade zinc oxide (containing by weight more than 80% of ZnO)
4	26090000	锡矿砂及其精矿	50	20	Tin ores and concentrates
5	28047000	磷	20	10	Phosphorus
6	29022000	苯	40	0	Benzene
7	31021000	尿素		260 元/吨	Urea
8	72021000	锰铁	20	5	Ferro-manganese
9	72022000	硅铁	25	0	Ferro-silicon
10	72023000	硅锰铁	20	5	Ferro-silico-manganese
11	72024000	铬铁	40	5	Ferro-chromium
12	74020000	未精炼铜,电解精炼用的铜阳极	30	10	Unrefined copper; copper anodes for electrolytic refining
13	74030000	未锻轧的精炼铜及铜合金	30	5	Refined copper and copper alloys, unwrought:
14	74040000	铜废碎料	30	10	Copper waste and scrap
15	74070000	铜条、杆、型材及异型材	30	0	Copper bars, rods and profiles
16	74080000	铜丝	30	0	Copper wire
17	74090000	铜板、片及带,厚度>0.15mm	30	0	Copper plates, sheets and strip, of a thickness exceeding 0.15mm
18	75020000	未锻轧镍	40	2	Unwrought nicklc
19	75089010	电镀用镍阳极	40	0	Electroplating anodes
20	76010000	未锻轧铝	30	5	Unwrought aluminium
21	76020000	铝废碎料	30	10	Aluminium waste and scrap
22	76040000	铝条、杆、型材及异型材	20	0	Aluminium bars, rods and profiles
23	76050000	铝丝	20	0	Aluminium wire
24	76060000	铝板、片及带,厚度>0.2mm	20	0	Aluminium plates, sheet and strip, of a thickness exceeding 0.2mm
25	79010000	未锻轧锌	20	0	Unwrought zinc
26	81101010	未锻轧锑	20	5	Unwrought antimony

表 8:纺织品出口暂定税率表

Table 8: Interim Duty Rate on Exported Textile

序号 No.	税则号列 Tariff Item	货 品 名 称	计量单位 Measuring unit	暂定税率(元/计量单位) Interim Duty Rate(RMB/ unit)	Article Description
1	61021000	毛制针织或钩编女式大衣、防风衣	件	0.3	Women's or girls' overcoats, etc, of wool…, knitted or crocheted
2	61022000	棉制针织或钩编女式大衣、防风衣	件	0.3	Women's or girls' overcoats, etc, of cotton, knitted or crocheted
3	61023000	化纤制针织或钩编女式大衣等	件	0.3	Women's or girls' overcoats, etc, of man-made fibres, knitted/crochetd
4	61029000	其他纺织材料制针织或钩编女式大衣、防风衣	件	0.3	Women's or girls' overcoats, etc, of other textiles, knitted/crochetd
5	61031900	其他纺织材料制针织或钩编男式西服套装	套	0.2	Men's or boys' suits of other textiles, nes, knitted or crocheted
6	61032100	毛制针织或钩编男式便服套装	套	0.2	Men's or boys' ensembles of wool or fine hair, knitted or crocheted
7	61032200	棉制针织或钩编男式便服套装	套	0.2	Men's or boys' ensembles of cotton, knitted or crocheted
8	61032300	合纤制针织或钩编男便服套装	套	0.2	Men's or boys' ensembles of synthetic fibres, knitted or crocheted
9	61032900	其他纺织材料制针织或钩编男式便服套装	套	0.2	Men's or boys' ensembles of other textiles, nes, knitted/crocheted
10	61034100	毛制针织或钩编男长裤、工装裤等	件	0.2	Men's or boys' trousers, etc, of wool …, knitted or crocheted
11	61034200	棉制针织或钩编男长裤、工装裤等	件	0.2	Men's or boys' trousers, etc, of cotton, knitted or crocheted
12	61034300	合纤制针织或钩编男长裤等	件	0.2	Men's or boys' trousers, etc, of synthetic fibres, knitted/crocheted
13	61034900	其他纺织材料制针织或钩编男长裤等	件	0.2	Men's or boys' trousers, etc, of other textiles, knitted/crocheted
14	61041200	棉制针织或钩编女式西服套装	套	0.2	Women's or girls' suits of cotton, knitted or crocheted
15	61041900	其他纺织材料制针织或钩编女式西服套装	套	0.2	Women's or girls' suits of other textiles, nes, knitted/crocheted
16	61042100	毛制针织或钩编女式便服套装	套	0.2	Women's or girls' ensembles, of wool …, knitted of crocheted
17	61042200	棉制针织或钩编女式便服套装	套	0.2	Women's or girls' ensembles, of cotton, knitted or crocheted
18	61042300	合纤制针织或钩编女便服套装	套	0.2	Women's or girls' ensembles, of synthetic, fibres, knitted/crocheted
19	61042900	其他纺织材料制针织式钩编女式便服套装	套	0.2	Women's or girls' ensembles, of other textiles, knitted/crocheted

序号 No.	税则号列 Tariff Item	货品名称	计量单位 Measuring unit	暂定税率（元/计量单位） Interim Duty Rate(RMB/ unit)	Article Description
20	61043100	毛制针织女式上衣	件	0.3	Women's or girls' jackets, of wool…, knitted or crocheted
21	61043200	棉制针织女式上衣	件	0.3	Women's or girls' jackets, of cotton, knitted or crocheted
22	61043300	合纤制针织女上衣	件	0.3	Women's or girls' jackets, of synthetic fibres, knitted/crocheted
23	61043900	其他纺织材料制针织女上衣	件	0.3	Woman's or girls' jackets, of other textiles, knitted or crocheted
24	61044100	毛制针织或钩编连衣裙	件	0.2	Dresses of wool or fine animal hair, knitted or crocheted
25	61044200	棉制针织或钩编连衣裙	件	0.2	Dresses of cotton, knitted or crocheted
26	61044300	合纤制针织或钩编连衣裙	件	0.2	Dresses of synthetic fibres, knitted or crocheted
27	61044400	人纤制针织或钩编连衣裙	件	0.2	Dresses of artificial fibres, knitted or crocheted
28	61044900	其他纺织材料制针织或钩编连衣裙	件	0.2	Dresses of other textile material, nes, knitted or crocheted
29	61045100	毛制针织或钩编裙子及裙裤	件	0.2	Skirts & divided skirts of wool or fine hair, knitted or crocheted
30	61045200	棉制针织裙子及裙裤	件	0.2	Skirts & divided skirts of cotton, knitted or crocheted
31	61045300	合纤制针织或钩编裙子及裙裤	件	0.2	Skirts & divided skirts of synthetic fibres, knitted or crocheted
32	61045900	其他纺织材料制针织或钩编裙子及裙裤	件	0.2	Skirts & divided skirts of other textiles, nes, knitted or crocheted
33	61046100	毛制针织或钩编女长裤、工装裤等	条	0.2	Women's or girls' trousers, etc, of wool…, knitted or crocheted
34	61046200	棉制针织或钩编女长裤、工装裤等	条	0.2	Women's or girls' trousers, etc, of cotton, knitted or crocheted
35	61046300	合纤制针织或钩编女长裤等	条	0.2	Women's or girls' trousers, etc, of synthetic, knitted or crocheted
36	61046900	其他纺织材料制针织或钩编长女裤等	条	0.2	Women's or girls' trousers, etc, of other textile, knitted/crocheted
37	61051000	棉制针织或钩编男衬衫	件	0.2	Men's or boys' shirts of cotton, knitted or crocheted
38	61052000	化纤制针织或钩编男衬衫	件	0.2	Men's or boys' shirts of man-made fibres, knitted or crocheted
39	61059000	其他纺织材料制针织或钩编男衬衫	件	0.2	Men's or boys' shirts of other textiles, nes, knitted or crocheted
40	61061000	棉制针织或钩编女衬衫	件	0.2	Women's or girls' blouses, etc, of cotton, knitted or crocheted

序号 No.	税则号列 Tariff Item	货品名称	计量单位 Measuring unit	暂定税率 (元/计量单位) Interim Duty Rate(RMB/ unit)	Article Description
41	61062000	化纤制针织或钩编女衬衫	件	0.2	Women's or girls' blouses, etc, of man-made fibres, knitted/crochetd
42	61069000	其他纺织材料制针织或钩编女衬衫	件	0.2	Women's or girls' blouses, etc, of other textiles, knitted/crocheted
43	61071100	棉制针织或钩编男内裤及三角裤	件	0.2	Men's or boys' underpants & briefs of cotton, knitted or crocheted
44	61071200	化纤制针织或钩编男内裤及三角裤	件	0.2	Men's or boys' underpants, etc, of man-made fibres, knitted/crochetd
45	61072100	棉制针织或钩编男长睡衣及睡衣裤	件(套)	0.2	Men's or boys' pyjamas of cotton, knitted or crocheted
46	61072200	化纤制针织或钩编男睡衣裤	件(套)	0.2	Men's or boys' pyjamas of man-made fibres, knitted or crocheted
47	61079100	棉制针织或钩编男浴衣、晨衣	件	0.2	Men's or boys' dressing gowns, etc, of cotton, knitted or crocheted
48	61079200	化纤制针织或钩编男浴衣、晨衣	件	0.2	Men's or boys' dressing gowns, of man-made fibres, knitted/crocheted
49	61081100	化纤制针织或钩编长衬裙及衬裙	件	0.2	Women's/girls' slips, etc, of man-made fibres, knitted or crocheted
50	61081910	棉制针织或钩编女式长衬裙及衬裙	件	0.2	Women's /girls' slips, etc, of cotton, knitted or crocheted
51	61082100	棉制针织或钩编女三角裤及短衬裤	件	0.2	Women's or girls' briefs & panties of cotton, knitted or crocheted
52	61082200	化纤制针织或钩编女三角裤及短衬裤	件	0.2	Women's or girls' briefs, etc, of man-made fibres, knitted/crocheted
53	61083100	棉制针织或钩编女睡衣及睡衣裤	件(套)	0.2	Women's or girls' nighties…, etc, of cotton, knitted or crocheted
54	61083200	化纤制针织或钩编女睡衣及睡衣裤	件(套)	0.2	Women's or girls' pyjamas, etc, of man-made fibres, knittd/crochetd
55	61089100	棉制针织或钩编女浴女、晨衣	件	0.2	Women's or girls' dressing gowns…, of cotton, knitted/crocheted
56	61089200	化纤制针织或钩编女浴衣、晨衣	件	0.2	Women's or girls' dressing gowns of man-made fibre, knitted/crochetd
57	61091000	棉制针织或钩编T恤衫、汗衫等	件	0.2	T-shirts, singlets & other vests, of cotton, knitted or crocheted
58	61099090	其他纺织材料制针织或钩编T恤衫、汗衫等	件	0.2	T-shirts, singlets, etc, of other textiles, nes, knitted/crocheted
59	61101100	羊毛制针织或钩编套头衫等	件	0.2	Jerseys, pullovers, etc, of wool, knitted or crocheted
60	61101920	兔毛制针织或钩编套头衫等	件	0.2	Jerseys, pullovers, etc, of rabbit or hare hair, knitted/crocheted

序号 No.	税则号列 Tariff Item	货品名称	计量单位 Measuring unit	暂定税率 (元/计量单位) Interim Duty Rate(RMB/ unit)	Article Description
61	61101990	其他毛制针织或钩编套头衫等	件	0.2	Jerseys, pullovers, etc, of other animal fine hair nes, knit/crochtd
62	61102000	棉制针织或钩编套头衫等	件	0.2	Jerseys, pullovers, etc, of cotton, knitted or crocheted
63	61103000	化纤制针织或钩编套头衫等	件	0.2	Jerseys, pullovers, etc, of man-made fibres, knitted or crocheted
64	61109090	其他纺织材料制针织或钩编套头衫等	件	0.2	Jerseys, pullovers, etc, of other textiles, knitted or crocheted
65	61121100	棉制针织或钩编运动服	套	0.2	Track - suits of cotton, knitted or crocheted
66	61121200	合纤制针织或钩编运动服	套	0.2	Track - suits of synthetic fibres, knitted or crocheted
67	61121900	其他纺织材料制针织或钩编运动服	套	0.2	Track - suits of other textiles, knitted or crocheted
68	61122090	其他纺织材料制针织或钩编滑雪服	套	0.2	Ski - suits, of other textile materials, knitted or crocheted
69	61130000	涂层经处理针织或钩编织物制服装	件	0.2	Garments of knitted or crocheted fabrics of 59.03,59.06,59.07
70	61141000	毛制针织或钩编的其他服装	件	0.2	Garments of wool or fine animal hair, knitted or crocheted, nes
71	61142000	棉制针织或钩编的其他服装	件	0.2	Garments of cotton, knitted or crocheted, nes
72	61143000	化纤制针织或钩编的其他服装	件	0.2	Garments of man-made fibres, knitted or crocheted, nes
73	61179000	其他针织或钩编衣着零件	千克	0.5	Parts of garments or clothing accessories, knitted or crocheted
74	62021100	毛制女式大衣、斗篷及类似品等	件	0.3	Women's or girls' overcoats, etc, of wool or fine animal hair
75	62021290	棉制女式大衣、斗篷及类似品等	件	0.3	Women's or girls' overcoats, etc, of cotton, nes
76	62021390	化纤制女式大衣、斗篷及类似品	件	0.3	Women's or girls' overcoats, etc, of man-made fibres, nes
77	62021900	其他纺织材料制女式大衣、斗篷及类似品	件	0.3	Women's or girls' overcoats, etc, of other textiles, nes
78	62029100	毛制女式带风帽防寒短上衣、防风衣	件	0.3	Women's or girls' anoraks, windcheaters, etc,of wool…
79	62029290	棉制女式带风帽防寒短上衣、防风衣	件	0.3	Women's or girls' anoraks, windcheaters…of cotton, nes
80	62029390	化纤制女式防风衣等	件	0.3	Women's or girls' anoraks, windcheaters… of man-made fibres, nes
81	62029900	其他纺织材料制防风衣、防风短上衣等	件	0.3	Women's or girls' anoraks, windcheaters, etc, of other textiles, nes
82	62031990	其他纺织材料制男式西服套装	套	0.2	Men's or boys' suits of other textiles, nes

序号 No.	税则号列 Tariff Item	货　品　名　称	计量单位 Measuring unit	暂定税率 (元/计量单位) Interim Duty Rate(RMB/ unit)	Article Description
83	62032100	毛制男式便服套装	套	0.2	Men's or boys' ensembles of wool or fine animal hair
84	62032200	棉制男式便服套装	套	0.2	Men's or boys' ensembles of cotton
85	62032300	合纤制男式便服套装	套	0.2	Men's or boys' ensembles of synthetic fibres
86	62032990	其他纺织材料制男式便服套装	套	0.2	Men's or boys' ensembles of other textiles, nes
87	62034100	毛制男式长裤、工装裤等	条	0.2	Men's or boys' trousers, breeches, etc, of wool or fine animal hair
88	62034210	棉制男式阿拉伯裤	条	0.2	Men's or boys' arabian trousers, breeches of cotton
89	62034290	棉制男式长裤、工装裤等	条	0.2	Men's or boys' trousers, breeches, nes, of cotton
90	62034310	合成纤维制男式阿拉伯裤	条	0.2	Men's or boys' arabian trousers, breeches of synthetic fibres
91	62034390	合纤制男式长裤、工装裤等	条	0.2	Men's or boys' trousers, breeches, nes, of synthetic fibres
92	62034990	其他纺织材料制男童裤、工装裤	条	0.2	Men's or boys' trousers, breeches, nes, of oth textile fibres
93	62041200	棉制女式西服套装	套	0.2	Women's or girls' suits of cotton
94	62041990	其他纺织材料制女式西服套装	套	0.2	Women's or girls' suits of other textiles, nes
95	62042100	毛制女式便服套装	套	0.2	Women's or girls' ensembles of wool or fine animal hair
96	62042200	棉制女式便服套装	套	0.2	Women's or girls' ensembles of cotton
97	62042300	合纤制女式便服套装	套	0.2	Women's or girls' ensembles of synthetic fibres
98	62042990	其他纺织材料制女式便服套装	套	0.2	Women's or girls' ensembles of other textile materials, nes
99	62043100	毛制女式上衣	件	0.3	Women's or girls' jackets & blazers of wool or fine animal hair
100	62043200	棉制女式上衣	件	0.2	Women's or girls' jackets & blazers of cotton
101	62043300	合纤制女式上衣	件	0.3	Women's or girls' jackets & blazers of synthetic fibres
102	62043990	其他纺织材料制女式上衣	件	0.3	Women's or girls' jackets of other textiles, nes
103	62044100	毛制连衣裙	件	0.2	Dresses of wool or fine animal hair
104	62044200	棉制连衣裙	件	0.2	Dresses of cotton
105	62044300	合纤制女式连衣裙	件	0.2	Dresses of synthetic fibres
106	62044400	人纤制女式连衣裙	件	0.2	Dresses of artificial fibres
107	62044990	其他纺织材料制连衣裙	件	0.2	Women's or girls' dresses of other textile materials, nes

序号 No.	税则号列 Tariff Item	货 品 名 称	计量单位 Measuring unit	暂定税率 (元/计量单位) Interim Duty Rate(RMB/ unit)	Article Description
108	62045100	毛制裙子及裙裤	件	0.2	Skirts & divided skirts of wool or fine animal hair
109	62045200	棉制裙子及裙裤	件	0.2	Skirts & divided skirts of cotton
110	62045300	合纤制裙子及裙裤	件	0.2	Skirts & divided skirts of synthetic fibres
111	62045990	其他纺织材料制裙子及裙裤	件	0.2	Women's or girls' skirts of other textile materials, nes
112	62046100	毛制女式长裤、工装裤等	条	0.2	Women's or girls' trousers/breeches, etc, of wool/fine animal hair
113	62046200	棉制女式长裤、工装裤等	条	0.2	Women's or girls' trousers, breeches, etc, of cotton
114	62046300	合纤制女式长裤、工装裤等	条	0.2	Women's or girls' trousers, breeches, etc, of synthetic fibres
115	62046900	其他纺织材料制女式长裤、工装裤等	条	0.2	Women's or girls' trousers, breeches, etc, of other textiles, nes
116	62052000	棉制男衬衫	件	0.2	Men's or boys' shirts of cotton
117	62053000	化纤制男衬衫	件	0.2	Men's or boys' shirts of man-made fibres
118	62059010	丝及绢丝制男衬衫	件	0.2	Men's or boys' shirts of silk or silk waste
119	62059090	其他纺织材料制男衬衫	件	0.2	Men's or boys' shirts of other textile materials, nes
120	62061000	丝及绢丝制女式衬衫	件	0.2	Women's or girls' blouses, shirts, etc, of silk or silk waste
121	62062000	毛制女衬衫	件	0.2	Women's or girls' blouses, shirts, etc, of wool or fine animal hair
122	62063000	棉制女衬衫	件	0.2	Women's or girls' blouses, shirts, etc, of cotton
123	62064000	化纤制女衬衫	件	0.2	Women's or girls' blouses, shirts, etc, of man-made fibres
124	62069000	其他纺织材料制女衬衫	件	0.2	Women's or girls' blouses, shirts, etc, of other textiles, nes
125	62071100	棉制男式内裤及三角裤	件	0.2	Men's or boys' underpants & briefs of cotton
126	62071920	化纤制男式内裤及三角裤	件	0.2	Men's or boys' underpants & briefs of man- made fibres
127	62072100	棉制男式长睡衣及睡衣裤	件(套)	0.2	Men's or boys' nightshirts & pyjamas of cotton
128	62072200	化纤制男式长睡衣及睡衣裤	件(套)	0.2	Men's or boys' nightshirts & pyjamas of man-made fibres
129	62079100	棉制男式浴衣、晨衣及类似品	件	0.2	Men's or boys' singlets, dressing gowns, etc, of cotton
130	62079200	化纤制男浴衣、晨衣及类似品	件	0.2	Men's or boys' singlets, dressing gowns, etc, of man-made fibres

序号 No.	税则号列 Tariff Item	货　品　名　称	计量单位 Measuring unit	暂定税率（元/计量单位）Interim Duty Rate(RMB/ unit)	Article Description
131	62081100	化纤制长衬裙及衬裙	件	0.2	Women's or girls' slips & petticoats of man-made fibres
132	62081920	棉制长衬裙及衬裙	件	0.2	Women's or girls' slips & petticoats of cotton
133	62082100	棉制女式睡衣及睡衣裤	件(套)	0.2	Women's or girls' nightdresses & pyjamas of cotton
134	62082200	化纤制女式睡衣及睡衣裤	件(套)	0.2	Women's or girls' nightdresses & pyjamas of man-made fibres
135	62089100	棉制女式背心、内衣、浴衣及类似品	件	0.2	Women's or girls' dressing gowns, panties, etc, of cotton
136	62089200	化纤制女式背心、内衣及类似品	件	0.2	Women's or girls' dressing gowns, panties, etc, of man-made fibres
137	62103000	用塑料、橡胶等处理的织物制女大衣等	件	0.3	Garments of 6202. 11 to 19, from fabrics of 59.03, 59.06 or 59.07
138	62104000	用塑料、橡胶等处理的织物制的其他男式服装	件	0.2	Men's or boys' garments of fabrics of 59.03, 59.06 or 59.07
139	62105000	用塑料、橡胶等处理的织物制的其他女式服装	件	0.2	Women's or girls' garments of fabrics of 59.03, 59.06 or 59.07
140	62112010	棉制滑雪服	套	0.2	Ski suits of cotton
141	62112090	其他纺织材料制滑雪服	套	0.2	Ski suits of other textile materials
142	62113100	毛制男式运动服及其他服装	件	0.2	Men's or boys' garments of wool or fine animal hair, nes
143	62113290	棉制男式运动服及其他服装	件	0.2	Men's or boys' garments of cotton, nes
144	62113390	化纤制男式运动服及其他服装	件	0.2	Men's or boys' garments of man-made fibres, nes
145	62114100	毛制女式运动服及其他服装	件	0.2	Women's or girls' garments of wool or fine animal hair
146	62114200	棉制女式运动服及其他服装	件	0.2	Women's or girls' garments of cotton, nes
147	62114300	化纤制女式运动服及其他服装	件	0.2	Women's or girls' garments of man-made fibres, nes
148	62179000	非针织非钩编服装或衣着零件	千克	0.5	Parts of garments or of clothing accessories, nes

表9:非全税目信息技术产品税率表

Table 9:Duty Rate on Specific Information Technology Products

序号 No.	ex	税则号列 Tariff Item	商品描述	Article Description	全税号税率 (%) M.F.N	EX项税率 (%) Ex.Rate
1	ex	70200019	用于插入熔化和氧化炉内以制备半导体晶片的石英反应管及夹持器	Quartz reactor tubes and holders designed for insertion into diffusion and oxidation furnaces for production of semiconductor wafers	10.5	0
2	ex	84199090	半导体工业用化学蒸镀装置的零件	Parts of chemical vapor deposition apparatus for semiconductor production	4	0
3	ex	84211990	半导体晶片加工用离心干燥器	Spin dryers for semiconductor wafer processing	10	0
4	ex	84622110	用于弯曲、折叠和矫直半导体引脚的数控机器	Machines for bending, folding and straightening semiconductor leads	9.7	0
5	ex	84622910	用于弯曲、折叠和矫直半导体引脚的非数控机器	Machines for bending, folding and straightening semiconductor leads	10	0
6	ex	85044091	自动数据处理设备机器及组件、电讯设备用的具有变流动能的半导体模块	Semicondutor modules with converting function for automatic data processing machines and units thereof, and telecommunication apparatus	10	0
7	ex	85044099	自动数据处理设备机器及组件、电讯设备用的其他静止变流器;ITA产品用的印刷电路组件,包括外接组件,如符合PCMCIA标准的卡	Other stastic converters for automatic data processing machines and units thereof, and telecommunication apparatus; Printed Circuit Assemblies for products falling within the ITA, including such assemblies for external connections such as cards that conform to the PCMCIA standard	10	0
8	ex	85181000	电讯用麦克风,频率范围在300赫兹到3.4千赫之间,直径不超过10毫米,高度不超过3毫米	Microphones having a frequency range of 300 Hz to 3,4 kHz with a diameter of not exceeding 10mm and a height not exceeding 3mm, for telecommunication use	10	0
9	ex	85184000	列入ITA的有线电话重复器用的电器扩音器	Electric amplifiers when used as repeaters in line telephony products falling within the ITA	12	0
10	ex	85189000	列入ITA的有线电话重复器用的电器扩音器的零件	Parts of electric amplifiers when used as repeaters in line telephony products falling within the ITA	10.5	0
11	ex	85279090	用于呼叫、提示和寻呼的便携式接收器	Portable receivers for calling, alerting or paging	9	0
12	ex	85291090	无线电话电报装置的天线	Aerials or antennae of a kind used with apparatus for radio-telephony and radio-telegraphy	2	0

序号 No.	ex	税则号列 Tariff Item	商品描述	Article Description	全税号税率 (%) M.F.N	EX 项税率 (%) Ex. Rate
13	ex	90309000	用于检测半导体晶片及器件的仪器的零件和附件;ITA 产品用的印刷电路组件,包括外接组件,如符合 PCMCIA 标准的卡	Parts and accessories of instruments and apparatus for measuring or checking semiconductor wafers or devices; Printed Circuit Assemblies for products falling within the ITA, including such assemblies for external connections such as cards that conform to the PCMCIA standard.	7	0

表 10：中华人民共和国进境物品进口税率表

Table 10: Duty Rate on Inward Articles of the People's Republic of China

税号 No.	物 品 名 称	税率 (%) Duty	Article Description
1	书报、刊物、教育专用电影片、幻灯片、原版录音带、录相带 金、银及其制品 食品、饮料 本表 2、3 税号及备注中所不包含的其他商品	10	Book, newspaper, journal, movie copy for educaiton, sliedo, cassette, video cassette, silver or article, food, beverage, and articles other than item 2,3
2	纺织品及其制成品 摄像机、摄录一体机、数码像机及其他电器用具 照像机、自行车、手表、钟表(含配件、附件) 化妆品	20	Textile product, camera, bicycles, watch, clock, cosmetic, video camera, digital camera and other electric products
3	烟、酒	50	Tobacco product, spirits, wine beer

注：

避孕用具和避孕药品，超过海关规定的自用合理数量部分按有关规定予以退运或按货物进口程序办理报关及验放手续

Note:

About contraceptive instruments and contraceptive medicines, the extra amount of which exceeding the reasonable self-using level set by customs regulations should be backtrack, or apply for customs registration and examination as required by goods-importing procedures.